Fairchild C-119 Flying Boxcar

ALSO BY SIMON D. BECK
AND FROM MCFARLAND

*Fairchild C-82 Packet: The Military
and Civil History* (2017)

*The Aircraft-Spotter's Film
and Television Companion* (2016)

Fairchild C-119 Flying Boxcar

The Complete Service Histories

Simon D. Beck

McFarland & Company, Inc., Publishers
Jefferson, North Carolina

ISBN (print) 978-1-4766-9607-2
ISBN (ebook) 978-1-4766-5367-9

LIBRARY OF CONGRESS CATALOGING DATA ARE AVAILABLE

Library of Congress Control Number 2024059941

© 2025 Simon D. Beck. All rights reserved

No part of this book may be reproduced or transmitted in any form or by any means, electronic or mechanical, including photocopying or recording, or by any information storage and retrieval system, without permission in writing from the publisher.

Front cover: ANG C-119G s/n 53-8087 (KF-190), August 1973 (Richard Vandervord)

Printed in the United States of America

*McFarland & Company, Inc., Publishers
Box 611, Jefferson, North Carolina 28640
www.mcfarlandpub.com*

Table of Contents

Acknowledgments vi
Preface 1
Abbreviations 3
Introduction 9

Section One—Technical Data 13
Section Two—Military Operators 41
Section Three—Civil Operators & Owners 135
Section Four—Service Histories 152

Appendix I: USAF & USMC Tail Codes 421
Appendix II: C-119 Retirement Bases & Disposals 423
Appendix III: Civil Registration Cross-Reference 430
Appendix IV: Civil Airtankers 432
Appendix V: Aircraft Losses 433
Appendix VI: Extant Airframes 439
Bibliography 443
Index 445

Acknowledgments

An undertaking such as this has required the expertise, assistance and kindness of many people in donating their time, memories and in many cases their photographic collections. The author would like to extend a sincere and grateful acknowledgment to the following persons and organizations, without whose help this book would not have been possible. In alphabetical order they are: Roger Bartels (C-119 pilot); Prof. Paul Bierman for C-119 Greenland operations; Chris Charland for his generous help with RCAF histories; Dave Ciocchi; Roy Davis of the AC-119 Association; Charlie De La Royere for his help with Belgian AF histories; Archie DiFante of the Air Force Historical Research Agency, Maxwell AFB; Lars Gleitsmann; Geoff Goodall for C-119 civil photos; Dave Gray for C-119 Bakalar AFB history; Roger Haneline (C-119 pilot) for technical help; Tammy Horton of the Air Force Historical Research Agency, Maxwell AFB, for her generous assistance with IARC Reels; Wayne Lo for the giving of his time on RoCAF Kaiser built C-119s; the late Chuck L. Lunsford for C-119 background; Peter Marson; National Air & Space Museum (NASM), Washington D.C., for help with their Fairchild Collection; National Archives and Records Administration (NARA), Washington D.C., for photographic assistance; Guy Nockels of Namib Film; Milo Peltzer for airtanker background; Jagan Pillarisetti for Indian AF histories; Michael Prophet for his generous help with photos; John S. Reffett; Wally Soplata for his father Walter's C-119 flight manual collection; Everett Sprous (AC-119K gunner [ret.]); Bob and Pam Thomas of the National Naval Aviation Museum (NNAM) for R4Q data and photos; Aad van der Voet; Richard Vandervord for his generous help with photos; Nick Veronico for help with photos; Capt. Bob West (C-119 pilot) for technical help; and Bill Zugelder and Kathy Snapp. Lastly a special thanks for the enduring patience of my wife Yee Miin and our two daughters Maegan and Ellie.

Preface

Fairchild C-119 Flying Boxcar: The Complete Service Histories is the accumulated result of a research project that began in late 2004 when I became interested in Fairchild aircraft cargo giants, namely the C-82, C-119 and C-123. Since then, I have collected photos, flight manuals, books, and original USAF publications and I have gathered multiple files of researched data from pilots and veteran service personnel who served on these aircraft around the globe.

It was during the later 2010s that I acquired all existing FAA C-119 aircraft files which then enabled me to form a complete picture of the Flying Boxcars' civil service.

In the early 2020s I was able to acquire the entire set of USAF Individual Aircraft Record Card (IARC) reels for the C-119. With 1,185 aircraft built and an average of up to 15 cards per aircraft to interpret, this represented a massive 17,500-plus cards to read and extract data from. The Covid-19 pandemic lockdowns in New Zealand beginning in March 2020, however, provided the perfect starting point for this endeavor, and by December of that year I had completed all IARC interpretations into the service histories that are now presented as the backbone of this book.

This resulting volume is designed primarily to be used as a technical "spot-check" reference manual and is not intended to be an anecdotal or historical narrative on the Flying Boxcars service career and contribution to aviation. For the first time a clear picture of the aircraft production run, service structure, unit assignments and final dispositions can be understood and referenced by historians, family of veterans who served on these aircraft or anyone with a keen interest in the Flying Boxcar.

Dates are presented as "Month dd yyyy" in the main body text and as "ddmmmyy" or "mmmyy" in the finer technical print.

Abbreviations

1Lt—1st Lieutenant—Military rank

2Lt—2nd Lieutenant—Military rank

A1C—Airman First Class—Military rank

A2C—Airman Second Class—Military rank

A3C—Airman Third Class—Military rank

AASBR—Arizona Aircraft Storage Branch—USAF organization

AB / ABS—Air Base—Military term

Acc—Accepted

ACO—Air Commando—USAF unit

ACS—Airways and Communications—USAF AACS unit

AD—Air Depot—Military term

ADC—Air Defense Command—USAF Major Command

ADE—Air Delivery—USAF unit

ADF—Air Defense (Fighter)—USAF unit

ADH—Advanced Headquarters—USAF Command

ADV—Air Division

AED—Air Engineering Development—USAF unit

AEMCO—Aircraft Engineering & Maintenance Co.—Civil contractor

AES—Aircraft Engineer Squadron—USMC

AEW—Airborne Early Warning & Control—USAF unit

AF—Air Field

AF—Air Force

AFA—Air Force Armament—USAF unit

AFB—Air Force Base

AFE—(U.S.) Air Force in Europe—USAF Major Command

AFM—Air Force Missile (Test Center)—USAF unit

AFR—Air Force Reserve—U.S. military branch

AFS—Air Force Station—USAF

AFT—Air (Force) Flight Test Center—USAF organization

AirFMFLANT—Air Fleet Marine Force, Atlantic—USMC

ALDIN—Allison Division Indianapolis—Civil contractor

AMB—Air Material Base—RCAF unit

AMC—Air Materiel Command—USAF Major Command

AMI—Aeronautica Militare Italiana

AML—Aeromedical (Light)—ANG unit

AMO—Airways and Communications, Mobile—USAF AACS unit

AMPAR—Air Materiel Command, Pacific Area—USAF

ANG—Air National Guard—U.S. military branch

ANGB—Air National Guard Base—ANG

AOM—Aeromedical (*see also:* AML)—ANG unit

AOT—Air Observers Training Center—USAF unit

APG—Air Proving Ground Command—USAF Major Command

(A)PGC—Air Proving Ground Command—USAF Major Command / unit

AR—Area—Military term

ARAp—Aircraft Registration Application—Civil

Abbreviations

ARC—Air Resupply and Communications—USAF ARCS unit

ARCS—Air Resupply and Communications Service—USAF organization

ARD—Air Research and Development Command—USAF Major Command

ARH—Air Refueling (Heavy)—USAF unit

ARM—Aerial Refueling, Medium—USAF unit

ARP—Aircraft Repair—USAF unit

ARS—Air Rescue

ARS—Air Reserve Station—USAF unit

ART—Air Reserve Training—USAF unit

AS—Air Station

ASD—Aircraft Storage & Disposition—USAF unit

ASL—Air Resupply—USAF ARCS unit

ASQ—Attack Squadron—VNAF unit

AT / ATS—Air Transport—USAF unit

ATC—Air Training Command—USAF Major Command

ATR—Aeromedical Transport—ANG unit

AU—Air University—USAF Major Command / unit

Aux—Auxiliary

Avble—Available

BAF—Belgian Air Force

BAR—Bureau of Aeronautics Representative—USN

BAS—Base Unit—Military term

BFL—Base Flight—Military term

Bku—Broken up

BLJ—Bombardment, Light, Jet—USAF unit

BMC—Birmingham Modification Center—Civil contractor

BOC—Brought on charge—Military term

BofS—Bill of Sale—Civil

BTA—Bombardment (Tactical)—USAF unit

CAC—Caribbean Air Command—USAF Major Command

CAM—Consolidated Aircraft Maintenance—USAF unit

Canx—Canceled

Capt—Captain—Military rank

CAR—Cambridge Research Center—USAF organization

CAT—Civil Air Transport, Inc.—Civil agency

CCT—Combat Crew Training—USAF unit

CE / CN—Center (CE from 1957)—Military term

CE&PE—Central Experimental & Proving Establishment—RCAF unit

CIA—Central Intelligence Agency—Civil agency

CJATC—Canadian Joint Air Training Centre—RCAF unit

CLM—Consolidated Maintenance—USAF unit

CM—Command—Military term

CMP—Composite—USAF unit

CMT—Cold Weather Materiel Test—USAF unit

CNC—Continental Air Command—USAF Major Command

CNR—Continental Air Command Reserve—USAF Major Command

COAAR—Continental AACS Area—USAF

CofA—Certificate of Airworthiness—Civil

CofR—Certificate of Registration—Civil

Com Flt—Communications Flight—RCAF unit

COS—Combat Support—USAF unit

Cpl—Corporal—Military rank

CRC—Control & Reporting Center—USAF organization

CTC—Chanute Technical Training Center—USAF organization

Cvtd—Converted

Cwo—Chief Warrant Officer—Military rank

CZ—(Panama) Canal Zone

d/b/a—Doing business as—Business term

del—Delivered—Military term

depl(s)—Deployment(s)—Military term

depmnt—Depot Maintenance

de-reg—De-registered—Civil term

Des—Designated

DFI—Dropped from inventory—USAF term

DOAOR—Department of the Army, Ops Range—U.S. Army organization

DoD—Department of Defense—U.S. Government

DP—Depot

DT—Detachment—Military term

DV—Division—Military term

DYNZZ—General Dynamics—Civil contractor

EI—Engineering & Installation—USAF AACS unit

ETAF—Ethiopian Air Force

FAA—Federal Aviation Administration—Civil agency

FASRON—Fleet Aircraft Service Squadron—USN

FDM / FLM—Field Maintenance—USAF unit

FEA—Far East Air Forces Command—USAF Major Command

FEAMCOM—Far East Air Materiel Command—USAF unit

FELFR—Far East Air Logistics Frontier—USAF

Ff—(date of) first flight

FI—Fighter-Interceptor—USAF unit

FMS—Foreign Military Sales—Military term

FRY—Ferrying—USAF unit

Ft—Feet—Military term

FT—Flight

FTA / FTN—Flying Training—USAF unit

FTG—Field Training—USAF unit

FTR—Flying Training Reserve (Center)—AFR unit

GP—Group—Military term

GSgt—Gunnery Sergeant—Military rank

GTT—Grupo de Transporte de Tropa—Brazilian AF unit

H&MS—Headquarters & Maintenance Squadron—USMC

H&P—Hawkins & Powers Aviation—Civil private co.

HACBA—Hayes Aircraft Corp. Birmingham, Alabama—Civil contractor

HAMRON—Headquarters & Maintenance Squadron—USMC

HEDRON—Headquarters Squadron—USMC

Hp—Horse power (engine power)

HQ—Headquarters—Military term

HQC—Headquarters Command—USAF Major Command

HQS—Headquarters Support—USAF unit

HVFS—Hemet Valley Flying Service—Civil private co.

IAF—Indian Air Force

IAP—International Airport—Civil

IARC—Individual Aircraft Record Card—USAF term

IM—Installation & Maintenance—USAF AACS unit

KTC—Keesler Technical Training Center—USAF organization

Lb—Pounds (Weight)

Lbst—Pounds Static (Thrust)

LOG—Air Force Logistics Command—USAF Major Command

LTC—Lowry Technical Training Center—USAF organization

MADDP—Manila Air Depot—USAF

MAI—Aircraft Maintenance—USAF unit

Maj—Major—Military rank

MAP—Military Assistance Program (1954–1982)—U.S. military term

MAP—Municipal Airport—Civil

MARS—Marine Aircraft Repair Squadron—USMC

MARTD—Marine Air Reserve Training Detachment—USMC

MASDC—Military Aircraft Storage & Disposition Center—USAF unit

MAT—Air Materiel—USAF unit

MATS—Military Air Transport Service—USAF Major Command

MAW—Marine Air Wing—USMC

MBA—Mobile Training Aids—USAF unit

MBT—Mobile Training—USAF unit

MCAF—Marine Corps. Air Field—USMC

MCAS—Marine Corps. Air Station—USMC

MCH—Mapping and Charting—USAF unit

MDAP—Military Defense Assistance Program (1949–1954)—U.S. military term

MDN—Depot Maintenance—USAF unit

MIDAR—Middleton Air Materiel Area—USAF

MIT—Military Training—USAF unit

MNT / mnt—Maintenance—USAF unit / term

MOAAR—Mobile Air Materiel Area—USAF

MOBAR—Mobile Air Materiel Area—USAF

MSgt—Master Sergeant—Military rank

MSU—Maintenance & Supply—USAF unit

MTG—Marine Training Group—USMC

MWSG—Marine Wing Support Group—USMC

NAATC—Naval Air Advanced Training Center—USN

NABTC—Naval Air Basic Training Center—USN

NAF—Naval Air Field—USN

NAMAR—Northern Air Materiel Area—USAF

NART—Naval Air Reserve Training—USN

NAS—Naval Air Station—USN

NAT—Navigator Training—USAF unit

NATC—Naval Air Test Center—USN

NEA(C)—Northeast Air Command—USAF Major Command

Nfd—No further details

NMPAR—Northern Air Materiel Area, Pacific—USAF

NPRO—Naval Plant Representative Office—USN

NTN—Navigator Training—USAF unit

Ntu—Not taken up

NWI—Northwest Industries Ltd.—Civil contractor

O&R—Overhaul & Repair—USMC

OGNAR—Ogden Air Materiel Area—USAF

OKLAR—Oklahoma Air Materiel Area—USAF

OP / OPS—Operational / Operations—Military term

OTC—Operational Test Center—USAF unit

OUT—Operational Training Unit—RCAF unit

PBL—Pilotless Bomb (Light)—USAF unit

PCN—Process Control Number—USAF-MASDC term

PDT—Parachute Development Test—USAF unit

PFC—Private First Class—Military rank

PT / PTS—Proof Test—USAF unit

PTN—Pilot Training—USAF unit

Pvt—Private—Military rank

RAE—Royal Aircraft Establishment—British RAF unit

RATO—Rocket Assisted Take-Off

RC—Reconnaissance—USAF unit

RCAF—Royal Canadian Air Force

RCM—Radar Countermeasures—USAF unit

RD—Repair Depot—RCAF unit

RDP—Rome Air Depot—USAF

REC—(Assigned to) Reclamation—Military term

redes—Redesignated

reg—Registration—Civil term

re-reg—Re-registered—Civil term

RES—Air Rescue—USAF unit / term

ret—Retired

retd—Retained

RFB—Aerial Refueling, Fighter-Bomber—USAF unit

RFC—(AF) Reserve Flying Center—AFR unit

RJAF—Royal Jordanian Air Force

RMAF—Royal Moroccan Air Force

RNAF—Royal Norwegian Air Force

ROD—Rome Air Development Center—USAF organization

ROMAR—Rome Air Materiel Area—USAF

rsvd—Reserved

RU—Requirements Unit—RCAF unit

SAAAR—San Antonio Air Materiel Area—USAF

SABBE—Sabena Airlines, Brussels, Belgium—Civil contractor

SAC—Strategic Air Command—USAF Major Command
SAL—Salvage
SAR—Search & Rescue—Military term
SBNAR—San Bernardino Air Materiel Area—USAF
Scr—Scrapped
SEA—South East Asia (Theatre)
Sgt—Sergeant—Military rank
SM—Strategic Reconnaissance (Medium)—USAF unit
SMAAR—Sacramento Air Materiel Area—USAF
SMPAR—Southern Pacific Air Materiel Area—USAF
SMT—Strategic Reconnaissance Test (Center)—USAF unit
SMW—Strategic (Medium) Weather—USAF unit
SO&ES—Station Operation & Engineer Squadron—USMC
SOC—Struck off charge—Military term
SOMAR—Southern Pacific Air Materiel Area—USAF
SOP—Special Operations—USAF unit
SPW—Special Weapons (Center)—USAF organization
SQ—Squadron—Military term
SRH—Strategic Reconnaissance (Heavy)—USAF unit
SRL—Strategic Reconnaissance (Light)—USAF unit
SSgt—Staff Sergeant—Military rank
STA—Strategic Aerospace—USAF unit
STM—Strategic Missile—USAF unit
STR—Strategic—USAF unit
SUT—Support—USAF unit
SWC—Special Weapons Center—USAF unit
TAC—Tactical Air Command—USAF Major Command
TAL—Tactical Airlift—USAF unit
TC—Air Training Command—USAF Major Command
TCA—Troop Carrier, Assault—USAF unit
TCG—Troop Carrier Group—USAF

TCSq—Troop Carrier Squadron—USAF unit
TCWG—Troop Carrier Wing—USAF
TES—Test Support—USAF unit
Tf—Transferred (to)
TFG—Tactical Fighter—USAF unit
TFW—Troop Carrier, Fixed-Wing—USAF unit
TNG—Training
TNP—Tactical Reconnaissance (Night-Photo)—USAF unit
TPJ—Tactical Reconnaissance (Photo, Jet)—USAF unit
TR—Tactical Reconnaissance—USAF unit
tranmnt—transient maintenance
TS—Tactical Support—USAF unit
TSD—Technical Support Depot—RCAF unit
TSgt—Technical Sergeant—Military rank
TSH—Air Transport (Heavy)—USAF unit
TSM—Air Transport, Medium—USAF unit
TSP—Air Transport—USAF unit
TSQ—Transport Squadron—VNAF unit
TST—Tactical Support—USAF unit
TSU—Technical Support Unit—RCAF unit
TTA—Technical Training—USAF unit
TTN—Technical Training—USAF unit
TWG—Tactical Wing—VNAF Wing
u/c—Undercarriage
UC / VC—Fleet Composite Squadron—USN
Unk—Unknown
USAF—United States Air Force
USAFE—United States Air Force in Europe—USAF Major Command
USMC—United States Marine Corps.—USMC
USN—United States Navy—USN
UT—Unit—Military term
VFR—Visual Flight Rules
VMGR—Marine Aerial Refuel Transport Squadron—USMC
VMR—Marine Transport Squadron—USMC
VNAF—South Vietnamese Air Force
VPAF—Vietnam People's Air Force
VR—Air Transport Squadron—USN

WDD—Wright (Air) Development Division—USAF organization

WEA—Weather—USAF unit

WER—Weather Reconnaissance—USAF unit

Wfu—withdrawn from use

WG—Wing—Military term

w/o—Written-off

WPAFB—Wright-Patterson Air Force Base—USAF

WRAAR—Warner-Robins Air Materiel Area—USAF

WRD DV—Wright (Air) Development Division—USAF organization

WRDCN/CE—Wright (Air) Development Center—USAF organization

Introduction

"The Flying Boxcar," "Twin Can Spam Can," "Pregnant P-38," "Dollar Nineteen," "Packplane," "Shadow" and "Stinger" are some of the names bestowed on the twin-boom Fairchild C-119 cargo and troop transport. The colloquial term "Flying Boxcar" became so synonymous with the aircraft that the USAF made it official when the new aircraft entered service in 1949, replacing its earlier sibling, the C-82A Packet. The C-119 bore the brunt of United States military transport logistics in the later years of aviation's piston-powered era, working tirelessly in freight transport, combat cargo drops, paratroop transport and aeromedical evacuation. In later years the Boxcar served as an armed gunship and even had some limited role in covert operations.

The predecessor to the C-119 was the Fairchild C-82A Packet of which 224 were built from 1944 to 1948. They remained in service until 1954 when sufficient numbers of C-119s were available to fill USAF troop carrier wings (USAF).

In terms of numbers built, 1,185 airframes were produced from 1948 to 1955, far surpassing any of the other larger category piston-powered transports built in the immediate years after World War II. When compared to the table below, it can be seen what an essential asset the C-119 became during the 1950s:

Introduction

1943	Lockheed C-69 / C-121	169
1944	Boeing C-97	888
1945	Douglas C-74	14
1946	Douglas C-118 / R6D	168
1947	Fairchild C-119 / R4Q	1185
1949	Fairchild C-123	307
1949	Douglas C-124	447
1949	Convair C-131 / T-29 / R4Y	512

The Boxcar served valiantly with U.S. forces from 1949 to 1975 in the Continental United States, Alaska, the Far East and Europe during the darkest days of the Cold War. They fought in two wars, firstly the Korean War, 1950–1953, and then the Vietnam War, 1968–1973. Service was also undertaken by the USAF Reserve, the Air National Guard and the United Sates Marine Corps. MAP aid straight from factory saw deliveries to Belgium, Canada, India and Italy during the 1950s. Many C-119s saw further service post–USAF through MAP aid to Brazil, Ethiopia, Jordan, Morocco, Norway, Taiwan and South Vietnam. Many were "loaned" by the U.S. to French forces during 1953–1954 and saw action at the infamous Battle of Dien Bien Phu—a now-historical event that marks one of the C-119's career highlights.

Not until the turboprop Lockheed C-130 Hercules entered service in the late 1950s was the C-119 surpassed in both numbers built and in mission diversity. Once front-line

Long lines of troop carrying C-119 Flying Boxcars were a common sight throughout the United States and in Europe during the 1950s (Nicholas A. Veronico Collection).

Introduction 11

C-130 squadron strength was achieved, the entire C-119 fleet was relegated to the USAF Reserve starting in 1957, so ending a nine-year leading role. The big jets arrived a short time later with the C-135, C-141 and C-5 making the heady days of C-119 service all but a distant memory.

Due to high operating costs, lack of spares and simply age, the C-119 Flying Boxcar had only limited service life in civilian clothing. Most were employed in fighting U.S. forest fires, converted to airtankers from the 1970s to the late 1980s. Some saw a colorful life in Alaska transporting freight and fish stocks to and from remote settlements and several ended their lives in this inhospitable part of the world as "wilderness wrecks" due to accidents and mishaps.

Today, the Flying Boxcar is remembered for its tough ruggedness and ability to get the job done in any part of the globe. It was a Cold War warrior that got on and got the mission completed without fuss or fanfare. The USAF Museum Program has done an outstanding job in preserving these aircraft, and today there are 42 surviving examples on public display at either USAF owned or privately operated museums within the United States. There are more than 30 more examples displayed around the world at various locations, mainly museums but also in public parks and on military parade grounds.

Many of the C-119C variant went to the Korean War like these 314th TCG examples at Yonpo, South Korea, in December 1950. Serials are 49-138 (10375), 49-129 (10366) and 49-139 (10376) (USAF).

The aeronautical legacy of the C-119 Flying Boxcar can best be attributed through the following accomplishments:

1. A ten-year front-line service across seventeen troop carrier wings with the USAF followed by a fifteen-year front-line service across fourteen wings with the USAF Reserve.
2. The contribution to the Korean War especially in the re-supply of troops during the Battle of Chosin Reservoir in late 1950.
3. The re-supply missions undertaken during the Battle of Dien Bien Phu in 1953–54 under French and American civilian flight crews.

Introduction

4. The use of the C-119J in the recovery of Corona spy satellites in the late 1950s and early 1960s.
5. As the definitive AC-119G and AC-119K gunship conversions for the Vietnam War.
6. The prominent use as USFS airtankers during the 1970s and 1980s.

C-119 Airtankers ready for action (Michael Prophet Collection).

Section One

Technical Data

Manufacturer: The Fairchild Engine and Airplane Corp., Aircraft Division, Hagerstown, Maryland, USA
Licensee Manufacturer: Kaiser Manufacturing Corp. (KMC), Willow Run, Ypsilanti, Michigan
Model Numbers: M-105A (XC-119A); M-107B (XC-120); M-110 (C-119); M-127 (R4Q-2); M-142 (YC-128A); M-156 (YC-128B); M-160 (YC-119H); M-203 (C-119J); M-484 (C-119K)
Designations: C-119 (USAF); R4Q (USMC); XC-120; YC-128
Official Name: Flying Boxcar
Other Names: Packplane (XC-120); Skyvan (YC-119H); Shadow (AC-119G); Stinger (AC-119K)
First Official Flight: December 17, 1947
Production Period: October 1948–December 1955
Production Contracts:

  W33-38 AC-19200
    Fairchild / November 25, 1947
    36 C-119B   48-319 / 48-329; 48-331 / 48-355
    1 XC-120   48-330
    18 C-119B   49-101 / 49-118
    81 C-119C   49-119 / 49-199
    8 R4Q-1   124324 / 124331
    53 C-119C   50-119 / 50-171
    31 R4Q-1   126574 / 126582; 128723 / 128744
    53 C-119C   51-2532 / 51-2584
    1 YC-119H   51-2585
    1 YC-119F   51-2586
    75 C-119CF 51-2587 / 51-2661
    56 C-119F   51-2662 / 51-2717 (18 MDAP)

  AF 33 (038)-18481
    Kaiser / December 20, 1950
    71 C-119F   51-8098 / 51-8168

  AF 33 (038)-18499
    Fairchild / December 22, 1950
    58 R4Q-2   131662 / 131719
    85 C-119F   51-7968 / 51-8052
    45 C-119G   51-8053 / 51-8097
    32 C-119F   (RCAF) 22104 / 22135
    41 C-119CF 51-8233 / 51-8273

3 C-119G　　51-17365 / 51-17367 (MDAP)
　　59 C-119G　　52-6000 / 52-6058 (MDAP)
　115 C-119G　　52-5840 / 52-5954
　　　2 C-119G　　52-9981, 52-9982

AF 33 (600)-22285
　Fairchild / August 2, 1952
　146 C-119G　　53-3136 / 53-3222; 53-7826 / 53-7884

AF 33 (600)-23919
　Fairchild / May 1, 1953
　　26 C-119G　　53-4637 / 53-4662 (MDAP)

AF 33 (600)-26021
　Fairchild / August 10, 1953
　　88 C-119G　　53-8069 / 53-8156

Total Built: 1,185—1,114 by Fairchild at Hagerstown
　　　　　　　　　　　　71 by Kaiser at Willow Run

Customer Deliveries:

USAF	C-119B / C / F / G	945
	XC-120	1
	YC-119H	1
USMC	R4Q-1 / -2	97
Belgium	C-119F / G	40
Canada	C-119F	35
India	C-119G	26
Italy	C-119G	40

U.S. Military Service Periods:

USAF	1949–1961
USAF Reserve	1954–1973
ANG	1958–1975
USMC	1950–1963
USMC Reserve	1961–1975

MAP / Foreign Operators: Belgium, Brazil, Canada, Ethiopia, France (USAF loans), India, Italy, Jordan, Morocco, Norway, Taiwan, United Nations, South Vietnam

Conflicts:
　The Korean War (1950–1953)
　The First Indochina War (1945–1955)
　The Suez Crisis (1956)
　The Lebanon Crisis (1958)
　The Congo Crisis (1960–1965)
　The Indo-Pakistani War (1965)
　The Vietnam War (1955–1975)

Description

　A twin-engine, twin-boom, high-mounted inverted gull wing, land monoplane of all metal construction with a tricycle undercarriage. The fuselage is of a semi-monocoque design which comprises frames, bulkheads, cross beams, stringers and skin all made up of 24ST and 75ST strength alclad of various thickness. The entire main wing from tip to tip is made up of three sections comprising the main center section and two outer wing sections. The main center section is of a cantilever structure of 24ST, 24SRT and 75ST strength alclad. The booms are made up of a semi-monocoque design with frames and skin made from 24ST alclad and stringers of 75ST alclad. A tricycle undercarriage design consists of two main landing struts with dual wheels and a steerable single nose wheel that was later upgraded to a dual wheel layout. Crew accommodation consists of five—Pilot (Aircraft Commander), Co-Pilot, Navigator, Radio Operator and Crew Chief / Flight Engineer.

Design Missions

Cargo Transport—wheeled vehicles up to 2½ ton trucks and artillery up to 75mm howitzers, cradled aircraft engines, propellers and a wide variety of other freight.

Troop / Paratroop Transport—42 troops with equipment can be carried with 20 folding seats on the port side and 22 seats on the starboard side fuselage. An additional 20 extra seats can be installed in the cargo hold center-line for a total of 62 troops or paratroops.

Aerial Equipment Drop—fuselage floor paratrainer doors can drop up to twenty 500 lb bundles in ten seconds via an internal roof mounted rail system. Large cargoes can be aerial delivered via parachute by removing the rear clamshell doors.

Aero-Medical Evacuation—35 litter patients can be carried in the cargo hold area and in emergency evacuation situations up to 76 troops may be carried—62 seated and 14 litter requirements.

Other C-119 missions include second-line unit support, satellite recovery, electronic warfare, VIP transport, covert reconnaissance and aerial gunship.

Dimensions / Areas

Overall Length:	86 feet 5 inches	26.36 meters
Overall Width:	109 feet 3 inches	33.30 meters
Overall Height:	26 feet 2 inches	7.99 meters
Fuselage Length:	60 feet 6 inches	18.45 meters
Fuselage Width:	11 feet 6 inches	3.50 meters
Fuselage Height:	13 feet 6 inches	4.11 meters
Cargo Hold Length:	36 feet 11 inches	11.25 meters
Cargo Hold Width:	9 feet 2 inches	2.79 meters
Cargo Hold Height:	8 feet	2.44 meters
Cargo Hold Deck Area:	353 sq. feet	32.80 sq. meters
Overall Wing Area:	1559.24 sq. feet	144.86 sq. meters

WEIGHTS

Design Gross Weight: 64,000 lb (29,029 kg)

	Empty	Basic Operating*	Fuel / Payload**	Max T/O***
C-119B:	37,691 lb (17,096 kg)	43,000 lb (19,504 kg)	25,000 lb (11,339 kg)	68,000 lb (30,844 kg)
C-119C:	39,942 lb (18,117 kg)	43,000 lb (19,504 kg)	30,150 lb (13,675 kg)	73,150 lb (33,180 kg)
R4Q-1:	39,942 lb (18,117 kg)	43,000 lb (19,504 kg)	30,150 lb (13,675 kg)	73,150 lb (33,180 kg)
C-119F:	39,809 lb (18,057 kg)	43,000 lb (19,504 kg)	34,000 lb (15,422 kg)	77,000 lb (34,926 kg)
R4Q-2:	39,809 lb (18,057 kg)	43,000 lb (19,504 kg)	34,000 lb (15,422 kg)	77,000 lb (34,926 kg)
C-119G:	39,982 lb (18,135 kg)	43,000 lb (19,504 kg)	34,000 lb (15,422 kg)	77,000 lb (34,926 kg)
---	---	---	---	---
XC-120: (Pack-Off)	37,572 lb (17,042 kg)	—	16,428 lb (Fuel) (7,451 kg)	54,000 lb (24,493 kg)
XC-120: (Pack-On)	—	—	20,000 lb (9,071 kg)	74,000 lb (33,565 kg)
YC-119H:	46,455 lb (21,071 kg)	—	30,000 lb (13,607 kg)	85,900 lb (38,963 kg)
---	---	---	---	---
AC-119G:	47,981 lb (21,763 kg)	51,226 lb (23,235 kg)	—	69,478 lb (31,514 kg)
AC-119K:	60,277 lb (27,341 kg)	68,450 lb (31,048 kg)	—	77,000 lb (34,926 kg)

* Average weight across all variants in C-119 flight manuals.
** Combined totals. Maximum payload limit dependent on minimum allowable fuel load and variant operating weight.
*** Maximum limit for a 100 fpm climb on one engine at sea level under standard atmosphere conditions.

ENGINES

Pratt & Whitney R-4360 Wasp Major
28 cylinder, 4 row, supercharged, air-cooled radial.
XC-119A: **R-4360-35** 3,250 hp (Dry)
C-119B: **R-4360-20** 3,250 hp (Dry)
C-119C / R4Q-1: **R-4360-20W / -20WA** 3,250 hp (Dry) / 3,500 hp (Wet)
- "A" denotes long rods.

Wright R-3350 Duplex Cyclone
18 cylinder, 2 row, turbo-compound (3x power recovery turbine), supercharged, air-cooled radial.
C-119F / C-119G: **R-3350-85** 3,250 hp (Dry) / 3,500 hp (Wet)
- Designated **-89** when modified with low-ratio turbine drives and armored exhaust hoods.
- Redesignated **-89A** when fitted with improved torque meter system.
- Redesignated **-89B** when fitted with manual spark control.

R4Q-2: **R-3350-30WA** 3,250 hp (Dry) / 3,500 hp (Wet)
- Designated **-36W** when modified with low-ratio turbine drives and armored exhaust hoods.

Propellers

Hamilton Standard
15ft diameter hydromatic, constant speed, full feathering, reversible pitch.

	Config.	Designation
XC-119A:	3 blades	—
C-119B:	4 blades	2H17Q3-26R
R4Q-1:	4 blades	2H17F3-24R
C-119C:	4 blades	2H17Q3-26R
		2J17G3-26R (from s/n: 51-2532)
C-119F:	4 blades	2J17G3-26R
C-119L:	3 blades	43H60-603

Aeroproducts
15 ft diameter constant speed, full feathering, reversible pitch.

	Config.	Designation
C-119G:	4 blades	A644FN-C1 / -C2

Fuel Capacity—Total Volume

Fuel Grade: 115/145 Octane (Alternate: 100/130)
Density: 6lb per U.S. Gal

	Tank Config.	Weight	Capacity
C-119B / C:	4 self-sealing tanks	16,428 lb	2,738 U.S. Gal
R4Q-1 / -2:	4 bladder tanks	17,412 lb	2,902 U.S. Gal
C-119F / G:	4 self-sealing tanks	16,224 lb	2,704 U.S. Gal

Oil Capacity

All variants have a two tanks, one in each nacelle behind the engine firewall and each with a total volume of 60 U.S. Gallons running grade 1100 oil.

Electrical System

The main electrical system is a 28 volt DC, one being supplied by two engine-driven 300 ampere generators. A 24 volt, 36 ampere storage battery is located under the rear cargo hold floor and a 28 volt, 175 ampere combustion engine-driven Auxiliary Power Plant is located in the top of the fuselage equipment area. The DC supply powers the engines, flaps, undercarriage, onboard lights and other equipment. For AC powered equipment such as flight instruments, autopilot and radios, 28 volt DC power is supplied to five 115 volt, 400 cycle single- and three-phase inverters. An external DC ground power receptacle is located on the right side of the aircraft. From production aircraft 51-2586 (10575) and up the undercarriage and wing flaps are not electrically but hydraulically actuated.

Hydraulic System

On production msn: 10301 / 10574 & 7001 / 7008, aircraft are equipped with a 980 to 1150 psi hydraulic system consisting of a 4.41 U.S. Gallon reservoir, electric pump, auto pressure switch and an accumulator. This system drives the nose wheel steering, main wheel brakes and elevator gust lock.

From production aircraft msn: 10575 and up, including all Kaiser builds, aircraft are equipped with a heavy duty 3,000 psi (high pressure) system for the undercarriage and flaps then the usual 980 psi (low pressure) system for the nose wheel steering, main wheel brakes and elevator gust lock. The high pressure system features a 9.3 U.S. Gallon reservoir oil tank.

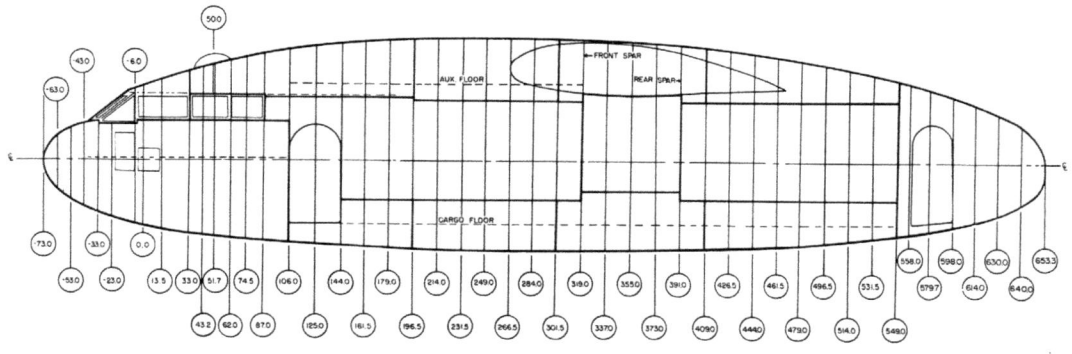

C-119C fuselage stations (USAF).

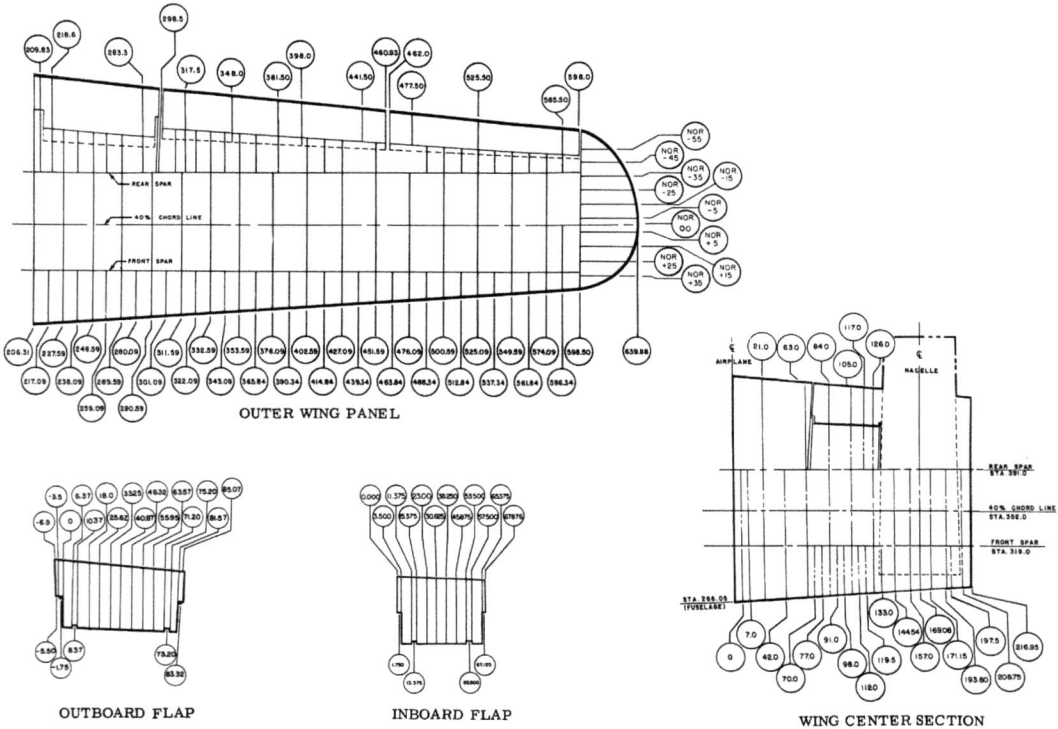

C-119C main wing stations (USAF).

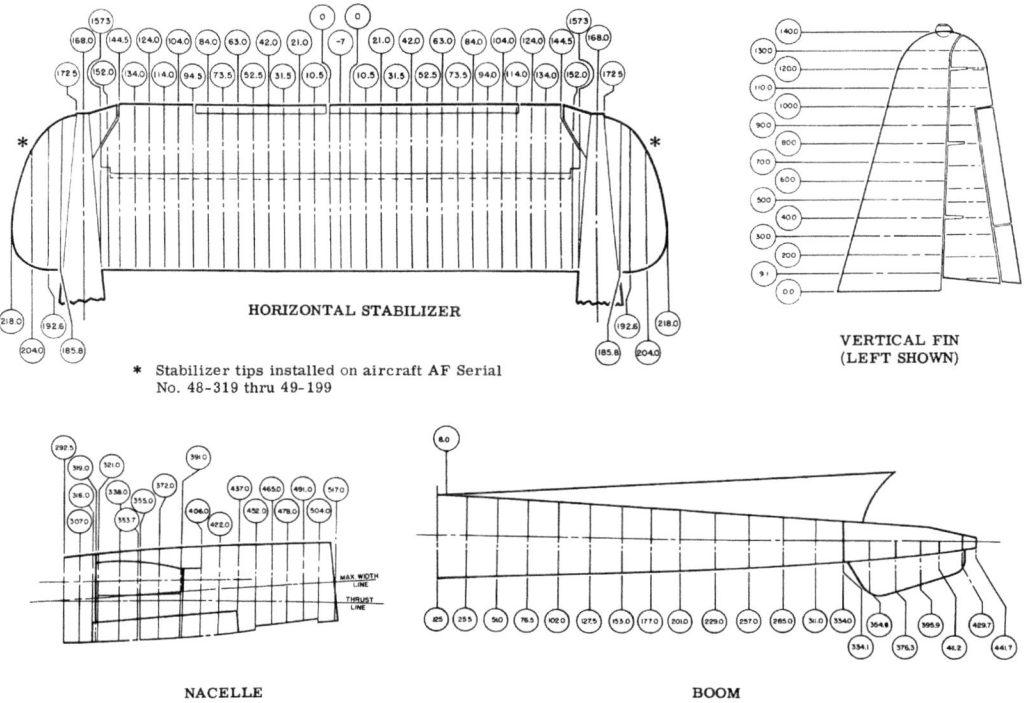

C-119C tail section stations (USAF).

Variants

Main Variants & Factory Prototypes

XC-119A

A single prototype begun as a feasibility project designated as the XC-82B (Model 105A) and using C-82A s/n 45-57769 (msn: 10139) as a base airframe to test a higher capacity and better performing version of the C-82 Packet. Engines were two 2,650hp P&W R-4360-35 Wasp Major radials and the flight-deck section was extensively redesigned to form a single-deck configuration. The XC-119A first flew on December 17, 1947, before delivery to the USAF in June 1948 redesignated as the C-119A. The ventral fins were later removed finalizing the initial C-119 tail layout.

C-119B

Fairchild Model M-110. The first main production variant powered by 3,250hp 28 cylinder P&W R-4360-20 radials with 4-bladed propellers that also included a 4 ft boom extension, strengthened booms, main wings and upgraded dual main wheel undercarriage with a steerable nose wheel. The C-119B also had a 14 inch wider fuselage to increase capacity and carry newer army equipment. The first C-119Bs entered regular service with the 314th TCG in late 1949. A total of 54 were delivered from June 1949 to March 1950 with 40 of these later upgraded to C-119C standard to include water injected radials and dorsal fins. Several of this batch remained in service with the USAF Reserve until 1967.

The prototype XC-119A was built from the 137th production C-82A s/n 45-57769 (10139) with more powerful radial engines and a radically redesigned flight-deck section (USAF).

The new XC-119A in-flight showing to good effect the new flight-deck design and longer engine nacelles for the R-4360 radials. This prototype retained the original fuselage, main wing, tail sections and undercarriage of the earlier C-82A Packet (USAF).

C-119B Production Summary

S/n	Qty	Delivered	Customer
48-319 / 48-329	11	02Jun49–28Dec49	USAF
48-331 / 48-355	25	20Dec49–04Apr50	USAF
49-101 / 49-118	18	22Dec49–13Mar50	USAF

C-119C & C-119CF

A major variant introduced in 1950 from the 56th airframe on the production line featured upgraded 3,500hp P&W R-4360-20WA water injected radial engines. A total of 303 were delivered to the USAF from April 1950 to May 1953, many being assigned to the Far East Air Forces Command during the Korean War. From s/n 50-119 dorsal fins were

Production C-119B s/n 48-332 (10314) sporting a new 14-inch wider cargo hold over the C-82A, twin main landing gear wheels and deleted ventral fins (USAF).

added and the small stabilizer tips deleted with all previous 81 C-119Cs getting the dorsal fins retro-fitted during regular maintenance.

The C-119CF was an unofficial designation to denote "C" configured aircraft which were fitted with "F" variant hydraulic actuated wing-flaps and undercarriage. While the air force officially retained the C-119C designation for these aircraft, ground and maintenance crews referred to them as the "CF" for servicing reasons. All 116 C-119C serialed 51-2587 thru 51-8273 were C-119CF configured. The last 41 serials, 51-8233 / 51-8273, were a Kaiser batch re-assigned to Fairchild when Kaiser had their contract numbers reduced. The last C-119C Boxcars were retired from the USAF Reserves in 1969.

Four C-119C, s/n 50-135, 50-139, 50-143, 50-147, were used to conduct tandem undercarriage tests during 1951. Each main gear strut was fitted with twin axles each with two wheels to determine rough and uneven field capabilities. The system was not developed and the Boxcars were retro-fitted in April 1953. Two C-119C, s/n 49-179, 51-2554, were used for RATO tests creating an extra 12,000 lb of thrust for short take-offs.

C-119C Production Summary

S/n	Qty	Delivered	Customer
49-119 / 49-199	81	04Apr50–08Dec50	USAF
50-119 / 50-171	53	09Dec50–04Jul51	USAF
51-2532 / 51-2584	53	20Jul51–05Jan52	USAF
51-2587 / 51-2661	75	27Mar52–09Jul52	USAF
51-8233 / 51-8273	41	04Feb53–04May53	USAF

C-119B converted to C-119C

Birmingham Modification Center & Fairchild / 1955–1959: 48-320, 48-322, 48-323, 48-325, 48-326, 48-327, 48-328, 48-329, 48-331, 48-332, 48-334, 48-335, 48-337, 48-339,

48-340, 48-341, 48-343, 48-344, 48-346, 48-347, 48-349, 48-351, 48-352, 48-354, 48-355, 49-101, 49-102, 49-104, 49-106, 49-107, 49-108, 49-109, 49-110, 49-111, 49-112, 49-113, 49-115, 49-116, 49-117, 49-118.

C-119G converted to C-119C
Temp cvtd 1964: 51-8060, 51-8106.

C-119CF s/n 51-8255 (10803) at Oakland, California, in July 1958. Note the standardized tail fittings but retention of the single nose-wheel (Nicholas A. Veronico Collection).

C-119B s/n 49-102 (10339) was one of forty of the original "B" variants converted to "C" standard with water-injected radials and dorsal fins but with the stabilizer tips retained (USAF).

YC-119D

Proposed 1951 service test variant featuring 3-wheel main undercarriage units with a detachable fuselage pod powered by P&W R-4360 radials. C-119C s/n 50-162 was earmarked but the project was canceled before conversion work began. Original designation was YC-128A.

YC-119E

Proposed 1951 version of the earlier YC-119D but with Wright R-3350 radials. C-119C s/n 50-171 and 51-2617 were to be converted but the project was canceled. Original designation was YC-128B.

YC-119F

The single unofficially designated YC-119F s/n 51-2586 (msn: 10575) was a working prototype for the 18 cylinder 3,500 hp Wright R-3350-85 Duplex-Cyclone radial engine which also featured three power recovery turbines, a two-stage supercharger and water injection. Hydraulically operated wing flaps and undercarriage was another new upgrade to this prototype which became available on January 7, 1952. After service test trials it entered regular USAF service upgraded to C-119G standard and was eventually disposed of via MAP to Taiwan in April 1969.

C-119F

This variant became the new production standard from July 1952 incorporating the powerplant and systems changes introduced on the YC-119F prototype. A total of 173, including 35 for the RCAF, were built from July 1952 to August 1953 by Fairchild with an additional 71 delivered by Kaiser-Frazer from their giant Willow Run plant from November 1952 to August 1953. Notable changes from previous variants are the now three, slightly narrower, exhaust stacks for the R-3350 radials with the stack pipes "biting" into the nacelle cowl flaps. The first deliveries retained the C-119C tail layout but from s/n 51-7968 (msn: 10707) in September 1952, ventral fins were introduced at factory and retro-fitted to all preceding C-119F and C-119C configured aircraft. Two hundred thirty-three C-119Fs were later upgraded to "G" standard including the 49 converted to the C-119J.

C-119F Production Summary

S/n	Qty	Delivered	Customer
51-2662 / 51-2686	25	17Jul52–29Aug52	USAF
51-2690 / 51-2707	18	28Sep52–14Oct52	Belgium
51-2708 / 51-2717	10	24Sep52–03Dec52	USAF
51-7968 / 51-8052	85	06Oct52–15Jun53	USAF
51-8098 / 51-8168	71	18Nov52–26Aug53	USAF
(RCAF) 22101 / 22135	35	27Aug52–27Aug53	Canada

C-119F s/n 51-8146 (KMC-149) seen at Oakland, California, in September 1956 after being upgraded to "G" standard with Aeroproducts propellers. This colorful livery was applied while the Boxcar served as a support aircraft to the USAF *Thunderbirds* aerobatic team (Nicholas A. Veronico Collection).

C-119G

The definitive variant of the Flying Boxcar series was the C-119G, the first of which was completed during November 1952 with deliveries starting in early 1953. The C-119G was identical to the earlier C-119F except for the use of Aeroproducts A644FN-C1 and later -C2 4-bladed propellers and some minor equipment upgrades. A total of 484 would be built up to December 1955 with 233 of the earlier C-119F variant also upgraded to this standard—the first completed on December 27, 1954. 88 C-119Gs, s/n 53-8069 / 53-8156, meant to be built by Kaiser-Frazer were instead built by Fairchild at Hagerstown over a twelve-month period starting in April 1954. The original Kaiser factory numbers, c/n KF-172 / KF-257, were retained but stamped onto Fairchild data-plates within the aircraft. Late production line "G" aircraft introduced the co-rotational (twin) nose wheel which was then retro-fitted to all earlier "C" and "G" variants already in service.

C-119G Production Summary

S/n	Qty	Delivered	Customer
51-8053 / 51-8097	45	28May53–26Aug53	USAF
51-17365 / 51-17367	3	08May53–06Jun53	Italy
52-5840 / 52-5954	115	02Sep53–18Mar54	USAF
52-6000 / 52-6058	59	19May53–18Mar54	Italy (37), Belgium (22)
52-9981, 52-9982	2	18Mar54	USAF
53-3136 / 53-3222	87	31Mar54–07Feb55	USAF
53-4637 / 53-4662	26	28Jan54–21Oct55	India
53-7826 / 53-7884	59	04Feb55–05Dec55	USAF
53-8069 / 53-8156	88	01Apr54–29Apr55	USAF

C-119F converted to C-119G

Birmingham Modification Center & Fairchild / 1954–1957: 51-2586, 51-2662, 51-2663, 51-2664, 51-2666, 51-2667, 51-2668, 51-2669, 51-2670, 51-2671, 51-2672, 51-2673, 51-2674, 51-2675, 51-2676, 51-2677, 51-2678, 51-2680, 51-2681, 51-2682, 51-2683, 51-2684, 51-2685, 51-2686, 51-2708, 51-2709, 51-2710, 51-2711, 51-2712, 51-2713, 51-2714, 51-2715, 51-2716, 51-7969, 51-7970, 51-7971, 51-7972, 51-7973, 51-7974, 51-7975, 51-7976, 51-7977, 51-7978, 51-7979, 51-7980, 51-7981, 51-7982, 51-7983, 51-7984, 51-7985, 51-7986, 51-7987, 51-7988, 51-7989, 51-7990, 51-7991, 51-7992, 51-7994, 51-7995, 51-7996, 51-7997, 51-7998, 51-7999, 51-8000, 51-8001, 51-8002, 51-8003, 51-8004, 51-8005, 51-8006, 51-8007, 51-8008, 51-8009, 51-8010, 51-8011, 51-8012, 51-8013, 51-8014, 51-8015, 51-8016, 51-8017, 51-8018, 51-8019, 51-8020, 51-8021, 51-8022, 51-8023, 51-8024, 51-8025, 51-8026, 51-8027, 51-8028, 51-8029, 51-8031, 51-8032, 51-8033, 51-8034, 51-8044, 51-8047, 51-8048, 51-8098, 51-8099, 51-8100, 51-8101, 51-8102, 51-8103, 51-8105, 51-8106, 51-8107, 51-8108, 51-8109, 51-8110, 51-8111, 51-8112, 51-8117, 51-8120, 51-8135, 51-8136, 51-8139, 51-8143, 51-8145, 51-8146, 51-8147, 51-8148, 51-8149, 51-8150, 51-8160, 51-8161, 51-8162, 51-8166, 51-8168.

Northwest Industries Ltd., Canada / 1955–1957: (RCAF) 22101 / 22135.

SABENA, Belgium / 1956, 1959: 51-2690 / 51-2707.

C-119F converted to C-119G standard as C-119J

1957–1958: 51-7968, 51-8030, 51-8035, 51-8036, 51-8037, 51-8038, 51-8039, 51-8040, 51-8041, 51-8042, 51-8043, 51-8045, 51-8046, 51-8049, 51-8050, 51-8051, 51-8052, 51-8113, 51-8114, 51-8115, 51-8116, 51-8119, 51-8121, 51-8122, 51-8123, 51-8124, 51-8125, 51-8126, 51-8127, 51-8128, 51-8129, 51-8130, 51-8131, 51-8132, 51-8134, 51-8137, 51-8138, 51-8140, 51-8141, 51-8144, 51-8152, 51-8154, 51-8155, 51-8156, 51-8157, 51-8158, 51-8159, 51-8164, 51-8167.

Section One—Technical Data

C-119G s/n 53-3139 (11150) in October 1959 while serving with the 434 TCWG. "139B" indicates this is the second C-119 with the nose code "139" in the same unit. The other was 51-8139 which would likely be coded "139A" (Nicholas A. Veronico Collection).

YC-119H Skyvan

Fairchild Model M-160. An experimental variant built from C-119C s/n 51-2585 (msn: 10574) which first flew on May 27, 1952. The design aim was to improve the performance and stability problems encountered in the standard C-119 design. The fuselage section remained unchanged, but the wings and tail areas were greatly redesigned with larger areas and the booms were lengthened. The engines were two 3,500 hp Wright R-3350-32W Duplex Cyclone radials and all fuel was now carried in two underwing external tanks. The new aircraft was to carry a 16,000 lb load up to a 1,000-mile radius. Flight tests proved positive with proposals for up to 195 C-119H Skyvans to be built at a new plant in Chicago. However, the Air Force canceled any further development and the single YC-119H was retired to Sheppard AFB in September 1954 for ground instructional duties.

YC-119H Skyvan s/n 51-2585 (10574) attempted to improve payload / range combinations plus improve the stability issues encountered with the standard Boxcar design. Note the extensively redesigned tail surfaces and wider wingspan (Gerald Balzer Collection, American Aviation Historical Society image archive).

YC-119H Basic Differences

	C-119C	YC-119H
Engines:	R-4360-20WA	R-3350-32WA
Propellers:	4-bladed	4-bladed
Overall Length:	86 ft 6 in	95 ft 8 in
Overall Wing Span:	109 ft 3 in	148 ft (36% greater area)
Overall Height:	26 ft 8 in	31 ft
Max Weight:	74,000 lb	86,000 lb

XC-120 Packplane

Fairchild Model M-107 with new c/n 8001. Prototype built from C-119B s/n 48-330 (msn: 10312) which features a detachable mission-interchangeable cargo pod, it first flew on August 11, 1950, before undertaking a much publicized flight test program. The airframe wing, tail and boom sections remained unchanged except for the addition of dorsal fins. The fuselage section was completely redesigned to feature a high-mounted flight-deck with the large bulbous pod attached underneath. The main undercarriage was also redesigned to feature 4-wheel long-tracked struts to keep the aircraft balanced on the ground, this necessitated a lengthening of the engine nacelles forward of the main wings. The detachable pod itself was supported on the ground by four castor wheels allowing quick removal or installation to the parent airframe. The engines were the standard C-119B R-4360-20 units. The XC-120 had its time in the lime-light due to is strange and unique appearance but was later declared obsolete and was assigned to reclamation in August 1954.

The one and only XC-120 Packplane s/n 48-330 (10312) with the multi-purpose detachable fuselage pod seemed a worthy concept on paper but proved impractical in reality (Gerald Balzer Collection, American Aviation Historical Society image archive).

R4Q-1

USMC version of the C-119B/C. The first eight delivered in early 1950 were built to C-119B standard but had a specially redesigned instrument panel suitable for USMC operations, bladder fuel tanks and new c/n: 7001 / 7008 were applied to this order. The remaining 31 were delivered from December 1951 thru March 1952 to C-119C standard with the -20W engines and dorsal fins. The last R4Q-1s were retired into NAS Litchfield Park by November 1959. The DoD Unified Designation System introduced in September 1962 redesignated the R4Q-1 as the C-119C but this was only a clerical application for the few

Section One—Technical Data

still stored at Litchfield Park and never used operationally. The R4Q-1 was retro-fitted with ventral fins, but most retained the single nose-wheel.

R4Q-1 Production Summary

BuNo.	Qty	Delivered	Customer
124324 / 124331	8	08Feb50–10Mar50	USMC
126574 / 126582	9	11Dec51–04Jan52	USMC
128723 / 128744	22	14Feb52–20Mar52	USMC

R4Q-1 BuNo. 124330 (7007) with VMR-252 at MCAS Cherry Point in April 1950. It was one of eight built for the USMC to C-119B standard without dorsal fins. National Archives (127-GR-1-3-508720).

A rare shot of an R4Q-1 with retro-fitted twin nose wheel and the AN/APS-42 nose radar fitted. This example is BuNo. 126581 (10550) with VMR-253 at MCAS Iwakuni, Japan, in August 1959 (Lawson Collection, National Naval Aviation Museum [NNAM]).

R4Q-2

Fairchild Model M-127. Introduced in 1953, the R4Q-2 was the USMC version of the C-119F with R-3350-85 engines and Hamilton-Standard propellers. A total of 58 were delivered in two batches and were redesignated as the C-119F in September 1962 under the DoD Unified Designation System. The R4Q-2 featured the AN/APS-42 weather radar in an extended nose, ventral fins and twin nose wheel but was never upgraded to "G" standard with Aeroproducts propellers. Bladder fuel tanks were retained, being the same capacity as the earlier R4Q-1.

R4Q-2 Production Summary

BuNo.	Qty	Delivered	Customer
131662 / 131719	58	04Mar53–10Jun53	USMC

R4Q-2 Packet BuNo. 131669 (10836) of VMR-353 landing at Wilmington, North Carolina, in late 1957 (Lawson Collection, National Naval Aviation Museum [NNAM]).

Variants by Conversion

EC-119A

"E" for Exempt prefix applied to the C-119A s/n 45-57769 (msn: 10139) in February 1949 for ECM test duties with Fairchild Aircraft. Wing-tip radomes were used in these tests which concluded in January 1951. The EC-119A was then retired to Chanute AFB and scrapped after December 1951.

EC-119B

Four C-119B airframes were used for test duties from 1950 to 1956, the "E" prefix for Exempt from normal duties. S/n 48-319, 48-320, 48-323, 48-339.

GC-119B

Redesignation of UC-119B s/n 48-333 as a ground instructional airframe at Sheppard AFB from late 1962.

JC-119B

Redesignation for s/n 48-339 (msn: 10321) after the "E" for Exempt was dropped in 1955 and replaced with "J" for Temporary Conversion. S/n 48-339 retained this designation until upgraded to JC-119C standard in 1956.

UC-119B

Utility designation applied to instructional airframe s/n 48-333 (msn: 10315) at Sheppard AFB in mid–1962. The more applicable "G" for Ground prefix was applied later that year.

EC-119C

Four Exempt status conversions for test duties from 1950 to 1956. S/n 48-320, 49-187, 50-135, 51-2582.

GC-119C

A single ground instructional airframe, s/n 49-120, briefly in use at Wright-Patterson AFB in late 1971.

JC-119C

Three C-119C conversions for temporary test duties from 1955 to 1961. S/n 48-339, 50-135, 50-142.

KC-119F

One C-119F, s/n 51-7968 (msn: 10707), was converted for aerial tanker trials from August 1953 to June 1954. The conversion was carried out by Fairchild and tests were undertaken at Wright-Patterson AFB, Ohio.

AC-119G Shadow

Twenty-six close fire support gunship conversions were made to selected C-119G Flying Boxcars during 1968 for use in the Vietnam War with the 17th SOP SQ / 14th SOP WG. Conversions were carried out by Fairchild-Hiller at their St. Augustine, Florida, facility under Phase III of the USAF Gunship Program Project *Combat Hornet*. The AC-119G was an all-weather, day or night mission capable gunship with an all up weight of 64,000 lbs fitted with a port side-firing weapons system consisting of:

- 4 General Electric GAU-2B/A 7.62mm Mini-Guns each installed in a MXU-470/A Module.

- Kearfott Products C707490010 Fire Control Computer.
- TRW Systems X24780 Fire Control Display.
- Chicago Aerial Industries 8053-100-1 Lead Computing Optical Gun-Sight.

- Electro-Optical Systems AN/AVG-3A Night Observation Scope (NOS).
- Electro-Optical Systems AVZ-8 Type 1A Illuminator (20kw 1.5 million candlelight).
- Fairchild-Hiller LAU-74/A Flare Launcher (24 Mk. 24 flares).
- Airesearch GTGE 85-127 60 kva Auxiliary Power Unit (APU).
- Smoke evacuation air scoops.
- Ceramic armor for crew protection.
- Updated flight, navigation and radio equipment.

The four mini-gun setup was originally fitted with the General Electric SUU-11 Gun Pod until replaced by the MXU-470/A Module in June 1969, this also applied to the AC-119K variant. The AC-119G requires a crew of eight: Pilot, Co-Pilot, Navigator, Flight

Engineer, NOS Operator, Illuminator Operator and two Gunners. Two AC-119Gs were lost to accidents with the remaining 24 transferred to the VNAF in August 1971, their final fates are unknown.

S/n	Mod No.	Cvtd	14 SOP WG	MAP S. Vietnam
52-5892	119	01Jul68	21Jan69	20Aug71
52-5898	106	11Jul68	27Dec69	24Aug71
52-5905	123	01Jul68	10Dec68	31Aug71
52-5907	109	01Jul68	20Dec68	W/o 11Oct69
52-5925	113A	01Jul68	07Jan69	27Aug71
52-5927	110	01Jul68	18Dec68	25Aug71
52-5938	116	01Jul68	15Jan69	04Sep71
52-5942	112	01Jul68	30Dec68	06Sep71
53-3136	111	01Jul68	10Dec68	07Sep71
53-3145	120	01Jul68	27Feb71	26Aug71
53-3170	121	01Jul68	29Jan69	08Sep71
53-3178	122S	01Jul68	06Dec68	21Aug71
53-3189	126	01Jul68	06Dec68	18Aug71
53-3192	125	01Jul68	05Dec68	19Aug71
53-3205	108	01Jul68	26Jan69	17Aug71
53-7833	102	13Jun68	16Apr71	30Aug71
53-7848	118S	01Jul68	04Feb69	02Sep71
53-7851	115	01Jul68	16Jan69	01Sep71
53-7852	114	01Jul68	01Jan69	30Aug71
53-8069	101	12Sep68	06Dec68	20Aug71
53-8089	103	03Jun68	01Feb71	28Aug71
53-8114	105	18Jun68	01Feb71	06Sep71
53-8115	107	14Jul68	08Mar71	03Sep71
53-8123	104	13Jun68	09Apr71	23Aug71
53-8131	124A	01Jul68	31Mar71	24Aug71
53-8155	117	01Jul68	31Dec68	W/o 28Apr70

AC-119G Shadow s/n 52-5927 (11106) in the distinctive SEA camouflage livery somewhere over South Vietnam (USAF).

EC-119G

One VC-119G s/n MM52-6031, and one C-119G s/n MM53-8146, converted to electronics configuration for the Italian AF in 1975 and 1976 respectively.

GC-119G

Redesignation of UC-119G ground instructional airframes s/n 53-3208 (msn: 11221) and s/n 53-8130 (msn: KF-233) at Sheppard AFB from late 1962, both assigned to reclamation in 1965.

JC-119G

Five C-119G temporary conversions from 1956 to 1964 for test duties. S/n 51-2586, 51-8040, 51-8046, 51-8098, 53-3136.

UC-119G

Utility designation applied to instructional airframes s/n 53-3208 (msn: 11221) and s/n 53-8130 (msn: KF-233) at Sheppard AFB in mid–1962. The more applicable "G" for Ground prefix was applied later that year.

VC-119G

One Italian AF C-119G s/n MM52-6031, converted to VIP configuration from 1960 to 1975, converted to an EC-119G in 1975.

C-119J

For aerial supply drops the rear clamshell doors on the C-119 were required to be removed as they could not be opened in-flight. This created a problem for crews dealing with the extreme cold and the aerodynamic drag that degraded aircraft performance. The solution was the Fairchild designed Flight-Operable Door System which was a wedge-shaped hood fitted in place of the clamshell doors that opened along the horizontal axis as opposed to the vertical on the clamshells. The system became known as the "beaver-tail doors" due to the flat wedge shape. On the ground, the entire door fitting could be opened vertically on hinges to allow the loading of bulk cargo, while in-flight the lower section of the hood fitting could be raised up internally without adding extra drag to the aircraft. The entire system was hydraulically actuated.

Fairchild designated the design as the Model 203 and the USAF as the C-119J. One C-119G s/n 53-3213 (msn: 11229), was bailed to Fairchild for the prototype modification in 1955. Once accepted a retrofit program was begun to modify the C-119 fleet under contract AF 36(6000)-2199. It soon became apparent the design was flawed when paratroops exiting aircraft were caught up in wind vortices causing jumpers to be injured, sometimes seriously, when they scraped the side of the hood structure. The program was halted at the 67th conversion with existing aircraft subsequently restricted from aerial drops of both paratroops and cargo, severely curtailing the mission of these C-119s. As a result, the C-119J fleet was slowly retired into storage from 1962 to 1964. 26 of these were delivered through MAP to Italy and 16 were later reinstated for unit support duties with the ADC.

The final totals are 49 C-119F and 18 C-119Gs converted with the F variants being upgraded to G standard prior to, or during, conversion to the C-119J. Final USAF dispositions are: 4 lost to accidents, 36 scrapped, 26 to Italy and 1 to museum status.

C-119J Conversion Summary

S/n	Cvtd Period	Disposition
C-119F		
51-7968	1957–1963	Italy spares 64
51-8030	1957–1964	Cvtd JC-119J 58-64. Scrapped
51-8035	1957–1964	Cvtd JC-119J 58-60. Italy spares 64
51-8036	1957–1968	W/o 08Aug68
51-8037	1957–1963	Museum status
51-8038	1957–1972	Scrapped
51-8039	1957–1972	Scrapped
51-8040	1959–1964	Cvtd JC-119J 57-59. Italy spares 64
51-8041	1957–1963	Scrapped
51-8042	1957–1963	Scrapped
51-8043	1957–1972	Scrapped
51-8045	1957–1972	Scrapped
51-8046	1959–1964	Cvtd JC-119J 57-59. MAP Italy 64
51-8049	1957–1972	Scrapped
51-8050	1957–1969	Scrapped
51-8051	1957–1959	W/o 07Mar59
51-8052	1957–1964	Scrapped
51-8113	1958–1963	MAP Italy 64
51-8114	1957–1962	Scrapped
51-8115	1957–1963	Scrapped
51-8116	1957–1964	Scrapped
51-8119	1957–1964	Scrapped
51-8121	1957–1963	MAP Italy 64
51-8122	1957–1958	W/o 14Nov58
51-8123	1957–1963	Scrapped.
51-8124	1957–1969	Scrapped
51-8125	1958–1963	MAP Italy 63
51-8126	1957–1964	Scrapped
51-8127	1957	W/o 24Oct57
51-8128	1957–1962	MAP Italy 64
51-8129	1958–1963	Scrapped
51-8130	1957–1963	MAP Italy 63
51-8131	1957–1963	Scrapped
51-8132	1957–1961	Scrapped
51-8134	1957–1962	Scrapped
51-8137	1957–1963	Scrapped
51-8138	1957–1963	Scrapped
51-8140	1957–1969	MAP Italy 70?
51-8141	1957–1969	Scrapped
51-8144	1957–1963	MAP Italy 64
51-8152	1957–1963	MAP Italy 64
51-8154	1957–1963	MAP Italy 63
51-8155	1957–1963	Scrapped
51-8156	1957–1963	MAP Italy 64
51-8157	1957–1969	Scrapped
51-8158	1957–1963	MAP Italy 64
51-8159	1957–1962	Italy spares 64
51-8164	1957–1964	Scrapped
51-8167	1957–1963	Scrapped
C-119G		
52-5845	1957–1972	Scrapped

Section One—Technical Data

S/n	Cvtd Period	Disposition
52-5849	1957–1963	MAP Italy 64
52-5851	1957–1962	MAP Italy 64
52-5866	1957–1963	MAP Italy 64
52-5875	1957–1963	MAP Italy spares 64
52-5877	1957–1972	Scrapped
52-5884	1957–1963	MAP Italy 64
52-5895	1957–1972	Scrapped
52-5896	1957–1963	MAP Italy 64
52-5897	1957–1963	MAP Italy 64
52-5903	1957–1972	Scrapped
52-5906	1957–1972	Scrapped
52-5947	1957–1962	MAP Italy 64
53-3213	1957–1963	Scrapped
53-7855	1959–1972	Scrapped
53-8098	1957–1963	MAP Italy 64
53-8101	1959–1972	Scrapped
53-8103	1957–1963	MAP Italy 64

The beaver-tail doors on the C-119J proved very successful for satellite recovery operations. Seen here is s/n 51-8041 (10919) which was used as a satellite recovery test aircraft at Edwards AFB (USAF).

EC-119J

Three Italian AF C-119J, s/n MM51-8030, MM52-5884, MM52-5896, each converted to electronics configuration in 1975, 1969 and 1973 respectively.

JC-119J

Four C-119J conversions for temporary test duties from 1957 to 1964. S/n 51-8030, 51-8035, 51-8040, 51-8046.

MC-119J

Some C-119Js served as aeromedical evacuation conversions with the ANG. Official records, albeit some are missing, list eight which might be the examples fitted with a self-contained aeromedical C-119 cargo hold module featuring insulation, sound proofing and specially fitted medical equipment. These conversions are listed as:

S/n	Cvtd Period	ANG Units
51-8124	1961–1963	183 AML SQ; 147 AML SQ
51-8155	1962–1963	145 AML SQ; 147 AML SQ
51-8157	1961–1962	140 AML SQ; 147 AML SQ
51-8158	1961–1963	150 AML SQ
51-8164	1961–1963	183 AML SQ; 147 AML SQ
51-8167	1961–1962	140 AML SQ; 147 AML SQ
53-7855	1961	102 AML SQ
53-8103	1961–1962	150 AML SQ

Photographic evidence shows at least four other ANG conversions to MC-119J standard: s/n 51-8121, 51-8123, 51-8129 and 51-8137. The MC-119J variant is usually identifiable from the Red Cross symbol on the vertical tails.

MC-119J s/n 51-8158 (KMC-161) conversion assigned to the 150th Aeromedical SQ with the New Jersey ANG from 1961 to 1963 (Nicholas A. Veronico Collection).

VC-119J

Four Italian AF C-119J, s/n MM51-8140, MM51-8144, MM51-8158, MM53-8103, converted to VIP configuration starting in 1969.

YC-119K

Fairchild-Hiller Model M-484. A Fairchild initiative to increase C-119 performance and reduce take-off distances through the use of two under-wing mounted General Electric J85-GE auxiliary jet engines of 2,850 lbst each. The jet-pod housings were inter-changeable with those used on the C-123K Provider. The prototype YC-119K s/n 53-3142 (msn: 11153) was completed in April 1967 and went on a demonstration tour in a colorful stars and stripes livery. Although showing promise the modification came too late in the C-119s career to warrant a fleet wide upgrade for the USAF. It did however find favor with the 26 high-weight AC-119K Stinger conversions and with several foreign air forces operating in hot-and-high conditions. The single YC-119K was retired to Davis-Monthan in March 1972 and eventually scrapped.

C-119K

Two foreign air forces opted for jet-augmented C-119K modifications which were carried out at Fairchild-Hiller's Crestview, Florida, facility in 1971-72. Jordan brought four in 1972 but one crashed on its delivery flight in October 1972. Ethiopia initially ordered 17 C-119Ks but this was reduced to ten.

AC-119K Stinger

26 close fire support gunship conversions were made to selected C-119G Flying Boxcars during 1968 for use in the Vietnam War with the 18th SOP SQ / 14th SOP WG. Conversions were carried out by Fairchild-Hiller at their St. Augustine, Florida, facility under Phase III of the USAF gunship program Project *Combat Hornet*. Service in the Vietnam Theatre began in late 1969 with six remaining in the U.S. for crew training. The AC-119K was an all-weather, day or night mission capable gunship upgraded to C-119K standard with two 2,850 lbst J-85-GE auxiliary jet engines for increased operating weights up to 80,400 lbs. All gunship equipment is the same as the AC-119G listed below but with the extra "K" variant fittings:

- 4 General Electric GAU-2B/A 7.62mm Mini-Guns each installed in a MXU-470/A Module.
- 2 General Electric M61 Vulcan 20mm Cannons.
- Kearfott Products C707490010 Fire Control Computer.
- TRW Systems X24780 Fire Control Display.
- Chicago Aerial Industries 8053-100-1 Lead Computing Optical Gun-Sight.
- Electro-Optical Systems AN/AVG-3A Night Observation Scope (NOS).
- Electro-Optical Systems AVZ-8 Type 1A Illuminator (20kw 1.5 million candlelight).
- Texas Instruments AN/AAD-4 Forward-Looking Infra-Red (FLIR).
- Sperry AN/ARQ-25 Ranging System.
- Itek APQ-25/26 Homing & Warning Radar.
- Motorola AN/APQ-133A Side-Looking Beacon Tracking Radar.
- Texas Instruments AN/APQ-135 Terrain Avoidance Radar.
- Marconi AN/APN-147 Doppler Radar.
- Fairchild-Hiller LAU-74/A Flare Launcher (24 Mk. 24 flares).
- Airesearch GTGE 85-127 60 kva Auxiliary Power Unit (APU).
- Smoke evacuation air scoops.
- Ceramic armor for crew protection.
- Updated flight, navigation and radio equipment.

Each aircraft required a crew of ten: pilot, co-pilot, navigator, flight engineer, nos operator, illuminator operator, flir operator and three gunners. 2 AC-119Ks were lost to accidents, 2 on combat missions with the remaining 22 transferred (8 of these from the U.S. which had been used as trainers and test aircraft), to the VNAF in November 1972.

Their ultimate fates are unknown although 53-7850 was found stored at Ho Chi Minh City Intl. Airport in the 2000s.

S/n	Mod No.	Cvtd	14 SOP WG	MAP S. Vietnam
52-5864	172	17Sep68	27Dec69	10Nov72
52-5889	176	07Oct68	23Oct69	10Nov72
52-5910	155	01Jul68	—	06Nov72
52-5911	156	01Jul68	—	06Nov72
52-5926	159	01Jul68	18Nov69	10Nov72
52-5935	174	02Oct68	18Nov69	W/o 06Jun70
52-5940	160	26Jul68	26Dec69	10Nov72
52-5945	175	02Oct68	26Dec69	10Nov72
52-9982	173	17Sep68	27Dec69	10Nov72
53-3154	167	12Aug68	27Dec69	10Nov72
53-3156	168	12Aug68	03Nov69	W/o 19Feb70
53-3187	153	01Jul68	—	31Jan73
53-3197	158	01Jul68	—	06Nov72
53-3211	161	26Jul68	18Nov69	31Jan73
53-7826	162	26Jul68	21Oct69	Combat loss 02May72
53-7830	163	26Jul68	27Dec69	10Nov72
53-7831	157	01Jul68	—	31Jan73
53-7839	152	01Jul68	—	31Jan73
53-7850	166	07Aug68	21Oct69	10Nov72
53-7854	170	30Aug68	18Nov69	Combat loss 23Aug72
53-7877	151	01Jul68	—	31Oct72
53-7879	169	30Aug68	21Oct69	10Nov72
53-7883	165	07Aug68	05Dec69	10Nov72
53-8121	171	10Sep68	21Oct69	10Nov72
53-8145	154	01Jul68	—	31Jan73
53-8148	164	26Jul68	18Nov69	10Nov72

An unidentified AC-119K taking off from Lockbourne AFB, Ohio, in December 1968. Note the AN/APQ-133 Side-Looking Radar protruding at the rear of the fuselage (National Archives KE-37970).

AC-119K Stinger s/n 52-5889 (11056) was shot down while in VNAF service on April 29, 1975, just a day before Saigon fell, ending the war in Vietnam (USAF).

ANG Rhode Island–based C-119L 53-3174 (11185) in October 1974. This aircraft became N55795 in civil service and was stolen on the night of November 20, 1980, from Kissimmee Airport in Florida, never to be seen again (Paul Goddard).

C-119L

The final variant in the C-119 series applied to 27 ANG C-119G aircraft—nine each with the 129 ACO GP at Hayward, 130 ACO GP Kanawha and the 143 SOP GP at Theodore F. Green Airport starting in 1973. When Aeroproducts stopped support of their 4-bladed propellers fitted to the C-119G variant, the USAF was forced to seek an alternative for the remaining ANG fleet. 3-bladed Hamilton Standard 43H60-603 hydromatic, full feathering, reverse pitch propellers were selected replacing the long troublesome Aeroproducts 4-bladed units. General Dynamics carried out the conversions for the ANG and the C-119L variant served until finally retired in September 1975.

ANG C-119L Conversions

53-3144, 53-3174, 53-3184, 53-3186, 53-3193, 53-3206, 53-3216, 53-7836, 53-7837, 53-7849, 53-7853, 53-7858, 53-7865, 53-7884, 53-8073, 53-8074, 53-8076, 53-8083, 53-8084, 53-8087, 53-8126, 53-8127, 53-8142, 53-8149, 53-8150, 53-8153, 53-8154.

Several USAF / VNAF AC-119G/K gunships still serving in SEA were similarly converted in their final years of service but did not get the "L" variant suffix. India and Taiwan converted numbers in their C-119G to C-119L standard, both countries still using their fleets into the 1980s, Taiwan until 1997.

A number civil Flying Boxcars were type certified as the C-119L with the Hawkins & Powers C-119G-3E variant, under FAA Type Certificate A24WE, being optional in either the Aeroproducts 4-bladed or Hamilton Standard 3-bladed configuration.

RC-119L

One experimental conversion was made in 1973 by Fairchild at Crestview to C-119G s/n 53-3181, then assigned to MASDC for reconnaissance research. It appears the testing was in aid of foreign users of the type and work was carried out with a camera fitted in the aft fuselage. It is thought Morocco, Taiwan and South Vietnam may have made use of RC-119G/L configured Flying Boxcars.

C-119 Jet-Pak

The first C-119 Jet-Pak configured Boxcars were for the Indian Air Force who needed the extra power for high altitude military operations in the northern regions of India. Civilian company Steward-Davis, Inc. supplied 26 3,400 lbst Jet-Pak 3402 (J34-WE) as kits for licensed installation in India by Hindustan Aircraft Ltd. (HAL).

A total of nine C-119 Jet-Pak modifications were made for service with U.S. civil operators from 1972 to 1987 for airtanker roles with Aero Union and Hemet Valley Fire Service. Other C-119 Jet-Paks were operated by Hawkins & Powers and owners in Alaska into the 1990s albeit on a limited basis.

C-119 Jet-Pak N1394N (10840) performing an engine run in Alaska during June 2003 (Michael Prophet Collection).

Type	Reg	msn / s/n	In Service
C-119C	N13742	10431 / 49-194	1976–1981
C-119C	N13743	10369 / 49-132	1974–1987
C-119C	N13744	10436 / 49-199	1972–1987
C-119C	N13745	10304 / 48-322	1974–1987
C-119C	N13746	10334 / 48-352	1976–1987
C-119F	N1394N	10840 / BuNo. 131673	1982–
C-119L	N8504X	KF-245 / 53-8142	1982–1987
C-119L	N8504Y	11232 / 53-3216	1982–
C-119L	N8504Z	11253 / 53-7836	1989–

C-119 STOLmaster

One former RCAF (s/n 22133) C-119G N383S (msn: 10992), was modified by Steward-Davis, Inc. at Long Beach California where the stock airframe arrived on May 9, 1967. U.S. aircraft dealer Aircraft International Associates owned the aircraft and contracted Steward-Davis for the conversion and development. The unique STOLmaster was a C-119 Jet-Pak variation intended for a range of power options so could be equipped with one, two or three of the Steward-Davis Jet-Pak 3402 pods housing the Westinghouse 24C-4D (J34-WE) jet-augmented engine. The aircraft had the Wright R-3350-89A piston radials as usual then civil operators had the choice of operating the STOLMaster with one (dorsal mounted), two (under-wing mounted) or three (dorsal and wing) of the Jet-Paks fitted. All jet pods were interchangeable and designed for quick installation or removal. The first flight of the conversion was in July 1967 followed by a sales tour of Central and South America. No further conversions were made and the single STOLmaster was sold and eventually became a C-119G-3E airtanker for the USFS.

The C-119 STOLmaster N383S (10992) seen here now converted as an airtanker for the USFS (Tanker 133) at Fox Field, California, in September 1977 (Geoff Goodall).

C-119G-3E

This variant is a specific version of the C-119 Jet-Pak operated under FAA Type Certificate A24WE held by Hawkins & Powers Aviation, Inc. of Wyoming. It applies only to the C-119G variant fitted with a J34WE-34 or -36 Jet-Pak and most were solely operated in airtanking roles during the annual fire season. The "3E" simply refers to "Three-Engines," these being the two Wright R3350s and the J34. The prototype was N8682 (msn: 10859) first converted in January 1972 and 11 other C-119Gs were converted by H&P many serving up to the type's withdrawal in 1987. In later years, N15501 (msn: 10955) was brought out of storage and restored to flying condition during 2003 for the filming of the motion picture *Flight of the Phoenix* (2004) in Namibia. Also, N8093 made the last flight of a C-119G-3E when it was transferred to the Hagertsown Aviation Museum during November 2008. Lastly, N961S is something of a mystery, it appears to have been converted with a Jet-Pak in 1980, designated as a C-119G-3E but operated under Aero Union TC A21WE.

Reg	msn / s/n	In Service
N15501	10955 / 22130	1981–1987
N15509	10775 / 22110	1976–1984
N383S	10992 / 22133	1976–1979
N3003	10737 / 22106	1980–1987
N3559	10870 / 22118	1975–1987
N3560	10957 / 22132	1973–1978
N3935	10824 / 22113	1972–1987
N48076	11005 / 52-5846	1976–1987
N5216R	10956 / 22131	1974–1987
N8093	10776 / 22111	1979–1987
N8682	10859 / 22115	1972–1981
N8832	10907 / 22123	1976–1985
N961S	10872 / 22120	1980–unk

Ill-fated Tanker 135 C-119G-3E N48076 (11005) crashed on September 16, 1987, while on fire-bombing runs in Shasta-Trinity National Park, California. The accident ended USFS C-119 airtanker operations (Michael Prophet Collection).

Section Two

Military Operators

Belgium

A total of 46 C-119 Flying Boxcars were operated by Belgium from 1952 to 1973. The first 18 were the C-119F variant delivered through MDAP from September 1952 to December 1952 followed by 22 C-119G through MDAP from August 1953 to May 1954. A further six C-119G were obtained from the USAF in July 1960. All aircraft were in service with No. 20 and No. 40 Squadrons / 15th Wing at Melsbroek AB, Belgium. The C-119Fs were returned to the USAF MAAG—Military Advisory Aid Group, in 1955 then upgraded to "G" standard by SABENA Airlines from 1956 with all but eight going back into Belgian service by 1959. Numbers were deployed to Africa during The Congo Crisis starting in 1960 to assist in the evacuation of foreign nationals from the newly independent country. Six aircraft were lost to accidents, one (s/n CP-36) while deployed to the Congo Crisis. Eight were transferred to Norway in 1955, two to Ethiopia in 1972 and two to civil owners. Two are preserved today and the remaining 26 scrapped at Koksijde AB by 1975.

18 New Built C-119F

BAF s/n	msn / USAF s/n	Call Sign	BOC-SOC	Disposition
CP-1	10681 / 51-2692	OT-CAA	25Sep52–12Sep55	To Norway
CP-2	10682 / 51-2693	OT-CAB	24Sep52–12Sep55	To Norway
CP-3	10683 / 51-2694	OT-CAC	24Sep52–08Feb72	To CP-14 59 / Scr
CP-4	10684 / 51-2695	OT-CAD	25Sep52–12Sep55	To Norway
CP-5	10685 / 51-2696	OT-CAE	25Sep52–22Oct75	To CP-11 59 / W/o
CP-6	10686 / 51-2697	OT-CAF	25Sep52–12Sep55	To Norway
CP-7	10687 / 51-2698	OT-CAG	25Sep52–12Sep55	To Norway
CP-8	10688 / 51-2699	OT-CAH	25Sep52–12Sep55	To Norway
CP-9	10689 / 51-2700	OT-CAI	28Oct52–22Oct75	Civil sale / UK
CP-10	10690 / 51-2701	OT-CAJ	28Oct52–22Oct75	Extant Belgium
CP-11	10691 / 51-2702	OT-CAK	28Oct52–12Sep55	To Norway
CP-12	10692 / 51-2703	OT-CAL	28Oct52–22Oct75	Scrapped
CP-13	10693 / 51-2704	OT-CAM	28Oct52–22Oct75	Scrapped
CP-14	10694 / 51-2705	OT-CAN	28Oct52–12Sep55	To Norway
CP-15	10695 / 51-2706	OT-CAO	28Oct52–22Oct75	Scrapped
CP-16	10696 / 51-2707	OT-CAP	28Oct52–22Oct75	Scrapped
CP-17	10679 / 51-2690	OT-CAQ	27Nov53–22Oct75	Scrapped
CP-18	10680 / 51-2691	OT-CAR	24Dec53–22Oct75	Scrapped

22 New Built C-119G

BAF s/n	msn / USAF s/n	Call Sign	BOC-SOC	Disposition
CP-19	11035 / 52-6034	OT-CAS	26Mar54–22Oct65	W/o 22Oct65
CP-20	11034 / 52-6033	OT-CAT	23Mar54–Jul74	Scrapped
CP-21	10952 / 52-6022	OT-CBA	04Aug53–Jul74	Civil sale / Extant
CP-22	10953 / 52-6023	OT-CBB	08Aug53–Jul74	Scrapped
CP-23	10951 / 52-6021	OT-CBC	09Sep53–12Dec61	W/o 12Dec61
CP-24	11077 / 52-6038	OT-CBD	17Dec53–Jul74	Scrapped
CP-25	11082 / 52-6043	OT-CBE	22Dec53–12Dec61	W/o 12Dec61
CP-26	11029 / 52-6028	OT-CBF	11Jan54–Jul74	Scrapped
CP-27	10997 / 52-6026	OT-CBG	18Jan54–Jul74	Scrapped
CP-28	11078 / 52-6039	OT-CBH	18Jan54–Jul74	Scrapped
CP-29	11086 / 52-6047	OT-CBI	19Jan54–03Jul72	To Ethiopia
CP-30	10998 / 52-6027	OT-CBJ	27Jan54–Jul74	Scrapped
CP-31	11036 / 52-6035	OT-CBK	09Feb54–Jul74	Scrapped
CP-32	11084 / 52-6045	OT-CBL	14Feb54–Jul74	Scrapped
CP-33	11033 / 52-6032	OT-CBM	14Feb54–Jul74	Scrapped
CP-34	11118 / 52-6050	OT-CBN	15Feb54–Jul74	Scrapped
CP-35	11120 / 52-6052	OT-CBO	17Feb54–Jul74	Scrapped
CP-36	11083 / 52-6044	OT-CBP	18Feb54–19Jul60	W/o 19Jul60
CP-37	11123 / 52-6055	OT-CBQ	18Feb54–03Jul72	To Ethiopia
CP-38	11119 / 52-6051	OT-CBR	21Feb54–Jul74	Scrapped
CP-39	11085 / 52-6046	OT-CBS	22Feb54–Jul74	Scrapped
CP-40	11146 / 52-6058	OT-CBT	09May54–Jul74	Scrapped

6 ex-USAF C-119G

BAF s/n	msn / USAF s/n	Call Sign	BOC-SOC	Disposition
CP-41	11246 / 53-7829	OT-CEA	17Jul60–22Oct75	Scrapped
CP-42	11260 / 53-7843	OT-CEB	17Jul60–22Oct75	Scrapped
CP-43	KF-241 / 53-8138	OT-CEC	17Jul60–22Oct75	Scrapped
CP-44	KF-244 / 53-8141	OT-CED	17Jul60–22Oct75	Scrapped
CP-45	KF-246 / 53-8143	OT-CEE	17Jul60–26Jun63	W/o 26Jun63
CP-46	KF-254 / 53-8151	OT-CEH	17Jul60–12Dec74	Extant Belgium

Belgian C-119G s/n CP-35 (11120) on public display during May 1972 at a UK airshow (Richard Vandervord).

Brazil

13 C-119Gs were acquired through MAP by the Forca Aerea Brasileira (Brazilian Air Force) and were BOC in batches beginning January 31, 1962. They were assigned to No. 2/1 Grupo de Transporte de Tropa (No. 2 Squadron / 1 Troop Transport Group) located at Aerea dos Afonsos AB in Rio de Janeiro. They served alongside the 12 Fairchild C-82A Packets of No. 1/1 Grupo de Transporte de Tropa (No. 1 Squadron / 1 Troop Transport Group) located at the same base for the carriage of cargo and paratroop brigades of the Brazilian army. C-119G s/n 2302 was lost in a training accident a few months into service and was replaced by an ex-USAF C-119G, FAB s/n 2312, in 1966 in order to retain the 12 aircraft squadron strength. With spares availability declining, increasing maintenance and the loss of two further airframes, s/n 2312 in 1973 and s/n 2307 in 1974 due to engine failures, it was decided to retire the fleet in favor of new C-130H Hercules transports. All aircraft were withdrawn from service in November 1974 with the last struck off in February 1975. Two were selected for preservation with the remaining eight scrapped.

FAB s/n	msn / USAF s/n	MAP Date	Disposition
2300	10959 / 51-8065	24Jan62	Scrapped 1975
2301	10960 / 51-8066	24Jan62	Scrapped 1975
2302	10961 / 51-8067	24Jan62	W/o 01Apr62
2303	10962 / 51-8068	24Jan62	Scrapped 1975
2304	10968 / 51-8074	24Jan62	Extant Brazil
2305	10970 / 51-8076	18Apr62	Extant Brazil
2306	10958 / 51-8064	02Jul62	Scrapped 1975
2307	10978 / 51-8084	03Aug62	W/o 27Jun74
2308	10969 / 51-8075	22Aug62	Scrapped 1975
2309	10971 / 51-8077	1962	Scrapped 1975
2310	10974 / 51-8080	1962	Scrapped 1975
2311	10986 / 51-8092	12Oct62	Scrapped 1975
2312	KF-231 / 53-8128	17Oct66	W/o 21Dec73

C-119G s/n FAB-2305 (10970) is one of two Flying Boxcars preserved by the Brazilian AF Museum in Rio de Janeiro as seen here in September 2022 (Richard Vandervord).

Canada

35 new built C-119F Flying Boxcars were operated by the Royal Canadian Air Force (RCAF) from 1952 to 1967 performing all usual mission requirements plus other tasks such as Arctic resupply along the DEW line, army co-op exercises and electronic warfare duties. The first three (s/n 22101 / 22103) were delivered in September 1952 and were the only RCAF Boxcars to have previous USAF serial numbers: 51-2687/ 51-2689. The two main RCAF units to operate the type were No. 435 *Chinthe* Squadron based at Edmonton, Alberta (later Namao, Alberta) and No. 436 *Elephant* Squadron based at Dorval, Quebec (later Downsview, Ontario). From December 1955 the entire fleet was upgraded to C-119G standard with new propellers, twin nose-wheels, ventral fins and the AN/APS-42 weather radar fitted in an extended nose. RCAF Boxcars played a key role in the 1956 Suez Crisis flying in United Nations peacekeeping markings and based out of Egypt. Four aircraft were lost to accidents in RCAF service but notably no loss of personnel was suffered. Fleet withdrawal of the remaining 31 Boxcars began in 1964 and all were in storage at Saskatoon by the end of 1965. The fleet was struck off charge in 1967.

RCAF s/n	msn	BOC-SOC	Disposition
22101	10676	08Sep52–09May67	U.S. civil N15505
22102	10677	08Sep52–02Mar67	Sold for spares
22103	10678	09Sep52–06Jun67	U.S. civil N8092
22104	10735	12Nov52–06Jun67	U.S. civil N3558
22105	10736	12Nov52–01Nov66	U.S. civil N15506
22106	10737	06Dec52–22Aug67	U.S. civil N3003
22107	10738	06Dec52–22Aug67	U.S. civil N966S
22108	10773	12Jan53–13Sep67	U.S. civil N5215R
22109	10774	12Jan53–01Dec66	U.S. civil N15504
22110	10775	12Jan53–13Sep67	U.S. civil N15509
22111	10776	12Jan53–06Jan67	U.S. civil N8093
22112	10823	18Mar53–05Apr67	U.S. civil N964S
22113	10824	18Mar53–05Apr67	U.S. civil N3935
22114	10825	18Mar53–01Feb67	U.S. civil N15502
22115	10859	24Apr53–09Apr67	U.S. civil N8682
22116	10860	24Apr53–07Jul67	U.S. civil N5217R
22117	10861	24Apr53–07Jul67	U.S. civil N15507
22118	10870	24Apr53–13Sep67	U.S. civil N3559
22119	10871	11May53–05Apr67	Sold for spares
22120	10872	11May53–07Jul67	U.S. civil N961S
22121	10905	16Jun53–07Jul67	U.S. civil N962S
22122	10906	16Jun53–09May67	U.S. civil N8091
22123	10907	16Jun53–22Aug67	U.S. civil N8832
22124	10908	16Jun53–27Apr56	W/o 19Mar56
22125	10942	10Jul53–20Jan58	W/o 13Jan58
22126	10943	10Jul53–13Sep67	U.S. civil N965S
22127	10944	10Jul53–20Sep60	W/o 11Aug60
22128	10945	10Jul53–04Mar59	W/o 13Dec58
22129	10954	31Jul53–22Aug67	U.S. civil N963S
22130	10955	31Jul53–06Jun67	U.S. civil N15501
22131	10956	17Aug53–13Sep67	U.S. civil N5216R
22132	10957	17Aug53–02Mar67	U.S. civil N3560
22133	10992	03Sep53–02Mar67	U.S. civil N383S
22134	10993	06Oct53–06Jun67	U.S. civil N15508
22135	10994	03Sep53–13Sep67	U.S. civil N8094

Section Two—Military Operators 45

RCAF C-119G s/n 22129 (10954) in full livery with the "lightning bolt" fuselage stripe. This Boxcar later became N963S, sold to Morocco as CN-AMO in 1968 (RCAF via Chris Charland).

ETAF C-119K s/n 917 (11171) seen here on a visit to The Netherlands in late 1972. Most of the Ethiopian Boxcar fleet were assigned to scrap by the late 1980s (George Kamp via Dick Lohuis).

Ethiopia

17 jet-assisted C-119K were ordered in 1970 by the Ethiopian Air Force (ETAF) for service with No. 1 Transport Squadron, but it appears the order was amended to eight in two batches—five, s/n 910/914, in 1970 from the Fairchild St. Augustine facility and three, s/n 915/917, in 1971 from the Fairchild Crestview facility. Two additional C-119Gs from Belgium were added to the fleet in 1972 for a total of ten aircraft. Ethiopian service continued until 1986 and by the late 1990s it is known that seven of the fleet were derelict at Debre Zeyit AB (now Bishoftu), Ethiopia and were later scrapped. USAF records show that the initial MAP deliveries from the U.S. were all of the C-119G type which is believed to be a clerical error.

ETAF s/n	msn / USAF s/n	MAP Date	Disposition
910	11261 / 53-7844	25Aug70	Derelict Asmara, Eritrea
911	11277 / 53-7856	26Aug70	Scrapped Debre Zeyit
912	11066 / 52-5899	12Sep70	Scrapped Debre Zeyit
913	11111 / 52-5932	24Nov70	
914	11252 / 53-7835	18Sep70	
915	11199 / 53-3188	11Jul71	Scrapped Debre Zeyit
916	KF-211 / 53-8108	06Aug71	Scrapped Debre Zeyit
917	11171 / 53-3160	20Aug71	Scrapped Debre Zeyit
918	11086 / 52-6047	29Sep72	Scrapped Debre Zeyit
919	11123 / 52-6055	29Sep72	Scrapped Debre Zeyit

France (French Indochina)

A total of eight C-119B and at least 55 C-119C were loaned from the USAF to the Armée de l'Air (French Air Force) from May 1953 through to September 1954 in support of French ground forces in Northern Vietnam during the First Indochina War.

Operation Squaw (06May53–16Jul53)

The U.S. Government initially loaned six C-119Cs to the French which would be piloted by American civilian crews from Civil Air Transport (CAT) based out of Taiwan and members of the French AF. The Boxcars came from the 483rd and 314th TCWG in Japan. Under the loan agreement the ownership was retained by the USAF but they carried French markings with French civil registrations applied as callsigns, these being: F-OYXA (537), F-OYXB (541), F-OYXC (543), F-OYXD (550), F-OYXE (581) and F-OYXF (583). Once type training had been completed in The Philippines the C-119s were based at Gia Lam AB outside Hanoi in Vietnam where resupply missions were flown to French soldiers. The six Boxcars clocked up 176 sorties flying a combined total of 517 hours.

Operation Ironage I thru V (05Dec53–12Mar54)

French and CAT aircrew training continued and C-119s were provided for the establishment of a base at Dien Bien Phu on the Vietnam–Laos border. *Ironage I* formally began on December 7, 1953, with 15 C-119s supplying tons of equipment, ammunition, food and barbed wire to the French encampment. Four more phases of this operation followed—*Ironage II* (22Dec53), *Ironage III* (09Jan54), *Ironage IV* (17Jan54) and *Ironage V* (01Feb54).

Operation Squaw II (12Mar54–07Sep54)

The largest operation in terms of acquired aircraft, in excess of 60 loaned C-119s for the continued resupply missions from Cat Bi AB, North Vietnam. In what by now had become the Battle of Dien Bien Phu, the French marked Flying Boxcars flew a total of 682 missions dropping a total of 5,000 tons of supplies and equipment, up to 2,000 tons more were dropped by C-47 aircraft. The C-119s and their aircrews bore the brunt of intense anti-aircraft fire on resupply drops, returning with chunks of wing, nacelles, fuselage and tail sections missing. Two C-119Cs were lost due to combat action during the battle—March 11, 1954, s/n 51-2546 was hit by shelling while on the ground and May 6, 1954, s/n 49-149 was shot down by AA fire. Another C-119C was lost on March 23, 1954, at Cat Bi due to pilot error. The USAF listed these losses as "lost to agency outside USAF." The Battle of Dien Bien Phu ended on May 7, 1954, marking the closure of the First Indochina War. The war weary Boxcars were returned to USAF service in September 1954 after completing further support sorties to isolated French units in Vietnam.

Loan periods as recorded in USAF records:

Code	msn / USAF s/n	USAF Loan Periods & Operation
106	10343 / 49-106	13Jun54–07Jul54.
108	10345 / 49-108	15Apr54–14May54 Squaw II.
109	10346 / 49-109	Ironage / 05Apr54–16May54 Squaw II.
110	10347 / 49-110	03May54–03Jul54.
112	10349 / 49-112	05Jun54–15Jul54.
114	10351 / 49-114	08–24Jun54.
116	10353 / 49-116	27May54–07Jun54.
118	10355 / 49-118	07–14May54.
121	10358 / 49-121	28Jun–27Jul54.
124	10361 / 49-124	21–26Jul54.
125	10362 / 49-125	Ironage / 11Apr54–26Jul54 Squaw II.
127	10364 / 49-127	07Aug54–01Sep54.
133	10370 / 49-133	12Apr54–18Jun54 Squaw II.
135	10372 / 49-135	19Jun54–10Jul54.
136	10373 / 49-136	Ironage / 12Mar54–17Apr54 Squaw II / 01Jul54–14Jul54.
138	10456 / 50-138	05Apr54–01Jun54 Squaw II.
139	10376 / 49-139	Ironage / 11Apr54–14May54 Squaw II / 10Jul54–Jul54.
139A	10457 / 50-139	24May54–02Jul54.
141	10378 / 49-141	24Jun54–24Aug54.
143	10380 / 49-143	Ironage / 09Apr54–24May54 Squaw II / 02Jul54–Jul54.
144	10381 / 49-144	10–26Apr54 Squaw II / 28May54–24Jul54.
147	10465 / 50-147	20Mar54–25Apr54 Squaw II.
149	10386 / 49-149	02–06May54 Squaw II.
151	10388 / 49-151	17–26Jul54.
152	10389 / 49-152	24Jun54–02Aug54.
154	10391 / 49-154	Ironage / 10–25Apr54 Squaw II / 01–07May54 / 26May54–07Aug54.
155	10392 / 49-155	17Mar54–22May54 Squaw II / 12Aug54–04Sep54.
164	10401 / 49-164	05Aug54–04Sep54.
165	10402 / 49-165	09–29Jun54 / 14–16Jul54.
167	10404 / 49-167	25Apr54–14May54 Squaw II / 23May54–08Jun54.
177	10414 / 49-177	30May54–17Jul54.
183	10420 / 49-183	30Apr54–03May54 Squaw II / 08May54–10Jun54.
184	10421 / 49-184	29Apr54–30May54 Squaw II.
185	10422 / 49-185	03May54–03Jul54 Squaw II.
186	10423 / 49-186	Ironage / 11–23Mar54 Squaw II.
187	10424 / 49-187	Ironage / Apr54–09May54 Squaw II.
532	10490 / 51-2532	05Apr54–12Jun54 Squaw II.
536	10494 / 51-2536	20Mar54–13May54 Squaw II.
537	10495 / 51-2537	03May53–26Jul53 Squaw I / 20Mar54–21Jun54 Squaw II / 16–26Jul54.
538	10496 / 51-2538	11Apr54–13May54 Squaw II.
540	10498 / 51-2540	24Aug54–04Sep54.
541	10499 / 51-2541	03May53–06Jul53 Squaw I / 26Jul53–03Aug53; 28–30Jun54 / 03–09Aug54.

Code	msn / USAF s/n	USAF Loan Periods & Operation
543	10501 / 51-2543	03May53–22Jun53, 08Jul53–03Aug53 Squaw I / 07Apr54–05Jun54 Squaw II / 02Jul54–04Aug54.
545	10503 / 51-2545	31Mar54–28May54 Squaw II / 28Jun54–24Jul54.
546	10504 / 51-2546	11Mar54 Squaw II.
550	10508 / 51-2550	03May53–30Jul53 Squaw I.
552	10510 / 51-2552	17Apr54–31May54 Squaw II.
557	10515 / 51-2557	17Apr54–03Jun54 Squaw II.
558	10516 / 51-2558	14Mar54–30Apr54 Squaw II / 03Jul54–01Aug54.
561	10519 / 51-2561	06Apr54–23May54 Squaw II.
562	10520 / 51-2562	05Apr54–07May54 Squaw II.
563	10521 / 51-2563	06–29Apr54 Squaw II.
564	10522 / 51-2564	17Apr54–14Jun54 Squaw II.
572	10530 / 51-2572	07May54–25Jun54.
575	10533 / 51-2575	17Apr54–24May54 Squaw II / 19–25Jun54.
576	10534 / 51-2576	26–27Jul54.
578	10536 / 51-2578	17Apr54–03Jun54 Squaw II.
581	10539 / 51-2581	03May53–24Jul53 Squaw I / 16Apr54–26Apr54 Squaw II.
583	10541 / 51-2583	03May53–08Jul53 Squaw I.

No loan allocation recorded in USAF records but photographic and pilot logbook evidence confirms they were in French service:

137	10374 / 49-137	Ironage / unk–10May54 Squaw II.
559	10517 / 51-2559	Ironage / Squaw II.
571	10529 / 51-2571	Ironage / Apr54–Jun54 Squaw II.
573	10531 / 51-2573	Ironage / Apr54–Jun54 Squaw II.

Unconfirmed identities:

131	10368 / 49-131	Ironage / Squaw II.
147	10384 / 49-147	Maybe assigned / 50-147 has a USAF loan allocation.
539	10497 / 51-2539	Squaw II.
555	10513 / 51-2555	Ironage / Squaw II.
577	10535 / 51-2577	Ironage / Apr54–Jun54 Squaw II.

Mistaken identities:

138	10375 / 49-138	49-138 w/o 1952 / 50-138 has a USAF loan allocation.
153	10390 / 49-153	49-153 w/o 1951 / Not 50-153, 51-8153 / 53-3153 to 483rd 1954.
163	10400 / 49-163	49-163 w/o 1950 / Not 50-163, 51-8163 / 53-3163 to 483rd 1954.

INDIA

A total of 79 C-119Gs were obtained by the Indian Air Force (IAF) through MAP deliveries from USAF stocks. 26, s/n IK441/IK466, were delivered new built from January 1954 to October 1955 as MDAP aid to No. 12 Squadron. 29, s/n BK502/BK530, were delivered ex-USAF from June to July 1960 to No. 19 Squadron followed by another 24, s/n BK962/BK985, from May to June 1963 for No. 48 Squadron to bolster army support of the border conflict with Communist China three, s/n BK986/BK988, additional deliveries from this batch were canceled. Unfortunately later IAF to USAF serial tie-ups are not known. IAF C-119s flew support missions under the United Nations Command during the Congo Crisis in 1960 and 1961. Border conflicts in India's northern regions

Section Two—Military Operators

with neighboring Communist China and Pakistan where high-altitude airstrips, up to 18,000 ft, and drop zones limited the Boxcars performance meant selected aircraft in the fleet would be upgraded with the Steward-Davis J3402 Jet-Pak. From 1960 U.S. company Steward-Davis, Inc. began a program with the IAF to develop a dorsal-mounted 3,400 lbst Westinghouse J-34 auxiliary jet engine to boost power. After a very successful test program Steward-Davis delivered 26 J3402 Jet-Pak kits to the IAF from February 1962 which were fitted in-country by Hindustan Aircraft Ltd. (HAL). Later C-119 *Jet-Packets* were fitted with the British 5,000 lbst thrust Bristol Orpheus 500 engine license built by HAL. It is not known how many units were finally fitted but they were a great success for the crews flying in the hazardous far north. The IAF had several C-119 combat losses during the 1965 Indo-Pakistani War when Pakistani F-86 Sabres destroyed three Boxcars on the ground during airstrikes during September. India also has one of the highest losses of life from C-119 operations with 12 accidents from 1959 to 1982, several of the worst accidents being 04Mar60-22 killed; 29Jan69-20 killed; 23Feb80-46 killed and 07Feb82-23 killed. After 1973 most IAF Boxcars had three-bladed C-119L propellers, the *Packet* name was retained throughout service with retirement taking place in 1984. The remaining 60 were scrapped, two still exist with another two believed to be in storage.

C-119G s/n IK442 (11102) of the Indian Air Force was one of 26 new builds delivered from January 1954. Many were upgraded with Jet-Paks and after 1973 three-bladed propellers (via Jagan Pillarisetti).

26 New Built C-119G

IAF s/n	msn / USAF s/n	MAP Date	Disposition
IK441	11101 / 53-4637	28Jan54	
IK442	11102 / 53-4638	28Jan54	
IK443	11224 / 53-4639	17Dec54	
IK444	11225 / 53-4640	17Dec54	Extant India
IK445	11226 / 53-4641	17Dec54	
IK446	11241 / 53-4642	04Feb55	
IK447	11242 / 53-4643	04Feb55	
IK448	11243 / 53-4644	15Feb55	
IK449	11244 / 53-4645	18Feb55	
IK450	11262 / 53-4646	31Aug55	Extant India
IK451	11263 / 53-4647	31Aug55	
IK452	11264 / 53-4648	31Aug55	

IAF s/n	msn / USAF s/n	MAP Date	Disposition
IK453	11265 / 53-4649	31Aug55	
IK454	11285 / 53-4650	14Sep55	
IK455	11286 / 53-4651	10Sep55	
IK456	11287 / 53-4652	14Sep55	
IK457	11288 / 53-4653	10Sep55	
IK458	11299 / 53-4654	06Oct55	
IK459	11300 / 53-4655	06Oct55	W/o 19Sep60
IK460	11301 / 53-4656	11Oct55	
IK461	11302 / 53-4657	12Oct55	
IK462	11309 / 53-4658	06Oct55	
IK463	11310 / 53-4659	06Oct55	
IK464	11311 / 53-4660	11Oct55	
IK465	11312 / 53-4661	11Oct55	
IK466	11313 / 53-4662	21Oct55	

29 ex-USAF C-119G. IAF s/n: BK502 / BK530
24 ex-USAF C-119G. IAF s/n: BK962 / BK985

Italy

A total of 71 Flying Boxcars were operated by the Aeronautica Militare Italiano (AMI) from 1953 to 1959. 40 new built C-119G were delivered through MDAP from May 1953 to March 1954. S/n 52-6007, 6008, 6009, 6017, 6019, 6020, 6024 were used at Olmsted AFB for pilot training July to September 1953. Five ex-United Nations C-119G were acquired in 1961 during the Congo Crisis and 21 ex-USAF C-119J were delivered from November 1963 to March 1964. An additional five ex-USAF airframes were acquired as spare parts sources. Assigned units were the 2nd and 98th Groups of the 46th Air Brigade (46th Transport Flight prior to 1954), based on the AMI side of Pisa Intl. Airport, Italy. The 50th Group was formed in 1964 for the C-119J variant until withdrawal in 1972 and five EC-119J electronics conversions were assigned to the 71st Group of the 14th Flight at Pratica di Mare, Italy from 1972 to 1979. A total of ten C-119 airframes were lost to accidents, four while on United Nations deployment to Africa during the Congo Crisis. Upon retirement the massive 55 remaining Boxcars were scrapped at Pisa from 1979 with some derelicts still noted into the 1980s.

40 New Built C-119G

AMI s/n	msn / USAF s/n	Codes*	MAP Date	Disposition
MM51-17365	10777 / 51-17365	46-9 / -26	08May53	Scrapped
MM51-17366	10778 / 51-17366	46-2 / -80	12May53	Scrapped
MM51-17367	10779 / 51-17367	46-4 / -27	02Jun53	Scrapped
MM52-6000	10780 / 52-6000	46-6 / -98	19May53	Scrapped
MM52-6001	10826 / 52-6001	46-13 / -33	27Jun53	Scrapped
MM52-6002	10827 / 52-6002	46-31	16Jun53	Scrapped
MM52-6003	10828 / 52-6003	46-5 / -34	29May53	Scrapped
MM52-6004	10862 / 52-6004	46-8 / -22	20Jun53	Scrapped
MM52-6005	10863 / 52-6005	46-3	29Jun53	W/o 30Mar63
MM52-6006	10864 / 52-6006	46-7 / -82	26Jun53	Scrapped
MM52-6007	10865 / 52-6007	46-37 / -87	29Jun53	Scrapped
MM52-6008	10866 / 52-6008	46-20 / -20	29Jun53	Scrapped
MM52-6009	10867 / 52-6009	46-24 / -24	10Jul53	Scrapped

Section Two—Military Operators

AMI s/n	msn / USAF s/n	Codes*	MAP Date	Disposition
MM52-6010	10868 / 52-6010	46-23 / -23	09Jul53	Scrapped
MM52-6011	10869 / 52-6011	46-15	16Jul53	W/o 15Feb61
MM52-6012	10909 / 52-5012	46-29 / -29	19Jun53	Scrapped
MM52-6013	10910 / 52-6013	46-16 / -37	10Jun53	Scrapped
MM52-6014	10911 / 52-6014	46-10	14Jun53	W/o 17Nov61
MM52-6015	10912 / 52-6015	46-19 / -99	15Jun53	Scrapped
MM52-6016	10946 / 52-6016	46-31 / -81	08Jan54	Scrapped
MM52-6017	10947 / 52-6017	46-18 / -39	10Aug53	Scrapped
MM52-6018	10948 / 52-6018	46-24 / -86	08Jul53	W/o 25Apr70
MM52-6019	10949 / 52-6019	46-21 / -21	07Aug53	Scrapped
MM52-6020	10950 / 52-6020	46-34 / -84	13Aug53	Extant Italy
MM52-6024	10995 / 52-6024	46-41 / -91	12Jan54	Scrapped
MM52-6025	10996 / 52-6025	46-36 / -36	22Jan54	Scrapped
MM52-6029	11030 / 52-6029	46-43 / -93	29Dec53	Extant Italy
MM52-6030	11031 / 52-6030	46-27 / -95	29Dec53	W/o 24Jan79
MM52-6031	11032 / 52-6031	46-30 / -30	29Dec53	**V, EC-119G**, scrapped
MM52-6036	11075 / 52-6036	46-11	04Dec53	W/o 20Apr64
MM52-6037	11076 / 52-6037	46-22	04Dec53	W/o 02Feb61
MM52-6040	11079 / 52-6040	46-25 / -25	29Dec53	Scrapped
MM52-6041	11080 / 52-6041	46-28 / -28	21Dec53	W/o 22Oct64
MM52-6042	11081 / 52-6042	46-32 / -32	04Jan54	Scrapped
MM52-6048	11116 / 52-6048	46-42 / -92	11Feb54	Scrapped
MM52-6049	11117 / 52-6049	46-33 / -83	08Feb54	Scrapped
MM52-6053	11121 / 52-6053	46-35 / -85	19Feb54	Scrapped
MM52-6054	11122 / 52-6054	46-38 / -88	19Feb54	Scrapped
MM52-6056	11144 / 52-6056	46-39 / -89	16Mar54	Scrapped
MM52-6057	11145 / 52-6057	46-40 / -90	16Mar54	Scrapped

5 ex-United Nations C-119G

AMI s/n	msn / USAF s/n	Codes*	MAP Date	Disposition
MM53-3200	11213 / 53-3200	46-48 / -38	1961	Extant Italy
MM53-3219	11235 / 53-3219	46-46 / -96	1961	To civil, scrapped
MM53-7828	11245 / 53-7828	46-47 / -97	1961	Scrapped
MM53-7845	11266 / 53-7845	46-44 / -94	1961	Scrapped
MM53-8146	KF-249 / 53-8146	46-45 / -35	1961	**EC-119G**, extant Italy

21 ex-USAF C-119J

AMI s/n	msn / USAF s/n	Codes*	MAP Date	Disposition
MM51-8046	10924 / 51-8046	-	28Jan64	W/o 14Dec64
MM51-8113	KMC-116 / 51-8113	46-49 / -69	31Jan64	Scrapped
MM51-8121	KMC-124 / 51-8121	46-50 / -50	15Jan64	Extant Italy
MM51-8125	KMC-128 / 51-8125	46-51 / -51	21Nov63	Scrapped
MM51-8128	KMC-131 / 51-8128	46-52 / -52	06Feb64	Scrapped
MM51-8130	KMC-133 / 51-8130	46-53 / -53	21Nov63	**EC-119J**, scrapped
MM51-8140	KMC-143 / 51-8140	46-54 / -54	1970?	**VC-119J**, scrapped
MM51-8144	KMC-147 / 51-8144	46-55 / -55	31Jan64	**VC-119J**, scrapped
MM51-8152	KMC-155 / 51-8152	46-56 / -56	19Feb64	Scrapped
MM51-8154	KMC-157 / 51-8154	46-57 / -57	05Nov63	W/o 26Jun69
MM51-8156	KMC-159 / 51-8156	46-58 / -58	14Feb64	Scrapped
MM51-8158	KMC-161 / 51-8158	46-62 / -62	14Feb64	**VC-119J**, scrapped
MM52-5849	11008 / 52-5849	46-59 / -59	20Jan64	Scrapped
MM52-5851	11010 / 52-5851	46-60 / -60	27Feb64	Scrapped
MM52-5866	11025 / 52-5866	46-61 / -61	27Feb64	Scrapped

AMI s/n	msn / USAF s/n	Codes*	MAP Date	Disposition
MM52-5884	11051 / 52-5884	46-63 / -63	03Mar64	EC-119J, scrapped
MM52-5896	11063 / 52-5896	46-64 / -64	19Feb64	EC-119J, scrapped
MM52-5897	11064 / 52-5897	46-65 / -65	21Feb64	Scrapped
MM52-5947	11134 / 52-5947	46-66 / -66	06Mar64	Scrapped
MM53-8098	KF-201 / 53-8098	46-67 / -67	03Mar64	Scrapped
MM53-8103	KF-206 / 53-8103	46-68 / -68	15Jan64	VC-119J, scrapped

Spares Airframes

—	10707 / 51-7968	—	28Jan64	
—	10913 / 51-8035	—	10Feb64	
—	10918 / 51-8040	—	10Feb64	
—	11042 / 52-5875	—	21Feb64	
—	KMC-162 / 51-8159	—	06Feb64	

* Codes are assignments pre, then post 1965.

Italian C-119G coded 46-24 (10867) was serialed MM52-6009 which is in small lettering on the tail. This aircraft, seen here in October 1977 at RAF Mildenhall, UK, was one of the 55 scrapped after retirement (Richard Vandervord).

Jordan

Four C-119K conversions were ordered by the Royal Jordanian Air Force (RJAF) in 1972. Conversion work was carried out at Fairchild Crestview, Florida, facility. S/n 118 crashed in France during delivery so only three C-119Ks actually served with the RJAF, albeit briefly as they were withdrawn in 1977.

RJAF s/n	msn / USAF s/n	MAP Date	Disposition
115	11022 / 52-5863	23Sep72	
116	11047 / 52-5880	23Sep72	
117	11097 / 52-5918	21Oct72	
118	11115 / 52-5936	21Oct72	W/o 27Oct72

Morocco

18 C-119 Flying Boxcars served the Royal Moroccan Air Force (RMAF) with the 1st Air Transport Squadron. The first six were C-119C types delivered through MAP from May 1962 to March 1963. An additional six C-119G were delivered from June 1966 to December 1966 through MAP from the USAF. The final six C-119G were acquired in December 1968 from a U.S. civilian source—Aircraft International Associates of Nebraska who had just purchased the entire RCAF C-119 fleet. An unconfirmed nineteenth is recorded as CN-AMS (msn: 10738), but it appears to have been canceled prior to delivery. CN-AMA and CN-AMD have been noted as RC-119C conversions. RMAF retirement took place in 1980 with up to four presently preserved at locations in Morocco.

RMAF s/n	msn / s/n	MAP Date	Disposition
CN-AMA	10374 / 49-137	05Nov62	
CN-AMB	10408 / 49-171	10May62	
CN-AMC	10424 / 49-187	01Mar63	
CN-AMD	10427 / 49-190	03Oct62	
CN-AME	10417 / 49-180	14Mar63	
CN-AMF	10420 / 49-183	29Oct62	
CN-AMG	11282 / 53-7861	22Jun66	W/o 14Mar69
CN-AMH	11283 / 53-7862	22Jun66	Extant Morocco
CN-AMI	11296 / 53-7871	22Jun66	
CN-AMJ	11314 / 53-7880	03Dec66	
CN-AMK	11278 / 53-7857	03Dec66	
CN-AML	KF-198 / 53-8095	25Nov66	
CN-AMM	10905 / (RCAF) 22121	23Dec68	
CN-AMN	10943 / (RCAF) 22126	23Dec68	Extant Morocco
CN-AMO	10954 / (RCAF) 22129	23Dec68	
CN-AMP	10823 / (RCAF) 22112	23Dec68	
CN-AMQ	10774 / (RCAF) 22109	23Dec68	
CN-AMR	10861 / (RCAF) 22117	23Dec68	

Moroccan AF C-119G s/n CN-AMH (11283) is seen here in 2010 as a gate guardian at Marrakesh Menara Airport, Marrakesh (Richard Vandervord).

Norway

Eight former Belgian C-119G were operated by No. 335 Squadron of the Royal Norwegian Air Force (RNAF) at Gardermoen AB Norway from July 1956 to July 1969. One aircraft was written off due to a misjudged landing with the remaining seven retired to the USAF at Davis-Monthan in 1969. The RNAF used three-letter identification codes and retained the original five-digit USAF serial on the tail.

S/n	msn / USAF s/n	Codes	BOC-SOC	Disposition
12692	10681 / 51-2692	BW-C	18Jul56–22May69	To USAF storage
12693	10682 / 51-2693	BW-B	26Jun56–07May69	To USAF storage
12695	10684 / 51-2695	BW-E	31Jul56–06Dec68	W/o 06Dec68
12697	10686 / 51-2697	BW-A	24Jun56–14Jul69	To USAF storage
12698	10687 / 51-2698	BW-F	08Aug56–22May69	To USAF storage
12699	10688 / 51-2699	BW-D	02Aug56–22May69	To USAF storage
12702	10691 / 51-2702	BW-H	31Aug56–Jun69	To USAF storage
12705	10694 / 51-2705	BW-G	07Sep56–Jun69	To USAF storage

Taiwan (Republic of China)

A total of 114 C-119G Flying Boxcars were operated by the Republic of China Air Force (RoCAF), unofficially known as the Taiwanese Air Force, through U.S. MAP deliveries. 16 were delivered in 1959; 12 in 1966; 43 in 1969; 35 in 1970; 3 in 1971 and 5 in 1973—these being acquired from the former VNAF fleet. The RoCAF s/n range was 3112 / 3224, the individual numbers appear to have been randomly applied to 113 aircraft with one likely acquired only for spares. Many RoCAF serials to original USAF serials are not known but it appears the 1959 batch had corresponding final two digits. Assigned units (& unit colors) were the 101st (yellow), 102nd (red) and the 103rd (blue) Troop Carrier Squadrons based at Pingtung AB. The 101st retained their Boxcars until 1986, the 102nd and 103rd until 1997. These squadrons operated under the 10th Transport Group / 6th Troop Carrier and Anti-Submarine Combined Wing. From 1973 large numbers of the RoCAF fleet received the C-119L 3-bladed propellers as blade replacements came round. Some Boxcars are also known to have been upgraded to RC-119L configuration and others as navigation trainers with extra dorsal astrodomes. Some were local AC-119G gunship conversions with a variety of armaments from six M3 Browning machine guns to two M61 Vulcans. 15 accidents have occurred, some with a significant loss of life. 12 have serials with no known, or confirmed USAF identity:

28Jul62	3163	Crashed into the sea.
07Apr69	3122	No details.
10Aug71	3149	Crashed into the sea.
20Nov71	3124	Crashed during exercise.
14Mar72	3117	Crashed Mt. Guanyin. 6 crew, 33 Special Forces passengers killed.
09May74	3128	Collided with 3166, Pintung AB. 2 killed.
09May74	3166	Collided with 3128, Pintung AB. 2 killed.
01Jul77	3205	Damaged beyond repair.
18Aug77	3221	Crashed on test flight.
28Aug79	3194	Crashed Mt. Daxue in bad weather. 5 crew, 15 passengers killed.
18Apr83	3182	Damaged beyond repair.
06Jun83	3197	Stalled after take-off due to engine fire. 38 fatalities (8 crew, 30 passengers) and 9 survivors.

Section Two—Military Operators

RoCAF C-119 withdrawal began in 1995 and an official final retirement ceremony was held on December 19, 1997, attended by Taiwanese Air Force officials, TV media and the public. The six C-119s present were s/n 3120, 3144, 3158, 3160, 3192, 3204.

RoCAF s/n	msn / USAF s/n	MAP Date	Disposition
3128 (?)	11107 / 52-5928	01Jul59	
3137	11124 / 52-5937	01Jul59	
3154 (?)	11141 / 52-5954	01Jul59	
3143	11154 / 53-3143	01Jul59	
3149 (?)	11160 / 53-3149	01Jul59	W/o 10Aug71
3153 (?)	11164 / 53-3153	01Jul59	
3155 (?)	11166 / 53-3155	01Jul59	
3163 (?)	11174 / 53-3163	01Jul59	W/o 28Jul62
3164	11175 / 53-3164	01Jul59	W/o 18Jul72
3165	11176 / 53-3165	01Jul59	
3169	11180 / 53-3169	01Jul59	
3171	11182 / 53-3171	01Jul59	
3172	11183 / 53-3172	01Jul59	
3176	11187 / 53-3176	01Jul59	
3177	11188 / 53-3177	01Jul59	
3175	11304 / 53-7875	01Jul59	
3129	11220 / 53-3207	22Apr66	
	11267 / 53-7846	21Mar66	
3131	11268 / 53-7847	20Apr66	
	11289 / 53-7864	21May66	
	11305 / 53-7876	21Apr66	
3133	KF-188 / 53-8085	21Apr66	
	KF-193 / 53-8090	23Apr66	
	KF-194 / 53-8091	20Apr66	
3141	KF-196 / 53-8093	21May66	
	KF-199 / 53-8096	21May66	
	KF-223 / 53-8120	21Apr66	
3139	KF-235 / 53-8132	21May66	
	10575 / 51-2586	10Apr69	
	10660 / 51-2671	03Dec69	
	10666 / 51-2677	03Feb69	
	10672 / 51-2683	06Mar69	
3125	10698 / 51-2709	06Mar69	
3174	10699 / 51-2710	02Dec69	
3162	10700 / 51-2711	08Apr69	
3112	10705 / 51-2716	03Feb69	
	10709 / 51-7970	02Dec69	
3158	10710 / 51-7971	03Feb69	Extant Taiwan
	10713 / 51-7974	03Dec69	
	10716 / 51-7977	06Mar69	
3123	10717 / 51-7978	06Mar69	
	10718 / 51-7979	08Apr69	
	10719 / 51-7980	03Feb69	
3160	10724 / 51-7985	21Apr69	Extant Taiwan
3119	10728 / 51-7989	06Mar69	
	10731 / 51-7992	08Apr69	
3181	10743 / 51-8000	02Dec69	W/o 01Jun96
3167	10750 / 51-8007	01Dec69	
	10756 / 51-8013	01Dec69	

Section Two—Military Operators

RoCAF s/n	msn / USAF s/n	MAP Date	Disposition
3147	10760 / 51-8017	08Apr69	
	10763 / 51-8020	01Dec69	
	10770 / 51-8027	03Dec69	
	10922 / 51-8044	08Apr69	
	10925 / 51-8047	10Apr69	
3157	10935 / 51-8057	10Apr69	
3151	10936 / 51-8058	08Apr69	
	10938 / 51-8060	06Mar69	
3120	10964 / 51-8070	06Mar69	Extant Taiwan
3144	10973 / 51-8079	06Mar69	Extant Taiwan
3116	10977 / 51-8083	20Feb69	
3126	10985 / 51-8091	06Mar69	
	10988 / 51-8094	21Apr69	
	11108 / 52-5929	06Mar69	
	11207 / 52-5923	08Apr69	
	KMC-113 / 51-8110	01Dec69	
	KMC-120 / 51-8117	01Dec69	
3152	KMC-139 / 51-8136	08Apr69	Extant Taiwan
3159	KMC-151 / 51-8148	21Apr69	
	KMC-163 / 51-8160	06Mar69	
	KMC-164 / 51-8161	03Feb69	
	KMC-169 / 51-8166	06Mar69	
3208	10658 / 51-2669	13Mar70	
	10670 / 51-2681	27Nov70	
	10704 / 51-2715	04Jan70	
3202	10706 / 51-8016	26Feb70	Extant Taiwan
3210	10712 / 51-7973	15Mar70	
3195	10720 / 51-7981	05Jan70	
3206	10729 / 51-7990	03Mar70	
3199	10739 / 51-7996	04Jan70	Extant S. Korea
	10745 / 51-8002	03Jan70	
3186	10747 / 51-8004	27Jan70	
	10748 / 51-8005	26Feb70	
	10751 / 51-8008	03Jan70	
	10754 / 51-8011	03Jan70	
	10758 / 51-8015	09Jan70	
	10761 / 51-8018	27Nov70	
3213	10768 / 51-8025	27Nov70	
3212	10857 / 51-8031	27Nov70	
3209	10926 / 51-8048	25Feb70	
	10934 / 51-8056	11Mar70	
3183	10965 / 51-8071	03Jan70	Extant Taiwan
3184	11003 / 52-5844	27Jan70	Extant Taiwan
3191	11007 / 52-5848	03Jan70	
	11020 / 52-5861	28Jan70	
3214	11028 / 52-5869	28Nov70	
3187	11037 / 52-5870	27Jan70	
3217	11103 / 52-5924	27Nov70	
	11138 / 52-5951	05Jan70	
	11139 / 52-5952	04Jan70	
3142	KMC-102 / 51-8099	28Jan70	
	KMC-105 / 51-8102	10Jan70	
3190	KMC-109 / 51-8106	05Jan70	Extant Taiwan

RoCAF s/n	msn / USAF s/n	MAP Date	Disposition
3192	KMC-123 / 51-8120	10Jan70	Extant Taiwan
3198	KMC-150 / 51-8147	09Jan70	
3204	KMC-153 / 51-8150	25Feb70	Extant Taiwan
	KF-208 / 53-8105	27Nov70	
3220	10723 / 51-7984	04Jan71	
	11015 / 52-5856	04Jan71	
3215	KF-222 / 53-8119	04Jan71	W/o 27Jan89

5 ex-VNAF C-119G

3140	11151 / 53-3140	73	
	11159 / 53-3148	25May73	
	11172 / 53-3161	26May73	
	11211 / 53-3198	25May73	
	KF-250 / 53-8147	26May73	

RoCAF C-119G s/n 3192 (KMC-123) is preserved at Kinmen National Park, Taiwan. The tiny nation today possesses four of the five Kaiser-built Flying Boxcars thought to remain in existence (Wayne Lo).

UNITED STATES OF AMERICA

Overview

A total of 1044 C-119 Flying Boxcars were operated by two main service branches of the United States armed forces—the United States Air Force (USAF) and United States Marine Corps (USMC) plus two air reserve components—the USAF Reserve (AFR) and Air National Guard (ANG). Overall U.S. military service was from December 1949 to September 1975.

947 C-119s were delivered to the USAF-941 for active duty plus three static test airframes, s/n 48-321, 53-3208, 53-8130. One C-119C, s/n 49-189, crashed during its delivery flight. The other two were the XC-120 and YC-119H prototypes. Post front-line service USAF C-119s were transferred to the AFR, ANG or exported to foreign military operators. 97 R4Q designated Boxcars were delivered to the USMC with some short, limited service by its associated service the United States Navy (USN).

With clamshell doors removed, the C-119 proved invaluable for the aerial resupply of troops on the ground as seen here with C-119B s/n 48-343 (10325) during the Korean War during 1951 (USAF).

United States Air Force (USAF)

A total of twelve troop carrier wings operated the C-119 in a front-line capacity under the USAF's Tactical Air Command (TAC) from 1949 to 1958. Regular joint USAF—U.S. Army exercises were held refining the art of battlefield paratroop delivery, aerial combat drops and cargo delivery to remote areas. TAC was a Major Command of the USAF tasked with overseeing tactical air power planning and utilization on a global scale. Numbers of C-119Bs were first delivered to the 314th Troop Carrier Wing at Smyrna AFB, Tennessee in December 1949. This wing would become the largest operator of front-line C-119s for the USAF both in the continental U.S. and initially in the Far East. The Korean War broke out in June 1950 seeing almost every C-119B and C-119C variant built to that point sent into the theater of conflict under USAF Major Command—Far East Air Forces Command (FEA). The C-119 squadrons, based mainly from Ashiya AB, Japan provided vital aerial links to the forces in South Korea for troop transport, base resupply, combat air-drops and personnel returning from the war zone. The 483rd TCWG C-119s would maintain a front-line presence at Ashiya AB up to late 1958. The Major Command—USAF in Europe (AFE) was established after World War II as a major force in the emerging Cold War era made widespread use of the C-119 by four troop carrier wings from 1951 to 1958. Resupply and paratroop missions and exercises with fellow NATO countries took place throughout Europe from the UK to as far south as the Mediterranean and North Africa.

54 C-119s were kept in front-line service in Europe until 1961 with the 10th, 11th, and 12th TCSq at Dreux AB in France.

USAF Front-line Unit Summary

Wing	SQ	Period	Qty	Remarks
Tactical Air Command (TAC)				
64 TCWG	17, 18, 35	1952, 1953–1954	70	Attached 63 TCWG 53-54
313 TCWG	29, 47, 48	1953–1955	91	
314 TCWG	37(A), 50, 61, 62	1949–1950, 1954–1957	232	To FEA 50-54
316 TCWG	36, 37, 75	1950–1954	182	To FEA 54-56
433 TCWG	67, 68, 69	1950–1951	48	To AFE 51
435 TCWG	76, 77, 78	1951–1952	81	To AFR 52
443 TCWG	309, 310, 343	1952–1953	51	
456 TCWG	744, 745, 746	1952–1956	107	Depl Japan 55-56
463 TCWG	16, 772, 773, 774, 775	1953–1957	93	
464 TCWG	776, 777, 778, 779	1953–1958	139	
465 TCWG	780, 781, 782	1953	77	To AFE 53
516 TCWG	345, 346, 347	1952–1953	48	
Far East Air Forces (FEA)				
314 TCWG	37(A), 50, 61, 62	1950–1954	152	
316 TCWG	36, 37, 75	1954–1956	91	
374 TCWG	21(A)	1950–1952, 1955–1956	15	
403 TCWG	63, 64, 65	1951–1953	78	To AFR 1953
483 TCWG	21(A), 36(A), 37(A), 75(A), 773(A), 815(A), 816(A), 817(A)	1952–1959	239	
USAF in Europe (AFE)				
60 TCWG	10, 11, 12	1952–1958	130	To 7305 AB GP 58-61
317 TCWG	39, 40, 41	1952–1958	162	
433 TCWG	67, 68, 69	1951–1952	48	To AFR 1955
465 TCWG	780, 781, 782	1953–1957	108	

As adequate numbers of the new C-119F / G began to be delivered to front-line TAC wings after the Korean War in 1953, some of the older C-119C types were reassigned second-line duties with Air Defense Command (ADC) and Strategic Air Command (SAC) as fighter and bomber unit support aircraft. Aerial mapping and chart work was carried out by C-119s assigned to the Military Air Transport Service (MATS) including assignments to the Airways & Air Communications Service (AACS) and Air Resupply & Communications Service (ARCS) of MATS. Air Proving Ground Command (APG) and Air Research and Development Command (ARD) used C-119s for R&D duties such as parachute drop tests and satellite recovery. Cold weather assignments for Arctic base resupply duties included assignments to Northeast Air Command (NEA) in Greenland and Canada. Alaskan Air Command (AAC) C-119s, based mainly at Ladd and Elmendorf AFB, kept bases in the northeastern Arctic reaches supplied.

The Vietnam War in Southeast Asia saw 52 C-119 Boxcars return to front-line USAF service from 1968 to 1973 as aerial AC-119 gunship platforms. 26 AC-119G served with the 17th SOP SQ and 18 AC-119K with the 18th SOP SQ of the 14th SOP WG based as detachments throughout South Vietnam and Thailand. The K variant was reassigned to the 56th SOP WG from 1971 until U.S. forces withdrew from the conflict in 1973. The gunship program in Vietnam was supported by the 1st ACO WG in Florida which used assigned AC-119s as training aircraft.

USAF Dispositions

The following figures relate solely to the 945 C-119s serving the USAF, AFRes and ANG and their status at the point of withdrawal from service. Across its 25 years of service up to 1975 a total of 134 Boxcars were lost to accidents (one while on loan to France) and another seven to acts of nature, in this case tornado damage. This represents an almost 15 percent loss of the entire fleet. The C-119 had a reputation for a high accident rate with many losses attributed to engine failure and subsequent loss of control. Many of the engine failures were catastrophic meaning the complete or sudden loss of an engine at critical moments like take-off resulting in airframe loss and fatalities. A total of 504 were assigned

Heavy cargo drops like vehicles and artillery required the use of multiple large parachutes (USAF).

C-119G s/n 51-8088 (10982) flying in formation with C-119G s/n 52-5880 (11047) (Nicholas A. Veronico Collection).

for scrap at retirement, 456 of these going to the MASDC "boneyard" at Davis-Monthan AFB in Arizona. A small number of MASDC residents avoided the furnaces going into civil ownership, museum care or to MAP customers. 273 Boxcars, mainly the C-119G variant, were assigned directly to MAP customers with large numbers going to allies of the then current SEA conflict—Taiwan and South Vietnam. Other C-119s disposed of from active service were three for civilian crash tests, 11 assigned museum status and five were assigned duties with the United Nations. Only six were lost in combat—two in the Korean War, two in the Vietnam War and two while on loan to the French in Indochina. Three C-119Cs were transferred out of service to unknown operators, one by civil commercial sale. No trace of the aircraft has been heard of since so these are likely transfers to agencies for various aeronautical tests.

Summary of USAF Final Dispositions (at point of retirement)

Combat Losses:	6 →	2 French in Laos
Airframe Losses:	140 →	1 French in Laos
Destroyed for Tests:	3	
Scrapped:	504 →	Incl. 12 S. Vietnam returned USAF
MAP Deliveries:	273	
Museum Status:	11 →	1 later lost to "Accident"
United Nations:	5	
Unknown Operators:	3	

The USAF unit histories for the C-119 are extensive and lengthy. The following layout has been adopted to bring some order to the various USAF commands and operational units:

1. Major Commands (MAJCOM) are listed as headings and are presented in alphabetical order.
2. Listed under each Major Command are the relevant Groups, Wings, Units and

C-119B s/n 49-111 (10348) after conversion to C-119C standard with dorsal fins and water-injected radials. "111" was one of the hundreds of Boxcars later scrapped in Tucson, Arizona (USAF).

Air Base Groups C-119 aircraft were assigned too. Squadrons have also been noted where applicable.
3. Assigned operating bases and C-119 serial numbers are listed under each unit followed by 2-digit year dates and relevant notes in brackets.

C-119J s/n 51-8039 (10917) assigned to the USAF's then secret 6593rd Test SQ based out of Hawaii for the aerial recovery of Corona spy satellites (USAF).

Air Defense Command (ADC)

Major Command established 21 March 1946 to oversee aerospace defense of the Continental United States. Became a sub-command of **Continental Air Command** (CNC) from 1 December 1948 to 1 July 1950. Re-established as a Major Command 1 January 1951.

1 Fighter (Air Defense) Group: Selfridge AFB, Michigan. 50-125 (56), 50-145 (55–56), 50-152 (55–56), 51-2578 (56).

1 Maintenance & Supply Group: Selfridge AFB, Michigan. 50-125 (57–58), 50-144 (57–58), 50-145 (56–59), 50-152 (56–57), 51-2555 (57–58), 51-2557 (57–59), 51-2567 (59), 51-2569 (57–58), 51-2578 (56–58), 53-8075 (59).

14 Consolidated Maintenance Squadron: Burlington IAP (Ethan Allen AFB), Vermont. 50-133 (57–58), 51-2543 (58), 51-2564 (57–58).

14 Fighter (Air Defense) Group: Burlington IAP (Ethan Allen AFB), Vermont. 50-133 (55–57), 51-2532 (55; 56), 51-2543 (55–56), 51-2564 (55–57).

15 Consolidated Maintenance Squadron: Niagara Falls ARS, New York. 48-322 (59).

15 Fighter (Air Defense) Group: Niagara Falls ARS, New York. 51-2541 (depl 56).

27 Advanced Headquarters Division: Norton AFB, California. 50-148 (54–56), 51-2535 (54–56).

29 Advanced Headquarters Division: Great Falls AFB, Montana (renamed Malmstrom AFB 01Oct55). 50-130 (54–56), 50-169 (54–56).

32 Advanced Headquarters Division: Hancock Field, New York (renamed Syracuse AFS 01Dec53), to Dobbins AFB, Georgia. 51-2543 (56–58), 51-2563 (58–59).

33 Fighter (Air Defense) Group: Otis AFB, Massachusetts. 49-197 (55–56), 50-157 (55–56), 50-170 (55–56), 51-2574 (55–56), 51-2660 (55–56).

33 Maintenance & Supply Group: Otis AFB, Massachusetts. 49-197 (56–57), 50-157 (56–57), 50-170 (56–57), 51-2574 (56–57), 51-2660 (56–unk).

34 Advanced Headquarters Division: Kirtland AFB, New Mexico. 50-125 (54), 50-158 (54–56).

35 Advanced Headquarters Division: Dobbins AFB, Georgia. 51-2563 (56–58).

52 Consolidated Maintenance Squadron: Suffolk County AFB, New York. 50-123 (57–59), 50-164 (57–58), 51-2547 (57–59), 51-2565 (57–58).

52 Fighter (Air Defense) Group: Suffolk County AFB, New York. 50-123 (55–57), 50-164 (55–57), 51-2547 (55–56), 51-2565 (56–57).

53 Air Materiel Squadron: Sioux City MAP, Iowa. 51-2587 (59–60).

53 Fighter (Air Defense) Group: Sioux City MAP, Iowa. 50-123 (depl 56), 50-136 (55–56), 50-140 (55–56), 51-2561 (55–56), 51-2658 (56–unk).

54 Fighter (Air Defense) Group: Greater Pittsburgh IAP, Pennsylvania. 50-119 (depl 56), 50-145 (57), 50-157 (56), 51-2581 (55–57), 53-7870 (55 mnt).

54 Fighter-Interceptor Squadron: Ellsworth AFB, South Dakota. 51-2558 (55).

56 Consolidated Maintenance Squadron: O'Hare IAP, Illinois. 49-199 (57–58), 50-121 (57–58), 51-2537 (58), 51-2541 (depl 57), 51-2659 (57–59), 51-8009 (59), 51-8021 (58).

56 Fighter (Air Defense) Group" O'Hare IAP, Illinois. 48-341 (depl 56), 49-125 (55–56), 49-199 (55–57), 50-121 (55–57), 51-2541 (depl 57), 51-2550 (55), 51-2571 (57), 51-2659 (55–57).

78 Fighter (Air Defense) Group: Hamilton AFB, California. 50-137 (55–58), 50-150 (55–58), 50-161 (depl 56), 51-2535 (56–58), 51-2572 (55–58), 51-2609 (55–58), 51-2625 (55–58). C-119J: 51-8036 (67–68), 51-8039 (66–69), 51-8045 (67–69), 51-8049 (66–69), 51-8050 (66–69), 51-8141 (69), 51-8157 (67–69), 52-5903 (67–69), 52-5906 (67–69), 53-8101 (67–69).

78 Maintenance & Supply Group: Hamilton AFB, California. 50-150 (58), 51-2535 (58), 51-2609 (58), 51-2625 (58), 51-8160 (58), 53-3168 (59).

79 Fighter (Air Defense) Group: Youngstown AFB, Ohio. 49-179 (57), 51-2557 (55–57).

82 Consolidated Maintenance Squadron: New Castle AFB, Delaware. 50-119 (57), 50-144 (57), 51-2569 (57).

82 Fighter (Air Defense) Group: New Castle AFB, Delaware. 50-119 (55–57), 50-144 (55–57), 51-2569 (56–57).

84 Fighter (Air Defense) Group: Geiger Field, Washington State. 50-148 (depl 57).

325 Fighter (Air Defense) Group: McChord AFB, Washington State. 50-141 (unk–58), 50-148 (56–58), 50-165 (55–57), 50-166 (56–58), 51-2575 (55–58), 51-2576 (55–58), 51-8039 (55), 51-8157 (56), 53-8077 (58).

325 Maintenance & Supply Group: McChord AFB, Washington State. 50-165 (57–58), 51-2575 (58), 51-2576 (58), 53-8090 (58).

327 Consolidated Maintenance Squadron: Truax AFB, Wisconsin. 51-2537 (57–58), 51-2541 (57–58), 51-2550 (57–58).

327 Fighter (Air Defense) Group: Truax AFB, Wisconsin (deployment to O'Hare IAP 55-56). 50-157 (56–57), 51-2537 (56–57), 51-2541 (56–57), 51-2550 (56–57).

328 Consolidated Maintenance Squadron: Richards-Gebaur AFB, Missouri. 49-199 (58), 50-130 (57–58), 50-134 (57), 50-136 (57), 50-140 (57), 50-148 (60), 50-150 (58), 50-158 (57), 51-2558 (57), 51-2561 (57), 51-2604 (57), 51-2659 (58), 51-8050 (58), 52-5860 (58), 52-5905 (60), 53-8076 (58).

328 Fighter (Air Defense) Group: Grandview AFB, Missouri (renamed Richards-Gebaur AFB 18Aug55). 49-146 (57), 50-130 (56–57), 50-134 (55–57), 50-136 (56–57), 50-140 (56–57), 50-158 (56–57), 50-161 (depl 56), 50-167 (56), 50-169 (56–57), 51-2558 (55–57), 51-2561 (56–57), 51-2604 (55–57).

329 Consolidated Maintenance Squadron: Stewart AFB, New York. 51-2550 (57).

329 Fighter (Air Defense) Group: Stewart AFB, New York. 50-121 (depl 56).

337 Consolidated Maintenance Squadron: Portland IAP, Oregon. 51-8069 (58–59).

337 Fighter (Air Defense) Group: Portland IAP, Oregon. 50-137 (depl 55-56), 50-166 (depl 56), 51-2625 (depl 56).

337 Fighter-Interceptor Squadron: Minneapolis-St. Paul IAP, Minnesota. 50-134 (54).

355 Fighter (Air Defense) Group: McGhee Tyson AFB, Tennessee. 50-123 (57), 50-164 (depl 57), 51-2565 (depl 57).

412 Consolidated Maintenance Squadron: Wurtsmith AFB, Michigan. 51-2555 (57).

412 Fighter (Air Defense) Group: Wurtsmith AFB, Michigan. 51-2555 (55–57).

414 Fighter (Air Defense) Group: Oxnard AFB, California. 49-119 (57).

432 Fighter-Interceptor Squadron: Truax AFB, Wisconsin. 49-199 (54).

444 Fighter-Interceptor Squadron: Charleston AFB, South Carolina. 51-2563 (55).

456 Fighter-Interceptor Squadron: Truax AFB, Wisconsin. 50-121 (54).

460 Fighter-Interceptor Squadron: Grand Forks AFB, North Dakota. C-119J: 53-8101 (depl 71).

475 Fighter (Air Defense) Group: Minneapolis-St. Paul IAP, Minnesota. 50-134 (unk–56), 51-2564 (depl 55–56).

500 Air Defense Group: Greater Pittsburgh IAP, Pennsylvania. 51-2581 (55).

501 Air Defense Group: O'Hare IAP, Illinois. 49-199 (54–55), 50-121 (54–55), 51-2659 (55).

502 Air Defense Group: Youngstown AFB, Ohio. 51-2557 (55).

514 Air Defense Group: Minneapolis-St. Paul IAP, Minnesota. 50-145 (55).

517 Air Defense Group: Burlington IAP (Ethan Allen AFB), Vermont. 50-133 (55), 51-2543 (55), 51-2564 (55).

518 Air Defense Group: Niagara Falls MAP, New York. 50-157 (55).

519 Air Defense Group: Suffolk County AFB, New York. 50-123 (54–55), 50-164 (54–55), 51-2547 (55), 51-2565 (56).

520 Air Defense Group: Truax AFB, Wisconsin. 51-2541 (55–56).

521 Air Defense Group: Sioux City MAP, Iowa. 50-136 (55), 50-140 (54–55), 51-2561 (55), 51-2658 (55–56).

525 Air Defense Group: New Castle AFB, Delaware. 50-119 (54–55), 50-144 (54–55).

527 Air Defense Group: Wurtsmith AFB, Michigan (63 FI SQ depl O'Hare IAP). 49-199 (55), 51-2555 (55).

528 Air Defense Group: Presque Isle AFB, Maine. 51-2543 (55), 51-2564 (55).

534 Air Defense Group: Kinross AFB, Michigan. 51-2550 (55).

539 Fighter-Interceptor Squadron: Stewart AFB, New York, to McGuire AFB, New Jersey (18Aug55). 51-2565 (56).

551 Airborne Early Warning and Control Wing: Otis AFB, Massachusetts. 49-197 (57–58), 50-119 (57–58), 50-157 (57–58), 50-170 (57–58), 51-2574 (57–58), 51-2581 (57–58).

551 Maintenance & Supply Group: Otis AFB, Massachusetts. 49-197 (57), 50-157 (57), 50-170 (57), 51-2574 (57).

564 Air Defense Group: Otis AFB, Massachusetts. 49-197 (54–55), 50-157 (55), 50-170 (54–55), 51-2574 (55).

566 Air Defense Group: Hamilton AFB, California. 50-137 (54–55), 50-150 (54–55), 51-2572 (55), 51-2609 (55), 51-2625 (55).

567 Air Defense Group: McChord AFB, Washington State. 50-141 (55–unk), 50-165 (54–55), 50-166 (54–56), 51-2572 (55), 51-2575 (55), 51-2576 (55).

575 Air Defense Group: Selfridge AFB, Michigan. 50-145 (54–55), 50-152 (54–55), 51-2555 (55).

4600 Air Base Group / Wing: Peterson AFB, Colorado. 49-120 (53–54), 49-179 (53), 50-167 (59), 51-7992 (59).

4600 Consolidated Maintenance Squadron: Peterson AFB, Colorado. 52-5906 (71).

4603 Air Base Group: Stewart AFB, New York. 50-131 (62), 51-2611 (66 sal), 51-7969 (69), 51-8038 (66–69), 51-8043 (66–69), 51-8124 (67–69), 51-8140 (67–69), 51-8141 (67–69), 52-5845 (67–69), 52-5877 (67–69), 52-5895 (67–69), 53-7855 (67–69).

4624 Air Base Squadron: Hancock Field, New York (renamed Syracuse AFS 01Dec53). 51-2543 (58).

4650 Combat Support Squadron: Richards-Gebaur AFB, Missouri. C-119J: 51-8038 (70–72), 51-8039 (70–72), 51-8043 (70–72), 51-8045 (70–72), 51-8049 (70–72), 52-5845 (70–72), 52-5877 (70–72), 52-5895 (70–72), 52-5903 (70–72), 52-5906 (70–72), 53-7855 (70–72), 53-8101 (70–72).

4676 Air Base Group: Richards-Gebaur AFB, Missouri. C-119J: 51-8038 (69–70), 51-8039 (69–70), 51-8043 (69–70), 51-8045 (69–70), 51-8049 (69–70), 52-5845 (69–70), 52-5877 (69–70), 52-5895 (69–70), 52-5903 (69–70), 52-5906 (69–70), 53-7855 (69–70), 53-8101 (69–70).

4676 Air Defense Group: Grandview AFB, Missouri (renamed Richards-Gebaur AFB 18Aug55). 50-134 (54–55), 50-136 (54–55), 51-2558 (55), 51-2604 (55).

4737 Air Base Wing: Pepperell AFB, Newfoundland, Canada. 51-2595 (58), 53-3210 (58).

4750 Training (Air Defense) Wing: Yuma AFB, Arizona (renamed Vincent AFB 13Oct56). 50-161 (54–59), 50-167 (54–59), 51-2535 (55), 51-2557 (55), 51-2557 (59), 51-2574

(55–56), 51-2609 (depl 56), 52-5907 (59), 53-8112 (59). Redesignated as 4750 ADF (Weapons) WG 01Sep54.

4756 Air Defense Wing: Moody AFB, Georgia, to Tyndall AFB, Florida. 50-161 (58), 50-167 (depl 58), 51-2543 (depl 56), 51-2569 (59), 51-8009 (57), 51-8057 (59).

Air Force Logistics Command (LOG)

Redesignated 1 April 1961 from **Air Materiel Command** (AMC). Maintains and manages all logistics and assets in the air force inventory. It over sees management of aircraft servicing, maintenance and Procurement. Headquartered at Wright-Patterson AFB, Ohio.

2704 Aircraft Storage & Disposition Group: *Operating Base:* Davis-Monthan AFB, Arizona. *Notes:* Designated as such 01Aug59. Previously designated **Arizona Aircraft Storage Branch** (AASBR). Redesignated **Military Aircraft Storage and Disposition Center** (MASDC) 01Feb65. See under: **Appendix II—USAF Retirement Bases & Disposals** for a complete listing of serial numbers.

2750 Air Base Wing: Wright-Patterson AFB, Ohio. 49-120 (69–71). Unit previously assigned to AMC.

2849 Air Base Group: Hill AFB, Utah. 51-8045 (72).

2850 Air Base Group: Brookley AFB, Alabama. 51-2574 (65).

2852 Air Base Wing: McClellan AFB, California. 50-171 (62).

2856 Air Base Wing: Griffiss AFB, New York. 49-101 (63).

Military Aircraft Storage and Disposition Center (MASDC): *Operating Base:* Davis-Monthan AFB, Arizona. *Notes:* Designated as such 01Feb65. Redesignated **Aerospace Maintenance and Regeneration Center** (AMARC) 01Oct85. Redesignated **309 Aerospace Maintenance and Regeneration Group** (AMARG) 02May07. See under: **Appendix II—USAF Retirement Bases & Disposals** for a complete listing of serial numbers.

Air Force Systems Command (SYS)

Redesignated 1 April 1961 from **Air Research and Development Command** (ARD).

3245 Air Base Wing: Laurence G. Hanscom Field, Massachusetts. C-119J: 51-8030 (62–64).

Control & Reporting Center Detachment: NAS Dallas, Texas. C-119J: 51-8039 (71).

Air Materiel Command (AMC)

Previously known as **Air Technical Service Command** (ATSC). Established 9 March 1946 incorporating R&D, procurement, supply and maintenance. The R&D section was separated 2 April 1950 as the **Air Research and Development Command** (ARDC).

2750 Air Base Group / Wing: *Operating Base:* Wright-Patterson AFB, Ohio. *Notes:* 48-321 for static tests only. Later reassigned to LOG. *C-119B Assignments:* 48-319 (53), 48-320 (49–51), 48-321 (49–unk), 48-322 (49), 48-323 (49–51; 56; 57-59), 48-325 (52), 48-330 (XC-120 51), 48-339 (50–51), 48-344 (58), 48-351 (56), 49-112 (56), 49-117 (55). *C-119C Assignments:* 49-124 (59), 49-131 (58), 49-146 (56), 49-162 (54), 49-187 (50; 55), 49-193 (60), 49-198 (50–51), 50-119 (58), 50-132 (57), 50-141 (58), 50-143 (53), 50-148 (56), 50-149 (55), 50-161 (57), 51-2535 (65), 51-2543 (56), 51-2547 (55), 51-2548 (57), 51-2555 (57), 51-2557 (58),

51-2564 (58), 51-2565 (57), 51-2566 (57), 51-2573 (57–58), 51-2578 (52). *C-119CF Assignments:* 51-2587 (59), 51-2616 (57). *C-119G Assignments:* 51-7979 (53), 51-8085 (54), 51-8090 (56), 53-8075 (54), 53-8083 (56), 53-8136 (56). *C-119J Assignment:* 51-8155 (57).

2845 Air Base Group: Griffiss AFB, New York. 51-8139 (55).

3110 Aircraft Maintenance Group: *Operating Base:* RAF Burtonwood, England. *C-119 Assignments:* 50-122 (55), 50-128 (56), 50-149 (56), 51-2612 (56), 51-2640 (56), 51-2647 (56), 52-5922 (56), 52-5923 (56), 53-3179 (55), 53-3180 (56), 53-3181 (56), 53-3182 (55), 53-3183 (56), 53-3184 (56), 53-3185 (56), 53-3186 (56), 53-3187 (56), 53-3188 (56), 53-3189 (56; 57), 53-3190 (56), 53-3191 (56), 53-3192 (56), 53-3193 (56), 53-3194 (56), 53-3195 (56), 53-3196 (56), 53-3197 (56), 53-3198 (56), 53-3199 (56), 53-3200 (56), 53-3202 (56), 53-3203 (56), 53-3204 (56), 53-3105 (56), 53-3206 (56), 53-3219 (56), 53-3220 (56), 53-3221 (56), 53-7826 (56), 53-7827 (56), 53-7828 (56), 53-7829 (56), 53-7830 (56), 53-7832 (56), 53-7833 (56), 53-7834 (56), 53-7835 (56), 53-7836 (56), 53-7837 (56), 53-7838 (56), 53-7839 (56), 53-7840 (56), 53-7842 (56), 53-7843 (56), 53-7844 (56), 53-7845 (56), 53-7846 (56), 53-7847 (56), 53-7848 (56), 53-7849 (56), 53-7850 (56), 53-7851 (56), 53-7852 (56), 53-7853 (56), 53-7853 (56), 53-7854 (56), 53-8104 (56), 53-8105 (56), 53-8106 (56), 53-8107 (56), 53-8108 (56), 53-8109 (56), 53-8110 (56), 53-8111 (56), 53-8112 (56), 53-8113 (56), 53-8115 (56), 53-8116 (56), 53-8117 (56), 53-8119 (56), 53-8120 (56), 53-8121 (56), 53-8122 (56), 53-8123 (56), 53-8124 (56), 53-8125 (56), 53-8126 (56), 53-8138 (56), 53-8139 (56), 53-8141 (56), 53-8142 (56), 53-8143 (56), 53-8144 (56), 53-8145 (56), 53-8146 (56), 53-8147 (56), 53-8148 (56), 53-8149 (56), 53-8150 (56), 53-8151 (56), 53-8152 (56), 53-8153 (56), 53-8154 (56), 53-8155 (56), 53-8156 (56).

3131 Aircraft Repair Squadron / Group: *Operating Base:* Chateauroux AB, France. *Notes:* Also designated 3131 MAI GP. Redes 3131 ARP SQ/GP 01Aug58. 51-2653, 51-8234 assigned to salvage. *C-119 Assignments:* 49-196 (58), 51-2533 (57), 51-2544 (56), 51-2589 (57–58), 51-2592 (57; 58-59), 51-2597 (56), 51-2600 (56), 51-2615 (56), 51-2617 (56; 57), 51-2623 (56–57), 51-2642 (57), 51-2646 (unk), 51-2653 (56), 51-8234 (56), 51-8262 (56), 53-3189 (57), 53-3196 (56), 53-3200 (56), 53-7827 (57), 53-7829 (56), 53-7833 (57), 53-7838 (57; 58), 53-7840 (57), 53-7843 (59), 53-7844 (56), 53-7852 (56), 53-8138 (59), 53-8154 (58).

3150 Aircraft Maintenance Group: Nouasseur AB, Morocco. 49-196 (56; 57), 51-2594 (57–58), 51-2600 (56), 51-8093 (56), 51-8244 (56), 52-5923 (56), 53-3189 (56).

Arizona Aircraft Storage Branch: Davis-Monthan AFB, Arizona. 50-169 (stor 58–59), 53-3145 (stor 58). Redesignated **2704 ASD GP** 01Aug59.

Depot Maintenance Centers: C-119 assignments under Air Materiel Command (AMC) and later carried over to Air Logistics Command (LOG).
- Air Materiel Command, Pacific Area (AMPAR). FEAMCOM / Tachikawa AB, Japan
- Middletown Air Materiel Area (MIDAR). Olmsted AFB, Pennsylvania
- Mobile Air Materiel Area (MOBAR / MOAAR). Brookley AFB, Alabama
- Northern Air Materiel Area (NAMAR). Miho AB & Tachikawa AB, Japan. Redesignated Northern Air Materiel Area, Pacific (NMPAR)
- Ogden Air Materiel Area (OGDAR). Hill AFB, Utah. Yakima Airport, Washington State
- Oklahoma Air Materiel Area (OKLAR). Tinker AFB, Oklahoma
- Rome Air Depot / Air Materiel Area (RDP DP / ROMAR). Griffiss AFB, New York

- Sacramento Air Materiel Area (SMAAR). McClellan AFB, California / Chu Lai S. Vietnam 69 53-7851
- San Antonio Air Materiel Area (SAAAR). Kelly AFB, Texas
- San Bernardino Air Materiel Area (SBNAR). Norton AFB, California
- Southern Pacific Air Materiel Area (SOMAR / SMPAR). Clark AB, The Philippines. Tainan, Taiwan
- Warner-Robins Air Materiel Area (WRAAR). Robins AFB, Georgia. Patrick AFB, Florida
- AFRMI—Malpensa, Italy
- MACVI—Vergiate Italy
- SABBE—Sabina Airlines, Brussels Belgium
- SASCD—Copenhagen, Denmark

Air Proving Ground Command (APG)

Major Command established for the testing of aircraft weapons systems first named as the **AAF Proving Ground Command** (PGC) 8 March 1946. Renamed **Air Proving Ground Command** (APGC) 10 July 1946. Renamed **Air Proving Ground** (APG), losing Major Command status under the Air Materiel Command (AMC) 1 June 1948. Renamed as **Air Proving Ground Command** (APGC) 20 December 1951.

3200 Aircraft Maintenance Wing: Eglin AFB, Florida. 48-330 (XC-120 53–54), 50-156 (53–54).

3200 Proof Test Group / Wing: Eglin AFB, Florida. 48-322 (49–50), 48-323 (49), 48-330 (XC-120 depl 51; 53), 48-333 (50–51), 49-193 (51), 50-142 (51; 52), 51-2555 (51–52).

3201 Air Base Group / Wing: Eglin AFB, Florida. 48-330 (XC-120 53–55), 48-333 (51–53), 48-342 (50–51), 50-142 (55), 50-156 (53–54), 52-5907 (55).

3203 Maintenance & Supply Group: Eglin AFB, Florida. 48-322 (49–50), 48-323 (49), 48-330 (XC-120 depl 51; 53), 48-333 (50–53), 48-342 (50–51), 49-179 (51), 49-193 (51), 51-2555 (51–52).

Air Force Operational Test Center: Eglin AFB, Florida. 48-339 (57).

Air Proving Ground Command Center: Eglin AFB, Florida. 48-339 (57–58), 49-137 (60), 51-8009 (59), 51-8079 (58), 51-8119 (62–63 stor), 52-5858 (58).

Air Research And Development Command (ARD)

Started as **Research and Development Command** on 23 January 1950 under AMC. Established as a separate Major Command 2 April 1950. Redesignated as such on 16 September 1950. Devoted to R&D of new weapons systems.

4401 Support Wing: 51-8113 (53).

4900 Air Base Group: Kirkland AFB, New Mexico. 49-187 (56), 50-140 (56), 50-167 (56), 51-2548 (55), 51-2567 (55), 51-2574 (56), 51-2577 (55), 51-2607 (56), 51-8105 (depl 55).

4901 Support Wing: Kirkland AFB, New Mexico. 49-101 (53), 50-135 (51), 51-8155 (53).

4925 Test Group: Kirtland AFB, New Mexico. 51-8105 (55).

6502 Parachute Development Test Group: NAS El Centro, California. 48-320 (52), 48-323 (51–52).

6510 Air Base Wing: Edwards AFB, California. 51-2586 (54), 51-8053 (53–54), 51-8098 (52–54).

6511 Aircraft Maintenance Group: NAS El Centro, California. 51-2586 (56–57).

6511 Parachute Development Test Group: NAS El Centro, California. 48-320 (52–55), 48-323 (52–54), 49-194 (53), 51-2586 (54–56), 51-8098 (54–57). Redesignated 6511 Test Group (Parachute) Nov54-55.

6515 Aircraft Maintenance Group: Edwards AFB, California. 50-142 (56–57), 51-2583 (55).

6520 Test Support Wing: Laurence G. Hanscom Field, Massachusetts. 51-7968 (54–55), 51-8113 (54), 51-8167 (54–55).

6530 Air Base Wing: Griffiss AFB, New York. 51-2575 (52).

6593 Test Squadron: Edwards AFB, California (01Aug58–01Nov59), to Hickham AFB, Hawaii (01Nov59–01Jul72). C-119J: 51-8037 (58–61), 51-8038 (58–61), 51-8039 (58–62), 51-8042 (58–61), 51-8043 (58–61), 51-8045 (58–61), 51-8049 (58–62), 51-8050 (58–61), 51-8115 (58–61). Used the C-119J 1958–1962 to recover classified spy satellites.

Air Engineering Development Division: WPAFB, Ohio. C-119J: 51-8030 (60–62), 51-8035 (60).

Air Force Armament Test Center: Eglin AFB, Florida. 50-142 (54–56), 51-8105 (55).

Air Force Flight Test Center: Edwards AFB, California. 50-142 (57), 51-2586 (57–61), 51-8039 (71), 51-8041 (58–61), 51-8050 (depl 58–59), 51-8051 (57), 51-8098 (57–58), 53-3142 (YC-119K 71–72), 53-3145 (AC-119G 68–69), 53-3187 (AC-119K 69–70), 53-8079 (58–59).

Air Force Missile Test Center: Patrick AFB, Florida. 49-194 (54), 51-8034 (56), 51-8135 (53–55), 51-8136 (53–55), 51-8142 (53–55), 51-8143 (53–55), 51-8149 (53–55), 51-8150 (53–55), 51-8151 (54).

Air Research & Development Command: Baltimore, Maryland (20Jan50–24Jan58), to Andrews AFB, Maryland (24Jan58–01Jul92). 50-122 (58–59), 51-8041 (58–59).

Cambridge Research Center: Laurence G. Hanscom Field, Massachusetts. 48-322 (58–59), 49-102 (55), 50-142 (56), 51-7968 (55), 51-8030 (57–60), 51-8035 (57–60), 51-8105 (55), 51-8106 (55–57), 51-8113 (55), 51-8139 (55–57), 51-8167 (55), 53-3195 (58).

Rome Air Development Center: Griffiss AFB, New York. 51-2575 (52).

Wright Air Development Center: WPAFB, Ohio. 48-320 (51–52), 48-323 (51), 48-330 (XC-120 51–53), 48-339 (51–57), 49-187 (51), 49-194 (52–53), 50-135 (51–57), 50-142 (60–61), 50-143 (52; 52–53), 51-2543 (51–52), 51-2578 (51–52), 51-2582 (52–53), 51-2585 (YC-119H 52–53), 51-7968 (KC-119F 53–54), 51-8040 (57–59), 51-8046 (57–59), 51-8053 (55), 51-8067 (54), 52-5858 (55), 52-5901 (55), 52-5907 (55), 52-5912 (55), 53-8131 (55). Renamed Wright Air Development Division 1959; Aeronautical Systems Division 1961.

Special Weapons Center: Kirtland AFB, New Mexico. 49-181 (58), 50-124 (57), 51-8103 (57–58), 51-8166 (58).

Air Training Command (TC / ATC)

Major Command established 1 July 1946 to oversee air force training duties. The abbreviation was reassigned the standard three letter code from TC to ATC in May58.

3310 Technical Training Wing: Scott AFB, Illinois. 49-124 (55), 49-155 (57), 50-123 (55), 50-135 (54), 50-139 (55), 50-151 (56), 51-2554 (54), 51-2579 (56), 51-7992 (56), 51-7999 (57), 51-8103 (55), 53-8084 (56), 53-8096 (56).

3320 Maintenance & Supply Group: Amarillo AFB, Texas. 49-136 (65), 49-139 (65), 49-155 (66), 49-162 (65), 49-182 (66). All assigned to reclamation.

3320 Technical Training Wing: Amarillo AFB, Texas. 48-341 (58), 49-133 (55–56), 49-159 (53–54), 49-162 (57), 49-187 (58), 50-156 (57), 50-158 (57), 51-2568 (56), 51-2625 (56), 51-7987 (58).

3345 Maintenance & Supply Group: Chanute AFB, Illinois. 53-7832 (ret 69).

3345 Technical Training Wing: Chanute AFB, Illinois. 45-57769 (51), 48-341 (50), 49-199 (58), 50-167 (56), 51-2550 (56), 51-2568 (57), 51-2607 (56), 51-2717 (57), 51-7997 (57), 51-8034 (56–57), 51-8048 (57), 51-8101 (56), 51-8104 (55–58), 51-8118 (55–58), 51-8123 (57–58), 51-8124 (57), 51-8125 (57), 51-8126 (57), 51-8127 (57), 51-8129 (57), 51-8131 (57–58), 51-8140 (57–58), 51-8141 (57), 51-8150 (56), 52-5844 (57), 52-5895 (57), 53-8127 (56–57), 53-8129 (56–57).

3380 Technical Training Wing: Keesler AFB, Mississippi. 49-126 (sal 53), 50-126 (58).

3415 Technical Training Wing: Lowry AFB, Colorado. 48-352 (58), 49-110 (58), 49-122 (55), 49-146 (58), 49-159 (56), 49-179 (56), 49-182 (56), 49-183 (57), 50-141 (56), 51-2545 (57), 51-2608 (57), 51-8111 (55), 52-5890 (58), 53-8081 (56), 53-8082 (56), 53-8083 (56).

3450 Technical Training Wing: Francis E. Warren AFB, Wyoming. 49-116 (56–57), 50-132 (54–55), 51-2550 (55).

3499 Mobile Training Aids Wing: Chanute AFB, Illinois. 51-8034 (55–56), 51-8044 (55–57), 51-8048 (55–57), 51-8113 (57–58), 51-8124 (57–58), 51-8125 (57–58), 51-8126 (57–58), 51-8127 (57), 51-8129 (57–58), 51-8134 (57–58), 51-8141 (57–58), 51-8154 (57–58), 51-8155 (57–58), 51-8157 (57–58), 51-8159 (57–58), 51-8167 (57–58), 53-8070 (56–57), 53-8073 (56–57), 53-8074 (56–57), 53-8075 (56–57), 53-8076 (56–57), 53-8078 (56–57), 53-8079 (56–57), 53-8081 (56–57), 53-8082 (56–57).

3500 Pilot Training Wing: Reese AFB, Texas. 49-165 (56), 51-2559 (59).

3510 Flying Training Wing: Randolph AFB, Texas. 50-141 (60), 51-8105 (55–56), 51-8134 (55), 51-8136 (55–56), 51-8146 (55), 51-8149 (55–56), 51-8150 (55–56), 53-8070 (54–56), 53-8072 (54–56), 53-8073 (54–56), 53-8074 (54–56), 53-8075 (54–56), 53-8076 (54–56), 53-8077 (54–56), 53-8078 (54–56), 53-8079 (54–56), 53-8080 (54–56), 53-8081 (54–56), 53-8082 (54–56), 53-8083 (54–56), 53-8084 (54–56), 53-8085 (54–56), 53-8086 (54–56), 53-8087 (54–56), 53-8088 (54–56), 53-8089 (54–56), 53-8090 (54–56), 53-8091 (54–56), 53-8092 (54–56), 53-8093 (54–56), 53-8094 (54–56), 53-8095 (54–56), 53-8127 (54–56), 53-8128 (54–56), 53-8129 (54–56). Redesignated as 3510 CCT WG Oct54.

3520 Combat Crew Training Wing: McConnell AFB, Kansas. 50-129 (57–58).

3530 Pilot Training Wing: Bryan AFB, Texas. 49-171 (57–58).

3535 Air Observers Training Wing: Mather AFB, California. 53-8080 (55).

3550 Flying Training Wing: Moody AFB, Georgia. 50-170 (60), 53-8073 (56), 53-8080 (54).

3560 Pilot Training Wing: Webb AFB, Texas. 50-166 (55), 50-167 (56), 51-8110 (59), 53-3195 (59).

Section Two—Military Operators

3565 Navigator Training Wing: James Connally AFB, Texas. 51-8127 (57).

3595 Combat Crew Training Wing: Nellis AFB, Nevada. 49-139 (57), 49-141 (55), 51-2572 (56), 51-8137 (57–58), 51-8146 (56–57), 51-8150 (56–57), 51-8158 (57–58).

3600 Combat Crew Training Wing: Luke AFB, Arizona. 50-169 (55–56), 51-2599 (58), 51-8111 (56), 51-8146 (55–56), 52-5880 (58), 53-8083 (55), 53-8127 (depl 55). 51-8146 as USAF *Thunderbirds* support aircraft.

3605 Air Observers Training Wing: Ellington AFB, Texas. 49-147 (56), 51-8147 (56), 53-3141 (56). Redesignated as 3605 NTN WG.

3615 Pilot Training Wing: Craig AFB, Alabama. 53-8087 (59).

3625 Combat Crew Training Wing: Tyndall AFB, Florida. 51-8002 (56).

3635 Combat Crew Training Wing: Stead AFB, Nevada. 51-2554 (56), 51-2578 (58), 51-2603 (55), 51-8103 (55). Redesignated as 3635 FTA WG 15Jul58.

3640 Pilot Training Wing: Laredo AFB, Texas. 50-139 (58).

3645 Combat Crew Training Wing: Laughlin AFB, Texas. 50-149 (55).

3700 Military Training Wing: Lackland AFB, Texas. 51-8134 (57), 51-8138 (57), 51-8164 (57), 51-8167 (57), 53-8083 (56–57), 53-8085 (56–57), 53-8086 (56–57), 53-8091 (56–57). Depl Sheppard Sep56.

3750 Maintenance & Supply Group: Sheppard AFB, Texas. 48-333 (62–63), 53-3208 (62–65), 53-8130 (62–65).

3750 Technical Training Wing: Sheppard AFB, Texas. 48-319 (54–56), 48-333 (53–62), 49-120 (depl 55), 49-148 (53–56), 49-159 (55), 49-187 (55), 49-194 (55), 49-195 (55), 51-2585 (YC-119H 54), 53-2308 (54–62), 53-8096 (56–57), 53-8130 (54–62).

Air Observers Training Center: Amarillo AFB, Texas. 51-2613 (59).

Chanute Technical Training Center: Chanute AFB, Illinois. 51-8087 (59), 51-8097 (59).

Keesler Technical Training Center: Keesler AFB, Mississippi. 51-2612 (60), 51-8262 (60).

Lowry Technical Training Center: Lowry AFB, Colorado. 51-2537 (58), 51-2538 (59), 51-2617 (59), 51-8010 (59).

Strategic Reconnaissance (Medium) Test Center: Sheppard AFB, Texas. 49-197 (59).

Air University (AU)

Established as a Major Command 29 November 1945. Named as such 12 March 1946 for educational, research and training duties.

3800 Air University Wing: Maxwell AFB, Alabama. 48-333 (52), 48-354 (56), 49-104 (58), 49-195 (52), 51-2553 (61), 51-2574 (56), 51-2578 (57), 51-2607 (56), 51-2643 (65), 51-8233 (58), 52-5895 (60), 53-3213 (56), 53-7855 (60), 53-8083 (54). Most in for mnt. Also designated as 3800 AB WG.

Alaskan Air Command (AAC)

Major Command established 18 December 1945 to oversee the air defense systems in Alaska. Also supported SAC operations. Previously a Numbered Air Force as the 11 Air Force.

5001 Air Base Wing: Ladd AFB, Alaska. 51-7985 (depl unk).

5001 Aircraft Maintenance Squadron: Ladd AFB, Alaska. 49-178 (56), 52-5918 (55).

5001 Composite Wing: Ladd AFB, Alaska. 49-190 (51), 51-2630 (52–53).

5039 Air Base Wing: Elmendorf AFB, Alaska. 49-119 (55–57), 49-122 (55–57), 49-124 (55–57), 49-125 (55–57), 49-127 (55–57), 49-128 (55), 49-131 (55–58), 49-133 (56–57), 49-154 (55–57), 49-177 (55–57).

5039 Air Transport Squadron: Elmendorf AFB, Alaska. 49-122 (57), 49-124 (57), 49-127 (57), 49-133 (57), 49-154 (57), 49-177 (57). Redesignated 5040 OP SQ 01Jun57 with records listing C-119 assignments, but all except 49-122 diverted to 5040 CLM GP 01Oct57.

5039 Aircraft Repair Squadron: Elmendorf AFB, Alaska. 49-119 (56), 49-121 (55; 56), 49-122 (56–57), 49-124 (57), 49-125 (56), 49-127 (56), 49-154 (56), 49-177 (56), 49-178 (56–57). Redesignated 5039 MAI (Aircraft Mnt) SQ then 5040 CLM GP 01Jun57.

5039 Maintenance Group: Elmendorf AFB, Alaska. 51-7985 (depl unk). Redesignated 5025 MNT GP.

5040 Consolidated Maintenance Group: Elmendorf AFB, Alaska. 49-124 (57), 49-127 (57), 49-133 (57), 49-154 (57), 49-177 (57), 53-8078 (61). Redesignated from 5039 ARP SQ 01Jun57.

5064 Cold Weather Materiel Testing Unit: Ladd AFB, Alaska. 48-333 (50–51), 48-339 (50–51), 51-2582 (52–53).

Caribbean Air Command (CAC)

Established as the **Panama Canal Air Force** 19Oct40. Redesignated as such 31Jul46 primarily tasked with the defense of the Panama Canal.

5700 Air Base Group: Albrook AFS, Panama Canal Zone. 51-7973 (54), 51-8064 (56), 51-8145 (58), 52-5903 (56), 52-5912 (58), 52-5947 (58).

Continental Air Command (CNC)

Established 1 December 1948 to oversee the United States Continental **Air Force Reserve** (AFR) and **Air National Guard** (ANG) components. It also oversaw **Air Defense Command** (ADC) and **Tactical Air Command** (TAC) within the Continental U.S. from 1 December 1948 to 1 July 1950. The command was inactivated 1 August 1968.

14 Air Force Headquarters: Robins AFB, Georgia. 51-8023 (59), 51-8116 (depl 59), 53-8073 (59).

445 Troop Carrier (Assault) Wing: Memphis MAP, Tennessee (Detachment). 50-141 (60), 51-2610 (60), 52-5936 (60), 53-8137 (60).

2253 Air Base Group: Greater Pittsburgh IAP ARS, Pennsylvania. 51-2657 (59).

2347 Air Base Group: Long Beach IAP, California. 50-126 (59), 51-8108 (60).

2370 Air Base Group: Long Beach IAP, California. 51-2609 (52–53).

2465 Air Base Group: Minneapolis-St. Paul IAP, Minnesota. 50-130 (58), 51-8037 (58), 51-8049 (58), 51-8076 (58), 51-8150 (59), 52-5892 (58), 53-3141 (58).

2500 Air Base Group / Wing: Mitchel AFB, New York. 48-322 (57–58), 48-323 (55–56), 48-335 (55–56), 48-344 (55), 48-346 (57), 49-102 (57–58), 49-104 (56), 49-106 (57),

49-107 (58), 49-108 (56–57), 49-117 (57–58), 50-121 (57), 51-2558 (58), 51-2649 (53), 51-8033 (56), 51-8136 (59), 51-8145 (59), 53-3211 (56), 53-7857 (59).

2578 Air Base Squadron / Group: Ellington AFB, Texas. 49-178 (58), 49-198 (59–60), 50-157 (58), 51-2569 (60), 51-2659 (60), 51-8028 (66–67), 52-5918 (58), 53-8081 (59).

2584 Air Base Squadron / Wing: Memphis MAP, Tennessee. 49-127 (59), 50-166 (58–59), 51-2540 (58–59), 51-2572 (58–59), 51-2580 (59), 51-2589 (59), 51-2600 (58–59), 51-8146 (58).

2585 Air Base Squadron: Miami IAP, Florida. 48-327 (depl 58), 51-2616 (60), 51-2631 (58), 52-5878 (59), 52-5909 (59), 52-5910 (59). Redesignated from 2585 RFC under CNR.

2589 Air Base Group: Dobbins ARB, Georgia. 51-2535 (59), 51-7971 (60).

2596 Flying Training Reserve Center: Hensley Field (NAS Dallas), Texas. 49-143 (52), 51-2628 (59).

Radar Countermeasures Squadron / Wing: Portland Intl., Oregon. 53-7846 (depl 63–64), 53-8105 (depl 63–64).

Continental Air Command Reserve (CNR)

Not technically a Major Command but more a sub-command of Continental Air Command (CNC) in use to distinguish Air Reserve units from Air National Guard units. To AFR 20Nov58.

2230 Air Force Reserve Flying Center: *Operating Base:* NAS New York, New York. *C-119 Service:* 17 1957. *C-119G Assignments:* 51-7981 (57), 51-8002 (57), 51-8076 (57), 51-8135 (57), 52-5851 (57), 52-5862 (depl 57), 52-5920 (57), 52-5933 (depl 57), 53-3215 (57), 53-3216 (57), 53-3217 (57), 53-7857 (57), 53-8094 (57), 53-8095 (57). *C-119J Assignments:* 51-8038 (57), 51-8052 (57), 51-8119 (57).

2231 Air Force Reserve Flying Center: *Operating Base:* Richard E. Byrd Airport (Byrd Field), Virginia. *C-119 Service:* 4 1957. *C-119C Assignments:* 49-125 (57), 50-128 (57), 50-154 (57), 51-2533 (57).

2233 Air Force Reserve Flying Center: *Operating Base:* Mitchel AFB, New York. *Notes:* Previously designated as 2233 CTR CN. Most C-119B cvtd to C-119C standard after assignment. *C-119 Service:* 57 1954–1958. *C-119B Assignments:* 48-320 (55–56; depl 57), 48-322 (54–56; 57–58), 48-323 (54–56), 48-326 (54–55; depl 56, 57), 48-327 (54–56), 48-328 (57), 48-329 (54–56), 48-331 (54–58), 48-332 (depl 56), 48-334 (54–56), 48-335 (54–56), 48-337 (54; 55–56), 48-340 (54–58), 48-341 (54–58), 48-343 (54–58), 48-344 (54–57), 48-346 (54–57), 48-347 (54–58), 48-349 (54–57), 48-351 (depl 56), 48-352 (54–55), 48-355 (54–55; depl 58), 49-101 (54–58), 49-102 (56–57), 49-104 (54–58), 49-106 (54–58), 49-107 (54–55; 56–58), 49-108 (56–58), 49-109 (54–58), 49-110 (54–58), 49-111 (54–58), 49-112 (54–58), 49-113 (54–58), 49-115 (54–58), 49-116 (54–58), 49-117 (56–57), 49-118 (depl 55; 56–58). *C-119C Assignments:* 49-119 (57–58), 49-122 (57–58), 49-127 (57–59), 49-131 (58), 49-170 (57–58), 49-182 (58), 50-124 (57–58), 50-128 (depl 57), 50-130 (58), 50-136 (58), 50-147 (58), 50-162 (58), 50-165 (58), 51-2558 (57–58), 51-2571 (58). *C-119CF Assignments:* 51-2595 (58), 51-2616 (58), 51-8233 (depl 57), 51-8237 (depl 57), 51-8238 (depl 57).

2234 Air Force Reserve Flying Center: *Operating Base:* Laurence G. Hanscom Field, Massachusetts. *Notes:* Detachment at Youngstown MAP, Ohio. *C-119 Service:* 21 1957–1959. *C-119CF Assignment:* 51-8238 (57). *C-119G Assignments:* 51-7981 (57–59), 51-8002 (57–59), 51-8106 (58–59), 51-8136 (57–59), 52-5856 (depl 57), 52-5858 (59), 53-3185

(58–59), 53-7857 (57–59), 53-8080 (57–59), 53-8094 (57–58), 53-8095 (57–59), 53-8109 (58–59), 53-8117 (58–59), 53-8121 (58–59), 53-8133 (57–59). *C-119J Assignments:* 51-8038 (57–58), 51-8052 (57–58), 51-8114 (depl 58), 51-8119 (57–59), 52-5851 (57–58; depl 58).

2235 Air Force Reserve Flying Center: *Operating Base:* Grenier AFB, New Hampshire. *C-119 Service:* 31 1957–1959. *C-119C Assignment:* 48-347 (depl 57-58). *C-119G Assignments:* 51-7975 (57–59), 51-7981 (57), 51-7982 (57–59), 51-7990 (57–59), 51-7994 (57–59), 51-8000 (57–59), 51-8002 (57), 51-8003 (57–59), 51-8006 (57–59), 51-8007 (57–59), 51-8008 (57–59), 51-8020 (57–59), 51-8066 (depl 57), 51-8076 (57), 51-8096 (depl 57), 51-8106 (57–58), 51-8136 (57), 52-5852 (57–58), 52-5872 (depl 57), 52-5873 (57–58), 52-5919 (57–59), 52-5920 (57), 52-5925 (depl 57), 52-5933 (depl 57), 53-3212 (57–59), 53-8086 (57–59), 53-8133 (57). *C-119J Assignments:* 51-8041 (unk–58), 51-8052 (58–59), 52-5895 (57–58).

2237 Air Force Reserve Flying Center: *Operating Base:* New Castle AFB, Delaware. NAS Willow Grove (20–21Jul58). *Notes:* Detachment at Niagara Falls IAP, New York. *C-119 Service:* 40 1957–1958. *C-119G Assignments:* 51-8033 (58), 51-8062 (57–58), 51-8065 (57–58), 51-8066 (58), 51-8068 (57–58), 51-8070 (57–58), 51-8094 (57–58), 51-8096 (58), 51-8102 (57–58), 51-8145 (57–58), 51-8147 (57–58), 52-5843 (57–58), 52-5847 (58), 52-5857 (57–58), 52-5871 (57–58), 52-5872 (58), 52-5873 (58), 52-5913 (58), 52-5915 (58), 52-5917 (57–58), 52-5922 (58), 52-5925 (58), 52-5931 (58), 52-5932 (57–58), 52-5935 (57–58), 52-5945 (57–58), 52-5948 (unk–58), 52-5950 (57–58), 52-5953 (57–58), 53-3182 (58), 53-3209 (58), 53-8097 (58), 53-8116 (58), 53-8123 (58). *C-119J Assignments:* 51-8039 (58), 51-8114 (57–58), 51-8122 (58), 51-8144 (57–58), 51-8152 (57–58), 52-5851 (58).

2242 Air Force Reserve Flying Center: *Operating Base:* Selfridge AFB, Michigan. *Notes:* Detachments at O'Hare IAP, Illinois and Niagara Falls IAP, New York. 51-8021 assigned for salvage. *C-119 Service:* 50 1957–1958. *C-119G Assignments:* 51-7978 (depl 57), 51-7984 (57–unk), 51-8005 (58), 51-8013 (58), 51-8021 (58), 51-8027 (58), 51-8047 (58), 51-8076 (57), 51-8087 (58), 51-8096 (57–58), 51-8097 (58), 51-8099 (58), 51-8112 (57–58), 51-8117 (58), 51-8135 (57–58), 51-8161 (58), 51-8162 (58), 52-5842 (57–58), 52-5844 (58), 52-5847 (57–58), 52-5855 (58), 52-5860 (58), 52-5861 (58), 52-5862 (57–58), 52-5872 (57–58), 52-5883 (58), 52-5901 (58), 52-5905 (58), 52-5907 (58), 52-5920 (57–58), 52-5925 (57–58), 52-5933 (57–58), 53-3161 (58), 53-3215 (57–58), 53-3216 (57–58), 53-3217 (57–58), 53-7856 (57–58), 53-8108 (58), 53-8125 (58). *C-119J Assignments:* 51-8039 (57–58), 51-8042 (58), 51-8049 (58), 51-8051 (57–58), 51-8066 (57–58), 51-8121 (58), 51-8122 (57–58), 52-5845 (unk–58), 52-5849 (57–58), 52-5877 (58), 53-8098 (58).

2252 Air Force Reserve Flying Center: *Operating Base:* Clinton County AFB, Ohio. *C-119 Service:* 43 1956–1959. *C-119C Assignments:* 48-320 (56–58), 48-323 (56–57), 48-327 (56–58), 48-329 (56–58), 48-334 (56–58), 48-337 (56–59), 49-124 (57–59), 49-125 (57–59), 49-129 (57–59), 49-156 (57–59), 49-157 (56–59), 49-167 (57–59), 49-171 (58–59), 49-177 (57–59), 49-178 (57–59), 49-180 (57–59), 50-120 (58–59), 50-128 (57–59), 50-133 (58–59), 50-135 (57–59), 50-138 (58–59), 50-142 (57–59), 50-154 (57–59), 51-2532 (57–59), 51-2533 (57–59), 51-2534 (56–59), 51-2539 (58–59), 51-2544 (57–59), 51-2549 (58–59), 51-2573 (57–59), 51-2582 (58–59). *C-119CF Assignments:* 51-2588 (57–59), 51-2591 (58–59), 51-2597 (56–59), 51-2605 (57), 51-2608 (56–59), 51-2614 (56–57), 51-2615 (57–59), 51-2620 (57–59), 51-2623 (58–59), 51-2626 (57–59), 51-8233 (56–57), 51-8238 (57).

2253 Air Force Reserve Flying Center: *Operating Base:* Greater Pittsburgh IAP Air Reserve Station, Pennsylvania. *Notes:* Most C-119B cvtd to C-119C standard after assignment. *C-119 Service:* 33 1954–1957. *C-119B Assignments:* 48-322 (56–57), 48-325 (54–57),

48-326 (55–57), 48-327 (depl 55, 56), 48-329 (depl 57), 48-331 (depl 56, 57), 48-332 (54–57), 48-334 (depl 55, 56, 57), 48-335 (56–57), 48-340 (depl 55), 48-341 (depl 56, 57), 48-343 (depl 55, 57), 48-344 (depl 56, 57), 48-346 (depl 56, 57), 48-349 (depl 55, 56, 57), 48-351 (54–57), 48-352 (55–57), 48-354 (56–57), 48-355 (depl 55; 55–58), 49-102 (55–56), 49-106 (depl 55–56), 49-107 (55), 49-108 (54–56), 49-109 (depl 56), 49-112 (depl 56), 49-113 (depl 56), 49-114 (54–55), 49-115 (depl 55), 49-116 (depl 56), 49-117 (54–56), 49-118 (54–56). *C-119C Assignments:* 49-157 (depl 57), 51-2534 (depl 57).

2255 Air Force Reserve Flying Center: *Operating Base:* Bradley IAP (Bradley ANGB), Connecticut. *C-119 Service:* 11 1958–1959. *C-119C Assignments:* 49-121 (58–59), 49-143 (58–59), 49-159 (58–59), 50-131 (58–59), 50-140 (58–59), 50-148 (58–59), 50-155 (58–59), 50-158 (59), 51-2536 (59), 51-2640 (58–59), 51-2545 (58–59).

2256 Air Force Reserve Flying Center: *Operating Base:* Niagara Falls MAP, New York. *C-119 Service:* 12 1957. *C-119G Assignments:* 51-8039 (57), 51-8066 (57), 51-8094 (57), 51-8096 (57), 52-5842 (57), 52-5847 (57), 52-5862 (57), 52-5872 (57), 52-5925 (57), 52-5933 (57), 52-5948 (57–unk). *C-119J Assignment:* 51-8122 (57).

2259 Air Force Reserve Flying Center / Detachment: *Operating Base:* Greater Pittsburgh IAP ARS, Pennsylvania. Andrews AFB, Maryland. *Notes:* Detachment at Youngstown MAP, Ohio. Deployment to Clinton, Ohio Aug57. *C-119 Service:* 49 1957–1958. *C-119B Assignment:* 48-349 (57–58). *C-119C Assignments:* 48-320 (58), 48-325 (57–58), 48-326 (57–58), 48-327 (58), 48-329 (58), 48-332 (57–58), 48-334 (58), 48-335 (57–58), 48-343 (58), 48-344 (57–58), 48-346 (57–58), 48-351 (57–58), 48-352 (57–58), 48-354 (57–58), 48-355 (58), 49-121 (58), 49-132 (58), 49-143 (58), 49-157 (depl 57-58), 49-159 (58), 49-178 (57), 49-187 (58), 50-131 (58), 50-140 (58), 50-148 (58), 50-155 (58), 50-158 (58), 51-2536 (58). *C-119CF Assignments:* 51-2598 (58), 51-2599 (57–58), 51-2605 (57–58), 51-2606 (57–58), 51-2612 (57–58), 51-2613 (57–58), 51-2614 (57–58), 51-2618 (58), 51-2619 (58), 51-2622 (57–58), 51-2630 (58), 51-2631 (57–58), 51-2632 (57–unk), 51-2640 (58), 51-2646 (57–unk), 51-2649 (58–unk), 51-2661 (57–unk), 51-8233 (57–58), 51-8237 (57–58), 51-8238 (57–58).

2346 Air Force Reserve Flying Center: *Operating Base:* Hamilton AFB, California. *Notes:* Detachments at Portland IAP, Oregon; McClellan AFB, California and Paine Field, Washington State. *C-119 Service:* 50 1958. *C-119G Assignments:* 51-7973 (58), 51-8059 (58), 51-8069 (58), 51-8087 (58–59), 51-8089 (58), 51-8098 (58), 52-5848 (58), 52-5890 (58), 52-5923 (58), 52-5927 (58), 52-5938 (58), 52-5939 (58), 52-5942 (58), 52-9981 (58), 52-9982 (58), 53-3142 (58), 53-3144 (58), 53-3146 (58), 53-3147 (58), 53-3156 (58), 53-3157 (58), 53-3160 (58), 53-3162 (58), 53-3166 (58), 53-3168 (58), 53-3170 (58), 53-3174 (58), 53-3183 (58), 53-3186 (58), 53-3190 (58), 53-3196 (58), 53-3198 (58), 53-3202 (58), 53-3203 (58), 53-3204 (58), 53-3205 (58), 53-3206 (58), 53-7876 (58), 53-7884 (58), 53-8071 (58), 53-8087 (58), 53-8089 (58), 53-8105 (58), 53-8110 (58), 53-8111 (58), 53-8114 (58), 53-8120 (58), 53-8124 (58), 53-8126 (58), 53-8131 (58).

2347 Air Force Reserve Flying Center: *Operating Base:* Long Beach AP, California. *Notes:* Detachment at Hill AFB, Utah. *C-119 Service:* 76 1958–1959. *C-119C Assignments:* 48-355 (depl 56), 50-137 (depl 56). *C-119G Assignments:* 51-7979 (57; 58-59), 51-7987 (58–59), 51-8011 (58–59), 51-8057 (58–59), 51-8060 (59), 51-8071 (59), 51-8072 (59), 51-8080 (58–59), 51-8081 (59), 51-8084 (59), 51-8088 (59), 51-8090 (59), 51-8092 (58–59), 51-8107 (58–59), 51-8160 (58–59), 51-8161 (58–59), 52-5840 (58–59), 52-5880 (59), 52-5898 (58–59), 52-5900 (58–59), 52-5929 (58–59), 52-5941 (58–59), 52-5942 (58), 52-5943 (59), 53-3145 (58–59), 53-3148 (58–59), 53-3151 (58–59), 53-3152 (58–59), 53-3158 (58–59), 53-3159 (58–59),

53-3160 (58), 53-3161 (58–59), 53-3166 (58), 53-3167 (58–59), 53-3168 (58), 53-3173 (58–59), 53-3174 (58), 53-3175 (58–59), 53-3179 (58–59), 53-3195 (58–59), 53-3196 (58), 53-3197 (58–59), 53-3198 (58), 53-3199 (58–59), 53-3201 (58–59), 53-3202 (58), 53-3203 (58), 53-3204 (58), 53-3205 (58), 53-3206 (58), 53-3207 (58–59), 53-7864 (58–59), 53-7865 (58–59), 53-7877 (58–59), 53-8072 (58–59), 53-8087 (58), 53-8090 (58–59), 53-8093 (58–59), 53-8096 (58–59), 53-8100 (58–59), 53-8104 (58–59), 53-8105 (58), 53-8106 (58–59), 53-8107 (58), 53-8110 (58), 53-8124 (58), 53-8128 (58–59), 53-8132 (58–59). *C-119J Assignments:* 51-8045 (58), 51-8132 (58–59), 52-5884 (58–59), 52-5903 (58–59), 52-5906 (58–59), 53-3213 (58–59).

2465 Air Force Reserve Flying Center: *Operating Base:* Minneapolis-St. Paul IAP ARS, Minnesota. *Notes:* 22 1957. *C-119G Assignments:* 51-7983 (depl 57), 51-7999 (depl 57), 51-8056 (57), 51-8074 (57), 51-8078 (57), 51-8150 (57), 51-8160 (57), 52-5841 (57), 52-5846 (57; depl 58), 52-5868 (57), 52-5874 (57), 52-5876 (57), 52-5889 (57), 52-5892 (57), 52-5900 (57), 52-5902 (57), 53-3141 (57), 53-8099 (57). *C-119J Assignments:* 51-7968 (57), 51-8037 (57), 51-8045 (57), 51-8130 (57).

2466 Air Force Reserve Flying Center: *Operating Base:* Bakalar AFB, Indiana. *Notes:* Detachments at Dress Memorial AP, Indiana and Scott AFB, Illinois. *C-119 Service:* 93 1957–1959. *C-119G Assignments:* 51-2662 (57–59), 51-2667 (57–unk), 51-2673 (57), 51-2675 (57), 51-2684 (57–59), 51-2685 (unk), 51-2710 (57), 51-2711 (depl 57), 51-2712 (57–unk), 51-7969 (depl 57), 51-7970 (depl 57; 57–59), 51-7976 (57–59), 51-7977 (57–59), 51-7978 (57–59), 51-7980 (depl 57), 51-7983 (57–59), 51-7986 (depl 57), 51-7988 (57–59), 51-7989 (57–59), 51-7991 (57–59), 51-7996 (57–59), 51-7997 (56–59), 51-7998 (57–58), 51-7999 (57–58), 51-8009 (57–59), 51-8010 (57–59), 51-8011 (depl 57; 57-58), 51-8013 (depl 57), 51-8014 (57–58), 51-8015 (57–59), 51-8018 (depl 57; 57–58), 51-8019 (depl 57; 57–59), 51-8021 (depl 57), 51-8022 (57–59), 51-8024 (57–58), 51-8027 (depl 57), 51-8028 (57–59), 51-8047 (depl 57), 51-8048 (depl 57), 51-8054 (58–59), 51-8061 (depl 57), 51-8063 (depl 57), 51-8073 (depl 57), 51-8074 (depl 57), 51-8076 (57), 51-8079 (57–59), 51-8100 (57–59), 51-8101 (57–59), 51-8117 (depl 57), 51-8135 (57), 51-8139 (depl 57; 57–59), 51-8161 (57–58), 51-8162 (57), 51-8168 (57–59), 52-5840 (57–58), 52-5845 (depl 57), 52-5846 (depl 57), 52-5850 (depl 57; 57–59), 52-5860 (depl 57), 52-5861 (depl 57), 52-5864 (depl 57; 57–59), 52-5868 (57), 52-5874 (57), 52-5876 (depl 57; 57), 52-5883 (depl 57), 52-5889 (57), 52-5892 (57), 52-5900 (57), 52-5902 (57), 52-5909 (58–59), 52-5912 (58–59), 52-5918 (depl 57), 52-5920 (57), 52-5947 (depl 57; C-119J 57–59), 52-5952 (57–59), 53-3136 (57–59), 53-3139 (depl 57; 57–59), 53-3140 (depl 57; 57–59), 53-3215 (57), 53-3216 (57), 53-3217 (57), 53-7855 (57–58), 53-7859 (57–59), 53-8079 (57), 53-8088 (57–59). *C-119J Assignments:* 51-8043 (57–58), 51-8050 (57–58), 51-8115 (57–58), 51-8132 (57–58), 51-8156 (57–59), 52-5877 (depl 57), 52-5897 (57–59), 53-8098 (depl 57).

2469 Air Force Reserve Flying Center: *Operating Base:* Scott AFB, Illinois. *Notes:* Detachment to Scott AFB, Illinois from 01Dec57. *C-119 Service:* 19 1957. *C-119G Assignments:* 51-2673 (57–unk), 51-2675 (57–unk), 51-2710 (57–unk), 51-7970 (57), 51-7983 (depl 57), 51-8011 (57), 51-8015 (57), 51-8018 (57), 51-8019 (57), 51-8139 (57), 51-8161 (57), 52-5840 (depl 57), 52-5850 (57), 52-5864 (57), 52-5947 (57), 53-3139 (57), 53-3140 (57). *C-119J Assignments:* 51-8043 (57), 51-8115 (depl 57).

2471 Air Reserve Training Center: *Operating Base:* O'Hare IAP ARS, Illinois. *C-119 Service:* 21 1957. *C-119C Assignment:* 49-149 (50). *C-119G Assignments:* 51-8013 (57), 51-8021 (57), 51-8027 (57), 51-8047 (57), 51-8087 (57), 51-8117 (57), 51-8162 (57), 52-5844 (57), 52-5845 (57), 52-5860 (57), 52-5861 (57), 52-5877 (57), 52-5883 (57), 52-5905 (57), 53-3137 (depl 57), 53-8098 (57). *C-119J Assignments:* 51-8042 (57), 51-8049 (57), 51-8121 (57), 51-8156 (depl 57).

2472 Air Reserve Training Center: *Operating Base:* Richards-Gebaur AFB, Missouri. *Notes:* Detachments at Tinker AFB, Oklahoma and Davis Field, Oklahoma. *C-119 Service:* 65 1957–1959. *C-119G Assignments:* 51-2670 (57–59), 51-2681 (58–unk), 51-2683 (58–59), 51-2684 (depl 57), 51-2686 (57–59), 51-2711 (57–unk), 51-7969 (57–59), 51-7974 (57–59), 51-7980 (57–59), 51-7986 (57–59), 51-7998 (58–59), 51-8016 (57–59), 51-8017 (57–59), 51-8018 (depl 58), 51-8024 (58–59), 51-8029 (57–59), 51-8031 (58), 51-8048 (57–59), 51-8061 (57–59), 51-8063 (57–59), 51-8073 (57–59), 51-8077 (57–58), 51-8078 (57), 51-8091 (57–59), 51-8093 (57–59), 51-8150 (57), 52-5863 (57–59), 52-5867 (57–59), 52-5868 (58), 52-5870 (58–59), 52-5876 (depl 57), 52-5902 (depl 57), 52-5916 (58–59), 52-5918 (57–59), 52-5924 (57–59), 52-5940 (58–59), 52-5951 (57–59), 53-3138 (57–59), 53-3180 (58), 53-3181 (58–59), 53-3184 (58), 53-3187 (58–59), 53-3188 (58–59), 53-3193 (58), 53-3194 (58–59), 53-3210 (58), 53-3211 (58), 53-7859 (depl 57–58), 53-8075 (57–59), 53-8076 (57–59), 53-8081 (57–59), 53-8091 (57–59), 53-8100 (57–58), 53-8101 (57–59), 53-8112 (58–59), 53-8115 (58–59), 53-8119 (58), 53-8122 (58–59), 53-8134 (57–59), 53-8135 (57–59), 53-8137 (58–59). *C-119J Assignments:* 51-8037 (57), 51-8128 (57–59), 52-5877 (58), 52-5896 (58–59).

2473 Air Force Reserve Flying Center: *Operating Base:* General Mitchell IAP-ARS, Wisconsin. *Notes:* Detachments at Minneapolis-St. Paul IAP, Minnesota and O'Hare IAP ARS, Illinois. *C-119 Service:* 56 1957–1959. *C-119G Assignments:* 51-7999 (58), 51-8012 (57–58), 51-8013 (57–58), 51-8014 (58), 51-8021 (57–58), 51-8027 (57–58), 51-8047 (57–58), 51-8053 (57–58), 51-8056 (57–59), 51-8074 (57–58), 51-8076 (57–58), 51-8078 (57–59), 51-8082 (57–58), 51-8087 (57–58), 51-8109 (57–58), 51-8117 (57–58), 51-8150 (57–58), 51-8160 (57–58), 51-8162 (57–58), 52-5841 (57–59), 52-5844 (57–58), 52-5846 (57–58), 52-5860 (57–58), 52-5861 (57–58), 52-5868 (57–58), 52-5874 (57–59), 52-5876 (57–58), 52-5883 (57–58), 52-5889 (57–59), 52-5892 (57–58), 52-5900 (57–58), 52-5902 (57–59), 52-5905 (57–58), 52-5911 (58–59), 52-5914 (58–59), 53-3137 (depl 57), 53-3141 (57–59), 53-3191 (58), 53-7858 (57–58), 53-8070 (57–58), 53-8074 (57–58), 53-8078 (57–58), 53-8079 (57–58), 53-8082 (57–58), 53-8099 (57–59), 53-8113 (58). *C-119J Assignments:* 51-7968 (57–58), 51-8037 (57–58), 51-8042 (57–58), 51-8045 (57–58), 51-8049 (57–58), 51-8121 (57–58), 51-8130 (57–58), 52-5845 (57–unk), 52-5877 (57–58), 53-8098 (57–58).

2559 Air Force Reserve Flying Center: *Operating Base:* Andrews AFB, Maryland. *C-119C Assignments:* 51-2545 (58), 51-8262 (58).

2577 Air Force Reserve Flying Center: *Operating Base:* Brooks AFB, Texas. *Notes:* Detachment 1 at NAS Dallas (Hensley Field), Texas. *C-119 Service:* 54 1957–1958. *C-119C Assignments:* 49-120 (58), 49-133 (57–60), 49-139 (58), 49-140 (58), 49-141 (58), 49-142 (58), 49-154 (57–58), 49-162 (57–58), 49-164 (58), 49-165 (58), 49-174 (57–58), 49-176 (58), 49-179 (58), 49-185 (57–58), 49-188 (58), 49-193 (58), 49-197 (58), 50-119 (58), 50-124 (57), 50-126 (58), 50-129 (58), 50-132 (58), 50-134 (57–58), 50-137 (58), 50-139 (58), 50-141 (58), 50-146 (depl 58), 50-157 (58), 50-168 (58), 50-170 (58), 51-2535 (58), 51-2548 (58), 51-2559 (57–58), 51-2561 (57–58), 51-2564 (58), 51-2566 (58), 51-2567 (58), 51-2577 (58), 51-2579 (depl 57–58), 51-2580 (58), 51-2584 (57–58). *C-119CF Assignments:* 51-2593 (58), 51-2596 (58), 51-2604 (57–58), 51-2607 (57–58), 51-2624 (58), 51-2627 (58), 51-2628 (58), 51-2629 (57–58), 51-2636 (58–59), 51-2639 (58–unk), 51-2643 (58–unk), 51-2654 (58–unk), 51-2655 (57–unk).

2578 Air Force Reserve Flying Center: *Operating Base:* Ellington AFB, Texas. *Notes:* Detachment at Alvin Callender Field, Louisiana. *C-119 Service:* 47 1958–unk. *C-119C Assignments:* 48-339 (58), 49-135 (58; depl 59), 49-136 (58), 49-137 (58), 49-139 (depl 58), 49-144 (58), 49-146 (58), 49-152 (58), 49-154 (58), 49-155 (58), 49-158 (58), 49-183

(58), 49-185 (depl 58), 49-190 (58), 49-194 (58), 49-198 (58), 50-119 (58), 50-134 (depl 58), 50-143 (58), 50-146 (58), 50-150 (58), 50-151 (58), 50-156 (58), 50-160 (58), 50-171 (58), 51-2538 (58), 51-2550 (58), 51-2553 (58), 51-2554 (58), 51-2562 (58), 51-2568 (58), 51-2574 (58), 51-2579 (57–58), 51-2581 (58). *C-119CF Assignments:* 51-2587 (58), 51-2594 (58), 51-2601 (57–58), 51-2603 (58), 51-2611 (58), 51-2617 (58), 51-2625 (58), 51-2633 (58–unk), 51-2634 (58–unk), 51-2641 (58–unk), 51-2644 (58–unk), 51-2651 (58–unk), 51-8244 (58).

2584 Air Force Reserve Flying Center: *Operating Base:* Memphis MAP, Tennessee. *C-119 Service:* 29 1958–1959. *C-119C Assignments:* 49-111 (depl 58), 49-121 (58), 49-137 (58), 50-120 (58), 50-131 (58), 50-133 (58), 50-140 (58), 50-166 (58), 51-2536 (57–58), 51-2540 (58), 51-2541 (58), 51-2550 (58), 51-2572 (55; 58), 51-2583 (58). *C-119CF Assignments:* 51-2589 (58), 51-2591 (58), 51-2592 (58), 51-2598 (58), 51-2600 (58), 51-2611 (58–59), 51-2618 (58), 51-2623 (58), 51-2628 (58), 51-2637 (58–unk), 51-2638 (58–unk), 51-2642 (58–unk), 51-2647 (58–unk), 51-2650 (58–unk). *C-119G Assignment:* 53-8123 (58).

2585 Air Force Reserve Flying Center: *Operating Base:* Miami IAP, Florida. *Notes:* Detachments at Donaldson AFB, South Carolina and Orlando, Florida. Deployment to Dobbins AFB Apr57. Redesignated as 2585 AB SQ under CNC 29Apr58. *C-119 Service:* 33 1957–1959. *C-119C Assignment:* 50-147 (58–59). *C-119G Assignments:* 51-7971 (57–58), 51-7972 (57–58), 51-8001 (57–58), 51-8004 (57–58), 51-8023 (57–58), 51-8025 (57–58), 51-8044 (57–58), 51-8055 (57–58), 51-8073 (depl 57), 51-8075 (57–58), 51-8083 (58), 51-8146 (57–58), 52-5856 (58), 52-5878 (58), 52-5887 (57–58), 52-5910 (58), 52-5926 (57–58), 52-5934 (57–58), 52-5936 (57–58), 52-5946 (57–58), 53-8069 (58), 53-8070 (57), 53-8073 (57–58), 53-8079 (57), 53-8083 (57–58), 53-8085 (57–58), 53-8127 (57–58), 53-8129 (57–58), 53-8136 (58). *C-119J Assignments:* 51-8036 (57–58), 51-8116 (57–58), 52-5895 (58).

2589 Air Force Reserve Flying Center: *Operating Base:* Dobbins AFB, Georgia. *Notes:* Detachment at Donaldson AFB, South Carolina. *C-119 Service:* 23 1955–1956; 1957–1958. *C-119C Assignments:* 49-143 (58), 49-159 (57–58), 49-187 (57–58), 50-136 (57–58), 50-140 (57–58), 50-148 (58), 50-155 (57–58), 50-158 (57–58), 50-170 (depl 58), 51-2536 (57), 51-2540 (57–58), 51-2545 (57–58). *C-119CF Assignments:* 51-2589 (depl 58), 51-2600 (57–58), 51-2601 (57), 51-2610 (57–58), 51-2611 (58), 51-2630 (57–58), 51-2640 (57–58), 51-2652 (57–unk), 51-2657 (57–unk). *C-119F Assignment:* 51-8099 (depl 55–56). *C-119G Assignment:* 51-8025 (depl 58).

2596 Air Force Reserve Flying Center: *Operating Base:* NAS Dallas (Hensley Field), Texas. *C-119 Service:* 10 1954; 1957. *C-119B Assignment:* 48-347 (depl 54). *C-119C Assignments:* 49-162 (57), 49-174 (57), 49-185 (57), 50-124 (57), 51-2559 (57), 51-2561 (57), 51-2579 (57). *C-119CF Assignments:* 51-2607 (57), 51-2656 (57–unk).

Far East Air Forces (FEA)

Major Command established as such on 1 January 1947 redesignated from the previous **Pacific Air Command, United States Army** first established after World War II. Redesignated as the **Pacific Air Forces** on 1 July 1957.

3 Bombardment (Light) Wing: Johnson AB, Japan. 53-7867 (55–57), 53-7878 (55–57), 53-7880 (56–57), 53-7883 (55–57).

8 Fighter-Bomber Wing: Itazuke AB, Japan. 51-2578 (depl 56), 52-5874 (depl 57), 53-7868 (55–57), 53-7873 (55–57).

18 Fighter-Bomber Wing: Kadena AB, Japan. 51-2537 (55), 51-2545 (55), 51-8066 (55), 53-3170 (55), 53-7869 (55–57), 53-7871 (56), 53-7874 (55–57).

Section Two—Military Operators

19 Bombardment (Medium) Wing: Andersen AFB, Guam. 48-322 (depl 53), 48-348 (depl 50), 49-162 (53), 50-138 (53), 51-2564 (53), 51-2583 (53).

24 Depot Maintenance Group: Clark AB, The Philippines. 48-348 (sal, 53), 51-2537 (53), 51-2541 (53), 51-2543 (53), 51-2550 (53), 51-2581 (53), 51-2583 (53).

35 Fighter-Interceptor Wing: Pusan East AB (K-9), South Korea (Dec50), to Johnson AB, Japan (25May51). 49-167 (depl 51), 51-2556 (depl 52).

49 Fighter-Bomber Wing: Miawa AB, Japan. 53-7870 (55–57), 53-7879 (55–57).

58 Fighter-Bomber Wing: Osan-Ni AB, South Korea. 53-7871 (55–56), 53-7872 (55), 53-7880 (depl 55–56).

67 Tactical Reconnaissance Wing: Itami AB, Japan. 52-5953 (56), 53-7872 (55–57), 53-7880 (55–56), 53-7881 (55–57).

75 Air Depot Wing: Miho AB, Japan. 48-328 (54).

314 Troop Carrier (Medium) Wing: *Group:* **314 Troop Carrier Group.** *Squadrons (Colors):* **37 TCSq (Attached)** (Yellow); **50 TCSq** (Red/White); **61 TCSq** (Green/White); **62 TCSq** (Blue/White). *Operating Base:* Ashiya AB, Japan (07Sep50–15Nov54). *Notes:* 37 TCSq at Komaki AB, Japan. Deployments to Clark AB The Philippines 1954 (French Indochina); Bofu, Japan and Tachikawa, Japan. *C-119 Service:* 152 1950–1954. *C-119B Assignments:* 48-324 (50), 48-325 (50–51), 48-325 (53–54), 48-327 (50–52; depl 53), 48-328 (50–53), 48-329 (50–53), 48-331 (50–53), 48-332 (50–54), 48-334 (50–52), 48-336 (50), 48-337 (50–52; 53), 48-340 (50; depl 54), 48-341 (50–52), 48-343 (50–52), 48-344 (50–51; depl 53–54), 48-345 (50–51), 48-346 (50–51), 48-347 (50–52), 48-348 (50–53), 48-349 (50–52; depl 54), 48-350 (50–51), 48-351 (50–52), 48-352 (50–52; depl 53–54), 48-353 (50), 48-354 (50–53; depl 53–54), 48-355 (50–52; depl 53–54), 49-101 (50–52; depl 54), 49-102 (50–52; depl 53–54), 49-103 (50–52), 49-104 (50–51), 49-105 (50), 49-106 (50–53; depl 54), 49-107 (50–52), 49-108 (50–52), 49-109 (50–52; depl 53–54), 49-110 (50–52; depl 53–54), 49-111 (50–52), 49-112 (50–53; depl 53–54), 49-113 (50–53), 49-114 (50–53), 49-115 (50–52), 49-116 (50–52), 49-117 (50–52), 49-118 (50–52; depl 54). *C-119C Assignments:* 49-119 (50–52; depl 54), 49-120 (50–52), 49-121 (50–54; 54), 49-122 (50–52; 54), 49-123 (50–51), 49-124 (50–53; depl 54), 49-125 (50–52; depl 54), 49-126 (50–52), 49-127 (50–53; 54), 49-128 (50–53; 54), 49-129 (50–53), 49-130 (50–51), 49-131 (50–52; depl 54), 49-132 (50–52), 49-133 (50–52; depl 54), 49-135 (50–54), 49-136 (50–54), 49-137 (50–54), 49-138 (50–52), 49-139 (50–52; depl 54), 49-140 (50–53; depl 54), 49-141 (50–53; depl 54), 49-142 (50–53), 49-143 (50–52; 53–54), 49-144 (50–52; depl 54), 49-145 (50), 49-146 (depl 50–51; 51–54), 49-147 (50–52; 54), 49-149 (50–52; depl 53–54), 49-150 (50–52), 49-151 (50–53; 54), 49-152 (50–52; 54), 49-153 (50–51), 49-154 (50–54), 49-155 (50–54), 49-156 (50–53), 49-157 (50–54), 49-158 (50–53), 49-159 (50–51), 49-160 (50–52), 49-161 (50–52), 49-162 (50–52), 49-164 (50–54), 49-165 (50–52; depl 54), 49-166 (50), 49-167 (50–54), 49-168 (50), 49-169 (50–51), 49-170 (50–52), 49-171 (50–52), 49-172 (50–51), 49-173 (50–52), 49-174 (50–52), 49-175 (50–52), 49-176 (50–54), 49-177 (50–53; depl 54), 49-178 (50–52), 49-180 (53–54), 49-182 (54), 49-185 (53–54), 50-138 (53–54), 51-2532 (52–54), 51-2537 (52–54), 51-2538 (54), 51-2540 (54), 51-2541 (52–54), 51-2543 (54), 51-2545 (52–54), 51-2546 (54), 51-2547 (54), 51-2548 (54), 51-2549 (52–54), 51-2550 (52–54), 51-2552 (53–54), 51-2553 (52–54), 51-2555 (depl 54), 51-2557 (depl 54), 51-2558 (52–54), 51-2559 (52–54), 51-2561 (54), 51-2562 (53–54), 51-2564 (depl 54), 51-2565 (54–55), 51-2569 (52–53), 51-2571 (depl 54), 51-2573 (52–54), 51-2574 (54), 51-2574 (depl 53–54), 51-2576 (depl 54), 51-2577 (52–54), 51-2581 (depl 54), 51-2583 (depl 54), 51-2584 (depl 54). *C-119G Assignments:* 51-8055 (depl 54), 51-8056 (depl 54), 51-8062 (depl 54), 51-8066 (depl 54),

51-8070 (depl 54), 51-8087 (depl 54), 51-8096 (depl 54), 52-5846 (depl 54), 52-5857 (depl 54), 52-5871 (depl 54), 52-5873 (depl 54), 52-5876 (depl 54), 52-5886 (depl 54), 52-5921 (depl 54), 52-5935 (depl 54).

316 Troop Carrier (Medium) Wing: *Group:* **316 Troop Carrier Group.** *Squadrons (Colors):* **36 TCSq** (Red/White); **37 TCSq** (Blue/White); **75 TCSq** (Green/White). *Operating Base:* Ashiya AB, Japan. *Notes:* Attached to **483 TCWG** 15Nov54–18Jun57. 51-8062 deployment to Andersen AFB, Guam May55. *C-119 Service:* 91 1954–1956. *C-119C Assignments:* 49-119 (54–55), 49-124 (54), 49-125 (54), 49-131 (54), 49-133 (54), 49-135 (54), 49-139 (54–55), 49-140 (54), 49-141 (54–55), 49-144 (54–55), 49-152 (54), 49-165 (54–55), 51-2532 (54), 51-2537 (54), 51-2541 (54–55), 51-2543 (54–55), 51-2547 (54–55), 51-2548 (54), 51-2550 (54–55), 51-2553 (54), 51-2555 (54–55), 51-2557 (54–55), 51-2558 (54–55), 51-2561 (54–55), 51-2564 (54–55), 51-2565 (55–56), 51-2571 (54), 51-2572 (54–55), 51-2573 (54), 51-2574 (54–55), 51-2575 (54–55), 51-2576 (54–55), 51-2581 (54–55), 51-2583 (54), 51-2584 (54). *C-119G Assignments:* 51-8055 (depl 54–55), 51-8056 (depl 54–55), 51-8062 (depl 54–55), 51-8065 (depl 55–56), 51-8066 (depl 54–55), 51-8068 (depl 55–56), 51-8070 (depl 54–56), 51-8074 (depl 54–55), 51-8076 (depl 55–56), 51-8085 (depl 54–55), 51-8087 (54–56), 51-8094 (depl 55–56), 51-8096 (depl 54–56), 52-5842 (depl 55, 55–56), 52-5843 (depl 54–56), 52-5846 (depl 54–56), 52-5852 (depl 54–56), 52-5857 (depl 54–56), 52-5861 (depl 55–56), 52-5862 (depl 55–56), 52-5868 (depl 55–56), 52-5871 (depl 54–56), 52-5872 (depl 54–56), 52-5873 (depl 54–56), 52-5874 (depl 55–56), 52-5876 (depl 54–56), 52-5881 (depl 54–55), 52-5883 (depl 55–56), 52-5885 (depl 54–56), 52-5886 (depl 54–55), 52-5887 (depl 55–56), 52-5889 (depl 55–56), 52-5892 (depl 55–56), 52-5900 (depl 55–56), 52-5902 (depl 54–56), 52-5905 (depl 55–56), 52-5917 (depl 55–56), 52-5919 (depl 55–56), 52-5920 (depl 54–56), 52-5921 (depl 54), 52-5932 (depl 55–56), 52-5933 (depl 55–56), 52-5934 (dep 55–56), 52-5935 (depl 54–56), 52-5936 (depl 55–56), 52-5944 (depl 55–56), 52-5945 (depl 55–56), 52-5946 (depl 55–56), 52-5948 (depl 55–56), 52-5950 (depl 55–56), 52-5953 (depl 55–56), 53-3166 (depl 55–56), 53-3175 (depl 55–56), 53-3177 (depl 55–56), 53-7875 (depl 56), 53-7876 (55–56).

319 Air Base Wing: Andersen AFB, Guam. 52-5944 (55), 53-3175 (55).

322 Troop Carrier (Medium) Squadron: Kadena AB, Japan. 49-129 (56–57), 49-157 (56).

374 Troop Carrier (Heavy) Wing: *Group:* **374 Troop Carrier Group.** *Squadron (Color):* **21 TCSq (Attached)** (Red, Yellow/Black, Yellow). *Operating Base:* Tachikawa AB, Japan. *Notes:* 21 TCSq at Itazuke AB briefly and "attached" to the **483 TCWG** from 18Sep56 but actually remained with the 374 TCWG. Deployment to Kanoya AB, Japan Oct50. *C-119 Service:* 15 1950–1952, 1955–1956. *C-119B Assignments:* 48-327 (52), 49-102 (depl 50-51), 49-104 (51–52). *C-119C Assignments:* 49-128 (depl 50-51), 49-129 (depl 51), 49-145 (50), 49-146 (50–51), 49-151 (depl 51), 49-153 (depl 50, 51), 49-154 (depl 50, 51), 49-171 (depl 51). *C-119G Assignments:* 53-3153 (depl 55), 53-7860 (55–56), 53-7862 (55–56), 53-7863 (55–56).

403 Troop Carrier (Medium) Wing: *Group:* **403 Troop Carrier Group.** *Squadrons:* **63 TCSq**; **64 TCSq**; **65 TCSq**. *Operating Base:* Ashiya AB, Japan. *Notes:* Wing reassigned to **Air Force Reserve** (AFR) 01Jan53. *C-119 Service:* 78 1951–1953. *C-119B Assignments:* 48-326 (51–52), 48-334 (52), 48-349 (depl 52), 49-101 (52), 49-102 (52), 49-103 (52), 49-107 (52), 49-108 (52), 49-109 (52), 49-110 (52), 49-111 (52–53), 49-114 (52), 49-115 (52–53), 49-116 (52), 49-117 (52–53), 49-118 (52). *C-119C Assignments:* 49-120 (depl 52), 49-122 (52), 49-125 (depl 52), 49-131 (52–53), 49-133 (52), 49-139 (52–53), 49-143 (52), 49-144 (52–53),

Section Two—Military Operators 81

49-147 (52), 49-152 (52), 49-154 (depl 52), 49-160 (52–53), 49-161 (52–53), 49-162 (52–53), 49-165 (52–53), 49-167 (depl 52), 49-170 (52–53), 49-171 (52–53), 49-174 (52–53), 49-178 (52–53), 49-181 (52–53), 49-182 (52–53), 49-184 (depl 52), 49-193 (52–53), 49-195 (52–53), 51-2532 (52), 51-2536 (52), 51-2537 (52), 51-2538 (52–53), 51-2539 (52–53), 51-2540 (52–53), 51-2541 (52), 51-2543 (52), 51-2545 (52), 51-2546 (52–53), 51-2548 (52), 51-2549 (52), 51-2550 (52), 50-2551 (52), 51-2552 (52–53), 51-2553 (52), 51-2555 (depl 52-53), 51-2556 (depl 52–53), 51-2557 (52–53), 51-2558 (52), 51-2559 (52), 51-2561 (52–53), 51-2562 (52–53), 51-2563 (52–53), 51-2565 (52–53), 51-2567 (depl 52-53), 51-2569 (52), 51-2571 (52–53), 51-2572 (52–53), 51-2573 (52), 51-2574 (52–53), 51-2576 (52–53), 51-2577 (52), 51-2578 (52–53), 51-2580 (depl 52–53), 51-2583 (52–53), 51-2584 (52–53).

421 Aerial Refueling (Medium) Squadron: Yokota AB, Japan. 53-7860 (56; 58), 53-7861 (55–unk), 53-7862 (56–unk), 53-7863 (56–58), 53-7882 (55–58).Further service under PAF.

437 Troop Carrier Wing: Tachikawa AB, Japan, to Brady Field, Japan. 49-153 (depl 51).

452 Bombardment (Light) Wing: Pusan East AB (K-9), South Korea. 49-150 (depl 51).

483 Troop Carrier (Medium) Wing: *Group:* 483 Troop Carrier Group. *Squadrons (Colors):* **21 TCSq (Attached)** Tachikawa AB Japan; **36 TCSq** (Red/White); **37 TCSq** (Blue/White); **75 TCSq** (Green/White); **773 TCSq**; **815 TCSq** (Red/White); **816 TCSq** (Green/White); **817 TCSq** (Blue/White). *Operating Base:* Ashiya AB, Japan. *Notes:* Deployments to Bofu, Japan; Tachikawa, Japan Sep-Dec56 and Clark AB, The Philippines. Deployment to unknown location at Akddoae Aug-Sep55. 52-5843 deployed to Albrook AFB, Panama CZ 57. *C-119 Service:* 239 1952–1959. *C-119B Assignments:* 48-322 (53–54), 48-325 (52–53; 54), 48-326 (52–53), 48-327 (53–54), 48-328 (53–54), 48-329 (53–54), 48-331 (53–54), 48-332 (54), 48-334 (52–54), 48-335 (53–54), 48-340 (53–54), 48-341 (53–54), 48-343 (53–54), 48-344 (53–54), 48-346 (53–54), 48-347 (53–54), 48-348 (53), 48-349 (53–54), 48-351 (53–54), 48-352 (53–54), 48-354 (53–55), 48-355 (53–54), 49-101 (54), 49-102 (53–54), 49-106 (54), 49-107 (52–54), 49-108 (53–54), 49-109 (53–54), 49-110 (53–54), 49-111 (53–54), 49-112 (53–54), 49-113 (54), 49-114 (54), 49-115 (53), 49-116 (53–54), 49-117 (53–54), 49-118 (53–54). *C-119C Assignments:* 49-119 (53–54), 49-120 (54), 49-121 (54), 49-122 (52–54), 49-124 (53–54), 49-125 (53–54), 49-127 (54), 49-128 (53–54), 49-131 (53–54), 49-132 (53–54), 49-133 (53–54), 49-135 (depl 53), 49-136 (depl 53, 54), 49-139 (53–54), 49-140 (54), 49-141 (54), 49-142 (54), 49-143 (53), 49-144 (53–54), 49-146 (depl 54), 49-147 (53–54), 49-149 (53–54), 49-151 (54), 49-152 (53–54), 49-155 (depl 53), 49-157 (54), 49-158 (54), 49-160 (53), 49-161 (53), 49-164 (depl 54), 49-165 (53–54), 49-170 (53), 49-171 (53–54), 49-174 (53–54), 49-177 (53–54), 49-178 (53), 49-180 (53), 49-181 (53), 49-182 (53–54), 49-183 (53–54), 49-184 (54), 49-185 (53), 49-186 (53–54), 49-187 (53–54), 49-188 (53–54), 49-190 (53), 49-191 (53–54), 49-193 (53), 49-195 (53), 49-196 (53), 50-138 (53; 54), 50-139 (54), 50-147 (53–54), 51-2532 (54), 51-2536 (52–55), 51-2537 (depl 53), 51-2538 (53–54), 51-2539 (53–55), 51-2540 (53–54), 51-2541 (depl 54), 51-2543 (52–54), 51-2546 (53–54), 51-2547 (53–54), 51-2548 (54), 51-2550 (depl 54), 51-2553 (depl 54), 51-2555 (depl 53; 54), 51-2556 (depl 53), 51-2557 (53–54), 51-2558 (54), 51-2559 (depl 54), 51-2561 (53–54), 51-2562 (54–55), 51-2563 (53–55), 51-2564 (53–54), 51-2565 (53–54), 51-2567 (depl 53), 51-2569 (54–55), 51-2571 (53–54), 51-2572 (53–54), 51-2573 (depl 54), 51-2574 (53–54), 51-2575 (53–54), 51-2576 (53–54), 51-2577 (depl 54), 51-2578 (53–54), 51-2580 (depl 53), 51-2581 (53–54), 51-2583 (53–54), 51-2584 (53–54). *C-119G Assignments:* 51-8055 (54–58), 51-8056 (54–57), 51-8062 (54–57), 51-8065

(54–57), 51-8066 (54–57), 51-8068 (54–57), 51-8070 (54–57), 51-8074 (54–57), 51-8076 (54–57), 51-8085 (54–55), 51-8087 (54–57), 51-8094 (54–57), 51-8096 (54–57), 52-5842 (54–57), 52-5843 (54–57), 52-5846 (54–57), 52-5847 (unk–57), 52-5852 (54–57), 52-5857 (54–57), 52-5861 (54–57), 52-5862 (54–57), 52-5868 (54–57), 52-5871 (54–57), 52-5872 (54–57), 52-5873 (54–57), 52-5874 (54–57), 52-5876 (54–57), 52-5881 (54–55), 52-5883 (54–57), 52-5885 (54–56), 52-5886 (54–55), 52-5887 (54–57), 52-5889 (55–57), 52-5891 (54–55), 52-5892 (54–57), 52-5900 (54–57), 52-5902 (54–57), 52-5905 (54–57), 52-5917 (54–57), 52-5919 (54–57), 52-5920 (54–57), 52-5921 (54), 52-5925 (54–57), 52-5926 (54–57), 52-5927 (54–58), 52-5928 (54–58), 52-5929 (54–58), 52-5930 (54–57), 52-5931 (54–58), 52-5832 (54–57), 52-5933 (54–57), 52-5934 (54–57), 52-5935 (54–57), 52-5936 (54–57), 52-5937 (54–58), 52-5938 (54–58), 52-5939 (54–58), 52-5940 (54–58), 52-5941 (54–58), 52-5942 (54–58), 52-5943 (54–58), 52-5944 (54–56), 52-5945 (54–57), 52-5946 (54–57), 52-5948 (54–57), 52-5950 (54–57), 52-5953 (54–57), 52-5954 (54–58), 52-9981 (54–58), 52-9982 (54–58), 53-3142 (54–58), 53-3143 (54–58), 53-3144 (54–58), 53-3145 (54–58), 53-3146 (54–58), 53-3147 (54–58), 53-3148 (54–58), 53-3149 (54–58), 53-3150 (54–56), 53-3151 (54–58), 53-3152 (54–58), 53-3153 (54–58), 53-3154 (54–59), 53-3155 (54–58), 53-3156 (54–58), 53-3157 (54–58), 53-3158 (54–58), 53-3159 (54–58), 53-3160 (54–58), 53-3161 (54–58), 53-3162 (54–58), 53-3163 (54–58), 53-3164 (54–58), 53-3165 (54–58), 53-3166 (54–58), 53-3167 (54–58), 53-3168 (54–58), 53-3169 (54–58), 53-3170 (54–58), 53-3171 (54–58), 53-3172 (54–58), 53-3173 (54–58), 53-3174 (54–58), 53-3175 (54–58), 53-3176 (54–58), 53-3177 (54–58), 53-7864 (55–58), 53-7875 (55–58), 53-7876 (56–58), 53-7879 (depl 56), 53-8087 (56–58), 53-8089 (depl 56; 56–58).

6002 Tactical Support Wing: Pusan East AB (K-9), South Korea. 49-166 (sal, 50).

6007 Composite Group: Yokota AB, Japan. 53-7860 (56–58). Further service under PAF.

6101 Air Base Wing: Komaki AB, Japan. 49-122 (52).

6106 Air Base Unit: Komaki AB, Japan. 48-324 (sal, 50).

6122 Air Base Group / Wing: Ashiya AB, Japan. 48-346 (51), 48-347 (52), 49-119 (51), 49-120 (51–52), 49-127 (52), 49-131 (51), 49-145 (sal, 50).

6149 Tactical Support Wing: Taegu AB (K-2), South Korea. 49-162 (50).

6200 Air Base Wing: Clark AB, The Philippines. 52-5939 (56), 53-7880 (56),

6200 Maintenance & Supply Group: Clark AB, The Philippines. 51-2556 (54).

6319 Air Base Wing: Andersen AFB, Guam. 48-337 (54–55), 49-111 (53), 53-3143 (54).

6332 Air Base Wing: Kadena AB, Japan. 51-2537 (54–55), 51-2545 (54–55).

6400 Field Maintenance Unit / Air Depot Wing: Taegu AB (K-2), South Korea, to Miho AB & Tachikawa AB, Japan. 48-328 (54), 48-337 (52–53), 48-346 (52), 48-354 (52), 48-355 (52), 49-104 (52–54), 49-120 (52), 49-131 (52), 49-137 (52), 49-147 (52), 49-149 (52), 49-162 (50), 49-164 (52–53), 49-165 (53).

6424 Air Depot Wing: Clark AB, The Philippines. 51-2536 (54), 51-2556 (54).

6480 Air Depot Wing: Tachikawa AB, Japan. 51-2552 (sal, 54).

6486 Air Base Wing: Hickham AFB, Hawaii. 51-2545 (55), 51-8087 (57), 52-5852 (57), 52-5874 (57), 52-5927 (58), 52-5932 (57), 52-5953 (55), 53-3177 (55). Reassigned to PAF Command Jul57.

Far East Air Logistics Frontier: Tachikawa AB, Japan. 53-7866 (55), 53-7877 (55).

Section Two—Military Operators 83

Far East Air Materiel Command: *Operating Base:* Tachikawa AB, Japan. *Notes:* Includes assignments to 13 ARP SQ for mnt. *C-119 Service:* 88 1950–1952. *C-119B Assignments:* 48-325 (51), 48-327 (51), 48-328 (51), 48-329 (51), 48-331 (51), 48-334 (50; 51), 48-337 (51–52), 48-341 (51), 48-343 (51), 48-344 (51), 48-346 (51–52), 48-347 (51), 48-348 (51), 48-349 (51), 48-351 (51), 48-352 (51), 48-354 (51–52), 48-355 (51), 49-101 (51), 49-102 (51), 49-103 (51), 49-104 (51), 49-106 (50; 51), 49-107 (51), 49-108 (51), 49-109 (51), 49-110 (51), 49-111 (51), 49-112 (51), 49-113 (51), 49-114 (51), 49-115 (51), 49-116 (51), 49-117 (51), 49-118 (51). *C-119C Assignments:* 49-119 (51), 49-121 (51), 49-122 (51), 49-124 (51), 49-125 (51), 49-126 (51), 49-127 (51), 49-128 (50; 51), 49-129 (51), 49-130 (51), 49-131 (51), 49-132 (51), 49-133 (51), 49-135 (51), 49-136 (51), 49-137 (50–51), 49-138 (50–51), 49-139 (51), 49-140 (51), 49-141 (51), 49-142 (51), 49-143 (51), 49-144 (51), 49-145 (50), 49-146 (50; 51), 49-147 (51; 51-52), 49-149 (51), 49-150 (51), 49-151 (51), 49-152 (51), 49-153 (50), 49-154 (50; 51), 49-155 (50–51), 49-156 (51), 49-157 (51), 49-158 (51), 49-159 (50–51), 49-160 (51), 49-161 (51), 49-162 (50–51; 51), 49-164 (51; 51-52), 49-165 (51), 49-167 (50–51; 51), 49-169 (51), 49-170 (51), 49-171 (51), 49-172 (51), 49-173 (51), 49-174 (51), 49-175 (51), 49-176 (51), 49-177 (50–51; 51), 49-178 (51).

Manila Air Depot: Clark AB, The Philippines. 53-7876 (55), 53-7877 (55). Maybe a forerunner of SOMAR / SMPAR.

Tachikawa Air Depot: Miho AB & Tachikawa AB, Japan. 48-328 (54–55).

Headquarters Command (HQC)

Established 15 December 1946 as **Bolling Field Command** for support of operations in the Washington D.C. area. Redesignated as such 17 March 1948 and not deactivated until 1 July 1976.

1001 Air Base Wing: Andrews AFB, Maryland. 48-347 (58), 48-351 (58), 51-2599 (58), 51-2626 (58), 51-8233 (58), 51-8262 (58), 52-5899 (57–58), 52-5952 (59).

1050 Air Base Wing: Andrews AFB, Maryland. 49-187 (51), 51-2542 (52).

1050 Maintenance & Supply Group: Andrews AFB, Maryland. 48-329 (50), 48-334 (50).

1100 Air Base Wing: Bolling AFB, District of Columbia. 48-333 (51).

1100 Maintenance & Supply Group: Bolling AFB, District of Columbia. 48-327 (57–58).

1100 Operations Group: Bolling AFB, District of Columbia. 51-2609 (57).

Military Air Transport Service (MATS)

Previously known as **Air Transport Command** (ATC). Established 1 June 1948 also incorporating the **Air Rescue Service** (ARS) and **Air Resupply & Communications Service** (ARCS). MATS was further organized into the **Military Air Transport Service, Continental Division** (MTC) and **Military Air Transport Service, Overseas Division** (MTO). Apart from freight logistics MATS also provided other services such as weather, photographic and aero medical duties.

7 Weather Group: Elmendorf AFB, Alaska. 49-178 (56–57).

53 Strategic (Medium) Weather Squadron: Kindley AFB, Bermuda Islands (21Feb51–05Nov53), RAF Burtonwood, England (07Nov53–25Apr59). 49-167 (55), 50-128 (54–56), 50-149 (56).

55 Strategic (Medium) Weather Squadron: McClellan AFB, California. 49-167 (54–55), 50-128 (56–57).

59 Weather Reconnaissance Squadron: Kindley AFB, Bermuda Islands. 49-167 (55–57), 50-128 (56).

62 Air Base Group: Larson AFB, Washington State. 50-148 (58).

63 Air Base Group: Donaldson AFB, South Carolina. 51-8160 (57).

80 Air Base Squadron: Dover AFB, Delaware. 51-2609 (53).

1300 Air Base Wing: Mountain Home AFB, Idaho, to Great Falls AFB, Montana. 51-7969 (52–53), 51-7970 (52–unk), 51-7971 (52–53), 51-7972 (52–53).

1371 Mapping and Charting Squadron: Palm Beach AFB, Florida. 50-149 (54–56), 50-151 (54–56), 50-153 (54–56).

1400 Field Maintenance Squadron: Keflavik Air Station, Iceland. 50-154 (55), 51-2544 (56–57), 51-2658 (55).

1401 Air Base Wing: Andrews AFB, Maryland. 48-341 (54–55), 49-167 (56), 50-149 (56), 51-8090 (53).

1405 Air Base Wing: Scott AFB, Illinois. 48-320 (58), 48-341 (58), 49-111 (59), 49-125 (57–58), 49-162 (59), 50-152 (59), 50-168 (58), 51-2583 (58), 51-2619 (58–59), 51-8018 (57–58), 52-5915 (58), 53-8122 (59).

1414 Air Base Group: Dhahran AB, Saudi Arabia. 50-133 (52).

1500 Operations Squadron: Hickham AFB, Hawaii. 48-326 (53–54), 48-331 (54), 48-355 (54), 49-118 (52), 49-147 (54), 49-156 (53; 53-54), 49-183 (55), 49-188 (54–55), 51-2537 (52), 51-8096 (54), 52-5929 (54), 52-5939 (54), 53-3177 (55). Later redesignated as 1500 FDM SQ then 1500 MAI SQ.

1501 Air Transport (Heavy) Wing: Travis AFB, California. 52-5848 (58), 53-8085 (58).

1502 Support Squadron: Kwajalein Island, Marshall Islands. 48-335 (50–51).

1503 Aircraft Maintenance Squadron: Tokyo IAP (Haneda AB), Japan. 49-119 (53–54).

1503 Air Transport (Heavy) Wing: Tachikawa AB, Japan. 50-143 (63).

1505 Air Base Group: Johnston Atoll, USA. 49-110 (53), 49-119 (53), 49-125 (53), 49-156 (53).

1509 Air Base Squadron: Johnston Atoll, USA. 49-104 (54), 49-122 (54), 49-183 (54–55), 52-5953 (55). Later redesignated as 1509 SUT SQ.

1600 Field Maintenance Squadron: Westover AFB, Massachusetts. 51-8088 (55).

1604 Air Base Wing: Kindley AFB, Bermuda Islands. 49-108 (60), 49-140 (60), 51-8041 (58), 51-8116 (60), 53-8078 (59).

1604 Air Materiel Squadron: Kindley AFB, Bermuda Islands. 49-179 (55), 51-2614 (55), 51-8053 (55), 52-5880 (55), 53-3192 (54), 53-7831 (55), 53-8149 (55), 53-8150 (55). Redesignated 1604 FDM SQ 05May55.

1604 Maintenance & Supply Squadron: Kindley AFB, Bermuda Islands. 50-151 (54).

1605 Air Base Wing: Lajes Field, Azores. 51-2628 (58), 53-7837 (61), 53-7850 (61).

1605 Field Maintenance Squadron: Lajes Field, Azores. 51-2657 (57–unk).

1607 Air Transport (Heavy) Wing: Dover AFB, Delaware. 48-329 (58), 49-107 (58), 50-162 (59), 51-2554 (59), 51-2644 (58), 52-5840 (58).

1607 Field Maintenance Squadron: Dover AFB, Delaware. 48-329 (58), 49-107 (55), 50-124 (54–55), 50-126 (54), 50-141 (54–55), 51-2616 (54–55). Later redesignated as 1607 MAI SQ.

1607 Maintenance & Supply Squadron: Dover AFB, Delaware. 51-2609 (53).

1608 Air Transport (Medium) Wing: Charleston AFB, South Carolina. 50-120 (58), 50-128 (59), 50-155 (59), 51-2595 (59), 51-2608 (60). Later redesignated 1608 TSH WG.

1608 Field Maintenance Squadron: Charleston AFB, South Carolina. 49-167 (56), 50-157 (56), 51-7987 (56), 51-8162 (55).

1610 Air Materiel Squadron: Grenier AFB, New Hampshire. 49-179 (55), 49-198 (54).

1611 Air Transport (Medium) Wing: McGuire AFB, New Jersey. 51-2580 (58–59), 51-8025 (58), 51-8089 (58), 52-5876 (58).

1611 Field Maintenance Squadron: McGuire AFB, New Jersey. 50-145 (58), 51-2547 (55–56), 53-8097 (56).

1631 Air Base Squadron: Prestwick, Scotland. 50-167 (53).

1631 Air Materiel Squadron: Prestwick, Scotland. 50-143 (55), 50-161 (53–54), 51-2533 (55–56), 51-2610 (55), 53-8140 (55–57).

1701 Air Base Group: Great Falls AFB, Montana. 51-2590 (sal, 52-53).

1707 Field Maintenance Squadron: Palm Beach AFB, Florida. 49-167 (55; 57), 50-154 (58).

1739 Ferrying Squadron: Amarillo AFB, Texas. 49-121 (54), 49-164 (56–57), 49-178 (57), 49-181 (54), 50-154 (52; 57), 51-2568 (53–54), 51-2579 (53–54), 51-2609 (52), 51-2619 (53), 51-8116 (57), 52-5950 (57), 53-8073 (57), 53-8096 (54–58), 53-8097 (54–58).

Airways & Air Communications Service (AACS)

1 Installation & Maintenance Squadron: Tinker AFB, Oklahoma. 49-164 (54–55), 51-2579 (54–55).

3 Airways & Communications, Mobile Squadron: Tinker AFB, Oklahoma. 50-153 (56–57), 51-2568 (54–58).

4 Installation & Maintenance Squadron: Erding AB, West Germany. 49-198 (54–55).

4 Airways & Communications, Mobile Squadron: Elmendorf AFB, Alaska. 49-121 (54–55).

1800 Airways & Communications Wing: Tinker AFB, Oklahoma. 49-164 (57), 50-151 (57), 51-2568 (depl 56), 51-2579 (57).

1807 Airways & Communications Wing: Erding AB, West Germany. 49-198 (54).

1855 Airways & Communications Flight: Elmendorf AFB, Alaska. 49-121 (55–58).

1857 Airways & Communications Flight: Rhein-Main AB, West Germany. 49-198 (57–58).

1881 Installation & Maintenance Squadron: Tinker AFB, Oklahoma. 49-164 (55–57), 50-151 (56–57), 51-2579 (55–57). Redesignated 1800 IM GP then EI GP 17Dec57.

1884 Airways & Communications Squadron: Erding AB, West Germany, to

Bordeaux AB France (16Apr56). 49-198 (55–57). Redesignated 1884 IM SQ Jun55; 1884 EI SQ Dec56.

Continental AACS Area: Tinker AFB, Oklahoma. 49-164 (57–58), 50-151 (57–58).

Air Resupply & Communications Service (ARCS)

580 Air Resupply & Communications Group / Wing: *Operating Bases:* Mountain Home AFB, Idaho (1Apr51–17Sep52). Wheelus AB, Libya (04Oct51). *Notes:* Redesignated as 580 ASL GP Sep53. Assigned to AFE while at Wheelus. *C-119 Service:* 14 1951–1956. *C-119 Assignments:* 50-142 (51; 51–52), 50-154 (51), 50-162 (51–55), 51-2534 (51–54), 51-2542 (51–53), 51-2544 (51–55), 51-2582 (53–55), 51-2605 (55–56), 51-2608 (55–56), 51-2612 (56), 51-2613 (55–56), 51-2614 (56), 51-2620 (55–56), 51-8237 (56).

581 Air Resupply & Communications Group / Wing: *Operating Bases:* Mountain Home AFB, Idaho (23Jul51–18Jul52). Clark AB, The Philippines (18Jul52–08Sep52). Kadena AB, Japan (08Sep53–01Sep56). *Notes:* USAF organization operated under MATS with its home base at Mountain Home AFB, Idaho. Redesignated as 581 ASL GP. Deployment to FEA for The Korean War. *C-119 Service:* 12 1950–1953. *C-119 Assignments:* 48-341 (depl 53), 49-129 (54–56), 49-156 (54–57), 49-157 (54–56), 49-158 (54–55), 50-142 (51), 50-154 (51–52), 51-2555 (51–53), 51-2556 (51–54), 51-2567 (51–54), 51-2578 (54–56), 51-2580 (51–55).

582 Air Resupply & Communications Group / Wing: *Operating Bases:* Mountain Home AFB, Idaho (24Sep52–01May53). Great Falls AFB, Montana (01May53–14Aug53). RAF Molesworth, England (21Feb54–25Oct56). *Notes:* Redesignated as 582 ASL GP Jun54. *C-119 Service:* 5 1953–1955. *C-119 Assignments:* 50-143 (53–55), 50-154 (53–55), 51-2533 (54–55), 51-2566 (53–55), 51-7972 (54).

1300 Air Base Wing: Mountain Home AFB, Idaho, to Great Falls AFB, Montana (21May53). 50-142 (52–53), 50-154 (52–53).

Northeast Air Command (NEA)

Established 01Oct50 for the defense of Greenland, Labrador and Newfoundland.

6511 Air Base Group: Narsarsuaq AB, Greenland. 52-5910 (unk–55), 52-5915 (55–unk).

6600 Air Base Group: Pepperrell AFB, Newfoundland, Canada. 51-8032 (53).

6602 Air Base Group: Ernest Harmon AFB, Newfoundland, Canada. 51-2611 (53).

6603 Air Base Group: Goose Bay AB, Labrador, Canada. 51-2533 (54), 51-2573 (52), 51-2589 (53), 51-2594 (53), 51-2595 (53), 51-2600 (53), 51-2617 (53), 51-2623 (53), 51-2630 (53), 51-2640 (53), 51-2648 (53), 51-8032 (54), 51-8106 (53).

6604 Air Base Wing: Pepperell AFB, Newfoundland, Canada. 51-8032 (55), 51-8048 (54), 53-3209 (55).

6605 Air Base Wing: Ernest Harmon AFB, Newfoundland, Canada. 49-198 (56), 51-8033 (57), 53-3209 (56), 53-3210 (56), 53-8113 (54).

6606 Air Base Wing: Goose Bay AFB, Labrador, Canada. 49-199 (54), 50-125 (54–55), 50-133 (55), 50-136 (54–55), 50-153 (54), 51-2709 (55), 51-2710 (55), 51-8031 (56), 51-8032 (55; 56), 51-8033 (54; 55), 51-8034 (55), 51-8044 (54), 51-8048 (54; 55), 51-8071 (55), 51-8088 (55), 52-5924 (56), 52-5952 (55), 53-3209 (56), 53-3210 (55), 53-3211 (55).

6607 Air Base Wing: Thule AB, Greenland. 53-3209 (55).

6611 Air Base Group: Narsarsuaq AB, Greenland. 51-8044 (54).

6612 Air Base Group: Thule AB, Greenland. 49-183 (52), 50-142 (52), 51-2536 (52), 51-2593 (53), 51-2641 (53), 51-8034 (53).

6614 Air Transport, Medium Squadron: Ernest Harmon AFB, Newfoundland, Canada. 51-8031 (53–57), 51-8032 (53–56), 51-8033 (53–57), 51-8034 (53–55), 51-8044 (54–55), 51-8048 (54–55), 51-8105 (52–unk), 51-8106 (53), 53-3209 (54–57), 53-3210 (54–57), 53-3211 (55–57).

6621 Air Base Squadron: Sondrestrom AB, Greenland. 51-8044 (54).

6622 Air Transport Squadron: Ernest Harmon AFB, Newfoundland, Canada. 51-8106 (52–53).

Pacific Air Forces (PAF)

Major Command designated as such 1 July 1957 to oversee the Asia-Pacific region. Previously named as **Far East Air Forces** (FEAF).

8 Tactical Fighter Wing: Itazuke AB, Japan. 53-7860 (59), 53-7869 (59), 53-7871 (59). For storage.

13 Air Division (Heavy) Headquarters / Division: Taipei, Taiwan. AC-119G: 52-5898 (71).

14 Special Operations Wing: *Squadrons:* **17 SOP SQ**; **18 SOP SQ**. *Operating Bases:* Nha Trang AB, South Vietnam (08Mar66–15Oct69). Phan Rang AB, South Vietnam (15Oct69–30Sep71). *Notes:* Detachments at Phu Cat, Da Nang, Tan Son Nhut and Tuy Hoa, South Vietnam & Nakhon Phanom and Udorn, Thailand. *C-119 Service:* 26 AC-119G & 18 AC-119K 1968–1971. *AC-119G Assignments:* 52-5892 (69–71), 52-5898 (69–71), 52-5905 (68–71), 52-5907 (68–69), 52-5925 (69–71), 52-5927 (68–71), 52-5938 (69–71), 52-5942 (68–71), 53-3136 (68–71), 53-3145 (71), 53-3170 (69–71), 53-3178 (68–71), 53-3189 (68–71), 53-3192 (68–71), 53-3205 (69–71), 53-7833 (71), 53-7848 (69–71), 53-7851 (69–71), 53-7852 (69–71), 53-8069 (68–71), 53-8089 (71), 53-8114 (71), 53-8115 (71), 53-8123 (71), 53-8131 (71), 53-8155 (68–70). *AC-119K Assignments:* 52-5864 (69–71), 52-5889 (69–71), 52-5926 (69–71), 52-5935 (69–70), 52-5940 (69–71), 52-5945 (69–71), 52-9982 (69–71), 53-3154 (69–71), 53-3156 (69–70), 53-3211 (69–71), 53-7826 (69–71), 53-7830 (69–71), 53-7850 (69–71), 53-7854 (69–71), 53-7879 (69–71), 53-7883 (69–71), 53-8121 (69–71), 53-8148 (69–71).

18 Tactical Fighter Wing: Kadena AB, Japan. 53-3136 (69), 53-3189 (69), 53-8069 (69).

56 Special Operations Wing: *Operating Base:* Da Nang AB, South Vietnam. *Notes:* Detachments at Phan Rang, South Vietnam & Nakhon Phanom Thailand. *C-119 Service:* 18 1971–1972. *AC-119K Assignments:* 52-5864 (71–72), 52-5889 (71–72), 52-5926 (71–72), 52-5940 (71–72), 52-5945 (71–72), 52-9982 (71–72), 53-3154 (71–72), 53-3187 (72–73), 53-3211 (71–72), 53-7826 (71–72), 53-7830 (71–72), 53-7850 (71–72), 53-7854 (71–72), 53-7877 (72), 53-7879 (71–72), 53-7883 (71–72), 53-8121 (71–72), 53-8148 (71–72).

313 Air Division (Heavy) Headquarters / Division: *Operating Base:* Kadena AB, Japan. *AC-119G Assignments:* 52-5892 (71), 52-5905 (71), 53-3170 (70), 53-3178 (71), 53-3189 (71), 53-3192 (70–71), 53-7848 (71), 53-8089 (71). *AC-119K Assignments:* 52-5864 (70), 52-5889 (70), 52-5926 (70), 52-5945 (71), 52-9982 (70), 53-3211 (70), 53-7854 (71), 53-7879 (71), 53-8121 (71).

366 Tactical Fighter Wing: Da Nang AB, South Vietnam. AC-119K: 52-5864 (71), 52-5889 (71).

377 Air Base Wing: Tan Son Nhut AB, South Vietnam. 53-3148 (73), 53-3157 (73), 53-3161 (73), 53-3175 (73), 53-3198 (73), 53-3202 (73), 53-3203 (73), 53-3218 (73), 53-3220 (73), 53-8147 (73)—10 returned to USAF from VNAF.

405 Fighter Wing: Clark AB, The Philippines. 51-7983 (73), 53-3148 (73), 53-3157 (73), 53-3161 (73), 53-3175 (73), 53-3194 (73), 53-3198 (73), 53-3202 (73), 53-3203 (73), 53-3218 (73), 53-3220 (73), 53-7842 (73), 53-8088 (73), 53-8104 (73), 53-8133 (73), 53-8147 (73)—16 stored for disposal.

432 Tactical Reconnaissance Wing: Udorn AB, Thailand. AC-119K: 52-5940 (72), 53-3211 (71).

460 Tactical Reconnaissance Wing: Tan Son Nhut AB, South Vietnam. AC-119K: 52-5889 (70), 52-5940 (70).

868 Tactical Missile Squadron: *Operating Base:* Tainan, Taiwan. *Notes:* 16 wfu and stored for MAP assignment to Taiwan. *C-119 Assignments:* 52-5928 (58–59), 52-5937 (58–59), 52-5954 (58–59), 53-3143 (58–59), 53-3149 (58–59), 53-3153 (58–59), 53-3155 (58–59), 53-3163 (58–59), 53-3164 (58–59), 53-3165 (58–59), 53-3169 (58–59), 53-3171 (58–59), 53-3172 (58–59), 53-3176 (58–59), 53-3177 (58–59), 53-7875 (58–59).

6041 Air Base Group: Johnson AB, Japan. 53-7862 (57–58), 53-7867 (57–60), 53-7871 (59), 53-7878 (57–60), 53-7880 (57–60), 53-7883 (57–60).

6102 Air Base Wing: Yokota AB, Japan. 53-7861 (58–60), 53-7862 (60), 53-7863 (58–60), 53-7872 (57–60), 53-7873 (60), 53-7878 (60), 53-7879 (60), 53-7881 (57–60), 53-7883 (60).

6139 Air Base Group: Misawa AB, Japan. 53-7860 (58–60), 53-7870 (57–60), 53-7879 (57–60).

6143 Air Base Wing: Itazuke AB, Japan. 53-7862 (58–60), 53-7868 (57–60), 53-7870 (60), 53-7873 (57–60).

6313 Air Base Wing: Kadena AB, Japan. 53-7869 (57–60), 53-7870 (60), 53-7871 (57–60), 53-7874 (57–60), 53-7882 (58–60).

6486 Air Base Wing: *See under:* **Far East Air Forces** (FEA)

6498 Air Base Wing: Da Nang AB, South Vietnam. AC-119K: 53-3187 (73), 53-3211 (72–73), 53-7831 (72–73), 53-7839 (72–73), 53-8145 (72–73). Transfers for MAP assignment to South Vietnam.

Strategic Air Command (SAC)

Major Command established 21 March 1946 to oversee the strategic bomber fleet with the ability to deploy to war globally at short notice. SAC also oversaw the aerial refueling fleet, atomic arsenal and strategic reconnaissance missions. The largest Major Command in the USAF and the most well-known as it was the frontline command against any nuclear strike during the Cold War Era.

2 Bombardment (Medium) Wing: Hunter AFB, Georgia. 49-190 (56), 50-129 (54–57), 50-131 (54–57), 51-2566 (depl 55, 56).

5 Air Base Group: Travis AFB, California. 49-120 (depl 57), 49-128 (53), 49-136 (54–57), 49-178 (53), 49-181 (53), 49-183 (55), 49-188 (54), 51-2559 (57), 51-2562 (57–58), 51-2567 (56–58), 51-2577 (54–55; 56–58), 51-7969 (58), 51-8140 (58).

5 Bombardment (Heavy) Wing: Travis AFB, California. 49-136 (55–56), 51-2567 (55–56).

5 Strategic Reconnaissance (Heavy) Wing: Travis AFB, California. 49-136 (54; 55), 49-183 (54–55), 51-2562 (55), 51-2567 (55).

6 Air Base Group: Walker AFB, New Mexico. 49-142 (59). Redesignated 6 COS GP Apr59.

6 Bombardment (Heavy) Wing: Walker AFB, New Mexico. 49-162 (54–55), 49-187 (54–56).

7 Air Base Group: Carswell AFB, Texas. 49-193 (57–58), 49-195 (57–58), 50-160 (57–58), 51-2539 (58).

7 Bombardment (Heavy) Wing: Carswell AFB, Texas. 48-325 (63–64), 49-137 (55–56), 49-170 (54–57), 49-181 (54–58).

8 Air Rescue Squadron: Camp Carson, Colorado. 49-179 (50).

9 Air Base Group: Mountain Home AFB, Idaho. 50-156 (56–57), 50-159 (56), 51-2548 (56–58), 51-2582 (56–58), 53-3183 (58).

9 Bombardment (Medium) Wing: Mountain Home AFB, Idaho. 50-132 (54–58), 51-2548 (depl 57–58), 51-2582 (55–56).

11 Air Base Group: Altus AFB, Oklahoma. 49-141 (57–58), 50-162 (57–58), 51-2539 (57–58).

11 Bombardment (Heavy) Wing: Carswell AFB, Texas. 49-193 (54–57), 49-195 (54–57), 50-160 (55–57).

12 Strategic Fighter Wing: Bergstrom AFB, Texas. 49-135 (54), 49-143 (54–57), 49-144 (55), 49-184 (54–57), 50-147 (55–57).

19 Bombardment (Medium) Wing: Pinecast AFB, Florida (11Jun54–01Jun56), to Homestead AFB, Florida (01Jun56–25Jul68). 50-160 (55), 51-2533 (55), 51-7989 (65).

22 Bombardment (Medium) Wing: March AFB, California. 49-135 (54–56), 49-165 (55), 51-2539 (55–56), 51-7989 (65).

26 Strategic Reconnaissance (Medium) Wing: Lockbourne AFB, Ohio. 51-2532 (55–57).

27 Strategic Fighter Wing: Bergstrom AFB, Texas. 49-151 (54–57), 49-155 (54), 50-120 (57), 50-139 (54–58), 50-160 (57). Redesignated 27 Fighter-Bomber WG to TAC 01Jul57.

28 Air Base Group / Wing: Ellsworth AFB, South Dakota (Named Rapid City AFB until Jun53). 49-136 (57–58), 49-159 (56–57), 49-165 (56–58), 49-182 (56), 51-2540 (56), 51-2554 (58), 51-2562 (56–57).

28 Bombardment (Heavy) Wing: Ellsworth AFB, South Dakota (Named as Rapid City AFB until Jun53). 49-159 (55–56), 49-182 (55–56).

28 Strategic Reconnaissance (Heavy) Wing: Ellsworth AFB, South Dakota (Named as Rapid City AFB until Jun53). 49-159 (54–55), 49-182 (54–55), 49-184 (51), 51-8079 (55).

31 Strategic Fighter Wing: Turner AFB, Georgia. 49-137 (54), 49-147 (54–55), 49-171 (unk–57), 49-174 (54–57), 49-176 (54–57), 49-180 (54–57), 50-124 (57), 50-147 (54), 50-154 (55–57). Redesignated as 31 Fighter-Bomber WG 01Apr57.

40 Bombardment (Medium) Wing: Smoky Hill AFB, Kansas. 49-144 (55–56), 49-183 (55), 51-2538 (55–56), 51-2553 (55–57).

42 Air Base Group: Loring AFB, Maine. 49-179 (57–58), 50-122 (57–58), 50-155 (57), 50-168 (56–58), 50-171 (58), 51-2554 (55).

42 Bombardment (Heavy) Wing: Limestone AFB, Maine (Renamed as Loring AFB, Maine 20Feb55). 49-162 (54), 49-179 (54–57), 50-122 (55–57), 50-155 (56), 50-171 (56–57), 51-2554 (53–55).

43 Bombardment (Medium) Wing: Davis-Monthan AFB, Arizona (17Nov47–15Mar60), to Andersen AFB, Guam (01Apr70–30Sep90). 49-165 (55–56), 51-2559 (56–57), 51-2562 (55–56), 51-2577 (56), 53-8089 (71). Redesignated as 43 STR WG 04Feb70.

44 Bombardment (Medium) Wing: Lake Charles AFB, Louisiana. 50-126 (54–57), 51-2584 (56), 51-2603 (54–57).

55 Strategic Reconnaissance (Medium) Wing: Forbes AFB, Kansas. 49-194 (55–56), 51-2571 (55).

68 Bombardment (Medium) Wing: Lake Charles AFB, Louisiana. 50-143 (55–56), 51-2607 (55–57), 51-2610 (unk–57).

70 Strategic Reconnaissance (Medium) Wing: Little Rock AFB, Arkansas. 49-151 (56), 49-152 (56–57), 50-143 (56–57).

72 Air Base Group: Ramey AFB, Puerto Rico. 50-147 (55).

72 Bombardment (Heavy) Wing: Ramey AFB, Puerto Rico. 50-164 (62), 51-2667 (68), 51-8092 (58).

72 Strategic Reconnaissance (Heavy) Wing: Ramey AFB, Puerto Rico. 49-137 (54–55), 50-147 (54–55).

90 Strategic Reconnaissance (Medium) Wing: Forbes AFB, Kansas. 49-146 (55–56), 51-2573 (55–56).

91 Strategic Reconnaissance (Medium) Wing: Lockbourne AFB, Ohio. 51-2554 (55–57), 51-2619 (55–57), 51-2624 (55–57).

92 Air Base Group: Fairchild AFB, Washington State. 49-136 (56), 49-155 (55), 50-156 (56–58), 51-2540 (56), 51-2580 (56–58), 51-2616 (56–58).

92 Bombardment (Heavy) Wing: Fairchild AFB, Washington State. 49-146 (55), 50-156 (54–56), 51-2580 (55–56), 51-8039 (72). Redesignated 92 STA WG 15Feb62.

93 Air Base Group: Castle AFB, California. 49-193 (58), 51-2583 (55).

93 Bombardment (Medium) Wing: Castle AFB, California. 51-2548 (55–56), 51-2583 (55–56).

95 Bombardment (Heavy) Wing: Biggs AFB, Texas. 49-120 (54–56), 49-142 (54–56).

95 Combat Support Group: Biggs AFB, Texas. 51-2594 (59), 53-8129 (59).

96 Air Base Group: Altus AFB, Oklahoma. 49-141 (56–57), 50-162 (56–57), 51-2539 (56–57), 51-2583 (56–57).

96 Bombardment (Medium) Wing: Altus AFB, Oklahoma. 50-162 (55–56), 51-2544 (55–56).

97 Bombardment (Medium) Wing: Biggs AFB, Texas. 49-139 (55–56), 49-158 (55–56), 51-2616 (53–54).

98 Bombardment (Medium) Wing: Lincoln AFB, Nebraska. 50-168 (55–56), 50-171 (55–56).

99 Bombardment (Heavy) Wing: Fairchild AFB, Washington State. 49-155 (55–56), 51-2540 (55–56).

Section Two—Military Operators

99 Strategic Reconnaissance (Heavy) Wing: Fairchild AFB, Washington State. 49-146 (54), 49-155 (54–55), 49-174 (55), 49-180 (55), 51-2540 (55).

301 Bombardment (Medium) Wing: Barksdale AFB, Louisiana. 50-154 (55), 51-2566 (55).

303 Bombardment (Medium) Wing: Davis-Monthan AFB, Arizona. 51-2559 (55–56), 51-2577 (55–56).

305 Bombardment (Medium) Wing: MacDill AFB, Florida, to Bunker Hill AFB, Indiana (from 01Jun59) (Renamed Grissom AFB 1968). 49-143 (54), 49-185 (54–57), 49-188 (54), 50-120 (55–57), 50-138 (54–57), 52-5850 (ret display 71). Redesignated 305 ARH WG 01Jan70.

306 Bombardment (Medium) Wing: MacDill AFB, Florida. 49-140 (54–57), 49-152 (54–56), 49-180 (57), 50-124 (55–57).

306 Combat Support Group: MacDill AFB, Florida. 51-2580 (60).

307 Bombardment (Medium) Wing: Lincoln AFB, Nebraska. 50-122 (55), 50-155 (54–56).

308 Bombardment (Medium) Wing: Hunter AFB, Georgia. 49-190 (54–57), 50-129 (56), 50-131 (56), 50-146 (54–56), 51-2566 (55–57).

310 Bombardment (Medium) Wing: Smoky Hill AFB, Kansas. 49-144 (55), 49-183 (55–56), 51-2538 (55; 56–57), 51-2571 (55–57), 51-2573 (56–57).

315 Air Base Group: Forbes AFB, Kansas. 49-194 (57).

320 Bombardment (Medium) Wing: March AFB, California. 49-132 (54–56), 49-141 (55–56).

321 Air Base Group: Pinecastle AFB, Florida. 49-120 (56).

321 Bombardment (Medium) Wing: Pinecastle AFB, Florida. 50-120 (55), 50-124 (54–55).

323 Fighter-Bomber Wing: Bunker Hill AFB, Indiana. 49-180 (57).

340 Air Base Group: Whiteman AFB, Missouri. 49-146 (57–58), 49-155 (56–58), 49-183 (57–58), 49-194 (56–58), 50-133 (59), 51-2540 (56–57), 51-2571 (58).

340 Bombardment (Medium) Wing: Whiteman AFB, Missouri. 49-144 (56–58), 49-146 (56–57), 49-183 (56–57; 57), 49-194 (56).

341 Strategic Missile Wing: Malmstrom AFB, Montana. 52-9982 (69).

376 Bombardment (Medium) Wing: Barksdale AFB, Louisiana. 51-2584 (55).

384 Bombardment (Medium) Wing: Little Rock AFB, Arkansas. 49-137 (56–57), 51-2584 (56–57; 57).

397 Bombardment (Heavy) Wing: Dow AFB, Maine. 53-8116 (65).

407 Air Base Group: Great Falls AFB, Montana (Renamed as Malmstrom AFB 01Oct55). 50-159 (55), 51-2616 (55).

407 Strategic Fighter Wing: Malmstrom AFB, Montana. 50-159 (depl 55; 56), 51-2616 (55–56).

484 Bombardment (Heavy) Wing: Turner AFB, Georgia. 50-147 (64).

506 Strategic Fighter Wing: Dow AFB, Maine (20Jan53), to Tinker AFB, Oklahoma (20Mar55). 49-147 (55–56), 49-151 (54), 49-188 (55–58), 50-146 (56–58), 51-2584 (56).

508 Strategic Fighter Wing: Turner AFB, Georgia. 49-171 (54–unk).

509 Bombardment (Medium) Wing: Walker AFB, New Mexico. 51-2536 (55–56), 51-2549 (unk–56).

801 Air Base Group: Lockbourne AFB, Ohio. 49-162 (55), 49-194 (55), 51-2532 (57), 51-2554 (57–58), 51-2619 (57–58), 51-2624 (57–58), 51-2660 (57), 51-8082 (59).

802 Air Base Group: Smoky Hill AFB, Kansas (Renamed Schilling AFB 16Mar57). 49-144 (55), 49-183 (55), 51-2538 (55; 57-58), 51-2540 (57), 51-2553 (57–58), 51-2571 (57–58), 51-2573 (57), 51-8081 (54).

803 Air Base Group: Davis-Monthan AFB, Arizona. 48-352 (53), 49-133 (53), 49-136 (56), 49-137 (59), 49-170 (57), 49-174 (54), 50-138 (58), 50-144 (59), 50-152 (57–58), 50-166 (57), 50-169 (57–58; 59), 51-2582 (56), 51-2583 (55), 51-8164 (54).

804 Air Base Group: Hunter AFB, Georgia. 49-190 (57–58), 50-129 (57–58), 50-131 (57–58), 51-2566 (57–58), 51-2631 (58–59).

805 Air Base Group: Barksdale AFB, Louisiana. 50-120 (56), 50-143 (55), 50-151 (55), 51-2584 (55; 57), 51-8110 (56).

806 Air Base Group: Lake Charles AFB, Louisiana. 50-126 (57–58), 50-143 (55), 51-2545 (57), 51-2584 (55–56), 51-2603 (57–58), 51-2607 (57), 51-2610 (57).

807 Air Base Group: March AFB, California. 49-135 (55–56), 49-139 (57), 50-146 (56), 51-2553 (55), 51-2564 (59), 51-2577 (55), 51-2584 (55), 51-2634 (53). 51-2564 detached to Little Rock AFB.

808 Air Base Group: Bergstrom AFB, Texas. 49-143 (57–58), 49-151 (55–57), 49-180 (55), 50-120 (57), 50-139 (57), 50-147 (57), 50-160 (57). Reassigned to TAC 01Jul57.

809 Air Base Group: MacDill AFB, Florida. 49-140 (57–58), 49-180 (58), 49-185 (57), 50-120 (57–58), 50-124 (57), 50-138 (57–58), 51-8112 (53).

810 Air Base Group: Biggs AFB, Texas. 49-120 (56–58), 49-132 (57–58), 49-135 (56–58), 49-139 (56–58), 49-142 (56–58), 49-158 (58), 50-119 (56–57), 50-153 (55), 50-167 (54), 51-2536 (58), 51-2547 (58), 51-2550 (56), 51-2574 (55), 53-3180 (58), 53-8077 (56).

812 Air Base Group: Walker AFB, New Mexico. 49-132 (58), 49-162 (55–57), 49-187 (55; 56–57), 51-2536 (56–57), 51-2549 (56–58), 51-2580 (56), 51-2583 (56–58).

813 Air Base Group: Pinecastle AFB, Florida. 49-120 (56).

814 Air Base Group: Fairchild AFB, Washington State, to Westover AFB, Massachusetts. 48-320 (58), 49-179 (57–58), 50-156 (56), 51-2539 (55), 51-2559 (55), 51-2573 (55), 51-2580 (56), 51-2584 (57), 51-2616 (56).

815 Air Base Group: Forbes AFB, Kansas. 50-157 (57), 51-2573 (56).

817 Air Base Group: Portsmouth IAP (Pease ANGB), New Hampshire. 49-183 (56), 51-2553 (57).

818 Air Base Group: Lincoln AFB, Nebraska. 49-140 (56), 49-185 (56), 51-2571 (56), 51-8027 (59).

819 Combat Support Group: Dyess AFB, Texas. 49-144 (60), 51-2563 (60), 51-8048 (60).

823 Combat Support Group: Homestead AFB, Florida. 50-126 (61), 50-164 (60), 51-2554 (61).

824 Air Base Group: Carswell AFB, Texas. 49-139 (57), 49-170 (54; 57), 49-187 (54), 49-193 (57), 49-195 (55; 57), 50-160 (57), 51-2545 (56–57), 51-2607 (57).

Section Two—Military Operators

825 Air Base Group: Little Rock AFB, Arkansas. 49-137 (57–58), 49-152 (57–58), 49-157 (59), 50-143 (57–58), 50-160 (57), 51-2571 (58), 51-2584 (57), 53-3140 (59).

3902 Air Base Wing: Offutt AFB, Nebraska. 49-108 (58), 50-158 (55), 50-162 (56), 51-8061 (55).

3904 Composite Wing: Camp Carson, Colorado, to Stead AFB, Nevada (Jan53). 49-179 (50–53), 51-2554 (51–53). Deployment to Peterson AFB, Colorado; RAF Mildenhall and RAF Burtonwood, England.

3906 Air Base Group: Sidi Slimane AB, Morocco. 49-196 (54–57).

3910 Air Base Group: RAF Mildenhall, England, to RAF Lakenheath, England (09May55). 49-191 (54–58).

3919 Air Base Group: RAF Fairford, England. 53-3181 (55), 53-3187 (55).

3920 Air Base Group: RAF Brize Norton, England. 51-8252 (54).

3960 Air Base Wing: Andersen AFB, Guam. 48-337 (55), 51-8083 (69), 52-5944 (55), 53-3136 (69), 53-3167 (68), 53-3175 (55), 53-7862 (56).

3970 Air Base Group: Torrejón AB, Spain. 53-8143 (58), 53-8149 (60).

4002 Air Base Squadron: Campbell AFB, Kentucky. 51-2715 (55).

4041 Air Base Wing: Bunker Hill AFB, Indiana. 48-334 (56).

4060 Air Base Group / Wing: Dow AFB, Maine. 49-131 (58), 49-147 (55), 53-8109 (58).

4061 Air Base Group: Malmstrom AFB, Montana. 50-165 (57), 51-2609 (rec 58).

4080 Air Base Group: Turner AFB, Georgia, to Laughlin AFB, Texas (01Apr57). 49-171 (57–58), 49-176 (57–58), 49-180 (57).

4080 Strategic Reconnaissance (Light) Wing: Turner AFB, Georgia (01May56–01Apr57), to Laughlin AFB, Texas (01Apr57–01Jul63), to Davis-Monthan AFB, Arizona (01Jul63–25Jun66). 49-176 (57), 49-180 (57), 51-8004 (63–64).

4081 Air Base Group: Ernest Harmon AFB, Newfoundland, Canada. 51-2593 (58), 53-7850 (61).

4082 Air Base Group: Goose Bay AB, Labrador, Canada. 51-2554 (57), 51-8112 (62).

4087 Air Transport, Medium Squadron: Ernest Harmon AFB, Newfoundland, Canada. 51-8031 (57–58), 51-8033 (57–58), 53-3209 (57–58), 53-3210 (57–58), 53-3211 (57–58).

4130 Combat Support Group: Bergstrom AFB, Texas. 49-140 (59).

4238 Air Base Wing: Barksdale AFB, Louisiana. 48-337 (58), 50-121 (59), 50-161 (58), 51-2571 (58), 53-7859 (58), 51-8244 (59–60).

4347 Combat Crew Training Wing: McConnell AFB, Kansas. 51-7974 (60).

Tactical Air Command (TAC)

Major Command established 21 March 1946 to oversee tactical air power planning and utilization on a global scale. Became a sub-command of **Continental Air Command** (CNC) from 1 December 1948 to 1 December 1950 then re-established as a Major Command.

1 Air Commando Wing: *Operating Bases:* Lockbourne AFB, Ohio, to Eglin 15Jul71, to Hurlburt, (from 30Oct72). *Notes:* Redesignated 1 SOP WG 08Jul68; 1 SOT GP 15Sep70.

Some 71 TCSq / 930 TCG C-119s deployed to Lockbourne AFB, Ohio 15Jun68–18Jun69 designated as 71 TCSq 01Jul67–15Jun68; 71 ACO SQ 15Jun68–08Jul68; 71 SOP SQ 08Jul68–18Jun69. *C-119 Assignments:* 22 C-119G, 26 AC-119G, 25 AC-119K 1968–1972. *C-119G Assignments:* 51-2708 (depl 68), 51-7976 (depl 68), 51-7983 (depl 68), 51-7988 (depl 68), 51-7996 (depl 68), 51-7997 (depl 68), 51-8009 (depl 68), 51-8010 (depl 68), 51-8024 (depl 69), 51-8063 (depl 68), 51-8100 (depl 68), 51-8101 (depl 68), 51-8168 (depl 68), 52-5846 (depl 68), 52-5929 (depl 68), 52-5952 (depl 68), 53-3137 (depl 68), 53-3147 (depl 68), 53-7849 (depl 63), 53-7851 (depl 63), 53-7859 (depl 63), 53-8154 (depl 68). *AC-119G Assignments:* 52-5892 (68), 52-5898 (69), 52-5905 (68), 52-5907 (68), 52-5925 (68), 52-5927 (68), 52-5938 (68), 52-5942 (68), 53-3136 (68), 53-3145 (69), 53-3170 (68), 53-3178 (68), 53-3189 (68), 53-3192 (68), 53-3205 (68), 53-7833 (69), 53-7848 (68), 53-7851 (68), 53-7852 (68), 53-8069 (68), 53-8089 (69), 53-8114 (69), 53-8115 (69), 53-8123 (69), 53-8131 (68; 69), 53-8155 (68). *AC-119K Assignments:* 52-5864 (69), 52-5889 (69), 52-5910 (69; 71–72), 52-5911 (69; 71–72), 52-5926 (69), 52-5935 (69), 52-5940 (69), 52-9982 (69), 53-3154 (69), 53-3156 (69), 53-3187 (71–72), 53-3197 (69; 71–72), 53-3211 (69), 53-7826 (69), 53-7830 (69), 53-7831 (69; 71–72), 53-7839 (69; 71–72), 53-7850 (69), 53-7854 (69), 53-7877 (71–72), 53-7879 (69), 53-7883 (69), 53-8121 (69), 53-8145 (69; 71-72), 53-8148 (69).

4 Fighter (Day) Wing: Seymour-Johnson AFB, North Carolina. 51-8005 (57–58), 52-5896 (57–58).

11 Pilotless Bomb (Light) Squadron: Patrick AFB, Florida. 49-194 (54–55).

17 Bombardment (Tactical) Wing: Eglin AFB, Florida. 51-7979 (57–58), 52-5866 (56–58), 52-5903 (57–58).

18 (Eighteenth) Air Force: Greenville AFB, South Carolina. 51-2536 (51).

27 Air Base Group: Bergstrom AFB, Texas. 49-143 (58), 50-139 (58), 51-8110 (58), 51-8111 (58–59), 53-8084 (58–59). Reassigned from SAC 01Jul57.

62 Troop Carrier (Heavy) Wing: *Group:* 62 Troop Carrier Group. *Operating Base:* Larson AFB, Washington State. *Notes:* C-119 not operated, brief assignments. *C-119 Assignments:* 51-2583 (55), 51-8108 (56), 51-8110 (depl 54; 54), 51-8120 (54–55).

63 Troop Carrier (Heavy) Wing: *Group:* 63 Troop Carrier Group. *Operating Base:* Donaldson AFB, South Carolina. *Notes:* Some C-119F cvtd to C-119G during service. *C-119 Service:* 15 1954–1957. *C-119B Assignment:* 48-327 (55). *C-119C Assignments:* 48-322 (57), 50-164 (56). *C-119F Assignments:* 51-7975 (depl 55), 51-7986 (depl 54), 51-7997 (60), 51-8030 (depl 54), 51-8130 (depl 54), 51-8147 (54–56), 51-8148 (54–55), 51-8160 (54–57), 51-8166 (54–55). *C-119G Assignments:* 51-8102 (55), 52-5907 (depl 56), 53-8090 (depl 56).

64 Troop Carrier (Medium) Wing: *Group:* 64 Troop Carrier Group. *Squadrons:* 17 TCSq; 18 TCSq; 35 TCSq. *Operating Base:* Greenville AFB, South Carolina. (Renamed Donaldson AFB 01Jul53 or 24Aug53). *Notes:* Attached to **63 TCWG** 15Oct53–01Mar54. Deployment to Seymour-Johnson AFB, North Carolina Apr54. *C-119 Service:* 70 1952, 1953–1954. *C-119C Assignments:* 49-194 (54), 50-142 (53–54), 51-2568 (53), 51-2579 (53). *C-119CF Assignments:* 51-2604 (53), 51-2606 (53), 51-2613 (53), 51-2616 (depl 53), 51-2617 (depl 53), 51-2621 (53), 51-2622 (53), 51-2623 (53), 51-2624 (53), 51-2629 (53), 51-2658 (53), 51-2659 (53), 51-2660 (53), 51-2661 (53). *C-119F Assignments:* 51-2680 (depl 52), 51-8043 (depl 53). *C-119G Assignments:* 51-8054 (53–54), 51-8055 (53–54), 51-8056 (53–54), 51-8057 (53–54), 51-8058 (53–54), 51-8059 (53–54), 51-8060 (53–54), 51-8061 (53–54), 51-8062 (53–54), 51-8063 (53–54), 51-8065 (53–54), 51-8066 (53–54), 51-8067 (53–54), 51-8074 (53–54),

51-8078 (53–54), 51-8081 (53–54), 51-8082 (53–54), 51-8083 (53–54), 51-8085 (53–54), 51-8094 (53–54), 51-8095 (53–54), 51-8096 (53–54), 51-8097 (53–54), 52-5845 (53–54), 52-5846 (53–54), 52-5848 (53–54), 52-5849 (53–54), 52-5850 (53–54), 52-5851 (53–54), 52-5852 (53–54), 52-5856 (53–54), 52-5857 (53–54), 52-5859 (53–54), 52-5860 (53–54), 52-5861 (53–54), 52-5862 (53–54), 52-5863 (53–54), 52-5864 (53–54), 52-5866 (53–54), 52-5867 (53–54), 52-5868 (53–54), 52-5869 (53–54), 52-5876 (53–54), 52-5877 (53–54), 52-5878 (53–54), 52-5917 (54), 52-5918 (54), 52-5919 (54), 52-5920 (54), 52-5921 (54).

69 Pilotless Bomb (Light) Squadron: Patrick AFB, Florida. 49-194 (54).

83 Fighter (Day) Wing: Seymour-Johnson AFB, North Carolina. 51-8005 (57), 52-5896 (57), 53-3213 (56–57).

312 Fighter-Bomber Wing: Clovis AFB, New Mexico (Renamed Cannon AFB Jul57). 51-8136 (56–57), 53-8072 (56–57), 53-8080 (56–57), 53-8093 (56–57).

313 Troop Carrier (Medium) Wing: *Group:* **313 Troop Carrier Group.** *Squadrons (Colors):* **29 TCSq** (Red); **47 TCSq** (Green); **48 TCSq** (Blue). *Operating Bases:* Mitchel AFB, New York (01Feb53–02Oct53). Sewart AFB, Tennessee (02Oct53–08Jun55). *Notes:* Group attached to **465 TCWG** 25Aug53–30Sep53 at Sewart. Deployments to Pope AFB, North Carolina Apr54; Sumpter Smith ANGB, Alabama Jul-Aug54 and Elmendorf AFB, Alaska Jan55. Some C-119F cvtd to C-119G during service. *C-119 Service:* 91 1953–1955. *C-119F Assignments:* 51-7969 (53–54), 51-7970 (53–54), 51-7971 (53–54), 51-7972 (54–55), 51-8016 (53–55), 51-8028 (53–55), 51-8029 (53–55), 51-8044 (53–54), 51-8047 (53–55), 51-8048 (53–54), 51-8110 (53–54), 51-8112 (53–54), 51-8113 (53–54), 51-8114 (53–54), 51-8117 (53–55), 51-8120 (53–54), 51-8145 (53–55), 51-8146 (53–55), 51-8147 (53–54), 51-8148 (53–54), 51-8151 (53–54), 51-8160 (53–54), 51-8161 (53–55), 51-8162 (53–55), 51-8166 (53–54), 51-8167 (53–54), 51-8168 (53–55). *C-119G Assignments:* 51-8060 (54–55), 51-8061 (54–55), 51-8063 (54–55), 51-8067 (54–55), 51-8068 (53–54), 51-8069 (53–55), 51-8070 (53–54), 51-8071 (53–55), 51-8072 (53–55), 51-8073 (53–55), 51-8075 (53–55), 51-8076 (53–55), 51-8077 (53–55), 51-8078 (54–55), 51-8079 (53–55), 51-8080 (53–55), 51-8081 (54–55), 51-8084 (53–55), 51-8086 (53), 51-8087 (53–54), 51-8088 (53–55), 51-8089 (53–55), 51-8090 (53–55), 51-8091 (53–55), 51-8092 (53–55), 51-8093 (53–55), 51-8095 (54–55), 52-5843 (53–54), 52-5844 (53–55), 52-5845 (54–55), 52-5850 (54–55), 52-5859 (54), 52-5860 (54–55), 52-5863 (54–55), 52-5864 (54–55), 52-5867 (54–55), 52-5877 (54–55), 52-5918 (54–55), 52-5924 (54–55), 52-5832 (54), 52-5933 (54), 52-5934 (54), 52-5935 (54), 52-5936 (54), 52-5944 (54), 52-5945 (54), 52-5946 (54), 52-5947 (54–55), 52-5948 (54), 52-5949 (54–55), 52-5950 (54), 52-5951 (54–55), 52-5952 (54–55), 52-5953 (54), 53-7857 (55), 53-7858 (55), 53-7859 (55), 53-8098 (54–55), 53-8099 (54–55; depl 55), 53-8100 (54–55), 53-8101 (54–55), 53-8102 (54), 53-8133 (54–55), 53-8134 (55).

314 Troop Carrier (Medium) Wing: *Group:* **314 Troop Carrier Group.** *Squadrons (Colors):* **37 TCSq (Attached)** (Yellow); **50 TCSq** (Red/White); **61 TCSq** (Green/White); **62 TCSq** (Blue/White). *Operating Base:* Smyrna AFB, Tennessee. (Renamed Sewart AFB 25Mar50). *Notes:* Deployments to Laurinburg-Maxton AP, North Carolina Apr50; Elmendorf AFB, Alaska Jan55; Ladd AFB, Alaska Jun-Aug55; Dreux AB, France Oct55–Apr56; Big Delta, Alaska Nov55–Jan56 and England AFB, Louisiana Nov56. Some C-119F cvtd to C-119G during service. See under: **Far East Air Forces** (FEA) for overseas assignments from 1950 to 1954. *C-119 Service:* 232 1949–1957. *C-119B Assignments:* 48-322 (52), 48-324 (49–50), 48-325 (49–50), 48-326 (49–51), 48-327 (49–50), 48-328 (49–50), 48-329 (49–50), 48-331 (49–50), 48-332 (49–50), 48-334 (49–50), 48-335 (49–51), 48-336 (50), 48-337 (50),

48-338 (50), 48-340 (50–52), 48-341 (50), 48-342 (50), 48-343 (50), 48-344 (50), 48-345 (50), 48-346 (50), 48-347 (50), 48-348 (50), 48-349 (50), 48-350 (50), 48-351 (50), 48-352 (50), 48-353 (50), 48-354 (50), 48-355 (50), 49-101 (49–50), 49-102 (49–50), 49-103 (49–50), 49-104 (49–50), 49-105 (50), 49-106 (50), 49-107 (50), 49-108 (50), 49-109 (50), 49-110 (50), 49-111 (50), 49-112 (50), 49-113 (50), 49-114 (50), 49-115 (50), 49-116 (50), 49-117 (50), 49-118 (50). ***C-119C Assignments:*** 49-119 (50), 49-120 (50), 49-121 (depl 50; 50), 49-122 (50), 49-123 (50), 49-122 (depl 50; 50), 49-125 (50), 49-126 (50), 49-127 (depl 50; 50), 49-128 (depl 50; 50), 49-129 (depl 50; 50), 49-130 (depl 50; 50), 49-131 (depl 50; 50), 49-132 (depl 50; 50), 49-133 (depl 50; 50), 49-134 (depl 50), 49-135 (50), 49-136 (50), 49-137 (50), 49-138 (50), 49-139 (50), 49-140 (50), 49-141 (50), 49-142 (50), 49-143 (50), 49-144 (50); 49-145 (50), 49-146 (50), 49-147 (50), 49-148 (50; depl 51), 49-149 (50), 49-150 (50), 49-151 (50), 49-152 (50), 49-153 (50), 49-154 (50), 49-155 (50), 49-156 (50), 49-157 (50), 49-158 (50), 49-159 (50), 49-160 (50), 49-161 (50), 49-162 (50), 49-163 (50), 49-164 (50), 49-165 (50), 49-166 (50), 49-167 (50), 49-168 (50), 49-169 (50), 49-170 (50), 49-171 (50), 49-172 (50), 49-173 (50), 49-174 (50), 49-175 (50), 49-176 (50), 49-177 (50), 49-178 (50), 49-180 (51–53), 49-181 (52; 56), 49-182 (del 51; 52), 49-184 (50–54), 49-185 (depl 51), 49-186 (depl 51), 49-190 (depl 51), 49-193 (depl 51), 49-194 (52), 49-195 (52), 49-199 (depl 56), 50-124 (56), 50-139 (52), 50-164 (56), 51-2538 (57), 51-2551 (depl 51). ***C-119F Assignments:*** 51-2663 (54–55), 51-2664 (54–55), 51-2665 (54–55), 51-2666 (54–unk), 51-2667 (54–unk), 51-2668 (54–unk), 51-2669 (54–unk), 51-2670 (54–unk), 51-2671 (54–unk), 51-2672 (54–unk), 51-2673 (54–unk), 51-2674 (54–unk), 51-2675 (54–unk), 51-2676 (54–unk), 51-2677 (54–unk), 51-2678 (54–unk), 51-2680 (54–55), 51-2681 (54–unk), 51-2682 (54–unk), 51-2683 (54–unk), 51-2684 (54–unk), 51-2685 (54–unk), 51-2686 (54–unk), 51-2708 (54–unk), 51-2709 (54–unk), 51-2710 (54–unk), 51-2711 (54–unk), 51-2717 (54–unk). 51-7969 (54–57), 51-7970 (54–56), 51-7971 (54–56), 51-8005 (depl 54), 51-8013 (54–57), 51-8014 (54–56), 51-8015 (54–57), 51-8017 (54–57), 51-8018 (54–57), 51-8019 (54–57), 51-8020 (54–57), 51-8021 (54–57), 51-8022 (54–57), 51-8023 (54–57), 51-8024 (54–57), 51-8025 (54–57), 51-8026 (54–56), 51-8027 (54–57), 51-8101 (54–55), 51-8102 (54–55), 51-8103 (54–55), 51-8104 (54–55), 51-8109 (54–55), 51-8111 (54–55), 51-8118 (54–55). ***C-119G Assignments:*** 51-7972 (55–57), 51-7989 (depl 56), 51-8016 (55–57), 51-8028 (55–57), 51-8029 (55–57), 51-8047 (55–57), 51-8060 (55), 51-8063 (55), 51-8067 (55–57), 51-8073 (55–57), 51-8075 (55–57), 51-8077 (55–57), 51-8078 (55–57), 51-8079 (depl 54; 55–57), 51-8081 (55), 51-8084 (55), 51-8089 (55), 51-8090 (55), 51-8091 (55–57), 51-8092 (55), 51-8093 (55–57), 51-8095 (55), 51-8117 (55–57), 51-8145 (55–57), 51-8147 (depl 56), 51-8148 (depl 56), 51-8161 (55–57), 51-8162 (55–57), 51-8168 (55–57), 52-5844 (55–57), 52-5845 (55–57), 52-5850 (56–57), 52-5860 (55–57), 52-5863 (56–57), 52-5864 (55–57), 52-5867 (56–57), 52-5877 (55–57), 52-5918 (55–57), 52-5924 (55–57), 52-5947 (56–57), 52-5951 (56–57), 52-5952 (56–57), 53-3215 (55–57), 53-7855 (55–57), 53-7856 (55–57), 53-7857 (55–57), 53-7858 (55–57), 53-7859 (55–57), 53-8098 (55–57), 53-8099 (55–57), 53-8100 (55–57), 53-8101 (55–56), 53-8133 (55–57), 53-8134 (55–57), 53-8135 (55–57). ***Under MAC:*** 53-8084 (ret 75 displayed at Little Rock AFB, not counted).

316 Troop Carrier (Medium) Wing: *Group:* 316 Troop Carrier Group. *Squadrons (Colors):* **36 TCSq** (Red/White); **37 TCSq** (Blue/White); **75 TCSq** (Green/White). *Operating Bases:* Smyrna AFB, Tennessee (from 04Nov49). (Renamed Sewart AFB 25Mar50). *Notes:* Deployments to Laurinburg-Maxton AP, North Carolina Aug51; Burlington IAP, Vermont Feb53; Pope AFB, North Carolina Apr54; Elmendorf AFB, Alaska Jul54 and Sumpter Smith ANGB, Alabama Aug54. See under: **Far East Air Forces** (FEA) for overseas assignments from 1954 to 1956. ***C-119 Service:*** 182 1950–1954. ***C-119B Assignments:*** 48-324 (depl 50), 48-325 (depl 50), 48-327 (depl 50), 48-328 (depl 50), 48-329 (depl 50), 48-331 (depl 50), 48-332 (depl 50), 48-334 (depl 50), 48-335 (depl 50), 48-336 (depl 50), 48-337 (depl 50), 48-340 (depl 50; 50–52), 48-342 (depl 50), 48-343 (depl 50), 48-344 (depl 50), 48-345 (depl 50), 48-346 (depl 50), 48-347 (depl 50), 48-348 (depl 50), 48-349 (depl 50), 48-350 (depl 50), 48-351 (depl 50), 48-352 (depl 50), 48-353 (depl 50), 48-354 (depl 50), 48-355 (depl 50), 49-101 (depl 50), 49-102 (depl 50), 49-103 (depl 50), 49-104 (depl 50), 49-105 (depl 50), 49-107 (depl 50), 49-108 (depl 50), 49-109 (depl 50), 49-110 (depl 50), 49-111 (depl 50), 49-112 (depl 50), 49-113 (depl 50), 49-114 (depl 50), 49-115 (depl 50), 49-116 (depl 50), 49-117 (depl 50), 49-118 (depl 50). ***C-119C Assignments:*** 49-121 (50), 49-122 (depl 50), 49-124 (50), 49-125 (50), 49-126 (50), 49-127 (50), 49-128 (50), 49-129 (50), 49-130 (50), 49-131 (50), 49-132 (50), 49-133 (50), 49-134 (50), 49-135 (50), 49-136 (50), 49-137 (50), 49-138 (50), 49-139 (50), 49-140 (50), 49-141 (50), 49-142 (50), 49-143 (50), 49-144 (50), 49-145 (50), 49-146 (50), 49-147 (50), 49-148 (50–52), 49-149 (50), 49-150 (50), 49-151 (50), 49-152 (50), 49-153 (50), 49-154 (50), 49-155 (50), 49-170 (depl 50), 49-171 (depl 50), 49-180 (50–51), 49-181 (50–52), 49-182 (5–52), 49-183 (50–52), 49-184 (50–52), 49-185 (50–52), 49-186 (50–52), 49-187 (50–52), 49-188 (50–52), 49-189 (50), 49-190 (50–52), 49-191 (50–52), 49-192 (50–51), 49-193 (50–52), 49-194 (50–52), 49-195 (50–52), 49-196 (50–52), 50-139 (51–52), 50-142 (53), 50-143 (51–53), 50-147 (51–52), 50-154 (depl 51), 51-2532 (51), 51-2533 (51), 51-2537 (51), 51-2556 (depl 51), 51-2567 (depl 51). ***C-119CF Assignments:*** 51-2603 (52), 51-2604 (52), 51-2605 (52), 51-2606 (52), 51-2607 (52), 51-2608 (52), 51-2609 (52), 51-2610 (52), 51-2611 (52), 51-2612 (52), 51-2613 (52), 51-2614 (52), 51-2615 (52), 51-2616 (52), 51-2617 (52), 51-2618 (52), 51-2619 (52), 51-2620 (52), 51-2621 (52), 51-2622 (52), 51-2623 (52), 51-2624 (52), 51-2625 (52), 51-2629 (depl 53). ***C-119F Assignments:*** 51-2663 (52–54), 51-2664 (52–54), 51-2665 (52–54), 51-2666 (52–54), 51-2667 (52–54), 51-2668 (52–54), 51-2669 (52–54), 51-2670 (52–54), 51-2671 (52–54), 51-2672 (52–54), 51-2673 (52–54), 51-2674 (52–54), 51-2675 (52–54), 51-2676 (52–54), 51-2677 (52–54), 51-2678 (52–54), 51-2679 (52–54), 51-2680 (52–54), 51-2681 (52–54), 51-2682 (52–54), 51-2683 (52–54), 51-2684 (52–54), 51-2685 (52–54), 51-2686 (52–54), 51-2708 (52–54), 51-2709 (52–54), 51-2710 (52–54), 51-2711 (52–54), 51-2717 (52–54), 51-7969 (54), 51-7970 (54), 51-8013 (52–54), 51-8014 (52–54), 51-8015 (52–54), 51-8017 (52–54), 51-8018 (52–54), 51-8019 (52–54), 51-8020 (52–54), 51-8021 (52–54), 51-8022 (52–54), 51-8023 (53–54), 51-8024 (53–54), 51-8025 (53–54), 51-8026 (53–54), 51-8027 (53–54), 51-8101 (53–54), 51-8102 (52–54), 51-8103 (52–54), 51-8104 (52–54), 51-8109 (52–54), 51-8111 (52–54), 51-8118 (52–54).

317 Tactical Airlift Wing: Lockbourne AFB, Ohio. 52-5906 (depl 70).

323 Fighter-Bomber Wing: Bunker Hill AFB, Indiana. 52-5851 (57), 52-5897 (56–57), 53-8094 (56–57), 53-8095 (56–57).

354 Fighter (Day) Wing: Myrtle Beach AFB, South Carolina. 48-328 (depl 60), 51-7984 (57), 51-8012 (59), 52-5895 (56–57).

363 Tactical Reconnaissance Wing: Shaw AFB, South Carolina. 51-7987 (depl 57), 51-8103 (depl 56), 52-5849 (56–57), 52-5875 (57–58), 52-5906 (56–57), 52-5918 (54), 53-8090 (56–57).

366 Fighter-Bomber Wing: Alexandria AFB, Louisiana (Renamed England AFB 23Jun55). 51-2714 (55–unk), 51-7987 (depl 55), 51-7992 (57), 51-8108 (depl 54), 51-8110 (55–57), 51-8135 (55–57), 51-8142 (55), 51-8166 (55–57), 53-8092 (depl 56).

401 Tactical Fighter Wing: England AFB, Louisiana. 51-8090 (sal 64).

405 Fighter-Bomber Wing: Langley AFB, Virginia. 51-2713 (depl 54), 51-8099 (depl 56; 57), 51-8107 (57), 51-8113 (55), 53-8128 (57).

419 Troop Carrier, Assault Fixed-Wing Group: *Wing:* **463 TCWG** (Attached Jul56). *Operating Base:* Ardmore AFB, Oklahoma. *C-119 Assignment:* 51-8101 (56).

433 Troop Carrier (Medium) Wing: *Group:* **433 Troop Carrier Group.** *Squadrons:* **67 TCSq**; **68 TCSq**; **69 TCSq.** *Operating Base:* Greenville AFB, South Carolina. *Notes:* See under: **USAF in Europe** (AFE) for overseas assignment. *C-119 Service:* 48 1950–1951. *C-119C Assignments:* 49-197 (50–51), 49-198 (50–51), 49-199 (50–51), 50-119 (50–51), 50-120 (51), 50-121 (51), 50-122 (51), 50-123 (51), 50-124 (51), 50-125 (51), 50-126 (51), 50-127 (51), 50-128 (51), 50-129 (51), 50-130 (51), 50-131 (51), 50-132 (51), 50-133 (51), 50-134 (51), 50-136 (51), 50-137 (51), 50-140 (51), 50-141 (51), 50-144 (51), 50-145 (51), 50-146 (51), 50-148 (51), 50-149 (51), 50-150 (51), 50-151 (51), 50-152 (51), 50-153 (51), 50-155 (51), 50-157 (51), 50-158 (51), 50-159 (51), 50-160 (51), 50-161 (51), 50-163 (51), 50-164 (51), 50-165 (51), 50-166 (51), 50-167 (51), 50-168 (51), 50-169 (51), 50-170 (51), 50-171 (51), 51-2535 (51).

435 Troop Carrier (Medium) Wing: *Group:* **435 Troop Carrier Group.** *Squadrons:* **76 TCSq**; **77 TCSq**; **78 TCSq.** *Operating Base:* Miami Intl. Airport, Florida. *Notes:* Deployments to Grenier AFB, New Hampshire Jan52 and Elmendorf AFB, Alaska Oct52. To **Air Force Reserve** (AFR) 01Dec52. *C-119 Service:* 81 1951–1952. *C-119B Assignments:* 48-340 (52). *C-119C Assignments:* 49-148 (52), 49-183 (52), 49-185 (52), 49-186 (52), 49-187 (52), 49-188 (52), 49-190 (52), 49-191 (52), 49-196 (52), 50-138 (52), 51-2532 (51–52), 51-2533 (51–54), 51-2536 (51–52), 51-2537 (51–52), 51-2538 (51–52), 51-2539 (51–52), 51-2540 (51–52), 51-2541 (51–52), 51-2543 (52), 51-2545 (51–52), 51-2546 (51–52), 51-2547 (51–52), 51-2548 (51–52), 51-2549 (51–52), 51-2550 (51–52), 51-2551 (51–52), 51-2552 (51–52), 51-2553 (51–52), 51-2557 (51–52), 51-2558 (51–52), 51-2559 (51–52), 51-2560 (51–52), 51-2561 (51–52), 51-2562 (51–52), 51-2563 (51–52), 51-2564 (51–52), 51-3565 (51–52), 51-2566 (51–52), 51-2568 (51–52), 51-2569 (51–52), 51-2570 (51–52), 51-2571 (51–52), 51-2572 (51–52), 51-2573 (51–52), 51-2574 (51–52), 51-2575 (51–52), 51-2576 (51–52), 51-2577 (51–52), 51-2578 (51–52), 51-2579 (51–52), 51-2581 (51–52), 51-2583 (52), 51-2584 (52). *C-119CF Assignments:* 51-2603 (52), 51-2604 (52), 51-2605 (52), 51-2606 (52), 51-2607 (52), 51-2608 (52), 51-2609 (52), 51-2610 (52), 51-2611 (52), 51-2612 (52), 51-2613 (52), 51-2614 (52), 51-2615 (52), 51-2616 (52), 51-2617 (52), 51-2618 (52), 51-2619 (52), 51-2620 (52), 51-2621 (52), 51-2622 (52), 51-2623 (52), 51-2624 (52), 51-2625 (52), 51-2658 (52), 51-2659 (52), 51-2660 (52), 51-2661 (52).

443 Troop Carrier (Medium) Wing: *Group:* **443 Troop Carrier Group.** *Squadrons:* **309 TCSq**; **310 TCSq**; **343 TCSq.** *Operating Base:* Greenville AFB, South Carolina. (Renamed Donaldson AFB 25Aug53). *Notes:* Deployment to Ladd AFB, Alaska Oct52. *C-119 Service:* 51 1952–1953. *C-119CF Assignments:* 51-2587 (52–53), 51-2588 (52–53), 51-2589 (52–53), 51-2590 (52), 51-2591 (52–53), 51-2592 (52–53), 51-2593 (52–53), 51-2594 (52–53), 51-2595 (52–53), 51-2596 (52–53), 51-2597 (52–53), 51-2598 (52–53), 51-2599 (52–53), 51-2600 (52–53), 51-2601 (52–53), 51-2602 (52–53), 51-2615 (53), 51-2616 (53), 51-2618 (53), 51-2626 (52–53), 51-2627 (52–53), 51-2628 (52–53), 51-2629 (52–53), 51-2630 (52), 51-2631 (52–53), 51-2632 (52–53), 51-2633 (52–53), 51-2634 (52–53), 51-2635 (52–53), 51-2636 (52–53), 51-2637 (52–53), 51-2638 (52–53), 51-2639 (52–53), 51-2640 (52–53), 51-2641 (52–53), 51-2642 (52–53), 51-2643 (52–53), 51-2644 (52–53), 51-2645 (52–53), 51-2646 (52–53), 51-2647 (52–53), 51-2648 (52–53), 51-2649 (52–53), 51-2650 (52–53), 51-2651 (52–53), 51-2652 (52–53), 51-2653 (52–53), 51-2654 (52–53), 51-2655 (52–53), 51-2656 (52–53), 51-2657 (52–53).

450 Fighter (Day) Wing: Foster AFB, Texas. 51-7985 (57–unk), 51-8103 (55–58), 51-8109 (55–57), 51-8111 (55–58), 51-8120 (55–58).

456 Troop Carrier (Medium) Wing: *Group:* 456 Troop Carrier Group. *Squadrons (Colors):* **744 TCSq**; **745 TCSq**; **746 TCSq**. *Operating Bases:* Miami IAP, Florida (01Dec52–25Jul53). Charleston AFB, South Carolina (25Jul53–16Oct55). Shiroi AB, Japan (10Nov55–10May56). Ardmore AFB, Oklahoma (25May56–09Jul56). *Notes:* Deployment to Langley AFB, Virginia Oct54. *C-119 Service:* 107 1952–1956. *C-119B Assignments:* 48-340 (52–53). *C-119C Assignments:* 49-148 (52–53), 49-180 (53), 49-183 (52–53), 49-185 (52–53), 49-186 (52–53), 49-187 (52–53), 49-188 (52–53), 49-190 (52–53), 49-191 (52–53), 49-196 (52–53), 50-138 (52–53), 50-139 (52–53), 50-142 (53), 50-143 (53), 50-147 (52–53), 51-2547 (52–53), 51-2564 (52–53), 51-2566 (52–53), 51-2568 (52–53), 51-2575 (52–53), 51-2579 (52–53), 51-2581 (52–53). *C-119CF Assignments:* 51-2603 (52–53), 51-2604 (52–53), 51-2605 (52–53), 51-2606 (52–53), 51-2607 (52–53), 51-2608 (52–53), 51-2609 (52), 51-2610 (52–53), 51-2611 (52–53), 51-2612 (52–53), 51-2613 (52–53), 51-2614 (52–53), 51-2615 (52–53), 51-2616 (52–53), 51-2617 (52–53), 51-2618 (52–53), 51-2619 (52–53), 51-2620 (52–53), 51-2621 (52–53), 51-2622 (52–53), 51-2623 (52–53), 51-2624 (52–53), 51-2625 (52–53), 51-2658 (52–53), 51-2659 (52–53), 51-2660 (52–53), 51-2661 (52–53). *C-119F Assignments:* 51-7968 (55–56), 51-8030 (53–56), 51-8035 (53–56), 51-8036 (53–56), 51-8037 (53–56), 51-8038 (53–56), 51-8039 (53–56), 51-8040 (53–56), 51-8041 (53–unk), 51-8042 (53–56), 51-8043 (53–56), 51-8045 (53–56), 51-8046 (53–56), 51-8049 (53–56), 51-8050 (53–56), 51-8051 (53–56), 51-8052 (53–56), 51-8101 (55–56), 51-8105 (54–55), 51-8106 (54–55), 51-8113 (55–56), 51-8114 (54–56), 51-8115 (53–56), 51-8116 (53–56), 51-8119 (53–56), 51-8121 (53–56), 51-8122 (53–56), 51-8123 (53–56), 51-8124 (53–56), 51-8125 (53–56), 51-8126 (53–56), 51-8127 (53–56), 51-8128 (53–56), 51-8129 (53–56), 51-8130 (53–56), 51-8131 (53–56), 51-8132 (53–56), 51-8133 (53–54), 51-8134 (53–56), 51-8137 (53–56), 51-8138 (53–56), 51-8139 (53–55), 51-8140 (53–56), 51-8141 (53–56), 51-8144 (53–56), 51-8152 (53–56), 51-8153 (53–54), 51-8154 (53–56), 51-8155 (53–56), 51-8156 (53–56), 51-8157 (53–56), 51-8158 (53–56), 51-8159 (53–56), 51-8163 (53), 51-8164 (53–56), 51-8165 (53–55), 51-8167 (55–56).

461 Bombardment (Tactical) Wing: Blytheville AFB, Arkansas. 50-168 (57–58), 51-7987 (57–58), 52-5884 (56–57), 53-8103 (56–58).

463 Troop Carrier (Medium) Wing: *Group:* 463 Troop Carrier Group. *Squadrons (Colors):* **16 TCSq**; **772 TCSq** (Red); **773 TCSq** (Yellow); **774 TCSq** (Green); **775 TCSq** (Blue). *Operating Bases:* Memphis MAP, Tennessee (16Jan53–01Sep53). Ardmore AFB, Oklahoma (01Sep53–15Jan59). Sewart AFB, Tennessee (15Jan59–01Jul63). *Notes:* Deployments to Charleston AFB, South Carolina Apr54; Gray AFB, Texas May54; Sewart AFB, Tennessee 55; North Auxiliary Airfield, South Carolina Apr55 and Apr & May56; Évreux-Fauville AB, France May55 and Laurinburg-Maxton AP, North Carolina Aug55. Some C-119F cvtd to C-119G during service. *C-119 Service:* 93 1953–1957. *C-119C Assignment:* 51-2563 (55). *C-119F Assignments:* 51-2662 (53–57), 51-2663 (55), 51-2664 (55), 51-2680 (55–unk), 51-2681 (55–unk), 51-2682 (55–unk), 51-2683 (55–unk), 51-2684 (55–unk), 51-2685 (55–unk), 51-2686 (depl 55), 51-2708 (depl 55), 51-2709 (depl 55), 51-2710 (depl 55), 51-2711 (depl 55; 57), 51-2712 (53–unk), 51-2713 (53–unk), 51-2714 (53–55), 51-2715 (53–55), 51-2716 (53–unk), 51-7969 (depl 55), 51-7970 (depl 55; 56–57), 51-7971 (depl 55; 56–57), 51-7972 (depl 55), 51-7973 (53–56), 51-7974 (53–56), 51-7975 (53–57), 51-7976 (53–57), 51-7977 (53–56), 51-7978 (53–56), 51-7979 (53–56), 51-7980 (53–57), 51-7981 (53–57), 51-7982 (53–56), 51-7983 (53–57), 51-7984 (53–56), 51-7985 (53–56), 51-7986 (53–57), 51-7987 (53–56), 51-7988 (53–57), 51-7989 (53–57), 51-7990 (53–57), 51-7991 (53–56), 51-7992 (53–56), 51-7993 (53–54), 51-7994 (53–57), 51-7995 (53–55),

51-7996 (53–56), 51-7997 (53–56), 51-7998 (53–56), 51-7999 (53–57), 51-8000 (53–57), 51-8001 (53–57), 51-8002 (53–57), 51-8003 (53–57), 51-8004 (53–57), 51-8005 (53–56), 51-8006 (53–56), 51-8007 (53–57), 51-8008 (53–56), 51-8009 (53–57), 51-8010 (53–57), 51-8011 (53–57), 51-8012 (53–56), 51-8014 (56–57), 51-8048 (54), 51-8099 (53–55), 51-8100 (53–56), 51-8101 (depl 56), 51-8107 (53–55), 51-8108 (53–56), 51-8110 (54–55), 51-8112 (54–57). *C-119G Assignments:* 51-8031 (58), 51-8047 (55), 51-8067 (55), 51-8075 (55), 51-8077 (55), 51-8078 (55), 51-8079 (55), 51-8091 (55), 51-8148 (55–57), 52-5850 (55–56), 52-5860 (55), 52-5951 (55–56), 52-5952 (55–56), 53-3212 (55–56), 53-3216 (55–57), 53-3217 (55–57), 53-7858 (55), 53-8099 (depl 55–56), 53-8134 (depl 55). *C-119J Assignments:* 51-8128 (58).

464 Troop Carrier Wing: *Group:* **464 Troop Carrier Group.** *Squadrons (Colors):* **776 TCSq** (Red); **777 TCSq** (Blue); **778 TCSq** (Green); **779 TCSq** (Yellow). *Operating Bases:* Lawson AFB, Georgia (01Feb53–21Sep54). Pope AFB, North Carolina (21Sep54–31Aug71). *Notes:* Deployments to Seymour Johnson AFB, North Carolina Apr54; limited deployment to Rhein-Main AB West Germany Oct54-Nov55; North Auxiliary Airfield, South Carolina Jun55 and U.S. Army Aberdeen Proving Ground, Maryland May56. Some were stationed at Sewart AFB, Tennessee Jun-Aug55 before the move to Pope. Early assignments were mainly for mnt. C-119C 49-110 was assigned Nov63-Jan64 but likely not in service. *C-119 Service:* 139 1953–1958. *C-119C Assignments:* 48-331 (56), 48-332 (57–58; 58), 48-340 (55; 59), 49-112 (56), 49-118 (56; 57), 49-122 (depl 63), 49-157 (57), 49-167 (56), 50-142 (54). *C-119CF Assignments:* 51-2596 (60), 51-2616 (59), 51-2627 (59). *C-119G Assignments:* 51-2664 (57–unk), 51-2673 (depl 67), 51-7973 (56–58), 51-7974 (56–57), 51-7981 (depl 56), 51-7994 (depl 58), 51-8005 (depl 57), 51-8012 (56–57), 51-8053 (54–57), 51-8054 (54–58), 51-8057 (54–58), 51-8058 (54–58), 51-8059 (54–58), 51-8060 (55–58), 51-8061 (55–57), 51-8063 (55–57), 51-8064 (unk–58), 51-8069 (55–58; 59–60), 51-8071 (55–58), 51-8072 (55–58), 51-8080 (55–58), 51-8081 (55–58), 51-8082 (depl 56), 51-8083 (54–58), 51-8084 (55–58), 51-8088 (55–58), 51-8089 (55–58), 51-8090 (55–58), 51-8092 (55–58), 51-8097 (54–58), 51-8102 (depl 55), 51-8108 (56–58), 51-8112 (depl 55), 51-8145 (55), 51-8146 (58), 51-8147 (depl 56), 51-8160 (depl 55), 51-8161 (depl 55; 55), 51-8162 (55), 51-8168 (55), 52-5840 (53–57), 52-5841 (53–57), 52-5842 (53–54), 52-5848 (54–58), 52-5849 (55–56), 52-5850 (depl 54), 52-5851 (54–56), 52-5852 (54), 52-5853 (53–54), 52-5854 (53–54), 52-5855 (53–58), 52-5856 (54–58), 52-5858 (unk–58), 52-5859 (54), 52-5863 (55–56), 52-5865 (55–58), 52-5866 (54–56), 52-5867 (55–56), 52-5868 (54), 52-5869 (54–58), 52-5870 (53–58), 52-5871 (53–54), 52-5872 (53–54), 52-5873 (53–54), 52-5874 (53–54), 52-5875 (53–57), 52-5878 (54–58), 52-5879 (53–54), 52-5880 (53–58), 52-5881 (53–54), 52-5882 (53–54), 52-5883 (53–54), 52-5884 (53–57), 52-5885 (53–54), 52-5886 (53–54), 52-5887 (53–54), 52-5888 (53–58), 52-5889 (53–54), 52-5890 (53–58), 52-5891 (53–54), 52-5892 (53–54), 52-5893 (53–54), 52-5894 (53–54), 52-5895 (53–56), 52-5896 (53–54), 52-5897 (53–56), 52-5898 (53–58), 52-5899 (unk–57), 52-5900 (53–54), 52-5901 (53–58), 52-5902 (53–54), 52-5903 (53–57), 52-5904 (53–54), 52-5905 (53–54), 52-5906 (53–54), 52-5907 (53–58), 52-5908 (53–54), 52-5909 (53–58), 52-5910 (53–58), 52-5911 (53–58), 52-5912 (53–58), 52-5913 (53–58), 52-5914 (53–58), 52-5915 (53–58), 52-5916 (54–58), 52-5918 (55), 52-5947 (55–56), 53-3136 (depl 57), 53-3137 (54–57), 53-3138 (54–57), 53-3139 (54–57), 53-3140 (54–57), 53-3141 (54–57), 53-3178 (54–58), 53-3207 (depl 56), 53-3213 (55–56), 53-3216 (depl 56-57), 53-7857 (55), 53-8069 (54–58), 53-8071 (54–58), 53-8100 (depl 56), 53-8101 (55), 53-8103 (54–56), 53-8131 (55–58), 53-8132 (depl 56), 53-8136 (55–58), 53-8137 (55–58).

465 Troop Carrier (Medium) Wing: *Group:* **465 Troop Carrier Group.** *Squadrons (Colors):* **780 TCSq** (Red); **781 TCSq** (Blue); **782 TCSq** (Green). *Operating Base:* Mitchel

AFB, New York (25Aug53–23Mar54). *Notes:* 465 TCG attached to 64 TCWG at Greenville (Donaldson) until 30Nov53. Deployment to Thule AB, Greenland Jun-Sep53. See under: **USAF in Europe** (AFE) for overseas assignment. *C-119 Service:* 77 1953. *C-119CF Assignments:* 51-2587 (53), 51-2588 (53), 51-2589 (53), 51-2591 (53), 51-2592 (53), 51-2593 (53), 51-2594 (53), 51-2595 (53), 51-2596 (53), 51-2597 (53), 51-2598 (53), 51-2599 (53), 51-2600 (53), 51-2601 (53), 51-2602 (53), 51-2606 (53), 51-2615 (53), 51-2616 (53), 51-2617 (53), 51-2618 (53), 51-2622 (53), 51-2623 (53), 51-2626 (53), 51-2627 (53), 51-2628 (53), 51-2629 (53), 51-2630 (53), 51-2631 (53), 51-2632 (53), 51-2633 (53), 51-2634 (53), 51-2635 (53), 51-2636 (53), 51-2637 (53), 51-2638 (53), 51-2639 (53), 51-2640 (53), 51-2641 (53), 51-2642 (53), 51-2643 (53), 51-2644 (53), 51-2645 (53), 51-2646 (53), 51-2647 (53), 51-2648 (53), 51-2649 (53), 51-2650 (53), 51-2651 (53), 51-2652 (53), 51-2653 (53), 51-2654 (53), 51-2655 (53), 51-2656 (53), 51-2657 (53), 51-2661 (53). *C-119F Assignments:* 51-7969 (depl 53), 51-7971 (depl 53), 51-8110 (depl unk), 51-8113 (depl 53). *C-119G Assignments:* 51-8068 (depl 53), 51-8069 (depl 53), 51-8070 (depl 53), 51-8071 (depl 53), 51-8072 (depl 53), 51-8073 (depl 53), 51-8076 (depl 53), 51-8079 (depl 53), 51-8080 (depl 53), 51-8086 (depl 53), 51-8087 (depl 53), 51-8088 (depl 53), 51-8089 (depl 53), 51-8090 (depl 53), 51-8091 (depl 53), 51-8092 (depl 53), 51-8093 (depl 53), 52-5843 (depl 53).

479 Fighter (Day) Wing: George AFB, California. 49-136 (57), 50-119 (56), 50-153 (56), 50-166 (57), 51-2567 (57), 51-2582 (56), 51-2628 (depl 53), 51-7989 (depl 54), 51-8002 (depl 54), 51-8105 (56–57), 51-8111 (depl 55-56), 51-8122 (depl 54), 51-8143 (55–57), 51-8148 (57), 51-8156 (depl 54), 53-8077 (56–57), 53-8092 (56–57).

506 Fighter-Bomber Wing: Tinker AFB, Oklahoma. 51-7992 (58–unk), 51-8120 (58), 51-8166 (58).

513 Troop Carrier (Fixed Wing) Group: *Operating Base:* Sewart AFB, Tennessee. *Notes:* Attached to the **314 TCWG** 08Nov55–15Apr66. *C-119 Service:* 8 1955–1956. *C-119F Assignments:* 51-2663 (55–unk), 51-2664 (55–unk), 51-2680 (depl 55–unk), 51-2681 (55–unk), 51-2682 (55–unk), 51-2683 (depl 55–unk), 51-2684 (depl 55–unk), 51-2685 (depl 55–unk).

514 Troop Carrier (Medium) Wing: *Operating Base:* Mitchel AFB, New York. *Notes:* Brief assignments as the Wing was inactivated 01Feb53. Wing re-activated under the **Air Force Reserve** (AFR) 01Apr53 with C-119s assigned from 1958 to 1970. *C-119 Service:* 9 1953. *C-119F Assignments:* 51-8016 (53), 51-8028 (52–53), 51-8029 (52–53), 51-8110 (53), 51-8112 (53), 51-8113 (53), 51-8114 (53), 51-8117 (53), 51-8120 (53).

516 Troop Carrier (Medium) Wing: *Group:* 516 Troop Carrier Group. *Squadrons:* **345 TCSq; 346 TCSq; 347 TCSq.** *Operating Base:* Memphis MAP, Tennessee. *C-119 Service:* 48 1952–1953. *C-119F Assignments:* 51-2662 (52–53), 51-2712 (52–53), 51-2713 (52–53), 51-2714 (52–53), 51-2715 (52–53), 51-2716 (52–53), 51-7975 (52–53), 51-7976 (52–53), 51-7977 (52–53), 51-7978 (52–53), 51-7979 (52–53), 51-7980 (52–53), 51-7981 (52–53), 51-7982 (52–53), 51-7983 (52–53), 51-7984 (52–53), 51-7985 (52–53), 51-7986 (52–53), 51-7987 (52–53), 51-7988 (52–53), 51-7989 (52–53), 51-7990 (52–53), 51-7991 (52–53), 51-7992 (52–53), 51-7993 (52–53), 51-7994 (52–53), 51-7995 (52–53), 51-7996 (52–53), 51-7997 (52–53), 51-7998 (52–53), 51-7999 (52–53), 51-8000 (52–53), 51-8001 (52–53), 51-8002 (52–53), 51-8003 (52–53), 51-8004 (52–53), 51-8005 (52–53), 51-8006 (52–53), 51-8007 (52–53), 51-8008 (52–53), 51-8009 (52–53), 51-8010 (52–53), 51-8011 (52–53), 51-8012 (52–53), 51-8099 (52–53), 51-8100 (52–53), 51-8107 (52–53), 51-8108 (52–53).

831 Air Base Group: George AFB, California. 51-8103 (58–59), 51-8105 (57–59), 51-8148 (57–58), 53-8077 (57–59), 53-8092 (57–59), 53-8110 (59), 53-8120 (58–59).

832 Air Base Group: Cannon AFB, New Mexico. 51-8136 (57), 53-8072 (57–58), 53-8080 (57), 53-8093 (57).

834 Air Base Group: England AFB, Louisiana. 51-7992 (57–58), 51-8110 (57–58), 51-8166 (57–58).

836 Air Base Group: Langley AFB, Virginia. 51-8055 (58), 51-8099 (57–58), 51-8107 (57–58), 51-8110 (59), 53-8084 (57–58), 53-8128 (57).

837 Air Base Group: Shaw AFB, South Carolina. 48-349 (58), 51-7978 (58), 51-8028 (58).

839 Air Base Group: Sewart AFB, Tennessee. 48-352 (59), 49-143 (59), 50-137 (60), 50-142 (58–59), 50-144 (59), 51-2548 (59), 51-2571 (58–59), 51-2578 (60), 51-2594 (58), 51-2599 (60–61), 51-2615 (60), 51-2628 (60), 51-2639 (61), 51-7997 (59).

4405 Operational Squadron: Langley AFB, Virginia. 51-8082 (57), 51-8099 (55–57), 51-8107 (55–57), 53-8084 (56–57), 53-8128 (56–57).

4410 Combat Crew Training Wing: *Operating Base:* Lockbourne AFB, Ohio. *Notes:* Redesignated as 4410 SOT GP 15Sep70. *C-119 Service:* 1 C-119G, 8 AC-119G, 26 AC-119K 1969–1971. *C-119G Assignment:* 51-8024 (69–70). *AC-119G Assignments:* 52-5898 (69), 53-3145 (69–71), 53-7833 (69–71), 53-8089 (69–71), 53-8114 (69–71), 53-8115 (69–71), 53-8123 (69–71), 53-8131 (69–71). *AC-119K Assignments:* 52-5864 (69), 52-5889 (69), 52-5910 (69–71), 52-5911 (69–71), 52-5926 (69), 52-5935 (69), 52-5940 (69), 52-5945 (69), 52-9982 (69), 53-3154 (69), 53-3156 (69), 53-3187 (70–71), 53-3197 (69–71), 53-3211 (69), 53-7826 (69), 53-7830 (69), 53-7831 (69–71), 53-7839 (69–71), 53-7850 (69), 53-7854 (69), 53-7877 (69–71), 53-7879 (69), 53-7883 (69), 53-8121 (69), 53-8145 (69–71), 53-8148 (69).

4411 Combat Crew Training Group: Shaw AFB, South Carolina. 51-2640 (63).

4413 Combat Crew Training Squadron: *Operating Base:* Lockbourne AFB, Ohio. *Notes:* Associate unit within the 1 ACO WG. *C-119 Service:* 3 C-119G, 7 AC-119G, 6 AC-119K 1968–1969. *C-119G Assignments:* 51-8024 (68–69), 51-8101 (depl 68), 52-5846 (depl 68). *AC-119G Assignments:* 52-5898 (68–69), 53-7833 (68–69), 53-8089 (68–69), 53-8114 (68–69), 53-8115 (68–69), 53-8123 (68–69), 53-8131 (68–69). *AC-119K Assignments:* 52-5910 (68–69), 52-5911 (68–69), 53-3197 (69), 53-7831 (68–69), 53-7839 (68–69), 53-8145 (68–69).

4415 Air Base Squadron / Group: Pope AFB, North Carolina. 48-322 (50–51), 51-7976 (54), 51-7987 (52), 51-8025 (53), 51-8084 (53), 51-8110 (unk), 52-5844 (53).

4440 Air Delivery Group: *Operating Base:* Langley AFB, Virginia. *Notes:* 83 aircraft for transfer to MAP operators or transfers from AFE to AFR. *C-119 Assignments:* 51-2592 (59), 51-2595 (58), 51-8037 (61), 51-8038 (61), 51-8042 (61), 51-8043 (61), 51-8045 (61), 51-8050 (61), 51-8115 (61), 52-5855 (60), 52-5865 (60), 52-5909 (60), 52-5912 (60), 53-3154 (59), 53-3189 (61), 53-3192 (61), 53-3209 (60), 53-3200 (60), 53-3212 (60), 53-3217 (60), 53-3218 (61), 53-3220 (61), 53-7826 (61), 53-7827 (61), 53-7830 (61), 53-7831 (61), 53-7832 (61), 53-7833 (61), 53-7834 (61), 53-7835 (61), 53-7836 (61), 53-7837 (61), 53-7838 (61), 53-7839 (61), 53-7840 (61), 53-7842 (61), 53-7844 (61), 53-7846 (61), 53-7847 (61), 53-7848 (61), 53-7849 (61), 53-7850 (61), 53-7851 (61), 53-7852 (61), 53-7853 (61), 53-7853 (61), 53-7854 (61), 53-7859 (60), 53-7861 (60), 53-7862 (60), 53-7869 (60), 53-7871 (60), 53-7873 (60), 53-7878 (60), 53-7879 (60), 53-7880 (60), 53-7882 (60), 53-7883 (60), 53-8070 (60), 53-8079 (60), 53-8081 (60), 53-8082 (60), 53-8092 (60), 53-8094 (60), 53-8097 (60), 53-8129 (60), 53-8134 (60), 53-8135 (60), 53-8136 (60), 53-8137 (60), 53-8139 (61), 53-8140 (61), 53-8142 (61), 53-8144

(61), 53-8145 (61), 53-8147 (61), 53-8148 (61), 53-8149 (61), 53-8150 (61), 53-8153 (61), 53-8154 (61), 53-8155 (61), 53-8156 (61).

4500 Air Base Wing
Langley AFB, Virginia. 50-130 (60), 50-162 (66), 51-2544 (64), 51-8065 (61), 52-5877 (71), 53-7839 (61).

4501 Headquarters Support Squadron
Donaldson AFB, South Carolina. 51-8082 (54–57), 51-8102 (55–57), 51-8147 (56–57), 52-5849 (54–55), 53-3207 (54–58), 53-3214 (54–55), 53-8101 (56–57), 53-8132 (54–58).

4510 Combat Crew Training Wing
Luke AFB, Arizona. 51-2567 (58), 51-8105 (58), 51-8110 (58–59), 51-8111 (59), 53-8084 (59).

4520 Combat Crew Training Wing
Nellis AFB, Nevada. 48-334 (59), 49-142 (60), 51-7985 (unk–59), 51-7992 (unk–59), 51-8120 (58–59), 51-8166 (58–59), 53-8137 (60).

United States Air Forces in Europe (AFE / USAFE)

Major Command established 7 August 1945 to oversee U.S. air power in post-war Europe.

10 Tactical Reconnaissance Wing: Spangdah AB, West Germany. 53-7834 (depl 59).

36 Fighter (Day) Wing: Bitburg AB, West Germany. 49-196 (55), 51-2629 (depl 57), 53-7833 (depl 57).

47 Bombardment (Light, Jet) Wing: RAF Sculthorpe, England. 51-8236 (55–58), 51-8244 (55–58), 51-8252 (unk), 51-8253 (55–unk), 51-8258 (55–unk), 51-8266 (55–unk), 51-8272 (55–unk).

48 Fighter-Bomber Wing: Chaumont-Semoutiers AB, France. 51-2597 (depl 56), 51-2622 (depl 57), 53-3195 (depl 57), 53-7828 (depl 58), 53-7838 (depl 59).

50 Fighter-Bomber Wing: Toul-Rosières AB, France. 53-8138 (depl 57), 53-8153 (depl 56).

59 Air Depot Wing: RAF Burtonwood, England. 49-179 (52), 51-8272 (53).

60 Troop Carrier Wing: *Squadrons (Colors):* **10 TCSq** (Red); **11 TCSq** (Green); **12 TCSq** (Blue). *Operating Bases:* Rhein-Main AB, West Germany (02Jun51–23Sep55). Dreux AB, France (23Sep55–25Sep58). *Notes:* The 10, 11, 12 TCSq were transferred to the 7305 AB GP (under the 322 ADV) at Dreux when the 60 TCWG was inactivated in 1958. C-119C deployment to RAF Sculthorpe, England 53-54. Three C-119s are listed as deployed to the 61 TCWG: 51-2612 (depl 55), 53-7828 (depl 55), 53-8148 (depl 55) at Dreux AB but this is likely a misprint on USAF IARC reels. *C-119 Service:* 130 1952–1958. *C-119C Assignments:* 49-178 (54), 49-181 (54), 49-190 (54), 49-196 (55), 50-155 (depl 53), 50-168 (depl 54), 51-2533 (depl 54), 51-2534 (depl 52–53), 51-2542 (depl 52, 53). *C-119CF Assignments:* 51-2603 (53–54; 55), 51-2604 (55), 51-2605 (53–55), 51-2607 (53–55), 51-2608 (53–55), 51-2609 (53–55), 51-2610 (53–unk), 51-2611 (53–56), 51-2612 (53–56), 51-2613 (53–55), 51-2614 (53–56), 51-2619 (53–55), 51-2620 (53–55), 51-2624 (53–55), 51-2625 (53–54), 51-2645 (depl 54), 51-2658 (unk–55), 51-2659 (54–55), 51-2660 (unk–55), 51-8233 (53–55), 51-8234 (53–55), 51-8235 (53), 51-8236 (53–55), 51-8237 (53–56), 51-8238 (53–55), 51-8239 (53–unk), 51-8240 (53–unk), 51-8241 (53), 51-8242 (53–unk), 51-8243 (53–unk), 51-8244 (53–55), 51-8245 (53–unk), 51-8246 (53–unk), 51-8247 (53–unk), 51-8248 (53–unk), 51-8249 (53–unk), 51-8250 (53–unk), 51-8251 (53–unk), 51-8252 (53–unk), 51-8253 (53–55), 51-8254 (53–unk), 51-8255

(53–55), 51-8256 (53–unk), 51-8257 (53–unk), 51-8258 (53–55), 51-8259 (53–unk), 51-8260 (53–unk), 51-8261 (53–unk), 51-8262 (53–55; depl 55), 51-8263 (53–unk), 51-8264 (53–unk), 51-8265 (53–unk), 51-8266 (53–55), 51-8267 (53–unk), 51-8268 (53–unk), 51-8269 (53–unk), 51-8270 (53–unk), 51-8271 (53–unk), 51-8272 (53–55), 51-8273 (53–54). *C-119F Assignment:* 51-2691 (depl 53). *C-119G Assignments:* 53-3185 (depl 55), 53-3189 (58), 53-3192 (58), 53-3195 (depl 55), 53-3200 (58), 53-3218 (55–58), 53-3219 (55–58), 53-3220 (55–58), 53-3221 (55–57), 53-3222 (55), 53-7826 (55–58), 53-7827 (55–58), 53-7828 (55–58), 53-7829 (55–58), 53-7830 (55–58), 53-7831 (55–58), 53-7832 (55–58), 53-7833 (55–58), 53-7834 (55–58), 53-7835 (55–58), 53-7836 (55–58), 53-7837 (55–58), 53-7838 (55–58), 53-7839 (55–58), 53-7840 (55–58), 53-7841 (55), 53-7842 (55–58), 53-7843 (55–58), 53-7844 (55–58), 53-7845 (55–58), 53-7846 (55–58), 53-7847 (55–58), 53-7848 (55–58), 53-7849 (55–58), 53-7850 (55–58), 53-7851 (55–58), 53-7852 (55–58), 53-7853 (55–58), 53-7853 (55–58), 53-7854 (55–58), 53-8120 (depl 55), 53-8138 (55–58), 53-8139 (55–58), 53-8140 (55; 57-58), 53-8141 (55–58), 53-8142 (55–58), 53-8143 (55–58), 53-8144 (55–58), 53-8145 (55–58), 53-8146 (55–58), 53-8147 (55–58), 53-8148 (55–58), 53-8149 (55–58), 53-8150 (55–58), 53-8151 (55–58), 53-8152 (55–58), 53-8153 (55–58), 53-8154 (55–58), 53-8155 (55–58), 53-8156 (55–58).

66 Tactical Reconnaissance Wing: Sembach AB, West Germany. 53-7836 (depl 57-58).

73 Air Depot Wing: Chateauroux AB, France. 50-119 (53), 50-134 (53), 50-149 (53–54), 51-2535 (53), 51-2614 (53), 51-2690 (52–53), 51-2691 (52–53). Redesignated as 7373 AB GP / WG 15Nov53.

80 Air Depot Wing: Nouasseur AB, Morocco. 51-8261 (53).

85 Air Depot Wing: Erding AB, West Germany. 51-2544 (53).

86 Fighter-Interceptor Wing: Landstuhl AB, West Germany. 51-2614 (56), 51-2646 (depl 54), 51-2648 (53–54), 53-3180 (56).

317 Air Base Group: Évreux-Fauville AB, France. 53-7829 (60), 53-7843 (60), 53-8138 (60), 53-8141 (60), 53-8143 (60), 53-8151 (60). Transition unit for MAP deliveries to Belgium.

317 Troop Carrier (Medium) Wing: *Group:* 317 Troop Carrier Group. *Squadrons (Colors):* **39 TCSq** (Yellow/Black); **40 TCSq** (Red/White); **41 TCSq** (Blue/White). *Operating Bases:* Rhein-Main AFB, West Germany (1952–1953). Neubiberg AB, West Germany (1953–1957). Évreux-Fauville AB, France (08Jul57–1958). *Notes:* Deployment to RAF Sculthorpe, England 52, 53. *C-119 Service:* 163 1952–1958. *C-119C Assignments:* 49-178 (54), 49-190 (54), 49-197 (52–54), 49-198 (52–54), 49-199 (52–54), 50-119 (52–54), 50-120 (52–55), 50-121 (52–54), 50-122 (52–54), 50-123 (53–54), 50-124 (52–54), 50-125 (52–54), 50-126 (52–54), 50-127 (52–53), 50-128 (52–54), 50-129 (52–54), 50-130 (52–54), 50-131 (52–54), 50-132 (52–54), 50-133 (52–55), 50-134 (52–54), 50-136 (52–54), 50-137 (52–54), 50-140 (52–54), 50-141 (52–54), 50-144 (52–54), 50-145 (52–54), 50-146 (52–54), 50-148 (53–54), 50-149 (52–54), 50-150 (52–54), 50-151 (52–54), 50-151 (52–54), 50-153 (52–54), 50-155 (52–54), 50-157 (52–54), 50-158 (52–54), 50-159 (52–54), 50-160 (52–54), 50-161 (52–53), 50-163 (52–54), 50-164 (52–54), 50-165 (52–54), 50-166 (52–54), 50-167 (52–53), 50-168 (52–54), 50-169 (52–54), 50-170 (52–54), 50-171 (52–54), 51-2535 (52–54). *C-119CF Assignments:* 51-2587 (57–58), 51-2588 (57), 51-2589 (57), 51-2591 (57–58), 51-2592 (57–58), 51-2593 (57–58), 51-2594 (57–58), 51-2595 (57–58), 51-2596 (57–58), 51-2598 (depl 55; 57-58), 51-2600 (57), 51-2601 (57), 51-2604 (53–55), 51-2608 (depl 56), 51-2609 (depl 54–unk), 51-2611

(57–58), 51-2613 (53), 51-2617 (57–58), 51-2618 (depl 55; 57–58), 51-2620 (depl 55), 51-2621 (53), 51-2622 (57), 51-2623 (57–58), 51-2624 (53), 51-2626 (57), 51-2627 (57–58), 51-2628 (57–58), 51-2629 (57), 51-2630 (57), 51-2631 (depl 57; 57), 51-2633 (57–58), 51-2634 (57–58), 51-2636 (57–58), 51-2637 (57–58), 51-2638 (57–58), 51-2639 (57–58), 51-2640 (57), 51-2641 (57–58), 51-2642 (57–58), 51-2643 (57–58), 51-2644 (57–58), 51-2646 (depl 53; 57), 51-2647 (57–58), 51-2649 (57–58), 51-2650 (57–58), 51-2651 (57–58), 51-2652 (57), 51-2654 (57–58), 51-2655 (57), 51-2656 (57), 51-2657 (57), 51-2658 (53–unk), 51-2659 (53–54), 51-2660 (53–unk), 51-8238 (depl 55), 51-8239 (depl 54), 51-8243 (depl 54), 51-8262 (57–58), 51-8264 (depl 54). *C-119G Assignments:* 51-7998 (54), 52-5922 (54–58), 52-5923 (54–58), 53-3179 (54–58), 53-3180 (54–58), 53-3181 (54–58), 53-3182 (54–58), 53-3183 (54–58), 53-3184 (54–58), 53-3185 (54–58), 53-3186 (54–58), 53-3187 (54–58), 53-3188 (54–58), 53-3189 (54–58), 53-3190 (54–58), 53-3191 (54–58), 53-3192 (54–58), 53-3193 (54–58), 53-3194 (54–58), 53-3195 (54–58), 53-3196 (54–58), 53-3197 (54–58), 53-3198 (54–58), 53-3199 (54–58), 53-3200 (54–58), 53-3201 (54–58), 53-3202 (54–58), 53-3203 (54–58), 53-3204 (54–58), 53-3205 (54–58), 53-3206 (54–58), 53-8104 (54–58), 53-8105 (54–58), 53-8106 (54–58), 53-8107 (54–58), 53-8108 (54–58), 53-8109 (54–58), 53-8110 (54–58), 53-8111 (54–58), 53-8112 (54–58), 53-8113 (54–58), 53-8114 (54–58), 53-8115 (54–58), 53-8116 (54–58), 53-8117 (54–58), 53-8118 (54–55), 53-8119 (54–58), 53-8120 (54–58), 53-8121 (54–58), 53-8122 (54–58), 53-8123 (54–58), 53-8124 (54–58), 53-8125 (54–58), 53-8126 (54–58).

388 Fighter-Bomber Wing: Étain-Rouvres AB, France. 53-7839 (depl 56).

406 Fighter-Interceptor Wing: RAF Manston, England. 51-2623 (depl 55), 53-8125 (depl 55).

433 Troop Carrier (Medium) Wing: *Group:* 433 Troop Carrier Group. *Squadrons:* **67** TCSq; **68** TCSq; **69** TCSq. *Operating Base:* Rhein-Main AB, West Germany. *Notes:* Wing stationed in West Germany 05Aug51–14Jul52. To **Air Force Reserve** (AFR) 18May55. *C-119 Service:* 48 1951–1952. *C-119C Assignments:* 49-197 (51–52), 49-198 (51–52), 49-199 (51–52), 50-119 (51–52), 50-120 (51–52), 50-121 (51–52), 50-122 (51–52), 50-123 (51–52), 50-124 (51–52), 50-125 (51–52), 50-126 (51–52), 50-127 (51–52), 50-128 (51–52), 50-129 (51–52), 50-130 (51–52), 50-131 (51–52), 50-132 (51–52), 50-133 (51–52), 50-134 (51–52), 50-136 (51–52), 50-137 (51–52), 50-140 (51–52), 50-141 (51–52), 50-144 (51–52), 50-145 (51–52), 50-146 (51–52), 50-148 (51–52), 50-149 (51–52), 50-150 (51–52), 50-151 (51–52), 50-152 (51–52), 50-153 (51–52), 50-155 (51–52), 50-157 (51–52), 50-158 (51–52), 50-159 (51–52), 50-160 (51–52), 50-161 (51–52), 50-163 (51–52), 50-164 (51–52), 50-165 (51–52), 50-166 (51–52), 50-167 (51–52), 50-168 (51–52), 50-169 (51–52), 50-170 (51–52), 50-171 (51–52), 51-2535 (51–52).

465 Troop Carrier Wing: *Group:* 465 Troop Carrier Group. *Squadrons (Colors):* **780** TCSq (Red); **781** TCSq (Blue) at Wiesbaden; **782** TCSq (Green). *Operating Bases:* Neubiberg AB, West Germany. Toul-Rosières AB, France (02Apr54–23May55). Évreux-Fauville AB, France (23May55–08Jul57). *Notes:* Deployments to Landstuhl West Germany. 781 TCSq stationed at Wiesbaden West Germany. *C-119 Service:* 108 1953–1957. *C-119C Assignment:* 49-191 (56). *C-119CF Assignments:* 51-2587 (53–57), 51-2588 (53–57), 51-2589 (53–57), 51-2591 (53–57), 51-2592 (53–57), 51-2593 (53–57), 51-2594 (53–57), 51-2595 (53–57), 51-2596 (53–57), 51-2597 (53–56), 51-2598 (53–57), 51-2599 (53–57), 51-2600 (53–57), 51-2601 (53–57), 51-2602 (53–57), 51-2606 (53–57), 51-2611 (56–57), 51-2612 (56–57), 51-2613 (57), 51-2615 (53–57), 51-2617 (53–57), 51-2618 (53–57), 51-2622 (53–57), 51-2623 (53–56), 51-2626 (53–57), 51-2627 (53–57), 51-2628 (53–57), 51-2629 (53–57), 51-2630 (53–57), 51-2631 (53–57), 51-2632 (53–57), 51-2633 (53–57), 51-2634

(53–57), 51-2636 (53–57), 51-2637 (53–57), 51-2638 (53–57), 51-2639 (53–57), 51-2640 (53–57), 51-2641 (53–57), 51-2642 (53–57), 51-2643 (53–57), 51-2644 (53–57), 51-2645 (53–55), 51-2646 (53–57), 51-2647 (53–57), 51-2648 (53–55), 51-2649 (53–57), 51-2650 (53–57), 51-2651 (53–57), 51-2652 (53–57), 51-2653 (53–56), 51-2654 (53–57), 51-2655 (53–57), 51-2656 (53–57), 51-2657 (53–57), 51-2661 (53–57), 51-8233 (55–56), 51-8234 (55), 51-8237 (56–57), 51-8238 (55–57), 51-8262 (55–57). *C-119G Assignments:* 52-5922 (depl 56–57), 52-5923 (depl 56–57), 53-3179 (depl 56–57), 53-3180 (depl 56–57), 53-3181 (depl 56–57), 53-3182 (depl 56–57), 53-3183 (depl 56–57), 53-3184 (depl 56–57), 53-3185 (depl 56–57), 53-3186 (depl 56–57), 53-3187 (depl 56–57), 53-3188 (depl 56–57), 53-3189 (depl 56–57), 53-3190 (depl 56–57), 53-3191 (depl 56–57), 53-3192 (depl 56–57), 53-3193 (depl 57), 53-3194 (depl 57), 53-3195 (depl 57), 53-3196 (depl 57), 53-3197 (depl 57), 53-3198 (depl 57), 53-3199 (depl 57), 53-3200 (depl 57), 53-3201 (depl 57), 53-3202 (depl 57), 53-3203 (depl 57), 53-3204 (depl 57), 53-3205 (depl 57), 53-3206 (depl 57), 53-8104 (depl 57), 53-8110 (depl 57), 53-8111 (depl 57), 53-8112 (depl 57), 53-8113 (depl 57), 53-8115 (depl 57), 53-8116 (depl 57), 53-8117 (depl 57), 53-8119 (depl 57), 53-8120 (depl 57), 53-8121 (depl 57), 53-8122 (depl 57), 53-8123 (depl 57), 53-8124 (depl 57), 53-8125 (depl 57), 53-8126 (depl 57).

1603 Air Transport Wing: Wheelus AB, Libya. 49-198 (55), 51-2613 (55), 51-2620 (55), 51-2653 (55).

7030 Support Group: Ramstein AB, West Germany. 53-7835 (58).

7100 Headquarters Support Wing: Wiesbaden AB, West Germany. 49-191 (56; 57), 49-196 (56), 50-130 (53), 50-141 (53), 50-144 (53), 50-169 (53–54), 51-2534 (53), 51-2544 (54), 51-2587 (57), 51-2620 (56), 51-2623 (56), 51-2636 (54), 51-2640 (54), 51-2647 (54), 53-3191 (55), 53-7849 (59).

7109 Support Wing: Wiesbaden AB, West Germany. 51-2599 (56).

7150 Air Base Group: Wiesbaden AB, West Germany. 50-129 (54), 50-130 (53), 50-131 (53), 50-144 (53).

7206 Support Group: Athens IAP, Greece. 53-3206 (57).

7207 Air Base Squadron: Aviano AB, Italy. 51-2596 (56), 51-2630 (56), 53-3205 (57), 53-8122 (57).

7216 Air Base Group: Adana AB, Turkey. 49-198 (56).

7244 Air Base Group: Dhahran AB, Saudi Arabia. 49-198 (54–55).

7272 Air Base Group / Wing: Wheelus AB, Libya. 51-2534 (56), 51-2598 (57), 51-2605 (56–57), 51-2608 (56), 51-2612 (56), 51-2614 (56), 51-2620 (56–57), 51-2628 (56), 51-2629 (57), 51-2640 (56), 51-8237 (56).

7280 Air Depot Wing: Nouasseur AB, Morocco. 50-164 (54), 51-2634 (54).

7280 Aircraft Maintenance Group: Nouasseur AB, Morocco. 53-3191 (55), 53-8119 (55).

7305 Air Base Group: *Operating Base:* Dreux AB, France. **Notes:** Acquired the 10, 11, 12 TCSq when the 60 TCWG was inactivated in 1958. The 7305 AB GP was under the 322 ADV. *C-119 Service:* 54 1958–1961. *C-119G Assignments:* 53-3189 (58–61), 53-3192 (59–61), 53-3200 (58–60), 53-3218 (58–61), 53-3219 (58–60), 53-3220 (58–61), 53-7826 (58–61), 53-7827 (58–61), 53-7828 (58–60), 53-7829 (58–60), 53-7830 (58–61), 53-7831 (58–61), 53-7832 (58–61), 53-7833 (58–61), 53-7834 (58–61), 53-7835 (58–61), 53-7836 (58–61), 53-7837 (58–61), 53-7838 (58–61), 53-7839 (58–61), 53-7840 (58–61), 53-7842 (58–61),

53-7843 (58–60), 53-7844 (58–61), 53-7845 (58–60), 53-7846 (58–61), 53-7847 (58–61), 53-7848 (58–61), 53-7849 (58–61), 53-7850 (58–61), 53-7851 (58–61), 53-7852 (58–61), 53-7853 (58–61), 53-7853 (58–61), 53-7854 (58–61), 53-8138 (58–60), 53-8139 (58–61), 53-8140 (58–61), 53-8141 (58–60), 53-8142 (58–61), 53-8143 (58–60), 53-8144 (58–61), 53-8145 (58–61), 53-8146 (58–60), 53-8147 (58–61), 53-8148 (58–61), 53-8149 (58–61), 53-8150 (58–61), 53-8151 (58–60), 53-8152 (58–60), 53-8153 (58–61), 53-8154 (58–61), 53-8155 (58–61), 53-8156 (58–61).

7305 Consolidated Aircraft Maintenance Squadron: Dreux AB, France. 53-3189 (61), 53-3192 (61), 53-7827 (61), 53-7832 (61), 53-7833 (61), 53-7834 (61), 53-7838 (61), 53-7839 (61), 53-7840 (61), 53-7842 (61), 53-7844 (61), 53-7846 (61), 53-7847 (61), 53-7848 (61), 53-7849 (61), 53-7851 (61), 53-7852 (61), 53-7853 (61), 53-7853 (61), 53-7854 (61), 53-8140 (61), 53-8144 (61), 53-8145 (61), 53-8147 (61), 53-8149 (61), 53-8153 (61), 53-8154 (61), 53-8155 (61), 53-8156 (61).

7310 Air Base Group: Rhein-Main AB, West Germany. 50-149 (56–57), 51-2614 (55), 51-2618 (56), 51-2633 (56), 51-2636 (56–57), 53-3221 (55–unk), 53-7829 (56; 59), 53-7837 (55–56), 53-8147 (depl 56), 53-8148 (depl 56), 53-8153 (56), 53-8155 (57).

7312 Air Base Group: Rhein-Main AB, West Germany. 51-2598 (56), 51-2631 (56).

7330 Flying Training Wing: Fürstenfeldbruck AB, West Germany. 51-2589 (depl 55), 51-2613 (depl 55), 51-2617 (depl 57).

7373 Air Base Group / Wing: Chateauroux AB, France. 50-146 (53), 50-150 (53), 50-163 (54), 50-164 (54), 50-170 (53–54), 51-2633 (54), 51-2654 (54), 51-8244 (53), 51-8246 (54), 51-8247 (53), 51-8268 (54). Redesignated from 73 AD WG 15Nov53. Also designated as 7373 AD WG.

7373 Aircraft Maintenance Group: Chateauroux AB, France. 49-196 (55), 50-168 (54), 50-170 (54), 51-2544 (55–56), 51-2597 (55), 51-2598 (unk), 51-2600 (55–56), 51-2617 (55–56), 51-2634 (54), 51-2637 (54), 51-2643 (56), 51-8251 (54), 51-8256 (54), 53-3180 (55), 53-3190 (55), 53-3197 (55), 53-7844 (55–56), 53-7852 (55), 53-8118 (55).

7413 Support Group: Bordeaux AB, France. 51-2631 (57).

7425 Air Base Group: Hahn AB, West Germany. 53-7829 (58; 59), 53-7852 (59), 53-8139 (58).

7559 Aircraft Maintenance Group: RAF Burtonwood, England. 49-178 (54–55), 50-122 (54–55), 50-128 (55), 50-132 (54), 50-143 (55), 50-153 (54), 50-154 (54), 50-155 (54), 50-158 (53), 50-159 (54–55), 50-160 (54–55), 50-168 (54–55), 50-171 (54–55), 51-2544 (53–54), 51-2611 (53), 51-2612 (55), 51-2615 (55), 51-8272 (53), 52-5922 (55).

7602 Support Wing: Seville, Spain. 51-2596 (57).

USAF Museum Allocations

Eleven C-119 Flying Boxcars were assigned directly to USAF museum status after retirement from active service for preservation in federal museums, base display exhibits at USAF field museums or loans to non-federal museums. Several C-119s owned by the USAF are on a loan basis to U.S. Army and Reserve museums. Some USAF field museums went into private (civil) ownership during the 1990s and these have been noted where possible. From the late 1980s the USAF Museum Program acquired 11 more Boxcars for preservation purposes.

Direct USAF Museum Status Assignments:

10446 / 51-0128 / C-119C	1966–1997	Florence Air & Missile Museum, Florence, South Carolina (Loan)
	1997–present	82 Airborne Division War Memorial Museum, Fort Bragg, North Carolina
10524 / 51-2566 / C-119C	1966–1986	CAF, Harlingen, Texas (Loan)
	1986–present	Museum of Aviation, Robins AFB, Georgia
10525 / 51-2567 / C-119C	1966–present	USAF History & Traditions Museum, Lackland AFB, Texas
10605 / 51-2616 / C-119C	1966–1979	Bradley Air Museum, Windsor Locks, Connecticut. (Loan) (Destroyed by tornado 1979)
10664 / 51-2675 / C-119G	1970–2012	Pate Museum of Transportation, Fort Worth, Texas (Loan) (To a private museum 2012, stored)
10767 / 51-8024 / C-119G	1970–present	Strategic Air Command & Aerospace Museum, Offutt AFB Nebraska. To Ashland, Nebraska 1997 (Privatized 1998)
10915 / 51-8037 / C-119J	1963–present	National Museum of the USAF, WPAFB, Dayton, Ohio
11005 / 52-5846 / C-119G	1971–1975	Michigan Military Museum, Saginaw, Michigan (Loan). To a civil operator 1975 as N48076
11009 / 52-5850 / C-119G	1971–present	Grissom Air Museum, Grissom AFB, Indiana (Privatized 1995)
KF-187 / 53-8084 / C-119G	1975–present	Little Rock AFB, Arkansas
KF-190 / 53-8087 / C-119G	1974–present	82 Airborne Division War Memorial Museum, Fort Bragg, North Carolina

USAF Acquisitions for Museum Status:

10736 / (RCAF) 22105 / C-119G	1988–2006	General Mitchell IAP ANGB, Wisconsin
	2006–present	Niagara Falls ARS, New York
10738 / (RCAF) 22107 / C-119G	1988–present	Hill Aerospace Museum, Hill AFB, Utah
10825 / (RCAF) 22114 / C-119G	1988–present	Aerospace Museum of California, McClellan AFB, California (Privatized 2001)
10906 / (RCAF) 22122 / C-119G	1988–present	March Field Air Museum, March AFB, California (Privatized 1996)
10993 / (RCAF) 22134 / C-119G	1988–present	Travis AFB Heritage Center, Travis AFB, California
11155 / 53-3144 / C-119G	1987–present	Hurlburt Field Memorial Air Park, Hurlburt Field, Florida

USAF Acquisitions for Museum Status (via USFS):

10334 / 48-0352 / C-119C	1992–2016	Edwards AFB Museum, California
	2016–present	Air Mobility Command Museum, Dover AFB, Delaware
10436 / 49-0199 / C-119C	1992–present	Castle Air Museum, Castle AFB, California (Privatized 1995)
10676 / (RCAF) 22101 / C-119G	1995–present	(U.S. Army) Don F. Pratt Memorial Museum, Fort Campbell, Kentucky.
10860 / (RCAF) 22116 / C-119G	1992–present	National Infantry Museum, Fort Benning, Georgia
10870 / (RCAF) 22118 / C-119G	1991–present	Air Mobility Command Museum, Dover AFB, Delaware

Section Two—Military Operators

Non-USAF Organizations

Aeroproducts Division, General Motors Corp.: The aircraft propeller division of General Motors Corp. located at Vandalia MAP, Ohio. Five C-119s were loaned in the early 1950s for Aeroproducts propeller development tests: s/n 48-319 (52–53), 49-187 (50–51), 52-6001 (53), 52-6002 (53), 52-5865 (unk) later used on the C-119G.

Aircraft Engineering & Maintenance Co. (AEMCO): A large aircraft maintenance and overhaul facility located at Oakland Intl. Airport, California, undertook multiple U.S. military contracts in the 1950s that included the C-82, C-54, C-74, F-100 and T-33. AEMCO serviced and upgraded numerous C-119s in the USAF fleet, mainly in the earlier years of service. The company was a subsidiary of Transocean Airlines which operated from 1946 to 1960.

Allison Division, General Motors Corp.: One C-119G, s/n 53-8088, bailment for test duties 1956–1957 at Indianapolis, Indiana.

Bell Aircraft Corp.: One C-119J, s/n 51-8040, assigned for test duties 1958–1959 at Niagara Falls, New York.

Bellanca Aircraft Corp.: One C-119G, s/n 52-5865, assigned in the mid–1950s for reasons unk at New Castle, Delaware.

Birmingham Modification Center (BMC): A major aviation engineering and modification hub made a significant contribution to the history of the C-119 Flying Boxcar. The parent company was Hayes Aircraft Corp. founded in 1951 which held the largest civilian contracts for USAF aircraft conversions and upgrades. BMC itself employed over 2,300 staff in a massive 1.38 million sq. ft. ten hangar complex at Birmingham IAP, Alabama. It is well known for its conversion upgrades of the B-25 and KB-50. Along with Fairchild at Hagerstown, BMC was the prime contractor involved in the 40 C-119B to C-119C conversions and 180 C-119F to C-119G conversions. They were also involved in the 1956 C-119 Standardization Program, plus most regular C-119 servicing and overhaul requirements throughout the 1950s and into the 1960s where work began to decline as the fleet was retired. The Hayes Dothan, Alabama, facility continued to undertake limited, specialized C-119 contracts into the 1970s.

China Airlines: A state-owned civilian airline of the Republic of China (Taiwan) carried out maintenance on two AC-119G, s/n 52-5898, 53-8069, and four AC-119K, s/n 52-5926, 53-3154, 53-3211, 53-8148, during 1971. Additional work on the standard C-119G for the USAF was also undertaken.

Civil Air Transport, Inc. (CAT): A civilian airline headquartered in Taipei, Taiwan formed in 1950 as a subsidiary company to Delaware corporation, Airdale Corp., which was secretly owned by the Central Intelligence Agency (CIA). CAT openly operated as a civilian airline but was also known to carry out covert operations throughout Southeast Asia. Staggered bailments of 40 USAF C-119Gs based in Japan were made to CAT between Nov54 to Dec56 for maintenance purposes. 51-8055 (22Nov54–unk), 51-8056 (20Jan55–unk), 51-8062 (21Jan55–unk), 51-8065 (27Sep55–13Nov55), 51-8066 (27Feb55–13May55), 51-8068 (19May55–29Jun55), 51-8070 (13May55–24Jun55), 51-8074 (21Jan55–09Mar55), 51-8087 (24Jun55–01Jul55), 51-8096 (23Aug55–01Nov55), 52-5843 (15Dec55–09Feb56), 52-5846 (19Oct55–17Dec55), 52-5852 (27Nov55–22Jan56), 52-5857 (27Jun55–07Sep55), 52-5862 (19Mar55–unk), 52-5871 (04Dec55–29Jan56), 52-5876 (17Aug55–26Oct55), 52-5885 (07Feb56–01Apr56), 52-5920 (20Oct55–29Jan56), 52-5926 (29Oct56–17Dec56),

52-5927 (29Jun55–19Oct55), 52-5928 (17Mar56–27May56), 52-5930 (19Feb56–23Apr56), 52-5935 (01Mar56–15May56), 52-5938 (30Dec55–21Feb56), 52-5939 (01Mar56–01May56), 52-5940 (07Jan56–02Mar56), 52-5943 (13Nov56–30Dec56), 52-5954 (28Feb56–26Mar56), 52-9981 (05Apr55–22May55), 52-9982 (16Jun55–unk), 53-3142 (22Feb55–14May55), 53-3149 (18Nov55–07Jan56), 53-3151 (23Sep55–18Nov55), 53-3155 (02Jul55–05Oct55), 53-3161 (30May55–01Jul55), 53-3163 (26May55–01Jul55), 53-3164 (17Apr55–28May55), 53-3167 (14May55–27Jun55), 53-3173 (03May55–17Jun55).

Department of the Army, Operations Range: Based at White Sands, New Mexico. One C-119C, s/n 50-0150, assigned 1960–1963 through AMC and stationed at Fort Monmouth, New Jersey.

Fairchild Aircraft Division: Apart from designing and manufacturing the 1,185 C-119 Flying Boxcars, Fairchild Aircraft, based at its Hagerstown Maryland factory, also provided the U.S. military with ongoing support and development projects including C-119 conversions such as the C-119J and K and the AC-119 gunship development. During the 1960s Fairchild shifted its support work from Hagerstown to plants in Florida. The St. Augustine facility converted the 52 AC-119 gunships and readied C-119s sold to foreign nations through MAP. Their Crestview (Bob Sikes AP) depot performed most of the C-119K conversions.

Flight Safety Foundation, Inc.: Civil entity whose Aviation Safety Engineering and Research Division conducted three full-scale crash tests on behalf of the USAF using C-119C airframes as test equipment. The objective was to test the USAF's CNU-103/E shipping containers durability in a real world crash environment. The C-119C airframes were s/n 48-334, 49-155, 49-181 and tests were carried out between Aug66 and Feb67 at a test site in Deer Valley, Arizona.

General Dynamics Aircraft Corp.: Held a civilian contract from 1973 to convert and service the 27 C-119Ls for the ANG.

General Motors Corp.: One C-119G, s/n 52-5865, assigned for a short period.

Hayes Aircraft Corp.: *See under:* **Birmingham Modification Center.**

Lear, Inc.: U.S. civilian aircraft manufacturer held a USAF contract for C-119 servicing in Alaska.

The Martin Aircraft Co.: One R4Q-1, BuNo. 128744, diverted from 16Feb52 to 18Jun52 for unknown project at Baltimore, Maryland, under AMC (USAF) and BAR (USN).

Royal Aircraft Establishment (RAE): Founded in 1904 in Farnborough, England and named as such in 1918 as a leading British aircraft Research and test facility. One C-119F, s/n 51-2611, was assigned for various duties 23Apr53-18Sep56. Another C-119CF, s/n 51-8254, was briefly assigned in early 1953.

Sabena Airlines: The national airline of Belgium based in Brussels. They were contracted, through AMC, to carry out routine maintenance and overhauls on the USAF C-119 fleet based in Europe during the 1950s. They were also contracted to convert the 18 Belgian C-119Fs to C-119G standard in 1956-57.

United States Air Force Reserve (AFR / AFRES)

The Air Force Reserve was restructured as a component of the USAF **Continental Air Command** (CNC) from December 1, 1948. The first C-119s transferred into the

Section Two—Military Operators

Reserves soon after The Korean War in 1954 and were early C-119B/C variants. When the new C-130 Hercules began arriving in front-line service from 1957 the AFRES then began receiving the entire 700 plus C-119 fleet. These were assigned to 27 Air Force Reserve Flying Centers at various bases throughout the continental United States—these units are listed under *Continental Air Command* in the USAF units section. As a result of air force restructuring, in November 1958 all C-119s were re-assigned into numbered wings—14 in total with two to four squadrons each. Squadrons were subordinate to Groups until April 14, 1959, when they reported directly to the Wing. At this time the AFRES almost exclusively operated the C-119 within its unit structure apart from a few wings which had the C-124 Globemaster II and the short lived C-123 Provider. From February 1963 Groups were again initiated and assigned to each Squadron. This allowed individual groups to be mobilized as opposed to the entire wing. CNC was inactivated in 1968 and the AFRES was established as an Air Force Separate Operating Agency on February 17, 1968. Troop Carrier Groups (TCG) were redesignated as Tactical Airlift Groups (TAL) from 1967 and some were redesignated Special Operations Groups (SOP) in 1969 for service in Vietnam. Low-rate retirement of the C-119Cs took place from around 1963 with the last of the later variant C-119Gs retired in 1973.

USAF Reserve Unit Summary

Wing	SQ (1963 Group)	Period
94 TCWG	731 (901), 732 (902)	1959–1966
302 TCWG	355 (906), 356 (907), 357 (908)	1959–1973
349 TCWG	312 (938), 313 (939), 314 (940), 97 (941)	1958–1969
403 TCWG	63 (927), 64 (928), 65 (929)	1958–1971
433 TCWG	67 (921), 68 (922), 69 (923)	1958–1971
434 TCWG	71 (930), 72 (931), 73 (932)	1959–1970
435 TCWG	76 (915), 77 (916), 78, 357 (908)	1958–1969
440 TCWG	95 (933), 96 (934)	1958–1971
442 TCWG	303, 304, 305	1959–1961
446 TCWG	357, 704 (924), 705 (925), 706 (926)	1958–1970
452 TCWG	728 (942), 729 (943), 730 (944), 733 (945)	1959–1969
459 TCWG	756 (909), 757 (910), 758 (911)	1958–1969

C-119G s/n 52-5952 (11139) in typical USAF Reserve livery with the "0-" extra zero on the serial and white upper fuselage surfaces. The black painted engine nacelles helped hide the extensive exhaust streaks the Flying Boxcar was known for (James R. Alvis).

Wing	SQ (1963 Group)	Period
512 TCWG	326 (912), 327 (913), 328 (914)	1958–1971
514 TCWG	335 (903), 336 (904), 337 (905)	1958–1971

Notes: 78 TCSq (435 TCWG) did not operate the C-119 after 08May61, to 917 TCG 1963 with the C-124. 442 TCWG was inactivated 1961. 357 TCSq initially at 446 TCWG, transferred to 302 TCWG 1961–1963 then 435 TCWG from 1963.

94 Troop Carrier (Medium) Wing: *Group / Squadrons:* **94 Troop Carrier Group** (14Jun52–14Apr59). **731 TCSq** / Laurence G. Hanscom Field, Massachusetts. **732 TCSq** / Grenier AFB, New Hampshire. *Operating Base:* Laurence G. Hanscom Field, Massachusetts. *C-119 Service:* 1959–1966. *C-119G Assignments:* 51-2676 (unk–63), 51-2715 (unk–63), 51-7975 (59–63), 51-7981 (59–63), 51-7982 (59–63), 51-7990 (59–63), 51-7994 (59–63), 51-8000 (59–63), 51-8002 (59–63), 51-8003 (59–63), 51-8004 (61–63), 51-8006 (59–63), 51-8007 (59–63), 51-8008 (59–63), 51-8020 (59–63), 51-8023 (61–63), 51-8052 (59–60; 63–64), 51-8103 (59–63), 51-8106 (59–63), 51-8110 (59–63), 51-8136 (59–63), 52-5852 (59–63), 52-5858 (59–62), 52-5919 (59–61), 53-3154 (depl 59), 53-3185 (59–63), 53-3212 (59–60), 53-7849 (61–63), 53-7857 (59–63), 53-7880 (60–63), 53-8077 (59–63), 53-8080 (59–63), 53-8086 (59–63), 53-8094 (59–60), 53-8095 (59–63), 53-8109 (59–63), 53-8117 (59–63), 53-8121 (59–63), 53-8133 (59–63), 53-8155 (61–63), 53-8156 (61–63). *C-119J Assignments:* 51-8114 (59), 51-8116 (depl 60), 51-8119 (59–60), 51-8144 (62–unk), 52-5896 (62–63). *C-119 Assignments from 11Feb63:*

  901 Troop Carrier Group / 731 Troop Carrier Squadron: Laurence G. Hanscom Field, Massachusetts. *C-119G:* 51-7981 (63–66), 51-8002 (63–66), 51-8004 (63–66), 51-8106 (63–66), 51-8136 (63–66), 53-3185 (63–64; 66), 53-7849 (63–66), 53-7857 (63–66), 53-7880 (63–66), 53-8077 (63–66), 53-8080 (63–66), 53-8095 (63–66), 53-8109 (63–66), 53-8117 (63–66), 53-8121 (63–66), 53-8133 (63–66).

  902 Troop Carrier Group / 732 Troop Carrier Squadron: Grenier AFB, New Hampshire. *C-119G:* 51-2676 (63–unk), 51-2715 (63–65), 51-7975 (63–65), 51-7982 (63–65), 51-7990 (63–65), 51-7994 (63), 51-8000 (63–65), 51-8003 (63–65), 51-8006 (63–65), 51-8007 (63–65), 51-8008 (63–65), 51-8020 (63–65), 51-8023 (63–65), 51-8103 (63–65), 51-8110 (63–65), 52-5852 (63), 53-8155 (63–65), 53-8156 (63–65). *C-119J:* 51-8119 (64).

302 Troop Carrier (Medium) Wing: *Group / Squadrons:* **302 Troop Carrier Group** (14Jun52–14Apr59). **355 TCSq** / Clinton County AFB, Ohio. **356 TCSq** / Clinton County AFB, Ohio. **357 TCSq** / Bates Field, Alabama. *Operating Bases:* Clinton County AFB, Ohio (14Jun52–02Aug71). Lockbourne AFB, Ohio (02Aug71–01Apr81). *Notes:* 357 TCSq transferred from the 446 TCWG to Bates Field 08May61, then the 357 TCSq / 908 TCG was transferred to the 435 TCWG 18Mar63. *C-119 Service:* 1959–1973. *C-119C Assignments:* 48-328 (62–63), 48-337 (59–61; 62–63), 49-124 (59–63), 49-125 (59–63), 49-129 (59–63), 49-156 (59–63), 49-157 (59–63), 49-167 (59–63), 49-171 (59–62), 49-177 (59–63), 49-178 (59–63), 49-180 (59–63), 49-196 (62–63), 50-120 (59–63), 50-128 (59–61), 50-133 (59–63), 50-135 (59–63), 50-138 (59–63), 50-142 (59–63), 50-154 (59–63), 51-2532 (59–63), 51-2533 (59–63), 51-2534 (59–63), 51-2539 (59–63), 51-2544 (59–63), 51-2549 (59–62), 51-2565 (60–63), 51-2573 (59–63), 51-2582 (59–63). *C-119CF Assignments:* 51-2588 (59–63), 51-2591 (59–63), 51-2592 (59–63), 51-2597 (59–63), 51-2608 (59–61), 51-2615 (59–63), 51-2620 (59–60), 51-2623 (59), 51-2626 (59–63), 51-2647 (unk–63). *C-119G Assignments:* 51-2677 (61–63), 51-7971 (61–63), 51-8044 (61–63), 51-8073 (61–63), 51-8083 (61–63), 52-5878 (61–63), 52-5919 (61–63), 52-5934 (61–63), 53-3220 (61–63), 53-7826 (61–63), 53-7830

(61–63), 53-7831 (61–63), 53-7837 (61–63), 53-7839 (61–63), 53-7850 (61–63), 53-7883 (61–63), 53-8085 (61–63), 53-8144 (61–63). *C-119 Assignments from 11Feb63:*

906 Troop Carrier Group / 355 Troop Carrier Squadron: Clinton County AFB, Ohio / Lockbourne AFB, Ohio (from 24Jul71). *C-119C:* 51-2532 (63–65), 51-2533 (63–65), 51-2534 (63–65), 51-2539 (63–65), 51-2565 (63), 51-2568 (65), 51-2573 (63–64), 51-2582 (63–65), 51-2584 (65). *C-119CF:* 51-2588 (63–65), 51-2591 (63–65), 51-2592 (63–65), 51-2595 (65), 51-2597 (63–65), 51-2615 (63–65), 51-2626 (63–65), 51-2632 (65), 51-2647 (63–65), 51-2655 (depl 65), 51-2656 (65), 51-8239 (63–65), 51-8247 (63–65), 51-8248 (65), 51-8251 (unk–65), 51-8260 (65), 51-8261 (65), 51-8267 (65), 51-8271 (unk–65). *C-119G:* 51-2666 (65–69), 51-2669 (70), 51-2670 (65–67), 51-2681 (70), 51-2714 (65–69), 51-2715 (65–70), 51-7973 (70), 51-7984 (70–71), 51-7990 (70), 51-8000 (65–69), 51-8007 (65–69), 51-8008 (65–70), 51-8016 (68–70), 51-8018 (70), 51-8020 (65–69), 51-8023 (65–70), 51-8025 (70), 51-8031 (70), 51-8056 (70), 51-8075 (62), 51-8071 (68–70), 51-8110 (65–69), 51-8111 (65–67), 51-8145 (70), 51-8150 (70), 52-5848 (68–70), 52-5856 (70–71), 52-5863 (70–71), 52-5869 (70), 52-5924 (70), 52-5936 (65–71), 52-5951 (69–70), 52-5952 (68–70), 53-3141 (69–72), 53-3162 (66), 53-3166 (70–72), 53-3179 (69–72), 53-3182 (69–71), 53-3185 (69–71), 53-3188 (71), 53-3194 (69), 53-3196 (69–71), 53-3201 (69–71), 53-7840 (69), 53-7842 (68–72), 53-8073 (65–69), 53-8077 (69–72), 53-8083 (65–69), 53-8099 (69–71; depl 72), 53-8105 (70), 53-8112 (69–72), 53-8117 (69–72), 53-8119 (70–71), 53-8122 (69–71), 53-8139 (65–71), 53-8156 (65–71).

907 Troop Carrier Group / 356 Troop Carrier Squadron: Clinton County AFB, Ohio / Lockbourne AFB, Ohio (from 24Jul71). *C-119C:* 48-328 (63), 48-337 (63), 49-124 (63–65), 49-125 (63–65), 49-129 (63–64), 49-154 (64–65), 49-156 (63–64), 49-157 (63–65), 49-167 (63–65), 49-177 (63–65), 49-178 (63–65), 49-196 (63–65), 50-120 (63–65), 50-128 (64–66), 50-133 (63–65), 50-135 (63–64), 50-138 (63–64), 50-142 (63–65), 50-151 (64–65), 50-154 (63–65), 51-2532 (65), 51-2533 (65), 51-2534 (65), 51-2539 (65), 51-2544 (63–64), 51-2547 (65), 51-2559 (65), 51-2561 (65–66), 51-2565 (63–65), 51-2568 (65), 51-2573 (64–65), 51-2582 (65), 51-2584 (65). *C-119CF:* 51-2588 (65–66), 51-2615 (65–66), 51-2626 (65–66), 51-2632 (65–66), 51-2638 (65–66), 51-2647 (65–66), 51-2655 (65–66), 51-2656 (65–66), 51-8239 (65–66), 51-8247 (65–66), 51-8248 (65–66), 51-8251 (65–66), 51-8260 (65–66), 51-8261 (65), 51-8267 (65–66), 51-8271 (65–66). *C-119G:* 51-7981 (66–70), 51-7996 (68–70), 51-8002 (66–70), 51-8004 (66–70), 51-8005 (69–70), 51-8011 (68–70), 51-8015 (69–70), 51-8048 (69–70), 51-8099 (69–70), 51-8102 (69–70), 51-8106 (66–70), 51-8120 (69–70), 51-8147 (69–70), 52-5844 (69–70), 52-5846 (70–71), 52-5861 (69–70), 52-5870 (69–70), 52-5880 (70–71; C-119K 72), 52-5900 (70–71), 52-5918 (66–71; C-119K 72), 52-5939 (66–71), 53-3141 (72), 53-3147 (68–72), 53-3151 (66–70), 53-3159 (66–68), 53-3162 (66–70), 53-3166 (72–73), 53-3168 (66–70), 53-3173 (66–72), 53-3180 (72–73), 53-3181 (69–73), 53-3183 (72–73), 53-3185 (71–72), 53-3194 (69–72), 53-3196 (71–73), 53-3201 (71–73), 53-7840 (69–72), 53-7842 (72), 53-7878 (68–73), 53-7882 (69–72), 53-8071 (66–72; 72), 53-8077 (72), 53-8080 (66–71), 53-8099 (71–72; 72–73), 53-8104 (66–72), 53-8106 (66–72), 53-8109 (66–72), 53-8112 (72), 53-8117 (72), 53-8122 (71–73). *C-119K:* 52-5863 (72), 52-5936 (72). *YC-119K:* 53-3142 (70).

908 Troop Carrier Group / 357 Troop Carrier Squadron: Bates Field, Alabama (Transferred to 435 TCWG 18Mar63). *C-119G:* 51-2677 (63), 51-7971 (63), 51-8044 (63), 51-8073 (63), 51-8083 (63), 53-3220 (63), 53-7826 (63), 53-7830 (63), 53-7831

(63), 53-7837 (63), 53-7839 (63), 53-7850 (63), 53-7883 (63), 53-8085 (63), 53-8144 (63).

349 Troop Carrier (Medium) Wing: *Group / Squadrons:* **349 Troop Carrier Group** (13Jun52–14Apr59). **312 TCSq** / Hamilton AFB, California. **313 TCSq** / Portland IAP, Oregon. **314 TCSq** / McClellan AFB, California. **97 TCSq** / Paine Field, Washington State. *Operating Base:* Hamilton AFB, California. *Notes:* 97 TCSq not attached until 25Mar58. *C-119 Service:* 1958–1969. *C-119G Assignments:* 51-2684 (62–63), 51-2716 (unk–63), 51-7973 (58–63), 51-8058 (58–63), 51-8059 (59–63), 51-8060 (59–63), 51-8064 (58–62), 51-8089 (58–63), 51-8098 (59–63), 51-8108 (59–63), 52-5848 (58–63), 52-5869 (59–63), 52-5890 (59–63), 52-5910 (61–63), 52-5923 (59–63), 52-5924 (61–63), 52-5927 (58–63), 52-5938 (59–63), 52-5939 (58–63), 52-5942 (58–63), 52-5951 (61–63), 52-9981 (58–63), 52-9982 (58–63), 53-3142 (58–63), 53-3144 (58–63), 53-3146 (58–63), 53-3147 (58–63), 53-3156 (58–63), 53-3157 (58–63), 53-3160 (58–63), 53-3162 (58–63), 53-3166 (58–63), 53-3168 (58–63), 53-3170 (58–63), 53-3174 (58–63), 53-3183 (58–63), 53-3186 (59–63), 53-3190 (58–63), 53-3196 (59–63), 53-3198 (59–63), 53-3202 (58–63), 53-3203 (58–63), 53-3204 (58–63), 53-3205 (58–63), 53-3206 (59–63), 53-7833 (61–63), 53-7835 (61–63), 53-7842 (61–63), 53-7846 (61–63), 53-7847 (61–63), 53-7869 (60–63), 53-7876 (58–63), 53-7878 (61–63), 53-7884 (58–63), 53-8071 (58–63), 53-8087 (58–63), 53-8089 (58–63), 53-8105 (58–63), 53-8110 (58–63), 53-8111 (58–63), 53-8112 (61–63), 53-8114 (58–63), 53-8115 (61–63), 53-8120 (59–63), 53-8124 (58–63), 53-8126 (58–63), 53-8131 (58–63). *C-119J Assignments:* 51-8138 (62–63), 51-8141 (62). *C-119 Assignments from 11Feb63:*

938 Troop Carrier Group / 312 Troop Carrier Squadron: Hamilton AFB, California. *C-119G:* 52-5910 (63–66), 52-5927 (63–67), 52-5938 (63–66), 52-5939 (63–66), 52-9981 (63–65), 53-3142 (63–66), 53-3147 (63–66), 53-3157 (63–66), 53-3162 (63–66), 53-3168 (63–66), 53-3170 (63–66), 53-3205 (63–67), 53-3206 (63–66), 53-7847 (63–66), 53-7876 (63–65), 53-8071 (63–66), 53-8110 (63–65), 53-8120 (63–66).

939 Troop Carrier Group / 313 Troop Carrier Squadron: Portland IAP, Oregon. *C-119G:* 51-2684 (63–68), 51-2713 (68–69), 51-2716 (63–69), 51-7973 (63–68), 51-8059 (63–68), 51-8089 (63–68), 52-5923 (63–68), 52-5951 (63–68), 53-3144 (63–68), 53-3166 (63–68), 53-3174 (63–68), 53-3196 (63–68), 53-3205 (67–68), 53-3206 (66–68), 53-7846 (63–66), 53-8105 (63–68), 53-8112 (63–68), 53-8114 (63–68), 53-8115 (63–68).

940 Troop Carrier Group / 314 Troop Carrier Squadron: McClellan AFB, California. *C-119G:* 51-8060 (63–65), 51-8098 (63–65), 52-5848 (63–65), 52-5869 (63–65), 52-5942 (63–65), 53-3146 (63–65), 53-3156 (63–65), 53-3160 (63–65), 53-3185 (64–66), 53-3204 (63–65), 53-7833 (63–65), 53-7835 (63–65), 53-7842 (63–65), 53-8087 (63–65), 53-8089 (63–65), 53-8124 (63–65), 53-8131 (63–65).

941 Troop Carrier Group / 97 Troop Carrier Squadron: Paine Field, Washington State. *C-119G:* 51-8058 (63–65), 51-8108 (63–65), 52-5890 (63–65), 52-5924 (63–65), 52-9982 (63–65), 53-3183 (63–65), 53-3186 (63–65), 53-3190 (63–65), 53-3198 (63–65), 53-3202 (63–65), 53-3203 (63–65), 53-7869 (63–65), 53-7878 (63–65), 53-7884 (63–65), 53-8111 (63–65), 53-8126 (63–65).

403 Troop Carrier (Medium) Wing: *Group / Squadrons:* **403 Troop Carrier Group** (01Jan53–14Apr59). **63 TCSq** / Selfridge AFB, Michigan. **64 TCSq** / O'Hare IAP-ARS, Illinois. **65 TCSq** / Davis Field, Oklahoma. *Operating Base:* Selfridge AFB, Michigan. *C-119 Service:* 1958–1971. *C-119G Assignments:* 51-2662 (unk–63), 51-2663 (unk–63), 51-2664 (unk–63), 51-2673 (unk–63), 51-2674 (unk–63), 51-2675 (unk–63), 51-2711 (unk–63), 51-7984 (unk–63), 51-7992 (61–63), 51-8005 (58–63), 51-8013 (59–63), 51-8018 (58–63), 51-8027 (58–63), 51-8031 (58–63), 51-8047 (58–63), 51-8051 (58–59), 51-8075 (61–62), 51-8077 (58–62), 51-8087 (59–61), 51-8097 (58–63), 51-8099 (58–63), 51-8112 (58–62), 51-8117 (59–63), 51-8135 (58–63), 51-8162 (58–63), 52-5844 (58–63), 52-5855 (58–60), 52-5860 (59–63), 52-5861 (58–63), 52-5868 (58–63), 52-5870 (61–63), 52-5883 (58–63), 52-5892 (58–63), 52-5901 (58–63), 52-5905 (58–63), 52-5907 (58–63), 52-5920 (58–63), 52-5940 (61–63), 53-3138 (61–63), 53-3178 (58–63), 53-3180 (58–63), 53-3184 (58–63), 53-3188 (61–63), 53-3193 (58–63), 53-3200 (58–60), 53-3211 (58–63), 53-3215 (58–63), 53-3216 (58–63), 53-3217 (58–60), 53-7838 (61–63), 53-7848 (61–63), 53-7851 (61–63), 53-7852 (61–63), 53-7856 (58–63), 53-7871 (60–63), 53-8108 (58–63), 53-8119 (58–63), 53-8125 (58–63), 53-8147 (61–63), 53-8153 (61–63). *C-119J Assignments:* 51-8121 (62–63), 51-8123 (62–63), 51-8129 (62–63), 51-8137 (62–63), 51-8140 (62–63), 52-5877 (58–61), 53-8098 (58–61). *C-119 Assignments from 11Feb63:*

927 Troop Carrier Group / 63 Troop Carrier Squadron: Selfridge AFB, Michigan. *C-119G:* 51-2662 (63–69), 51-2663 (63–69), 51-2664 (63–69), 51-2674 (63–69), 51-2675 (63–69), 51-2683 (68–69), 51-7984 (63–69), 51-8005 (63–69; 69), 51-8060 (68–69), 51-8070 (68–69), 51-8097 (63–69), 51-8099 (63–69), 51-8135 (63–69), 51-8160 (68–69), 51-8166 (68–69), 52-5864 (67–68), 52-5920 (63), 52-5929 (68–69), 52-9982 (68), 53-3157 (66–67), 53-3215 (63), 53-3216 (63–69), 53-7856 (63–67), 53-7871 (63–69), 53-8108 (63–69), 53-8125 (63–69), 53-8131 (68), 53-8147 (63–67).

928 Troop Carrier Group / 64 Troop Carrier Squadron: O'Hare IAP-ARS, Illinois. *C-119C:* 48-327 (66–67). *C-119G:* 51-2586 (68–69), 51-2671 (67–69), 51-7974 (69), 51-8005 (69), 51-8013 (63–69), 51-8025 (70), 51-8027 (63–69), 51-8047 (63–69), 51-8057 (68–69), 51-8098 (68), 51-8117 (63–69), 51-8162 (63–69), 52-5844 (63–69), 52-5850 (70–71), 52-5860 (63), 52-5861 (63–69), 52-5870 (63–69), 52-5883 (63–69), 52-5901 (63–69), 52-5905 (63–68), 52-5907 (63–68), 53-3137 (68–71), 53-3160 (71), 53-3188 (63–71), 53-3190 (68–70), 53-7851 (63–68), 53-7852 (63–68), 53-8108 (69–71), 53-8125 (69–71), 53-8153 (63–69).

929 Troop Carrier Group / 65 Troop Carrier Squadron: Davis Field, Oklahoma. *C-119G:* 51-2673 (63–65), 51-2711 (63–65), 51-7992 (63–65), 51-7992 (63–65), 51-8018 (63–65), 51-8031 (63–65), 52-5868 (63–65), 52-5892 (63–65), 52-5940 (63–65), 53-3138 (63), 53-3178 (63–65), 53-3180 (63–65), 53-3184 (63–65), 53-3193 (63–65), 53-3211 (63–65), 53-7838 (63–65), 53-7848 (63–65), 53-8119 (63–65).

433 Troop Carrier (Medium) Wing: *Group / Squadrons:* **433 Troop Carrier Group** (18May55–14Apr59). **67 TCSq** / Brooks AFB, Texas. **68 TCSq** / Kelly AFB, Texas. **69 TCSq** / NAS Dallas, Texas. *Operating Base:* Brooks AFB, Texas (18May55–14Apr59). Kelly AFB, Texas (01Nov60–). *Notes:* 67 TCSq transferred to Kelly 01Nov60. 68 TCSq transferred to Kelly 21May60. 69 TCSq at NAS Dallas (Hensley Field) transferred to Carswell 03Mar63. *C-119 Service:* 1958–1971. *C-119C Assignments:* 49-120 (58–63), 49-133 (60–63), 49-139 (58–63), 49-140 (58–63), 49-141 (58–63), 49-142 (58–63), 49-154 (58–61), 49-162 (58–63), 49-164 (58–63), 49-165 (58–63), 49-174 (58–63), 49-176 (58–63), 49-179 (58–61; 62–63), 49-185 (58–63), 49-188 (58–63), 49-191 (62–63), 49-193 (58–63), 49-197 (58–63), 50-119 (60–63), 50-123 (59–63), 50-126 (58–63), 50-129 (58–63), 50-132 (58–63), 50-134 (58–63), 59-137 (58–63), 50-139 (58–63), 50-141 (58–63), 50-157 (58–63), 50-168 (58–63), 50-170 (58–63), 51-2535 (58–63), 51-2547 (59–63), 51-2548 (58–63), 51-2559 (58–63), 51-2561 (58–63), 51-2564 (58–63), 51-2566 (58–63), 51-2567 (58–63), 51-2577 (58–63), 51-2580 (59–63), 51-2584 (58–63). *C-119CF Assignments:* 51-2593 (58–63), 51-2596 (58–63), 51-2604

(58–63), 51-2607 (58–63), 51-2624 (58–59), 51-2627 (58–63), 51-2628 (58–63), 51-2629 (58–63), 51-2636 (59–63), 51-2639 (unk–61), 51-2643 (unk–63), 51-2654 (unk–63), 51-2655 (unk–63), 51-2656 (unk–63), 51-8236 (58–61). *C-119 Assignments from 17Jan63:*

    921 Troop Carrier Group / 67 Troop Carrier Squadron: Kelly AFB, Texas. *C-119C:* 49-139 (63–65), 49-140 (63–64), 49-141 (63–65), 49-142 (63–65), 49-165 (63–65), 49-176 (63–65), 49-188 (63–65), 49-191 (63), 49-193 (63–65), 49-197 (63–65), 50-119 (63–65), 50-123 (63–65), 50-126 (63–65), 50-129 (63–65), 50-132 (63–64; 65), 50-139 (63), 51-2548 (65–66), 51-2564 (65–66), 51-2566 (65–66), 51-2567 (65–66), 51-2577 (65–66), 51-2580 (65–66). *C-119CF:* 51-2593 (65–66), 51-2596 (65–66), 51-2607 (65–66), 51-2627 (65–66), 51-2628 (65–66), 51-2629 (65–66), 51-2636 (63–66), 51-2643 (63–66), 51-2654 (63–66), 51-2658 (65–66).

    922 Troop Carrier Group / 68 Troop Carrier Squadron: Kelly AFB, Texas. *C-119C:* 49-141 (65), 49-142 (65), 49-164 (63–65), 49-165 (65), 49-176 (65), 49-179 (63–65), 49-197 (65), 50-119 (65), 50-123 (65), 50-126 (65), 50-129 (65), 50-132 (64–65), 50-137 (63), 50-141 (63–65), 50-157 (63–65), 50-168 (63–65), 50-170 (63–65), 51-2548 (63–65), 51-2564 (63–65), 51-2566 (63–65), 51-2567 (63–65), 51-2577 (63–65), 51-2580 (63–65). *C-19CF:* 51-2593 (63–65), 51-2596 (63–65), 51-2627 (63–65), 51-2628 (63–65). *C-119G:* 51-2673 (65–68), 51-2675 (depl 70), 51-2711 (65–69), 51-7983 (69), 51-7992 (65–69), 51-8017 (68–69), 51-8018 (65–70), 51-8031 (65–70), 51-8089 (68–71), 52-5856 (67–70), 52-5868 (65–71), 52-5892 (65–68), 52-5924 (68–70), 52-5940 (65–68), 53-3144 (68–71), 53-3166 (69–70), 53-3167 (66–67), 53-3174 (69–71), 53-3178 (65–68), 53-3180 (65–70), 53-3183 (68–70), 53-3184 (65–71), 53-3186 (68–71), 53-3193 (65–71), 53-3211 (65–68), 53-7838 (65–66), 53-7848 (65–68), 53-8105 (69–70), 53-8111 (68–71), 53-8119 (65–70).

    923 Troop Carrier Group / 69 Troop Carrier Squadron: NAS Dallas, Texas / Carswell AFB, Texas (from 03Mar63). *C-119C:* 49-120 (63), 49-133 (63–64), 49-162 (63–65), 49-174 (63–65), 49-185 (63–65), 50-134 (63–65), 51-2535 (63–65), 51-2547 (63–65), 51-2559 (63–65), 51-2561 (63–65), 51-2584 (63–65). *C-119CF:* 51-2604 (63–65), 51-2607 (63–65), 51-2629 (63–65), 51-2655 (63–65), 51-2656 (63–65), 51-2658 (63–65), 51-8260 (unk–65).

434 Troop Carrier (Medium) Group: *Group / Squadrons:* 434 Troop Carrier Group (01Feb53–14Apr59). **71 TCSq** / Atterbury AFB, Indiana. **72 TCSq** / Atterbury AFB, Indiana. **73 TCSq** / Atterbury AFB, Indiana. *Operating Bases:* Atterbury AFB, Indiana (01Feb53–31Dec69). (Renamed Bakalar AFB 13Nov54). Grissom AFB, Indiana (15Jan70–). *Notes:* 73 TCSq transferred to Scott AFB, Illinois 16Nov57. Some 71 TCSq C-119s deployed to Lockbourne AFB, Ohio with the 1 ACO WG 68-69. *C-119 Service:* 1959–1970. *C-119G Assignments:* 51-2662 (59–unk), 51-2667 (59–63), 51-2671 (unk–63), 51-2672 (unk–63), 51-2678 (unk–63), 51-2682 (unk–63), 51-2684 (59–62), 51-2685 (unk–63), 51-2708 (unk–63), 51-2709 (unk–63), 51-2710 (unk–63), 51-7970 (59–63), 51-7976 (59–63), 51-7977 (59–63), 51-7978 (59–63), 51-7983 (59–63), 51-7988 (59–63), 51-7989 (59–63), 51-7991 (59–63), 51-7996 (59–63), 51-7997 (59–63), 51-7998 (61–63), 51-8009 (59–63), 51-8010 (59–63), 51-8015 (59–63), 51-8019 (69–63), 51-8022 (59–63), 51-8024 (61–63), 51-8028 (59–63), 51-8054 (59–63), 51-8063 (61–63), 51-8074 (depl 60), 51-8079 (59–63), 51-8091 (61–63), 51-8100 (59–63), 51-8101 (59–61), 51-8139 (59–61), 51-8168 (59–63), 52-5850 (59–63), 52-5856 (61–63), 52-5864 (59–63), 52-5909 (59–60), 52-5912 (59–60), 52-5952 (59–63), 53-3137 (59–63), 53-3139 (59–63), 53-3140 (59–63), 53-7844 (61–63), 53-7854 (61–63), 53-7859 (59–60), 53-7873 (60–63), 53-8069 (61–63), 53-8088 (59–63), 53-8125 (depl

60–61), 53-8144 (61), 53-8145 (61–63), 53-8149 (61–63), 53-8154 (61–63). *C-119J Assignments:* 51-8037 (61–63), 51-8038 (61–63), 51-8039 (62–63), 51-8041 (61–63), 51-8042 (61–63), 51-8043 (61–63), 51-8045 (61–63), 51-8049 (62–63), 51-8050 (61–63), 51-8115 (61–63), 51-8156 (59–61), 52-5897 (59–61), 52-5947 (59–61). *C-119 Assignments from 11Feb63:*

930 Troop Carrier Group / 71 Troop Carrier Squadron: Bakalar AFB, Indiana / Grissom AFB, Indiana (from 15Jan70). *C-119C:* 51-2575 (rec 65). *C-119G:* 51-2662 (69–70), 51-2663 (69–70), 51-2664 (69–70), 51-2674 (69–70), 51-2675 (69–70), 51-2684 (69–70), 51-2685 (69–70), 51-2708 (67–68), 51-7976 (63–68), 51-7983 (63–68), 51-7988 (63–68), 51-7991 (69–70), 51-7996 (63–68), 51-7997 (63–68), 51-7998 (69–70), 51-8009 (63–68), 51-8010 (63–68), 51-8022 (69–70), 51-8024 (63–68), 51-8028 (63–66), 51-8054 (69–70), 51-8063 (63–68), 51-8072 (69), 51-8100 (63–68), 51-8135 (69–70), 51-8168 (63–69), 52-5846 (69–70), 52-5863 (69–70), 52-5869 (69–70), 52-5880 (69–70), 52-5900 (69–70), 52-5929 (68), 52-5938 (68), 52-5952 (68), 53-3137 (68), 53-3147 (68), 53-3167 (66), 53-3197 (68), 53-3201 (68), 53-7844 (63–67), 53-8069 (63–68), 53-8088 (63–67), 53-8154 (63–68). *C-119J:* 51-8042 (63), 51-8043 (63), 51-8045 (63).

931 Troop Carrier Group / 72 Troop Carrier Squadron: Bakalar AFB, Indiana / Grissom AFB, Indiana (from 15Jan70). *C-119G:* 51-2682 (63–68), 51-2684 (69), 51-2685 (63–69), 51-2709 (63–69), 51-7977 (63–69), 51-7978 (63–69), 51-7989 (63–69), 51-7991 (63–69), 51-7998 (63–69), 51-8022 (63–69), 51-8054 (63–69), 51-8072 (69), 51-8079 (63–69), 51-8091 (63–69), 51-8168 (depl 68–69), 52-5863 (68–69), 52-5880 (68–69), 52-5929 (68), 52-5951 (68–69), 52-5952 (63–68), 53-3137 (63–68), 53-3157 (depl 63), 53-3196 (68–69), 53-3197 (68), 53-3201 (68–69), 53-7873 (63–67), 53-8112 (68–69), 53-8148 (68), 53-8149 (63–69), 53-8154 (68–69). *C-119J:* 51-8037 (63), 51-8038 (63), 51-8041 (63), 51-8115 (63), 52-5900 (69).

932 Troop Carrier Group / 73 Troop Carrier Squadron: Scott AFB, Illinois. *C-119G:* 51-2667 (63–66), 51-2671 (63–67), 51-2672 (63–66), 51-2678 (63–67), 51-2678 (63–67), 51-2708 (63–67), 51-2710 (63–66), 51-7970 (63–66), 51-8015 (63–67), 51-8019 (63–66), 51-8136 (66; 68–69), 52-5850 (63–67), 52-5856 (63–67), 52-5864 (63–67), 53-3139 (63–66), 53-3140 (63–66), 53-8145 (63–66). *C-119J:* 51-8039 (63), 51-8049 (63), 51-8050 (63), 53-7854 (63–66).

435 Troop Carrier (Medium) Wing: *Group / Squadrons:* 435 Troop Carrier Group (01Dec52–14Apr59). **76 TCSq** / Miami IAP, Florida. **77 TCSq** / Donaldson AFB, South Carolina. **78 TCSq** / Bates Field, Alabama. *Operating Bases:* Miami IAP, Florida (01Dec52–25Jul60). Homestead AFB, Florida (25Jul60–01Dec65). *Notes:* 76 TCSq transferred to Homestead 25Jul60. 77 TCSq transferred to Carswell 01Apr63. 78 TCSq transferred to Barksdale 08May61. 357 TCSq / 908 TCG transferred from 302 TCWG 18Mar63. *C-119 Service:* 1958–1969. *C-119C Assignments:* 49-117 (61), 49-136 (61), 49-137 (61), 49-139 (depl 59), 49-183 (61), 49-184 (61), 49-190 (61), 49-191 (61), 49-194 (61), 49-196 (61), 51-2541 (depl 59). *C-119CF Assignment:* 51-2616 (depl 60). *C-119G Assignments:* 51-2664 (59–unk), 51-2666 (unk–63), 51-2669 (unk–63), 51-2670 (unk–63), 51-2714 (62–63), 51-7971 (58–61), 51-7972 (58–63), 51-7992 (59–61), 51-8001 (58–61), 51-8004 (58–61), 51-8023 (58–61), 51-8025 (58–61), 51-8044 (59–61), 51-8055 (58–61), 51-8067 (58–61), 51-8075 (58–61), 51-8083 (58–61), 51-8087 (61–63), 51-8105 (59–60), 51-8111 (59–63), 51-8146 (59–63), 52-5856 (58–61), 52-5878 (58–61), 52-5887 (59–63), 52-5910 (58–61), 52-5926 (58–63), 52-5934 (58–61), 52-5936 (58–63), 52-5946 (58–63), 53-3188 (depl 60), 53-3220 (61), 53-7826 (61), 53-7830 (61), 53-7831 (61), 53-7837 (61), 53-7839 (61), 53-7850 (61), 53-7883 (60–61),

53-8069 (58–61), 53-8073 (58–63), 53-8076 (61–63), 53-8083 (58–63), 53-8085 (58–61), 53-8127 (58–63), 53-8129 (58–60), 53-8136 (58–60), 53-8139 (61–63). **C-119J Assignments:** 51-8046 (60–61; 62), 51-8052 (60–61), 51-8114 (59–61; 62), 51-8116 (58–61), 51-8119 (60; 62), 52-5851 (unk–61), 52-5895 (58–61). *C-119 Assignments from 17Jan63:*

915 Troop Carrier Group / 76 Troop Carrier Squadron: Homestead AFB, Florida. **C-119G:** 51-2666 (63–65), 51-2669 (63–65), 51-2670 (63–65), 51-2676 (unk–65; 66), 51-2714 (63–65), 51-7972 (63–65), 51-8087 (63–65), 51-8111 (63–65), 51-8146 (63–65), 51-8160 (depl 65–66), 52-5887 (63), 52-5926 (63–65), 52-5936 (63–65), 52-5946 (63–65), 53-3220 (66–67), 53-7839 (depl 66), 53-7883 (depl 65–66), 53-8073 (63–65), 53-8076 (63–65), 53-8083 (63–65), 53-8127 (63–65), 53-8139 (63–65).

916 Troop Carrier Group / 77 Troop Carrier Squadron: Carswell AFB, Texas (from 01Apr63). **C-119CF:** 51-2632 (65).

917 Troop Carrier Group / 78 Troop Carrier Squadron: Barksdale AFB, Louisiana. Did not operate the C-119 after 08May61.

908 Troop Carrier Group / 357 Troop Carrier Squadron: Bates Field, Alabama (Transferred from 302 TCWG 18Mar63) / Brookley AFB, Alabama (from 01Oct64). **C-119G:** 51-2677 (63–69), 51-2681 (68–69), 51-2716 (69), 51-7971 (63–69), 51-7980 (68–69), 51-7983 (68–69), 51-8015 (67–69), 51-8044 (63–69), 51-8048 (68–69), 51-8058 (68–69), 51-8073 (63–69), 51-8083 (63–69), 51-8107 (66–69), 51-8161 (68–69), 52-5878 (63), 52-5890 (68–69), 52-5919 (63–66), 52-5923 (68–69), 52-5927 (67–68), 52-5934 (63–69), 53-3206 (68–69), 53-3220 (63–66; 67-68), 53-7826 (63–68), 53-7830 (63–68), 53-7831 (63–68), 53-7837 (63–69), 53-7839 (63–68), 53-7850 (63–68), 53-7883 (63–68), 53-8085 (63–66), 53-8144 (63).

440 Troop Carrier (Medium) Wing: *Group / Squadrons:* 440 Troop Carrier Group (15Jun52–14Apr59). **95 TCSq** / General Mitchell IAP-ARS, Wisconsin. **96 TCSq** / Minneapolis-St. Paul IAP-ARS, Michigan. *Operating Base:* General Mitchell IAP-ARS, Wisconsin. *Notes:* 97 TCSq attached until 25Mar58 then transferred to the 349 TCWG. Some (52–5846) 934 TAL GP aircraft deployed to 4413 CCT WG (1 ACO WG) Lockbourne AFB, Ohio May68. *C-119 Service:* 1958–1971. **C-119G Assignments:** 51-2680 (unk–63), 51-2712 (unk–63), 51-7980 (depl 61), 51-7999 (58–63), 51-8012 (58–63), 51-8014 (58–63), 51-8025 (61–63), 51-8053 (58–63), 51-8055 (61–63), 51-8056 (59–63), 51-8074 (59–61), 51-8076 (58–62), 51-8078 (59–63), 51-8082 (58–63), 51-8101 (61–63), 51-8109 (59–63), 51-8150 (58–63), 52-5841 (59–63), 52-5846 (59–63), 52-5874 (59–63), 52-5876 (58–63), 52-5889 (59–63), 52-5902 (59–63), 52-5911 (59–63), 52-5914 (59–63), 53-3141 (59–63), 53-3189 (61–63), 53-3191 (58–63), 53-3194 (61–63), 53-7827 (61–63), 53-7834 (61–63), 53-7840 (61–63), 53-7858 (58–63), 53-7879 (60–63), 53-8070 (58–60), 53-8074 (58–63), 53-8075 (61–63), 53-8078 (58–63), 53-8079 (59–60), 53-8082 (58–60), 53-8099 (59–63), 53-8113 (58–63), 53-8122 (61–63). **C-119J Assignments:** 51-7968 (58–61; 62–63), 51-8130 (58–61; 62–63), 51-8156 (62–unk). *C-119 Assignments from 11Feb63:*

933 Troop Carrier Group / 95 Troop Carrier Squadron: General Mitchell IAP-ARS, Wisconsin. **C-119G:** 51-2668 (68–70), 51-2680 (63–65), 51-2708 (67), 51-2712 (63–71), 51-7986 (68), 51-7997 (68–70), 51-7999 (63–70), 51-8012 (63–70), 51-8014 (63–71), 51-8025 (70), 51-8053 (63–70), 51-8055 (63–70), 51-8082 (63–70), 51-8101 (63–71), 51-8109 (63–68), 52-5840 (68–70), 52-5850 (70), 52-5874 (70), 52-5914 (70), 53-3191 (63–71), 53-3206 (69–71), 53-7827 (63), 53-7834 (63), 53-7858 (63–71), 53-7876 (65–66), 53-7879 (63–68), 53-8074 (63–68), 53-8075 (63–70), 53-8078 (63), 53-8110 (66–71), 53-8113 (63–71).

Section Two—Military Operators

934 Troop Carrier Group / 96 Troop Carrier Squadron: Minneapolis-St. Paul IAP-ARS, Michigan. *C-119G:* 51-7979 (68–69), 51-7984 (69–70), 51-8025 (63–70), 51-8056 (63–70), 51-8062 (68–69), 51-8078 (63–69), 51-8150 (63–70), 52-5841 (63), 52-5846 (63–68), 52-5850 (67–70), 52-5874 (63–70), 52-5876 (63–70), 52-5889 (63–68), 52-5902 (63), 52-5911 (63–68), 52-5914 (63–70), 52-5917 (68–69), 53-3141 (63–69), 53-3179 (69), 53-3189 (63–68), 53-3194 (63–69), 53-7840 (63–69), 53-8077 (69), 53-8099 (63–69), 53-8117 (69), 53-8120 (66), 53-8122 (63–69).

442 Troop Carrier (Medium) Wing: *Group / Squadrons:* **442 Troop Carrier Group** (15Jun52–14Apr59). **303 TCSq** / Richards-Gebaur AFB, Missouri. **304 TCSq** / Richards-Gebaur AFB, Missouri. **305 TCSq** / Tinker AFB, Oklahoma. *Operating Base:* Richards-Gebaur AFB, Missouri. *Notes:* Inactivated from the AFR in 1961. *C-119 Service:* 45 1959–1961. *C-119G Assignments:* 51-2670 (59–unk), 51-2671 (unk), 51-2683 (59–unk), 51-2686 (59–unk), 51-7969 (59–61), 51-7974 (59–61), 51-7980 (59–61), 51-7986 (59–61), 51-7998 (59–61), 51-8016 (59–61), 51-8017 (59–61), 51-8024 (59–61), 51-8029 (59–61), 51-8048 (59–61), 51-8061 (59–61), 51-8063 (59–61), 51-8073 (59–61), 51-8091 (59–61), 51-8093 (59–61), 52-5863 (59–61), 52-5867 (59–61), 52-5870 (59–61), 52-5916 (59–61), 52-5918 (59–61), 52-5924 (59–61), 52-5940 (59–61), 52-5951 (59–61), 53-3138 (59–61), 53-3181 (59–61), 53-3187 (59–61), 53-3188 (59–61), 53-3194 (59–61), 53-7878 (60–61), 53-8075 (59–61), 53-8076 (59–61), 53-8081 (59–60), 53-8091 (59–61), 53-8112 (59–61), 53-8115 (59–61), 53-8122 (59–61), 53-8134 (59–60), 53-8135 (59–60), 53-8137 (59–60). *C-119J Assignments:* 51-8128 (59–61), 53-8101 (59–61).

446 Troop Carrier (Medium) Wing: *Group / Squadrons:* **446 Troop Carrier Group** (25May55–14Apr59). **357 TCSq** / Alvin Callender Field (NAS New Orleans), Louisiana. **704 TCSq** / Ellington AFB, Texas. **705 TCSq** / Ellington AFB, Texas. **706 TCSq** / Barksdale AFB, Louisiana. *Operating Base:* Ellington AFB, Texas. *Notes:* 357 TCSq at NAS New Orleans (Alvin Callender Field) then transferred to Bates Field (302 TCWG) from 08May61. 706 TCSq at Barksdale until transferred to NAS New Orleans 08May61. *C-119 Service:* 1958–1970. *C-119C Assignments:* 48-339 (58–61), 49-117 (58–61), 49-135 (58–61), 49-136 (58–61), 49-137 (58–61), 49-144 (58–61), 49-146 (58–61), 49-151 (58–61), 49-152 (58–59), 49-155 (58–61), 49-158 (58–61), 49-181 (58–61), 49-183 (58–61), 49-184 (59–61), 49-190 (58–61), 49-191 (58–61), 49-194 (58–61), 49-196 (58–61), 49-198 (58–62), 49-199 (58–61), 50-121 (59–63), 50-125 (59–63), 50-128 (62–63), 50-143 (58–61; 62–63), 50-144 (58–63), 50-145 (59–61), 50-146 (58–63), 50-147 (59–63), 50-150 (58–60), 50-151 (58–63), 50-152 (59–63), 50-156 (58–63), 50-160 (58–61), 50-161 (59–63), 50-164 (58–62), 50-167 (59–63), 50-169 (59–63), 50-171 (58–63), 51-2537 (58–63), 51-2538 (58–63), 51-2541 (58–63), 51-2543 (58–63), 51-2549 (62), 51-2550 (58–63), 51-2553 (58–63), 51-2554 (58–63), 51-2555 (58–63), 51-2557 (59–63), 51-2561 (depl 58), 51-2562 (58–63), 51-2563 (59–63), 51-2568 (58–63), 51-2569 (59–61), 51-2574 (58–63), 51-2575 (58–63), 51-2576 (58–63), 51-2578 (58–63), 51-2579 (58–63), 51-2581 (58–63), 51-2583 (58–63). *C-119CF Assignments:* 51-2587 (58–63), 51-2594 (58–63), 51-2601 (58–61), 51-2603 (58–63), 51-2610 (58–61), 51-2617 (58–63), 51-2625 (58–63), 51-2633 (unk–63), 51-2634 (unk–63), 51-2641 (unk–63), 51-2644 (unk–63), 51-2651 (unk–63), 51-2652 (unk–63), 51-2659 (59–61), 51-2660 (unk–63), 51-8244 (58–59). *C-119 Assignments from 17Jan63:*

924 Troop Carrier Group / 704 Troop Carrier Squadron: Ellington AFB, Texas. *C-119C:* 50-128 (63–64), 50-146 (63–65), 50-152 (63–65), 50-161 (63–65), 50-167 (63–65), 50-171 (63–65), 51-2537 (63–65), 51-2543 (63–65), 51-2557 (63–65), 51-2568 (63–65), 51-2575 (63–65), 51-2578 (63–65). *C-119CF:* 51-2594 (63–65),

51-2603 (63–65), 51-8250 (unk–65), 51-8253 (unk–64), 51-8266 (unk–65), 51-8272 (unk–65). *C-119G:* 51-2681 (65–68), 51-2686 (65–68), 51-7980 (68), 51-7987 (65–68), 51-8011 (65), 51-8016 (65–68), 51-8017 (68), 51-8048 (65–68), 51-8058 (68), 51-8071 (66–68), 51-8111 (65), 51-8160 (65–67), 51-8161 (65; 68), 52-5948 (65–67), 52-5869 (65–68), 52-5890 (65–68), 52-9982 (65–68), 53-3183 (68), 53-3183 (72), 53-3186 (65–68), 53-3198 (65–67), 53-3202 (65–67), 53-7884 (65–68), 53-8111 (65–68).

925 Troop Carrier Group / 705 Troop Carrier Squadrons: Ellington AFB, Texas. *C-119C:* 49-154 (63–64), 50-121 (63–65), 50-125 (63–65), 50-143 (63–65), 50-144 (63–65), 50-147 (63–64), 50-156 (63–65), 50-169 (63–65), 51-2538 (63–65), 51-2541 (63–65), 51-2553 (63–65), 51-2554 (63–65), 51-2563 (63–65), 51-2576 (63–65). *C-119CF:* 51-2617 (63–65), 51-2633 (63–65), 51-2634 (63–65), 51-2652 (63–65). *C-119G:* 51-2683 (65–68), 51-7980 (65–68), 51-8011 (65–68), 51-8017 (65–68), 51-8058 (65–68), 51-8088 (65–68), 51-8108 (65–68), 51-8160 (67–68), 51-8161 (65–68), 52-5848 (67–68), 52-5924 (65–68), 53-3146 (65–67), 53-3183 (65–68), 53-3203 (65–67), 53-7835 (65–67), 53-7842 (65–68), 53-7878 (65–68), 53-8124 (65–67).

926 Troop Carrier Group / 706 Troop Carrier Squadron: NAS New Orleans, Louisiana. *C-119C:* 50-151 (63–64), 51-2540 (63–65), 51-2543 (65), 51-2550 (63–65), 51-2555 (63–64), 51-2562 (63–65), 51-2572 (63–65), 51-2574 (63–65), 51-2579 (63–65), 51-2581 (63–65), 51-2583 (63–65). *C-119CF:* 51-2587 (63–65), 51-2594 (65), 51-2603 (65), 51-2617 (65), 51-2625 (63–65), 51-2633 (65), 51-2634 (65), 51-2641 (63–65), 51-2644 (63–65), 51-2651 (63–65), 51-2660 (63–65), 51-8246 (unk–65), 51-8250 (65), 51-8253 (64–65), 51-8258 (unk–65), 51-8266 (65), 51-8272 (65). *C-119G:* 51-2669 (65–70), 51-2676 (65–66; 66–69), 51-2681 (69–70), 51-7972 (65–69), 51-7973 (68–70), 51-7975 (65–69), 51-7982 (65–69), 51-7990 (65–70), 51-8003 (65–69), 51-8006 (65–70), 51-8010 (68–69), 51-8044 (69), 51-8058 (69), 51-8059 (68–69), 51-8073 (69–70), 51-8087 (65–70), 51-8103 (65–69), 51-8108 (68), 51-8146 (65–69), 52-5923 (69), 52-5926 (65–68), 52-5946 (65–70), 53-7884 (68–69), 53-8076 (65–69), 53-8127 (65–69), 53-8155 (65–68).

452 Troop Carrier (Medium) Wing: *Group / Squadrons:* 452 Troop Carrier Group (13Jun52–14Apr59). **728 TCSq** / Long Beach MAP, California. **729 TCSq** / Long Beach MAP, California. **730 TCSq** / Long Beach MAP, California. **733 TCSq** / Hill AFB, Utah. *Operating Bases:* Long Beach MAP, California (13Jun52–14Oct60). March AFB, California (14Oct60–01Jan72). *Notes:* 728, 729, 730 TCSq transferred to March with the Wing 14Oct60. *C-119 Service:* 1959–1969. *C-119G Assignments:* 51-2586 (61–63), 51-2668 (unk–63), 51-2681 (unk–63), 51-2683 (unk–63), 51-2686 (unk–63), 51-2714 (unk–62), 51-7969 (61–63), 51-7974 (61–63), 51-7979 (59–63), 51-7980 (61–63), 51-7986 (61–63), 51-7987 (59–63), 51-8011 (59–63), 51-8016 (61–63), 51-8017 (61–63), 51-8029 (61–63), 51-8048 (61–63), 51-8057 (59–63), 51-8071 (59–63), 51-8072 (59–63), 51-8080 (59–62), 51-8081 (59–63), 51-8084 (59–62), 51-8088 (59–63), 51-8090 (59–63), 51-8092 (59–62), 51-8107 (59–63), 51-8139 (61–63), 51-8160 (59–63), 51-8161 (59–63), 52-5840 (59–63), 52-5880 (59–63), 52-5898 (59–63), 52-5900 (59–63), 52-5916 (61–63), 52-5918 (61–63), 52-5929 (59–63), 52-5941 (59–63), 52-5943 (59–63), 53-3145 (59–63), 53-3148 (59–63), 53-3151 (59–63), 53-3152 (59–63), 53-3158 (59–63), 53-3159 (59–63), 53-3161 (59–63), 53-3167 (59–63), 53-3173 (59–63), 53-3175 (59–63), 53-3179 (59–63), 53-3195 (59–63), 53-3197 (59–63), 53-3199 (59–63), 53-3201 (59–63), 53-3207 (59–63), 53-3218 (61–63), 53-7836 (61–63), 53-7857 (depl 61), 53-7861 (60–63), 53-7862 (60–63), 53-7864 (59–63), 53-7865 (59–63), 53-7877 (59–63), 53-8072 (59–62), 53-8090 (59–63), 53-8093 (59–63), 53-8096 (59–63),

53-8104 (59–63), 53-8106 (59–63), 53-8128 (59–63), 53-8132 (59–63), 53-8142 (61–63), 53-8148 (61–63), 53-8150 (61–63). *C-119J Assignments:* 51-8035 (60–61; 62), 51-8132 (59–61), 52-5884 (59–61), 52-5903 (59–61), 52-5906 (59–61), 52-5947 (unk–62), 53-3213 (59–61). *C-119 Assignments from 17Jan63: (11Feb63)?*

942 Troop Carrier Group / 728 Troop Carrier Squadron: March AFB, California. *C-119G:* 51-2586 (63–65), 51-2681 (63–65), 51-2683 (65), 51-2686 (65), 51-2713 (63–65), 51-7974 (65–66), 51-7980 (63–65), 51-7986 (63–65), 51-7987 (65), 51-8011 (65), 51-8016 (65), 51-8017 (65), 51-8048 (65), 51-8057 (63–65), 51-8071 (65–66), 51-8088 (63–65), 51-8160 (65), 51-8161 (63–65), 52-5929 (63–65), 53-3136 (64), 53-3145 (63–65), 53-3152 (63–65), 53-3158 (63–65), 53-3175 (63–65), 53-3179 (63–65), 53-3197 (63–65), 53-3199 (63), 53-7862 (depl 65–66), 53-8142 (63–65).

943 Troop Carrier Group / 729 Troop Carrier Squadron: March AFB, California. *C-119G:* 51-2684 (68–69), 51-2686 (63–65), 51-2713 (69), 51-7969 (63–69), 51-7974 (63–65; 66–69; 69), 51-7987 (63–65), 51-8029 (63–69), 51-8048 (63–65), 51-8072 (63–69), 51-8081 (63–69), 51-8090 (63–64), 51-8107 (63–66), 51-8139 (68), 51-8160 (63–65), 52-5898 (65–68), 52-5900 (63–69), 52-5910 (67–68), 52-5941 (63), 53-3136 (64–68), 53-3145 (65–68), 53-3152 (65–69), 53-3158 (65–69), 53-3166 (68–69), 53-3174 (68–69), 53-3179 (68–69), 53-7836 (63–68), 53-7862 (63–66), 53-7865 (63–68), 53-7877 (68), 53-8105 (68–69), 53-8142 (65–69), 53-8150 (63–69).

944 Troop Carrier Group / 730 Troop Carrier Squadron: March AFB, California. *C-119G:* 51-2668 (63–68), 51-2683 (63–65), 51-2713 (65–68; depl 68), 51-7979 (63–68), 51-8011 (63–65), 51-8016 (63–65), 51-8017 (63–65), 51-8071 (63–65), 51-8139 (63–68), 52-5840 (63–68), 52-5880 (63–68), 52-5898 (63–65), 52-5910 (66–67), 52-5929 (65–68), 52-5938 (66–68), 52-5943 (63), 53-3147 (66–68), 53-3175 (65–67), 53-3179 (65–68), 53-3195 (63–66), 53-3197 (65–68), 53-3201 (63–68), 53-3218 (63–67), 53-7861 (63–66), 53-7877 (63–68), 53-8148 (63–68).

945 Troop Carrier Group / 733 Troop Carrier Squadron: Hill AFB, Utah. *C-119G:* 51-8139 (depl 66), 52-5916 (63), 52-5918 (63–66), 53-3148 (63–66), 53-3151 (63–66), 53-3159 (63–66), 53-3161 (63–66), 53-3167 (63–66), 53-3173 (63–66), 53-3207 (63–66), 53-7864 (63–66), 53-8090 (63–66), 53-8093 (63–66), 53-8096 (63–66), 53-8104 (63–66), 53-8106 (63–66), 53-8128 (63–66), 53-8132 (63–66).

459 Troop Carrier (Medium) Wing: *Group / Squadrons:* **459 Troop Carrier Group** (26Jan55–14Apr59). **756 TCSq** / Andrews AFB, Maryland. **757 TCSq** / Youngstown MAP-ARS, Ohio. **758 TCSq** / Greater Pittsburgh IAP ARS, Pennsylvania. *Operating Base:* Andrews AFB, Maryland. *C-119 Service:* 1958–1969. *C-119C Assignments:* 48-320 (58–61), 48-325 (58–63), 48-326 (58–61), 48-327 (58–63), 48-329 (58–61), 48-332 (58–61), 48-334 (58–63), 48-335 (58–61), 48-343 (58–61), 48-344 (58–61), 48-349 (58–61), 48-351 (58–61), 48-352 (58–63), 48-354 (58–61), 48-355 (58–61), 49-132 (58–63), 49-143 (58), 49-144 (61–63), 49-146 (61), 49-151 (61–63), 49-155 (61–63), 49-158 (61–63), 49-181 (61–63), 49-187 (58–63), 49-194 (62–63), 49-198 (62–63), 49-199 (61–63), 50-158 (59), 51-2536 (58–59), 51-2540 (59–63), 51-2565 (60), 51-2572 (59–63). *C-119CF Assignments:* 51-2598 (58–63), 51-2599 (58–63), 51-2605 (58–61), 51-2606 (58–63), 51-2612 (58–63), 51-2613 (58–63), 51-2614 (58–63), 51-2618 (58–62), 51-2619 (59–63), 51-2620 (60–63), 51-2622 (58–63), 51-2630 (58–63), 51-2631 (59–63), 51-2637 (unk–63), 51-2642 (unk–63), 51-2646 (unk–63), 51-2649 (unk–63), 51-2650 (unk–63), 51-2661 (unk–63), 51-8233 (59–63), 51-8237 (58–63), 51-8262 (59–63). *C-119 Assignments from 17Jan63:*

909 Troop Carrier Group / 756 Troop Carrier (Medium) Squadron: Andrews

AFB, Maryland. *C-119CF:* 51-2599 (63–unk), 51-2604 (65–66), 51-2612 (63–unk), 51-2613 (63–66), 51-2614 (63–66), 51-2630 (63–64), 51-2632 (63–64), 51-2661 (63–66), 51-8233 (63–66), 51-8236 (63–66), 51-8237 (63–66), 51-8240 (unk–66), 51-8242 (unk–66), 51-8243 (unk–66), 51-8245 (unk–66), 51-8254 (unk–66), 51-8255 (unk–66), 51-8257 (unk–66), 51-8262 (63).

910 Troop Carrier Group / 757 Troop Carrier (Medium) Squadron: Youngstown MAP ARS, Ohio. *C-119CF:* 51-2597 (66), 51-2598 (63–66), 51-2606 (63–66), 51-2619 (63–66), 51-2620 (63–66), 51-2622 (63–67), 51-2631 (63–66), 51-2632 (64–65), 51-2637 (63–66), 51-2642 (63–66), 51-2646 (63–66), 51-2649 (63–66), 51-2650 (63–66), 51-2652 (65–66), 51-8236 (unk–65), 51-8238 (unk–66), 51-8239 (66), 51-8247 (66), 51-8249 (unk–66), 51-8251 (66), 51-8260 (66–67), 51-8264 (unk–67), 51-8265 (unk–66), 51-8267 (66), 51-8268 (unk–67), 51-8269 (unk–66), 51-8271 (66–67). *C-119G:* 51-2667 (66–68), 51-2670 (67–69), 51-2672 (66–69), 51-2678 (67–69), 51-2686 (68), 51-2710 (66–69), 51-7970 (66–69), 51-7987 (68–69), 51-7988 (68–69), 51-8001 (68–69), 51-8033 (68–69), 51-8063 (68–69), 51-8088 (68–69), 51-8107 (69), 51-8111 (67–69), 52-5869 (68–69), 53-3139 (66–67), 53-3140 (66–67), 53-3148 (66–67), 53-3161 (66–67), 53-3185 (66–69), 53-7849 (66–69), 53-7854 (66–68), 53-8077 (66–69), 53-8117 (66–69), 53-8121 (66–68), 53-8133 (66–67), 53-8145 (66–68), 53-8154 (69).

911 Troop Carrier Group / 758 Troop Carrier (Medium) Squadron: Greater Pittsburgh IAP ARS, Pennsylvania. *C-119C:* 48-320 (63–66), 48-322 (63–67), 48-325 (63), 48-327 (63–66), 48-334 (63–66), 48-335 (63–66), 48-349 (63–66), 48-352 (63–66), 49-132 (63–67), 49-144 (63–65), 49-151 (63–66), 49-155 (63–66), 49-158 (63–66), 49-181 (63–66), 49-194 (63–67), 49-198 (63–65), 49-199 (63–67), 50-123 (65–66), 50-168 (65–66), 51-2540 (63), 51-2572 (63).

512 Troop Carrier (Medium) Wing: *Group / Squadrons:* **512 Troop Carrier Group** (14Jun52–14Apr59). **326 TCSq** / New Castle County Airport, Delaware. **327 TCSq** / New Castle County Airport, Delaware. **328 TCSq** / Paine AFB, Washington State. *Operating Bases:* New Castle County Airport, Delaware (14Jun52–20Jul58). NAS Willow Grove, Pennsylvania (20Jul58–08Jan65). Carswell AFB, Texas (08Jan65–29Jun71). *Notes:* 326, 327 TCSq transferred to NAS Willow Grove 20Jul58. 328 TCSq transferred to Niagara Falls 25Mar58.: *C-119 Service:* 1958–1971. *C-119G Assignments:* 51-2664 (unk–59), 51-7975 (59), 51-7985 (59–63), 51-7990 (59), 51-7994 (59), 51-8001 (61–63), 51-8008 (depl 59), 51-8020 (depl 59), 51-8025 (depl 59), 51-8033 (58–63), 51-8061 (61–63), 51-8062 (58–63), 51-8065 (58–62), 51-8066 (58–61), 51-8067 (61), 51-8068 (58–61), 51-8069 (59–63), 51-8070 (58–63), 51-8093 (61–63), 51-8094 (58–63), 51-8096 (58–63), 51-8102 (59–63), 51-8106 (depl 59), 51-8120 (59–63), 51-8145 (58–63), 51-8147 (58–63), 51-8148 (58–63), 51-8166 (59–63), 52-5842 (58–63), 52-4843 (58–63), 52-5847 (58–63), 52-5857 (58–63), 52-5862 (58–63), 52-5863 (61–63), 52-5865 (58–60), 52-5867 (61–63), 52-5871 (58–63), 52-5872 (58–63), 52-5873 (59–63), 52-5888 (58–63), 52-5899 (59–63), 52-5913 (58–63), 52-5915 (58–63), 52-5917 (58–63), 52-5922 (58–63), 52-5925 (58–63), 52-5931 (58–63), 52-5932 (58–63), 52-5933 (58–63), 52-5935 (58–63), 52-5945 (59–63), 52-5948 (58–63), 52-5950 (58–63), 52-5953 (58–63), 53-3154 (59–63), 53-3181 (61–63), 53-3182 (59–63), 53-3187 (61–63), 53-3192 (61–63), 53-3209 (58–60), 53-7832 (61–63), 53-7853 (61–63), 53-7882 (60–63), 53-8073 (depl 59), 53-8083 (depl 59), 53-8084 (59–63), 53-8091 (61–63), 53-8092 (59–60), 53-8097 (58–60), 54-8116 (58–63), 53-8123 (58–63), 53-8140 (61). *C-119J Assignments:* 51-8046 (59–60), 51-8113 (62–63), 51-8114 (58–59), 51-8116 (depl 59; 62–64), 51-8126 (62–64), 51-8154 (62–63), 52-5851 (58–unk). **C-119 Assignments from 11Feb63:**

912 Troop Carrier Group / 326 Troop Carrier Squadron: NAS Willow Grove, Pennsylvania. *C-119G:* 51-8001 (63–68), 51-8033 (63–68), 51-8062 (63–68), 51-8069 (63–68), 51-8070 (63–68), 51-8136 (66–68), 51-8145 (63–68), 51-8148 (63–68), 51-8166 (63–68), 52-5863 (63–68), 52-5871 (63–68), 52-5888 (63), 52-5899 (63–64), 52-5917 (63–68), 52-5935 (63–68), 52-5950 (depl 66–67), 52-9981 (65–68), 53-3187 (63–68), 53-3192 (63), 53-7847 (66), 53-8084 (63–68), 53-8116 (63–65), 53-8123 (63–68).

913 Troop Carrier Group / 327 Troop Carrier Squadron: NAS Willow Grove, Pennsylvania. *C-119G:* 51-2681 (70), 51-7984 (70), 51-7985 (63–69), 51-8006 (70–71), 51-8015 (69), 51-8048 (69), 51-8069 (68–69), 51-8073 (70–71), 51-8087 (70–71), 51-8094 (63–69), 51-8097 (69–70), 51-8099 (69), 51-8100 (68), 51-8102 (63–69), 51-8120 (63–69), 51-8147 (63–69), 51-8148 (68–69), 51-8166 (depl 66–67), 52-5843 (63), 52-5857 (63), 52-5871 (68–69), 52-5874 (70–71), 52-5876 (70), 52-5899 (64–69), 52-5914 (70–71), 52-5932 (63–69), 52-5935 (68), 52-5945 (63–68), 52-5946 (70–71), 52-5950 (63–69), 52-5953 (63–69), 52-9981 (68–69), 53-3154 (63–68), 53-3181 (63–69), 53-3182 (63–69), 53-3192 (63–68), 53-7832 (63–69), 53-7882 (63–69), 53-8084 (68–69). *YC-119K:* 53-3142 (67).

914 Troop Carrier Group / 328 Troop Carrier Squadron: Niagara Falls IAP ARS, New York. *C-119G:* 51-2678 (67), 51-8009 (68–71), 51-8061 (63–70), 51-8093 (63–71), 51-8096 (63–68), 52-5842 (63), 52-5847 (63–70), 52-5862 (63–71), 52-5872 (63–70), 52-5873 (63–70), 52-5913 (63–70), 52-5915 (63–71), 52-5922 (63–70), 52-5925 (63–68), 52-5931 (63–71), 52-5933 (63–70), 52-5948 (63–70), 53-3170 (66–68), 53-3204 (68–71), 53-7853 (63–71; depl 71), 53-7869 (68–70), 53-8087 (68–69), 53-8091 (63–66), 53-8126 (68–71).

514 Troop Carrier (Medium) Wing: *Group / Squadrons:* **514 Troop Carrier Group** (01Apr53–14Apr59). **335 TCSq** / Mitchel AFB, New York. **336 TCSq** / Mitchel AFB, New York. **337 TCSq** / Mitchel AFB, New York. *Operating Bases:* Mitchel AFB, New York (01Apr53–15Mar61). McGuire AFB, New Jersey (15Mar61–). *Notes:* 335 TCSq transferred to McGuire 15Mar61. 336 TCSq transferred to Stewart 15Mar61. 337 TCSq transferred to Bradley 08Jul58. Some 903 TCG aircraft deployed to Myrtle Beach South Carolina Jul67. 1 C-119G (53-7833) deployed to (CNC) Boston ANG 21May67. 1 903 TCG C-119G (53-8126) deployed to Westover AFB, Massachusetts Apr66. *C-119 Service:* 1958–1971. *C-119C Assignments:* 48-322 (59–61), 48-328 (60–61), 48-331 (58–61), 48-340 (58–61), 48-341 (58–61), 48-347 (58–61), 49-101 (58–63), 49-104 (58–63), 49-106 (58–63), 49-107 (58–63), 49-108 (58–63), 49-109 (58–63), 49-110 (58–63), 49-111 (58–63), 49-112 (58–63), 49-113 (58–63), 49-115 (58–63), 49-116 (58–63), 49-117 (61–63), 49-118 (58–63), 49-119 (58–unk), 49-121 (59–63), 49-122 (58–63), 49-127 (59–63), 49-131 (58–63), 49-135 (61–63), 49-136 (62–63), 49-143 (59–61), 49-159 (59–63), 49-170 (58–63), 49-182 (58–63), 49-184 (61–63), 50-124 (58–63), 50-130 (58–63), 50-131 (59–63), 50-136 (58–63), 50-140 (59–63), 50-148 (59–63), 50-155 (59–63), 50-158 (59–63), 50-162 (58–63), 50-165 (58–63), 50-166 (59–63), 51-2536 (59–63), 51-2545 (59–63), 51-2558 (58–63), 51-2571 (59–63). *C-119CF Assignments:* 51-2589 (59–63), 51-2595 (58–63), 51-2600 (59–63), 51-2611 (59–63), 51-2616 (58–63), 51-2623 (59–63), 51-2638 (unk–63), 51-2640 (59–63), 51-2657 (unk–63). *C-119 Assignments from 17Jan63:*

903 Troop Carrier Group / 335 Troop Carrier Squadron: McGuire AFB, New Jersey. *C-119C:* 49-127 (63), 49-131 (63–64), 49-135 (63–65), 49-159 (63–66), 49-170 (63–66), 49-182 (63–66), 49-184 (63–66), 50-124 (63–65), 50-130 (63–66), 50-131 (63–66), 50-136 (63–66), 50-140 (63–66), 50-148 (63–66), 50-155 (63–66), 50-158 (63–66), 50-162 (63–66), 50-165 (63–66), 50-166 (63–66), 51-2558 (65–66), 51-2571

(65–66), 51-2580 (66). *C-119CF:* 51-2616 (depl 65–66). *C-119G:* 51-2586 (67–68), 51-7986 (66–68), 51-8057 (66–68), 51-8060 (66–68), 51-8098 (66–68), 52-5942 (67–68), 53-3156 (67–68), 53-3160 (67–71), 53-3190 (67–68), 53-3204 (66–68), 53-7833 (67–68), 53-7869 (66–68), 53-8087 (66–68), 53-8089 (67–68), 53-8126 (66–68), 53-8131 (66–68).

904 Troop Carrier Group / 336 Troop Carrier Squadron: Stewart AFB, New York. *C-119C:* 49-101 (63–65), 49-104 (63–65), 49-106 (63–65), 49-107 (63–65), 49-108 (63–65), 49-109 (63–65), 49-110 (63), 49-111 (63–65), 49-112 (63–65), 49-113 (63–65), 49-115 (63–65), 49-116 (63–65), 49-117 (63–65), 49-118 (63–65), 49-121 (63–65), 49-122 (63–65), 49-136 (63–65). *C-119G:* 51-2586 (65–67), 51-7986 (65–66), 51-8057 (65–66), 51-8060 (65–66), 51-8098 (65–66), 52-5942 (65–67), 53-3156 (65–67), 53-3160 (65–67), 53-3190 (65–67), 53-3204 (65–66), 53-7833 (65–67), 53-7869 (65–66), 53-8087 (65–66), 53-8089 (65–67), 53-8126 (65–66), 53-8131 (65–66).

905 Troop Carrier Group / 337 Troop Carrier Squadron: Bradley IAP, Connecticut. *C-119C:* 51-2536 (63–65), 51-2545 (63–65), 51-2558 (63–65), 51-2571 (63–65). *C-119CF:* 51-2589 (63–66), 51-2595 (63–65), 51-2599 (unk–66), 51-2600 (63–66), 51-2611 (63–66), 51-2612 (unk–66), 51-2616 (63–66), 51-2623 (63–66), 51-2638 (63–65), 51-2640 (63), 51-2657 (63–66), 51-8248 (unk–65), 51-8252 (unk–66), 51-8261 (unk–65), 51-8267 (unk–65), 51-8270 (unk–66).

Air National Guard (ANG)

The ANG is a federal military reserve force of the USAF that normally operates under the jurisdiction of U.S. state governors but can be federalized as an active part of the USAF if called on in an emergency. 21 ANG groups and squadrons operated the C-119C/G/J variants from 1958 to 1975. A large portion of the ANG units were aeromedical squadrons assigned the C-119J withdrawn early by the USAF due to design flaws with the redesigned cargo doors. At least a dozen C-119s were further modified as the MC-119J which featured a cargo hold capsule that housed ready-to-go medical facilities. The final three operational Boxcar squadrons in 1973, the 129th, 130th and 143rd, converted to the 3-bladed propeller C-119L which provided the last U.S. military service of the type before they too were retired in September, 1975.

C-119CF s/n 51-8236 (10784) was assigned to the 187th Aeromedical SQ (red cross emblem) with the Wyoming ANG from 1961 to 1963. The "0-18236" prefixed zero in the tail number was used by the USAF to denote aircraft more than ten years old (USAF).

102 Aeromedical (Light) Squadron: New York ANGB, New York. C-119J: 51-8036 (58–64), 51-8040 (59–63), 51-8121 (58–62), 51-8122 (58), 51-8144 (58–62), 51-8152 (58–63), 52-5845 (58–63), 52-5849 (58–63), 53-3213 (61–63), 53-7855 (58–63), 53-8098 (61–63), 53-8101 (61–63).

106 Aeromedical (Light) Group: New York ANGB, New York. C-119J: 53-3213 (63), 53-8098 (63).

106 Tactical Reconnaissance (Photo, Jet) Squadron: Birmingham MAP, Alabama. C-119J: 51-8116 (58).

112 Consolidated Maintenance Squadron: Greater Pittsburgh IAP, Pennsylvania. C-119J: 52-5903 (61).

124 Consolidated Maintenance Squadron: Boise ANGB (Gowen Field), Idaho. C-119C: 51-2537 (59). C-119G: 52-9981 (58), 53-3186 (58).

125 Fighter-Interceptor Squadron: Tulsa MAP, Oklahoma. C-119G: 51-8031 (58).

129 Air Commando Group: Hayward Airport, California. C-119CF: 51-2601 (63–69), 51-2605 (63–69), 51-2608 (63–69), 51-2610 (63–69), 51-2618 (63–69), 51-8256 (64–69), 51-8259 (unk–69), 51-8263 (unk–69). C-119G: 53-3184 (71), 53-3186 (71), 53-3193 (71), 53-3216 (69–73 L), 53-7836 (68–75 L), 53-7865 (68–75 L), 53-8074 (68–75 L), 53-8076 (69–75 L), 53-8127 (69–75 L), 53-8142 (69–75 L), 53-8150 (69–75 L), 53-8153 (69–75 L). Redesignated as 129 SOP GP 11Sep68. Deployment to Aviano AB Italy Oct74. 9 cvtd to C-119L 1973.

130 Air Commando Group: Kanawha Airport, West Virginia. C-119C: 48-326 (63), 48-332 (63), 48-343 (63), 48-351 (63), 48-354 (63), 48-355 (63), 49-119 (unk–69), 49-120 (63–69), 49-127 (63–69), 49-129 (64–69), 49-131 (64–69), 49-133 (64–69), 49-143 (63), 49-156 (64–69), 49-191 (63–69). C-119G: 53-7832 (69), 53-7837 (69–75 L), 53-7849 (69–75 L), 53-7884 (69–75 L), 53-8073 (69–75 L), 53-8083 (69–75 L), 53-8084 (69–75 L), 53-8087 (69–74 L), 53-8149 (69–75 L), 53-8154 (69–75 L). Redesignated as 130 SOP GP 1968. 9 cvtd to C-119L 1973.

137 Aeromedical (Light) Squadron: Westchester County Airport, New York. C-119C: 48-320 (61–63), 48-322 (61–63), 48-328 (61–62), 48-331 (61–63), 48-335 (61–63), 48-337 (61–62), 48-339 (61–62), 48-340 (61–63), 48-341 (61–63), 48-347 (61–63). C-119J: 53-8101 (61).

140 Aeromedical (Light) Squadron: General Spaatz Field, Pennsylvania, to Olmsted AFB, Pennsylvania (01Feb61). C-119J: 51-7968 (61–62), 51-8035 (61–62), 51-8123 (58–62), 51-8128 (61–62), 51-8130 (61–62), 51-8131 (58–62), 51-8156 (61–62), 51-8157 (58–62), 51-8159 (58–62), 51-8167 (58–62).

143 Special Operations Group: Theodore F. Green AP, Rhode Island. C-119G: 53-3144 (71–73 L), 53-3174 (71–75 L), 53-3184 (71–75 L), 53-3186 (71–75 L), 53-3193 (71–75 L), 53-3206 (71–75 L), 53-7853 (71–75 L), 53-7858 (71–75 L), 53-8126 (71–75 L). All 9 cvtd to C-119L 1973.

144 Consolidated Maintenance Squadron: Fresno ANGB, California. C-119CF: 51-2614 (58).

145 Aeromedical (Light) Squadron: Akron-Canton Airport, Ohio (17Mar56–01Jul61), to Clinton County AFB, Ohio (01Jul61–01Oct71). C-119J: 51-8116 (61–62), 51-8129 (58–62), 51-8134 (58–61), 51-8154 (58–62), 51-8155 (58–62), 52-5875 (58–63). Redesignated as 145 ATR SQ 01Jul61.

145 Consolidated Maintenance Squadron: Douglas IAP, North Carolina. C-119J: 51-8052 (61).

147 Aeromedical (Light) Squadron: Greater Pittsburgh IAP, Pennsylvania. C-119J: 51-8124 (62–63), 51-8131 (62–63), 51-8134 (61–62), 51-8155 (62–63), 51-8157 (62–63), 51-8164 (62–64), 51-8167 (62–63), 52-5851 (61–62), 52-5895 (61–64), 52-5896 (61–62), 52-5897 (61–63), 52-5903 (61–63), 53-8103 (62–63).

150 Aeromedical (Light) Squadron: Newark Airport, New Jersey. C-119J: 51-8113 (58–62), 51-8125 (58–63), 51-8126 (58–62), 51-8158 (58–63), 52-5866 (58–63), 52-5877 (61–64), 52-5884 (61–63), 52-5906 (61–62), 52-5947 (61–unk), 53-8103 (58–62).

153 Consolidated Maintenance Squadron: Cheyenne MAP, Wyoming. C-119J: 52-5895 (61), 52-5896 (61).

156 Aeromedical (Light) Squadron: Douglas IAP, North Carolina. C-119C: 49-136 (61–62), 49-137 (61–62), 49-179 (61–62), 49-183 (61–62), 49-190 (61–62), 49-191 (61–62), 49-194 (61–62), 49-196 (61–62), 50-128 (61–62), 50-143 (61–62). C-119J: 51-8046 (61), 51-8052 (61), 51-8114 (61), 51-8115 (61).

167 Aeromedical (Light) Squadron: Martinsburg (Shepherd Field), West Virginia. C-119C: 48-326 (61–63), 48-329 (61–63), 48-332 (61–63), 48-343 (61–63), 48-344 (61–63), 48-351 (61–63), 48-354 (61–63), 48-355 (61–63), 49-143 (61–63), 49-154 (61–63). C-119J: 51-8035 (61), 52-5877 (61), 52-5897 (61).

183 Aeromedical (Light) Squadron: Hawkins Field, Mississippi. C-119J: 51-8046 (61–62), 51-8052 (61–63), 51-8114 (61–62), 51-8116 (61), 51-8119 (61–62), 51-8124 (58–62), 51-8137 (58–62), 51-8138 (57–62), 51-8140 (58–62), 51-8141 (58–62), 51-8164 (57–62).

187 Aeromedical (Light) Squadron: Cheyenne MAP, Wyoming. C-119CF: 51-2601 (61–63), 51-2605 (61–63), 51-2608 (61–63), 51-2610 (61–63), 51-2618 (depl 61; 62–63), 51-2625 (58), 51-8236 (61–63). C-119J: 51-8119 (61), 51-8128 (61), 52-5851 (61), 52-5895 (61), 52-5896 (61).

United States Marine Corps (USMC) & Navy (USN)

A total of 97 C-119s, designated as R4Q by the Navy, were operated by the USMC then USMC Reserve from 1950 to 1975. Officially retaining the earlier C-82 name *Packet* they filled a tactical assault and aerial resupply transport role. The first eight were built to C-119B standard as the R4Q-1 and delivered to VMR-252 at MCAS Cherry Point North Carolina during 1950. The following 33 deliveries of the R4Q-1 were built to C-119C standard serving with VMR-153 also at Cherry Point and VMR-253 at MCAS El Toro California. Aircraft from these units served during the Korean War in support of the Marine ground forces. VMR-253 would retain a presence in the Far East based in Japan until 1959 exclusively operating the R4Q-1 type. The 58 C-119F configured R4Q-2 variant was introduced into USMC service in 1953 with VMR-153, -252 and -353. VMR-253 and -352 received some examples from 1959 to 1961 when the R4Q-1 was retired from service. VMR-353 was the only USMC squadron still using the R4Q-2 when they were redesignated C-119F in September 1962. Three USMC Reserve Training Detachment (MARTD) squadrons began receiving the last 19 serving C-119Fs in 1961, the type staying in Reserve service up to July 1975 when the few remaining examples were retired to MASDC at Davis-Monthan AFB.

The USN only saw limited use of the R4Q Packets, fleet tactical support squadron VR-24 operating four aircraft briefly from 1960 to 1962.

USMC Unit Summary

SQ	Period	Qty
VMR-153	1952–1959	41
VMR-252	1950–1961	71
VMR-253	1952–1961	39
VMR-352	1959–1961	5
VMR-353	1953–1963	47

MARTD Unit Summary

SQ	Period	Qty
VMR-216	1962–1972	17
VMR-222	1961–1969	12
VMR-234	1961–1975	17

USMC Dispositions

11 USMC Boxcars were lost to accidents during its 25 year service period. A total of 70 were simply scrapped—55 at the Navy's storage facility Litchfield Park, one was scrapped at Dothan, Alabama, after retirement with another 14 at MASDC in Tucson. 16 found their way into civilian service either as registered flying aircraft or as spare parts sources.

R4Q-2 BuNo. 131696 (10881) of VMR-352 MCAS El Toro in December 1960 just several months before it was retired and scrapped at NAF Litchfield Park, Arizona (Lawson Collection, National Naval Aviation Museum [NNAM]).

R4Q-2 BuNo. 131678 (10845) of VMR-252 at Oklahoma City in 1956 (Lawson Collection, National Naval Aviation Museum [NNAM]).

128 Section Two—Military Operators

United States Marine Corps

Aircraft Engineer Squadron 12 (AES-12): MCAS Quantico, Virginia. R4Q-1: 126575 (53).

Air Fleet Marine Force, Atlantic (AirFMFLANT): MCAS Cherry Point, North Carolina. R4Q-1: 126578 (52), 128739 (52), 128740 (52), 128741 (52), 128742 (52), 128743 (52). R4Q-2: 131718 (depl 53), 131719 (depl 53–54).

Headquarters Squadron 2 (HEDRON-2): MCAS Cherry Point, North Carolina. R4Q-1: 126578 (53).

Headquarters Squadron 25 (HEDRON-25): MCAS El Toro, California. R4Q-1: 126575 (54–55), 126576 (54), 126580 (54), 128737 (54–55). Redesignated as H&MS-25 09Apr54.

Headquarters Squadron 27 (HEDRON-27): MCAS Cherry Point, North Carolina. R4Q-2: 131709 (54), 131711 (54), 131712 (54–55), 131719 (54). Also designated **Marine Wing Support Group** (MWSG-27).

Headquarters Squadron 37 (HEDRON-37): MCAS Miami, Florida. R4Q-2: 131718 (54). Also designated **Marine Wing Support Group** (MWSG-37).

Headquarters & Maintenance Squadron, 1st Marine Air Wing (H&MS MAW-1): Pohang Airbase (K-3), South Korea. R4Q-1: 126576 (54–55).

Headquarters & Maintenance Squadron 11 (HAMRON-11): MCAS Edenton, North Carolina. R4Q-2: 131711 (53), 131712 (53).

Headquarters & Maintenance Squadron 13 (H&MS-13): MCAS Kaneohe, Hawaii. R4Q-1: 124330 (54).

Headquarters & Maintenance Squadron 14 (HAMRON-14): NAS Roosevelt Roads, Puerto Rico, to MCAS Edenton, North Carolina. R4Q-1: 126578 (53–54), 128743 (53–54). R4Q-2: 131709 (53), 131710 (53).

Headquarters & Maintenance Squadron 14 (H&MS-14): MCAS Edenton, North Carolina. R4Q-2: 131665 (depl 54–55).

Headquarters & Maintenance Squadron 15 (HAMRON-15): MCAS El Toro, California. R4Q-1: 128737 (depl 53), 128739 (depl 53).

Headquarters & Maintenance Squadron 24 (HAMRON-24): MCAS Cherry Point, North Carolina. R4Q-2: 131706 (53–55), 131707 (53–55).

Headquarters & Maintenance Squadron 25 (H&MS-25): MCAS El Toro, California. R4Q-1: 124328 (54), 124330 (54), 126578 (54–55), 126580 (54), 126581 (depl 55), 128724 (55), 128726 (55), 128730 (55), 128737 (55), 128739 (54–55), 128743 (54–55).

Headquarters & Maintenance Squadron 26 (HAMRON-26): MCAS Cherry Point, North Carolina. R4Q-1: 128743 (53). R4Q-2: 131710 (53–54).

Headquarters & Maintenance Squadron 26 (H&MS-26): MCAF New River, Jacksonville, Florida. R4Q-2: 131710 (54–55).

Headquarters & Maintenance Squadron 31 (HAMRON-31): MCAS Miami, Florida. R4Q-2: 131713 (53–54), 131715 (53–55).

Headquarters & Maintenance Squadron 31 (H&MS-31): MCAS Miami, Florida. R4Q-2: 131713 (54–55).

Headquarters & Maintenance Squadron 32 (HAMRON-32): MCAS Miami, Florida. R4Q-2: 131716 (53–54), 131717 (53–54).

Headquarters & Maintenance Squadron 32 (H&MS-32): MCAS Cherry Point, North Carolina. R4Q-2: 131711 (54–55), 131713 (55), 131716 (54), 131717 (54–55).

Marine Aircraft Repair Squadron 27 (MARS-27): MCAS Cherry Point, North Carolina. R4Q-1: 126578 (53). R4Q-2: 131683 (63), 131709 (53–54), 131712 (55), 131719 (54–55).

Marine Aircraft Repair Squadron 37 (MARS-37): MCAS Miami, Florida, to MCAS El Toro, California. R4Q-1: 128724 (55–56). R4Q-2: 131708 (54–55).

Marine Training Group 10 (MTG-10): MCAS El Toro, California. R4Q-1: 124328 (54), 124330 (53–54), 126574 (53), 126575 (53–54), 126576 (53–54), 126577 (53), 126579 (53), 126580 (53–54), 126581 (53; depl 53), 126582 (53), 128737 (53–54), 128738 (53), 128739 (53–54), 128740 (53), 128741 (53).

Marine Training Group 20 (MTG-20): MCAS Cherry Point, North Carolina. R4Q-2: 131711 (53–54), 131712 (53–54), 131714 (53), 131718 (53–54), 131719 (53–54).

Station Operation & Engineer Squadron (SO&ES): MCAS Cherry Point, North Carolina. R4Q-1: 128736 (58–59). R4Q-2: 131691 (63–65), 131706 (62–63).

VMR-153: *Operating Base:* MCAS Cherry Point, North Carolina. *Notes:* 2 R4Q-2 to **Detachment** at NAS Port Lyautey, Morocco Nov56–Jan57. *R4Q Service:* 41 1952–1959. *R4Q-1 Assignments:* 124326 (52–53), 124327 (52–53), 124328 (52–53), 124329 (52–53), 124330 (52–53), 124331 (52). *R4Q-2 Assignments:* 131662 (53–59), 131663 (53), 131664 (53–58), 131665 (53–58), 131666 (53–57), 131667 (53–59), 131668 (53–59), 131669 (53–57), 131670 (53–59), 131671 (53–59), 131672 (53–57), 131673 (53–58), 131674 (53–57), 131675 (53–59), 131676 (53–58), 131678 (depl 58–59), 131679 (depl 56, 58–59), 131683 (57–58), 131685 (59), 131687 (58–59), 131688 (57–59), 131689 (58–59), 131695 (57–59), 131696 (58–59), 131700 (58–59), 131701 (56–58), 131706 (55–56; 58–59), 131707 (56–59), 131708 (58–59), 131709 (54–56), 131710 (55–59), 131712 (58–59), 131714 (53–58), 131716 (54–56), 131717 (55–56).

VMR-252: *Operating Base:* MCAS Cherry Point, North Carolina. *Notes:* 128742 detached to MCAF New River, Jacksonville, Florida Jul53–Oct54. 12 R4Q-2 to **Detachment** at NAS Port Lyautey, Morocco Aug56–Jan61. Later redesignated as VMGR-252. *R4Q Service:* 71 1950–1961. *R4Q-1 Assignments:* 124324 (50–52), 124325 (50–51), 124326 (50–52), 124327 (50–52), 124328 (50–52), 124329 (50–52), 124330 (50–52), 124331 (50–52), 126574 (52–53), 126575 (52–53), 126576 (52–53), 126577 (52–53), 126578 (52–53), 126579 (52–53), 126580 (52–53), 126581 (52–53), 128737 (52–53), 128738 (52–53), 128739 (52–53), 128740 (52–53), 128741 (52–53), 128742 (52–54), 128743 (52–53). *R4Q-2 Assignments:* 131662 (59–61), 131664 (58–59), 131665 (58–61), 131667 (59–61), 131668 (59–62), 131669 (60–61), 131672 (60–61), 131673 (58–61), 131674 (59–61), 131675 (59–61), 131677 (53–58), 131678 (53–61), 131679 (53–61), 131680 (53–61), 131681 (53–55), 131682 (53–57; 60–61), 131683 (53–57; depl 61), 131684 (53–61), 131685 (53–59), 131686 (53–57), 131687 (53–58; 60–61), 131688 (53–57; 59), 131689 (53–58), 131690 (58–59), 131691 (53–58; 60–61), 131692 (53–57), 131694 (58–60), 131695 (61), 131696 (57–58), 131697 (56–59), 131698 (56–61), 131699 (58–60), 131700 (56–58), 131701 (58–61), 131702 (depl 56–57; 60–61), 131704 (57–61), 131705 (57–59), 131706 (59–61), 131707 (55–56; 59–61), 131708 (55–56; 59), 131710 (59–60), 131711 (53; 55–61), 131712 (53; 55–58; 61), 131713 (60–61), 131714 (58–61), 131715 (58–61), 131717 (56–58; depl 60–61), 131719 (55–61).

VMR-253: *Operating Bases:* MCAS El Toro, California (1952–1953). MCAF Itami, Japan (1953–1955). MCAF Iwakuni, Japan (1955–1961). *R4Q Service:* 39 1952–1961. *R4Q-1*

Assignments: 126574 (53–59), 126575 (55–59), 126576 (55–58), 126577 (53–59), 126578 (55–59), 126579 (53–54), 126580 (54–59), 126581 (53–59), 126582 (51–53; 57–59), 128723 (52–54; 59), 128724 (52–55; 56–59), 128725 (52–59), 128726 (52–53; 55–56), 128727 (52–53), 128728 (52–59), 128729 (52–53), 128730 (52–53; 55–59), 128731 (52; 54–58), 128732 (52–54), 128733 (52–59), 128734 (52–54), 128735 (52–53), 128736 (52–54), 128737 (55–59), 128738 (53–59), 128740 (53–59), 128741 (53–58), 128743 (55–59), 128744 (53–56). *R4Q-2 Assignments:* 131664 (59–61), 131666 (59–61), 131670 (59–62), 131671 (59–61), 131677 (59–61), 131690 (59–61), 131693 (59–61), 131700 (59–61), 131708 (59–61), 131718 (59–61).

VMR-352: *Operating Base:* MCAS El Toro, California. *R4Q Service:* 5 1959–1961. *R4Q-2 Assignments:* 131685 (59–61), 131688 (59–61), 131695 (59–61), 131696 (59–61), 131712 (59–61).

VMR-353: *Operating Bases:* MCAS Miami, Florida (–31Aug59). MCAS Cherry Point, North Carolina (31Aug59–). *R4Q Service:* 47 1953–1963. *R4Q-2 Assignments:* 131662 (61–63), 131666 (57–59), 131669 (57–60; 61-62), 131672 (57–60), 131673 (61–62), 131674 (57–59), 131675 (61–62), 131676 (58–62), 131677 (58–59), 131678 (61–63), 131679 (61–62), 131680 (61–62), 131682 (57–60), 131683 (58–63), 131686 (57–61), 131687 (59–60; 61–62), 131689 (59–61), 131690 (53–58), 131691 (58–60), 131692 (57–61), 131693 (53–59), 131694 (53–58; 60–63), 131695 (53–57; 61–63), 131696 (53–56), 131697 (53–56), 131698 (53–56; 61–63), 131699 (53–58; 60–62), 131700 (53–56), 131701 (53–56; 61–63), 131702 (53–60; 61–62), 131703 (53–56), 131704 (53–57; 61–62), 131705 (53–56; 59–61), 131706 (56–58; 61–62), 131707 (61–63), 131708 (53–54; 57–58), 131709 (57–61), 131710 (60–62), 131711 (61–63), 131712 (61–63), 131713 (55–60), 131714 (61), 131715 (55–58; 61–62), 131716 (56), 131717 (58–62), 131718 (54–59), 131719 (61–63).

VMGR-152: *Operating Base:* MCAF Iwakuni, Japan. *R4Q-2 Assignment:* 131670 (62).

Two USMC Reserve C-119F Packets, BuNo. 131689 (10856) and BuNo. 131679 (10846), of VMR-234 NAS Glenview during July 1972 (National Archives 127-GG-112-A149867).

Section Two—Military Operators 131

Marine Air Reserve Training Detachments (MARTD)

VMR-216: *Operating Bases:* NAS Seattle, Washington (1962–28Apr70). NAS Whidbey Island, Washington (28Apr70–1972). *R4Q Service:* 17 1962–1972. *R4Q-2 Assignments:* 131664 (66–72), 131669 (70–72), 131670 (62–64; 70–72), 131673 (62–63; 69–72), 131677 (65–67), 131679 (62–65), 131688 (67–72), 131689 (64–66), 131690 (62–63; 67–70), 131691 (68–70), 131695 (67–72), 131699 (64–72), 131700 (63–unk), 131704 (62–64), 131708 (65–67), 131715 (64–66; 69–72), 131717 (66–72).

VMR-222: *Operating Base:* NAS Grosse Ile, Michigan. *R4Q Service:* 12 1961–1969. *R4Q-2 Assignments:* 131664 (63–unk; depl 69), 131669 (65–67), 131670 (64–69), 131673 (63–68), 131689 (61–64), 131690 (63–65), 131699 (62–64), 131700 (68–69), 131706 (66–69), 131708 (61–65; 68), 131717 (62), 131718 (61–63; 66–69).

VMR-234: *Operating Bases:* NAS Minneapolis, Minnesota (1947–11Feb70). (Renamed NAS Twin Cities 01Jul63). NAS Glenview, Illinois (11Feb70–1975). *R4Q Service:* 17 1961–1975. *R4Q-2 Assignments:* 131664 (61–63), 131669 (62–65; 67–69; 72–73), 131670 (depl 70; 72–73), 131677 (61–65; 68–75), 131679 (65–74), 131688 (depl 70; 72–73), 131689 (66–72), 131690 (61–62; 65–67; 70–74), 131691 (65–67; depl 70; 72–75), 131695 (depl 70; 72–73), 131700 (61–63; 66–67; 69–75), 131704 (64–67), 131706 (63–66; 69–75), 131708 (68–75), 131715 (62–64; 67–69), 131717 (62–66; depl 70), 131718 (63–66).

U.S. Navy

5th Naval District: NAS Norfolk, Virginia. R4Q-1: 124329 (58–59), 128723 (56–58).

6th Naval District: NAS Jacksonville, Florida. R4Q-1: 128735 (58–59).

11th Naval District: NAS North Island, California. R4Q-1: 128727 (58–59).

12th Naval District: NAS Alameda, California. R4Q-1: 128729 (58–59).

Fleet Aircraft Service Squadron 117 (FASRON-117): NAS Barber's Point, Hawaii. R4Q-1: 124330 (54).

Naval Air Advanced Training Center (NAATC): NAS Corpus Christi, Texas. R4Q-1: 124329 (56–58), 128739 (58–59), 128742 (55–56).

Naval Air Basic Training Center (NABTC): NAS Pensacola, Florida. R4Q-1: 124330 (57–59), 128739 (55–57).

Naval Air Test Center (NATC): NAS Patuxent River, Maryland. R4Q-1: 124324 (50), 124326 (50), 126574 (51–52), 126582 (54), 128735 (57). R4Q-2: 131665 (depl 54), 131700 (56).

UC-1: NAS Barber's Point, Hawaii. R4Q-2: 131714 (61–63).

VR-24: NAS Port Lyautey, Morocco. R4Q-2: 131665 (61–62), 131668 (60–62), 131691 (61–62), 131713 (61–62).

USMC Museum Allocations

10846 / 131679	1981–present	(U.S. Army) Don F. Pratt Memorial Museum, Fort Campbell, Kentucky
10893 / 131708	1989–2002	Marine Corps Air-Ground Museum, Quantico, Virginia (on loan at MCAS El Toro California)
	2002–present	Flying Leatherneck Aviation Museum, MCAS Miramar, California

South Vietnam

A total of 35 C-119G Flying Boxcars were operated by the Republic of Vietnam Air Force (VNAF) during the Vietnam War with initial deliveries of 18 from the USAF starting in January 1968 to the 413th Transport Squadron (*Red Dragons*) at Tan Son Nhut AB near Saigon. Four more were delivered over the next two years and as the Nixon Administration's Vietnamization policy advanced another 13 during 1972. A total of 16 are known to have been transferred back to the USAF in early 1973 with some then going to Taiwan. The remainder were earmarked for use as night patrol RC-119Gs under the 720th Reconnaissance Squadron but the sensing equipment was not delivered due to budget cuts. They were then assigned to supplement the newly formed C-130 Hercules squadrons. At least one C-119G s/n 53-8080 (msn: KF-183) is known to have been used by the newly formed Vietnam Peoples' Air Force (VPAF) after the fall of South Vietnam in April 1975.

As part of the Vietnamization program the USAF transferred the remaining 24 AC-119G gunships to the 819th Attack Squadron (*Black Dragons*) at Tan Son Nhut, the VNAF now trained and equipped enough to handle more sophisticated air operations. A further 22 AC-119K gunships were then transferred from November 1972 to the 821st Attack Squadron (*White Dragons*) at Tan Son Nhut AB with a detachment based at Da Nang.

The ultimate fate of the VNAF C/AC-119 fleet still in service when Saigon fell on April 30, 1975, are largely unknown. It is known that in the final days three AC-119s were damaged on the ground at Tan Son Nhut by shelling and one, s/n 52-5889, was lost in combat on the night of April 29, 1975. 37 AC-119G / K gunships plus eight standard C-119Gs are known to have been seized by the victorious North Vietnamese the following day but three AC-119s did escape to Thailand. Decades later a single AC-119K, s/n 53-7850, was discovered stored in Vietnam, it is to be restored for museum display.

413 Transport Squadron—35 C-119G

VNAF s/n	msn / USAF s/n	MAP Date	Disposition
517983	10722 / 51-7983	11Aug69	Back to USAF 04Jan73
533139	11150 / 53-3139	28Mar68	
533140	11151 / 53-3140	08Mar68	Taiwan 73
533141	11152 / 53-3141	08Jul72	
533146	11157 / 53-3146	27Jan68	Combat loss 18Feb68
533147	11158 / 53-3147	03Jul72	
533148	11159 / 53-3148	19Mar68	Back to USAF 15Jan73
533157	11168 / 53-3157	25Mar68	Back to USAF 06Feb73
533161	11172 / 53-3161	12Feb68	Back to USAF 12Jan73
533167	11178 / 53-3167	25Mar68	
533173	11184 / 53-3173	31Jul72	
533175	11186 / 53-3175	06Feb68	Back to USAF 07Jan73
533185	11196 / 53-3185	03Jul72	
533194	11205 / 53-3194	23Jun72	Back to USAF 08Mar73
533198	11211 / 53-3198	27Jan68	Back to USAF 11Jan73
533202	11215 / 53-3202	14Mar68	Back to USAF 05Jan73
533203	11216 / 53-3203	29Mar68	Back to USAF 06Feb73
533218	11234 / 53-3218	25Mar68	Back to USAF 11Jan73
533220	11236 / 53-3220	28Jul68	Back to USAF 10Jan73
537840	11257 / 53-7840	01Jul72	
537842	11259 / 53-7842	01Jul72	Back to USAF 09Mar73
537873	11298 / 53-7873	25Mar68	

Section Two—Military Operators 133

538077	KF-180 / 53-8077	31Jul72	
538080	KF-183 / 53-8080	08Sep71	VPAF 918 TS Reg 75
538088	KF-191 / 53-8088	24Mar68	Back to USAF 10Jan73
538104	KF-207 / 53-8104	03Jul72	Back to USAF 11Mar73
538104	KF-209 / 53-8106	01Jul72	
538109	KF-212 / 53-8109	08Jul72	
538112	KF-215 / 53-8112	16Jul72	
538117	KF-220 / 53-8117	31Jul72	
538124	KF-227 / 53-8124	22Mar68	
538133	KF-236 / 53-8133	25Mar68	Back to USAF 04Jan73
538139	KF-242 / 53-8139	08Sep71	
538147	KF-250 / 53-8147	26Mar68	Back to USAF 05Jan73
538156	KF-259 / 53-8156	02Sep71	

819 Attack Squadron—24 AC-119G

VNAF s/n	msn / USAF s/n	MAP Date	Disposition
525892	11059 / 52-5892	20Aug71	
525898	11065 / 52-5898	24Aug71	
525905	11072 / 52-5905	31Aug71	
525925	11104 / 52-5925	27Aug71	
525927	11106 / 52-5927	25Aug71	
525938	11125 / 52-5938	04Sep71	
525942	11129 / 52-5942	06Sep71	
533136	11147 / 53-3136	07Sep71	
533145	11156 / 53-3145	26Aug71	
533170	11181 / 53-3170	08Sep71	
533178	11189 / 53-3178	21Aug71	
533189	11200 / 53-3189	18Aug71	
533192	11203 / 53-3192	19Aug71	
533205	11218 / 53-3205	17Aug71	
537833	11250 / 53-7833	30Aug71	
537848	11269 / 53-7848	02Sep71	
537851	11272 / 53-7851	01Sep71	
537852	11273 / 53-7852	30Aug71	
538069	KF-172 / 53-8069	20Aug71	
538089	KF-192 / 53-8089	28Aug71	
538114	KF-217 / 53-8114	06Sep71	
538115	KF-218 / 53-8115	03Sep71	
538123	KF-226 / 53-8123	23Aug71	
538131	KF-234 / 53-8131	24Aug71	

821 Attack Squadron—22 AC-119K

VNAF s/n	msn / USAF s/n	MAP Date	Disposition
525864	11023 / 52-5864	10Nov72	
525889	11056 / 52-5889	10Nov72	Combat loss 29Apr75
525910	11089 / 52-5910	06Nov72	
525911	11090 / 52-5911	06Nov72	
525926	11105 / 52-5926	10Nov72	
525940	11127 / 52-5940	10Nov72	
525945	11132 / 52-5945	10Nov72	
529982	11143 / 52-9982	10Nov72	
533154	11165 / 53-3154	10Nov72	
533187	11198 / 53-3187	31Jan73	
533197	11210 / 53-3197	06Nov72	

VNAF s/n	msn / USAF s/n	MAP Date	Disposition
533211	11227 / 53-3211	31Jan73	
537830	11247 / 53-7830	10Nov72	
537831	11248 / 53-7831	31Jan73	
537839	11256 / 53-7839	31Jan73	W/o 01Mar73
537850	11271 / 53-7850	10Nov72	Extant Vietnam
537877	11306 / 53-7877	31Oct72	
537879	11308 / 53-7879	10Nov72	
537883	11317 / 53-7883	10Nov72	
538121	KF-224 / 53-8121	10Nov72	
538145	KF-248 / 53-8145	31Jan73	
538148	KF-251 / 53-8148	10Nov72	

Section Three

Civil Operators & Owners

Primarily listed in this section are civil commercial operators, such as airlines and freight companies, who flew the C-119 Flying Boxcar as a means of business and revenue. Also listed are aircraft dealers, scrap dealers, financial institutions if named on FAA Bills of Sale, civilian public and private museums, private individuals and any aircraft used for commercial gimmick purposes such as restaurants and motels etc. Private individuals are listed by surname but companies that carry the owners name are listed as worded in the title.

Equatorial Guinea

BATA International Airlines / Equatorial Guinea

African operator acquired one ex-BAF C-119G as 3C-ABA (10689 / 51-2700) for cargo operations. The aircraft was flown to the UK from Belgium but never left Manston Airport in Great Britain. Ownership was from 1976 to 1984.

Great Britain

Aces High Ltd. / Surrey

Well-known warbird operator. Owned one C-119G as G-BLSW (10689 / 51-2700) from 1984 to 1985.

The Wings Museum / Surrey

Owns the nose section of C-119G (10689 / 51-2700) acquired for display in 2007. It has since been sold to an unknown buyer.

Italy

Ditellandia Air Park / Castel Volturno

One ex-AMI C-119J (11235 / MM53-3219) from 1994 to 2004. Aircraft presumed scrapped after the park was closed.

Friulano Aero Club / Campoformido

One C-119G (11030 / MM52-6029) on display from 1978 to 2004.

Piana delle Orme Museum / Latina

One static display C-119G (KF249 / MM53-8146).

Pisa International Airport / Pisa

One static display C-119G (11213 / MM53-3200).

REPUBLIC OF CYPRUS

One unknown operator registered a C-119F (10885 / N3267U) as 5B-CFG sometime after 1988 for United Nations lease operations in Africa. The aircraft was abandoned in Kenya around 1996 with the reg never apparently marked on the aircraft.

UNITED STATES OF AMERICA

There were a total of 65 C-119 Flying Boxcars registered with U.S. N-number registrations which were operated in a variety of commercial freight, agricultural, fire-bombing and other roles. The following list breaks down the initial civilian purchaser against the immediate former military operator:

Purchaser	Qty	Former Military
Aircraft International Associates	29	RCAF C-119G.
Dross Metals	18	8 USMC, 10 USAF.
Aero Union Corp.	11	C-119C.
Starbird, Inc.	4	C-119L.
Jack R. Munson	1	USAF C-119G, sold to H&P Aviation.
City of Pueblo	1	USMC C-119F for museum display.
BATA Intl.	1	Belgian C-119G.

Out of the total of 65, only 35 were actually operated in civil operations on a regular basis under a series of FAA restricted type certificates. 24 were briefly flown or registered only as a spares source and the remaining six were exported, in this case to the Moroccan Air Force. Notably, the former RCAF fleet played a major role in North American C-119 civil service, many taking up USFS airtanker assignments.

Summary of U.S. Civil Registered C-119 Aircraft

msn / s/n	Type	Reg	Service	Current Status
10304 / 48-322	C-119C Jet-Pak	N13745	1967–1987	Stored
10334 / 48-352	C-119C Jet-Pak	N13746	1967–1987	Museum
10369 / 49-132	C-119C Jet-Pak	N13743	1967–1987	Museum
10370 / 49-133	C-119C	N9955F	-	Broken up spares
10431 / 49-194	C-119C Jet-Pak	N13742	1967–1981	W/o 08Jul81
10436 / 49-199	C-119C Jet-Pak	N13744	1967–1987	Museum
10594 / 51-2605	C-119C	N9966F	-	Broken up spares
10597 / 51-2608	C-119C	N9959F	-	Broken up spares
10599 / 51-2610	C-119C	N9961F	-	Broken up spares
10607 / 51-2618	C-119C	N9960F	-	Broken up spares
10676 / (RCAF) 22101	C-119G	N15505	-	Broken up spares
10678 / (RCAF) 22103	C-119G	N8092	-	Spares
10689 / 51-2700	C-119G	N2700	1975–1987	Broken up stored
10735 / (RCAF) 22104	C-119G	N3558	-	Scrapped
10736 / (RCAF) 22105	C-119G	N15506	-	Museum
10737 / (RCAF) 22106	C-119G-3E	N3003	1967–1993	Derelict
10738 / (RCAF) 22107	C-119G	N966S	1967–1988	Museum

Section Three—Civil Operators & Owners

msn / s/n	Type	Reg	Service	Current Status
10773 / (RCAF) 22108	C-119G	N5215R	-	Museum
10774 / (RCAF) 22109	C-119G	N15504	-	Exported Morocco
10775 / (RCAF) 22110	C-119G-3E	N15509	1967–1984	W/o 21Apr84
10776 / (RCAF) 22111	C-119G-3E	N8093	1967–1997	Museum
10811 / 51-8263	C-119C	N9956F	-	Broken up spares
10823 / (RCAF) 22112	C-119G	N964S	-	Exported Morocco
10824 / (RCAF) 22113	C-119G-3E	N3935	1967–1987	Museum
10825 / (RCAF) 22114	C-119G	N15502	-	Museum
10831 / BuNo. 131664	C-119F	N131DM	-	Broken up spares
10836 / BuNo. 131669	C-119F	N13626	1980–1983	W/o 08May83
10840 / BuNo. 131673	C-119F Jet-Pak	N1394N	1980–present	Stored Alaska
10844 / BuNo. 131677	C-119F	N49543, N175ML	1980–1994	Museum
10855 / BuNo. 131688	C-119F	N99574	-	Museum
10859 / (RCAF) 22115	C-119G-3E	N8682	1967–1981	W/o 27Jun81
10860 / (RCAF) 22116	C-119G	N5217R	-	Museum
10861 / (RCAF) 22117	C-119G	N15507	-	Exported Morocco
10870 / (RCAF) 22118	C-119G-3E	N3559	1967–1987	Museum
10872 / (RCAF) 22120	C-119G-3E	N961S	1967–1987	Unknown
10875 / BuNo. 131690	C-119F	N4234S	1980–1982	Scrapped
10880 / BuNo. 131695	C-119F	N8501W	1980–present	Stored Alaska
10885 / BuNo. 131700	C-119F	N3267U	1980–1996	Scrapped
10893 / BuNo. 131708	C-119F	N7051U	-	Museum
10905 / (RCAF) 22121	C-119G	N962S	-	Exported Morocco
10906 / (RCAF) 22122	C-119G	N8091	-	Museum
10907 / (RCAF) 22123	C-119G-3E	N8832	1966–1985	W/o 17Aug85
10943 / (RCAF) 22126	C-119G	N965S	-	Exported Morocco
10954 / (RCAF) 22129	C-119G	N963S	-	Exported Morocco
10955 / (RCAF) 22130	C-119G-3E	N15501	1967–present	Museum flyable
10956 / (RCAF) 22131	C-119G-3E	N5216R	1967–1987	Museum
10957 / (RCAF) 22132	C-119G-3E	N3560	1967–1978	W/o 10Jun78
10992 / (RCAF) 22133	C-119G-3E	N383S	1967–1979	W/o 08Jun79
10993 / (RCAF) 22134	C-119G	N15508	1967–1987	Museum
10994 / (RCAF) 22135	C-119G	N8094	-	Derelict
11005 / 52-5846	C-119G-3E	N48076	1975–1987	W/o 16Sep87
11155 / 53-3144	C-119L	N37484, N8512N	1980–1987	Museum
11185 / 53-3174	C-119L	N55795	1980	Unknown
11219 / 53-3206	C-119L	N4999K, N90268	1978–1980	W/o 05Jul80
11232 / 53-3216	C-119L Jet-Pak	N8504Y	1979–unk	Unknown
11253 / 53-7836	C-119L Jet-Pak	N8504Z	1979–present	Stored Alaska
11270 / 53-7849	C-119L	N1040E	1978–1979	W/o 08Jul79
11318 / 53-7884	C-119L	N37483, N8512K	1979–1981	W/o 01Oct81
KF-176 / 53-8073	C-119L	N9027K	1978–unk	Museum
KF-179 / 53-8076	C-119L	N8505A	1979–1992	Derelict
KF-230 / 53-8127	C-119L	N4999N, N90269	1978–1979	W/o 06Jul79
KF-245 / 53-8142	C-119L Jet-Pak	N8504X	1979–1987	W/o 13May87
KF-253 / 53-8150	C-119L	N37636	1979–1993	Derelict
KF-256 / 53-8153	C-119L	N8504W	1979–1981	W/o 07Sep81
KF-257 / 53-8154	C-119L	N90267	1978–1988	Broken up spares

The following companies are known U.S. civil and commercial operators of the C-119 as recorded in FAA files. Also included are non-direct commercial ownership entities as aircraft dealers, financial institutions, private enterprises and individuals and non-federal museums. Not included are federal museums or entities that have a C-119 on a "loan basis" from the military. For these listings see: USAF or USMC Museum Allocations under United States of America in Military Operators.

ABBAS International, Inc. / Dallas, Texas

Operated one C-119F as N3267U (10885) from 1983 to 1987.

Aero Union Corp. / Chico, California

Originally founded in 1959 as Western Air Industries, the name was changed to Aero Union Corp. in 1961 and went on to become the largest U.S. airtanker operator leading the way in airtanker conversions, operations and training. Five C-119C Boxcars were purchased in 1967 and converted to airtankers with FAA Type Certificate A21WE being issued to the company for fire-bombing duties, six additional C variants were later purchased as spares. All C-119 aircraft were sold to **Hemet Valley Flying Service Co.** by 1976.

N13742	10431 / 49-194	1967–1976	C-119C.
N13743	10369 / 49-132	1967–1975	C-119C Jet-Pak.
N13744	10436 / 49-199	1967–1976	C-119C Jet-Pak.
N13745	10304 / 48-322	1967–1975	C-119C Jet-Pak.
N13746	10334 / 48-352	1967–1975	C-119C.
—	10370 / 49-133	1969–1976	Spares airframe.
—	10811 / 51-8263	1969–1976	Spares airframe.
—	10597 / 51-2608	1969–1976	Spares airframe.
—	10607 / 51-2618	1969–1976	Spares airframe.
—	10599 / 51-2610	1969–1976	Spares airframe.
—	10594 / 51-2505	1969–1976	Spares airframe.

AFSCO, Inc.

Aaron Ferer & Sons Co. Major U.S. scrap processing company. *See under:* **Aircraft International Associates**

Aircraft International Associates / Omaha, Nebraska

An aircraft dealership initially set up in Dallas, Texas, as a joint venture partnership between three companies: **AFSCO, Inc.** (Harvey D. Ferer); **Frank Shelley Electronics, Inc.** (G. Wayne Shelley) and **Renstrom Enterprises** (Carl W. Renstrom). They purchased the 31 retired RCAF Boxcars during 1966–1967 for resale to commercial operators. RCAF records show initial sales to partner Frank Shelley Electronics, Inc. in 1966 but FAA documents show sales corrected to Aircraft International Associates as a whole during 1967. Most of the fleet was ferried into storage in Lincoln, Nebraska, except for N383S, which was converted into the unique C-119G STOLmaster in 1967. Two ex-RCAF airframes (22102, 22119) appear to have been purchased but broken up for spares; one (22104) was also scrapped in Canada; six were exported to the Moroccan AF and 22 had been sold to Hawkins & Powers Aviation by 1975 concluding Aircraft International's involvement with the C-119.

N15501	10955 / (RCAF) 22130	1967–1975	To H&P Aviation.
N15502	10825 / (RCAF) 22114	1967–1975	To H&P Aviation.
N15504	10774 / (RCAF) 22109	1967–1968	Exported to Morocco.
N15505	10676 / (RCAF) 22101	1967–1975	To H&P Aviation.
N15506	10736 / (RCAF) 22105	1966–1975	To H&P Aviation.
N15507	10861 / (RCAF) 22117	1967–1968	Exported to Morocco.
N15508	10993 / (RCAF) 22134	1967–1975	To H&P Aviation.
N15509	10775 / (RCAF) 22109	1967–1975	To H&P Aviation.
N383S	10992 / (RCAF) 22133	1967–1975	To H&P Aviation.

N3003	10737 / (RCAF) 22106	1967–1975	To H&P Aviation.
N3558	10735 / (RCAF) 22104	1967–1975	Scrapped.
N3559	10870 / (RCAF) 22118	1967–1975	To H&P Aviation.
N3560	10957 / (RCAF) 22132	1967–1972	To H&P Aviation.
N3935	10824 / (RCAF) 22113	1967–1972	To H&P Aviation.
N5215R	10773 / (RCAF) 22108	1967–1975	To H&P Aviation.
N5216R	10956 / (RCAF) 22131	1967–1974	To H&P Aviation.
N5217R	10860 / (RCAF) 22116	1967–1975	To H&P Aviation.
N8091	10906 / (RCAF) 22122	1967–1975	To H&P Aviation.
N8092	10678 / (RCAF) 22103	1967–1975	To H&P Aviation.
N8093	10776 / (RCAF) 22111	1967–1975	To H&P Aviation.
N8094	10994 / (RCAF) 22135	1967–1975	To H&P Aviation.
N8682	10859 / (RCAF) 22115	1967–1971	To H&P Aviation.
N8832	10907 / (RCAF) 22123	1966–1975	To H&P Aviation.
N961S	10872 / (RCAF) 22120	1967–1975	To H&P Aviation.
N962S	10905 / (RCAF) 22121	1967–1968	Exported to Morocco.
N963S	10954 / (RCAF) 22129	1967–1968	Exported to Morocco.
N964S	10823 / (RCAF) 22112	1967–1968	Exported to Morocco.
N965S	10943 / (RCAF) 22126	1967–1968	Exported to Morocco.
N966S	10738 / (RCAF) 22107	1967–1975	To H&P Aviation.
—	10677 / (RCAF) 22102	1967	Broken up spares.
—	10871 / (RCAF) 22119	1967	Broken up spares.

Alaska Aircraft Leasing, Inc. / Anchorage, Alaska

An Alaska based aircraft leasing company that went bankrupt in 1992.

N8504X	KF-245 / 53-8142	1987	C-119L Jet-Pak. W/o 13May87.
N8504Z	11253 / 53-7836	1987–1992	C-119L Jet-Pak.
N8505A	KF-179 / 53-8076	1987–1992	
N9027K	KF-176 / 53-8073	1987–1992	

Alaska Aviation Museum / Anchorage, Alaska

One C-119G N9027K (KF-176) stored for future display.

Alaska Commercial Fishing and Agriculture Bank / Anchorage, Alaska

Lending bank who repossessed three C-119s from J.D. **Gifford & Associates, Inc.** and two from **Starbird, Inc.** after these companies failed. All aircraft were registered in the banks name for resale.

N8504X	KF-245 / 53-8142	1984–1985	To Northern Pacific Transport.
N8504Y	11232 / 53-3216	1984–1985	To Stebbins & Ambler.
N8505A	KF-179 / 53-8076	1984–1985	To Northern Pacific Transport.
N9027K	KF-176 / 53-8073	1984–1985	To Northern Pacific Transport.
N90267	KF-257 / 53-8154	1984–1985 &	To Stebbins & Ambler.
		1989–1990s	To H&P Aviation.

American Air Freight Co. / Laredo, Texas

Owned one C-119L N37484 (11155) from 1986 to 1987.

Anchorage Flight, Inc. / Anchorage, Alaska

No details.

N8504Z	11253 / 53-7836	1992–1994	C-119L Jet-Pak.
N9027K	KF-176 / 53-8073	1992–1993	

Arbor Air, Inc. / Ann Arbor, Michigan

President was **William Waara** who was also involved with **Michigan Aerial Applicator**. The vice-president was apparently **Louis P. Minkoff**. *See also:* **Michigan Aerial Applicator, Louis P. Minkoff, William Waara**

N37483	11318 / 53-7884	1980–1981
N37484	11155 / 53-3144	1980–1981

Atterbury-Bakalar Air Museum / Columbus, Indiana

One C-119G N3003 (10737) acquired in 2019 from Greybull for static display outdoors.

B&G Industries, LLC / Greybull, Wyoming

Aircraft structural repair, fabrication and servicing company based at **H&P Aviation's** former facility at South Big Horn County Airport near Greybull, Wyoming. One C-119G N3003 (10737) was acquired from auction in 2006 and sold to a museum in 2019.

Ball, Gerald C. / Anchorage, Alaska

No details. One C-119G N8504Z (11253) from 1985 to 1986.

Battle Mountain Air Museum / Battle Mountain Airport, Nevada

One C-119G (10956 / N5216R) for static display from 1991 to 2019. Parts from an abandoned **H&P Aviation** C-119G N8832 were used to help in the static restoration. The museum closed at an unknown point and N5216R became derelict until acquired by **Rolling Boxcar, Inc.** in 2019.

Brady, David / Cartersville, Georgia

Warbird operator. One C-119L N37636 (KF-253) from 1989 to 1990.

Brooks Fuel / Fairbanks, Alaska

One C-119L N9027K (KF-176) from 1996 to 2017.

Bud's Flying Service Ltd. / Rising City, Nebraska

Operated by Alvin Gruenewald. Bud's held FAA Type Certificate A35CE for their single C-119L N37484 (11155) registered under their sister company **El Marc Air** from 1981 to 1982 and directly under Bud's from 1982 to 1983. The aircraft was managed and maintained at Laredo Airport in Texas by Gene Williams, owner of **Texas Aerial Applicators, Inc.** *See also:* **El Marc Air**

The City of Pueblo / Pueblo, Colorado

See under: **Pueblo Weisbrod Aircraft Museum**

Central Air Service, Inc. / Tucson, Arizona

Owned C-119L N37484 (11155) circa 1987 before it was acquired by the USAF Museum.

Chandalar Development Associates / Bellevue, Washington

One C-119F as N13626 (10836) from 1981 to 1982.

Classic Air Transport, Inc. / Ridgewood, New Jersey

One C-119L N9027K (KF-176) from 1993 to 1996.

Comutair / Gering, Nebraska

Operated one C-119F as N3267U (10885) from 1987. Used for a period by a Cyprus operator who leased it to the United Nations in Africa.

Consolidated Aviation Enterprises, Inc.
See under: **Downey, John P.**

Cottington, Jim [sic]
Real name Howard F. Cottingham from Dallas, Texas. Veteran C-119 pilot and Dallas businessman. Appears to have been acting on behalf of La Mesa Air Charter Service of Laredo, Texas, for the purchase of C-119G as N37483 (11318). "Jim Cottington" appears on the FAA BofS document presumably when he picked up the aircraft off El Marc Air in Nebraska.

D&G, Inc. / Greybull, Wyoming
Created as an aircraft leasing company by Hawkins & Powers Aviation, Inc. founders Dan Hawkins & Gene Powers and run in conjunction with their H&P Aviation operations. *See also:* **Hawkins & Powers Aviation, Inc.**

N3003	10737 / (RCAF) 22106	1993–2005
N37636	KF-253 / 53-8150	1993–2005
N8093	10776 / (RCAF) 22111	1993–2005
N8094	10994 / (RCAF) 22135	1993–2005
N8505A	KF-179 / 53-8076	1993–2005

Delta Associates, Inc. / Anchorage, Alaska
Company appears to have become **Delta Leasing** during 1987 and was run by Terry Luther and Ted King. Delta leased to Alaskan air freight operators such as **Stebbins & Ambler Air Transport**.

N1394N	10840 / BuNo. 131673	1985–1989
N8501W	10880 / BuNo. 131695	1985–1998

D.M.I. Aviation, Inc.
See under: **Dross Metals, Inc.**

Downey, John P. (d/b/a Consolidated Aviation Enterprises, Inc.) / East Middlebury, Vermont
One C-119G (10689) purchased from Aces High Ltd. The aircraft was registered in the U.S. in 1985 as N2700 but it remained in Great Britain eventually becoming derelict until broken up in 1994.

Dross Metals, Inc. / Tucson, Arizona
One of the largest civilian aviation-based companies located around the perimeter of the giant military AMARC facility who acquire ex-military aircraft for recycling, restoration or parts. Dross Metals purchased and sold 18 C-119 Boxcars from 1978 to 1987. As well as providing civil conversions, upgrades and parts they held FAA Type Certificate A8NW pertaining to the commercial operation of former USMC C-119Fs. Dross was renamed as **D.M.I. Aviation, Inc.** in 1982 and survives today as ARMair on the same site.

N131DM	10831 / BuNo. 131664	1981–2013	Broken up as a workshop.
N1394N	10840 / BuNo. 131673	1980–1982	To Elling Halvorson.
N13626	10836 / BuNo. 131669	1980–1981	To Chandalar Dev. Assoc.
N3267U	10885 / BuNo. 131700	1980–1983 & 1987	To ABBAS Intl. To Comutair.
N37483	11318 / 53-7884	1978–1979	To Michigan Aerial Applicator.
N37484	11155 / 53-3144	1978–1979	To Michigan Aerial Applicator.
N37636	KF-253 / 53-8150	1979–1980 & 1982–1986	To Raven Intl. To Mike Ivers.

N49543	10844 / BuNo. 131677	1984	To Marine Lumber.
N55795	11185 / 53-3174	1980	To Juan Perez.
—	10837 / BuNo. 131670	1981	Scrapped.
—	10875 / BuNo. 131690	1980–1981	To Tobacco Road Farms as N4234S.
—	10880 / BuNo. 131695	1980–1981	To Pacific Intl. Foods as N8501W.
—	10884 / BuNo. 131699	1980	Scrapped.
—	10893 / BuNo. 131708	1980–1988	To H&P Aviation as N7051U.
—	10900 / BuNo. 131715	1980	Noted derelict 2011, scrapped.
—	11148 / 53-3137	1979	Scrapped.
—	11232 / 53-3216	1979–1980	To J.D. Gifford as N8504Y.
—	11253 / 53-7836	1979–1980	To J.D. Gifford as N8504Z.
—	11270 / 53-7849	1978–1979	To Lambeth Aircraft as N1040E.
—	KF-179 / 53-8076	1979–1980	To J.D. Gifford as N8505A.
—	KF-245 / 53-8142	1979–1980	To J.D. Gifford as N8504X.
—	KF-256 / 53-8153	1979–1980	To J.D. Gifford as N8504W.

Elling Halvorson, Inc. / Redmond, Washington State

A joint venture company.

N1394N	10840 / BuNo. 131673	1982–1985
N8501W	10880 / BuNo. 131695	1982–1985

El Marc Air / Columbus, Nebraska

Air freight company operated by Alvin Gruenewald who also operated **Bud's Flying Service Ltd.** to which N37484 was conveyed in 1982. *See also:* **Bud's Flying Service Ltd.**

N37483	11318 / 53-7884	1981
N37484	11155 / 53-3144	1981–1982

Everts Air Fuel, Inc. / Fairbanks, Alaska

One C-119L Jet-Pak N8504Z (11253) from 1994. Presently stored at Fairbanks Intl. Airport.

F. H. Chew Sales Co. / Phoenix, Arizona

No details. Possibly an aircraft broker for **Aero Union Corp.**

N13742	10431 / 49-194	1967
N13743	10369 / 49-132	1967
N13744	10436 / 49-199	1967

The First National Bank of Anchorage / Anchorage, Alaska

Lending bank who repossessed three C-119s from **Alaska Aircraft Leasing, Inc.** The aircraft were registered in the banks name until they could be resold.

N8504Z	11253 / 53-7836	1992	To H&P Aviation.
N8505A	KF-179 / 53-8076	1992	To H&P Aviation.
N9027K	KF-176 / 53-8073	1992	To Anchorage Flight.

Flying J Ranch Airpark / Pima, Arizona

Derelict C-119G (KF-177) acquired from Tucson and stored on private property. *See also:* **Jenkins, Howard E.**

Frank Shelley Electronics, Inc. / Los Angeles, California

The C-119G STOLmaster N383S (10992) was registered during 1967. *See also:* **Aircraft International Associates**

Grantham, William A. & Sergio A. Tomassoni / Buckeye, Arizona

One C-119G-3E N15509 (10775) from 1977 to 1980. Company later renamed as T&G Aviation, Inc. moving to Chandler, Arizona. *See also:* **Tomassoni, Sergio A. & T&G Aviation, Inc.**

Hagerstown Aviation Museum / Hagerstown, Maryland

One C-119G (10776 / N8093) for static display from 2006.

Hamilton, Rudy / Shageluk, Alaska

Private Shageluk resident who purchased the fuselage of N8504X (KF-245) for use as a workshop after its 1987 crash landing.

Hawkins & Powers Aviation, Inc. / Greybull, Wyoming

Often abbreviated as: H&P. Established in 1969 by Dan Hawkins and Gene Powers, this company became a legend among the aviation community for their wide use of piston-engined aircraft used in freight, agricultural, fisheries and fire-bombing. In particular, H&P are well known for their use of the C-119G Flying Boxcar with a total of 28 of the type registered in their name—22 of these were acquired in a single deal for **Aircraft International Associates** former RCAF C-119G fleet. The C-119G-3E Jet-Pak configuration was an H&P innovation based on the Steward-Davis Jet-Pak with FAA Type Certificate A24WE issued to the company in 1972 for operation. 13 were flown in this configuration mostly for firefighting. Being the most prolific C-119 operator also meant H&P suffered the most losses with five C-119G-3Es being lost to accidents in their 17 years of operation. The last crash, N48076 on 15Sep87, finished C-119 operations in the fire-bombing role due to structural concerns. H&P's Boxcar fleet was put out to pasture with several later becoming involved in the *U.S. Air Tanker Scandal* of the 1990s. The company closed for business in 2005, their remaining aviation assets were auctioned off in 2006 by **The Pride Capital Group, LLC**. *See also:* **D&G, Inc.**

N15501	10955 / (RCAF) 22130	1975–2005	C-119G-3E.
N15502	10825 / (RCAF) 22114	1975–1988	Spares airframe.
N15505	10676 / (RCAF) 22101	1975–1995	Spares airframe.
N15506	10736 / (RCAF) 22105	1975–1988	Spares airframe.
N15508	10993 / (RCAF) 22134	1975–1988	Not cvtd.
N15509	10775 / (RCAF) 22110	1975	C-119G-3E.
N383S	10992 / (RCAF) 22133	1975–1979	C-119G-3E. W/o 08Jun79.
N3003	10737 / (RCAF) 22106	1975–1993	C-119G-3E.
N3559	10870 / (RCAF) 22118	1975–1991	C-119G-3E.
N3560	10957 / (RCAF) 22132	1972–1978	C-119G-3E. W/o 10Jun78.
N3935	10824 / (RCAF) 22113	1972–2006	C-119G-3E.
N37636	KF-253 / 53-8150	1990–1993	Stored.
N48076	11005 / 52-5846	1975–1987	C-119G-3E. W/o 16Sep87.
N5215R	10773 / (RCAF) 22108	1975–unk	Spares airframe.
N5216R	10956 / (RCAF) 22131	1974–1991	C-119G-3E.
N5217R	10860 / (RCAF) 22116	1975–1992	Spares airframe.
N7051U	10893 / BuNo. 131708	1988–1989	Stored.
N8091	10906 / (RCAF) 22122	1975–1988	Spares airframe.
N8092	10678 / (RCAF) 22103	1975–1991	Spares airframe.
N8093	10776 / (RCAF) 22111	1975–1993	C-119G-3E.
N8094	10994 / (RCAF) 22135	1975–1993	Spares airframe.
N8504Z	11253 / 53-7836	1992	C-119L Jet-Pak. Stored.
N8505A	KF-179 / 53-8076	1992–1993	Stored.
N8682	10859 / (RCAF) 22115	1971–1981	C-119G-3E. W/o 27Jun81.

N8832	10907 / (RCAF) 22123	1975–1985	C-119G-3E. W/o 17Aug85.
N961S	10872 / (RCAF) 22120	1975–unk	C-119G-3E.
N966S	10738 / (RCAF) 22107	1975–1988	Not cvtd.
—	KF-257 / 53-8154	1990s–2006	Ex-N90267. Stored.

Heims Seafoods, Inc. / Myrtle Point, Oregon

Two C-119s acquired for the transport of sea foods.

N3003	10737 / (RCAF) 22106	1980–1982
N8093	10776 / (RCAF) 22111	1980

Hemet Valley Flying Service Co. (HVFS) / Hemet, California

Founded as WHV Flying Service in 1952, the small crop-dusting business expanded into airtanking operations during 1957. HVFS acquired five C-119C airtankers from **Aero Union Corp.** starting in 1975 putting all into fire service with Jet-Paks under FAA Type Certificate A21WE. The purchase included all spares and six spare C-119C stored at Tucson. They were issued with N-numbers in the hope of restoring them to airworthy condition but were soon assessed as beyond economic repair and instead dismantled and shipped to Hemet on trucks. The C-119s provided excellent firefighting service until withdrawn from use in 1987 and stored. HVFS ceased operations in 1996.

N13742	10431 / 49-194	1976–1981	C-119C Jet-Pak. W/o 08Jul81.
N13743	10369 / 49-132	1975–1991	C-119C Jet-Pak.
N13744	10436 / 49-199	1976–1992	C-119C Jet-Pak.
N13745	10304 / 48-322	1975–1992	C-119C Jet-Pak.
N13746	10334 / 48-352	1975–1992	C-119C Jet-Pak.
N9955F	10370 / 49-133	1976–1980s	Spares airframe.
N9956F	10811 / 51-8263	1976–1980s	Spares airframe.
N9959F	10597 / 51-2608	1976–1980s	Spares airframe.
N9960F	10607 / 51-2618	1976–1980s	Spares airframe.
N9961F	10599 / 51-2610	1976–1980s	Spares airframe.
N9966F	10594 / 51-2505	1976–1980s	Spares airframe.

Ivers, Mike / Yakutat, Alaska

One C-119L N37636 (KF253) from 1986 to 1989.

J. D. Gifford & Associates, Inc. / Anchorage, Alaska

Also known as **Gifford Aviation, Inc.**, an Alaskan air freight operator. They appear to have leased at least one of **Starbird, Inc.'s** C-119s.

N8504W	KF-256 / 53-8153	1980–1981	W/o 07Sep81.
N8504X	KF-245 / 53-8142	1980–1984	C-119L Jet-Pak.
N8504Y	11232 / 53-3216	1980–1984	C-119L Jet-Pak.
N8504Z	11253 / 53-7836	1980–1985	
N8505A	KF-179 / 53-8076	1980–1984	

Jenkins, Howard E. (Flying J Ranch Airpark) / Pima, Arizona

A local resident from Safford Arizona purchased C-119G KF-177 (53-8074) from a Tucson scrapyard in 1990 for display on his private ranch airstrip that is sign-posted as having being established in 1973.

Kolar, Inc. / Tucson, Arizona

Local AMARC scrap and aircraft dealer. At the time of this writing at least 70 C-119s are known to have been acquired from MASDC Jan-Apr76.

51-2662, 51-2663, 51-2664, 51-2668, 51-2670, 51-2674, 51-2676, 51-2678, 51-2684, 51-2685, 51-2692, 51-2693, 51-2699, 51-2702, 51-7972, 51-7975, 51-7982, 51-7988, 51-7991, 51-7997, 51-7998, 51-7999, 51-8003, 51-8012, 51-8022, 51-8023, 51-8053, 51-8054, 51-8055, 51-8059, 51-8061, 51-8063, 51-8069, 51-8072, 51-8078, 51-8082, 51-8097, 51-8103, 51-8107, 51-8111, 51-8135, 51-8145, 51-8146, 51-8168, 52-5840, 52-5847, 52-5871, 52-5872, 52-5873, 52-5876, 52-5883, 52-5901, 52-5913, 52-5922, 52-5933, 52-5948, 52-5950, 52-5953, 52-9981, 53-3151, 53-3162, 53-3168, 53-3190, 53-7869, 53-8075. BuNo. 131662, 131675, 131678, 131689, 131717.

La Mesa Air Charter Service / Laredo, Texas
See under: **Cottington, Jim**

Lambeth Aircraft Corp. / Phoenix, Arizona
Acquired C-119L N1040E (11270) in 1979, however it crashed outside Phoenix, Arizona, within weeks of being purchased.

Lauridsen, Hans O. / Phoenix, Arizona
Private warbird collector. Purchased N15501 (10955) Aug 2006 for display in the **Lauridsen Aviation Museum** at Buckeye Municipal Airport, Arizona.

Lee-Argyle Corp. / Miami, Florida
The C-119G STOLmaster N383S (10992) was registered to this company during 1969 probably in connection with Aircraft International's Latin American sales tour.

Marine Lumber, Inc. / Nantucket, Massachusetts
Building company acquired one C-119F N175ML (10844) from 1984 to 1994 for the carriage of staff and materials to and from Nantucket Island.

Michigan Aerial Applicator / Dothan, Alabama
President was one **William Waara** who also ran **Arbor Air**. Two C-119L aircraft were owned from 1979 to 1980. N37483 was to become N8512K and N37484 to N8512N but these were ntu. Ownership of both were transferred to Waara's other company **Arbor Air, Inc.** in 1980. *See also:* **Arbor Air, Inc., Louis P. Minkoff, William Waara**

N37483	11318 / 53-7884	1979–1980
N37484	11155 / 53-3144	1979–1980

Mid Atlantic Air Museum / Reading, Pennsylvania
One C-119F (10844 / BuNo. 131677) for static display from 1994.

Milestones of Flight Air Museum / Fox Field, California
Located at Gen William J. Fox Airfield. One C-119C (10304 / N13745) for static display. The aircraft is in dire need of restoration and is the oldest C-119 airframe in existence.

Minkoff, Louis P. / Ypsilanti, Michigan
Vice-president of **Arbor Air, Inc.** and business partner of **William A. Waara**. Minkoff is named on a BofS document as having purchased C-119L N37484 (11155) in Dec82 from **Bud's Flying Service Ltd.** *See also:* **Arbor Air Inc., Michigan Aerial Applicator, William Waara**

Munson, Jack R. / Muncie, Indiana
One C-119G (11005) in 1975, no registration, sold to **H&P Aviation, Inc.**

Museum of Flight & Aerial Firefighting / Greybull, Wyoming

Founded in 1987 adjacent the **H&P Aviation** facility for the preservation of local aircraft history. Two ex-H&P C-119G aircraft on display.

N3935	10824 / (RCAF) 22113	2006–present	Marked as: "N5216R / 136"
N5215R	10773 / (RCAF) 22108	unk–present	Marked as: "06"

National Warplane Museum / Geneseo, New York

One C-119G (10678 / N8092) for static display since 1991.

Northern Pacific Transport, Inc. / Anchorage, Alaska

Air freight service. Also leased C-119 N3003 (10737) during May83.

N8504X	KF-245 / 53-8142	1985–1987	C-119L Jet-Pak.
N8504Z	11253 / 53-7836	1986–1987	
N8505A	KF-179 / 53-8076	1985–1987	
N9027K	KF-176 / 53-8073	1985–1987	

Pacific International Foods, Inc. / Kenai, Alaska

A major food company that operated one C-119L N8501W (10880) during 1981 under FAA Type Certificate A6NW. The aircraft was repossessed by **Seattle-First National Bank** within months of commencing operations.

Palm Springs Air Museum / Palm Springs, California

Static display of forward fuselage C-119L 53-8154 (KF-257).

Perez, Juan / Miami, Florida & Williams, Emmett B.
(Williams Auto Parts) / Lavonia, Georgia

Acquired one C-119L as N55795 (11185) in 1980. Aircraft disappeared after an FAA Notice of Seizure was issued to the aircraft for unlawful operation.

Pima Air & Space Museum / Tucson, Arizona

Two static displays. C-119C (10394 / 49-0157) was acquired from MASDC in 1967 and a former civil C-119C (10369 / N13743) was acquired in HVFS livery with Jet-Pak.

Powers, D.A. / Greybull, Wyoming

Co-owner of **H&P Aviation, Inc.** Briefly owned one C-119G N8093 (10776) in 1991.

The Pride Capital Group, LLC / Deerfield, Illinois

Financial company which handled the closure and auctioneering of assets from **Hawkins & Powers Aviation** during 2005–2006.

N15501	10955 / (RCAF) 22130	2005–2006	To Hans O. Lauridsen.
N3003	10737 / (RCAF) 22106	2005–2006	To B&G Industries, LLC.
N37636	KF-253 / 53-8150	2005–2006	To Sheppard Trucking.
N8093	10776 / (RCAF) 22111	2005–2006	To Hagerstown Aviation Museum.
N8094	10994 / (RCAF) 22135	2005–2006	To Sheppard Trucking.
N8505A	KF-179 / 53-8076	2005–2006	To Sheppard Trucking.

Pueblo Weisbrod Aircraft Museum / Pueblo, Colorado

An aviation museum owned by The City of Pueblo and located at Pueblo Memorial Airport. Founded in the mid–1970s by the then Pueblo City Manager Fred Weisbrod, it acquired a former USMC C-119F (10855 / BuNo. 131688) in 1977 for static display. Briefly held civil reg N99574.

Rachanski, Ed / Nevada

Private individual who restored the forward fuselage of C-119L N90267 (KF-257).

Raven International, Inc. / Huntington, West Virginia
No details. One C-119L N37636 (KF-253) from 1980 to 1982.

Reffett, John S. & Luther, Terrence E. / Anchorage, Alaska
Co-ownership of C-119F Jet-Pak N1394N (10840) from 1989 just before it became stranded on Kodiak Island not to be rescued until 2002. Luther had been a co-partner in **Delta Leasing**, the C-119's previously registered owner. In 2015, **John S. Reffett** became the sole registered owner when Luther sold his half of the aircraft. Reffett also acquired C-119F N8501W (10880) as sole owner in 1998. Both aircraft are presently stored at Palmer, Alaska. *See also:* **Delta Associates, Inc.**

Renstrom, Carl W. / Omaha, Nebraska
The C-119G STOLmaster N383S (10992) registered during 1969–1970. *See also:* **Aircraft International Associates**; **Ren-Aire Aviation, Inc.**

Ren-Aire Aviation, Inc. / Omaha, Nebraska
Carl W. Renstrom's aviation company. The C-119G STOLmaster N383S (10992) was registered to this company from 1970 to 1975, probably as a separate interest from **Aircraft International Associates**. *See also:* **Aircraft International Associates**

Renstrom Enterprises
See under: **Aircraft International Associates**

Rolling Boxcar, Inc. / Alaska
A private partnership formed to restore C-119G N5216R (10956) as a mobile display unit in support of Alaska's military history. Airframe was acquired in early 2019.

Salamatof Seafoods, Inc. / Kenai, Alaska
Transportation of seafoods in the Alaskan region.

N8093	10776 / (RCAF) 22111	1979
N8682	10859 / (RCAF) 22115	1979

Seattle-First National Bank / Olympia, Washington State
Lending bank who repossessed N8501W (10880) from **Pacific Intl. Foods, Inc.** in 1981. The aircraft was registered in the bank's name until resold to **Elling Halvorson, Inc.** in 1982.

Sheppard Jr., Harold d/b/a Sheppard Trucking / Riverton, Wyoming
Three C-119s purchased at the H&P Aviation auction in 2006 but all airframes remain derelict at South Big Horn County Airport near Greybull. Ex-N13745 acquired in 2016.

N37636	KF-253 / 53-8150	2006–present
N8094	10994 / (RCAF) 22135	2006–present
N8505A	KF-179 / 53-8076	2006–present
N13745	10304 / 48-322	2016–present

Southwestern Alloys Corp. / Tucson, Arizona
Local Tucson scrap and aircraft dealer founded after World War II for smelting of scrapped military airframes. Renamed as **National Aircraft, Inc.** in 1986. The business was closed and the yard cleared by April 2012.

10686 / 51-2697	09Jul76	Scrapped.
10687 / 51-2698	30Jun76	Scrapped.
10694 / 51-2705	13Jul76	Scrapped.
10844 / BuNo. 131677	1980–1984	To D.M.I. Aviation, Inc.

10876 / BuNo. 131691	1980–2003	To Fox film studios 2003.
10891 / BuNo. 131706	1980–2003	To Fox film studios 2003.
10900 / BuNo. 131715		Scrapped, noted on site 2010.
KF-174 / 53-8071		Scrapped, noted on site 2010.

Starbird, Inc. / Reno, Nevada

Appears to be an air freight company originally located near Everett, Washington State. They held FAA Type Certificate A5NW and in addition to their own C-119s converted J.D. **Gifford Associates, Inc.** fleet during 1982. They seem to have gone out of business by 1984 when their lending bank repossessed their two remaining aircraft.

N9027K	KF-176 / 53-8073	1978–1984	
N90267	KF-257 / 53-8154	1978–1984	
N90268	11219 / 53-3206	1978–1980	W/o 05Jul80.
N90269	KF-230 / 53-8127	1978–1979	W/o 06Jul79.

Stebbins & Ambler Air Transport / Anchorage, Alaska

A trading name for a joint partnership between Stebbins Community Association and the Village of Ambler. N90267 was broken up for parts and de-registered at some point during 1988 only to be repossessed again by the companies lender in 1989. Another Boxcar, C-119F Jet-Pak N1394N (10840) was leased from **Delta Leasing** 1987–1989.

N8504Y	11232 / 53-3216	1985–1990s	C-119L Jet-Pak.
N90267	KF-257 / 53-8154	1985–1989	

Supra International, Inc. / Fairbanks, Alaska

One C-119F as N13626 (10836) from 1982 until it was w/o in 1984.

T&G Aviation, Inc. / Chandler, Arizona

Renamed by Tomassoni & Grantham. One C-119G-3E N15509 (10775) from 1980 until it was written-off in Alaska in 1984. *See also:* **Grantham, William A. & Sergio A. Tomassoni**.

Texas Aerial Applicators, Inc. / Laredo, Texas

Purchased C-119L N37484 (11155) in 1983 which company owner Gene D. Williams had been maintaining on behalf of **Bud's Flying Service Ltd.** since late 1982. Aircraft was sold in 1986.

Tobacco Road Farms, Inc. / Ronda, North Carolina

No details on the company itself but C-119F N4234S (10875) was registered under Tobacco Road Farms in April 1981. The aircraft was flown to Colombia but impounded in Bogota by Colombian AF authorities in 1982 marked as "N4234C." It was later scrapped along with the large numbers of other impounded aircraft stored at the airport.

Tomassoni, Sergio A. d/b/a Sergio Aviation / Buckeye, Arizona

One C-119G-3E N15509 (10775) from 1975 to 1977. Tomassoni went into partnership with William A. Grantham in 1977 renaming the company. *See also:* **Grantham, William A. & Sergio A. Tomassoni**

20th Century Fox Film Corp. / Century City, California

Major U.S. film studio acquired for filming purposes several derelict USMC Boxcars for *The Flight of the Phoenix* (1965) and the later remake *Flight of the Phoenix* (2004). All were scrapped after filming.

10549 / BuNo. 126580	1965		Fabricated by Allied Aircraft, Inc. Tucson.
10876 / BuNo. 131691	2003–2004		
10885 / BuNo. 131700	2003–2004		Ex-N3267U, located in Kenya.
10891 / BuNo. 131706	2003–2004		

United Nations (UN) / New York, New York

An intergovernmental organization first founded in 1945 to maintain international peace and security. Five C-119Gs were operated for missions in The Congo during the conflict there in the early 1960s. All aircraft were later transferred to the Italian AF.

UNO-101	11213 / 53-3200	1960–1961
UNO-102	11235 / 53-3219	1960–1961
UNO-103	11245 / 53-7828	1960–1961
UNO-104	11266 / 53-7845	1960–1961
UNO-105	KF-249 / 53-8146	1960–1961

U.S. Veterans Museum / Grandbury, Texas

Former USAF museum owned C-119G (10664 / 51-2675) acquired by the museum in 2012. Disassembled and moved by road to the museum in 2012 but appears to remain in storage.

Waara, William

President of both **Michigan Aerial Applicator** and **Arbor Air, Inc.** He was also the holder for FAA Type Certificate A32CE pertaining to mods carried out on C-119L N37483 (11318). *See also:* **Arbor Air Inc.**, **Michigan Aerial Applicator**, **Louis P. Minkoff**

Williams, Emmett B. (Williams Auto Parts) / Lavonia, Georgia

See under: **Perez, Juan**

C-119G-3E N15509 (10775) of T&G Aviation at Chandler, Arizona, in October 1979. This Boxcar slid off the Tobin Creek Mine Airstrip in Alaska during take-off on April 21, 1984. The wreckage remains at the site today (Geoff Goodall).

C-119G-3E N8093 (10776) retains the original RCAF lightning bolt livery from the 1960s in this photograph taken at Greybull, Wyoming, in 1989. Flying as Tanker 140 this aircraft had a film appearance in *Always* (1989) and is today preserved at the Hagerstown Aviation Museum (Geoff Goodall).

Flight-deck on C-119L Jet-Pak N8504Z (11253) in August 2022 (Michael Prophet Collection).

Section Two—Military Operators

C-119F N4234S (10875) is seen here in Bogota, Colombia, during November 1982 after being impounded by local authorities for reasons unknown. It was later scrapped along with other "derelicts" confiscated by the Colombians (Richard Vandervord).

C-119F N3267U (10885) had a colorful civil life flying humanitarian aid throughout African countries in the 1980s and is seen here at Stansted Airport, UK, in December 1987. It was later abandoned in Kenya until revived in Namibia for a starring role in *Flight of the Phoenix* (2004) (Richard Vandervord).

The oldest surviving Boxcar in the world is this C-119C Jet-Pak N13745 (10304) which served as Airtanker 82 with Hemet Valley Flying Service as seen here in October 1979. Hemet Valley operated five C-119C fire-bombers from 1975 to 1987 (Richard Vandervord).

Section Four

Service Histories

This section presents the complete service histories for all 1,185 C-119 aircraft built. A table titled C-119 Production Summary is included below to assist in easier referencing of any particular serial number, construction number, type or customer.

The entire production run is presented in consecutive order of factory construction number (or what is also known as the manufacturer's serial number [msn]), followed by the customer serial number then aircraft designation at the time of completion.

Aircraft Available (Avble), Accepted (Acc) and Delivery (del) dates are included for each entry. Aircraft changes of designation, serial number or civil registrations are presented in bold text for easier reference.

U.S. Military Units—USAF, AFRES, USMC and ANG unit histories and dates listed here are interpreted entirely from archival USAF and Naval IARC reels and are presented as they are found in these reels. It should be noted many of the dates of assignment will be "late recordings" and can be running behind by up to a month or more in some cases.

Presentation of these units are in the following order—the unit; with the major command in brackets; operating base, assignment date, and any additional notes as required such as squadron assignments or maintenance (mnt), etc. For clarity, unit assignments are separated by use of hyphenation throughout the text string. Deployments from the parent unit are listed with "depl" and non-deployment, or temporary assignments, are initialed with a "to" in order to signify the assignment is in fact temporary, not a permanent move.

Regular, or scheduled, depot maintenance areas, U.S. and foreign, are not listed unless they fall between unit assignments. A complete list of U.S. and foreign depot maintenance areas are however listed under the relevant area in the USAF Military Operators section.

Assignments for irregular maintenance are not listed unless they are, (1) numbered units; (2) assigned where there is a designation change or; (3) where the assignment carries a significant connection to projects or special tests.

Foreign Operator Units—Historically military unit and aircraft service histories of non-U.S. C-119 military operators has been difficult to trace for aviation historians. Canada and several European nations do have ample histories and records as has been presented here but many second-hand, Far and Middle East nation military unit histories, including dispositions, remain unanswered. USAF IARC reels do record foreign dispositions which has been very helpful in accurately recording numbers transferred to particular air forces.

The Mutual Defense Assistance Program (MDAP) was established on October 6, 1949, and was renamed as the Military Assistance Program (MAP) on August 26, 1954.

Section Four—Service Histories

Most C-119 dispositions to foreign nations, going into the 1970s, were through what is short termed in records as "MAP" programs.

Civil Histories—C-119 civilian service was limited but well recorded through Federal Aviation Administration (FAA) records. Owner name or company name are presented first followed by the date of purchase according to the Bill of Sale (BofS) document. Aircraft Registration Application (ARAp) dates are also included where possible. USFS airtanker numbers are included in bold where applicable.

C-119 Production Summary

msn	s/n	Type	Qty	Customer
10301 / 10311	48-319 / 48-329	C-119B	11	USAF
10312	48-330	XC-120	1	USAF
10313 / 10337	48-331 / 48-355	C-119B	25	USAF
10338 / 10346	49-101 / 49-109	C-119B	9	USAF
7001 / 7008	BuNo. 124324 / 124331	R4Q-1	8	USMC
10347 / 10355	49-110 / 49-118	C-119B	9	USAF
10356 / 10436	49-119 / 49-199	C-119C	81	USAF
10437 / 10489	50-119 / 50-171	C-119C	53	USAF
10490 / 10542	51-2532 / 51-2584	C-119C	53	USAF
10543 / 10551	BuNo. 126574 / 126582	R4Q-1	9	USMC
10552 / 10573	BuNo. 128723 / 128744	R4Q-1	22	USMC
10574	51-2585	YC-119H	1	USAF
10575	51-2586	YC-119F	1	USAF
10576 / 10650	51-2587 / 51-2661	C-119CF	75	USAF
10651 / 10675	51-2662 / 51-2686	C-119F	25	USAF
10676 / 10678	(RCAF) 22101 / 22103	C-119F	3	Canada
10679 / 10696	51-2690 / 51-2707	C-119F	18	Belgium
10697 / 10705	51-2708 / 51-2716	C-119F	9	USAF
10706	51-8016	C-119F	1	USAF
10707 / 10734	51-7968 / 51-7995	C-119F	28	USAF
10735 / 10738	(RCAF) 22104 / 22107	C-119F	4	Canada
10739 / 10758	51-7996 / 51-8015	C-119F	20	USAF
10759	51-2717	C-119F	1	USAF
10760 / 10772	51-8017 / 51-8029	C-119F	13	USAF
10773 / 10776	(RCAF) 22108 / 22111	C-119F	4	Canada
10777 / 10779	51-17365 / 51-17367	C-119G	3	Italy
10780	52-6000	C-119G	1	Italy
10781 / 10821	51-8233 / 51-8273	C-119CF	41	USAF
10822	51-8030	C-119F	1	USAF
10823 / 10825	(RCAF) 22112 / 22114	C-119F	3	Canada
10826 / 10828	52-6001 / 52-6003	C-119G	3	Italy
10829 / 10856	BuNo. 131662 / 131689	R4Q-2	28	USMC
10857, 10858	51-8031, 51-8032	C-119F	2	USAF
10859 / 10861	(RCAF) 22115 / 22117	C-119F	3	Canada
10862 / 10869	52-6004 / 52-6011	C-119G	8	Italy
10870 / 10872	(RCAF) 22118 / 22120	C-119F	3	Canada
10873, 10874	51-8033, 51-8034	C-119F	2	USAF
10875 / 10904	BuNo. 131690 / 131719	R4Q-2	30	USMC
10905 / 10908	(RCAF) 22121 / 22124	C-119F	4	Canada
10909 / 10912	52-6012 / 52-6015	C-119G	4	Italy
10913 / 10930	51-8035 / 51-8052	C-119F	18	USAF
10931 / 10941	51-8053 / 51-8063	C-119G	11	USAF
10942 / 10945	(RCAF) 22125 / 22128	C-119F	4	Canada

msn	s/n	Type	Qty	Customer
10946 / 10950	52-6016 / 52-6020	C-119G	5	Italy
10951 / 10953	52-6021 / 52-6023	C-119G	3	Belgium
10954 / 10957	(RCAF) 22129 / 22132	C-119F	4	Canada
10958 / 10991	51-8064 / 51-8097	C-119G	34	USAF
10992 / 10994	(RCAF) 22133 / 22135	C-119F	3	Canada
10995, 10996	52-6024, 52-6025	C-119G	2	Italy
10997, 10998	52-6026, 52-6027	C-119G	2	Belgium
10999 / 11028	52-5840 / 52-5869	C-119G	30	USAF
11029	52-6028	C-119G	1	Belgium
11030 / 11032	52-6029 / 52-6031	C-119G	3	Italy
11033 / 11036	52-6032 / 52-6035	C-119G	4	Belgium
11037 / 11074	52-5870 / 52-5907	C-119G	38	USAF
11075, 11076	52-6036, 52-6037	C-119G	2	Italy
11077, 11078	52-6038, 52-6039	C-119G	2	Belgium
11079 / 11081	52-6040 / 52-6042	C-119G	3	Italy
11082 / 11086	52-6043 / 52-6047	C-119G	5	Belgium
11087 / 11100	52-5908 / 52-5921	C-119G	14	USAF
11101, 11102	53-4637, 53-4638	C-119G	2	India
11103 / 11115	52-5924 / 52-5936	C-119G	13	USAF
11116, 11117	52-6048, 52-6049	C-119G	2	Italy
11118 / 11120	52-6050 / 52-6052	C-119G	3	Belgium
11121, 11122	52-6053, 52-6054	C-119G	2	Italy
11123	52-6055	C-119G	1	Belgium
11124 / 11141	52-5937 / 52-5954	C-119G	18	USAF
11142, 11143	52-9981, 52-9982	C-119G	2	USAF
11144, 11145	52-6056, 52-6057	C-119G	2	Italy
11146	52-6058	C-119G	1	Belgium
11147 / 11205	53-3136 / 53-3194	C-119G	59	USAF
11206, 11207	52-5922, 52-5923	C-119G	2	USAF
11208 / 11223	53-3195 / 53-3210	C-119G	16	USAF
11224 / 11226	53-4639 / 53-4641	C-119G	3	India
11227 / 11238	53-3211 / 53-3222	C-119G	12	USAF
11239, 11240	53-7826, 53-7827	C-119G	2	USAF
11241 / 11244	53-4642 / 53-4645	C-119G	4	India
11245 / 11261	53-7828 / 53-7844	C-119G	17	USAF
11262 / 11265	53-4646 / 53-4649	C-119G	4	India
11266 / 11284	53-7845 / 53-7863	C-119G	19	USAF
11285 / 11288	53-4650 / 53-4653	C-119G	4	India
11289 / 11298	53-7864 / 53-7873	C-119G	10	USAF
11299 / 11302	53-4654 / 53-4657	C-119G	4	India
11303 / 11308	53-7874 / 53-7879	C-119G	6	USAF
11309 / 11313	53-4658 / 53-4662	C-119G	5	India
11314 / 11318	53-7880 / 53-7884	C-119G	5	USAF
KMC-101 / KMC-171	51-8098 / 51-8168	C-119F	71	USAF
KF-172 / KF-259	53-8069 / 53-8156	C-119G	88	USAF

10139 / 45-57769 / XC-119A-FA

C-119 prototype. C-82A cvtd on production line as **XB-82B** with fuselage redesign and upgraded radial engines. Ff: 17Dec47 as **XC-119A** (**Model 105A**) prototype; retained by Fairchild (AMC) Hagerstown for flight tests – Avble & Acc for USAF service: 24May48; Del: 14Jun48 to USAF, redes **C-119A**; cvtd **EC-119A** 11Feb49 for ECM tests with Fairchild through AMC – Wfu to 3345 TTN WG (TC) Chanute 18Jan51 as mock-up airframe – DFI: 02Dec51 due to abnormal deterioration in use – Final Disposition: Scrapped.

Section Four—Service Histories

10301 / 48-319 / C-119B-FA

First production aircraft. Avble & Acc: 26May49; Del: 02Jun49 to USAF – bailment Fairchild (AMC) Hagerstown for test duties – bailment Aeroproducts Division (AMC) Vandalia 11Sep52; cvtd **EC-119B** 27Oct52. Forced landing due to mechanical failure 7 miles E of Bremen Ohio 17Sep53, no fatalities but aircraft substantially damaged – 2750 AB WG (AMC) WPAFB 18Sep53, probably didn't enter service due to accident – Middletown ARP (AMC) Olmsted 23Sep53 for repairs – Wfu to 3750 TTN WG (TC) Sheppard 09Jul54 for ground instruction – DFI: 24Jul56 – Final Disposition: Scrapped.

10302 / 48-320 / C-119B-FA

Avble: 15Apr49; Acc: 06May49; Del: 27Oct49 to USAF – 2750 AB WG (AMC) WPAFB 28Oct49; cvtd **EC-119B** 24Jul50; bailment Fairchild (AMC) Hagerstown for mods 28Feb51 – WRDCN (ARD) WPAFB 17Apr51 – 6502 PDT GP (ARD) El Centro 07Jul52 – 6511 PDT GP (AMC) El Centro 27Aug52; bailment BMC (AMC) Birmingham 04Nov53 for mods – 2233 RFC (CNR) Mitchel 29Jul55; to BMC (AMC) Birmingham 16Jan56, cvtd **EC-119C** 16Mar56, cvtd **C-119C** 12Apr56 – 2252 RFC (CNR) Clinton County 04Oct56; depl 2233 RFC (CNR) Mitchel 23Sep57 – 2259 RFC DT (CNR) Greater Pittsburgh 21Feb58 – re-serialed **48-0320** 1958; to 1405 AB WG (MATS) Scott 13Apr58 mnt; to 814 AB GP (SAC) Westover 19Oct58 mnt – 758 TCSq (459 TCG) (AFR) Greater Pittsburgh 19Dec58 – 137 AML SQ (NY-ANG) Westchester 02May61 – 911 TCG (459 TCWG) (AFR) Greater Pittsburgh 10Mar63 – Wfu to MASDC (LOG) Davis-Monthan 08Jul66 for storage – DFI: 02May68 – Final Disposition: Scrapped.

10303 / 48-321 / C-119B-FA

Avble: 31Mar49; Acc: 28Apr49; Del: 29Apr49 to USAF – 2750 AB GP (AMC) WPAFB 29Apr49, static test airframe – Final Disposition: Scrapped.

10304 / 48-322 / C-119B-FA

Avble: 02May49; Acc: 30Jun49; Del: 25Oct49 to USAF – 2750 AB WG (AMC) WPAFB 26Oct49 – 3203 MSU GP & 3200 PT GP (APG) Eglin 02Nov49 – 4415 AB SQ (TAC) Pope 01Dec50 – MIDAR (AMC) Olmsted 14Jan51 depot mnt – 314 TCG (TAC) Sewart 21Apr52 – SMAAR (AMC) McClellan 24Dec52 mnt – 483 TCWG (FEA) Ashiya AB Japan 03Feb53; depl 19 Bomb WG (FEA) Andersen 11Feb53; depl Bofu Japan 24May53 – bailment BMC (AMC) Birmingham 18May54 mods – 2233 RFC (CNR) Mitchel 16Sep54; to BMC (AMC) Birmingham 28Nov55, cvtd **C-119C** 06Feb56 – 2253 RFC (CNR) Greater Pittsburgh 29Feb56; to 63 TCWG (TAC) Donaldson 16Feb57 mnt – 2233 RFC (CNR) Mitchel 22Nov57; to 2500 AB WG (CNC) Mitchel 13Dec57; re-serialed **48-0322** 1958 – CAR Center (ARD) Hanscom 01Nov58 – 514 TCWG (AFR) Mitchel 22Jan59; to 15 CLM SQ (ADC) Niagara Falls 21May59 mnt – 137 AML SQ (NY-ANG) Westchester 30Mar61 – 911 TCG (459 TCWG) (AFR) Greater Pittsburgh 10Mar63 – Wfu to MASDC (LOG) Davis-Monthan 07Jan67 for storage – REC: 22Jun67 – BofS Aero Union Corp. Chico California 14Nov67; ARAp **N13745** 24Jan68; DFI: 02May68; cvtd for fire-bombing with fire retardant tank and avionics upgrade 26Jun71, Tanker **C14**; cvtd **C-119C Jet-Pak** (J34-WE) 10Apr74 – BofS Hemet Valley Flying Service Co. Hemet California 18Feb75; Tanker **82**; recovered elevator 12Apr82; wfu and stored 1987 – Donated via U.S. Forest Service circa 1992, listed as "gave to U.S. Govt for free," to Milestones of Flight Air Museum Fox Field California for static display – N13745 de-reg 01Sep95. The museum closed in 2015, the C-119 apparently acquired by Sheppard Trucking of Wyoming via Auction 01Feb16 – Final Disposition: Extant.

10305 / 48-323 / C-119B-FA

Avble: 17May49; Acc: 15Aug49; Del: 25Oct49 to USAF – 3203 MSU GP & 3200 PT GP (APG) Eglin 28Oct49 – 2750 AB WG (AMC) WPAFB 08Nov49; cvtd **EC-119B** 15Nov49; to 2500 AB GP (CNC) Mitchel 28Aug50 mnt – WRDCN (ARD) WPAFB 01Apr51 – 6502 PDT GP (ARD) El Centro 22Sep51; to WRAAR (AMC) Robins 04Mar52 for project – 6511 PDT GP (AMC) El Centro 01Aug52; to SBNAR (AMC) Norton 06Oct52 & 22Apr53 – bailment BMC (AMC) Birmingham 14Apr54 mods – 2233 RFC (CNR) Mitchel 26Oct54, cvtd **C-119B** 26Apr55; to Fairchild (AMC) Hagerstown 21Sep55, cvtd **C-119C** 06Dec55; to 2500 AB WG (CNC) Mitchel 16Dec55 – 2252 RFC (CNR) Clinton 03Oct56; to 2750 AB WG (AMC) WPAFB 08Dec56 – 2750 AB WG (AMC) WPAFB 27Dec57; re-serialed **48-0323** 1958. Crashed at WPAFB Ohio 20Oct59, cause and fatalities unk – DFI: 31Oct59 – Final Disposition: Accident.

10306 / 48-324 / C-119B-FA

Avble: 26Aug49; Acc: 22Sep49; bailment Fairchild (AMC) Hagerstown 28Nov49 for tests; Del: 16Dec49 to USAF – 314 TCG (CNC) Smyrna 23Dec49; depl Laurinburg-Maxton Airport North Carolina 20Apr50; depl 316 TCG (CNC) Sewart 21Jul50 – 314 TCG (FEA) Komaki AB Japan 10Sep50; assigned 50 TCSq; to Ashyia AB Japan 18Sep50. Damaged beyond repair while parked due to a ground collision when T-33A, s/n: 49-976, suffered a landing accident at Komaki AB Japan 25Sep50. No fatalities but both aircraft were w/o – 6106 AB UT (FEA) Komaki AB Japan 30Sep50 likely for parts salvage – DFI: 06Dec50 – Final Disposition: Accident.

10307 / 48-325 / C-119B-FA

Avble: 22Jul49; Acc: 22Sep49; bailment Fairchild (AMC) Hagerstown for tests; Del: 20Dec49 to USAF – 314 TCG (CNC) Smyrna 22Dec49; depl Laurinburg-Maxton Airport North Carolina 20Apr50; depl 316 TCG (CNC) Sewart 21Jul50 – 314 TCG (FEA) Komaki AB Japan 10Sep50; to Ashyia AB Japan 18Sep50; to 13 ARP SQ (FEAMCOM) Tachikawa AB Japan 17Aug51 mods – MIDAR (AMC) Olmsted 17Dec51 mnt – 2750 AB WG (AMC) WPAFB 15Feb52 – 483 TCWG (FEA) Ashyia AB Japan 25Feb52 – 314 TCG (FEA) Ashiya AB Japan 27May53 – 483 TCWG (FEA) Ashiya AB Japan 12Jul54 – 2253 RFC (CNR) Greater Pittsburgh 23Sep54; to 2500 AB WG (CNC) Mitchel 13Feb55; to BMC (AMC) Birmingham 28Oct55, cvtd **C-119C** 29Nov55; to 2500 AB WG (CNC) Mitchel 15Jul57 mnt – 2259 RFC DT (CNR) Greater Pittsburgh 01Dec57; re-serialed **48-0325** 1958 – 758 TCSq (459 TCG) (AFR) Greater Pittsburgh 19Dec58; 911 TCG (459 TCWG) (AFR) Greater Pittsburgh 10Mar63 – 7 Bomb WG (SAC) Carswell 08Jul63. Listed as w/o in accident 09Jul63, location, cause and fatalities unk – Final Disposition: Accident.

10308 / 48-326 / C-119B-FA

Avble: 17Jun49; Acc: 15Aug49; Del: 27Oct49 to USAF – retained Fairchild (AMC) Hagerstown for tests – 314 TCG (CNC) Smyrna 23Dec49 – 403 TCWG (FEA) Ashyia AB Japan 15Aug51 – 483 TCWG (FEA) Ashiya AB Japan 01Jan52; depl Bofu Japan 26May53 – bailment BMC (AMC) Birmingham 01Mar53 mods – 1500 OP SQ (MATS) Hickam 10Nov53 – bailment BMC (AMC) Birmingham 24Mar54 mods – 2233 RFC CN (CNR) Mitchel 21Aug54 – 2253 RFC (CNR) Greater Pittsburgh 17Apr55; to BMC (AMC) Birmingham 12Mar56, cvtd **C-119C** 14May56; depl 2233 RFC (CNR) Mitchel 10Aug56 & 23Sep57 –

2259 RFC DT (CNR) Greater Pittsburgh 01Dec57; re-serialed **48-0326** 1958 – 758 TCM SQ (459 TCG) (AFR) Greater Pittsburgh 19Dec58 – 167 AML SQ (WV-ANG) Martinsburg 25Apr61 – 130 ACO GP (WV-ANG) Kanawha 15Aug63 – Wfu to 2704 ASD GP (LOG) Davis-Monthan 13Nov63 for storage – REC: 26Jun64 – Final Disposition: Scrapped.

10309 / 48-327 / C-119B-FA

Avble: 17Jun49; Acc: 15Aug49; bailment Fairchild (AMC) Hagerstown 28Nov49 for tests; Del: 17Dec49 to USAF – 314 TCG (CNC) Smyrna 23Dec49; depl Laurinburg-Maxton Airport North Carolina 20Apr50; depl 316 TCG (CNC) Sewart 21Jul50 – 314 TCG (FEA) Komaki AB Japan 10Sep50; to Ashyia AB Japan 18Sep50; to 13 ARP SQ (FEAMCOM) Tachikawa AB Japan 22Aug51 mods – BMC (AMC) Birmingham 28Aug52 – 374 TCWG (FEA) Tachikawa AB Japan 02Sep52 – MIDAR (AMC) Olmsted 24Sep52 mnt – 483 TCWG (FEA) Ashyia AB Japan 08Apr53; depl 314 TCWG (FEA) Ashyia AB Japan 24May53 – 2233 RFC (CNR) Mitchel 29Oct54; to 63 TCWG (TAC) Donaldson 01Aug55 mnt; depl 2253 RFC (CNR) Greater Pittsburgh 19Aug55; to BMC (AMC) Birmingham 15Sep55, cvtd **C-119C** 29Nov55; depl 2253 RFC (CNR) Greater Pittsburgh 14Jun56 – 2252 RFC (CNR) Clinton 03Oct56; to 1100 MSU GP (HQC) Bolling 31Jul57 mnt – 2259 RFC DT (CNR) Greater Pittsburgh 19Jan58; to 2500 AB WG (CNC) Mitchel 14Feb58 mnt; re-serialed **48-0327** 1958; depl 2585 AB SQ (CNC) Miami 01Jun58 mnt – 758 TCSq (459 TCWG) (AFR) Greater Pittsburgh 19Dec58; 911 TCG (459 TCWG) (AFR) Greater Pittsburgh 17Jan63 – 928 TCG (403 TCWG) (AFR) O'Hare 26Sep66. Wfu into storage at O'Hare – REC: 01Oct67 – Final Disposition: Scrapped.

10310 / 48-328 / C-119B-FA

Avble: 06Sep49; Acc: 22Sep49; bailment Fairchild (AMC) Hagerstown 28Nov49 for tests; Del: 17Dec49 to USAF – 314 TCG (CNC) Smyrna 22Dec49; depl Laurinburg-Maxton Airport North Carolina 20Apr50; depl 316 TCG (CNC) Sewart 21Jul50 – 314 TCG (FEA) Komaki AB Japan 10Sep50; to Ashyia AB Japan 18Sep50; to 13 ARP SQ (FEAMCOM) Tachikawa AB Japan 15Aug51 mods – bailment BMC (AMC) Birmingham 12Nov52 mods – 483 TCWG (FEA) Ashyia Japan 08Aug53; depl Bofu Japan 17Nov53 – 75 AD WG (FEA) Miho AB Japan 01Mar54 mnt – 6400 AD WG (FEA) Miho AB Japan 25Jun54 – TAD DP (FEA) Miho AB Japan 25Nov54 mnt – NAMAR (AMC) Miho AB & Tachikawa AB Japan 01Oct55 – 2233 RFC (CNR) Mitchel 21Mar57; to 2500 AB WG (CNC) Mitchel 05Apr57 mnt – MIDAR (AMC) Olmsted 08Apr57; re-serialed **48-0328** 1958; cvtd **C-119C** 27Nov59 – 514 TCWG (AFR) Mitchel 24Mar60; depl 354 Fighter WG (TAC) Myrtle Beach 01Sep60 – 137 AML SQ (NY-ANG) Westchester 10Mar61 – 302 TCWG (AFR) Clinton 03Oct62; 907 TCG (302 TCWG) (AFR) Clinton 11Feb63 – Wfu to 2704 ASD GP (LOG) Davis-Monthan 01Mar63 for storage – REC: 27Jul64 – Final Disposition: Scrapped.

10311 / 48-329 / C-119B-FA

Avble: 22Aug49; Acc: 22Sep49; bailment Fairchild (AMC) Hagerstown 28Nov49 for tests; Del: 28Dec49 to USAF – 314 TCG (CNC) Smyrna 29Dec49; to 1050 MSU GP (HQC) Andrews 28Feb50 mnt; depl Laurinburg-Maxton Airport North Carolina 20Apr50; depl 316 TCG (CNC) Sewart 21Jul50 – 314 TCG (FEA) Komaki AB Japan 10Sep50; to Ashyia AB Japan 30Sep50; to 13 ARP SQ (FEAMCOM) Tachikawa AB Japan 30Oct51 for depot mods – bailment BMC (AMC) Birmingham 15Aug52 mods – 483 TCWG (FEA) Ashyia AB Japan 25May53 – 2233 RFC (CNR) Mitchel 06Sep54; bailment BMC (AMC) Birmingham 24May55; to Fairchild (AMC) Hagerstown 28Sep55, cvtd **C-119C** 18Nov55 – 2252 RFC

(CNR) Clinton 05Oct56; depl 2253 RFC (CNR) Greater 15Jun57; to 1607 FLM SQ (MATS) Dover 10Jan58 mnt; to 1607 TSH WG (MATS) Dover 11Feb58 mnt; re-serialed **48-0329** 1958 – 2259 RFC DT (CNR) Greater 04Apr58 – 758 TCM SQ (AFR) Greater Pittsburgh 19Dec58 – 167 AML SQ (WV-ANG) Martinsburg 08Apr61 – Wfu to 2704 ASD GP (LOG) Davis-Monthan 13Nov63 for storage – REC: 26Jun64 – Final Disposition: Scrapped.

10312 / 48-330 / XC-120-FA Packplane

Assigned new msn: **8001**. Avble, Acc & Del: 21Dec50 to USAF. Experimental prototype with detachable fuselage cargo pod. Bailment Fairchild (AMC) Hagerstown 25Dec50 for flight tests – 2750 AB WG (AMC) WPAFB 11Jan51 – WRDCN (ARD) WPAFB 16May51; depl 3203 MSU GP & 3200 PT WG (APG) Eglin 21May51 for tests; bailment Fairchild (AMC) Hagerstown 30Jun52 likely for mods – assigned variously 3200 PT GP, 3203 MSU GP, 3200 MNT WG & 3201 AB WG (APG) Eglin from 16May53 for experimental test duties. Wfu as obsolete – REC: 18Aug54 – DFI: 30Jan55 – Final Disposition: Scrapped.

10313 / 48-331 / C-119B-FA

Avble: 20Jul49; Acc: 15Aug49; bailment Fairchild (AMC) Hagerstown 28Nov49 for tests; Del: 20Dec49 to USAF – 314 TCG (CNC) Smyrna 22Dec49; depl Laurinburg-Maxton Airport North Carolina 20Apr50; depl 316 TCG (CNC) Sewart 21Jul50 – 314 TCG (FEA) Komaki AB Japan 10Sep50; to Ashyia AB Japan 18Sep50; to 13 ARP SQ (FEAMCOM) Tachikawa AB Japan 18Aug51 mods – bailment BMC (AMC) Birmingham 13Apr53 mods – 483 TCWG (FEA) Ashyia AB Japan 06Nov53; to 1500 OP SQ (MATS) Hickam 18Jan54 mnt – 2233 RFC (CNR) Mitchel 30Aug54; to BMC (AMC) Birmingham 26Oct55, cvtd **C-119C** 06Dec55; depl 2253 RFC (CNR) Greater Pittsburgh 14Jun56; to 464 TCWG (TAC) Pope 27Aug56 mnt; depl 2253 RFC (CNR) Greater Pittsburgh 14Jun57; re-serialed **48-0331** 1958 – 514 TCWG (AFR) Mitchel 19Dec58 – 137 AML SQ (NY-ANG) Westchester 30Mar61 – Wfu to 2704 ASD GP (LOG) Davis-Monthan 30Mar63 for storage – REC: 27Jul64 – Final Disposition: Scrapped.

10314 / 48-332 / C-119B-FA

Avble: 22Jul49; Acc: 23Sep49; bailment Fairchild (AMC) Hagerstown 28Nov49 for tests; Del: 20Dec49 to USAF – 314 TCG (CNC) Smyrna 22Dec49; depl Laurinburg-Maxton Airport North Carolina 20Apr50; depl 316 TCG (CNC) Sewart 21Jul50 – 314 TCG (FEA) Komaki AB Japan 10Sep50; to Ashyia AB Japan 18Sep50; to 13 ARP SQ (FEAMCOM) Tachikawa AB Japan 02Aug51 mods – 483 TCWG (FEA) Ashyia AB Japan 30Jun54 – 2253 RFC (CNR) Greater Pittsburgh 25Sep54; to Fairchild (AMC) Hagerstown 22Mar55, cvtd **C-119C** 26Apr55; depl 2233 RFC (CNR) Mitchel 02Jul56 – 2259 RFC DT (CNR) Greater Pittsburgh 01Dec57; to 464 TCWG (TAC) Pope 01Dec57 & 15Apr58 mnt; to 2500 AB WG (CNC) Mitchel 13Feb58 mnt; re-serialed **48-0332** 1958 – 758 TCSq (459 TCG) (AFR) Greater Pittsburgh 19Dec58; to 2500 AB WG (CNC) Mitchel 22Oct60 mnt – 167 AML SQ (WV-ANG) Martinsburg 08Apr61 – 130 ACO GP (ANG) Kanawha 05Sep63 – Wfu to 2704 ASD GP (LOG) Davis-Monthan 17Nov63 for storage – REC: 26Jun64 – Final Disposition: Scrapped.

10315 / 48-333 / C-119B-FA

Avble: 16Aug49; Acc: 23Sep49; bailment Fairchild (AMC) Hagerstown 28Nov49 for tests; Del: 10Jan50 to USAF – assigned variously 3203 MSU GP, 3200 PT GP & 3201 AB WG (APG) Eglin from 13Jan50 for experimental duties; to 5064 CMT UT (AAC) Ladd 30Nov50; to 1100 AB WG (HQC) Bolling 05Jun51 mnt; to 3800 AU WG (AU) Maxwell 21Oct52 mnt – 3750 TTN WG (TC) Sheppard 14Apr53, to ground instruction; re-serialed

48-0333 1958 – 3750 MSU GP (ATC) Sheppard cvtd **UC-119B** 30Jun62; cvtd **GC-119B** 30Nov62 – REC: 13May63 – Final Disposition: Scrapped.

10316 / 48-334 / C-119B-FA

Avble: 25Aug49; Acc: 24Sep49; bailment Fairchild (AMC) Hagerstown 28Nov49 for tests; Del: 28Dec49 to USAF – 314 TCG (CNC) Smyrna 29Dec49; to 1050 MSU GP (HQC) Andrews 28Feb50 mnt; depl Laurinburg-Maxton Airport North Carolina 20Apr50; depl 316 TCG (CNC) Sewart 21Jul50 – 314 TCG (FEA) Ashyia AB Japan 09Nov50; to 13 ARP SQ (FEAMCOM) Tachikawa AB Japan 09Nov50 mods; to 37 TCSq (Attached) (FEA) Komaki AB Japan 01Feb51; to 13 ARP SQ (FEAMCOM) Tachikawa AB Japan 10Nov51 – 403 TCWG (FEA) Ashyia AB Japan 28Apr52; to 6400 AD WG (FEA) Tachikawa AB Japan 16Sep52 mnt – 483 TCWG (FEA) Ashyia AB Japan 01Nov52 – 2233 RFC (CNR) Mitchel 05Sep54; depl 2253 RFC (CNR) Greater Pittsburgh 10Aug55; to Fairchild (AMC) Hagerstown 20Mar56, cvtd **C-119C** 24May56; depl 2253 RFC (CNR) Greater Pittsburgh 14Jun56; to 464 TCWG (TAC) Pope 03Jul56 mnt – 2252 RFC (CNR) Clinton 05Oct56; depl 2253 RFC (CNR) Greater Pittsburgh 15Jun57; to 4041 AB WG (SAC) Bunker 03Sep57 mnt; re-serialed **48-0334** 1958 – 2259 RFC DT (CNR) Greater Pittsburgh 25Apr58 – 758 TCSq (459 TCG) (AFR) Greater Pittsburgh 19Dec58; to 4520 CCT WG (TAC) Nellis 28Sep59 mnt; 911 TCG (459 TCWG) (AFR) Greater Pittsburgh 22Feb63. Wfu and assigned 16Aug66 for USAF allotted destructive tests at Deer Valley Arizona – Final Disposition: Tested to Destruction.

10317 / 48-335 / C-119B-FA

Avble: 01Sep49; Acc: 24Sep49; bailment Fairchild (AMC) Hagerstown 28Nov49 for tests; Del: 29Dec49 to USAF – 314 TCG (CNC) Smyrna 29Dec49; depl Laurinburg-Maxton Airport North Carolina 20Apr50; depl 316 TCG (CNC) Sewart 21Jul50 – 1502 SUT SQ (MATS) Kwajalein Is. Marshall Islands 03Oct50 mnt – OGNAR (AMC) Hill 12Apr51 depot mnt – bailment BMC (AMC) Birmingham 22Sep53 mods – 483 TCWG (FEA) Ashyia AB Japan 21Dec53 – 2233 RFC (CNR) Mitchel 20Sep54; to 2500 AB WG (CNC) Mitchel 26Sep55 – BMC (AMC) Birmingham 16Apr56, cvtd **C-119C** 29Jul56 – 2253 RFC (CNR) Greater Pittsburgh 16Aug56 – 2259 RFC DT (CNR) Greater Pittsburgh 01Dec57; re-serialed **48-0335** 1958 – 758 TCSq (459 TCG) (AFR) Greater Pittsburgh 19Dec58 – 137 AML SQ (NY-ANG) Westchester 07Apr61 – 911 TCG (459 TCWG) (AFR) Greater Pittsburgh 10Feb63. Wfu to storage at Greater Pittsburgh – REC: 12Nov66 – Final Disposition: Scrapped.

10318 / 48-336 / C-119B-FA

Avble: 01Sep49; Acc: 24Sep49; bailment Fairchild (AMC) Hagerstown 01Dec49 for tests; Del: 16Jan50 to USAF – 314 TCG (CNC) Smyrna 18Jan50; depl Laurinburg-Maxton Airport North Carolina 24Apr50; depl 316 TCG (CNC) Sewart 13Jul50 – 314 TCG (FEA) Komaki AB Japan 10Sep50; to Ashyia AB Japan 18Sep50. Abandoned on ground at Yonpo Airfield (K-27) North Korea 16Dec50 due to advancing enemy troops, presumed destroyed – DFI: 18Dec50 – Final Disposition: Combat Loss.

10319 / 48-337 / C-119B-FA

Avble: 12Sep49; Acc: 24Sep49; Del: 24Jan50 to USAF – 314 TCG (CNC) Smyrna 26Jan50; depl Laurinburg-Maxton Airport North Carolina 24Apr50; depl 316 TCG (CNC) Sewart 13Jul50 – 314 TCG (FEA) Komaki AB Japan 10Sep50; to Ashyia AB Japan 18Sep50; to 13 ARP SQ (FEAMCOM) Tachikawa AB Japan 29Jul51, 24Aug51 & 01Oct51 mods – 6400 AD WG (FEA) Tachikawa AB Japan 31Jan52 mnt & mods – 314 TCWG (FEA) Ashiya AB Japan

24Apr53 – bailment BMC (AMC) Birmingham 10Dec53 mods – 483 TCWG (FEA) Ashiya AB Japan 01Jun54 – 2233 RFC (CNR) Mitchel 06Sep54 – 6319 AB WG (FEA) Andersen 18Sep54 mnt – 3960 AB WG (SAC) Andersen 10Apr55 mnt – 2233 RFC (CNR) Mitchel 20Oct55; to BMC (AMC) Birmingham, cvtd **C-119C** 22Mar56 – 2252 RFC (CNR) Clinton 03Oct56; re-serialed **48-0337** 1958; to 4238 AB GP (SAC) Barksdale 11Sep58 mnt – 302 TCWG (AFR) Clinton 19Mar59 – 137 AML SQ (NY-ANG) Westchester 26May61 – 302 TCWG (AFR) Clinton 03Oct62; 907 TCG (302 TCWG) (AFR) Clinton 11Feb63 – Wfu to 2704 ASD GP (LOG) Davis-Monthan 05Jul63 for storage – REC: 27Jul64 – Final Disposition: Scrapped.

10320 / 48-338 / C-119B-FA

Avble: 12Sep49; Acc: 26Sep49; retained Fairchild (AMC) Hagerstown for tests; Del: 24Jan50 to USAF – 314 TCG (CNC) Smyrna 26Jan50; assigned 61 TCSq; depl Laurinburg-Maxton Airport North Carolina 24Apr50. Made a forced landing due to a no. 2 engine failure near Decatur AFB Alabama 29Jun50. Pilot: James W. Bonner, no fatalities but two of the five crew were injured and the aircraft damaged beyond repair – REC: 30Jun50, noted as being put in storage for parts – Final Disposition: Accident.

10321 / 48-339 / C-119B-FA

Avble: 15Sep49; Acc: 26Sep49; Del: 06Feb50 to USAF – 2750 AB WG (AMC) WPAFB 06Feb50, cvtd **EC-119B** 27Jun50; to 5064 CMT UT (AAC) Ladd 05Oct50 – WRDCN (ARD) WPAFB 01Apr51; bailment BMC (AMC) Birmingham 14Jul54 mods; bailment Fairchild (AMC) Hagerstown 06Sep55, redes **JC-119B** 01Dec55, cvtd **JC-119C** 23Mar56 – OTC (APG) Eglin 27Nov57 – APG Center (APG) Eglin 01Dec57; re-serialed **48-0339** 1958 – Hayes (AMC) Birmingham, cvtd **C-119C** 06Jun58 – 2578 RFC (CNR) Ellington 13Sep58 – 446 TCWG (AFR) Ellington 20Sep58 – 137 AML SQ (NY-ANG) Westchester 08Apr61. Listed as being w/o in a ground accident 01Nov62, location, cause and fatalities unk – Final Disposition: Accident.

10322 / 48-340 / C-119B-FA

Avble: 20Sep49; Acc: 27Sep49; Del: 09Feb50 to USAF – 314 TCG (CNC) Smyrna 09Feb50; depl Laurinburg-Maxton Airport North Carolina 24Apr50; depl 316 TCG (CNC) Sewart 13Jul50 – 314 TCG (FEA) Komaki AB Japan 10Sep50 – 316 & 314 TCG (TAC) Sewart 19Sep50; depl to Laurinburg-Maxton Airport North Carolina 20Aug51 – 435 TCWG (TAC) Miami 07Aug52 – 456 TCWG (TAC) Miami 03Dec52 – 483 TCWG (FEA) Ashiya AB Japan 22Feb53; depl 314 TCWG (FEA) Ashiya AB Japan 13Jan54 – 2233 RFC (CNR) Mitchel 27Aug54; depl 2253 RFC (CNR) Greater 15Aug55; to 464 TCWG (TAC) Pope 16Oct55 mnt; to BMC (AMC) Birmingham 09Nov55, cvtd **C-119C** 06Dec55; re-serialed **48-0340** 1958; 514 TCWG (AFR) Mitchel 19Dec58; to 464 TCWG (TAC) Pope 28Oct59 mnt – 137 AML SQ (NY-ANG) Westchester 17Mar61 – Wfu to 2704 ASD GP (LOG) Davis-Monthan 26Apr63 for storage – REC: 27Jul64 – Final Disposition: Scrapped.

10323 / 48-341 / C-119B-FA

Avble: 23Sep49; Acc: 27Sep49; Del: 18Feb50 to USAF – 314 TCWG (CNC) Smyrna 23Feb50; to 3345 TTN WG (TC) Chanute 01Jun50 for ground instruction – 314 TCWG (FEA) Ashiya AB Japan 30Sep50; to 13 ARP SQ (FEAMCOM) Tachikawa AB Japan 01Jun51 mods – bailment BMC (AMC) Birmingham 26Aug52 mnt – MIDAR (AMC) Olmsted 02Oct52 mnt – 483 TCWG (FEA) Ashiya AB Japan 07May53; depl 581 ARC WG (FEA) Clark AB The Philippines 24Aug53; depl Bofu Japan 07Nov53 – 2233 RFC (CNR) Mitchel 02Oct54; to 1401 AB WG (MATS) Andrews 23Nov54 mnt; bailment BMC

Section Four—Service Histories

(AMC) Birmingham 21Jun55; bailment Fairchild (AMC) Hagerstown 27Aug55, cvtd **C-119C** 13Sep55; depl 2253 RFC (CNR) Greater Pittsburgh 14Jun56; depl 56 Fighter GP (ADC) O'Hare 15Jul56; depl 2253 RFC (CNR) Greater Pittsburgh 14Jun57; to 1405 AB WG (MATS) Scott 10Feb58 mnt; re-serialed **48-0341** 1958; to 3320 TTA WG (ATC) Amarillo 28Sep58 mnt – 514 TCWG (AFR) Mitchel 19Dec58 – 137 AML SQ (NY-ANG) Westchester 10Mar61 – Wfu to 2704 ASD GP (LOG) Davis-Monthan 12Apr63 for storage – REC: 27Jul64 – Final Disposition: Scrapped.

10324 / 48-342 / C-119B-FA

Avble & Acc: 27Sep49; retained Fairchild (AMC) Hagerstown for tests; Del: 09Feb50 to USAF – 314 TCG (CNC) Smyrna 09Feb50; depl Laurinburg-Maxton Airport North Carolina 24Apr50; depls 316 TCG (CNC) Sewart 13Jul50 & 19Sep50 – 3203 MSU GP & 3201 AB GP (APG) Eglin 09Feb51 for test duties. Made a belly crash landing on a road due to engine failure 8 miles W El Paso Texas 15Jul51. Pilot: Howard F. Mason, no fatalities but 2 injured – REC: 18Jul51 – DFI: 23Nov51 – Final Disposition: Accident.

10325 / 48-343 / C-119B-FA

Avble & Acc: 27Sep49; Del: 18Feb50 to USAF – 314 TCG (CNC) Smyrna 23Feb50; depl Laurinburg-Maxton Airport North Carolina 24Apr50; depl 316 TCG (CNC) Sewart 13Jul50 – 314 TCG (FEA) Komaki AB Japan 10Sep50; to Ashiya AB Japan 18Sep50; to 13 ARP SQ (FEAMCOM) Tachikawa AB Japan 11Jul51 mods – bailment BMC (AMC) Birmingham 12Nov52 for mods – 483 TCWG (FEA) Ashiya AB Japan 09Jan53; depl Bofu Japan 17Nov53 – 2233 RFC (CNR) Mitchel 26Aug54; depl 2253 RFC (AMC) Greater Pittsburgh 14Aug55; to BMC (AMC) Birmingham 10Oct55, cvtd **C-119C** 29Nov55; to 2253 RFC (CNR) Greater Pittsburgh 21Jun57 – 2259 RFC DT (CNR) Greater Pittsburgh 07Feb58; re-serialed **48-0343** 1958 – 758 TCSq (459 TCG) (AFR) Greater Pittsburgh 19Dec58 – 167 AML SQ (WV-ANG) Martinsburg 25Apr61 – 130 ACO GP (ANG) Kanawha 16Nov63 – Wfu to 2704 ASD GP (LOG) Davis-Monthan 16Nov63 for storage – REC: 26Jun64 – Final Disposition: Scrapped.

10326 / 48-344 / C-119B-FA

Avble: 04Oct49; Acc: 20Oct49; Del: 18Feb50 to USAF – 314 TCG (CNC) Smyrna 23Feb50; depl Laurinburg-Maxton Airport North Carolina 24Apr50; depl 316 TCG (CNC) Sewart 13Jul50 – 314 TCG (FEA) Komaki AB Japan 10Sep50; to Ashiya AB Japan 18Sep50; to 13 ARP SQ (FEAMCOM) Tachikawa AB Japan 05Jul51 mods – MIDAR (AMC) Olmsted 17Dec51 mnt – 483 TCWG (FEA) Ashiya AB Japan 13Mar53; depl 314 TCWG (FEA) Ashiya AB Japan 01Sep53 – 2233 RFC (CNR) Mitchel 05Sep54; to 2500 AB WG (CNC) Mitchel 29Nov55; to BMC (AMC) Birmingham 22Mar56, cvtd **C-119C** 16May56; depl 2253 RFC (CNR) Greater Pittsburgh 14Jun56, 14Jun57 & 05Nov57 – 2259 RFC DT (CNR) Greater Pittsburgh 01Dec57; to 2750 AB WG (AMC) WPAFB 20Feb58 & 16Jun58 mnt; re-serialed **48-0344** 1958 – 758 TCSq (459 TCG) (AFR) Greater Pittsburgh 19Dec58 – 167 AML SQ (WV-ANG) Martinsburg 08Apr61 – Wfu to 2704 ASD GP (LOG) Davis-Monthan 13May63 for storage – REC: 27Jul64 – Final Disposition: Scrapped.

10327 / 48-345 / C-119B-FA

Avble: 06Oct49; Acc: 20Oct49; retained Fairchild (AMC) Hagerstown for tests; Del: 03Mar50 to USAF – 314 TCG (CNC) Smyrna 06Mar50; depl Laurinburg-Maxton Airport North Carolina 26Apr50; depl 316 TCG (CNC) Sewart 13Jul50 – 314 TCG (FEA) Komaki

AB Japan 10Sep50; to Ashyia AB Japan 18Sep50. Lost during a combat mission 2 miles E Hongcheon S. Korea 29Mar51 due to structural failure of the right engine which broke away from its mounts. Listed as making a belly crash landing. Crew were Maj Maurice Johnson (pilot), 2Lt Kenneth E. Becker, SSgt Reybur J. Grott, Sgt John C. Kmiec and F.N.U. Neylon. Maj Johnson was killed in the crash with the other 4 crew injured – DFI: 25Apr51 – Final Disposition: Accident.

10328 / 48-346 / C-119B-FA

Avble: 04Oct49; Acc: 20Oct49; Del: 03Mar50 to USAF – 314 TCG (CNC) Smyrna 07Mar50; depl Laurinburg-Maxton Airport North Carolina 24Apr50; depl 316 TCG (CNC) Sewart 13Jul50 – 314 TCG (FEA) Komaki AB Japan 10Sep50; to Ashyia AB Japan 18Sep50 – 6122 AB GP (FEA) Ashyia AB Japan 20May51 – 13 ARP SQ (FEAMCOM) Tachikawa AB Japan 22Jul51 mnt – 6400 AD WG (FEA) Tachikawa AB Japan 31Jan52 mnt – bailment BMC (AMC) Birmingham 09Apr52 mods – 483 TCWG (FEA) Ashyia AB Japan 14May53 – 2233 RFC (CNR) Mitchel 29Oct54; bailment BMC (AMC) Birmingham 17Mar55; to Fairchild (AMC) Hagerstown 29Jul55, cvtd **C-119C** 15Sep55; depl 2253 RFC (CNR) Greater Pittsburgh 14Jun56, 15Jun57 & 05Nov57; to 2500 AB WG (CNC) Mitchel 24Jul57 mnt – 2259 RFC DT (CNR) Greater Pittsburgh 01Dec57; re-serialed **48-0346** 1958; to AEMCO (AMC) Oakland 28Jun58 mnt. Crashed and burned on a test flight after overhaul out of Oakland IAP California 05Feb59, 3 civilian crew were killed – Final Disposition: Accident.

10329 / 48-347 / C-119B-FA

Avble: 10Oct49; Acc: 24Oct49; Del: 22Mar50 to USAF; briefly retained Fairchild (AMC) Hagerstown for project duties – 314 TCG (CNC) Smyrna 24Mar50; depl Laurinburg-Maxton Airport North Carolina 20Apr50; depl 316 TCG (CNC) Sewart 13Jul50 – 314 TCG (FEA) Komaki AB Japan 10Sep50; to Ashyia AB Japan 18Sep50; to 13 ARP SQ (FEAMCOM) Tachikawa AB Japan 20Aug51 mods; to 6122 AB WG (FEA) Ashyia AB Japan 05Apr52 – bailment BMC (AMC) Birmingham 19Nov52 mods – 483 TCWG (FEA) Ashyia AB Japan 10Sep53 – bailment BMC (AMC) Birmingham 02Mar54 mods – MIDAR (AMC) Olmsted 22Apr54 mnt – 2233 RFC CN (CNR) Mitchel 03Nov54; depl 2596 RFC (CNR) NAS Dallas 19Nov54; to BMC (AMC) Birmingham 28Dec55, cvtd **C-119C** 27Feb56; depl 2235 RFC (CNR) Grenier 12Oct57; re-serialed **48-0347** 1958; to 1001 AB WG (HQC) Andrews 20Aug58 mnt – 514 TCWG (AFR) Mitchel 19Dec58 – 137 AML SQ (NY-ANG) Westchester 21Apr61 – Wfu to 2704 ASD GP (LOG) Davis-Monthan 30Mar63 for storage – REC: 27Jul64 – Final Disposition: Scrapped.

10330 / 48-348 / C-119B-FA

Avble: 14Oct49; Acc: 24Oct49; Del: 22Mar50 to USAF; briefly retained Fairchild (AMC) Hagerstown for project duties – 314 TCG (CNC) Smyrna 24Mar50; depl Laurinburg-Maxton Airport North Carolina 20Apr50; depl 316 TCG (CNC) Sewart 13Jul50 – 314 TCG (FEA) Komaki AB Japan 10Sep50; to Ashyia AB Japan 18Sep50; depl 19 Bomb WG (FEA) Andersen 20Nov50; to 13 ARP SQ (FEAMCOM) Tachikawa AB Japan 18Aug51 for depot mods – 483 TCWG (FEA) Ashyia 27Feb53, assigned 815 TCSq. Ditched at sea due to engine failure 3 miles W of Laiya Philippine Islands 23Aug53. Pilot: Harry A. Witt and crew rescued. To 24 MND GP (FEA) Clark 28Aug53 for REC & SAL – DFI: 09Sep53 – Final Disposition: Accident.

Section Four—Service Histories

10331 / 48-349 / C-119B-FA

Avble: 17Oct49; Acc: 24Oct49; Del: 25Mar50 to USAF; briefly retained Fairchild (AMC) Hagerstown for project duties – 314 TCG (CNC) Smyrna 24Mar50; depl Laurinburg-Maxton Airport North Carolina 20Apr50; depl 316 TCG (CNC) Sewart 13Jul50 – 314 TCG (FEA) Komaki AB Japan 10Sep50; to Ashiya AB Japan 18Sep50; to 13 ARP SQ (FEAMCOM) Tachikawa AB Japan 28Jul51 mods; depl 403 TCWG (FEA) Ashiya AB Japan 10May52 – bailment BMC (AMC) Birmingham 28Aug52 mods – 483 TCWG (FEA) Ashiya AB Japan 27May53; depl 314 TCWG (FEA) Ashiya AB Japan 16Apr54 – 2233 RFC (CNR) Mitchel 02Nov54; depl 2253 RFC (CNR) Greater Pittsburgh 12Oct55, 12Jul56 & 12Oct57 – 2259 RFC DT (CNR) Greater Pittsburgh 01Dec57; to Hayes (AMC) Birmingham cvtd **C-119C** 20Mar58; re-serialed **48-0349** 1958; to 837 AB GP (TAC) Shaw 07Nov58 mnt – 758 TCSq (459 TCG) (AFR) Greater Pittsburgh 19Dec58; 911 TCG (459 TCWG) (AFR) Greater Pittsburgh 17Jan63 – Wfu to MASDC (LOG) Davis-Monthan 07Jul66 for storage – REC: 31May67 – DFI: 02May68 – Final Disposition: Scrapped.

10332 / 48-350 / C-119B-FA

Avble: 19Oct49; Acc: 27Oct49; Del: 29Mar50 to USAF; briefly retained Fairchild (AMC) Hagerstown for project duties – 314 TCG (CNC) Sewart 30Mar50; depl Laurinburg-Maxton Airport North Carolina 20Apr50. Taxiing incident at Sewart AFB 07Jun50. Depl 316 TCG (CNC) Sewart 13Jul50. Had a minor mechanical failure 08Sep50 35 miles NW NAS Kwajalein Marshall Islands – 314 TCG (FEA) Komaki AB Japan 10Sep50; to Ashyia AB Japan 18Sep50. Suffered a taxiing accident while parked at NAS Agana Guam 14Oct50. Lost on a combat mission over Korea 03Jun51 due to friendly fire 1.9 miles NE Inje S. Korea. The aircraft (with C-119C s/n: 49-123) was approaching a drop zone area when fired on by U.S. artillery. One of the aircraft was hit in the tail, lost control and collided with the other. 4 crew killed were: 1Lt Eric W. Anderson, 1Lt John M. Gilbert, SSgt Houston N. Rich and Sgt Floyd N. Alexander – DFI: 24Jul51 – Final Disposition: Accident.

10333 / 48-351 / C-119B-FA

Avble: 25Oct49; Acc: 26Oct49; Del: 27Mar50 to USAF; briefly retained Fairchild (AMC) Hagerstown for project duties – 314 TCG (CNC) Sewart 28Mar50; depl Laurinburg-Maxton Airport North Carolina 20Apr50; depl 316 TCG (CNC) Sewart 13Jul50 – 314 TCG (FEA) Komaki AB Japan 10Sep50; to Ashiya AB Japan 18Sep50; to 13 ARP SQ (FEAMCOM) Tachikawa AB Japan 05Aug51 mods – bailment BMC (AMC) Birmingham 10Nov52 mods – 483 TCWG (FEA) Ashiya AB Japan 02Jan53 – 2253 RFC (CNR) Greater Pittsburgh 05Oct54; to BMC (AMC) Birmingham 28Aug55, cvtd **C-119C** 15Nov55; to 2750 AB WG (AMC) WPAFB 02Mar56 mnt; depl 2233 RFC (CNR) Mitchel 30Jun56 – 2259 RFC DT (CNR) Greater Pittsburgh 01Dec57; to 1001 AB WG (HQC) Andrews 04Feb58; re-serialed **48-0351** 1958 – 758 TCSq (459 TCG) (AFR) Greater Pittsburgh 19Dec58 – 167 AML SQ (WV-ANG) Martinsburg 08Apr61 – 130 ACO GP (ANG) Kanawha 20Sep63 – Wfu to 2704 ASD GP (LOG) Davis-Monthan 13Nov63 for storage – REC: 26Jun64 – Final Disposition: Scrapped.

10334 / 48-352 / C-119B-FA

Avble & Acc: 25Oct49; Del: 29Mar50 to USAF; briefly retained Fairchild Aircraft (AMC) Hagerstown for project duties – 314 TCG (CNC) Sewart 30Mar50; depl Laurinburg-Maxton Airport North Carolina 24Apr50; depl 316 TCG (CNC) Sewart 13Jul50 – 314 TCG (FEA) Komaki AB Japan 10Sep50; to Ashiya AB Japan 18Sep50; to 13 ARP SQ (FEAMCOM) Tachikawa AB Japan 26Jul51 mods – bailment BMC (AMC) Birmingham

20Nov52 mods; to 803 AB GP (SAC) Davis-Monthan 10Jan53 mnt – 483 TCWG (FEA) Ashiya AB Japan 01Sep53; depl 314 TCWG (FEA) Ashiya AB Japan 24Oct53 – 2233 RFC (CNR) Mitchel 27Oct54 – 2253 RFC (CNR) Greater Pittsburgh 16Apr55; to BMC (AMC) Birmingham 15Dec55, cvtd **C-119C** 23Feb56 – 2259 RFC DT (CNR) Greater Pittsburgh 01Dec57; re-serialed **48-0352** 1958; to 3415 TTA WG (ATC) Lowry 28Oct58 mnt – 758 TCSq (459 TCG) (AFR) Greater Pittsburgh 19Dec58; to 839 AB GP (TAC) Sewart 23Jun59 mnt; 911 TCG (459 TCWG) (AFR) Greater Pittsburgh 17Jan63 – Wfu to MASDC (LOG) Davis-Monthan 03Dec66 for storage – REC: 22Jun67 – BofS Aero Union Corp. Chico California 14Nov67; ARAp **N13746** 24Jan68 – BofS Hemet Valley Flying Service Co. Hemet California 18Feb75; cvtd **C-119C Jet-Pak** (J34-WE) with water and chemical tank installation 01Jun76 for airtanker duties, Tanker **87**; cvtd with hopper installation for seed dropping 08Oct77-10Jan78; recovered elevator 26Apr82; cvtd seed dropping 04-16Oct85; TV appearance *MacGyver* episode *The Heist* (S.1 Ep.5) Nov85 involved a stunt dropping a car out the back of the aircraft in-flight. Wfu and stored 1987 – To USAF Museum Program and placed on static display at the Air Force Flight Test Museum Edwards AFB California. N13746 de-reg 28Sep92 – Air Mobility Command Museum Dover AFB Delaware 2016. Dismantled and transported from Edwards to Dover 16Dec16 on a C-5M Galaxy (s/n 86-0017) for a full restoration – Final Disposition: Extant.

10335 / 48-353 / C-119B-FA

Avble & Acc: 26Oct49; retained Fairchild (AMC) Hagerstown for project duties; Del: 04Apr50 to USAF – 314 TCG (CNC) Sewart 06Apr50; depl Laurinburg-Maxton Airport North Carolina 20Apr50; depl 316 TCG (CNC) Sewart 13Jul50 – 314 TCG (FEA) Komaki AB Japan 10Sep50; assigned 62 TCSq; to Ashyia AB Japan 18Sep50. Flew into terrain near Ashiya AB Japan 13Sep50. Pilot Ellis S. O'Connell and crew killed – REC: 30Sep50 – DFI: 06Dec50 – Final Disposition: Accident.

10336 / 48-354 / C-119B-FA

Avble: 28Oct49; Acc: 31Oct49; Del: 04Apr50 to USAF; briefly retained Fairchild (AMC) Hagerstown for project duties – 314 TCG (CNC) Sewart 06Apr50; depl Laurinburg-Maxton Airport North Carolina 20Apr50; depl 316 TCG (CNC) Sewart 13Jul50 – 314 TCG (FEA) Komaki AB Japan 10Sep50; to Ashiya AB Japan 18Sep50; to 13 ARP SQ (FEAMCOM) Tachikawa AB Japan 02Aug51 & 26Nov51 mods; to 6400 AD WG (FEA) Tachikawa AB Japan 31Jan52 – bailment BMC (AMC) Birmingham 13Apr53 mods – 483 TCWG (FEA) Ashiya AB Japan 12Nov53; depl 314 TCWG (FEA) Ashiya AB Japan 13Dec53 – MIDAR (AMC) Olmsted 27Apr55, cvtd **C-119C** 04Jan56 – 2253 RFC (CNR) Greater Pittsburgh 10Feb56; to 3800 AB WG (AU) Maxwell 09Dec56 mnt – 2259 RFC DT (CNR) Greater Pittsburgh 01Dec57; re-serialed **48-0354** 1958 – 758 TCM SQ (459 TCG) (AFR) Greater Pittsburgh 19Dec58 – 167 AML SQ (WV-ANG) Martinsburg 08Apr61 – 130 ACO GP (ANG) Kanawha 26Sep63 – Wfu to 2704 ASD GP (LOG) Davis-Monthan 13Nov63 for storage – REC: 26Jun64 – Final Disposition: Scrapped.

10337 / 48-355 / C-119B-FA

Avble & Acc: 31Oct49; Del: 04Apr50 to USAF; briefly retained Fairchild (AMC) Hagerstown for project duties – 314 TCG (CNC) Sewart 06Apr50; depl Laurinburg-Maxton Airport North Carolina 20Apr50; depl 316 TCG (CNC) Sewart 21Jul50 – 314 TCG (FEA) Komaki AB Japan 10Sep50; to Ashiya AB Japan 18Sep50; to 13 ARP SQ (FEAMCOM) Tachikawa AB Japan 23Aug51 mods – 6400 AD WG (FEA) Tachikawa AB Japan

11Apr52 – bailment BMC (AMC) Birmingham 14Jun52 mods – 483 TCWG (FEA) Ashiya AB Japan 09Sep53; depl 314 TCWG (FEA) Ashiya AB Japan 24Sep53; to 1500 MAI SQ (MATS) Hickam Sep54 mnt – 2233 RFC (CNR) Mitchel 03Sep54; bailment BMC (AMC) Birmingham 19May55; depl 2253 RFC (CNR) Greater Pittsburgh 18Aug55 – 2253 RFC (CNR) Greater Pittsburgh 24Oct55; cvtd **C-119C** 10Sep56; depl 2347 RFC (CNR) Long Beach 02Dec56; depl 2233 RFC (CNR) Mitchel 10Jan58 – 2259 RFC (CNR) Greater Pittsburgh 11Feb58; re-serialed **48-0355** 1958 – 758 TCSq (459 TCG) (AFR) Greater Pittsburgh 19Dec58 – 167 AML SQ (WV-ANG) Martinsburg 25Apr61 – 130 ACO GP (ANG) Kanawha 05Sep63 – Wfu to 2704 ASD GP (LOG) Davis-Monthan 16Nov63 for storage – REC: 26Jun64 – Final Disposition: Scrapped.

10338 / 49-101 / C-119B-10-FA

Avble: 16Nov49; Acc: 20Dec49; Del: 22Dec49 to USAF – 314 TCG (CNC) Smyrna 29Dec49; depl Laurinburg-Maxton Airport North Carolina 20Apr50; depl 316 TCG (CNC) Sewart 13Jul50 – 314 TCG (FEA) Komaki AB Japan 10Sep50; to Ashiya AB Japan 18Sep50; depl 37 TCSq (314 TCG) Komaki AB Japan 08Nov50; to 13 ARP SQ (FEAMCOM) Tachikawa AB Japan 23Aug51 mods – 403 TCWG (FEA) Ashiya AB Japan 28Apr52 – bailment BMC (AMC) Birmingham 07Nov52 mods – 4901 SUT WG (ARD) Kirkland 18Sep53 mnt – OGDAR (AMC) Hill 29Sep53 mnt – 483 TCWG (FEA) Ashiya AB Japan 02Feb54; depl 314 TCWG (FEA) Ashiya AB Japan 08Feb54 – 2233 RFC (CNR) Mitchel 10Oct54; to Fairchild (AMC) Hagerstown 27Jul55, cvtd **C-119C** 02Sep55; re-serialed **49-0101** 1958 – 514 TCWG (AFR) Mitchel 19Dec58, assigned 336 TCSq Stewart 18Mar61; 904 TCG (514 TCWG) (AFR) Stewart 31May63; to 2856 AB WG (LOG) Griffiss 14Aug63 – Wfu to MASDC (LOG) Davis-Monthan 06Jun65 for storage – REC: 21Jun66 – Final Disposition: Scrapped.

10339 / 49-102 / C-119B-10-FA

Avble: 21Nov49; Acc: 20Dec49; Del: 28Dec49 to USAF – 314 TCG (CNC) Smyrna 30Dec49; depl Laurinburg-Maxton Airport North Carolina 20Apr50; depl 316 TCG (CNC) Sewart 21Jul50 – 314 TCG (FEA) Ashiya AB Japan 30Sep50; depl 37 TCSq (314 TCG) Komaki AB Japan 08Nov50; depl 374 TCWG (FEA) Tachikawa AB Japan 14Dec50; to 13 ARP SQ (FEAMCOM) Tachikawa AB Japan 30Jul51 mods; depl Komaki AB Japan 30Aug51 – 403 TCWG (FEA) Ashiya AB Japan 28Apr52 – bailment BMC (AMC) Birmingham 19Aug52 mods – 483 TCWG (FEA) Ashiya AB Japan 14Jul53; depl 314 TCWG (FEA) Ashiya AB Japan 23Jul53 – SMAAR (AMC) McClellan 09Oct54 mnt – 2253 RFC (CNR) Greater Pittsburgh 31Mar55; to CAR Center (ARD) Hanscom 03Sep55 mnt – BMC (AMC) Birmingham 10Feb56, cvtd **C-119C** 02Apr56 – 2233 RFC (CNR) Mitchel 16Apr56 – 2500 AB WG (CNC) Mitchel 26Mar57; re-serialed **49-0102** 1958. Listed as having ground accident 16Sep58, nfd – Final Disposition: Accident.

10340 / 49-103 / C-119B-10-FA

Avble: 23Nov49; Acc: 20Dec49; Del: 28Dec49 to USAF – 314 TCG (CNC) Smyrna 29Dec49; depl Laurinburg-Maxton Airport North Carolina 20Apr50; depl 316 TCG (CNC) Sewart 13Jul50 – 314 TCG (FEA) Komaki AB Japan 10Sep50, assigned 37 TCSq; to Ashiya AB Japan 18Sep50; depl Komaki AB Japan 29Nov50; to 13 ARP SQ (FEAMCOM) Tachikawa AB Japan 30Jun51 mods – 403 TCWG (FEA) Ashiya AB Japan 28Apr52, assigned 63 TCSq. Made a forced landing due to engine failure 2 miles SE Niigata AB Japan 25Jul52, Pilot: Jack Parker, no fatalities – DFI: 12Aug52 – Final Disposition: Accident.

10341 / 49-104 / C-119B-10-FA

Avble: 28Nov49; Acc: 21Dec49; Del: 29Dec49 to USAF – 314 TCG (CNC) Smyrna 29Dec49; depl 316 TCG (CNC) Sewart 13Jul50 – 314 TCG (FEA) Komaki AB Japan 14Sep50; to Ashiya AB Japan 18Sep50; depl 37 TCSQ (314 TCG) Komaki AB Japan 08Nov50; to 13 ARP SQ (FEAMCOM) Tachikawa AB Japan 30Jun51 mods – 374 TCWG (FEA) Tachikawa AB Japan 24Sep51; to 13 ARP SQ (FEAMCOM) Tachikawa AB Japan 06Oct51 mods – 6400 AD WG (FEA) 31Jan52 for storage and mnt – bailment BMC (AMC) Birmingham 09Mar54; to 1509 AB SQ (MATS) Johnston 19Mar54 mnt – 2233 RFC (CNR) Mitchel 10Oct54; to BMC (AMC) Birmingham 28Nov55, cvtd **C-119C** 09Feb56; to 2500 AB WG (CNC) Mitchel 18May56 mnt; re-serialed **49-0104** 1958 – 3800 AB WG (AU) Maxwell 06Oct58 – 514 TCWG (AFR) Mitchel 19Dec58, assigned 336 TCSq Stewart 18Mar61; 904 TCG (514 TCWG) (AFR) Stewart 31May63 – Wfu to MASDC (LOG) Davis-Monthan 06Jun65 for storage – REC: 21Jun66 – Final Disposition: Scrapped.

10342 / 49-105 / C-119B-10-FA

Avble: 30Nov49; Acc: 24Dec49; Del: 11Jan50 to USAF – 314 TCG (CNC) Smyrna 16Jan50; depl Laurinburg-Maxton Airport North Carolina 20Apr50; depl 316 TCG (CNC) Sewart 13Jul50 & 23Aug50 – 314 TCG (FEA) Ashiya AB Japan 30Sep50, assigned 37 TCSq. Crashed returning from a combat mission during an instrument landing due to weather conditions at Ashiya AB Japan 12Oct50. Pilot: Norman E. Morrison, 3 of the 4 crew killed, 4 passengers survived – DFI: 06Dec50 – Final Disposition: Accident.

10343 / 49-106 / C-119B-10-FA

Avble: 02Dec49; Acc: 24Dec49; Del: 16Jan50 to USAF – 314 TCG (CNC) Smyrna 18Jan50; depl Laurinburg-Maxton Airport North Carolina 20Apr50 – 314 TCG (FEA) Komaki AB Japan 10Sep50; to Ashiya AB Japan 18Sep50; to 13 ARP SQ (FEAMCOM) Tachikawa AB Japan 02Nov50 & 01Aug51; depl Bofu Japan 18May53 – bailment BMC (AMC) Birmingham 13Nov53 – 483 TCWG (FEA) Ashiya AB Japan 08Apr54; depl 314 TCWG (FEA) Ashiya AB Japan 22Apr54; loan to French Armée de l'Air Indochina 13Jun54-07Jul54, French markings applied – 2233 RFC (CNR) Mitchel 28Oct54; depl 2253 RFC (CNR) Greater Pittsburgh 17Apr55 & 30Jul55; to BMC (AMC) Birmingham 28Feb56, cvtd **C-119C** 18Apr56; to 2500 AB WG (CNC) Mitchel 26Mar57; re-serialed **49-0106** 1958 – 514 TCWG (AFR) Mitchel 19Dec58, assigned 336 TCSq Stewart 15Mar61; 904 TCG (514 TCWG) (AFR) Stewart 31May63 – Wfu to MASDC (LOG) Davis-Monthan 15May65 for storage – REC: 21Jun66 – Final Disposition: Scrapped.

10344 / 49-107 / C-119B-10-FA

Avble: 08Dec49; Acc: 28Dec49; Del: 16Jan50 to USAF – 314 TCG (CNC) Smyrna 18Jan50; depl Laurinburg-Maxton Airport North Carolina 24Apr50; depl 316 TCG (CNC) Sewart 13Jul50 – 314 TCG (FEA) Komaki AB Japan 10Sep50; to Ashiya AB Japan 18Sep50; depl 37 TCSQ (314 TCG) Komaki AB Japan 08Nov50; to 13 ARP SQ (FEAMCOM) Tachikawa AB Japan 24Aug51 – 403 TCWG (FEA) Ashiya AB Japan 28Apr52 – 483 TCWG (FEA) Ashiya AB Japan 01Nov52; depl Bofu Japan 25May53 – 2233 RFC (CNR) Mitchel 09Sep54; to 1607 FDM SQ (MATS) Dover 03Jan55 – 2253 RFC (CNR) Greater Pittsburgh 16Aug55 – WRAAR (AMC) Robins 26Aug55 mnt – BMC (AMC) Birmingham 25Apr56, cvtd **C-119C** 29Jul56 – 2233 RFC (CNR) Mitchel 07Aug56; re-serialed **49-0107** 1958; to 2500 AB WG (CNC) Mitchel 05May58 mnt; to 1607 TSH WG (MATS) Dover 28Jul58 mnt – 514 TCWG (AFR) Mitchel 19Dec58, assigned 336 TCSq Stewart 15Mar61; 904 TCG (514

TCWG) (AFR) Stewart 31May63 – Wfu to MASDC (LOG) Davis-Monthan 12Apr65 for storage – REC: 21Jun66 – Final Disposition: Scrapped.

10345 / 49-108 / C-119B-10-FA

Avble: 12Dec49; Acc: 28Dec49; Del: 25Jan50 to USAF – 314 TCG (CNC) Smyrna 27Jan50; depl Laurinburg-Maxton Airport North Carolina 24Apr50; depl 316 TCG (CNC) Sewart 13Jul50 – 314 TCG (FEA) Ashiya AB Japan 30Sep50; depl 37 TCSQ (314 TCG) Komaki AB Japan 08Nov50; to 13 ARP SQ (FEAMCOM) Tachikawa AB Japan 28Jul51 – 403 TCWG (FEA) Ashiya AB Japan 28Apr52 – bailment BMC (AMC) Birmingham 01Nov52 – 483 TCWG (FEA) Ashiya AB Japan 02Jan53; loan to French Armée de l'Air Indochina 15Apr54-14May54, French markings applied – 2253 RFC (CNR) Greater Pittsburgh 05Oct54; to BMC (AMC) Birmingham 15Sep55, cvtd **C-119C** 29Nov55 – 2233 RFC (CNR) Mitchel 30Jun56; to 2500 AB WG (CNC) Mitchel 11Oct56; re-serialed **49-0108** 1958; to 3902 AB WG (SAC) Offutt 06Mar58 mnt – 514 TCWG (AFR) Mitchel 19Dec58; to 1604 AB WG (MATS) Kindley 21Nov60 mnt; assigned 336 TCSq Stewart 15Mar61; 904 TCG (514 TCWG) (AFR) Stewart 31May63 – Wfu to MASDC (LOG) Davis-Monthan 15May65 for storage – REC: 21Jun66 – Final Disposition: Scrapped.

10346 / 49-109 / C-119B-10-FA

Avble: 13Dec49; Acc: 28Dec49; Del: 16Jan50 to USAF – 314 TCG (CNC) Smyrna 18Jan50; depl Laurinburg-Maxton Airport North Carolina 24Apr50; depl 316 TCG (CNC) Sewart 13Jul50 – 314 TCG (FEA) Komaki AB Japan 14Sep50; to Ashiya AB Japan 18Sep50; depl 37 TCSQ (314 TCG) Komaki AB Japan 08Nov50; to 13 ARP SQ (FEAMCOM) Tachikawa AB Japan 20Jul51 – 403 TCWG (FEA) Ashiya 28Apr52 – bailment BMC (AMC) Birmingham 02Sep52 – 483 TCWG (FEA) Ashiya AB Japan 26May53; depl 314 TCWG (FEA) Ashiya AB & Tachikawa AB Japan 08Jun53; loan to French Armée de l'Air Indochina early54-16May54, French markings applied – 2233 RFC (CNR) Mitchel 03Dec54; to BMC (AMC) Birmingham 01Oct55, cvtd **C-119C** 29Nov55; depl 2253 RFC (CNR) Greater Pittsburgh 14Jun56; re-serialed **49-0109** 1958 – 514 TCWG (AFR) Mitchel 19Dec58, assigned 336 TCSq Stewart 21Mar61; 904 TCG (514 TCWG) (AFR) Stewart 31May63 – Wfu to MASDC (LOG) Davis-Monthan 20May65 for storage – REC: 21Jun66 – Final Disposition: Scrapped.

7001 / BuNo. 124324 / R4Q-1

Avble: 19Dec49; Acc: 29Dec49; Del: 09Feb50 to USMC – NATC NAS Patuxent River 08Feb50 for tests – VMR-252 MCAS Cherry Point 01Sep50 – O&R MCAS Cherry Point 18Feb52 mnt – O&R NAS San Diego 13Aug53 mnt – O&R NAS Corpus Christi 25Jan56 mnt – Wfu to NAF Litchfield Park 29Apr56 for storage – DFI: 13May60 – Final Disposition: Scrapped.

7002 / BuNo. 124325 / R4Q-1

Avble: 20Dec49; Acc: 01Feb50; Del: 20Feb50 to USMC – VMR-252 MCAS Cherry Point 01Mar50. Crashed near Eastville Virginia due to a structural failure while flying in bad weather from MCAS Cherry Point to New York-Floyd Bennet Field 07Feb51, the 9 fatalities were: Capt Kenneth H. Dieffenbach, Capt Edward E. Rost, MSgt Ralph H. Alderman, Sr., MSgt Robert C. Holly, MSgt Robert L. Jonasson, SSgt James E. Thomas, TSgt Walter J. Hoover, TSgt Melvin E. Rainey and Cpl Leonard McDonald – DFI: 08Mar51 – Final Disposition: Accident.

7003 / BuNo. 124326 / R4Q-1

Avble: 29Dec49; Acc: 27Jan50; Del: 08Feb50 to USMC – NATC NAS Patuxent River 08Feb50 for tests – VMR-252 MCAS Cherry Point 06Jul50 – VMR-153 MCAS Cherry Point 18Jun52 – O&R MCAS Cherry Point 30Apr53 mnt – O&R NAS San Diego 21Nov53 mnt – O&R NAS Corpus Christi 13Apr55 mnt – O&R NAS Jacksonville 19Jul57 mnt – Wfu to NAF Litchfield Park 18May58 for storage – DFI: 13May60 – Final Disposition: Scrapped.

7004 / BuNo. 124327 / R4Q-1

Avble: 31Dec49; Acc: 09Feb50; Del: 10Mar50 to USMC – VMR-252 MCAS Cherry Point 14Mar50 – VMR-153 MCAS Cherry Point 18Jun52 – O&R MCAS Cherry Point 30Apr53 mnt – Pool Maintenance MCAS El Toro 15Dec53 mnt – O&R NAS San Diego 11Feb54 mnt – O&R NAS Corpus Christi 29Mar55 mnt – Wfu to NAF Litchfield Park 13Aug57 for storage – DFI: 13May60 – Final Disposition: Scrapped.

7005 / BuNo. 124328 / R4Q-1

Avble: 06Jan50; Acc: 07Feb50; Del: 20Feb50 to USMC – VMR-252 MCAS Cherry Point 01Mar50 – VMR-153 MCAS Cherry Point 18Jun52 – O&R MCAS Cherry Point 30Apr53 mnt – MTG-10 MCAS El Toro 05Jan54 – H&MS-25 MCAS El Toro 13Mar54 mnt – O&R NAS San Diego 17Aug54 mnt – O&R NAS Corpus Christi 08Jun55 mnt – Wfu to NAF Litchfield Park 29Aug57 for storage – DFI: 13May60 – Final Disposition: Scrapped.

7006 / BuNo. 124329 / R4Q-1

Avble: 16Jan50; Acc: 08Feb50; Del: 10Mar50 to USMC – VMR-252 MCAS Cherry Point 14Mar50 – VMR-153 MCAS Cherry Point 18Jun52 – O&R MCAS Cherry Point 30Apr53 mnt – O&R NAS San Diego 01Nov54 mnt – O&R NAS Corpus Christi 02Dec54 mnt – NAATC NAS Corpus Christi 08Apr56 – O&R NAS Corpus Christi 08Apr58 – 5 Naval District NAS Norfolk 30Nov58 – Wfu to NAF Litchfield Park 15Jul59 for storage – DFI: 13May60 – Final Disposition: Scrapped.

7007 / BuNo. 124330 / R4Q-1

Avble: 16Jan50; Acc: 31Jan50; Del: 20Feb50 to USMC – VMR-252 MCAS Cherry Point 01Mar50 – VMR-153 MCAS Cherry Point 18Jun52 – O&R MCAS Cherry Point 30Apr53 mnt – O&R NAS San Diego 27Oct53 mnt – MTG-10 MCAS El Toro 29Dec53 – H&MS-13 MCAS Kaneohe 20Feb54 – FASRON-117 NAS Barber's Point 30Apr54 – H&MS-25 MCAS El Toro 27Jul54 – O&R NAS San Diego 17Aug54 mnt – O&R NAS Corpus Christi 29Oct54 mnt – NABTC NAS Pensacola 27Nov57 – Wfu to NAF Litchfield Park 29May59 for storage – DFI: 13May60 – Final Disposition: Scrapped.

7008 / BuNo. 124331 / R4Q-1

Avble: 18Jan50; Acc: 20Feb50; Del: 10Mar50 to USMC – VMR-252 MCAS Cherry Point 14Mar50 – VMR-153 MCAS Cherry Point 18Jun52. Crashed after take-off 1.9 miles from MCAS Cherry Point 27Jun52, the pilot radioed an emergency return to the base before the aircraft lost control. The 5 fatalities were: Capt John W. Godfrey (student pilot), Capt Robert L. McCartney, Jr., (student pilot), 1Lt Benjamin A. Phipps (pilot), TSgt James B. Merritt (crew chief) and Cpl James M. Coker (assist crew chief) – DFI: 01Jul52 – Final Disposition: Accident.

Section Four—Service Histories

10347 / 49-110 / C-119B-12-FA

Avble: 24Jan50; Acc: 18Feb50; Del: 23Feb50 to USAF – 314 TCG (CNC) Smyrna 28Feb50; depl Laurinburg-Maxton Airport North Carolina 24Apr50; depl 316 TCG (CNC) Sewart 13Jul50 – 314 TCG (FEA) Komaki AB Japan 14Sep50; to Ashiya AB Japan 18Sep50; depl 37 TCSQ (314 TCG) Komaki AB Japan 08Nov50; to 13 ARP SQ (FEAMCOM) Tachikawa AB Japan 25Jul51 – 403 TCWG (FEA) Ashiya 28Apr52 – bailment BMC (AMC) Birmingham 09Dec52 – 483 TCWG (FEA) Ashiya AB Japan 10Sep53; to 1505 AB GP (MATS) Johnston 20Sep53 mnt; depl 314 TCWG (FEA) Ashiya AB Japan 21Oct53; loan to French Armée de l'Air Indochina 03May54-03Jul54, French markings applied – 2233 RFC (CNR) Mitchel 15Sep54; to Fairchild (AMC) Hagerstown 21Sep55, cvtd **C-119C** 04Nov55; re-serialed **49-0110** 1958; to 3415 TTA WG (ATC) Lowry 25Aug58 – 514 TCWG (AFR) Mitchel 19Dec58, assigned 336 TCSq Stewart 25Apr61; 904 TCG (514 TCWG) (AFR) Stewart 05Nov63 – 464 TCWG (TAC) Pope 12Nov63, likely no active service – REC: 02Jan64 – Final Disposition: Scrapped.

10348 / 49-111 / C-119B-12-FA

Avble: 24Jan50; Acc: 17Feb50; Del: 23Feb50 to USAF – 314 TCG (CNC) Smyrna 27Feb50; depl Laurinburg-Maxton Airport North Carolina 24Apr50; depl 316 TCG (CNC) Sewart 13Jul50 – 314 TCG (FEA) Komaki AB Japan 14Sep50; to Ashiya AB Japan 18Sep50; depl 37 TCSQ (314 TCG) Komaki AB Japan 08Nov50; to 13 ARP SQ (FEAMCOM) Tachikawa AB Japan 22Aug51 – 403 TCWG (FEA) Ashiya 26Apr52 – 483 TCWG (FEA) Ashiya AB Japan 01Jan53; bailment BMC (AMC) Birmingham 11Aug53; to 6319 AB WG (FEA) Andersen 15Aug53 mnt – 2233 RFC (CNR) Mitchel 27Oct54; to Fairchild (AMC) Hagerstown 28Sep55, cvtd **C-119C** 24Oct55; re-serialed **49-0111** 1958; depl 2584 RFC (CNR) Memphis 19Jun58 – 514 TCWG (AFR) Mitchel 19Dec58; to 1405 AB WG (MATS) Scott 11Jun59 mnt; assigned 336 TCSq Stewart 15Mar61; 904 TCG (514 TCWG) (AFR) Stewart 31May63 – Wfu to MASDC (LOG) Davis-Monthan 12Apr65 for storage – REC: 21Jun66 – Final Disposition: Scrapped.

10349 / 49-112 / C-119B-12-FA

Avble: 3Jan50; Acc: 24Feb50; Del: 09Mar50 to USAF – 314 TCG (CNC) Smyrna 10Mar50; depl Laurinburg-Maxton Airport North Carolina 20Apr50; depl 316 TCG (CNC) Sewart 02Aug50 – 314 TCG (FEA) Komaki AB Japan 10Sep50; to Ashiya AB Japan 18Sep50; depl 37 TCSQ (314 TCG) Komaki AB Japan 08Nov50; to 13 ARP SQ (FEAMCOM) Tachikawa AB Japan 22Aug51 – bailment BMC (AMC) Birmingham 13Apr53 – 483 TCWG (FEA) Ashiya AB Japan 02Nov53; depl 314 TCWG (FEA) Ashiya 13Dec53; loan to French Armée de l'Air Indochina 05Jun54-15Jul54, French markings applied – 2233 RFC (CNR) Mitchel 23Sep54; to BMC (AMC) Birmingham 05Aug55, cvtd **C-119C** 28Nov55; depl 2253 RFC (CNR) Greater Pittsburgh 14Jun56; to 464 TCWG (TAC) Pope 08Aug56 mnt; to 2750 AB WG (AMC) WPAFB 03Oct56 mnt; re-serialed **49-0112** 1958; AEMCO (AMC) Oakland IAP 19Jun58 – 514 TCWG (AFR) Mitchel 19Dec58, assigned 336 TCSq Stewart 15Mar61; 904 TCG (514 TCWG) (AFR) Stewart 31May63 – Wfu to MASDC (LOG) Davis-Monthan 12Jun65 for storage – REC: 21Jun66 – Final Disposition: Scrapped.

10350 / 49-113 / C-119B-12-FA

Avble: 3Jan50; Acc: 22Feb50; Del: 03Mar50 to USAF – 314 TCG (CNC) Smyrna 07Mar50; depl Laurinburg-Maxton Airport North Carolina 24Apr50; depl 316 TCG (CNC) Sewart 13Jul50 – 314 TCG (FEA) Komaki AB Japan 10Sep50; to Ashiya AB Japan

18Sep50; depl 37 TCSQ (314 TCG) Komaki AB Japan 08Nov50; to 13 ARP SQ (FEAMCOM) Tachikawa AB Japan 07Jun51 – bailment BMC (AMC) Birmingham 09Feb53 – 483 TCWG (FEA) Ashiya AB Japan 26Feb54 – 2233 RFC (CNR) Mitchel 18Nov54; to Fairchild (AMC) Hagerstown 01Aug55, cvtd **C-119C** 08Sep55; depl 2253 RFC (CNR) Greater Pittsburgh 14Jun56; re-serialed **49-0113** 1958 – 514 TCWG (AFR) Mitchel 19Dec58, assigned 336 TCSq Stewart 15Mar61; 904 TCG (514 TCWG) (AFR) Stewart 31May63 – Wfu to MASDC (LOG) Davis-Monthan 06Jun65 for storage – REC: 21Jun66 – Final Disposition: Scrapped.

10351 / 49-114 / C-119B-12-FA

Avble: 02Feb50; Acc: 24Feb50; Del: 06Mar50 to USAF – 314 TCG (CNC) Smyrna 09Mar50; depl Laurinburg-Maxton Airport North Carolina 20Apr50; depl 316 TCG (CNC) Sewart 13Jul50 – 314 TCG (FEA) Komaki AB Japan 14Sep50; to Ashiya AB Japan 18Sep50; depl 37 TCSQ (314 TCG) Komaki AB Japan 08Nov50; to 13 ARP SQ (FEAMCOM) Tachikawa AB Japan 03Aug51; depl 403 TCWG (FEA) Ashiya AB Japan 19May52; depl Tachikawa AB Japan 23Jun53 – bailment BMC (AMC) Birmingham 30Aug53 – 483 TCWG (FEA) Ashiya AB Japan 02Feb54; loan to French Armée de l'Air Indochina 08-24Jun54, French markings applied – 2253 RFC (CNR) Greater Pittsburgh 08Sep54. Crashed during landing 08Mar55 at Greater Pittsburgh, no fatalities – DFI: 14Mar55 – Final Disposition: Accident.

10352 / 49-115 / C-119B-12-FA

Avble: 07Feb50; Acc: 24Feb50; Del: 09Mar50 to USAF – 314 TCG (CNC) Smyrna 10Mar50; depl Laurinburg-Maxton Airport North Carolina 20Apr50; depl 316 TCG (CNC) Sewart 13Jul50 – 314 TCG (FEA) Ashiya AB Japan 30Sep50; depl 37 TCSQ (314 TCG) Komaki AB Japan 08Nov50; to 13 ARP SQ (FEAMCOM) Tachikawa AB Japan 08Jul51 – 403 TCWG (FEA) Ashiya AB Japan 28Apr52 – 483 TCWG (FEA) Ashiya AB Japan 01Jan53 – bailment BMC (AMC) Birmingham 04Jul53 – 2233 RFC (CNR) Mitchel 03Sep54; depl 2253 RFC (CNR) Greater Pittsburgh 16Aug55; to BMC (AMC) Birmingham 06Sep55, cvtd **C-119C** 29Nov55; re-serialed **49-0115** 1958 – 514 TCWG (AFR) Mitchel 19Dec58, assigned 336 TCSq Stewart 18Mar61; 904 TCG (514 TCWG) (AFR) Stewart 31May63 – Wfu to MASDC (LOG) Davis-Monthan 15May65 for storage – REC: 21Jun66 – Final Disposition: Scrapped.

10353 / 49-116 / C-119B-12-FA

Avble: 09Feb50; Acc: 27Feb50; Del: 10Mar50 to USAF – 314 TCG (CNC) Smyrna 10Mar50; depl Laurinburg-Maxton Airport North Carolina 26Apr50; depl 316 TCG (CNC) Sewart 13Jul50 – 314 TCG (FEA) Ashiya AB Japan 30Sep50; depl 37 TCSQ (314 TCG) Komaki AB Japan 08Nov50; to 13 ARP SQ (FEAMCOM) Tachikawa AB Japan 07Aug51 – 403 TCWG (FEA) Ashiya AB Japan 28Apr52 – bailment BMC (AMC) Birmingham 26Aug52 – 483 TCWG (FEA) Ashiya AB Japan 29Apr53; loan to French Armée de l'Air Indochina 27May54-07Jun54, French markings applied – 2233 RFC (CNR) Mitchel 03Oct54; to Fairchild (AMC) Hagerstown 29Jul55, cvtd **C-119C** 07Sep55; depl 2253 RFC (CNR) Greater Pittsburgh 14Jun56; to 3450 TTA WG (TC) Francis E. Warren 30Sep56; re-serialed **49-0116** 1958 – 514 TCWG (AFR) Mitchel 19Dec58, assigned 336 TCSq Stewart 15Mar61; 904 TCG (514 TCWG) (AFR) Stewart 31May63 – Wfu to MASDC (LOG) Davis-Monthan 20May65 for storage – REC: 21Jun66 – Final Disposition: Scrapped.

10354 / 49-117 / C-119B-12-FA

Avble: 13Feb50; Acc: 01Mar50; Del: 09Mar50 to USAF – 314 TCG (CNC) Smyrna 10Mar50; depl Laurinburg-Maxton Airport North Carolina 20Apr50; depl 316 TCG

(CNC) Sewart 13Jul50 – 314 TCG (FEA) Komaki AB Japan 10Sep50; to Ashiya AB Japan 18Sep50; depl 37 TCSQ (314 TCG) Komaki AB Japan 08Nov50; to 13 ARP SQ (FEAM-COM) Tachikawa AB Japan 08Aug51 – 403 TCWG (FEA) Ashiya AB Japan 28Apr52 – 483 TCWG (FEA) Ashiya AB Japan 01Jan53 – 2253 RFC (CNR) Greater Pittsburgh 11Sep54; to 2750 AB WG (AMC) WPAFB 16Oct55 mnt; to BMC (AMC) Birmingham 18Jan56, cvtd **C-119C** 15Mar56 – 2233 RFC (CNR) Mitchel 23Mar56 – 2500 AB WG (CNC) Mitchel 27Nov57 mnt; re-serialed **49-0117** 1958 – Hayes (AMC) Birmingham 11Jul58 – 446 TCWG (AFR) Ellington 11Nov58; assigned 706 TCSq Barksdale 12Apr59 – 78 TCSq (435 TCWG) (AFR) Barksdale 08May61 – 336 TCSq (514 TCG) (AFR) Stewart 13May61; 904 TCG (514 TCWG) (AFR) Stewart 31May63 – Wfu to MASDC (LOG) Davis-Monthan 12Apr65 for storage – REC: 21Jun66 – Final Disposition: Scrapped.

10355 / 49-118 / C-119B-12-FA

Avble: 17Feb50; Acc: 07Mar50; Del: 13Mar50 to USAF – 314 TCG (CNC) Smyrna 15Mar50; depl Laurinburg-Maxton Airport North Carolina 20Apr50; depl 316 TCG (CNC) Sewart 13Jul50 – 314 TCG (FEA) Komaki AB Japan 14Sep50; to Ashiya AB Japan 18Sep50; depl 37 TCSQ (314 TCG) Komaki AB Japan 08Nov50; to 13 ARP SQ (FEAMCOM) Tachikawa AB Japan 05Jul51 – 403 TCWG (FEA) Ashiya AB Japan 28Apr52 – MIDAR (AMC) Olmsted 22Sep52 – 1500 OP SQ (MATS) Hickham 06Oct52 mnt – BMC (AMC) Birmingham 09Dec52 – 483 TCWG (FEA) Ashiya AB Japan 01Jan53; depl 314 TCWG (FEA) Ashiya AB Japan 08Mar54; loan to French Armée de l'Air Indochina 07-14May54, French markings applied – 2253 RFC (CNR) Greater Pittsburgh 23Sep54; depl 2233 RFC (CNR) Mitchel 14Jul55; to BMC (AMC) Birmingham 24Oct56, cvtd **C-119C** 06Dec56; to 464 TCWG (TAC) Pope 03Jul56 mnt – 2233 RFC (CNR) Mitchel 12Sep56; to 464 TCWG (TAC) Pope 28Aug57 mnt; re-serialed **49-0118** 1958 – 514 TCWG (AFR) Mitchel 19Dec58, assigned 336 TCSq Stewart 15Mar61; 904 TCG (514 TCWG) (AFR) Stewart 31May63 – Wfu to MASDC (LOG) Davis-Monthan 06Jun65 for storage – REC: 21Jun66 – Final Disposition: Scrapped.

10356 / 49-119 / C-119C-12-FA

Avble: 27Feb50; Acc: 30Mar50; Del: 04Apr50 to USAF – 314 TCG (CNC) Sewart 06Apr50; depl Laurinburg-Maxton Airport North Carolina 20Apr50 – 314 TCG (FEA) Komaki AB Japan 10Sep50; to Ashiya AB Japan 18Sep50; to 13 ARP SQ (FEAMCOM) Tachikawa AB Japan 04Jun51; depl 6122 AB WG (FEA) Ashiya AB Japan 24Nov51 – MIDAR (AMC) Olmsted 25Sep52 – BMC (AMC) Birmingham 14Jan53 – 483 TCWG (FEA) Ashiya AB Japan 05Aug53; to 1505 AB GP (MATS) Johnston 15Dec53 mnt; to 1503 MAI SQ (MATS) Tokyo IAP Japan 20DSec53 mnt; depl 314 TCWG (FEA) Ashiya AB Japan 29Jan54 – 316 TCWG (FEA) Ashiya AB Japan 15Nov54 – MIDAR (AMC) Olmsted 07Jan55 – bailment Fairchild (AMC) Hagerstown 10Oct55 – 5039 AB WG (AAC) Elmendorf 19Nov55; to 5039 MAI SQ (AAC) Elmendorf 09Oct56 – BMC (AMC) Birmingham 23Jan57 – 2233 RFC (CNR) Mitchel 29May57; to 414 Fighter GP (ADC) Oxnard 03Jul57 mnt; re-serialed **49-0119** 1958 – 514 TCWG (AFR) Mitchel 19Dec58, assigned 336 TCSq Stewart 15Mar61 – 130 ACO GP (ANG) Kanawha date unk – Wfu to MASDC (LOG) Davis-Monthan 11Mar69 for storage, PCN: **CJ292** – DFI: 26Mar69 – Final Disposition: Scrapped.

10357 / 49-120 / C-119C-12-FA

Avble: 27Feb50; Acc: 27Mar50; Del: 29Mar50 to USAF – 314 TCG (CNC) Sewart 30Mar50; depl Laurinburg-Maxton Airport North Carolina 20Apr50 – 314 TCG (FEA) Komaki AB Japan 10Sep50; to Ashiya AB Japan 18Sep50; depl 6122 AB WG (FEA) Ashiya

AB Japan 31Oct51; depl 403 TCWG (FEA) Ashiya AB Japan 14Apr52 – 6400 AD WG (FEA) Tachikawa AB Japan 13Jun52 mnt – OGDAR (AMC) Hill 25Sep52 mnt – 4600 AB GP (ADC) Peterson 23Sep53 mnt – BMC (AMC) Birmingham 18Feb54 mnt – 483 TCWG (FEA) Ashiya AB Japan 03Jun54 – 95 Bomb WG (SAC) Biggs 14Sep54; depl 3750 TTN WG (TC) Sheppard 11Mar55 – 813 AB GP (SAC) Pinecastle 31May56 – 321 AB GP (SAC) Pinecastle 01Jun56 – 810 AB GP (SAC) Biggs 26Jun56; depl 5 AB GP (SAC) Travis 25Mar57; re-serialed **49-0120** 1958 – 2577 RFC DT (CNR) NAS Dallas 04Dec58 – 69 TCSq (433 TCG) (AFR) NAS Dallas 20Sep58; 923 TCG (433 TCWG) (AFR) NAS Dallas 17Jan63; to Carswell 02Mar63 – 130 ACO GP (ANG) Kanawha 19Dec63 – 2750 AB WG (LOG) WPAFB 15Jan69, cvtd **GC-119C** 01Feb71 – DFI: 04Nov71 – Final Disposition: Scrapped.

10358 / 49-121 / C-119C-12-FA

Avble: 02Mar50; Acc: 10Apr50; Del: 12Apr50 to USAF – 316 TCG (CNC) Sewart 12Apr50; depl 314 TCG (CNC) Laurinburg-Maxton Airport North Carolina 20Apr50 – 314 TCG (CNC) Sewart 13Jul50 – 314 TCG (FEA) Komaki AB Japan 10Sep50; to Ashiya AB Japan 18Sep50; to 13 ARP SQ (FEAMCOM) Tachikawa AB Japan 03Jul51; depl Tachikawa AB Japan 23Jun53 – 483 TCWG (FEA) Ashiya AB Japan 04Feb54; to 1739 FRY SQ (MATS) Amarillo 25Mar54 mnt – 314 TCWG (FEA) Ashiya AB Japan 19Apr54; loan to French Armée de l'Air Indochina 28Jun54-27Jul54, French markings applied; depl Clark AB The Philippines 24Aug54 – 4 AMO SQ (MATS) Elmendorf 27Oct54 – 1855 ACS FT (MATS) Elmendorf 20Mar55; to 5039 ARP SQ (AAC) Elmendorf 13Oct55 & 25May56 – 2584 RFC (CNR) Memphis 14Feb58; re-serialed **49-0121** 1958 – Hayes (AMC) Birmingham 02May58 – 2259 RFC (CNR) Andrews 04Aug58 – 2255 RFC (CNR) Bradley 16Oct58 – 337 TCSq (514 TCG) (AFR) Bradley 12Mar59; 336 TCSq (514 TCG) (AFR) Stewart 20Mar61; 904 TCG (514 TCWG) (AFR) Stewart 31May63 – Wfu to MASDC (LOG) Davis-Monthan 12Jun65 for storage – DFI: 21Jun66 – Final Disposition: Scrapped.

10359 / 49-122 / C-119C-12-FA

Avble: 04Mar50; Acc: 31Mar50; Del: 08Apr50 to USAF – 314 TCG (CNC) Sewart 11Apr50; depl Laurinburg-Maxton Airport North Carolina 20Apr50 – depl 316 TCG (CNC) Sewart 17Aug50 – 314 TCG (FEA) Komaki AB Japan 10Sep50; to Ashiya AB Japan 18Sep50; depl 37 TCSq (314 TCG) Komaki AB Japan 08Nov50; to 13 ARP SQ (FEAMCOM) Tachikawa AB Japan 12Aug51; to 6101 AB WG (FEA) Komaki AB Japan 13Feb52 – 403 TCWG (FEA) Ashiya AB Japan 28Apr52 – 483 TCWG (FEA) Ashiya AB Japan 01Nov52; to 1509 AB SQ (MATS) Johnston 01Apr54 – 314 TCWG (FEA) Ashiya AB Japan 02Aug54; depl Clark AB The Philippines 27Aug54 – MIDAR (AMC) Olmsted 27Oct54 – Fairchild (AMC) Hagerstown 09Apr55 – 3415 TTN WG (TC) Lowry 10Aug55 mnt – 5039 AB WG (AAC) Elmendorf 30Aug55; to 5039 MAI SQ (AAC) Elmendorf 19Nov56 – 5039 AT SQ (AAC) Elmendorf 01Jun57 – 2233 RFC (CNR) Mitchel 19Aug57; re-serialed **49-0122** 1958 – 514 TCWG (AFR) Mitchel 19Dec58, assigned 336 TCSq Stewart 15Mar61; depl 464 TCWG (TAC) Pope 16Feb63; 904 TCG (514 TCWG) (AFR) Stewart 31May63 – Wfu to MASDC (LOG) Davis-Monthan 29Apr65 for storage – DFI: 21Jun66 – Final Disposition: Scrapped.

10360 / 49-123 / C-119C-12-FA

Avble: 01Mar50; Acc: 03Apr50; Del: 08Apr50 to USAF; briefly retained Fairchild (AMC) Hagerstown for project duties – 314 TCG (CNC) Sewart 11Apr50; depl Laurinburg-Maxton Airport North Carolina 24Apr50; Had a minor take-off incident at Pope AFB 18Aug50 – 314 TCG (FEA) Ashyia AB Japan 30Sep50. Lost on a combat mission over

Korea 03Jun51 due to friendly fire 1.9 miles NE Inje S. Korea. The aircraft (along with C-119B s/n: 48-350) was approaching a drop zone area when fired on by U.S. artillery. One of the aircraft was hit in the tail, lost control and collided with the other. 7 crew killed were: 1Lt James W. Bonner, 1Lt David L. Cook (pilot), 1Lt Bradley B. Irish, 2Lt Harry H, Sherman, Jr., SSgt Carl J. Dorsey, SSgt Winfred D. Morgan and Cpl Jack A. Beck – DFI: 24Jul51 – Final Disposition: Accident.

10361 / 49-124 / C-119C-12-FA

Avble: 06Mar50; Acc: 03Apr50; Del: 12Apr50 to USAF – 316 TCG (CNC) Sewart 12Apr50; depl 314 TCG (CNC) Laurinburg-Maxton Airport North Carolina 20Apr50 – 314 TCG (CNC) Sewart 13Jul50 – 314 TCG (FEA) Komaki AB Japan 10Sep50; to Ashiya AB Japan 18Sep50; to 13 ARP SQ (FEAMCOM) Tachikawa AB Japan 22Jul51 – BMC (AMC) Birmingham 29Jul53 – 483 TCWG (FEA) Ashiya AB Japan 07Dec53 – depl 314 TCWG (FEA) Ashiya AB Japan 01Jun54; depl 314 TCWG Clark AB The Philippines 07Jul54; loan to French Armée de l'Air Indochina 21-26Jul54, French markings applied – 316 TCWG (FEA) Ashiya AB Japan 15Nov54 – MIDAR (AMC) Olmsted 08Dec54 – bailment Fairchild (AMC) Hagerstown 01Aug55 – 5039 AB WG (AAC) Elmendorf 09Sep55; to 3310 TTN WG (TC) Scott 09Sep55 mnt; to 5039 MAI SQ (AAC) Elmendorf 21Jan57 – 5039 AT SQ (AAC) Elmendorf 01Jun57 – 5040 CLM GP (AAC) Elmendorf 01Oct57 – 2252 RFC (CNR) Clinton 01Dec57; re-serialed **49-0124** 1958 – 302 TCWG (AFR) Clinton 19Mar59; to 2750 AB WG (AMC) WPAFB 23May59; 907 TCG (302 TCWG) (AFR) Clinton 11Feb63 – Wfu to MASDC (LOG) Davis-Monthan 09Oct65 for storage – DFI: 02May68 – Final Disposition: Scrapped.

10362 / 49-125 / C-119C-13-FA

Avble: 21Mar50; Acc: 10Apr50; Del: 04May50 to USAF; briefly retained Fairchild (AMC) Hagerstown for project duties – 316 TCG (CNC) Sewart 05May50 – 314 TCG (CNC) Sewart 13Jul50 – 314 TCG (FEA) Komaki AB Japan 10Sep50; to Ashiya AB Japan 18Sep50; to 13 ARP SQ (FEAMCOM) Tachikawa AB Japan 22Jul51; depl 403 TCWG (FEA) Ashiya AB Japan 25Apr52 – bailment BMC (AMC) Birmingham 20Aug52 – 483 TCWG (FEA) Ashiya AB Japan 17Jul53; to 1505 AB GP (MATS) Johnston 03Aug53 mnt; depl 314 TCWG (FEA) Ashiya AB Japan 11Aug53; loan to French Armée de l'Air Indochina early54-26Jul54, French markings applied; depl 314 TCWG Clark AB The Philippines 26Jul54 – 316 TCWG (FEA) Ashiya AB Japan 15Nov54 – MIDAR (AMC) Olmsted 01Dec54 – bailment Fairchild (AMC) Hagerstown 28Aug55 mods – 5039 AB WG (AAC) Elmendorf 22Sep55; to 56 Fighter GP (ADC) O'Hare IAP 26Sep55 mnt; to 5039 MAI SQ (AAC) Elmendorf 24Sep56 – Hayes (AMC) Birmingham 06Jan57 – 2231 RFC (CNR) Byrd Field 23Sep57 – 2252 RFC (CNR) Clinton 27Sep57; to 1405 AB WG (MATS) Scott 06Dec57 mnt; re-serialed **49-0125** 1958 – Hayes (AMC) Birmingham 02Mar59 – 302 TCWG (AFR) Clinton 17Apr59; 907 TCG (AFR) Clinton 11Feb63 – Wfu to MASDC (LOG) Davis-Monthan 28Sep65 for storage – DFI: 02May68 – Final Disposition: Scrapped.

10363 / 49-126 / C-119C-13-FA

Avble: 18Mar50; Acc: 13Apr50; Del: 06May50 to USAF; briefly retained Fairchild (AMC) Hagerstown for project duties – 316 TCG (CNC) Sewart 10May50 – 314 TCG (CNC) Sewart 13Jul50 – 314 TCG (FEA) Komaki AB Japan 10Sep50; to Ashiya AB Japan 18Sep50; to 13 ARP SQ (FEAMCOM) Tachikawa AB Japan 22Jun51 & 09Dec51 – MIDAR (AMC) Olmsted 02Jan52 mnt – bailment BMC (AMC) Birmingham 07May52. Crashed-landed after engine failure 0.3 miles W Jackson-Hawkins Field Missouri 14Feb53. To 3380

TTN WG (TC) Keesler 15Feb53 assessed as damaged beyond repair – REC: 28Feb53 – DFI: 09Sep54 – Final Disposition: Accident.

10364 / 49-127 / C-119C-13-FA

Avble: 15Mar50; Acc: 11Apr50; Del: 19Apr50 to USAF; briefly retained Fairchild (AMC) Hagerstown for project duties – 316 TCG (CNC) Sewart 18Apr50; depl 314 TCG (CNC) Laurinburg-Maxton Airport North Carolina 24Apr50 – 314 TCG (CNC) Sewart 13Jul50 – 314 TCG (FEA) Komaki AB Japan 10Sep50; to Ashiya AB Japan 18Sep50; to 13 ARP SQ (FEAMCOM) Tachikawa AB Japan 27May51; depl 6122 AB WG (FEA) Ashiya AB Japan 09Jan52 – bailment BMC (AMC) Birmingham 22Nov53 – 483 TCWG (FEA) Ashiya AB Japan 08Apr54 – 314 TCWG (FEA) Ashiya AB Japan 23Jul54; depl 314 TCWG Clark AB The Philippines 06Aug54; loan to French Armée de l'Air Indochina 07Aug54–01Sep54, French markings applied – MIDAR (AMC) Olmsted 28Oct54 mnt – bailment Fairchild (AMC) Hagerstown 05Jun55 – 5039 AB WG (AAC) Elmendorf 08Jul55; to 5039 MAI SQ (AAC) Elmendorf 25Apr56 & 25Jun56 – 5039 AT SQ (AAC) Elmendorf 01Jun57 – 5040 CLM GP (AAC) Elmendorf 01Oct57 – 2233 RFC (CNR) Mitchel 09Nov57; re-serialed **49-0127** 1958 – Hayes (AMC) Birmingham 16Oct58 – 514 TCWG (AFR) Mitchel 23Jan59; to 2584 AB SQ (CNC) Memphis 28Jun59 mnt; 903 TCG (514 TCWG) McGuire 31May63 – 130 ACO GP (ANG) Kanawha 30Dec63 – Wfu to MASDC (LOG) Davis-Monthan 28Mar69 for storage – DFI: 09Apr69 – Final Disposition: Scrapped.

10365 / 49-128 / C-119C-13-FA

Avble: 21Mar50; Acc: 12Apr50; Del: 19Apr50 to USAF; briefly retained Fairchild (AMC) Hagerstown for project duties – 316 TCG (CNC) Sewart 18Apr50; depl 314 TCG (CNC) Laurinburg-Maxton Airport North Carolina 24Apr50 – 314 TCG (CNC) Sewart 13Jul50 – 314 TCG (FEA) Komaki AB Japan 10Sep50; to Ashiya AB Japan 18Sep50; to 13 ARP SQ (FEAMCOM) Tachikawa AB Japan 02Oct50 & 24Jul51; depl 374 TCWG (FEA) Tachikawa AB Japan 18Oct50 – 483 TCWG (FEA) Ashiya AB Japan 20Feb53; depl Bofu Japan 24May53; to 5 AB GP (SAC) Travis 12Dec53 – 314 TCWG (FEA) Ashiya AB Japan 27Jun54; depl 314 TCWG Clark AB The Philippines Jul54 – MIDAR (AMC) Olmsted 10Nov54 mnt – bailment Fairchild (AMC) Hagerstown 27Jul55 mods – 5039 AB WG (AAC) Elmendorf 23Aug55. Crashed on take-off at Sparrevohn AFS Alaska 27Dec55, 8 onboard no fatalities, aircraft damaged beyond repair – DFI: 11Jan56 – Final Disposition: Accident.

10366 / 49-129 / C-119C-13-FA

Avble: 24Mar50; Acc: 18Apr50; Del: 25Apr50 to USAF; briefly retained Fairchild (AMC) Hagerstown for project duties – 316 TCG (CNC) Sewart 26Apr50; depl 314 TCG (CNC) Laurinburg-Maxton Airport North Carolina 04May50 – 314 TCG (CNC) Sewart 13Jul50 – 314 TCG (FEA) Komaki AB Japan 10Sep50; to Ashiya AB Japan 18Sep50; to 13 ARP SQ (FEAMCOM) Tachikawa AB Japan 04Apr51; depl 374 TCWG (FEA) Tachikawa AB Japan 17Apr51 – bailment BMC (AMC) Birmingham 12Dec53 – 581 ASL GP (FEA) Clark AB The Philippines 03Jun54 – SMPAR (AMC) Clark AB The Philippines 30Sep56 – 322 TCSq (FEA) Kadena AB Japan 07Dec56 – 2252 RFC (CNR) Clinton 28Dec57; re-serialed **49-0129** 1958 – 302 TCWG (AFR) Clinton 19Mar59; 907 TCG (302 TCWG) Clinton 11Feb63 – 130 ACO GP (ANG) Kanawha 21Jan64 – Wfu to MASDC (LOG) Davis-Monthan 18Jan69 for storage – DFI: 03Feb69 – Final Disposition: Scrapped.

10367 / 49-130 / C-119C-13-FA

Avble: 24Mar50; Acc: 12Apr50; Del: 19Apr50 to USAF; briefly retained Fairchild (AMC) Hagerstown for project duties – 316 TCG (CNC) Sewart 18Apr50; depl 314 TCG Laurinburg-Maxton Airport North Carolina 24Apr50 – 314 TCG (CNC) Sewart 13Jul50 – 314 TCG (FEA) Komaki AB Japan 10Sep50, assigned 61 TCSq; to Ashyia AB Japan 18Sep50. Had a landing accident due to mechanical failure at Ashyia AB 24Jun51. To 13 ARP SQ (FEAMCOM) Tachikawa AB Japan 21Aug51 mods. Crashed at sea due to engine failure 08Dec51. Pilot: Raymond T. Frazier, crew bailed out but three were killed with one listed as missing and one rescued – DFI: 03Apr52 – Final Disposition: Accident.

10368 / 49-131 / C-119C-13-FA

Avble: 28Mar50; Acc: 17Apr50; Del: 20Apr50 to USAF; briefly retained Fairchild (AMC) Hagerstown for project duties – 316 TCG (CNC) Sewart 21Apr50; depl 314 TCG Laurinburg-Maxton Airport North Carolina 24Apr50 – 314 TCG (CNC) Sewart 13Jul50 – 314 TCG (FEA) Komaki AB Japan 10Sep50; to Ashyia AB Japan 18Sep50; to 13 ARP SQ (FEAMCOM) Tachikawa AB Japan 30Jul51 & 19Dec51; to 6122 AB WG (FEA) Ashyia AB Japan 29Aug51; to 6400 AD WG (FEA) Tachikawa AB Japan 31Jan52 – 403 TCWG (FEA) Ashyia AB Japan 12May52 – 483 TCWG (FEA) Ashyia AB Japan 01Jan53; depl 314 TCWG (FEA) Ashyia AB Japan 03Jun54 – 316 TCWG (FEA) Ashyia AB Japan 15Nov54 – MIDAR (AMC) Olmsted 20Dec54 – bailment Fairchild (AMC) Hagerstown 07Sep55 – 5039 AB WG (AAC) Elmendorf 31Oct55 – **RECORDS MISSING** – Re-serialed **49-0131** 1958 – 2233 RFC (CNR) Mitchel 01Apr58; to 2750 AB WG (AMC) WPAFB 22Sep58 mnt; to 4060 AB GP (SAC) Dow 02Dec58 mnt – 514 TCWG (AFR) Mitchel 19Dec58, assigned 335 TCSq McGuire 15Mar61; 903 TCG (514 TCWG) (AFR) McGuire 31May63 – 130 ACO GP (ANG) Kanawha 30Dec64 – Wfu to MASDC (LOG) Davis-Monthan 18Jan69 for storage – DFI: 03Feb69 – Final Disposition: Scrapped.

10369 / 49-132 / C-119C-13-FA

Avble: 29Mar50; Acc: 17Apr50; Del: 20Apr50 to USAF; briefly retained Fairchild (AMC) Hagerstown for project duties – 316 TCG (CNC) Sewart 21Apr50; depl 314 TCG Laurinburg-Maxton Airport North Carolina 24Apr50 – 314 TCG (CNC) Sewart 13Jul50 – 314 TCG (FEA) Komaki AB Japan 10Sep50; to Ashyia AB Japan 18Sep50; to 13 ARP SQ (FEAMCOM) Tachikawa AB Japan 12Aug51 – BMC (AMC) Birmingham 16Aug52 – 483 TCWG (FEA) Ashyia AB Japan 06Aug53; depl Bofu Japan 17Nov53 – 320 Bomb WG (SAC) March 09Nov54 – OGDAR (AMC) Hill 22Jun56 – 810 AB GP (SAC) Biggs 21Nov57; to 812 AB GP (SAC) Walker 10Feb58; re-serialed **49-0132** 1958 – 2259 RFC DT (CNR) Youngstown 09Apr58; to Greater Pittsburgh 29May58 – 758 TCSq (459 TCG) (AFR) Greater Pittsburgh 19Dec58; 911 TCG (459 TCWG) (AFR) Greater Pittsburgh 17Jan63 – Wfu to MASDC (LOG) Davis-Monthan 07Jan67 for storage – DFI: 22Jun67 – BofS F.H. Chew Sales Co. Phoenix Arizona 20Dec67 – BofS Aero Union Corp. Chico California 20Dec67; ARAp **N13743** 24Jan68; cvtd to airtanker duties Jun71 with avionics upgrade and fire retardant tank installation, Tanker **E12**; cvtd **C-119C Jet-Pak** (J34-WE) 21May74 – BofS Hemet Valley Flying Service Co. Hemet California 18Feb75; Tanker **81**; took part in motion picture filming at California City Airport Feb82, title unk; recovered elevator 27Mar82; wfu and stored 1987 – To U.S. Forest Service 1991 and donated to Pima Air & Space Museum Tucson Arizona. Displayed today in original HVFS livery – Reg N13743 canx 18Dec12 – Final Disposition: Extant.

10370 / 49-133 / C-119C-13-FA

Avble: 03Apr50; Acc: 18Apr50; Del: 26Apr50 to USAF; briefly retained Fairchild (AMC) Hagerstown for project duties – 316 TCG (CNC) Sewart 26Apr50; depl 314 TCG Laurinburg-Maxton Airport North Carolina 02May50 – 314 TCG (CNC) Sewart 13Jul50 – 314 TCG (FEA) Komaki AB Japan 10Sep50; to Ashyia AB Japan 18Sep50; to 13 ARP SQ (FEAMCOM) Tachikawa AB Japan 04Jul51 – 403 TCWG (FEA) Ashyia AB Japan 30May52 – BMC (AMC) Birmingham 16Aug52; to 803 AB WG (SAC) Davis-Monthan 20Jan53 mnt – 483 TCWG (FEA) Ashyia AB Japan 05Nov53; loan to French Armée de l'Air Indochina 12Apr54-18Jun54, French markings applied; depl 314 TCWG (FEA) Clark AB The Philippines Aug54 – 316 TCWG (FEA) Ashyia AB Japan 15Nov54 – MIDAR (AMC) Olmsted 20Dec54 – bailment Fairchild (AMC) Hagerstown 20Sep55 – 3320 TTN WG (TC) Amarillo 24Oct55 mnt – 5039 AB WG (AAC) Elmendorf 21Feb56 – 5039 AT SQ (AAC) Elmendorf 01Jun57 – 5040 CLM GP (AAC) Elmendorf 01Oct57 – 2577 RFC DT (CNR) NAS Dallas 05Dec57; re-serialed **49-0133** 1958 – 69 TCSq (433 TCG) (AFR) NAS Dallas 21Jun60; 923 TCG (433 TCWG) (AFR) NAS Dallas 17Jan63; to Carswell 02Mar63 – 130 ACO GP (ANG) Kanawha 27Feb64 – Wfu to MASDC (LOG) Davis-Monthan 28Mar69 for storage – DFI: 09Apr69 – Aero Union Corp. Chico California 1969, spares airframe – Hemet Valley Flying Service Co. Hemet California CofR **N9955F** 15Oct76, spares airframe, bku 1980s – Reg N9955F canx 02May11 – Final Disposition: Scrapped.

10371 / 49-134 / C-119C-13-FA

Avble: 04Apr50; Acc: 21Apr50; Del: 26Apr50 to USAF; briefly retained Fairchild (AMC) Hagerstown for project duties – 316 TCG (CNC) Sewart 26Apr50, assigned 37 TCSq; depl 314 TCG (CNC) Laurinburg-Maxton Airport North Carolina 02May50. Crashed due to a no. 2 engine fire 1.5 miles SE Hendersonville Tennessee 28Jun50. Pilot Arthur R. Busch, one of 32 aircraft undertaking night paratroop exercises near Fort Campbell. When the fire started the aircraft turned for an emergency landing with the 26 paratroops onboard then ordered to bail out. The aircraft crashed a short time later some 20 miles from base killing all four crew members. Notable as the first C-119 accident and the first fatal C-119 accident – REC: 30Jun50 – DFI: 31Jul50 – Final Disposition: Accident.

10372 / 49-135 / C-119C-13-FA

Avble: 04Apr50; Acc: 20Apr50; Del: 04May50 to USAF; briefly retained Fairchild (AMC) Hagerstown for project duties – 316 TCG (CNC) Sewart 05May50 – 314 TCG (CNC) Sewart 13Jul50 – 314 TCG (FEA) Komaki AB Japan 10Sep50; to Ashyia AB Japan 18Sep50; to 13 ARP SQ (FEAMCOM) Tachikawa AB Japan 25May51; depl 483 TCWG (FEA) Ashiya AB Japan 18Feb53; depl Bofu Japan 19May53; loan to French Armée de l'Air Indochina 19Jun54-10Jul54, French markings applied; depl Clark AB The Philippines 09Aug54 – 316 TCWG (FEA) Ashyia AB Japan 15Nov54 – 12 Fighter WG (SAC) Bergstrom 24Dec54 – 22 Bomb WG (SAC) March 29Dec54; to 807 AB GP (SAC) March 10Nov55 – 810 AB GP (SAC) Biggs 15Aug56; re-serialed **49-0135** 1958 – 2578 RFC (CNR) Ellington 20Aug58 – 446 TCWG (AFR) Ellington 20Sep58; assigned 706 TCSq (AFR) Barksdale 14Mar59; to 2578 AB GP (CNC) Ellington 11Jul59 – 514 TCWG (AFR) Mitchel; assigned 335 TCSq McGuire 06May61; 903 TCG (514 TCWG) (AFR) McGuire 31May63 – Wfu to MASDC (LOG) Davis-Monthan 08Dec65 for storage – DFI: 02May68 – Final Disposition: Scrapped.

Section Four—Service Histories

10373 / 49-136 / C-119C-13-FA

Avble: 10Apr50; Acc: 25Apr50; Del: 04May50 to USAF; briefly retained Fairchild (AMC) Hagerstown for project duties – 316 TCG (CNC) Sewart 05May50 – 314 TCG (CNC) Sewart 13Jul50 – 314 TCG (FEA) Komaki AB Japan 10Sep50; to Ashyia AB Japan 18Sep50; to 13 ARP SQ (FEAMCOM) Tachikawa AB Japan 21Jun51; depl 483 TCWG (FEA) Ashiya AB Japan 23Feb53 & 17Apr54; loan to French Armée de l'Air Indochina early54-14Jul54, French markings applied – Assigned variously 5 AB WG, 5 SRH WG, 5 Bomb WG (SAC) Travis from 05Nov54; to 803 AB GP (SAC) Davis-Monthan 12Mar56; to 92 AB GP (SAC) Fairchild 03Dec56; to 479 Fighter WG (TAC) George 01Mar57 – 28 AB WG (SAC) Ellsworth 08Apr57 – 2578 RFC (CNR) Ellington 04Jan58; re-serialed **49-0136** 1958 – 446 TCWG (AFR) Ellington 20Sep58; assigned 706 TCSq (AFR) Barksdale 12Apr59 – 78 TCSq (435 TCWG) (AFR) Barksdale 08May61 – 156 AML SQ (NC-ANG) Douglas 13Nov61 – 514 TCWG (AFR) Stewart; assigned 336 TCSq 01Nov62; 904 TCG (514 TCWG) Stewart 31May63 – Wfu to 3320 MSU GP (ATC) Amarillo 11Jun65 – REC: 14Jun65 – Final Disposition: Scrapped.

10374 / 49-137 / C-119C-13-FA

Avble: 12Apr50; Acc: 01May50; Del: 11May50 to USAF; briefly retained Fairchild (AMC) Hagerstown for project duties – 316 TCG (CNC) Sewart 12May50 – 314 TCG (CNC) Sewart 13Jul50 – 314 TCG (FEA) Komaki AB Japan 10Sep50; to Ashyia AB Japan 18Sep50; to 13 ARP SQ (FEAMCOM) Tachikawa AB Japan 08Nov51; to 6400 AD WG (FEA) Tachikawa AB Japan 31Jan52; loan to French Armée de l'Air Indochina early54-10May54, French markings applied – BMC (AMC) Birmingham 10May54 – 31 Fighter WG (SAC) Turner 14Sep54 – 72 SRH WG (SAC) Ramey 29Nov54 – Fairchild (AMC) Hagerstown 24Sep55 – 7 Bomb WG (SAC) Carswell 03Nov55 – 384 Bomb WG (SAC) Little Rock 09May56 – 825 AB GP (SAC) Little Rock 06Aug57; re-serialed **49-0137** 1958 – 2584 RFC (CNR) Memphis 24Apr58 – 2578 RFC (CNR) Ellington 13Sep58 – 446 TCWG (AFR) Ellington 20Sep58; assigned 706 TCSq Barksdale 21Jul59; to 803 COS GP (SAC) Davis-Monthan 11Aug59 mnt; to PGC Center (APG) Eglin 28Sep60 mnt – 78 TCSq (435 TCWG) (AFR) Barksdale 08May61 – 156 AML SQ (NC-ANG) Douglas 29Apr61 – Fairchild (LOG) St. Augustine 12Jul62 – MAP Morocco 05Nov62, RMAF s/n: **CN-AMA**, 1 AT SQ – Final Disposition: Unknown.

10375 / 49-138 / C-119C-13-FA

Avble: 17Apr50; Acc: 05May50; Del: 15May50 to USAF; briefly retained Fairchild (AMC) Hagerstown for project duties – 316 TCG (CNC) Sewart 17May50 – 314 TCWG (CNC) Sewart 13Jul50 – 314 TCG (FEA) Komaki AB Japan 10Sep50, assigned 62 TCSq; to Ashiya AB Japan 18Sep50. Had a minor landing accident due to a mechanical failure at Ashyia AB Japan 07Dec50. To 13 ARP SQ (FEAMCOM) Tachikawa AB Japan 01Dec50 for repairs and mnt. Crashed into terrain 1 mile NE Ashyia AB Japan 24Jul52. Pilot: John E. O'Donnell, Jr., plus 4 other crew killed. The aircraft came down on a Japanese beer hall building also killing 2 civilians on the ground – DFI: 24Jul52 – Final Disposition: Accident.

10376 / 49-139 / C-119C-13-FA

Avble: 18Apr50; Acc: 01May50; Del: 11May50 to USAF; briefly retained Fairchild (AMC) Hagerstown for project duties – 316 TCG (CNC) Sewart 12May50 – 314 TCG (CNC) Sewart 13Jul50 – 314 TCG (FEA) Komaki AB Japan 10Sep50; to Ashyia AB Japan 18Sep50; to 13 ARP SQ (FEAMCOM) Tachikawa AB Japan 11Jul51 – 403 TCWG (FEA) Ashiya AB Japan 05May52 – 483 TCWG (FEA) Ashiya AB Japan 06Feb53; loan to French Armée de l'Air Indochina early54-Jul54, French markings applied; depl 314 TCWG (FEA) Ashiya AB

Japan 04Apr54; depl 314 TCWG (FEA) Clark AB The Philippines Jul54 – 316 TCWG (FEA) Ashiya AB Japan 15Nov54 – 97 Bomb WG (SAC) Biggs 06Jan55 – 810 AB GP (SAC) Biggs 25Jul56; to 807 AB GP (SAC) March 21Jan57; to 824 AB GP (SAC) Carswell 11Feb57 mnt; to 3595 CCT WG (TC) Nellis 27Mar57 mnt – 2577 RFC (CNR) Brooks 22Mar58; re-serialed **49-0139** 1958; depl 2578 RFC (CNR) Ellington 01May58 – 433 TCWG (AFR) Brooks 18Sep58; depl 435 TCWG (AFR) Miami IAP 21Jun59; to Kelly 31Aug60; 921 TCG (433 TCWG) Kelly 31May63 – Wfu to 3320 MSU GP (ATC) Amarillo 01Oct65 – REC: 19Oct65 – Final Disposition: Scrapped.

10377 / 49-140 / C-119C-14-FA

Avble: 26Apr50; Acc: 13May50; Del: 20May50 to USAF; briefly retained Fairchild (AMC) Hagerstown for project duties – 316 TCG (CNC) Sewart 22May50 – 314 TCG (CNC) Sewart 13Jul50 – 314 TCG (FEA) Komaki AB Japan 10Sep50; to Ashyia AB Japan 18Sep50; to 13 ARP SQ (FEAMCOM) Tachikawa AB Japan 07Aug51 – BMC (AMC) Birmingham 28Jan53 – 483 TCWG (FEA) Ashyia AB Japan 25May54; depl 314 TCWG (FEA) Ashyia AB Japan 26Jun54; depl 314 TCWG (FEA) Clark AB The Philippines 01Jul54 – 316 TCWG (FEA) Ashyia AB Japan 15Nov54 – 306 Bomb WG (SAC) MacDill 22Nov54; to 818 AB GP (SAC) Lincoln 21Jul56 – 809 AB GP (SAC) MacDill 06Aug57 – 2577 RFC (CNR) Brooks 25Apr58; re-serialed **49-0140** 1958 – 433 TCWG (AFR) Brooks 18Sep58; to 4130 COS GP (SAC) Bergstrom 07Feb59 mnt; to 1604 AB WG (MATS) Kindley 24Jul60 mnt; to Kelly 31Aug60; 921 TCG (433 TCWG) (AFR) Kelly 31May63 – Wfu to REC: 01Sep64 – Final Disposition: Scrapped.

10378 / 49-141 / C-119C-14-FA

Avble: 26Apr50; Acc: 16May50; Del: 02Jun50 to USAF; briefly retained Fairchild (AMC) Hagerstown for project duties – 316 TCG (CNC) Sewart 05Jun50 – 314 TCG (CNC) Sewart 13Jul50 – 314 TCG (FEA) Komaki AB Japan 10Sep50; to Ashyia AB Japan 18Sep50; to 13 ARP SQ (FEAMCOM) Tachikawa AB Japan 30Jun51 – BMC (AMC) Birmingham 26Sep53 – 483 TCWG (FEA) Ashyia AB Japan 05Feb54; depl 314 TCWG (FEA) Ashyia AB Japan 16Apr54; loan to French Armée de l'Air Indochina 24Jun54–24Aug54, French markings applied – 316 TCWG (FEA) Ashyia AB Japan 15Nov54 – 320 Bomb WG (SAC) March 11Jan55; to 3595 CCT WG (TC) Nellis 27Mar55 mnt – 96 AB GP (SAC) Altus 27Jul56 – 11 AB GP (SAC) Altus 13Dec57; re-serialed **49-0141** 1958 – 2577 RFC (CNR) Brooks 17Jun58 – 433 TCWG (AFR) Brooks 18Sep58; to Kelly 31Aug60; 921 TCG (433 TCWG) (AFR) Kelly 31May63; 922 TCG (433 TCWG) (AFR) Kelly 10Oct65 – Wfu to MASDC (LOG) Davis-Monthan 29Oct65 for storage – DFI: 02May68 – Final Disposition: Scrapped.

10379 / 49-142 / C-119C-14-FA

Avble: 28Apr50; Acc: 23May50; Del: 02Jun50 to USAF; briefly retained Fairchild (AMC) Hagerstown for project duties – 316 TCG (CNC) Sewart 05Jun50 – 314 TCG (CNC) Sewart 13Jul50 – 314 TCG (FEA) Komaki AB Japan 10Sep50; to Ashyia AB Japan 18Sep50; to 13 ARP SQ (FEAMCOM) Tachikawa AB Japan 08Jul51 – BMC (AMC) Birmingham 23Nov53 – 483 TCWG (FEA) Ashyia AB Japan 05May54 – 95 Bomb WG (SAC) Biggs 03Sep54 – 810 AB GP (SAC) Biggs 25Jul56; re-serialed **49-0142** 1958 – 2577 RFC (CNR) Brooks 26Jun58 – 433 TCWG (AFR) Brooks 18Sep58; to 6 AB GP (SAC) Walker 22Mar59; to 4520 CCT WG (TAC) Nellis 19Jun60; to Kelly 31Aug60; 921 TCG (433 TCWG) (AFR) Kelly 31May63; 922 TCG (433 TCWG) (AFR) Kelly 10Oct65 – Wfu to MASDC (LOG) Davis-Monthan 29Oct65 for storage – DFI: 02May68 – Final Disposition: Scrapped.

Section Four—Service Histories

10380 / 49-143 / C-119C-14-FA

Avble: 01May50; Acc: 18May50; Del: 27May50 to USAF; briefly retained Fairchild (AMC) Hagerstown for project duties – 316 TCG (CNC) Sewart 29May50 – 314 TCG (CNC) Sewart 13Jul50 – 314 TCG (FEA) Komaki AB Japan 10Sep50; to Ashyia AB Japan 18Sep50; to 13 ARP SQ (FEAMCOM) Tachikawa AB Japan 25Jul51 – 403 TCWG (FEA) Ashiya AB Japan 04Jun52 – MIDAR (AMC) Olmsted 22Sep52; to 2596 FTR CN (CNC) Hensley 04Oct52 mnt – BMC (AMC) Birmingham 06Jan53 – 483 TCWG (FEA) Ashiya AB Japan 29Sep53 – 314 TCWG (FEA) Ashiya AB Japan 21Oct53; loan to French Armée de l'Air Indochina early54-Jul54, French markings applied – 305 Bomb WG (SAC) MacDill 10Nov54 – 12 Fighter WG (SAC) Bergstrom 13Nov54 – 808 AB GP (SAC) Bergstrom 01Jul57 – 27 AB GP (TAC) Bergstrom 01Feb58; re-serialed **49-0143** 1958 – 2589 RFC (CNR) Dobbins 11Apr58 – 2259 RFC DT (CNR) Greater Pittsburgh 02Jul58 – 758 TCSq (459 TCG) (AFR) Greater Pittsburgh 19Dec58 – 2255 RFC (AFR) Bradley 21Dec58 – 337 TCSq (514 TCG) (AFR) Bradley 19Mar59; to 839 AB GP (TAC) Sewart 27Apr59 mnt; assigned 337 TCSq McGuire 23Mar61 – 167 AML SQ (WV-ANG) Martinsburg 31May61 – 130 ACO GP (ANG) Kanawha 25Oct63 – Wfu to 2704 ASD GP (LOG) Davis-Monthan 21Nov63 for storage – DFI: 26Jun64 – Final Disposition: Scrapped.

10381 / 49-144 / C-119C-14-FA

Avble: 03May50; Acc: 25May50; Del: 13Jun50 to USAF; briefly retained Fairchild (AMC) Hagerstown for project duties – 316 TCG (CNC) Sewart 13Jun50 – 314 TCG (CNC) Sewart 13Jul50 – 314 TCG (FEA) Komaki AB Japan 10Sep50; to Ashyia AB Japan 18Sep50; to 13 ARP SQ (FEAMCOM) Tachikawa AB Japan 28Jul51 – 403 TCWG (FEA) Ashiya AB Japan 23May52 – 483 TCWG (FEA) Ashiya AB Japan 01Jan53; depl 314 TCWG (FEA) Ashiya AB Japan 07Feb54; loan to French Armée de l'Air Indochina 10Apr54-24Jul54, French markings applied – 316 TCWG (FEA) Ashiya AB Japan 15Nov54 – 40 Bomb WG (SAC) Smoky Hill 28Feb55 – to 12 Fighter WG (SAC) Bergstrom 04Apr55; to 802 AB GP (SAC) Smoky Hill Jun55; to 310 Bomb WG (SAC) Smoky Hill 01Jul55 – 340 AB GP (SAC) Whiteman dates unk 56-58 – 2578 RFC (CNR) Ellington 27Jan58; re-serialed **49-0144** 1958 – 446 TCWG (AFR) Ellington 20Sep58; assigned 706 TCSq Barksdale 18Apr59; to 819 COS GP (SAC) Dyess 25Apr60 mnt – 758 TCSq (459 TCG) (AFR) Greater Pittsburgh 29Apr61; 911 TCSq (459 TCWG) (AFR) Greater Pittsburgh 17Jan63 – Wfu to MASDC (LOG) Davis-Monthan 20Oct65 for storage – DFI: 02May68 – Final Disposition: Scrapped.

10382 / 49-145 / C-119C-14-FA

Avble: 06May50; Acc: 18May50; Del: 27May50 to USAF; briefly retained Fairchild (AMC) Hagerstown for project duties – 316 TCG (CNC) Sewart 29May50 – 314 TCG (CNC) Sewart 13Jul50 – 374 TCWG (FEA) Tachikawa AB Japan 02Aug50 for project duties – FEAMCOM (FEA) 22Aug50 – 314 TCG (FEA) Ashiya AB Japan 18Sep50, assigned 62 TCSq. Made a forced landing due to mechanical failure at Ashyia AB Japan 14Sep50. Pilot James M. Inks, no fatalities but aircraft damaged beyond repair – 6122 AB GP (FEA) Ashiya AB Japan 18Sep50 likely for SAL – REC: 09Oct50 – DFI: 08Nov50 – Final Disposition: Accident.

10383 / 49-146 / C-119C-14-FA

Avble: 10May50; Acc: 25May50; Del: 13Jun50 to USAF; briefly retained Fairchild (AMC) Hagerstown for project duties – 316 TCG (CNC) Sewart 13Jun50 – 314 TCG (CNC) Sewart 13Jul50 – 374 TCWG (FEA) Tachikawa AB Japan 02Aug50; to FEAMCOM (FEA) 21Aug50; depl 314 TCWG (FEA) Ashiya AB Japan 18Sep50; assigned 21 TCSq (374 TCWG)

Itazuke AB Japan 23Jan51 – 13 ARP SQ (FEAMCOM) Tachikawa AB Japan 05Mar51 – 314 TCWG (FEA) Ashiya AB Japan 10May51; to 13 ARP SQ (FEAMCOM) Tachikawa AB Japan 15Aug51; depl 483 TCWG (FEA) Ashiya AB Japan 21May54; depl Clark AB The Philippines 09Jul54 – 99 SRH WG (SAC) Fairchild 12Oct54 – SMAAR (AMC) McClellan 03Nov54 mnt – 92 Bomb WG (SAC) Fairchild 27Mar55 – OKLAR (AMC) Tinker 04Apr55 mnt – 90 SM WG (SAC) Forbes 13Apr55; to 2750 AB WG (AMC) WPAFB 02Apr56 – 340 Bomb WG (SAC) Whiteman 04May56 – 328 Fighter GP (ADC) Grandview 02Jan57 mnt – 340 AB GP (SAC) Whiteman 17Feb57; re-serialed **49-0146** 1958; to 3415 TTA WG (TC) Lowry 12Mar58 mnt – 2578 RFC (CNR) Ellington 20May58 – 446 TCWG (AFR) Ellington 20Sep58; assigned 706 TCSq Barksdale 23Apr59 – 758 TCSq (459 TCG) (AFR) Greater Pittsburgh 29Apr61 – Wfu to REC: 16Nov61 – Final Disposition: Scrapped.

10384 / 49-147 / C-119C-14-FA

Avble: 12May50; Acc: 25May50; Del: 13Jun50 to USAF; briefly retained Fairchild (AMC) Hagerstown for project duties – 316 TCG (CNC) Sewart 13Jun50 – 314 TCG (CNC) Sewart 13Jul50 – 314 TCG (FEA) Komaki AB Japan 10Sep50; to Ashyia AB Japan 18Sep50; to 13 ARP SQ (FEAMCOM) Tachikawa AB Japan 01Aug51 & 22Dec51; to 6400 AD WG (FEA) Tachikawa AB Japan 31Jan52 – 403 TCWG (FEA) Ashiya AB Japan 16May52 – MIDAR (AMC) Olmsted 22Sep52 mnt and storage – OKLAR (AMC) Tinker 15Jun53 – 483 TCWG (FEA) Ashiya AB Japan 01Jul53 – 314 TCWG (FEA) Ashiya AB Japan 15Jun54; depl Clark AB The Philippines 01Jul54 – 1500 FDM SQ (MATS) Hickham 17Nov54 mnt – 31 Fighter WG (SAC) Turner 03Dec54 – 4060 AB GP (SAC) Dow 22Mar55 – BMC (AMC) Birmingham 06Apr55 – 506 Fighter WG (SAC) Tinker 21Jun55; to 3605 AOT WG (TC) Ellington 02Apr56 mnt. Wfu to SAL reasons unk – REC: 19Apr56 – Final Disposition: Scrapped.

10385 / 49-148 / C-119C-14-FA

Avble: 16May50; Acc: 26May50; Del: 15Jun50 to USAF; briefly retained Fairchild (AMC) Hagerstown for project duties – 314 TCG (CNC) Sewart 15Jun50 – 316 TCG (CNC) Sewart 19Sep50; depl 314 TCG (TAC) Sewart 05Jun51; depl Laurinburg-Maxton Airport North Carolina 24Aug51 – 435 TCWG (TAC) Miami IAP 05May52; depl Elmendorf 24Oct52 – 456 TCWG (TAC) Miami IAP 01Dec52 – Wfu to 3750 TTN WG (TC) Sheppard 25Feb53, assigned Class 01Z (ground instruction). To Class 26 (ground instruction) 24Jul56 – Final Disposition: Scrapped.

10386 / 49-149 / C-119C-14-FA

Avble: 18May50; Acc: 02Jun50; Del: 16Jun50 to USAF; briefly retained Fairchild (AMC) Hagerstown for project duties – 316 TCG (CNC) Sewart 16Jun50; briefly to 2471 ART Center (CNR) O'Hare 13Jul50 mnt – 314 TCG (CNC) Sewart 13Jul50 – 314 TCG (FEA) Komaki AB Japan 10Sep50; to Ashyia AB Japan 02Aug50; to 13 ARP SQ (FEAMCOM) Tachikawa AB Japan 10Sep51; to 6400 AD WG (FEA) Tachikawa AB Japan 31Jan52 – BMC (AMC) Birmingham Jun52 – 483 TCWG (FEA) Ashiya AB Japan 13May53; depl 314 TCWG (FEA) Ashiya AB Japan 26May53; loan to French Armée de l'Air Indochina 02May54, French markings applied. Shot down 06May54 during the final stages of the Battle of Dien Bien Phu in northern Vietnam. While making a supply drop with five other Boxcars, it was struck by 37mm ground fire disabling the port engine and seriously damaging the horizontal stabilizer. The aircraft flew another 40 minutes before crashing and burning in jungle just short of an airstrip in Laos. Killed were American pilots James B. "Earthquake McGoon" McGovern, Jr., and Wallace Buford plus 2 French paratroopers. A Malay paratrooper and French 2Lt Jean

Arlaux survived, but the Malay soldier died some time later. Listed as officially transferred out of USAF 10May54 – DFI: 17Feb55 – Final Disposition: Combat Loss.

10387 / 49-150 / C-119C-14-FA

Avble: 20May50; Acc: 06Jun50; Del: 16Jun50 to USAF; briefly retained Fairchild Aircraft (AMC) Hagerstown for project duties – 316 TCG (CNC) Sewart 16Jun50 – 314 TCG (CNC) Sewart 13Jul50 – 314 TCG (FEA) Komaki AB Japan 10Sep50, assigned 62 TCSq; to Ashyia AB Japan 18Sep50; to 13 ARP SQ (FEAMCOM) Tachikawa AB Japan 12Aug51; depl 452 BL WG (FEA) Pusan East AB (K-9) S. Korea 01Sep51. Minor mechanical failure 50 miles E Oshima Japan 09Dec51. To 13 ARP SQ (FEAMCOM) Tachikawa AB Japan 09Dec51 mnt; to MIDAR (AMC) Olmstead 24Dec51 mnt. Made a forced landing due to mechanical (engine) failure 12 miles NE Hachita, New Mexico 04Jan52. Pilot: Thomas E. Shaffmaster, no fatalities but aircraft damaged beyond repair – DFI: 19Mar52 – Final Disposition: Accident.

10388 / 49-151 / C-119C-14-FA

Avble: 24May50; Acc: 07Jun50; Del: 16Jun50 to USAF; briefly retained Fairchild (AMC) Hagerstown for project duties – 316 TCG (CNC) Sewart 16Jun50 – 314 TCG (CNC) Sewart 13Jul50 – 314 TCG (FEA) Komaki AB Japan 10Sep50; to Ashyia AB Japan 18Sep50; depl 21 TCSq (374 TCWG) (FEA) Itazuke AB Japan 23Jan51; depl 374 TCWG (FEA) Tachikawa AB Japan 24Jan51; to 13 ARP SQ (FEAMCOM) Tachikawa AB Japan 18Aug51 – BMC (AMC) Birmingham 07Feb53 – 483 TCWG (FEA) Ashiya AB Japan 13Apr54 – 314 TCWG (FEA) Ashiya AB Japan 09May54; loan to French Armée de l'Air Indochina 17-26Jul54, French markings applied; depl Clark AB The Philippines 21Aug54 – 506 Fighter WG (SAC) Dow 29Oct54 – SMAAR (AMC) McClellan 30Oct54 mnt – Assigned variously 27 Fighter WG & 808 AB GP (SAC) Bergstrom from 19Nov54; to 70 SM WG (SAC) Little Rock 18Aug56 – MIDAR (AMC) Olmsted 06Jun57; re-serialed **49-0151** 1958 – 446 TCWG (AFR) Ellington 15Jun58; assigned 706 TCSq Barksdale 31Jul59 – 758 TCSq (459 TCG) (AFR) Greater Pittsburgh 14Apr61; 911 TCG (459 TCWG) (AFR) Greater Pittsburgh 17Jan63 – Wfu to MASDC (LOG) Davis-Monthan 11Nov66 for storage – DFI: 02May68 – Final Disposition: Scrapped.

10389 / 49-152 / C-119C-14-FA

Avble: 31May50; Acc: 09Jun50; Del: 20Jun50 to USAF; briefly retained Fairchild (AMC) Hagerstown for project duties – 316 TCG (CNC) Sewart 21Jun50 – 314 TCG (CNC) Sewart 13Jul50 – 314 TCG (FEA) Komaki AB Japan 10Sep50; to Ashyia AB Japan 18Sep50; to 13 ARP SQ (FEAMCOM) Tachikawa AB Japan 19Jun51 – 403 TCWG (FEA) Ashiya AB Japan 23May52 – BMC (AMC) Birmingham 16Aug52 – 483 TCWG (FEA) Ashiya AB Japan 11Feb53 – 314 TCWG (FEA) Ashiya AB Japan 14Jun54; loan to French Armée de l'Air Indochina 24Jun54-02Aug54, French markings applied; depl Clark AB The Philippines 02Aug54 – 316 TCWG (FEA) Ashiya AB Japan 15Nov54 – 306 Bomb WG (SAC) MacDill 24Nov54 – 70 SM WG (SAC) Little Rock 20Mar56 – 825 AB WG (SAC) Little Rock 06Aug57 – 2578 RFC (CNR) Ellington 27Jan58; to re-serialed **49-0152** 1958 – 446 TCWG (AFR) Ellington 20Sep58; assigned 706 TCSq Barksdale 14Mar59. Ground accident 06Jun59 – DFI: 31Jul59 – Final Disposition: Accident.

10390 / 49-153 / C-119C-14-FA

Avble: 01Jun50; Acc: 14Jun50; Del: 24Jun50 to USAF; briefly retained Fairchild Aircraft (AMC) Hagerstown for project duties – 316 TCG (CNC) Sewart 26Jun50 – 314 TCG

(CNC) Sewart 13Jul50; depl 374 TCWG (FEA) Tachikawa AB Japan 02Aug50 project duties; to FEAMCOM (FEA) Tachikawa AB Japan 22Aug50 – 314 TCWG (FEA) Ashiya AB Japan 18Sep50; depl 21 TCSq (374 TCWG) (FEA) Itazuke AB Japan 23Jan51, to 374 TCWG (FEA) Tachikawa AB Japan 24Jan51; depl 437 TCWG (FEA) Tachikawa AB Japan 29Mar51; to Brady Field Japan 31Mar51. Had a structural failure 10 miles S Atsugi AB Japan 24May51. Pilot: George W. Blank and five other crew bailed out and were rescued. The neatly trimmed aircraft reportedly flew on and was subsequently required to be shot down by American fighter aircraft while still over the sea – DFI: 24Jul51 – Final Disposition: Accident.

10391 / 49-154 / C-119C-14-FA

Avble: 06Jun50; Acc: 14Jun50; Del: 24Jun50 to USAF; briefly retained Fairchild Aircraft (AMC) Hagerstown for project duties – 316 TCG (CNC) Sewart 26Jun50 – 314 TCG (CNC) Sewart 13Jul50; depl 374 TCWG (FEA) Tachikawa AB Japan 02Aug50 project duties; to FEAMCOM (FEA) Tachikawa AB Japan 21Aug50; depl 374 TCWG (FEA) Kanoya AB Japan 18Oct50 – 314 TCWG (FEA) Ashiya AB Japan 02Dec50; depl 21 TCSq (374 TCWG) (FEA) Itazuke AB Japan 23Jan51, multiple depls 374 TCWG (FEA) Tachikawa AB Japan from 24Jan51; to 13 ARP SQ (FEAMCOM) Tachikawa AB Japan 24May51; depl 403 TCWG (FEA) Ashiya AB Japan 16Jun52; depl Bofu Japan 18May53; loan to French Armée de l'Air Indochina early54-07Aug54, French markings applied – BMC (AMC) Birmingham 15Oct54; briefly to WRAAR (AMC) Robbins 13Jan55 – MIDAR (AMC) Olmsted 03Feb55 – bailment Fairchild (AMC) Hagerstown 19May55 – 5039 AB WG (AAC) Elmendorf 06Aug55; to 5039 ARP SQ (AAC) Elmendorf 07Mar56; depl 463 TCWG (TAC) Ardmore 23Mar56 – 5039 AT SQ (AAC) Elmendorf 01Jun57 – 5040 CLM GP (AAC) Elmendorf 01Oct57 – 2577 RFC DT (CNR) NAS Dallas 01Dec57; re-serialed **49-0154** 1958; depl 2578 RFC (CNR) Ellington 25May58 – 69 TCSq (433 TCWG) (AFR) NAS Dallas 20Sep58 – 167 AML SQ (WV-ANG) Martinsburg 05May61 – 925 TCG (446 TCWG) (AFR) Ellington 27May63 – 907 TCG (302 TCWG) (AFR) Clinton 18Mar64 – Wfu to MASDC (LOG) Davis-Monthan 28Sep65 for storage – DFI: 02May68 – Final Disposition: Scrapped.

10392 / 49-155 / C-119C-15-FA

Avble: 07Jun50; Acc: 21Jun50; Del: 26Jun50 to USAF; briefly retained Fairchild Aircraft (AMC) Hagerstown for project duties – 316 TCG (CNC) Sewart 26Jun50 – 314 TCG (CNC) Sewart 13Jul50 – 314 TCG (FEA) Komaki AB Japan 10Sep50; to Ashiya AB Japan 18Sep50; to 13 ARP SQ (FEAMCOM) Tachikawa AB Japan 04Dec50; depl 483 TCWG (FEA) Ashiya AB Japan 26May53; loan to French Armée de l'Air Indochina 17Mar54-04Sep54, French markings applied; depl Clark AB The Philippines 11Aug54 – 27 Fighter WG (SAC) Bergstrom 27Oct54 – 99 SRH WG (SAC) Fairchild 10Nov54 – BMC (AMC) Birmingham 14Aug55 – 99 Bomb WG (SAC) Fairchild 18Nov55 – 92 AB WG (SAC) Fairchild 04Sep56 – 340 AB GP (SAC) Whiteman 26Sep56; to 3310 TTA WG (TC) Scott 05Feb57; re-serialed **49-0155** 1958 – 2578 RFC (CNR) Ellington 29Apr58 – 446 TCWG (AFR) Ellington 20Sep58; assigned 706 TCSq Barksdale 21May59 – 758 TCSq (459 TCWG) (AFR) Greater Pittsburgh 29Apr61; 911 TCG (459 TCWG) (AFR) Greater Pittsburgh 63 – Wfu to 3320 MSU GP (ATC) Amarillo 16Aug66 – REC: 17Aug66 – Final Disposition: Scrapped.

10393 / 49-156 / C-119C-15-FA

Avble: 10Jun50; Acc: 22Jun50; Del: 14Jul50 to USAF – 314 TCWG (CNC) Sewart 19Jul50 – 314 TCG (FEA) Komaki AB Japan 10Sep50; to Ashyia AB Japan 18Sep50; to 13 ARP SQ (FEAMCOM) Tachikawa AB Japan 06Jul51; depl Bofu Japan 18May53 – BMC

(AMC) Birmingham 03Sep53; to 1500 OP SQ (MATS) Hickham 17Sep53 & 27Dec53 mnt; to 1505 AB GP (MATS) Johnston 16Dec53 mnt – 581 ASL GP (FEA) Clark AB The Philippines 24Aug54 – SMPAR (AMC) Clark AB The Philippines 07Jun57 – SMAAR (AMC) McClellan 29Dec57 – 2252 RFC (CNR) Clinton 23Dec57; re-serialed **49-0156** 1958 – 302 TCWG (AFR) Clinton 19Mar59; 907 TCG (302 TCWG) (AFR) Clinton 11Feb63 – 130 ACO GP (ANG) Kanawha 21Jan64 – Wfu to MASDC (LOG) Davis-Monthan 28Mar69 for storage – DFI: 09Apr69 – Final Disposition: Scrapped.

10394 / 49-157 / C-119C-15-FA

Avble: 13Jun50; Acc: 23Jun50; Del: 14Jul50 to USAF – 314 TCWG (CNC) Sewart 19Jul50 – 314 TCG (FEA) Komaki AB Japan 10Sep50; to Ashyia AB Japan 18Sep50; to 13 ARP SQ (FEAMCOM) Tachikawa AB Japan 01Jun51; depl Bofu Japan 18May53 – BMC (AMC) Birmingham 18Jan54 – 483 TCWG (FEA) Ashyia AB Japan 08Jul54 – 581 ASL GP (FEA) Clark AB The Philippines 18Jul54 – 322 TCSq (FEA) Kadena AB Japan 18Sep56 – 2252 RFC (CNR) Clinton 09Dec56; depl 2253 RFC (CNR) Greater Pittsburgh 13Jun57; to 464 TCWG (TAC) Pope 29Jul57 mnt; depl 2259 RFC (CNR) Greater Pittsburgh 01Dec57; re-serialed **49-0157** 1958; to 825 AB GP (SAC) Little Rock 03Jun59 mnt – 302 TCWG (AFR) Clinton 19Mar59; 907 TCG (302 TCWG) (AFR) Clinton 11Feb63 – Wfu to MASDC (LOG) Davis-Monthan 09Oct65 for storage, PCN: **CJ112**. Assigned Museum status 07Jun67 – Displayed at the Pima Air & Space Museum Tucson Arizona – Final Disposition: Extant.

10395 / 49-158 / C-119C-15-FA

Avble: 15Jun50; Acc: 28Jun50; Del: 14Jul50 to USAF – 314 TCWG (CNC) Sewart 19Jul50 – 314 TCG (FEA) Komaki AB Japan 10Sep50; to Ashyia AB Japan 18Sep50; to 13 ARP SQ (FEAMCOM) Tachikawa AB Japan 12Jun51; depl Bofu Japan 18May53 – BMC (AMC) Birmingham 05Dec53 – 483 TCWG (FEA) Ashyia AB Japan 19Mar54 – 581 ASL GP (FEA) Clark AB The Philippines 19May54 – 97 Bomb WG (SAC) Biggs 01Feb55 – 810 AB GP (SAC) Biggs 25Nov56; re-serialed **49-0158** 1958; 2578 RFC (CNR) Ellington 19Jul58 – 446 TCWG (AFR) Ellington 20Sep58; assigned 706 TCSq Barksdale 1959 – 758 TCSq (459 TCWG) (AFR) Greater Pittsburgh 06May61; 911 TCG (459 TCWG) (AFR) Greater Pittsburgh 17Jan63. Wfu and assigned 16Aug66 for USAF allotted destructive tests at Deer Valley Arizona – Final Disposition: Tested to Destruction.

10396 / 49-159 / C-119C-15-FA

Avble: 17Jun50; Acc: 30Jun50; Del: 14Jul50 to USAF – 314 TCWG (CNC) Sewart 19Jul50 – 314 TCG (FEA) Komaki AB Japan 10Sep50; to Ashyia AB Japan 18Sep50; to 13 ARP SQ (FEAMCOM) Tachikawa AB Japan 03Nov50 – OGDAR (AMC) Hill 03Feb51 – BMC (AMC) Birmingham 30Jul53 – 3320 TTN WG (TC) Amarillo 07Nov53 mnt – 28 SRH WG (SAC) Ellsworth 31Aug54; to 3750 TTN WG (TC) Sheppard 18Apr55 – 28 Bomb WG (SAC) Ellsworth 01Oct55; to 3415 TTA WG (TC) Lowry 21Mar56 mnt – 28 AB GP (SAC) Ellsworth 25Jul56 – 2589 RFC (CNR) Dobbins 06Dec57; re-serialed **49-0159** 1958; 2259 RFC DT (CNR) Greater Pittsburgh 02Jul58 – 2255 RFC (CNR) Bradley 22Nov58 – 337 TCSq (514 TCWG) (AFR) Bradley 19Mar59; 903 TCG (514 TCWG) (AFR) McGuire 31May63 – Wfu to MASDC (LOG) Davis-Monthan 06Apr66 for storage – DFI: 03May68 – Final Disposition: Scrapped.

10397 / 49-160 / C-119C-15-FA

Avble: 21Jun50; Acc: 30Jun50; Del: 15Jul50 to USAF – 314 TCWG (CNC) Sewart 18Jul50 – 314 TCG (FEA) Komaki AB Japan 10Sep50; to Ashyia AB Japan 18Sep50; to 13 ARP SQ

(FEAMCOM) Tachikawa AB Japan 03Aug51 – 403 TCWG (FEA) Ashiya AB Japan 14May52 – 483 TCWG (FEA) Ashiya AB Japan 01Jan53. Left engine exploded and fell off in-flight 6.3 miles SE Daejeon S. Korea 11Mar53, the right engine also failed. Pilot: Lt Sheldon L. McConnell and 3 other crew bailed out. Once on the ground co-pilot 2Lt James W. Patton hiked to nearby Daejeon to find a telephone to report the crash. USAF lists as a crash while on a combat mission from Japan to S. Korea – DFI: 27Apr54 – Final Disposition: Accident.

10398 / 49-161 / C-119C-15-FA

Avble: 24Jun50; Acc: 12Jul50; Del: 17Jul50 to USAF – 314 TCWG (CNC) Sewart 21Jul50 – 314 TCG (FEA) Komaki AB Japan 10Sep50; to Ashyia AB Japan 18Sep50; to 13 ARP SQ (FEAMCOM) Tachikawa AB Japan 06Jun51 – 403 TCWG (FEA) Ashiya AB Japan 06Jun52 – 483 TCWG (FEA) Ashiya AB Japan 01Jan53. Crashed over water 15 miles NNE Ashiya AB Japan 23Jun53 while on a combat mission. 4 crew and 3 passengers killed – DFI: 05Aug54 – Final Disposition: Accident.

10399 / 49-162 / C-119C-15-FA

Avble: 27Jun50; Acc: 14Jul50; Del: 18Jul50 to USAF – 314 TCWG (CNC) Sewart 21Jul50 – 314 TCG (FEA) Komaki AB Japan 10Sep50; to Ashyia AB Japan 18Sep50; to 6149 TST WG (FEA) Taegu AB (K-2) S. Korea 10Oct50; to 6400 FDM UT (FEA) Taegu AB (K-2) S. Korea 18Oct50; to 13 ARP SQ (FEAMCOM) Tachikawa AB Japan 05Nov50 & 25Jul51; 403 TCWG (FEA) Ashiya AB Japan 26May52 – 19 Bomb WG (FEA) Andersen 23Jan53 – OGDAR (AMC) Hill 02Mar53 – BMC (AMC) Birmingham 30Jul53 – 42 Bomb WG (SAC) Limestone 31Jul54 – 6 Bomb WG (SAC) Walker 18Nov54; to 2750 AB WG (AMC) WPAFB 19Nov54 – 801 AB GP (SAC) Lockbourne 12Dec55 – 812 AB GP (SAC) Walker 19Dec55 – 2596 RFC (CNR) NAS Dallas 31Oct57 – 3320 TTA WG (TC) Amarillo 04Nov57 mnt – 2577 RFC DT (CNR) NAS Dallas 19Nov57; re-serialed **49-0162** 1958 – 69 TCSq (433 TCWG) (AFR) NAS Dallas 20Sep58; to 1405 AB GP (MATS) Scott 16Aug59 mnt; 923 TCG (433 TCWG) (AFR) NAS Dallas 17Jan63; to Carswell 31May63 – Wfu to 3320 MSU GP (ATC) Amarillo 17Sep65 – REC: 21Sep65 – Final Disposition: Scrapped.

10400 / 49-163 / C-119C-15-FA

Avble: 29Jun50; Acc: 13Jul50; Del: 18Jul50 to USAF – 314 TCG (CNC) Sewart 21Jul50; assigned 50 TCSq. Had a minor taxiing accident at Campbell AFB Kentucky 24Aug50. Suffered a catastrophic structural failure during a routine navigation training flight 7 miles NE of Parsons Tennessee 01Sep50. Pilot: George A. Tschappat, seven of the nine crew onboard were killed apparently while bailing out. The IARC lists the accident as being at Ashiya AB Japan, but this may be due to that assignment being printed to the card the day before the accident occurred and not corrected – REC: 30Sep50 – DFI: 19Oct50 – Final Disposition: Accident.

10401 / 49-164 / C-119C-15-FA

Avble: 06Jul50; Acc: 14Jul50; Del: 19Jul50 to USAF – 314 TCWG (CNC) Sewart 21Jul50 – 314 TCG (FEA) Komaki AB Japan 10Sep50; to Ashyia AB Japan 18Sep50; to 13 ARP SQ (FEAMCOM) Tachikawa AB Japan 26Jul51 & 17Nov51; to 6400 AD WG (FEA) Tachikawa AB Japan 31Jan52; depl 483 TCWG (FEA) Ashiya AB Japan 26Jun54; depl Clark AB The Philippines 04Aug54; loan to French Armée de l'Air Indochina 05Aug54-04Sep54, French markings applied – 1 IM SQ (MATS) Tinker 27Oct54 – 1881 IM SQ (MATS) Tinker 16Mar55; to 3415 TTN WG (TC) Lowry 07Jul55 mnt; to 1739 FRY SQ (MATS) Amarillo

17Dec56 – 1800 ACS WG (MATS) Tinker 28Oct57 – COAAR (MATS) Tinker 01Nov57; re-serialed **49-0164** 1958 – 2577 RFC (CNR) Brooks 02Jun58 – 433 TCWG (AFR) Brooks 18Sep58; to Kelly 31Aug60; 922 TCG (433 TCWG) (AFR) Kelly 31May63 – Wfu to MASDC (LOG) Davis-Monthan 22Oct65 for storage – DFI: 03May68 – Final Disposition: Scrapped.

10402 / 49-165 / C-119C-15-FA

Avble: 07Jul50; Acc: 18Jul50; Del: 19Jul50 to USAF – 314 TCWG (CNC) Sewart 21Jul50 – 314 TCG (FEA) Komaki AB Japan 10Sep50; to Ashyia AB Japan 18Sep50; to 13 ARP SQ (FEAMCOM) Tachikawa AB Japan 02Jun51 – 403 TCWG (FEA) Ashiya AB Japan 16May52 – 483 TCWG (FEA) Ashiya AB Japan 01Jan53; to 6400 AD WG (FEA) Tachikawa AB Japan 10Feb53 mnt; loan to French Armée de l'Air Indochina 09Jun54-16Jul54, French markings applied; depl 314 TCWG (FEA) Ashiya AB Japan 02Jul54 – 316 TCWG (FEA) Ashiya AB Japan 15Nov54 – 22 Bomb WG (SAC) March 19Jan55 – SMAAR (AMC) McClellan 27Jan55 – 43 Bomb WG (SAC) Davis-Monthan 27Mar55; to 3500 PTN WG (TC) Reese 21Feb56 mnt – 28 AB GP (SAC) Ellsworth 16Jul56; re-serialed **49-0165** 1958 – 2577 RFC (CNR) Brooks 25Jun58 – 433 TCWG (AFR) Brooks 18Sep58; to Kelly 31Aug60; 921 TCG (433 TCWG) (AFR) Kelly 31May63 – 922 TCG (433 TCWG) (AFR) Kelly 10Oct65 – Wfu to MASDC (LOG) Davis-Monthan 22Oct65 for storage – DFI: 03May68 – Final Disposition: Scrapped.

10403 / 49-166 / C-119C-15-FA

Avble: 11Jul50; Acc: 21Jul50; Del: 22Jul50 to USAF – 314 TCG (CNC) Sewart 25Jul50 – 314 TCG (FEA) Komaki AB Japan 10Sep50; assigned 50 TCSq.; to Ashiya AB Japan 18Sep50. Crash landed after a combat mission at Pusan East AB (K-9) S. Korea due to a non-combat related mechanical failure 09Nov50. Pilot Charles E. Turnispeed, no fatalities but aircraft damaged beyond repair. 6002 TS WG (FEA) Pusan East AB (K-9) S. Korea 12Nov50 likely for parts salvage, derelict fuselage cvtd to an officers' snack bar at Pusan – REC: 03Dec50 – DFI: 26Dec50 – Final Disposition: Accident.

10404 / 49-167 / C-119C-15-FA

Avble: 13Jul50; Acc: 22Jul50; Del: 25Jul50 to USAF – 314 TCWG (CNC) Sewart 26Jul50 – 314 TCG (FEA) Komaki AB Japan 10Sep50; to Ashyia AB Japan 18Sep50; to 13 ARP SQ (FEAMCOM) Tachikawa AB Japan 17Dec50 & 07Jun51; depl 35 Fighter WG (FEA) Pusan East AB (K-9) S. Korea 07Jan51; depl 403 TCWG (FEA) Ashiya AB Japan 29Apr52; depl Bofu Japan 18May53; loan to French Armée de l'Air Indochina 25Apr54-08Jun54, French markings applied – BMC (AMC) Birmingham 30Jun54 – 55 SMW SQ (MATS) McClellan 11Nov54 – 53 WER SQ (MATS) Kindley 27Feb55 – 59 WER FT (MATS) Kindley 17May55; to 1707 FDM SQ (MATS) Palm Beach 12Aug55; to 464 TCWG (TAC) Pope 14Mar56 mnt; to 1401 AB WG (MATS) Andrews 01May56, 08Aug56, 06Sep56; to 1608 FDM SQ (MATS) Charleston 23Aug56; to 1707 FDM SQ (MATS) Palm Beach 02Jan57 – 2252 RFC (CNR) Clinton 19Jul57; re-serialed **49-0167** 1958 – 302 TCWG (AFR) Clinton 19Mar59; 907 TCG (302 TCWG) (AFR) Clinton 11Feb63 – Wfu to MASDC (LOG) Davis-Monthan 07Nov65 for storage – DFI: 03May68 – Final Disposition: Scrapped.

10405 / 49-168 / C-119C-15-FA

Avble: 15Jul50; Acc: 22Jul50; Del: 25Jul50 to USAF – 314 TCWG (CNC) Sewart 26Jul50 – 314 TCG (FEA) Komaki AB Japan 10Sep50; to Ashiya AB Japan 18Sep50. Lost on a combat mission over Korea 23Mar51 due to enemy action, five crew bailed out, one survived the crash and two were killed – DFI: 04Apr51 – Final Disposition: Combat Loss.

10406 / 49-169 / C-119C-15-FA

Avble: 19Jul50; Acc: 08Aug; Del: 14Aug50 to USAF – 314 TCG (CNC) Sewart 16Aug50 – 314 TCG (CNC) Komaki AB Japan 10Sep50, assigned 50 TCSq; to Ashyia AB Japan 18Sep50; to 13 ARP SQ (FEAMCOM) Tachikawa AB Japan 24Jul51. Crashed into treetops about 3.7 miles from Runway 19 at Tachikawa AB Japan 27Sep51. 5 crew killed were: 1Lt Lynuel Bevers (pilot), 2Lt Joseph E. Binns, 2Lt Eugene H. Class, TSgt Jack Davis and Sgt Ernest D. Carrara plus 9 passengers were also killed – DFI: 23Nov51 – Final Disposition: Accident.

10407 / 49-170 / C-119C-16-FA

Avble: 21Jul50; Acc: 31Jul50; Del: 04Aug50 to USAF – 314 TCWG (CNC) Sewart 05Aug50; depl 316 TCG (CNC) Sewart 17Aug50 – 314 TCG (FEA) Komaki AB Japan 14Sep50; to Ashyia AB Japan 18Sep50; assigned 37 TCSq Komaki AB Japan 08Nov50; to 13 ARP SQ (FEAMCOM) Tachikawa AB Japan 13Jul51; depl Komaki AB Japan 13Jul51 – 403 TCWG (FEA) Ashiya AB Japan 28Apr52 – 483 TCWG (FEA) Ashiya AB Japan 01Jan53 – BMC (AMC) Birmingham 09Dec53 – 824 AB GP (SAC) Carswell 20May54 – 7 Bomb WG (SAC) Carswell 30May54 – 803 AB GO (SAC) Davis-Monthan 20Aug57 – 824 AB GP (SAC) Carswell 27Aug57 – 2233 RFC (CNR) Mitchel 14Nov57; re-serialed **49-0170** 1958 – 514 TCWG (AFR) Mitchel 19Dec58; to McGuire 18Mar61; 903 TCG (514 TCWG) (AFR) McGuire 31May63 – Wfu to MASDC (LOG) Davis-Monthan 25May66 for storage – DFI: 03May68 – Final Disposition: Scrapped.

10408 / 49-171 / C-119C-16-FA

Avble: 25Jul50; Acc: 08Aug50; Del: 10Aug50 to USAF – 314 TCWG (CNC) Sewart 11Aug50; depl 316 TCG (CNC) Sewart 17Aug50 – 314 TCG (FEA) Komaki AB Japan 10Sep50; to Ashyia AB Japan 18Sep50; depl 374 TCWG (FEA) Tachikawa AB Japan 06Feb51; to 13 ARP SQ (FEAMCOM) Tachikawa AB Japan 02Jun51; depl Komaki AB Japan 31Aug51 – 403 TCWG (FEA) Ashiya AB Japan 28Apr52 – 483 TCWG (FEA) Ashiya AB Japan 01Jan53 – BMC (AMC) Birmingham 19Jan54 – 508 Fighter WG (SAC) Turner 26Aug54 – **RECORDS MISSING** – 31 Fighter WG (SAC) Turner date unk – 4080 AB GP (SAC) Laughlin 15Apr57; to 3530 PTN WG (TC) Bryan 30Oct57 mnt; re-serialed **49-0171** 1958–2252 RFC (CNR) Clinton 30Mar58 – 302 TCWG (AFR) Clinton 19Mar59 – MAP Morocco 10May62, RMAF s/n: **CN-AMB**, 1 AT SQ – Final Disposition: Unknown.

10409 / 49-172 / C-119C-16-FA

Avble: 27Jul50; Acc: 15Aug50; Del: 24Aug50 to USAF – 314 TCG (CNC) Sewart 25Aug50 – 314 TCG (FEA) Komaki AB Japan 10Sep50; to Ashyia AB Japan 18Sep50; to 13 ARP SQ (FEAMCOM) Tachikawa AB Japan 19Jun51. Suffered a structural failure, likely due to prop blade failure, during a supply mission to Seoul AB (K-16) crashing into the sea 20 miles SE of Pohang Dong AB (K-3) S. Korea 16Jan52. 3 crew were killed: Capt Joseph R. Hewitt, 1Lt Robert K. Bancker (pilot), TSgt William A. Metcalfe and 2 survived: 1Lt Charles S. Aldrich and Cpl Billy J. Robinson – DFI: 19Mar52 – Final Disposition: Accident.

10410 / 49-173 / C-119C-16-FA

Avble: 28Jul50; Acc: 09Aug50; Del: 26Aug50 to USAF – 314 TCG (CNC) Sewart 30Aug50 – 314 TCG (CNC) Komaki AB Japan 10Sep50, assigned 61 TCSq; to Ashyia AB Japan 18Sep50; to 13 ARP SQ (FEAMCOM) Tachikawa AB Japan 08Jul51. Crashed into terrain during a night approach 8 miles NW Ashiya AB Japan 29Apr52. Pilot: Morford W. Locke plus 3 other crew killed – DFI: 25Jun52 – Final Disposition: Accident.

10411 / 49-174 / C-119C-16-FA

　　Avble: 16Aug50; Acc: 01Sep50; Del: 02Sep50 to USAF – 314 TCWG (CNC) Sewart 10Sep50 – 314 TCG (FEA) Ashyia AB Japan 19Sep50; assigned 37 TCSq Komaki AB Japan 08Nov50; to 13 ARP SQ (FEAMCOM) Tachikawa AB Japan 20Jul51; depl Komaki AB Japan 25Jul51 – 403 TCWG (FEA) Ashyia AB Japan 28Apr52 – 483 TCWG (FEA) Ashiya AB Japan 01Jan53 – BMC (AMC) Birmingham 18Jan54 – 31 Fighter WG (SAC) Turner 18Jun54; to 803 AB GP (SAC) Davis-Monthan 04Nov54; to 99 SRH WG (SAC) Fairchild 29Mar55 – 2596 RFC (CNR) NAS Dallas 27Nov57 – 2577 RFC DT (CNR) NAS Dallas 01Dec57; re-serialed **49-0174** 1958 – 69 TCSq (433 TCWG) (AFR) NAS Dallas 20Sep58; 923 TCG (433 TCWG) (AFR) NAS Dallas 17Jan63; to Carswell 02Mar63 – Wfu to MASDC (LOG) Davis-Monthan 21Sep65 for storage – DFI: 21Jun66 – Final Disposition: Scrapped.

10412 / 49-175 / C-119C-16-FA

　　Avble: 22Aug50; Acc: 30Aug50; Del: 02Sep50 to USAF – 314 TCWG (CNC) Sewart 02Sep50 – 314 TCG (FEA) Komaki AB Japan 10Sep50, assigned 50 TCSq; to Ashiya AB Japan 18Sep50; to 13 ARP SQ (FEAMCOM) Tachikawa AB Japan 13Jul51. Had a severe landing accident at Iwakuni MCAS Japan 09Oct52 damaging the airframe beyond repair. Pilot: Donald P. Leme and crew survived – REC: 09Nov52 – DFI: 10Dec52 – Final Disposition: Accident.

10413 / 49-176 / C-119C-16-FA

　　Avble: 22Aug50; Acc: 31Aug50; Del: 02Sep50 to USAF – 314 TCWG (CNC) Sewart 02Sep50 – 314 TCG (FEA) Komaki AB Japan 10Sep50; to Ashiya AB Japan 18Sep50; to 13 ARP SQ (FEAMCOM) Tachikawa AB Japan 04Jul51; depl Bofu Japan 18May53 – BMC (AMC) Birmingham 18Jan54 – 31 Fighter WG (SAC) Turner 17Jun54 – 4080 SRL WG (SAC) Turner 25Mar57 – 4080 AB GP (SAC) Turner 01Apr57; to Laughlin 22Apr57; re-serialed **49-0176** 1958 – 2577 RFC (CNR) Brooks 21Apr58 – 433 TCWG (AFR) Brooks 18Sep58; to Kelly 31Aug60; 921 TCG (433 TCWG) (AFR) Kelly 31May63; 922 TCG (433 TCWG) (AFR) Kelly 10Oct65 – Wfu to MASDC (LOG) Davis-Monthan 22Oct65 for storage – DFI: 03May68 – Final Disposition: Scrapped.

10414 / 49-177 / C-119C-16-FA

　　Avble: 15Aug50; Acc: 03Sep50; Del: 04Sep50 to USAF – 314 TCWG (CNC) Sewart 10Sep50 – 314 TCG (FEA) Ashyia AB Japan 19Sep50; to 13 ARP SQ (FEAMCOM) Tachikawa AB Japan 02Nov50 & 26May51 – 483 TCWG (FEA) Ashyia AB Japan 23Feb53; depl Bofu Japan 24May53; depl 314 TCWG (FEA) Ashyia AB Japan 16Mar54; loan to French Armée de l'Air Indochina 30May54-17Jul54, French markings applied – MIDAR (AMC) Olmsted 28Oct54 – bailment Fairchild (AMC) Hagerstown 19May55 – 5039 AB WG (AAC) Elmendorf 13Jun55; to 5039 MAI SQ (AAC) Elmendorf 24Nov56 – 5039 AT SQ (AAC) Elmendorf 01Jun57 – 5040 CLM GP (AAC) Elmendorf 01Oct57 – 2252 RFC (CNR) Clinton 16Dec57; re-serialed **49-0177** 1958 – 302 TCWG (AFR) Clinton 19Mar59; 907 TCG (302 TCWG) (AFR) Clinton 11Feb63 – Wfu to MASDC (LOG) Davis-Monthan 09Oct65 for storage – DFI: 03May68 – Final Disposition: Scrapped.

10415 / 49-178 / C-119C-16-FA

　　Avble: 28Aug50; Acc: 06Sep50; Del: 06Sep50 to USAF – 314 TCWG (CNC) Sewart 10Sep50 – 314 TCG (FEA) Ashyia AB Japan 19Sep50; assigned 37 TCSq Komaki AB Japan 08Nov50; to 13 ARP SQ (FEAMCOM) Tachikawa AB Japan 08Jun51; depl Komaki AB Japan 13Jun51 – 403 TCWG (FEA) Ashyia AB Japan 28Apr52 – 483 TCWG (FEA) Ashiya

AB Japan 01Jan53 – 5 AB GP (SAC) Travis 22Nov53 – BMC (AMC) Birmingham 26Nov53 – 60 TCWG (AFE) Rhein-Main AB W. Germany 29Apr54 – 317 TCWG (AFE) Neubiberg AB W. Germany 06May54 – 7559 MAI GP (AFE) RAF Burtonwood England 09Dec54 – MIDAR (AMC) Olmsted 01Oct55 – 7 WEA GP (MATS) Elmendorf 10Apr56; to 5001 MAI SQ (AAC) Ladd 17Nov56; to 5039 MAI SQ (AAC) Elmendorf 29Nov56 – 1739 FRY SQ (MATS) Amarillo 05Jul57 – 2259 RFC (CNR) Andrews 09Jul57 – 2252 RFC (CNR) Clinton 11Jul57; re-serialed **49-0178** 1958; depl 2578 AB GP (CNC) Ellington 11Dec58 – Hayes (AMC) Birmingham 18Mar59 – 302 TCWG (AFR) Clinton Sep59; 907 TCG (302 TCWG) (AFR) Clinton 11Feb63 – Wfu to MASDC (LOG) Davis-Monthan 09Oct65 for storage – DFI: 03May68 – Final Disposition: Scrapped.

10416 / 49-179 / C-119C-16-FA

Avble: 30Aug50; Acc: 06Sep50; Del: 06Sep50 to USAF – 8 ARS SQ (SAC) Camp Carson 17Oct50 – 3904 CMP WG (SAC) Camp Carson 01Nov50; to 3203 MSU GP (APG) Eglin 23Mar51 mnt; depl Peterson 26Mar52; to RAF Mildenhall England (SAC) 15Aug52; to 59 AD WG (AFE) RAF Burtonwood England 05Sep52 mnt; to 4600 AB GP (ADC) Peterson 15Jan53 mnt; to Stead 20May53 – OGDAR (AMC) Hill 20Jul53 – BMC (AMC) Birmingham 09Jan54 – 42 Bomb WG(SAC) Limestone 07Feb54; to 1610 MAT SQ (MATS) Grenier 14Feb55; to 1604 MAT SQ (MATS) Kindley 20Feb55; to 3415 TTA WG (TC) Lowry 28Sep56 mnt – 42 AB GP (SAC) Loring 01Feb57; to 79 Fighter GP (ADC) Youngstown 27Mar57 mnt; to 814 AB GP (SAC) Westover 19Dec57; re-serialed **49-0179** 1958 – 2577 RFC (CNR) Brooks 29Jun58 – 433 TCWG (AFR) Brooks 18Sep58; to Kelly 31Aug60 – 156 AML SQ (NC-ANG) Douglas 18Apr61 – 433 TCG (AFR) Kelly 12Oct62; 922 TCG (433 TCWG) Kelly 31May63 – Wfu to MASDC (LOG) Davis-Monthan 29Oct65 for storage – DFI: 03May68 – Final Disposition: Scrapped.

10417 / 49-180 / C-119C-16-FA

Avble: 05Sep50; Acc: 26Oct50; Del: 30Oct50 to USAF – 316 TCG (CNC) Sewart 01Nov50 – 314 TCWG (TAC) Sewart 27Jun51 – 456 TCWG (TAC) Miami IAP 15Jan53 – 483 TCWG (FEA) Ashiya AB Japan 14Feb53 – 314 TCWG (FEA) Ashiya AB Japan 20Mar53 – BMC (AMC) Birmingham 03Feb54 mnt – 31 Fighter WG (SAC) Turner 16Jul54; to 99 SRH WG (SAC) Fairchild 24Apr55; to 808 AB GP (SAC) Bergstrom 07Jul55 – 4080 SRL WG (SAC) Turner 01Apr57; to 323 Fighter-Bomber WG (SAC) Bunker Hill 10Apr57 mnt – 4080 AB GP (SAC) Laughlin 13Apr57; to 306 Bomb WG (SAC) MacDill 23Jul57 – 2252 RFC (CNR) Clinton 31Oct57; re-serialed **49-0180** 1958; to 809 AB GP (SAC) MacDill 31Mar58 – 302 TCWG (AFR) Clinton 19Mar59 – Fairchild (LOG) St. Augustine 07Jan63 – MAP Morocco 14Mar63, RMAF s/n: **CN-AME**, 1 AT SQ – Final Disposition: Unknown.

10418 / 49-181 / C-119C-16-FA

Avble: 06Sep50; Acc: 20Sep50; Del: 22Sep50 to USAF – 316 TCG (CNC) Sewart 24Sep50; depl Laurinburg-Maxton Airport North Carolina 24Aug51; 314 TCWG (TAC) Sewart 14May52 – 403 TCWG (FEA) Ashiya AB Japan 14Jul52 – 483 TCWG (FEA) Ashiya AB Japan 01Jan53 – 5 AB GP (SAC) Travis 30Nov53 mnt – BMC (AMC) Birmingham 04Dec53 mnt – 60 TCWG (AFE) Rhein-Main AB W. Germany 15Apr54 – 1739 FRY SQ (MATS) Amarillo 30Apr54 – 7 Bomb WG (SAC) Carswell 22Jun54; to 314 TCWG (TAC) Sewart 12Mar56 mnt; re-serialed **49-0181** 1958 – SPW Center (ARD) Kirtland 27Jun58 – 446 TCWG (AFR) Ellington 05Dec58; assigned 706 TCSq Barksdale 22Apr59 – 758 TCSq

(459 TCWG) (AFR) Greater Pittsburgh 29Apr61; 911 TCG (459 TCWG) (AFR) Greater Pittsburgh 31May63. Wfu and assigned 26Nov66 for USAF allotted destructive tests at Deer Valley Arizona – Final Disposition: Tested to Destruction.

10419 / 49-182 / C-119C-16-FA

Avble: 11Sep50; Acc: 25Sep50; Del: 26Sep50 to USAF – 316 TCG (CNC) Sewart 26Sep50; depl 314 TCWG (TAC) Sewart 09Jul51; depl Laurinburg-Maxton Airport North Carolina 20Aug51 – 314 TCWG (TAC) Sewart 15May52 – 403 TCWG (FEA) Ashiya AB Japan 14Jul52 – 483 TCWG (FEA) Ashiya AB Japan 01Jan53 – 314 TCWG (FEA) Ashiya AB Japan 16Apr54 – BMC (AMC) Birmingham 11May54 mnt – 28 SRH WG (SAC) Ellsworth 19Aug54 – 28 Bomb WG (SAC) Ellsworth 05Oct55; to 3415 TTA WG (TC) Lowry 28Feb56 mnt – 28 AB GP (SAC) Ellsworth 25Jul56; re-serialed **49-0182** 1958–2233 RFC (CNR) Mitchel 30Mar58 – 514 TCWG (AFR) Mitchel 19Dec58; to McGuire 10May61; 903 TCG (514 TCWG) (AFR) McGuire 31May63 – Fairchild (LOG) St.. Augustine 1966 – Wfu to 3320 MSU GP (ATC) Amarillo & REC: 21Jun66 – Final Disposition: Scrapped.

10420 / 49-183 / C-119C-16-FA

Avble: 12Sep50; Acc: 26Sep50; Del: 28Sep50 to USAF – 316 TCG (CNC) Sewart 28Sep50; depl Laurinburg-Maxton Airport North Carolina 24Aug51 – 435 TCWG (TAC) Miami IAP 04Aug52; to 6612 AB GP (NEA) Thule AB Greenland 17Aug52 mnt; depl Elmendorf 24Oct52 – 456 TCWG (TAC) Miami 02Dec52 – BMC (AMC) Birmingham 16Apr53 mnt – 483 TCWG (FEA) Ashiya AB Japan 21Nov53; loan to French Armée de l'Air Indochina 30Apr54-10Jun54, French markings applied – 5 SRH WG (SAC) Travis 29Oct54; to 1509 SUT SQ (MATS) Johnston 03Nov54 mnt; to 1500 FDM SQ (MATS) Hickham 06Feb55 mnt; to 5 AB GP (SAC) Travis 06Feb55 – bailment Fairchild (AMC) Hagerstown 11Apr55 – 802 AB GP (SAC) Smoky Hill 05May55 – 40 Bomb WG (SAC) Smoky Hill 01Jul55 – 310 Bomb WG (SAC) Smoky Hill 04Oct55 – 340 Bomb WG (SAC) Whiteman 09May56; to 817 AB GP (SAC) Portsmouth 19Nov56 – 3415 TTA WG (TC) Lowry 02Feb57 mnt – 340 AB GP (SAC) Whiteman 30Mar57; to 340 Bomb WG (SAC) Whiteman 21Aug57; re-serialed **49-0183** 1958 – OGDAR (AMC) Hill 14Mar58 – 2578 RFC (CNR) Ellington 18Jun58 – 446 TCWG (AFR) Ellington 20Sep58; assigned 706 TCSq Barksdale 09Feb59 – 78 TCSq (435 TCWG) (AFR) Barskdale 06May61 – 156 AML SQ (NC-ANG) Douglas 06May61 – Fairchild (LOG) St.. Augustine 13Jun62 – MAP Morocco 29Oct62, RMAF s/n: **CN-AMF**, 1 AT SQ – Final Disposition: Unknown.

10421 / 49-184 / C-119C-16-FA

Avble: 16Sep50; Acc: 30Sep50; Del: 04Oct50 to USAF – Assigned variously 314 TCWG & 316 TCG (CNC) Sewart from 08Oct50; depl Shaw 08Oct50; to 28 SRH WG (SAC) Rapid City 20Sep51; depl Laurinburg-Maxton Airport North Carolina 20Aug51; depl 403 TCWG (FEA) Ashiya AB Japan 14Aug52; loan to French Armée de l'Air Indochina 29Apr54-30May54, French markings applied – 483 TCWG (FEA) Ashiya AB Japan 22Jul54 – BMC (AMC) Birmingham 27Oct54 mnt – 12 Fighter WG (SAC) Bergstrom 16Nov54 – MIDAR (AMC) Olmsted 10Jun57; re-serialed **49-0184** 1958 – 706 TCSq (446 TCWG) (AFR) Barksdale 12Jun59 – 78 TCSq (435 TCWG) (AFR) Barksdale 08May61 – 514 TCWG (AFR) Mitchel 13May61, to McGuire 13May61; 903 TCG (514 TCWG) (AFR) McGuire 31May63 – Wfu to MASDC (LOG) Davis-Monthan 06Apr66 – DFI: 03May68 – Final Disposition: Scrapped.

10422 / 49-185 / C-119C-17-FA

Avble: 19Sep50; Acc: 05Oct50; Del: 07Oct50 to USAF – 316 TCG (CNC) Sewart 10Oct50; depl 314 TCWG (TAC) Sewart 24Aug51 – 435 TCWG (TAC) Miami 12Aug52; depl Elmendorf 24Oct52 – 456 TCWG (TAC) Miami 02Dec52 – 483 TCWG (FEA) Ashiya AB Japan 16Feb53 – 314 TCWG (FEA) Ashiya AB Japan 15Apr53; loan to French Armée de l'Air Indochina 03May54-03Jul54, French markings applied – BMC (AMC) Birmingham 23Jul54 mnt – 305 Bomb WG (SAC) MacDill 07Dec54; to 818 AB GP (SAC) Lincoln 17Jul56 – 809 AB GP (SAC) MacDill 16Sep57 – 2596 RFC (CNR) NAS Dallas 11Oct57 – 2577 RFC DT (CNR) NAS Dallas 01Dec57; re-serialed **49-0185** 1958; depl 2578 RFC (CNR) Ellington 03May58 – 69 TCSq (433 TCWG) (AFR) NAS Dallas 18Sep58; 923 TCG (433 TCWG) (AFR) NAS Dallas 17Jan63; to Carswell 31May63 – Wfu to MASDC (LOG) Davis-Monthan 21Sep65 for storage – REC: 21Jun66 – Final Disposition: Scrapped.

10423 / 49-186 / C-119C-17-FA

Avble: 26Sep50; Acc: 25Oct50; Del: 30Oct50 to USAF – 316 TCG (CNC) Sewart 31Oct50; depl 314 TCWG (TAC) Sewart 09Jul51 – 435 TCWG (TAC) Miami 02Sep52; depl Elmendorf 24Oct52 – 456 TCWG (TAC) Miami 02Dec52 – BMC (AMC) Birmingham 01May53 mnt – 483 TCWG (FEA) Ashiya AB Japan 07Dec53; loan to French Armée de l'Air Indochina early54, French markings applied. Crashed on take-off at Cat Bi Airfield N. Vietnam 23Mar54 when the French co-pilot raised the u/c before the aircraft became airborne. The aircraft settled onto the runway damaging the entire fuselage underside, the 2,000 lbs of fuel and 8,000 lbs of napalm onboard did not ignite. Listed as officially transferred out of USAF 11Mar54 – DFI: 17Feb55 – Final Disposition: Accident.

10424 / 49-187 / C-119C-17-FA

Avble: 26Sep50; Acc: 10Oct50; Del: 14Oct50 to USAF – 316 TCG (CNC) Sewart 16Oct50; to 2750 AB WG (AMC) WPAFB 19Oct50, cvtd **EC-119C** 25Oct50 for experimental duties; bailment Aeroproducts Division (AMC) Vandalia 16Oct50; to WRDCN (ARD) WPAFB 18Jul51, cvtd **C-119C** 19Jul51; depl Laurinburg-Maxton Airport North Carolina 20Aug51; to 1050 AB WG (HQC) Andrews 26Dec51 mnt – 435 TCWG (TAC) Miami 05May52 – 456 TCWG (TAC) Miami IAP 01Dec52 – 483 TCWG (FEA) Ashiya AB Japan 16Feb53; depl Bofu Japan 08Nov53; loan to French Armée de l'Air Indochina early54-09May54, French markings applied – BMC (AMC) Birmingham 11May54 mnt – 824 AB GP (SAC) Carswell 01Sep54 – 6 Bomb WG (SAC) Walker 06Oct54; to 3750 TTN WG (TC) Sheppard 27Mar55; to 812 AB GP (SAC) Walker 04Nov55; to 2750 AB WG (AMC) WPAFB 04Nov55; to 812 AB GP (SAC) Walker 22Nov55; to 4900 AB GP (ARD) Kirtland 15Feb56 – 812 AB GP (SAC) Walker 25Jul56 – 2589 RFC (CNR) Dobbins 08Nov57; to 3320 TTA WG (TC) Amarillo 58; re-serialed **49-0187** 1958 – 2259 RFC DT (CNR) Greater Pittsburgh 12Jul58 – 758 TCSq (459 TCWG) Greater Pittsburgh 19Dec58 – Fairchild (LOG) St. Augustine – MAP Morocco 01Mar63, RMAF s/n: **CN-AMC**, 1 AT SQ – Final Disposition: Unknown.

10425 / 49-188 / C-119C-17-FA

Avble: 28Sep50; Acc: 21Oct50; Del: 26Oct50 to USAF – 316 TCG (CNC) Sewart 29Oct50; depl Laurinburg-Maxton Airport North Carolina 20Aug51 – 435 TCWG (TAC) Miami 04Aug52; depl Elmendorf 24Oct52 – 456 TCWG (TAC) Miami 02Dec52 – BMC (AMC) Birmingham 16Apr53 mnt – 483 TCWG (FEA) Ashiya AB Japan 16Sep53 – 5 AB GP (SAC) Travis 17Nov54 – 305 Bomb WG (SAC) MacDill 29Nov54 – 1500 FDM SQ

(MATS) Hickham 06Dec54 mnt; 506 Fighter WG (SAC) Dow 16Jan55; to Tinker 20Mar55; re-serialed **49-0188** 1958 – 2577 RFC (CNR) Brooks 26Jun58 – 433 TCWG (AFR) Brooks 18Sep58; to Kelly 31Aug60; 921 TCG (433 TCWG) (AFR) Kelly 31May63 – Wfu to MASDC (LOG) Davis-Monthan 30Sep65 for storage – DFI: 03May68 – Final Disposition: Scrapped.

10426 / 49-189 / C-119C-17-FA

Avble: 03Oct50; Acc: 28Oct50; Del: 02Nov50 to USAF – Crew bailed out due to fuel starvation due to weather conditions 7 miles WNW Griffiss AFB New York 02Nov50. Pilot: George P. Postor, no fatalities. Appears to have been on its delivery flight, which would make this aircraft the shortest serving Boxcar in USAF history! Assigned 316 TCG (CNC) Sewart 07Nov50 with 36 TCSq – REC: 08Nov50 – DFI: 19Dec50 – Final Disposition: Accident.

10427 / 49-190 / C-119C-17-FA

Avble: 04Oct50; Acc: 28Oct50; Del: 02Nov50 to USAF – 316 TCG (CNC) Sewart 07Nov50; to 5001 CMP WG (AAC) Ladd 25Apr51 mnt – depl 314 TCWG (TAC) Sewart 05Oct51 – 435 TCWG (TAC) Miami 05May52; depl Elmendorf 31Oct52 – 456 TCWG (TAC) Miami 02Dec52 – 483 TCWG (FEA) Ashiya AB Japan 16Feb53 – BMC (AMC) Birmingham 22Nov53 mnt – 60 TCWG (AFE) Rhein-Main AB W. Germany 15Apr54 – 317 TCWG (AFE) Neubiberg AB W. Germany 30Apr54 – 308 Bomb WG (SAC) Hunter 19Dec54; to 2 Bomb WG (SAC) Hunter 21Aug56 – 804 AB GP (SAC) Hunter 05Aug57; re-serialed **49-0190** 1958 – 2578 RFC (CNR) Ellington 08Jun58 – 446 TCWG (AFR) Ellington 20Sep58; assigned 706 TCSq Barksdale 06Oct60 – 78 TCSq (435 TCWG) (AFR) Barksdale 08May61 – 156 AML SQ (NC-ANG) Douglas 06May61 – Fairchild (LOG) St. Augustine 13Jun62 – MAP Morocco 03Oct62, RMAF s/n: **CN-AMD,** 1 AT SQ – Final Disposition: Unknown.

10428 / 49-191 / C-119C-17-FA

Avble: 06Oct50; Acc: 31Oct50; Del: 07Nov50 to USAF – 316 TCG (CNC) Sewart 09Nov50; depl Laurinburg-Maxton Airport North Carolina 24Aug51 – 435 TCWG (TAC) Miami 11Apr52; depl Elmendorf 24Oct52 – 456 TCWG (TAC) Miami 02Dec52 – 483 TCWG (FEA) Ashiya AB Japan 17Feb53; depl Bofu Japan 17Nov53 – BMC (AMC) Birmingham 28Jan54 mnt – 3910 AB GP (SAC) RAF Mildenhall England 29Jul54; to RAF Lakenheath England 09May55; to 465 TCWG (AFE) Evreux-Fauville AB France 26Jan56 mnt; to 7100 SUT WG (AFE) Wiesbaden AB W. Germany 30Jul56 & 21Jan57 mnt; re-serialed **49-0191** 1958 – 2578 RFC (CNR) Ellington 10Feb58 – 446 TCWG (AFR) Ellington 20Sep58; assigned 706 TCSq Barksdale Mar59 – 78 TCSq (435 TCWG) (AFR) Barksdale 08May61 – 156 AML SQ (NC-ANG) Douglas 29Apr61 – Fairchild (LOG) St. Augustine 13Jun62 – 433 TCWG (AFR) Kelly 17Aug62; 921 TCG (433 TCWG) (AFR) Kelly 31May63 – 130 ACO GP (ANG) Kanawha 18Dec63 – Wfu to MASDC (LOG) Davis-Monthan 28Mar69 for storage – DFI: 09Apr69 – Final Disposition: Scrapped.

10429 / 49-192 / C-119C-17-FA

Avble: 10Oct50; Acc: 01Nov50; Del: 17Nov50 to USAF – 316 TCG (CNC) Sewart 20Nov50, assigned 36 TCSq. Suffered a structural failure while flying in a storm 12 miles SE Jackson Tennessee 04Jun51. Pilot: Kenneth W. Christensen and six other crew onboard, four were killed but three successfully managed to bail out – REC: 05Jun51 – DFI: 25Jul51 – Final Disposition: Accident.

10430 / 49-193 / C-119C-17-FA

Avble: 16Oct50; Acc: 07Nov50; Del: 21Nov50 to USAF – 316 TCG (CNC) Sewart 21Nov50; depl 3203 MSU GP & 3200 PTS WG (APG) Eglin from 04Feb51; depl 314 TCWG (TAC) Sewart 01Jun51 & 15Nov51; depl Laurinburg-Maxton Airport North Carolina 24Aug51 – 403 TCWG (FEA) Ashiya AB Japan 14Jul52 – 483 TCWG (FEA) Ashiya AB Japan 01Jan53; depl Bofu Japan 17Nov53 – BMC (AMC) Birmingham 12Dec53 – 11 Bomb WG (SAC) Carswell 09Jun54 – 824 AB GP (SAC) Carswell 06Aug57 – 7 AB GP (SAC) Carswell 01Dec57; re-serialed **49-0193** 1958 – to 93 AB GP (SAC) Castle 03May58 mnt – 2577 RFC (CNR) Brooks 18Jun58 – 433 TCWG (AFR) Brooks 18Sep58; to 2750 AB WG (AMC) WPAFB 18Jun60; to Kelly 31Aug60; 921 TCG (433 TCWG) (AFR) Kelly 31May63 – Wfu to MASDC (LOG) Davis-Monthan 30Sep65 for storage – DFI: 03May68 – Final Disposition: Scrapped.

10431 / 49-194 / C-119C-17-FA

Avble: 18Oct50; Acc: 02Nov50; Del: 07Nov50 to USAF – 316 TCG (CNC) Sewart 09Nov50; depl Laurinburg-Maxton Airport North Carolina 20Aug51 – 314 TCWG (TAC) Sewart 08Jul52 – WRDCN (ARD) WPAFB 15Oct52 – 6511 PDT GP (ARD) El Centro 14Jan53 for test duties – SBNAR (AMC) Norton 10Jun53 – 64 TCWG (TAC) Donaldson 05Jan54 – 69 PBL SQ (TAC) Patrick 19Apr54; to AFM Center (ARD) Patrick18May54 mnt – 11 PBL SQ (TAC) Orlando 08Sep54 – BMC (AMC) Birmingham 01Feb55 – 55 SM WG (SAC) Forbes 12May55; to 3750 TTN WG (TC) Sheppard 01Jun55; to 801 AB GP (SAC) Lockbourne 07Sep55 – **RECORDS MISSING** – 340 Bomb WG (SAC) Whiteman 17Jul56 – 340 AB GP (SAC) Whiteman 31Oct56; to 315 AB GP (SAC) Forbes 09May57; re-serialed **49-0194** 1958 – 2578 RFC (CNR) Ellington 31Jan58 – 446 TCWG (AFR) Ellington 20Sep58; assigned 706 TCSq Barksdale Apr59 – 78 TCSq (43 TCWG) (AFR) Barksdale 08May61 – 156 AML SQ (NC-ANG) Douglas 06May61 – 758 TCSq (459 TCWG) (AFR) Greater Pittsburgh 14Nov62; 911 TCG (459 TCWG) (AFR) Greater Pittsburgh 17Jan63 – Wfu to MASDC (LOG) Davis-Monthan 07Jan67 for storage – DFI: 22Jun67 – BofS F.H. Chew Sales Co. Phoenix Arizona 20Dec67 – BofS Aero Union Corp. Chico California 20Dec67; ARAp **N13742** 24Jan68; cvtd to airtanker duties Mar70, Tanker **85** – BofS Hemet Valley Flying Service Co. Hemet California 09Feb76; cvtd **C-119C Jet-Pak** (J34-WE) circa 1976, Tanker **88**. Crashed due to a fatal structural failure of the left wing during a fire-bombing run in Los Padres National Forest California 08Jul81; killed in the crash were pilot Louis Remschner (52yrs) and co-pilot Ted Sveum (25yrs). It was later found the failure of a control cable lead to control surface flutter that overstressed the airframe. Remains of the aircraft can still be found at the site today – N13742 de-reg 15Jun84 – Final Disposition: Accident.

10432 / 49-195 / C-119C-17-FA

Avble: 21Oct50; Acc: 14Nov50; Del: 05Dec50 to USAF – 316 TCG (TAC) Sewart 05Dec50; depl Laurinburg-Maxton Airport North Carolina 20Aug51; to 3800 AU WG (AU) Maxwell 12May52 mnt – 314 TCWG (TAC) Sewart 14May52 – 403 TCWG (FEA) Ashiya AB Japan 14Jul52 – 483 TCWG (FEA) Ashiya AB Japan 01Jan53; depl Bofu Japan 24May53 – BMC (AMC) Birmingham 12Dec53 – 11 Bomb WG (SAC) Carswell 04Jun54; to 3750 TTN WG (TC) Sheppard 07May55; to 824 AB GP (SAC) Carswell 05Jun55 – 824 AB GP (SAC) Carswell 06Aug57 – 7 AB GP (SAC) Carswell 01Dec57; re-serialed **49-0195** 1958. While on a flight from Sheppard AFB to Carswell AFB 27Mar58, collided midair over Bridgeport Texas with C-124C s/n: 52-0981. The 3 crew on the C-119C were killed along with all 15 personnel onboard the C-124C – Final Disposition: Accident.

10433 / 49-196 / C-119C-17-FA

Avble: 25Oct50; Acc: 28Nov50; Del: 06Dec50 to USAF – 316 TCG (TAC) Sewart 06Dec50; depl Laurinburg-Maxton Airport North Carolina 24Aug51 – 435 TCWG (TAC) Miami 02Sep52; depl Elmendorf 24Oct52 – 456 TCWG (TAC) Miami 02Dec52 – 483 TCWG (FEA) Ashiya AB Japan 16Feb53 – BMC (AMC) Birmingham 27Aug53 mnt – 3906 AB GP (SAC) Sidi Slimane AB Morocco 19Aug54; to 60 TCWG (AFE) Rhein-Main AB W. Germany 17Mar55 & 22May55 mnt; to 36 Fighter WG (AFE) Bitburg AB W. Germany 14Dec55 mnt; to 7373 MAI GP (AFE) Chateauroux AB, France 20Dec55 mnt; to 3150 MAI GP (AMC) Nouasseur AB Morocco 11Jul56 & 11Feb57; to 7100 SUT WG (AFE) Wiesbaden AB W. Germany 08Dec56 mnt – SABBE (AMC) Brussels 06Aug57; re-serialed **49-0196** 1958 – 3131 ARP SQ (AMC) Chateauroux AB France 01Aug58 – 446 TCWG (AFR) Ellington 17Oct58; assigned 706 TCSq Barksdale 12Jun59 – 78 TCSq (435 TCWG) (AFR) Barksdale 08May61 – 156 AML SQ (NC-ANG) Douglas 29Apr61 – 302 TCWG (AFR) Clinton 21Dec62; 907 TCG (302 TCWG) (AFR) Clinton 11Feb63 – Wfu to MASDC (LOG) Davis-Monthan 28Sep65 for storage – DFI: 03May68 – Final Disposition: Scrapped.

10434 / 49-197 / C-119C-17-FA

Avble: 30Oct50; Acc: 16Nov50; Del: 21Nov50 to USAF – 433 TCWG (TAC) Greenville 21Nov50 – 433 TCWG (AFE) Rhein-Main AB W. Germany 01Aug51 – 317 TCWG (AFE) Rhein-Main AB W. Germany 14Jul52; depl RAF Sculthorpe England 15Feb52; to Neubiberg AB W. Germany 21Mar53 – 564 ADF GP (ADC) Otis 28Nov54 – 33 Fighter GP (ADC) Otis 18Aug55 – 33 MSU GP (ADC) Otis 18Oct56 – 551 MSU GP (ADC) Otis 18Aug57 – 551 AEW WG (ADC) Otis 01Sep57; re-serialed **49-0197** 1958 – 2577 RFC (CNR) Brooks 02Mar58 – 433 TCWG (AFR) Brooks 18Sep58; to SMT Center (ATC) Sheppard 21Mar59; to Kelly 31Aug60; 921 TCG (433 TCWG) (AFR) Kelly 31May63; 922 TCG (433 TCWG) (AFR) Kelly 10Oct65 – Wfu to MASDC (LOG) Davis-Monthan 22Oct65 for storage – DFI: 03May68 – Final Disposition: Scrapped.

10435 / 49-198 / C-119C-17-FA

Avble: 04Nov50; Acc: 17Nov50; Del: 22Nov50 to USAF – 433 TCWG (TAC) Greenville 21Nov50; to 2750 AB WG (AMC) WPAFB 29Dec50 – 433 TCWG (AFE) Rhein-Main AB W. Germany 01Aug51 – 317 TCWG (AFE) Rhein-Main AB W. Germany 14Jul52; to Neubiberg AB W. Germany 25Mar53; depl RAF Sculthorpe England 27Mar53 & 30Jul53 – BMC (AMC) Birmingham 05May54; to 1610 MSU SQ (MATS) Grenier 14May54 mnt – 1807 ACS WG (MATS) Erding AB W. Germany14Sep54 – 4 IM SQ (MATS) Erding AB W. Germany 30Sep54; to 7244 AB GP (AFE) Dhahran AB Saudi Arabia 13Dec54 – 1884 ACS SQ (MATS) Erding AB W. Germany 31Mar55; to 1603 TSP WG (AFE) Wheelus AB Libya 23May55; to Bordeaux AB France 16Apr56; to 6605 AB WG (NEA) Ernest Harmon Canada 01Jun56 mnt; to 7216 AB SQ (AFE) Adana AB Turkey 15Aug56 mnt – 1857 ACS FT (MATS) Rhein-Main AB W. Germany 08Dec57; re-serialed **49-0198** 1958–2578 RFC (CNR) Ellington 10Jan58 – 446 TCWG (AFR) Ellington 20Sep58; assigned 706 TCSq Barksdale 20Mar59; to 2578 AB GP (CNC) Ellington 09Aug59 – 758 TCSq (459 TCWG) (AFR) Greater Pittsburgh 06May62; 911 TCG (459 TCWG) (AFR) Greater Pittsburgh 31May63 – Wfu to MASDC (LOG) Davis-Monthan 20Oct65 for storage – DFI: 03May68 – Final Disposition: Scrapped.

10436 / 49-199 / C-119C-17-FA

Avble: 11Nov50; Acc: 30Nov50; Del: 08Dec50 to USAF – 433 TCWG (TAC) Greenville 08Dec50 – 433 TCWG (AFE) Rhein-Main AB W. Germany 01Aug51 – 317 TCWG (AFE)

Rhein-Main AB W. Germany 14Jul52; to Neubiberg AB W. Germany 15May53 – 6606 AB WG (NEA) Goose Bay Canada 27Nov54 mnt – 432 Fighter SQ (520 ADF GP) (ADC) Truax 24Nov54 – 501 ADF GP (ADC) O'Hare 04Dec54; assigned 42 & 62 Fighter SQ – 527 ADF GP (ADC) Wurtsmith 23Jun55; assigned 63 Fighter SQ O'Hare (depl) 14Jul55 – 56 Fighter GP (ADC) O'Hare 01Nov55; to 314 TCWG (TAC) Sewart 22Jun56 mnt – 56 CLM SQ (ADC) O'Hare IAP 08Sep57; to 3345 TTA WG (TC) Chanute 25Jan58; re-serialed **49-0199** 1958; to 328 CLM SQ (ADC) Richards 16Jul58 – 446 TCWG (AFR) Ellington 05Nov58; assigned 706 TCSq Barksdale 14Mar59 – 758 TCSq (459 TCWG) (AFR) Greater Pittsburgh 29Apr61; 911 TCG (459 TCWG) (AFR) Greater Pittsburgh 17Jan63 – Wfu to MASDC (LOG) Davis-Monthan 07Jan67 for storage – DFI: 22Jun67 – BofS F.H. Chew Sales Co. Phoenix Arizona 20Dec67 – BofS Aero Union Corp. Chico California 20Dec67; ARAp **N13744** 24Jan68; cvtd **C-119C Jet-Pak** (J34-WE) with fire retardant tank installation 12Jun72, Tanker **E13** – BofS Hemet Valley Flying Service Co. Hemet California 09Feb76; Tanker **86**; took part in motion picture filming at California City Airport Jul78, title unk; recovered elevator 13Aug81; tail control surfaces recovered and serviced 13Mar87; wfu and stored 1987 – To USAF Museum Program and placed on static display at Castle AFB Museum California correctly marked as: "9199" – N13744 de-reg 17Sep92 – Final Disposition: Extant.

10437 / 50-119 / C-119C-18-FA

Avble: 16Nov50; Acc: 07Dec50; Del: 09Dec50 to USAF – 433 TCWG (TAC) Greenville 08Dec50 – 433 TCWG (AFE) Rhein-Main AB W. Germany 01Aug51 – 317 TCWG (AFE) Rhein-Main AB W. Germany 14Jul52; to Neubiberg AB W. Germany 23Mar53; depl RAF Sculthorpe England 11Aug53; to 73 AD WG (AFE) Chateauroux AB France 01Sep53 – 525 ADF Group (ADC) New Castle 24Nov54, assigned 96 Fighter SQ – BMC (AMC) Birmingham 08Jun55 – 82 Fighter GP (ADC) New Castle 20Aug55; depl 54 Fighter GP (ADC) Greater Pittsburgh 26Mar56; to 479 Fighter WG (TAC) George 08Jul56 mnt; to 810 AB GP (SAC) Biggs 10Dec56 mnt – 82 CLM SQ (ADC) New Castle 08Jul57 – 551 AEW WG (ADC) Otis 21Nov57; re-serialed **50-0119** 1958; to 2750 AB WG (AMC) WPAFB 10Feb58 mnt – 2577 RFC (CNR) Brooks 22Mar58; depl 2578 RFC (CNR) Ellington 01May58 – 433 TCWG (AFR) Brooks 18Sep58; to Kelly 31Aug60; 921 TCG (433 TCWG) (AFR) Kelly 31May63; 922 TCG (433 TCWG) (AFR) Kelly 10Oct65 – Wfu to MASDC (LOG) Davis-Monthan 22Oct65 for storage – DFI: 03May68 – Final Disposition: Scrapped.

10438 / 50-120 / C-119C-18-FA

Avble: 18Nov50; Acc: 30Dec50; Del: 09Jan51 to USAF – 433 TCWG (TAC) Greenville 09Jan51 – 433 TCWG (AFE) Rhein-Main AB W. Germany 01Aug51 – 317 TCWG (AFE) Rhein-Main AB W. Germany 14Jul52; depl RAF Sculthorpe England 15Dec52; to Neubiberg AB W. Germany 13Apr53 – 321 Bomb WG (SAC) Pinecastle 04Jan55 – bailment Fairchild (AMC) Hagerstown 15Aug55 – 305 Bomb WG (SAC) MacDill 21Sep55; to 805 AB GP (SAC) Barksdale 27Sep56; to 808 AB GP (SAC) Bergstrom 13Apr57; to 27 Fighter WG (SAC) Bergstrom 23May57 – 809 AB GP (SAC) MacDill 06Aug57; re-serialed **50-0120** 1958; to 1608 TSM WG (MATS) Charleston 06Mar58 mnt – 2584 RFC (CNR) Memphis 30Apr58 – 2252 RFC (CNR) Clinton 06Aug58 – 302 TCWG (AFR) Clinton 19Mar59; 907 TCG (302 TCWG) (AFR) Clinton 11Feb63 – Wfu to MASDC (LOG) Davis-Monthan 06Nov65 for storage – DFI: 03May68 – Final Disposition: Scrapped.

10439 / 50-121 / C-119C-18-FA

Avble: 22Nov50; Acc: 23Dec50; Del: 03Jan51 to USAF – 433 TCWG (TAC) Greenville 03Jan51 – 433 TCWG (AFE) Rhein-Main AB W. Germany 01Aug51 – 317 TCWG (AFE) Rhein-Main AB W. Germany 14Jul52; depl RAF Sculthorpe England 27Apr53; to Neubiberg AB W. Germany 14Jun53 – 456 Fighter SQ (520 ADF GP) (ADC) Truax 27Nov54 – 501 ADF GP (ADC) O'Hare 06Dec54, assigned 62 Fighter SQ 17Feb55 – BMC (AMC) Birmingham 04Oct55 – 56 Fighter GP (ADC) O'Hare 29Dec55; depl 329 Fighter GP (ADC) Stewart 13Dec56; to 2500 AB WG (CNC) Mitchel 01Jan57 mnt – 56 CLM SQ (ADC) O'Hare 08Sep57; re-serialed **50-0121** 1958 – Hayes (AMC) Birmingham 30Sep58 – 446 TCWG (AFR) Ellington 05Jan59; to 4238 COS GP (SAC) Barksdale 07Feb59 mnt; 925 TCG (446 TCWG) (AFR) Ellington 31May63 – Wfu to MASDC (LOG) Davis-Monthan 18Sep65 for storage – REC: 21Jun66 – Final Disposition: Scrapped.

10440 / 50-122 / C-119C-18-FA

Avble: 27Nov50; Acc: 13Dec50; Del: 04Jan51 to USAF – 433 TCWG (TAC) Greenville 04Jan51 – 433 TCWG (AFE) Rhein-Main AB W. Germany 01Aug51 – 317 TCWG (AFE) Rhein-Main AB W. Germany 14Jul52; to Neubiberg AB W. Germany 18Mar53; depl RAF Sculthorpe England 22 Sep53 – 7559 MAI GP (AFE) RAF Burtonwood England 10Dec54 – 3110 MAI GP (AMC) RAF Burtonwood England 01Jan55 – 307 Bomb WG (SAC) Lincoln 02Feb55 – 42 Bomb WG (SAC) Loring 09May55 – 42 AB GP (SAC) Loring 01Feb57; re-serialed **50-0122** 1958 – ARD CM (ARD) Andrews 25Jul58 – Fairchild (AMC) Hagerstown 02Dec59; to Mitchel 25Apr60 for test duties – Wfu to 2704 ASD GP (LOG) Davis-Monthan 19Jun63 for storage – REC: 27Jul64 – Final Disposition: Scrapped.

10441 / 50-123 / C-119C-18-FA

Avble: 30Nov50; Acc: 28Dec50; Del: 09Jan51 to USAF – 433 TCWG (TAC) Greenville 09Jan51 – 433 TCWG (AFE) Rhein-Main AB W. Germany 01Aug51 – 317 TCWG (AFE) Rhein-Main AB W. Germany 14Jul52; depl RAF Sculthorpe England 20Feb53; to Neubiberg AB W. Germany 15May53 – 519 ADF GP (ADC) Suffolk 10Dec54; to 3310 TTN WG (TC) Scott 26Feb55 mnt – 52 Fighter GP (ADC) Suffolk 18Aug55; depl 53 Fighter GP (ADC) Sioux City 06Mar56 – 52 CLM SQ (ADC) Suffolk 13Feb57; to 355 Fighter GP (ADC) McGhee Tyson 21Apr57; re-serialed **50-0123** 1958 – 433 TCWG (AFR) Brooks 15Jan59; to Kelly 31Aug60; 921 TCG (433 TCWG) (AFR) Kelly 31May63; 922 TCG (433 TCWG) (AFR) Kelly 10Oct65 – 911 TCG (459 TCWG) (AFR) Greater Pittsburgh 19Oct65 – Wfu to MASDC (LOG) Davis-Monthan 06Jul66 for storage – DFI: 03May68 – Final Disposition: Scrapped.

10442 / 50-124 / C-119C-18-FA

Avble: 04Dec50; Acc: 30Dec50; Del: 09Jan51 to USAF – 433 TCWG (TAC) Greenville 09Jan51 – 433 TCWG (AFE) Rhein-Main AB W. Germany 01Aug51 – 317 TCWG (AFE) Rhein-Main AB W. Germany 14Jul52; depl RAF Sculthorpe England 06Nov52 & 24Mar53; to Neubiberg AB W. Germany 15May53 – 321 Bomb WG (SAC) Pinecastle 02Nov54; to 1607 FDM SQ (MATS) Dover 06Dec54 – 306 Bomb WG (SAC) MacDill 20Sep55; to 314 TCWG (TAC) Sewart 16Mar56 mnt; to SPW Center (ARD) Kirtland 22Apr57; to 31 Fighter-Bomber WG (SAC) Turner 17Jun57 mnt – 809 AB GP (SAC) MacDill 06Aug57 – 2596 RFC (CNR) NAS Dallas 18Nov57 – 2577 RFC DT (CNR) NAS Dallas 01Dec57 – 2233 RFC (CNR) Mitchel 07Dec57; re-serialed **50-0124** 1958 – 514 TCWG (AFR) Mitchel 19Dec58; to McGuire 15Mar61; 903 TCG (514 TCWG) (AFR) McGuire 31May63 – Wfu to MASDC (LOG) Davis-Monthan 14Jul65 for storage – REC: 21Jun66 – Final Disposition: Scrapped.

10443 / 50-125 / C-119C-18-FA

Avble: 06Dec50; Acc: 28Dec50; Del: 09Jan51 to USAF – 433 TCWG (TAC) Greenville 09Jan51 – 433 TCWG (AFE) Rhein-Main AB W. Germany 01Aug51 – 317 TCWG (AFE) Rhein-Main AB W. Germany 14Jul52; to Neubiberg AB W. Germany 23Mar53; depl RAF Sculthorpe England 16Sep53 – 34 ADH DV (ADC) Kirtland 27Oct54 – 6606 AB WG (NEA) Goose Bay Canada 02Nov54 – BMC (AMC) Birmingham 06Nov55 – 1 Fighter GP (ADC) Selfridge 12Feb56 – MOBAR (AMC) Brookley 27Sep56 – 1 MSU GP (ADC) Selfridge 05Jan57; re-serialed **50-0125** 1958 – Hayes (AMC) Birmingham 25Nov58 – 446 TCWG (AFR) Ellington 28Feb59; 925 TCG (446 TCWG) (AFR) Ellington 31May63 – Wfu to MASDC (LOG) Davis-Monthan 18Sep65 for storage – REC: 21Jun66 – Final Disposition: Scrapped.

10444 / 50-126 / C-119C-18-FA

Avble: 08Dec50; Acc: 28Dec50; Del: 09Jan51 to USAF – 433 TCWG (TAC) Greenville 09Jan51 – 433 TCWG (AFE) Rhein-Main AB W. Germany 01Aug51 – 317 TCWG (AFE) Rhein-Main AB W. Germany 14Jul52; depl RAF Sculthorpe England 27Jan53; to Neubiberg AB W. Germany 18Mar53 – 44 Bomb WG (SAC) Lake Charles 03Nov54 – 806 AB GP (SAC) Lake Charles 05Aug57; re-serialed **50-0126** 1958 – 2577 RFC (CNR) Brooks 13Jun58; to 3380 TTA GP (ATC) Keesler 07Aug58 mnt – 433 TCWG (AFR) Brooks 18Sep58; to 2347 AB GP (CNC) Long Beach 04Jan59 mnt; to Kelly 31Aug60; to 823 COS GP (SAC) Homestead 20Jan61 mnt; 921 TCG (433 TCWG) (AFR) Kelly 31May63; 922 TCG (433 TCWG) (AFR) Kelly 10Oct65 – Wfu to MASDC (LOG) Davis-Monthan 22Oct65 for storage – DFI: 03May68 – Final Disposition: Scrapped.

10445 / 50-127 / C-119C-18-FA

Avble: 12Dec50; Acc: 27Feb51; Del: 06Mar51 to USAF – 433 TCWG (TAC) Greenville 06Mar51 – 433 TCWG (AFE) Rhein-Main AB W. Germany 01Aug51 – 317 TCWG (AFE) Rhein-Main AB W. Germany 14Jul52, assigned 39 TCSq. Crashed due to ground collision during approach in a snow storm 15 miles N Bitburg AB West Germany 10Feb53. Pilot: Louis D. Rassi and 4 other crew killed in crash – REC: 10Feb53 – Final Disposition: Accident.

10446 / 50-128 / C-119C-18-FA

Avble: 16Dec50; Acc: 24Feb51; Del: 06Mar51 to USAF – 433 TCWG (TAC) Greenville 06Mar51 – 433 TCWG (AFE) Rhein-Main AB W. Germany 01Aug51 – 317 TCWG (AFE) Rhein-Main AB W. Germany 14Jul52; to Neubiberg AB W. Germany 21Mar53; depl RAF Sculthorpe England 28Aug53 & 04Sep53 – BMC (AMC) Birmingham 05May54; to 1607 MAI SQ (MATS) Dover 54 – 53 SMW SQ (MATS) RAF Burtonwood England 03Dec54; to 7559 MAI GP (AFE) RAF Burtonwood England 27Jul55 & 10Nov55 mnt; to 3110 MAI GP (AMC) RAF Burtonwood England 01Jan56 – Fairchild (AMC) Hagerstown 25Apr56 – MIDAR (AMC) Olmsted 15May56 – 55 WER SQ (MATS) McClellan 15Sep56; to 59 WER SQ (MATS) Kindley 06Nov56 – 2231 RFC (CNR) Byrd 11Jun57 – 2252 RFC (CNR) Clinton 09Aug57; depl 2233 RFC (CNR) Mitchel 24Sep57; re-serialed **50-0128** 1958 – 302 TCWG (AFR) Clinton 19Mar59; to 1608 TSH WG (MATS) Charleston 25Apr59 – 156 AML SQ (NC-ANG) Douglas 25Apr61 – 446 TCWG (AFR) Ellington 18Dec62; 924 TCG (446 TCWG) (AFR) Ellington 31May63 – 907 TCG (302 TCWG) (AFR) Clinton 29May64 – Wfu 15Nov66 to museum / school status. Loan to the Florence Air & Missile Museum Florence S. Carolina 1966–1997. Displayed from 1997 at 82 Airborne Division War Memorial Museum Fort Bragg North Carolina – Final Disposition: Extant.

10447 / 50-129 / C-119C-18-FA

Avble: 19Dec50; Acc: 23Feb51; Del: 06Mar51 to USAF – 433 TCWG (TAC) Greenville 06Mar51 – 433 TCWG (AFE) Rhein-Main AB W. Germany 01Aug51 – 317 TCWG (AFE) Rhein-Main AB W. Germany 14Jul52; depl RAF Sculthorpe England 16Mar53; to Neubiberg AB W. Germany 15May53; to 7150 AB GP (AFE) Wiesbaden AB W. Germany 05Feb54 – 2 Bomb WG (SAC) Hunter 15Dec54; to 308 Bomb WG (SAC) Hunter 01Aug56 & 06Nov56 – 804 AB GP (SAC) Hunter 05Aug57; to 3520 CCT WG (TC) McConnell 22Dec57 mnt; re-serialed **50-0129** 1958 – 2577 RFC (CNR) Brooks 25Mar58 – 433 TCWG (AFR) Brooks 18Sep58; to Kelly 31Aug60; 921 TCG (433 TCWG) (AFR) Kelly 31May63; 922 TCG (433 TCWG) Kelly 10Oct65 – Wfu to MASDC (LOG) Davis-Monthan 29Oct65 for storage – DFI: 03May68 – Final Disposition: Scrapped.

10448 / 50-130 / C-119C-18-FA

Avble: 22Dec50; Acc: 27Feb51; Del: 06Mar51 to USAF – 433 TCWG (TAC) Greenville 06Mar51 – 433 TCWG (AFE) Rhein-Main AB W. Germany 01Aug51 – 317 TCWG (AFE) Rhein-Main AB W. Germany 14Jul52; to Neubiberg AB W. Germany 21Mar53; depls RAF Sculthorpe England 31Mar53 & 31Aug53; to 7150 AB GP (AFE) Wiesbaden AB W. Germany 06Nov53; to 7100 HQS WG (AFE) Wiesbaden AB W. Germany 15Nov53 – BMC (AMC) Birmingham 29Jul54 – 29 ADH DV (ADC) Great Falls 16Nov54 – 328 Fighter GP (ADC) Grandview 22Sep56 – 328 CLM SQ (ADC) Richards 01Aug57; re-serialed **50-0130** 1958 – 2233 RFC (CNR) Mitchel 01Apr58 – 2465 AB GP (CNC) Minneapolis-St. Paul 13Aug58 mnt – 514 TCWG (AFR) Mitchel 23Dec58; to 4500 AB WG (TAC) Langley 04Aug60; to McGuire 14Mar61; 903 TCG (514 TCWG) (AFR) McGuire 31May63 – Wfu to MASDC (LOG) Davis-Monthan 11Jun66 for storage – DFI: 03May68 – Final Disposition: Scrapped.

10449 / 50-131 / C-119C-18-FA

Avble: 23Dec50; Acc: 28Feb51; Del: 08Mar51 to USAF – 433 TCWG (TAC) Greenville 07Mar51 – 433 TCWG (AFE) Rhein-Main AB W. Germany 01Aug51 – 317 TCWG (AFE) Rhein-Main AB W. Germany 14Jul52; to Neubiberg AB W. Germany 19Mar53; depl RAF Sculthorpe England 13Jun53; to 7150 AB GP (AFE) Wiesbaden AB W. Germany 22Sep53 – 2 Bomb WG (SAC) Hunter 04Jan54; to 308 Bomb WG (SAC) Hunter 01Aug56 – 804 AB GP (SAC) Hunter 05Aug57; re-serialed **50-0131** 1958–2584 RFC (CNR) Memphis 09Jan58 – 2259 RFC DT (CNR) Greater Pittsburgh 16Jul58 – 2255 RFC (AFR) Bradley 22Nov58 – 337 TCSq (514 TCWG) (AFR) Bradley 19Mar59; 514 TCWG (AFR) McGuire 11May61; to 4603 AB GP (ADC) Stewart 21Sep62; 903 TCG (514 TCWG) (AFR) McGuire 31May63 – Wfu to MASDC (LOG) Davis-Monthan 24Jun66 for storage – DFI: 03May68 – Final Disposition: Scrapped.

10450 / 50-132 / C-119C-19-FA

Avble: 29Dec50; Acc: 01Mar51; Del: 16Mar51 to USAF – 433 TCWG (TAC) Greenville 16Mar51 – 433 TCWG (AFE) Rhein-Main AB W. Germany 01Aug51 – 317 TCWG (AFE) Rhein-Main AB W. Germany 14Jul52; depl RAF Sculthorpe England 04Feb53 & 06May53; to Neubiberg AB W. Germany 08Jun53 – 7559 MAI GP (AFE) RAF Burtonwood England 12Sep54 – 9 Bomb WG (SAC) Mt. Home 17Oct54; to 3450 TTN WG (TC) Francis E. Warren 17Oct54 mnt; to 2750 AB WG (AMC) WPAFB 22Apr57 mnt; re-serialed **50-0132** 1958 – 2577 RFC (CNR) Brooks 18Apr58 – 433 TCWG (AFR) Brooks 18Sep58; to Kelly 31Aug60; 922 TCG (433 TCWG) (AFR) Kelly 31May63; 921 TCG (433 TCWG) (AFR) Kelly 01Sep64;

922 TCG (433 TCWG) (AFR) Kelly 10Oct65 – Wfu to MASDC (LOG) Davis-Monthan 29Oct65 for storage – DFI: 03May68 – Final Disposition: Scrapped.

10451 / 50-133 / C-119C-19-FA

Avble: 02Jan51; Acc: 07Mar51; Del: 15Mar51 to USAF – 433 TCWG (TAC) Greenville 15Mar51 – 433 TCWG (AFE) Rhein-Main AB W. Germany 01Aug51; to 1414 AB GP (MATS) Dhahran AB Saudi Arabia 27Feb52 mnt – 317 TCWG (AFE) Rhein-Main AB W. Germany 14Jul52; depls RAF Sculthorpe England 19Dec52 & 07Jul53; to Neubiberg AB W. Germany 24Mar53 – 37 Fighter SQ (517 ADF GP) (ADC) Burlington 04Jan55; to 6606 AB WG (NEA) Goose Bay Canada 09Jan55 mnt; 517 ADF GP (ADC) Burlington 27Apr55 – 14 Fighter GP (ADC) Burlington 55 – 14 CLM SQ (ADC) Burlington 19Aug57; re-serialed **50-0133** 1958 – 2584 RFC (CNR) Memphis 25Apr58 – 2252 RFC (CNR) Clinton 12Aug58 – 302 TCWG (AFR) Clinton 19Mar59; to 340 COS GP (SAC) Whiteman 24Jul59 mnt; 907 TCG (302 TCWG) (AFR) Clinton 11Feb63 – Wfu 15Dec65 to museum / school status. No record can be found of 50-0133 at any museum, likely bku for school instructional use – Final Disposition: Scrapped.

10452 / 50-134 / C-119C-19-FA

Avble: 04Jan51; Acc: 09Mar51; Del: 21Mar51 to USAF – 433 TCWG (TAC) Greenville 21Mar51 – 433 TCWG (AFE) Rhein-Main AB W. Germany 01Aug51 – 317 TCWG (AFE) Rhein-Main AB W. Germany 14Jul52; to Neubiberg AB W. Germany 15May53; to 73 AD WG (AFE) Chateauroux AB France 23Sep53 mnt – 337 Fighter SQ (514 ADF WG) (ADC) Minneapolis-St. Paul 15Oct54 – 4676 ADF GP (ADC) Grandview 25Oct54 – 328 Fighter GP (ADC) Grandview 21Jul55; to 475 Fighter GP (ADC) Minneapolis-St. Paul date unk – BMC (AMC) Birmingham 23May57 – 328 CLM SQ (ADC) Richards 28Aug57 – 2577 RFC DT (CNR) NAS Dallas 18Nov57; re-serialed **50-0134** 1958; depl 2578 RFC (CNR) Ellington 03May58 – 69 TCSq (433 TCWG) (AFR) NAS Dallas 29Sep58; 923 TCG (433 TCWG) (AFR) NAS Dallas 17Jan63; to Carswell 31May63 – Wfu to MASDC (LOG) Davis-Monthan 21Sep65 for storage – REC: 21Jun66 – Final Disposition: Scrapped.

10453 / 50-135 / C-119C-19-FA

Avble: 28Feb51; Acc: 30Apr51; Del: 05May51 to USAF – WRDCN (ARD) WPAFB 05May51, cvtd **EC-119C** 04Jul51 for tandem u/c tests; to 4901 SUT WG (ARD) Kirtland 11Nov51; multiple bailments Fairchild (AMC) Hagerstown from 27Feb53; to 3310 TTN WG (TC) Scott 26Jun54 mnt; to BMC (AMC) Birmingham 26Sep54; redes **JC-119C** 01Dec55; cvtd **C-119C** 17Jun57 – 2252 RFC (CNR) Clinton 28Jun57; re-serialed **50-0135** 1958 – Hayes (AMC) Birmingham 20Feb59 – 302 TCWG (AFR) Clinton 04Apr59; 907 TCG (302 TCWG) (AFR) Clinton 11Feb63. Collided mid-air when C-119C s/n: 50-0138 struck C-119C s/n: 50-0135 6.3 miles NE Clinton County AFB Ohio 18Apr64 while in a formation approach to land in scattered cloud and low light conditions. Both aircraft crashed with the loss of all 9 lives on 50-0135 and 8 of the 10 on 50-0138. The 9 fatalities onboard 50-0135 were the 5 aircrew: Maj Francis J. Brock (mission inspector), Maj James A. Hopkins (pilot), Capt Ernest B. Milligan (co-pilot), Capt Robert L. Timmons (navigator), SSgt Richard F. Davis (flight engineer) and 4 passengers: Lt Col Ray J. Glaze (paratrooper), 1Lt Donald B. Becker (paratrooper), SSgt William Cornell (paratrooper) and SSgt Joseph T. Kelly (paratrooper). The aircraft were part of a nine ship formation in three groups of three returning from a night paradrop mission that was aborted due to deteriorating weather – DFI: 27Apr64 – Final Disposition: Accident.

Section Four—Service Histories

10454 / 50-136 / C-119C-19-FA

Avble: 19Feb51; Acc: 13Mar51; Del: 21Mar51 to USAF – 433 TCWG (TAC) Greenville 21Mar51 – 433 TCWG (AFE) Rhein-Main AB W. Germany 01Aug51 – 317 TCWG (AFE) Rhein-Main AB W. Germany 14Jul52; to Neubiberg AB W. Germany 25Mar53; depls RAF Sculthorpe England 03Jun53 & 13Aug53 – 326 Fighter SQ (4676 ADF GP) (ADC) Grandview 12Oct54; 6606 AB WG (NEA) Goose Bay Canada 11Nov54 mnt; 4676 ADF GP (ADC) Grandview 07Feb55 – 521 ADF GP (ADC) Sioux City 27Feb55 – 53 Fighter GP (ADC) Sioux City 18Aug55 – 328 Fighter GP (ADC) Grandview 05Oct56 – 328 CLM SQ (ADC) Richards 01Aug57 – 2589 RFC (CNR) Dobbins 02Dec57; re-serialed **50-0136** 1958 – Hayes (AMC) Birmingham 26May58 – 2233 RFC (CNR) Mitchel 27Aug58 – 514 TCWG (AFR) Mitchel 19Dec58; to McGuire 18Mar61; 903 TCG (514 TCWG) (AFR) McGuire 31May63 – Wfu to MASDC (LOG) Davis-Monthan 20Jun66 for storage – DFI: 03May68 – Final Disposition: Scrapped.

10455 / 50-137 / C-119C-19-FA

Avble: 20Feb51; Acc: 09Mar51; Del: 21Mar51 to USAF – 433 TCWG (TAC) Greenville 21Mar51 – 433 TCWG (AFE) Rhein-Main AB W. Germany 01Aug51 – 317 TCWG (AFE) Rhein-Main AB W. Germany 14Jul52; depl RAF Sculthorpe England 27Feb53; to Neubiberg AB W. Germany 31Mar53 – 566 ADF GP (ADC) Hamilton 29Oct54 – BMC (AMC) Birmingham 15Jun55 – 78 Fighter GP (ADC) Hamilton 30Aug55; depl 337 Fighter GP (ADC) Portland 16Dec55; depl 2347 RFC (CNR) Long Beach 20Mar56; re-serialed **59-0137** 1958 – 2577 RFC (CNR) Brooks 17Mar58 – 433 TCWG (AFR) Brooks 18Sep58; to 839 AB GP (TAC) Sewart 19Jan60 mnt; to Kelly 31Aug60; 922 TCG (433 TCWG) (AFR) Kelly 31May63. Listed as crashed 30Oct63, nfd – Final Disposition: Accident.

10456 / 50-138 / C-119C-19-FA

Avble: 22Feb51; Acc: 14Mar51; Del: 02Apr51 to USAF – Retained on bailment Fairchild (AMC) Hagerstown 02Apr51; to MIDAR (AMC) Olmsted 19Feb52 – 435 TCWG (TAC) Miami 05Feb52; depl Elmendorf 24Oct52 – 456 TCWG (TAC) Miami 02Dec52 – 483 TCWG (FEA) Ashiya AB Japan 16Feb53; to 19 Bomb WG (FEA) Andersen 18Mar53 – 314 TCWG (FEA) Ashiya AB Japan 28Mar53; loan to French Armée de l'Air Indochina 05Apr54-01Jun54, French markings applied – 483 TCWG (FEA) Ashiya AB Japan 25Jun54 – BMC (AMC) Birmingham 06Jul54 – 305 Bomb WG (SAC) MacDill 02Dec54 – Hayes (AMC) Birmingham 16Aug57 – 809 AB GP (SAC) MacDill 13Nov57; re-serialed **59-0138** 1958 – 2252 RFC (CNR) Clinton 11Apr58; to 803 AB GP (SAC) Davis-Monthan 23May58 mnt – 302 TCWG (AFR) Clinton 19Mar59; 907 TCG (302 TCWG) (AFR) Clinton 11Feb63. Collided mid-air when C-119C s/n: 50-0138 struck C-119C s/n: 50-0135 6.3 miles NE Clinton County AFB Ohio 18Apr64 while in a formation approach to land in scattered cloud and low light conditions. Both aircraft crashed with the loss of all 9 lives on 50-0135 and 8 of the 10 on 50-0138. The 8 fatalities onboard 50-0138 were the 4 aircrew: Lt Col Richard M. Griswold (pilot), Maj Stanley H. Heismann (co-pilot), Capt Woodson B. Gudgell (navigator), SSgt Clyde Grimes (radio operator) and 4 passengers: Lt Col Samuel W. Sardis (paratrooper), Capt Calvin F. Kemp (paratrooper), Corp Peter A. Weart (paratrooper) and P1C James W. Kremer (paratrooper). The two survivors were airman SSgt Bill Zugelder (flight engineer) and passenger Sgt William Kremer (paratrooper), both seriously injured after parachuting to the ground. The aircraft were part of a nine ship formation in three groups of three returning from a night paradrop mission that was aborted due to deteriorating weather – DFI: 27Apr64 – Final Disposition: Accident.

10457 / 50-139 / C-119C-19-FA

Avble: 29Mar51; Acc: 30Apr51; Del: 07May51 to USAF – 316 TCG (TAC) Sewart 07May51; depl Laurinburg-Maxton Airport North Carolina 24Aug51 – 314 TCWG (TAC) Sewart 12Aug52 – 456 TCWG (TAC) Miami 29Dec52 – BMC (AMC) Birmingham 16Apr53 – 483 TCWG (FEA) Ashiya AB Japan 01Feb54; loan to French Armée de l'Air Indochina 24May54-02Jul54, French markings applied – 27 Fighter WG (SAC) Bergstrom 03Nov54; to 3310 TTN WG (TC) Scott 05Jan55 mnt; to 808 AB GP (SAC) Bergstorm 01Jul57; re-serialed **59-0139** 1958 – 27 AB GP (TAC) Bergstrom 01Feb58 – 2577 RFC (CNR) Brooks 09Jun58 – 3640 PTN WG (ATC) Laredo 01Sep58 – 433 TCWG (AFR) Brooks 19Oct58; to Kelly 31Aug60; 921 TCG (433 TCWG) (AFR) Kelly 31May63 – Wfu to 2704 ASD GP (LOG) Davis-Monthan 17Jun63 for storage – REC: 27Jul64 – Final Disposition: Scrapped.

10458 / 50-140 / C-119C-19-FA

Avble: 28Feb51; Acc: 15Mar51; Del: 22Mar51 to USAF – 433 TCWG (TAC) Greenville 22Mar51 – 433 TCWG (AFE) Rhein-Main AB W. Germany 01Aug51 – 317 TCWG (AFE) Rhein-Main AB W. Germany 14Jul52; to Neubiberg AB W. Germany 25Mar53; depl RAF Sculthorpe England 29May53 – 521 ADF GP (ADC) Sioux City 18Oct54 – 53 Fighter GP (ADC) Sioux City 18Aug55; to 4900 AB GP (ARD) Kirtland 09Jul56 mnt – 328 Fighter GP (ADC) Grandview 27Oct56 – 328 CLM SQ (ADC) Richards 01Aug57 – 2589 RFC (CNR) Dobbins 29Nov57; re-serialed **59-0140** 1958 – 2584 RFC (CNR) Memphis 23Jan58 – 2259 RFC DT (CNR) Greater Pittsburgh 17Jul58 – 2255 RFC (CNR) Bradley 17Oct58 – Hayes (AMC) Birmingham 19Mar59 – 337 TCSq (514 TCWG) (AFR) Bradley 18Jun59; 514 TCWG (AFR) McGuire 15Mar61; 903 TCG (514 TCWG) (AFR) McGuire 31May63 – Wfu to MASDC (LOG) Davis-Monthan 13May66 for storage – DFI: 03May68 – Final Disposition: Scrapped.

10459 / 50-141 / C-119C-19-FA

Avble: 02Mar51; Acc: 17Mar51; Del: 18Apr51 to USAF – 433 TCWG (TAC) Greenville 19Apr51 – 433 TCWG (AFE) Rhein-Main AB W. Germany 01Aug51 – 317 TCWG (AFE) Rhein-Main AB W. Germany 14Jul52; depl RAF Sculthorpe England 16Jan53; to Neubiberg AB W. Germany 15May53; to 7100 HQS WG (AFE) Wiesbaden AB W. Germany 25Nov53 – 1607 FDM SQ (MATS) Dover 01Dec54 mnt – 567 ADF GP (ADC) McChord 13Feb55 – **RECORDS MISSING** – 325 Fighter GP (ADC) McChord date unk; to 3415 TTA WG (TC) Lowry 03Dec56 mnt; re-serialed **59-0141** 1958 – 2577 RFC (CNR) Brooks 07Mar58; to 2750 AB WG (AMC) WPAFB 08May58 mnt – 433 TCWG (AFR) Brooks 18Sep58; to 3510 FTA WG (ATC) Randolph 12Mar60 mnt; to 445 TCA DT (CNC) Memphis 30Apr60 mnt; to Kelly 31Aug60; 922 TCG (433 TCWG) (AFR) Kelly 31May63 – Wfu to MASDC (LOG) Davis-Monthan 14Nov65 for storage – DFI: 03May68 – Final Disposition: Scrapped.

10460 / 50-142 / C-119C-19-FA

Avble: 07Mar51; Acc: 16Mar51; Del: 17Apr51 to USAF – Fairchild (AMC) Hagerstown 17Apr51 – 580 ARC WG (MATS) Mt. Home 14Jun51 – 581 ARC WG (MATS) Mt. Home 18Sep51; to 3200 PTS WG (APG) Eglin 15Oct51 test duties – 580 ARC WG (MATS) Mt. Home 28Dec51 – 1300 AB WG (MATS) Mt. Home 07Feb52; to 3200 PTS WG (APG) Eglin 19Mar52 test duties; to 6612 AB GP (NEA) Thule AB Greenland 23Jul52 mnt – 316 TCG (TAC) Sewart 23Jan53; depl Burlington IAP 06Feb53 – 456 TCWG (TAC) Miami IAP 10Mar53 – 64 TCWG (TAC) Greenville 25Jun53 – 464 TCWG (TAC) Lawson 05Feb54; to Pope 07Sep54 – AFA Center (ARD) Eglin 14Sep54 experimental duties; to 3201 AB WG (APG) Eglin 26Apr55 & 25Jul55 mnt; cvtd **JC-119C** 01Dec55 – CAR Center (ARD)

Hanscom 22Jan56 – 6515 MAI GP (ARD) Edwards 14May56; to BMC (AMC) Birmingham 13Jul56, cvtd **C-119C** 24Sep56 – AFT Center (ARD) Edwards 01Apr57; cvtd **JC-119C** 01May57 – 2252 RFC (CNR) Clinton 01Jul57; re-serialed **59-0142** 1958; to 839 AB GP (TAC) Sewart 06Dec58 mnt – Hayes (AMC) Birmingham 18Mar59 – 302 TCWG (AFR) Clinton 04May59; cvtd **C-119C** 20May59; WDD (ARD) WPAFB 22Aug60; cvtd **JC-119C** 23Aug60; to MIDAR (AMC) Olmsted 23Sep60; cvtd **C-119C** 02Mar61; 907 TCG (AFR) Clinton 11Feb63 – Wfu to MASDC (LOG) Davis-Monthan 06Nov65 for storage – DFI: 03May68 – Final Disposition: Scrapped.

10461 / 50-143 / C-119C-19-FA

Avble: 24Apr51; Acc: 15May51; Del: 22May51 to USAF – 316 TCG (TAC) Sewart 22May51; depl Laurinburg-Maxton Airport North Carolina 24Aug51; to WRDCN (ARD) WPAFB 10Sep52 & 01Dec52 – bailment Fairchild (AMC) Hagerstown 20Feb53 – 456 TCWG (TAC) Miami 03Apr53; to 2750 AB WG (AMC) WPAFB 13Apr53 mnt – 582 ARC WG (MATS) Great Falls 22May53; to RAF Molesworth England 23Jun54; to 7559 MAI GP (AFE) RAF Burtonwood England 16Feb55 mnt – 1631 MAT SQ (MATS) Prestwick Scotland 22Jun55 mnt – 805 AB GP (SAC) Barksdale 21Aug55 – 806 AB GP (SAC) Lake Charles 22Sep55 – bailment Fairchild (AMC) Hagerstown 27Oct55 – 68 Bomb WG (SAC) Lake Charles 29Dec55 – 70 SM WG (SAC) Little Rock 01Mar56 – BMC (AMC) Birmingham 11Jul56 – 825 AB GP (SAC) Little Rock 06Aug57; re-serialed **50-0143** 1958 – 2578 RFC (CNR) Ellington 01Feb58 – 446 TCWG (AFR) Ellington 20Sep58 – 156 AML SQ (NC-ANG) Douglas 06May61 – 446 TCWG (AFR) Ellington 21Oct62; depl 1503 TSH WG (MATS) Tachikawa AB Japan 13Mar63; 925 TCG (446 TCWG) (AFR) Ellington 31May63 – Wfu to MASDC (LOG) Davis-Monthan 18Sep65 for storage – REC: 21Jun66 – Final Disposition: Scrapped.

10462 / 50-144 / C-119C-19-FA

Avble: 10Mar51; Acc: 14Apr51; Del: 20Apr51 to USAF – 433 TCWG (TAC) Greenville 20Apr51 – 433 TCWG (AFE) Rhein-Main AB W. Germany 01Aug51 – 317 TCWG (AFE) Rhein-Main AB W. Germany 14Jul52; to Neubiberg AB W. Germany 25Mar53; to 7150 AB GP (AFE) Wiesbaden AB W. Germany Nov53; to 7100 HQS WG (AFE) Wiesbaden AB W. Germany Nov53 – 332 Fighter SQ (525 ADF GP) (ADC) New Castle 01Dec54; 525 ADF GP (ADC) New Castle 02Dec54 – BMC (AMC) Birmingham 17Aug55 – 82 Fighter GP (ADC) New Castle 09Nov55 – 82 CLM SQ (ADC) New Castle 08Jul57 – 1 MSU GP (ADC) Selfridge 25Nov57; to 839 AB GP (TAC) Sewart 02Dec57 mnt; re-serialed **50-0144** 1958 – 446 TCWG (AFR) Ellington 17Oct58; to 803 COS GP (SAC) Davis-Monthan 12Mar59 mnt; 925 TCG (446 TCWG) (AFR) Ellington 31May63 – Wfu to MASDC (LOG) Davis-Monthan 18Sep65 for storage – REC: 21Jun66 – Final Disposition: Scrapped.

10463 / 50-145 / C-119C-19-FA

Avble: 14Mar51; Acc: 09Apr51; Del: 20Apr51 to USAF – 433 TCWG (TAC) Greenville 20Apr51 – 433 TCWG (AFE) Rhein-Main AB W. Germany 01Aug51 – 317 TCWG (AFE) Rhein-Main AB W. Germany 14Jul52; to Neubiberg AB W. Germany 13May53; depl RAF Sculthorpe England 24Jun53 – 575 ADF GP (ADC) Selfridge 22Nov54; depl 514 ADF GP (ADC) Minneapolis-St. Paul 10Mar55 – 1 Fighter GP (ADC) Selfridge 18Aug55 – 1 MSU GP (ADC) Selfridge 18Oct56; to 54 Fighter GP (ADC) Greater Pittsburgh 08Jan57; re-serialed **50-0145** 1958; to 1611 FDM SQ (MATS) McGuire 10Jan58 mnt – Hayes (AMC) Birmingham 29Jul59 – 446 TCWG (AFR) Ellington 17Nov59. Shortly after take-off from Whitehorse AP Yukon Canada 23Nov61 a seized brake drum caught fire forcing the aircraft to be abandoned. 4 of the 10 personnel onboard bailed out but the other 6 were killed when the aircraft subsequently crashed 24 miles S Whitehorse. 3 of the 4 that bailed out were rescued the following day but the fourth was never found even after an extensive search. Those onboard were the 4 aircrew: Capt Wayne D. Sager (pilot – killed in crash), Capt Milton L. Kimey (co-pilot killed in crash), Capt Alan J. White (navigator – killed in crash), MSgt Roger J. Forstner (mnt chief – bailed out, missing) and 6 passengers: SSgt Lonnie Z. Foreman (killed in crash), SSgt Leroy Cotton (bailed out, rescued), SSgt Clyde Nicholas (bailed out, rescued), A1C Jean R. Conklin (bailed out, rescued), A1C William J. Murphy (killed in crash) and A1C David L. Paul (killed in crash). The aircraft had been in Alaska taking part in military exercises at the time. Wreckage can still be found at the crash site today – DFI: 18Jan62 – Final Disposition: Accident.

10464 / 50-146 / C-119C-19-FA

Avble: 16Mar51; Acc: 12Apr51; Del: 18Apr51 to USAF – 433 TCWG (TAC) Greenville 19Apr51 – 433 TCWG (AFE) Rhein-Main AB W. Germany 01Aug51 – 317 TCWG (AFE) Rhein-Main AB W. Germany 14Jul52; to Neubiberg AB W. Germany 21Mar53; depl RAF Sculthorpe England 23Apr53; to 7373 AB GP (AFE) Chateauroux AB France 7Nov53 – 308 Bomb WG (SAC) Hunter 20Dec54; to 807 AB GP (SAC) March 15Jun56 – 506 Fighter WG (SAC) Tinker 10Aug56; re-serialed **50-0146** 1958 – 2578 RFC (CNR) Ellington 26May58; depl 2577 RFC DT (CNR) NAS Dallas 27Jun58 – 446 TCWG (AFR) Ellington 20Sep58; 924 TCG (446 TCWG) (AFR) Ellington 31May63 – Wfu to MASDC (LOG) Davis-Monthan 24Aug65 for storage – REC: 21Jun66 – Final Disposition: Scrapped.

10465 / 50-147 / C-119C-20-FA

Avble: 17May51; Acc: 31May51; Del: 12Jun51 to USAF – 316 TCG (TAC) Sewart 11Jun51 – 456 TCWG (TAC) Miami IAP 29Dec52 – 483 TCWG (FEA) Ashiya AB Japan 18Feb53; loan to French Armée de l'Air Indochina 20Mar54-25Apr54, French markings applied – BMC (AMC) Birmingham 02Jun54 – 31 Fighter WG (SAC) Turner 14Sep54 – WRAAR (AMC) Robins 10Nov54 – 72 SRH WG (SAC) Ramey AFB Puerto Rico 06Dec54 – 72 AB GP (SAC) Ramey AFB Puerto Rico Oct55 – 12 Fighter WG (SAC) Bergstrom 17Oct55 – BMC (AMC) Birmingham 24May57; to 808 AB GP (TAC) Bergstrom 30Aug57; re-serialed **50-0147** 1958 – 2233 RFC (CNR) Mitchel 12Apr58 – 2585 RFC (CNR) Miami 29Apr58 – Hayes (AMC) Birmingham 27Jan59 – 446 TCWG (AFR) Ellington 07Aug59; briefly assigned 357 TCSq NAS New Orleans 05Nov60; 925 TCG (446 TCWG) (AFR) Ellington 31May63 – 484 Bomb WG (SAC) Turner 19May64. Listed as having accident 20May64, nfd – Final Disposition: Accident.

10466 / 50-148 / C-119C-20-FA

Avble: 22Mar51; Acc: 12Apr51; Del: 18Apr51 to USAF – 433 TCWG (TAC) Greenville 18Apr51 – 433 TCWG (AFE) Rhein-Main AB W. Germany 01Aug51 – 317 TCWG (AFE) Rhein-Main AB W. Germany 14Jul52; to Neubiberg AB W. Germany 24Mar53; depl RAF Sculthorpe England 14May53 – 27 ADH DV (ADC) Norton 09Nov54 – Fairchild (AMC) Hagerstown 15Mar56 – 325 Fighter GP (ADC) McChord 13Apr56; to 2750 AB WG (AMC) WPAFB 13Apr56 mnt; depl 84 Fighter GP (ADC) Geiger Field 19Sep57; re-serialed **50-0148** 1958; to 62 AB GP (MATS) Larson 17Jan58 mnt – 2589 RFC (CNR) Dobbins 08Apr58 – 2259 RFC DT (CNR) Greater Pittsburgh 12Jul58 – 2255 RFC (CNR) Bradley 17Oct58 – Hayes (AMC) Birmingham 16Jan59 – 337 TCSq (514 TCWG) (AFR) Bradley 11Apr59; to 328 CLM SQ (ADC) Richards 23May60 mnt; 514 TCWG (AFR) McGuire 22Mar61; 903

TCG (514 TCWG) (AFR) McGuire 31May63 – Wfu to MASDC (LOG) Davis-Monthan 24Jun66 for storage – DFI: 03May68 – Final Disposition: Scrapped.

10467 / 50-149 / C-119C-20-FA

Avble: 28Mar51; Acc: 18Apr51; Del: 23Apr51 to USAF – 433 TCWG (TAC) Greenville 23Apr51 – 433 TCWG (AFE) Rhein-Main AB W. Germany 01Aug51 – 317 TCWG (AFE) Rhein-Main AB W. Germany 14Jul52; to Neubiberg AB W. Germany 15May53; depls RAF Sculthorpe England 08Jun53 & 16Jul53; to 73 AD WG (AFE) Chateauroux AB France 02Nov53 – BMC (AMC) Birmingham 13May54 – 1371 MCH SQ (MATS) Palm Beach 18Sep54; to 3645 CCT WG (TC) Laughlin 22Feb55 mnt; to 2750 AB WG (AMC) WPAFB 15Mar55 mnt – 53 WER SQ (MATS) RAF Burtonwood England 28Apr56; to 1401 AB WG (MATS) Andrews 01May56; to 3110 MAI GP (AMC) RAF Burtonwood England 05Jun56. Struck a radio mast on Mt. Feldberg W. Germany 30Nov56 while approaching Wiesbaden AB. Pilot Lt James L. Jarrett ordered the other nine crew and passengers to bail out then flew the crippled aircraft for 20 miles to Rhein-Main AB where a forced landing was made – 7310 AB GP (AFE) Rhein-Main AB W. Germany 01Dec56, damaged beyond repair, to salvage – REC: 18Jan57 – Final Disposition: Accident.

10468 / 50-150 / C-119C-20-FA

Avble: 29Mar51; Acc: 14Apr51; Del: 20Apr51 to USAF – 433 TCWG (TAC) Greenville 20Apr51 – 433 TCWG (AFE) Rhein-Main AB W. Germany 01Aug51 – 317 TCWG (AFE) Rhein-Main AB W. Germany 14Jul52; depl RAF Sculthorpe England 22Dec52; to Neubiberg AB W. Germany 21Mar53; to 7373 AB GP (AFE) Chateauroux AB France 18Dec53 – 566 ADF GP (ADC) Hamilton 15Oct54 – 78 Fighter GP (ADC) Hamilton 18Aug55; re-serialed **50-0150** 1958 – 328 CLM SQ (ADC) Richards 29Apr58 – 78 MSU GP (ADC) Hamilton 01May58 – 2578 RFC (CNR) Ellington 07Jul58 – 446 TCWG (AFR) Ellington 20Sep58 – DOAOR (AMC) Ft Monmo 08Dec60 – Wfu to 2704 ASD GP (LOG) Davis-Monthan 04Feb63 for storage – REC: 27Jul64 – Final Disposition: Scrapped.

10469 / 50-151 / C-119C-20-FA

Avble: 03Apr51; Acc: 18Apr51; Del: 02May51 to USAF – 433 TCWG (TAC) Greenville 02May51 – 433 TCWG (AFE) Rhein-Main AB W. Germany 01Aug51 – 317 TCWG (AFE) Rhein-Main AB W. Germany 14Jul52; depls RAF Sculthorpe England 31Jan53 & 27Jul53; to Neubiberg AB W. Germany 19Mar53; to 1604 MSU SQ (MATS) Kindley 14Mar54 mnt – 1371 MCH SQ (MATS) Palm Beach 02Nov54; to 805 AB GP (SAC) Barksdale 18Oct55 mnt – 1881 IM GP (MATS) Tinker 06Mar56; to 3310 TTN WG (TC) Scott 21Nov56 – 1800 ASC WG (MATS) Tinker 28Oct57 – COAAR (MATS) Tinker 01Nov57; re-serialed **50-0151** 1958 – 2578 RFC DT (CNR) Alvin Callender Field 09Aug58 – 357 TCSq (446 TCG) (AFR) Alvin Callender Field 20Sep58; to NAS New Orleans 01Dec58; 706 TCSq (446 TCWG) (AFR) NAS New Orleans 08May61; 926 TCG (446 TCWG) (AFR) NAS New Orleans 31May63 – 907 TCG (302 TCWG) (AFR) Clinton 15May64 – Wfu to MASDC (LOG) Davis-Monthan 04Dec65 for storage – DFI: 03May68 – Final Disposition: Scrapped.

10470 / 50-152 / C-119C-20-FA

Avble: 03Apr51; Acc: 25Apr51; Del: 07May51 to USAF – 433 TCWG (TAC) Greenville 08May51 – 433 TCWG (AFE) Rhein-Main AB W. Germany 01Aug51 – 317 TCWG (AFE) Rhein-Main AB W. Germany 14Jul52; to Neubiberg AB W. Germany 21Mar53; depl RAF Sculthorpe England 12May53 – 575 ADF GP (ADC) Selfridge 29Nov54 – 1

Fighter GP (ADC) Selfridge 18Aug55 – 1 MSU GP (ADC) Selfridge 18Oct56; to 803 AB GP (SAC) Davis-Monthan 24Nov57 mnt; re-serialed **50-0152** 1958 – Hayes (AMC) Birmingham 30Dec58 – 446 TCWG (AFR) Ellington 15Apr59; to 1405 AB WG (MATS) Scott 05Jun59 mnt; 924 TCG (446 TCWG) (AFR) Ellington 31May63 – Wfu to MASDC (LOG) Davis-Monthan 26Aug65 for storage – REC: 21Jun66 – Final Disposition: Scrapped.

10471 / 50-153 / C-119C-20-FA

Avble: 07Apr51; Acc: 01May51; Del: 07May51 to USAF – 433 TCWG (TAC) Greenville 08May51 – 433 TCWG (AFE) Rhein-Main AB W. Germany 01Aug51 – 317 TCWG (AFE) Rhein-Main AB W. Germany 14Jul52; depl RAF Sculthorpe England 15Dec52; to Neubiberg AB W. Germany 15May53; to 7559 MND GP (AFE) RAF Burtonwood England 01Mar54 mnt – 6606 AB WG (NEA) Goose Bay Canada 30Oct54 mnt – 1371 MCH SQ (MATS) Palm Beach 08Nov54; to 810 AB GP (SAC) Biggs 24Aug55 mnt – 3 AMO SQ (MATS) Tinker 11Feb56; to 479 Figher WG (TAC) George 21Mar56 mnt – OKLAR (AMC) Tinker 28Aug57; re-serialed **50-0153** 1958. Listed as having an accident, nfd – REC: 31Dec58 – Final Disposition: Accident.

10472 / 50-154 / C-119C-20-FA

Avble: 07Apr51; Acc: 01May51; Del: 16May51 to USAF – 580 ARC WG (MATS) Mt. Home 16May51 – 581 ARC WG (MATS) Mt. Home 26Aug51; depl 316 TCG (TAC) Sewart 10Oct51 – 1300 AB WG (MATS) Mt. Home 07Feb52; to 1739 FRY SQ (MATS) Amarillo 19Nov52; to Great Falls 21May53 – 582 ARC WG (MATS) Great Falls 16Jun53; to Molesworth 29Jan54; to 7559 MAI GP (AFE) RAF Burtonwood England 08Jun54 – 301 Bomb WG (SAC) Barksdale 12Jun55; to 1400 FDM SQ (MATS) Keflavik AS Iceland 23Jun55 mnt – 31 Fighter WG (SAC) Turner 19Oct55 – 1739 FRY SQ (MATS) Amarillo 12Sep57 – 2231 RFC (CNR) Byrd 26Sep57 – 2252 RFC (CNR) Clinton 27Sep57; re-serialed **50-0154** 1958; to 1707 TSH WG (MATS) Palm Beach 08Jul58 mnt – Hayes (AMC) Birmingham 18Mar59 – 302 TCWG (AFR) Clinton 04May59; 907 TCG (302 TCWG) (AFR) Clinton 11Feb63 – Wfu to MASDC (LOG) Davis-Monthan 28Dec65 for storage – DFI: 03May68 – Final Disposition: Scrapped.

10473 / 50-155 / C-119C-20-FA

Avble: 11Apr51; Acc: 01May51; Del: 16May51 to USAF – 433 TCWG (TAC) Greenville 16May51 – 433 TCWG (AFE) Rhein-Main AB W. Germany 01Aug51 – 317 TCWG (AFE) Rhein-Main AB W. Germany 14Jul52; to Neubiberg AB W. Germany 15Apr53; depl RAF Sculthorpe England 09Sep53; depl 60 TCWG (AFE) Rhein-Main AB W. Germany 24Nov53; to 7559 MAI GP (AFE) RAF Burtonwood England 14Dec54 – 307 Bomb WG (SAC) Lincoln 15Dec54 – 42 Bomb WG (SAC) Loring 15May56 – BMC (AMC) Birmingham 02Nov56 – 42 AB GP (SAC) Loring 26Feb57 – 2589 RFC (CNR) Dobbins 25Nov57; re-serialed **50-0155** 1958 – 2259 RFC DT (CNR) Greater Pittsburgh 27Jul58 – 2255 RFC (CNR) Bradley 29Sep58 – 337 TCSq (514 TCG) (AFR) Bradley 19Mar59; to 1608 TSH WG (MATS) Charleston 23Sep59; 514 TCWG (AFR) Mitchel 14Mar61; to McGuire 15Mar61; 903 TCG (514 TCWG) (AFR) McGuire 31May63 – Wfu to MASDC (LOG) Davis-Monthan 13May66 for storage – DFI: 06May68 – Final Disposition: Scrapped.

10474 / 50-156 / C-119C-20-FA

Avble: 13Apr51; Acc: 01May51; Del: 01May51 to USAF – bailment Fairchild (AMC) Hagerstown 01May51 – Multiple assignments 3201 AB WG & 3200 MAI WG (APG) Eglin from 24Jul53 for experimental duties – 92 Bomb WG (SAC) Fairchild 02Oct54 – 92 AB GP (SAC) Fairchild 04Sep56; to 814 AB GP (SAC) Fairchild 56; to 9 AB GP (SAC) Mt. Home

20Dec56; to 3320 TTA WG (TC) Amarillo 27Jun57 mnt; re-serialed **50-0156** 1958 – 2578 RFC (CNR) Ellington 22Apr58 – 446 TCWG (AFR) Ellington 20Sep58; 925 TCG (446 TCWG) (AFR) Ellington 31May63 – Wfu to MASDC (LOG) Davis-Monthan 18Sep65 for storage – DFI: 21Jun66 – Final Disposition: Scrapped.

10475 / 50-157 / C-119C-20-FA

Avble: 18Apr51; Acc: 10May51; Del: 25May51 to USAF – 433 TCWG (TAC) Greenville 24May51 – 433 TCWG (AFE) Rhein-Main AB W. Germany 01Aug51 – 317 TCWG (AFE) Rhein-Main AB W. Germany 14Jul52; to Neubiberg AB W. Germany 12May53; depl RAF Sculthorpe England 31Jul53 – BMC (AMC) Birmingham 26Oct54 – 47 Fighter SQ (518 ADF GP) (ADC) Niagara Falls MAP 31Mar55 – 564 ADF GP (ADC) Otis 06Apr55 – bailment Fairchild (AMC) Hagerstown 15Aug55 – 33 Fighter GP (ADC) Otis 21Sep55; depl 54 Fighter GP (ADC) Greater Pittsburgh 23Feb56; to 1608 FDM SQ (MATS) Charleston 02Apr56 mnt – 33 MSU GP (ADC) Otis 18Oct56; depl 327 Fighter GP (ADC) Truax 02Dec56; to 815 AB GP (SAC) Forbes 25Mar57 mnt – 551 MSU GP (ADC) Otis 18Aug57 – 551 AEW WG (ADC) Otis 01Sep57; re-serialed **50-0157** 1958 – 2577 RFC (CNR) Brooks 28Mar58 – 433 TCWG (AFR) Brooks 18Sep58; to 2578 AB GP (CNC) Ellington 25Sep58 mnt; to Kelly 31Aug60; 922 TCG (433 TCWG) (AFR) Kelly 31May63 – Wfu to MASDC (LOG) Davis-Monthan 22Oct65 for storage – DFI: 06May68 – Final Disposition: Scrapped.

10476 / 50-158 / C-119C-20-FA

Avble: 19Apr51; Acc: 08May51; Del: 09May51 to USAF – 433 TCWG (TAC) Greenville 08May51 – 433 TCWG (AFE) Rhein-Main AB W. Germany 01Aug51 – 317 TCWG (AFE) Rhein-Main AB W. Germany 14Jul52; depl RAF Sculthorpe England 15Feb52; to Neubiberg AB W. Germany 25Mar53; to 7559 MND GP (AFE) RAF Burtonwood England 21Sep53 – 34 ADH DV (ADC) Kirtland 08Oct54; to 3902 AB WG (SAC) Offutt 05Aug55 mnt – 328 Fighter GP (ADC) Grandview 07Aug56; to 3320 TTA WG (TC) Amarillo 06Feb57 mnt – 328 CLM SQ (ADC) Richards 01Aug57 – 2589 RFC (CNR) Dobbins 18Dec57; re-serialed **50-0158** 1958 – 2259 RFC DT (CNR) Greater Pittsburgh 12Jul58 – Hayes (AMC) Birmingham 28Oct58 – 758 TCSq (459 TCG) (AFR) Greater Pittsburgh 06Jan59 – 2255 RFC (AFR) Bradley 23Jan59 – 337 TCSq (514 TCWG) (AFR) Bradley 19Mar59; 514 TCWG (AFR) McGuire 22Mar61; 903 TCG (514 TCWG) (AFR) McGuire 31May63 – Wfu to MASDC (LOG) Davis-Monthan 11Jun66 for storage – DFI: 06May68 – Final Disposition: Scrapped.

10477 / 50-159 / C-119C-20-FA

Avble: 23Apr51; Acc: 19May51; Del: 02Jun51 to USAF – 433 TCWG (TAC) Greenville 01Jun51 – 433 TCWG (AFE) Rhein-Main AB W. Germany 01Aug51 – 317 TCWG (AFE) Rhein-Main AB W. Germany 14Jul52; depls RAF Sculthorpe England 31Dec52 & 06Apr53; to Neubiberg AB W. Germany 21Mar53 – 7559 MAI GP (AFE) RAF Burtonwood England 13Dec54 – 407 AB GP (SAC) Great Falls 17Oct55; depl 407 Fighter WG (SAC) Malmstrom 18Oct55 – bailment Fairchild (AMC) Hagerstown 21Dec55 – 407 Fighter WG (SAC) Malmstrom 31Jan56 – BMC (AMC) Birmingham 24Apr56 – 9 AB GP (SAC) Mt. Home 10Jul56. Crashed and burned after striking telephone poles on a ranch 5 miles SW Federal Wyoming 27Jul56. 4 crew onboard were all killed – Final Disposition: Accident.

10478 / 50-160 / C-119C-20-FA

Avble: 14May51; Acc: 14Jun51; Del: 28Jun51 to USAF – 433 TCWG (TAC) Greenville 28Jun51 – 433 TCWG (AFE) Rhein-Main AB W. Germany 01Aug51 – 317 TCWG (AFE)

Rhein-Main AB W. Germany 14Jul52; depl RAF Sculthorpe England 22Jan53; to Neubiberg AB W. Germany 25Mar53 – 7559 MAI GP (AFE) RAF Burtonwood England 15Dec54 – 19 Bomb WG (SAC) Pinecast 28Sep55 – 11 Bomb WG (SAC) Carswell 07Oct55; to 808 AB GP (SAC) Bergstrom 01Apr57; to 27 Fighter WG (SAC) Bergstrom 23May57; to 824 AB GP (SAC) Carswell 18Jun57 – 824 AB GP (SAC) Carswell 06Aug57 – 825 AB GP (SAC) Little Rock 27Nov57 – 7 AB GP (SAC) Carswell 04Dec57; re-serialed **50-0160** 1958 – 2578 RFC (CNR) Ellington 26Jul58 – 446 TCWG (CNR) Ellington 20Sep58; assigned 357 TCSq (446 TCG) (AFR) Alvin Callender Field 25Sep58; to NAS New Orleans 01Dec58; 706 TCSq (446 TCWG) (AFR) NAS New Orleans 08May61. Lost to operator outside USAF 02Nov61 – Final Disposition: Unknown.

10479 / 50-161 / C-119C-20-FA

Avble: 14May51; Acc: 06Jun51; Del: 15Jun51 to USAF – 433 TCWG (TAC) Greenville 15Jun51 – 433 TCWG (AFE) Rhein-Main AB W. Germany 01Aug51 – 317 TCWG (AFE) Rhein-Main AB W. Germany 14Jul52; depls RAF Sculthorpe England 14Jan53 & 30Apr53; to Neubiberg AB W. Germany 25Mar53 – BMC (AMC) Birmingham 20Nov53; to 1631 AB SQ (MATS) Prestwick Scotland 30Nov53 mnt – 4750 TNG GP (ADC) Yuma 02Aug54; depl Youngstown date unk; depl 328 Fighter GP (ADC) Grandview 16Sep56; depl 78 Fighter GP (ADC) Hamilton 21Nov56; to 2750 AB WG (AMC) WPAFB 08Jan57 mnt; re-serialed **50-0161** 1958; to 4238 AB GP (SAC) Barksdale 10Jul58; to 4756 ADF WG (ADC) Tyndall 26Jul58 – 446 TCWG (AFR) Ellington 02Apr59; 924 TCG (446 TCWG) (AFR) Ellington 31May63 – Wfu to MASDC (LOG) Davis-Monthan 26Aug65 for storage – DFI: 21Jun66 – Final Disposition: Scrapped.

10480 / 50-162 / C-119C-21-FA

Rsvd YC-119D prototype canx. Avble: 16May51; Acc: 06Jun51; Del: 15Jun51 to USAF – 580 ARC WG (MATS) Mt. Home 15Jun51; to Wheelus AB Libya 04Oct51 – 96 Bomb WG (SAC) Altus 10Sep55 – 3902 AB WG (SAC) Offutt 24Jun56 – 96 AB GP (SAC) Altus 25Jul56 – 11 AB GP (SAC) Altus 13Dec57; re-serialed **50-0162** 1958 – 2233 RFC (CNR) Mitchel 24Jun58 – 514 TCWG (AFR) Mitchel 19Dec58; to 1607 TSH WG (MATS) Dover 04Apr59; to McGuire 15May61; 903 TCG (514 TCWG) (AFR) McGuire 31May63 – 4500 AB WG (TAC) Langley 02Feb66. Listed as having an accident 25Apr66, nfd – Final Disposition: Accident.

10481 / 50-163 / C-119C-21-FA

Avble: 21May51; Acc: 28Jun51; Del: 04Jul51 to USAF – 433 TCWG (TAC) Greenville 03Jul51 – 433 TCWG (AFE) Rhein-Main AB W. Germany 01Aug51 – 317 TCWG (AFE) Rhein-Main AB W. Germany 14Jul52; depl RAF Sculthorpe England 25Feb53; to Neubiberg AB W. Germany 27Mar53 – BMC (AMC) Birmingham 07Feb54 – 7373 AD WG (AFE) Chateauroux AB France 08Feb54. Suffered an inflight engine failure 4 miles S Caceres Spain 09Feb54. Pilot: Thomas G. Johnson plus 7 other crew bailed out, the aircraft came down in a prairie and burned – Final Disposition: Accident.

10482 / 50-164 / C-119C-21-FA

Avble: 25May51; Acc: 12Jun51; Del: 04Jul51 to USAF – 433 TCWG (TAC) Greenville 03Jul51 – 433 TCWG (AFE) Rhein-Main AB W. Germany 01Aug51 – 317 TCWG (AFE) Rhein-Main AB W. Germany 14Jul52; depl RAF Sculthorpe England 19Mar53; to Neubiberg AB W. Germany 30Apr53; to 7280 AD WG (AFE) Nouasseur AB Morocco 11Jan54; to 7373 AD WG (AFE) Chateauroux AB France 24Jan54 – 519 ADF GP (ADC) Suffolk 10Dec54

– OGDAR (AMC) Hill 26May55 – 52 Fighter GP (ADC) Suffolk 05Sep55; to 63 TCWG (TAC) Donaldson 15Mar56 mnt; to 314 TCWG (TAC) Sewart 12Nov56 mnt – 52 CLM SQ (ADC) Suffolk 13Feb57; depl 355 Fighter GP (ADC) McGhee Tyson 22Apr57; re-serialed **50-0164** 1958 – Hayes (AMC) Birmingham 18Aug58 – 446 TCWG (AFR) Ellington 21Nov58; to 823 COS GP (SAC) Homestead 27Aug60 mnt – 72 Bomb WG (SAC) Ramey AFB Puerto Rico 22Jun62. Lost due to ground accident 25Jun62, nfd – Final Disposition: Accident.

10483 / 50-165 / C-119C-21-FA

Avble: 12Jun51; Acc: 12Jun51; Del: 04Jul51 to USAF – 433 TCWG (TAC) Greenville 03Jul51 – 433 TCWG (AFE) Rhein-Main AB W. Germany 01Aug51 – 317 TCWG (AFE) Rhein-Main AB W. Germany 14Jul52; depls RAF Sculthorpe England 15Dec52, 04May53 & 18Aug53; to Neubiberg AB W. Germany 15May53 – 567 ADF GP (ADC) McChord 10Oct54 – MOBAR (AMC) Brookley 04May55 mnt – BMC (AMC) Birmingham date unk – 325 Fighter GP (ADC) McChord 03Sep55; to 4061 AB GP (SAC) Malmstrom 11Jul57 mnt – 325 MSU GP (ADC) McChord 30JApr57; re-serialed **50-0165** 1958 – 2233 RFC (CNR) Mitchel 25Jun58 – 514 TCWG (AFR) Mitchel 19Dec58; to McGuire 15Mar61; 903 TCG (514 TCWG) (AFR) McGuire 31May63 – Wfu to MASDC (LOG) Davis-Monthan 13May66 for storage – DFI: 06May68 – Final Disposition: Scrapped.

10484 / 50-166 / C-119C-21-FA

Avble: 01Jun51; Acc: 14Jun51; Del: 04Jul51 to USAF – 433 TCWG (TAC) Greenville 03Jul51 – 433 TCWG (AFE) Rhein-Main AB W. Germany 01Aug51 – 317 TCWG (AFE) Rhein-Main AB W. Germany 14Jul52; to Neubiberg AB W. Germany 15May53 – 567 ADF GP (ADC) McChord 04Oct54; to 3560 PTN WG (TC) Webb 12Jun55 mnt – 325 Fighter GP (ADC) McChord 18Aug56; depl 337 Fighter GP (ADC) Portland 27Nov56; to 803 AB GP (SAC) Davis-Monthan 04Mar57 mnt; to 479 Fighter WG (TAC) George 29Jul57 mnt; re-serialed **50-0166** 1958 – 2584 RFC (CNR) Memphis 05Mar58 – 2584 AB SQ (CNC) Memphis 19Dec58 – 514 TCWG (AFR) Mitchel 09Jun59; to McGuire 15Mar61; 903 TCG (514 TCWG) (AFR) McGuire 31May63 – Wfu to MASDC (LOG) Davis-Monthan 11Jun66 for storage – DFI: 06May68 – Final Disposition: Scrapped.

10485 / 50-167 / C-119C-21-FA

Avble: 05Jun51; Acc: 28Jun51; Del: 04Jul51 to USAF – 433 TCWG (TAC) Greenville 04Jul51 – 433 TCWG (AFE) Rhein-Main AB W. Germany 01Aug51 – 317 TCWG (AFE) Rhein-Main AB W. Germany 14Jul52; to Neubiberg AB W. Germany 15May53; depl RAF Sculthorpe England 04Sep53 – BMC (AMC) Birmingham 20Nov53; to 1631 AB SQ (MATS) Prestwick Scotland 25Nov53 mnt – 4750 TNG GP (ADC) Yuma 08Aug54; to 810 AB GP (SAC) Biggs 04Nov54; to 3345 TTA WG (TC) Chanute 11Jan56 mnt; depl 328 Fighter GP (ADC) Grandview 06Mar56; to 4900 AB GP (ARD) Kirtland 16Mar56 mnt; to 3560 PTN WG (TC) Webb 15Jun56 mnt; re-serialed **50-0167** 1958; depl 4756 ADF WG (ADC) Tyndall 03Jun58; to 4600 AB WG (ADC) Peterson 16Jan59 – Hayes (AMC) Birmingham 11May59 – 446 TCWG (AFR) Ellington 16Sep59; 924 TCG (446 TCWG) (AFR) Ellington 31May63 – Wfu to MASDC (LOG) Davis-Monthan 26Aug65 for storage – DFI: 21Jun66 – Final Disposition: Scrapped.

10486 / 50-168 / C-119C-21-FA

Avble: 04Jun51; Acc: 28Jun51; Del: 04Jul51 to USAF – 433 TCWG (TAC) Greenville 04Jul51 – 433 TCWG (AFE) Rhein-Main AB W. Germany 01Aug51 – 317 TCWG (AFE)

Rhein-Main AB W. Germany 14Jul52; to Neubiberg AB W. Germany 15May53; depl 60 TCWG (AFE) Rhein-Main AB W. Germany 31Jul54; to 7373 MAI GP (AFE) Chateauroux AB France 07Sep54 – 7559 MAI GP (AFE) RAF Burtonwood England 15Dec54 mnt – 98 Bomb WG (SAC) Lincoln 23Nov55 – 42 AB GP (SAC) Loring 12May56; to 461 BTA WG (TAC) Blytheville 26Nov57 mnt; re-serialed **50-0168** 1958 – 2577 RFC (CNR) Brooks 28Mar58; to 1405 AB WG (MATS) Scott 29Mar58 mnt – 433 TCWG (AFR) Brooks 18Sep58; to Kelly 31Aug60; 922 TCG (433 TCWG) (AFR) Kelly 31May63 – 911 TCG (459 TCWG) (AFR) Greater Pittsburgh 19Oct65 – Wfu to MASDC (LOG) Davis-Monthan 07Jul66 for storage – DFI: 06May68 – Final Disposition: Scrapped.

10487 / 50-169 / C-119C-21-FA

Avble: 22Jun51; Acc: 09Jul51; Del: 10Jul51 to USAF – 433 TCWG (TAC) Greenville 10Jul51 – 433 TCWG (AFE) Rhein-Main AB W. Germany 01Aug51 – 317 TCWG (AFE) Rhein-Main AB W. Germany 14Jul52; depl RAF Sculthorpe England 03Mar53; to Neubiberg AB W. Germany 06Apr53; to 7100 HQS WG (AFE) Wiesbaden AB W. Germany 28Dec53 – 29 ADH DV (ADC) Great Falls 07Oct54; to 3600 CCT WG (TC) Luke 08Dec55 mnt – 328 Fighter GP (ADC) Grandview 21Aug56 – 803 AB GP (SAC) Davis-Monthan 10Aug57 mnt; re-serialed **50-0169** 1958 – AASBR (AMC) Davis-Monthan 23Oct58 for storage – Hayes (AMC) Birmingham 28Jan59 – 446 TCWG (AFR) Ellington 27Apr59; to 803 COS GP (SAC) Davis-Monthan 06Jun59 mnt; 925 TCG (446 TCWG) (AFR) Ellington 31May63 – Wfu to MASDC (LOG) Davis-Monthan 18Sep65 for storage – DFI: 06May68 – Final Disposition: Scrapped.

10488 / 50-170 / C-119C-21-FA

Avble: 22Jun51; Acc: 03Jul51; Del: 10Jul51 to USAF – 433 TCWG (TAC) Greenville 10Jul51 – 433 TCWG (AFE) Rhein-Main AB W. Germany 01Aug51 – 317 TCWG (AFE) Rhein-Main AB W. Germany 14Jul52; depl RAF Sculthorpe England 16Dec52; to Neubiberg AB W. Germany 19Mar53; to 7373 AB GP (AFE) Chateauroux AB France 17Dec53; to 7373 MAI GP (AFE) Chateauroux AB France 01Mar54 – 564 ADF GP (ADC) Otis 28Nov54 – BMC (AMC) Birmingham 14Jul55 – 33 Fighter GP (ADC) Otis 02Nov55 – 33 MSU GP (ADC) Otis 18Oct56 – 551 MSU GP (ADC) Otis 18Aug57 – 551 AEW WG (ADC) Otis 01Sep57; re-serialed **50-0170** 1958 – 2577 RFC (CNR) Brooks 01Apr58; depl 2589 RFC (CNR) Dobbins 02Mar58 – OKLAR (AMC) Tinker 19Aug58 – 433 TCWG (AFR) Brooks 08Nov58; to Kelly 26Oct60; to 3550 FTA WG (ATC) Moody 14Nov60 mnt; 922 TCG (433 TCWG) (AFR) Kelly 31May63 – Wfu to MASDC (LOG) Davis-Monthan 22Oct65 for storage – DFI: 06May68 – Final Disposition: Scrapped.

10489 / 50-171 / C-119C-21-FA

Rsvd YC-119E prototype canx. Avble: 11Jun51; Acc: 28Jun51; Del: 04Jul51 to USAF – 433 TCWG (TAC) Greenville 04Jul51 – 433 TCWG (AFE) Rhein-Main AB W. Germany 01Aug51 – 317 TCWG (AFE) Rhein-Main AB W. Germany 14Jul52; to Neubiberg AB W. Germany 21Mar53; depl RAF Sculthorpe England 28May53 – 7559 MAI GP (AFE) RAF Burtonwood England 14Dec54 – 98 Bomb WG (SAC) Lincoln 23Nov55 – 42 Bomb WG (SAC) Loring 01May56 – BMC (AMC) Birmingham 29Dec57; re-serialed **50-0171** 1958 – 42 AB GP (SAC) Loring 06Apr58 – 2578 RFC (CNR) Ellington 13Jun58 – 446 TCWG (AFR) Ellington 20Sep58; to 2852 AB WG (LOG) McClellan 27Sep62 mnt; 924 TCG (446 TCWG) (AFR) Ellington 31May63 – Wfu to MASDC (LOG) Davis-Monthan 26Aug65 for storage – DFI: 21Jun66 – Final Disposition: Scrapped.

Section Four—Service Histories

10490 / 51-2532 / C-119C-22-FA

Avble: 27Jun51; Acc: 26Jul51; Del: 20Jul51 to USAF – 316 TCG (TAC) Sewart 20Jul51; depl Laurinburg-Maxton Airport North Carolina 24Aug51 – 435 TCWG (TAC) Miami 10Sep51; depl Grenier New Hampshire 13Jan52 – 403 TCWG (FEA) Ashiya AB Japan 01Mar52 – 314 TCWG (FEA) Ashiya AB Japan 07Oct52; loan to French Armée de l'Air Indochina 05Apr54-12Jun54, French markings applied – 483 TCWG (FEA) Ashiya AB Japan 14Nov54 – 316 TCWG (FEA) Ashiya AB Japan 15Nov54 – BMC (AMC) Birmingham Nov54 – 26 SM WG (SAC) Lockbourne 23Feb55; to 14 Fighter GP (ADC) Burlington 24Dec55 & 16Jan56 mnt – 801 AB GP (SAC) Lockbourne 01Feb57 – 2252 RFC (CNR) Clinton 30Sep57 – 302 TCWG (AFR) Clinton 19Mar59; 906 TCG (302 TCWG) (AFR) Clinton 11Feb63; 907 TCG (302 TCWG) (AFR) Clinton 05Oct65 – Wfu to MASDC (LOG) Davis-Monthan 06Nov65 for storage – DFI: 06May68 – Final Disposition: Scrapped.

10491 / 51-2533 / C-119C-22-FA

Avble: 26Jun51; Acc: 12Jul51; Del: 17Jul51 to USAF – 316 TCG (TAC) Sewart 16Jul51; depl Laurinburg-Maxton Airport North Carolina 20Aug51 – 435 TCWG (TAC) Miami 10Sep51; depl Grenier New Hampshire 03Jan52; depl Elmendorf Alaska 24Oct52 – 582 ARC GP (MATS) RAF Molesworth England 29Jan54; to 6603 AB GP (NEA) Goose Bay Canada 12Feb54 mnt; depl 60 TCWG (AFE) Rhein-Main AB W. Germany 27Mar54 – 19 Bomb WG (SAC) Pinecast 26Sep55 – 1631 MAT SQ (MATS) Prestwick Scotland 14Nov55 – SABBE (AMC) Brussels Belgium 15Sep56 – 3131 MAI GP (AMC) Chateauroux AB France 20May57 – 2231 RFC (CNR) Byrd 23Jun57 – 2252 RFC (CNR) Clinton 03Aug57 – 302 TCWG (AFR) Clinton 19Mar59; 906 TCG (302 TCWG) (AFR) Clinton 11Feb63; 907 TCG (302 TCWG) (AFR) Clinton 15Nov65 – Wfu to MASDC (LOG) Davis-Monthan 29Dec65 for storage – DFI: 06May68 – Final Disposition: Scrapped.

10492 / 51-2534 / C-119C-22-FA

Avble: 30Jun51; Acc: 14Jul51; Del: 21Jul51 to USAF – 580 ARC WG (MATS) Mt. Home 20Jul51; to Wheelus AB Libya 09Oct51; depl 60 TCWG (AFE) Rhein-Main AB W. Germany 23Dec52; to 7100 SUT WG (AFE) Wiesbaden AB W. Germany 08Oct53 – Assigned SAC Jun54 – Sep56 units unk – 7272 AB GP (AFE) Wheelus AB Libya 15Sep56 – 2252 RFC (CNR) Clinton 07Nov56; depl 2253 RFC (CNR) Greater Pittsburgh 15Jun57 – 302 TCWG (AFR) Clinton 19Mar59; 906 TCG (301 TCWG) (AFR) Clinton 11Feb63; 907 TCG (302 TCWG) (AFR) Clinton 05Oct65 – Wfu to MASDC (LOG) Davis-Monthan 29Dec65 for storage – DFI: 06May68 – Final Disposition: Scrapped.

10493 / 51-2535 / C-119C-22-FA

Avble: 29Jun51; Acc: 12Jul51; Del: 16Jul51 to USAF – 433 TCWG (TAC) Greenville 16Jul51 – 433 TCWG (AFE) Rhein-Main AB W. Germany 01Aug51 – 317 TCWG (AFE) Rhein-Main AB W. Germany 14Jul52; to Neubiberg AB W. Germany 21Mar53; depl RAF Sculthorpe England 14Jul53; to 73 AD WG (AFE) Chateauroux AB France 06Feb53 – **RECORDS INDECIPHERABLE** – 27 ADH DV (ADC) Norton 54; to 4750 ADF WG (ADC) Yuma 01Dec55 – Fairchild (AMC) Hagerstown 07Feb56 – 78 Fighter GP (ADC) Hamilton 10Mar56 – 78 MSU GP (ADC) Hamilton 30Apr58 – 2577 RFC DT (CNR) NAS Dallas 20Jun58 – 69 TCSq (433 TCWG) (AFR) NAS Dallas 20Sep58; to 2589 AB GP (CNC) Dobbins 30May59 mnt; 923 TCG (433 TCWG) (AFR) NAS Dallas 17Jan63; to Carswell 31May63 – 2750 AB WG (LOG) WPAFB 06Jul65 – REC: 08Dec65 – Final Disposition: Scrapped.

10494 / 51-2536 / C-119C-22-FA

Avble: 30Jun51; Acc: 12Jul51; Del: 16Jul51 to USAF – 18 AF (TAC) Greenville 16Jul51 – 435 TCWG (TAC) Miami 03Aug51; to 6612 AB GP (NEA) Thule AB Greenland 23Jul52 mnt – 403 TCWG (FEA) Ashiya AB Japan 14Sep52 – 483 TCWG (FEA) Ashiya AB Japan 01Nov52; loan to French Armée de l'Air Indochina 20Mar54-13May54, French markings applied; to 6424 AD WG (FEA) Clark AB The Philippines 10Jun54 mnt – 509 Bomb WG (SAC) Walker 15Feb55 – 812 AB GP (SAC) Walker 25Jul56 – 2589 RFC (CNR) Dobbins 22Nov57 – 2584 RFC (CNR) Memphis 20Dec57 – 810 AB GP (SAC) Biggs 17Apr58 – Hayes (AMC) Birmingham 08May58 – 2259 RFC (CNR) Andrews 25Aug58 – 459 TCWG (AFR) Andrews 04Nov58 – 2255 RFC (CNR) Bradley 18Dec59 – 337 TCSq (514 TCWG) (AFR) Bradley 19Mar59; 905 TCG (514 TCWG) (AFR) Bradley 17Jan63 – Wfu to MASDC (LOG) Davis-Monthan 09Oct65 for storage – DFI: 06May68 – Final Disposition: Scrapped.

10495 / 51-2537 / C-119C-22-FA

Avble: 03Jul51; Acc: 12Jul51; Del: 19Jul51 to USAF – 316 TCG (TAC) Sewart 18Jul51; depl Laurinburg- Maxton Airport North Carolina 20Aug51 – 435 TCWG (TAC) Miami 11Sep51; depl Grenier New Hampshire 03Feb52 – 403 TCWG (FEA) Ashiya AB Japan 01Aug52; to 1500 OPS SQ (MATS) Hickham 17Aug52 mnt – 314 TCWG (FEA) Ashiya AB Japan 23Aug52; loan to French Armée de l'Air Indochina 03May–26Jul53, French markings applied callsign: F-OYXA; to 24 MND GP (FEA) Clark AB The Philippines 26Jul53; depl 483 TCWG (FEA) Ashiya AB Japan 02Nov53; loan to French Armée de l'Air Indochina 20Mar54-26Jul54, French markings applied; depl Clark AB The Philippines 26Jul54 – 316 TCWG (FEA) Ashiya AB Japan 54 – 6332 AB WG (FEA) Kadena AB Japan 54 – 18 Fighter-Bomber WG (FEA) Kadena AB Japan 04May55 – BMC (AMC) Birmingham 30Jun55 – 327 Fighter GP (ADC) Truax 12Oct56; depl O'Hare Michigan date unk – 327 CLM SQ (ADC) Truax 08Nov57 – 56 CLM SQ (ADC) O'Hare 27Feb58 – Hayes (AMC) Birmingham 31Jul58 – 446 TCWG (AFR) Ellington 24Oct58; to 124 CLM SQ (ID-ANG) Boise 26Jun59 mnt; to LTC (ATC) Lowry 24Jul58 mnt; 924 TCG (446 TCWG) (AFR) Ellington 31May63 – Wfu to MASDC (LOG) Davis-Monthan 24Aug65 for storage – REC: 21Jun66 – Final Disposition: Scrapped.

10496 / 51-2538 / C-119C-22-FA

Avble: 06Jul51; Acc: 14Jul51; Del: 18Jul51 to USAF – 435 TCWG (TAC) Miami 19Jul51 – 403 TCWG (FEA) Ashiya AB Japan 21Jul52 – 483 TCWG (FEA) Ashiya AB Japan 01Jan53; depl Bofu Japan 19Nov53; loan to French Armée de l'Air Indochina 11Apr54-13May54, French markings applied – 314 TCWG (FEA) Ashiya AB Japan 07Jul54 – **RECORDS INDECIPHERABLE** – BMC (AMC) Birmingham date unk – 310 Bomb WG (SAC) Smoky Hill 16Mar55 – 40 Bomb WG (SAC) Smoky Hill 31Mar55; to 802 AB GP (SAC) Smoky Hill 02Jun55 – bailment Fairchild (AMC) Hagerstown 28Aug56 – 310 Bomb WG (SAC) Smoky Hill 05Oct56 – 802 AB GP (SAC) Smoky Hill 01Feb57; to 314 TCWG (TAC) Sewart 01May57 mnt – 2578 RFC (CNR) Ellington 20May58 – 446 TCWG (AFR) Ellington 20Sep58; to LTC (ATC) Lowry 26Jul59 mnt; 925 TCG (446 TCWG) (AFR) Ellington 31May63 – Wfu to MASDC (LOG) Davis-Monthan 18Sep65 for storage – REC: 21Jun66 – Final Disposition: Scrapped.

10497 / 51-2539 / C-119C-22-FA

Avble: 07Jul51; Acc: 17Jul51; Del: 18Jul51 to USAF – 435 TCWG (TAC) Miami 19Jul51; depl Grenier New Hampshire 12Jan52 – 403 TCWG (FEA) Ashiya AB Japan 31Jul52 – 483

TCWG (FEA) Ashiya AB Japan 01Jan53; depl Clark AB The Philippines 16Apr54 – 814 AB GP (SAC) Fairchild 20Feb55 – 22 Bomb WG (SAC) March 20Feb55 – 96 AB GP (SAC) Altus 18Jul56 – 11 AB GP (SAC) Altus 13Dec57 – 7 AB GP (SAC) Carswell 04Apr58 – 2252 RFC (CNR) Clinton 02Jul58 – 302 TCWG (AFR) Clinton 19Mar59; to MAP operator unk 20Apr62; 906 TCG (302 TCWG) (AFR) Clinton 31May63; 907 TCG (302 TCWG) (AFR) Clinton 05Oct65 – Wfu to MASDC (LOG) Davis-Monthan 04Dec65 for storage – DFI: 06May68 – Final Disposition: Scrapped.

10498 / 51-2540 / C-119C-22-FA

Avble: 10Jul51; Acc: 26Jul51; Del: 30Jul51 to USAF – 435 TCWG (TAC) Miami 30Jul51; depl Grenier New Hampshire 19Jan52 – 403 TCWG (FEA) Ashiya AB Japan 13Aug52 – 483 TCWG (FEA) Ashiya AB Japan 01Jan53; depl Clark AB The Philippines 17Apr54 – 314 TCWG (FEA) Ashiya AB Japan 06Jul54; depl Clark AB The Philippines 23Aug54; loan to French Armée de l'Air Indochina 24Aug54-04Sep54, French markings applied – **RECORDS INDECIPHERABLE** – BMC (AMC) Birmingham 54-55 – 99 SRH WG (SAC) Fairchild 02Mar55 – 99 Bomb WG (SAC) Fairchild 01Oct55; to 28 AB GP (SAC) Ellsworth 01Mar56 – 92 AB GP (SAC) Fairchild 04Sep56 – 340 AB GP (SAC) Whiteman 26Sep56 – Hayes (AMC) Birmingham 08Jul57 – 802 AB GP (SAC) Schilling 19Oct57 – 2589 RFC (CNR) Dobbins 18Dec57 – 2584 RFC (CNR) Memphis 29Oct58 – 2584 AB SQ (CNC) Memphis 19Dec58 – 758 TCSq (459 TCWG) (AFR) Greater Pittsburgh 17Mar59; 911 TCG (459 TCWG) (AFR) Greater Pittsburgh 17Jan63 – 926 TCG (446 TCWG) (AFR) NAS New Orleans 02Mar63 – Wfu to MASDC (LOG) Davis-Monthan 28Aug65 for storage – REC: 21Jun66 – Final Disposition: Scrapped.

10499 / 51-2541 / C-119C-22-FA

Avble: 11Jul51; Acc: 28Jul51; Del: 07Aug61 to USAF – 435 TCWG (TAC) Miami 07Aug51; depl Grenier New Hampshire 20Jan52 – 403 TCWG (FEA) Ashiya AB Japan 52 – 314 TCWG (FEA) Ashiya AB Japan 04Nov52; loan to French Armée de l'Air Indochina 03May–03Aug53, French markings applied callsign: F-OYXB; to 24 MND GP (FEA) Clark AB The Philippines 06Jul53 mnt; depl 483 TCWG (FEA) Clark AB The Philippines 14Apr54; loan to French Armée de l'Air Indochina 28Jun54-09Aug54, French markings applied; depl Clark AB The Philippines 09Aug54 – 316 TCWG (FEA) Ashiya AB Japan 54-55 – 456 Fighter SQ (520 ADF GP) (ADC) Truax 22May55; 520 ADF GP (ADC) Truax 14Jun55 – 327 Fighter GP (ADC) Truax 06Sep56; depl 15 Fighter GP (ADC) Niagara Falls 10Jul56; depl 56 Fighter GP (ADC) O'Hare 05Aug57; depl 56 CLM SQ (ADC) O'Hare 08Sep57 – 327 CLM SQ (ADC) Truax 08Nov57 – 2584 RFC (CNR) Memphis 05Feb58 – 446 TCWG (AFR) Ellington 12Oct58; depl 435 TCWG (AFR) Miami 21Jun59; 925 TCG (446 TCWG) (AFR) Ellington 31May63 – Wfu to MASDC (LOG) Davis-Monthan 19Sep65 for storage – REC: 21Jun66 – Final Disposition: Scrapped.

10500 / 51-2542 / C-119C-22-FA

Avble: 12Jul51; Acc: 02Aug51; Del: 14Aug51 to USAF – 580 ARC WG (MATS) Mt. Home 13Aug51; to 1050 AB WG (HQC) Andrews 19Mar52 mnt; to Wheelus AB Libya date unk; depls 60 TCWG (AFE) Rhein-Main AB W. Germany 15Dec52, 10May53, 05Jun53 & 25Jul53. Crashed in the Libyan Desert 30 miles S Gharyan Libya 09Aug53 after running out of fuel on a flight from Udine Airfield to Wheelus AB. 6 crew and 18 passengers were later rescued – DFI: 26Oct53 – Final Disposition: Accident.

10501 / 51-2543 / C-119C-22-FA

Avble: 17Jul51; Acc: 11Aug51; Del: 18Aug51 to USAF – WRDCN (ARD) WPAFB 17Aug51 for experimental duties – 435 TCWG (TAC) Miami IAP 26Feb52 – 403 TCWG (FEA) Ashiya AB Japan 21Jul52 – 483 TCWG (FEA) Ashiya AB Japan 01Nov52; loan to French Armée de l'Air Indochina 03May–03Aug53, French markings applied callsign: F-OYXC; to 24 MND GP (FEA) Clark AB The Philippines 22Jun53 mnt; depl Bofu Japan 07Nov53; loan to French Armée de l'Air Indochina 07Apr54-04Aug54, French markings applied – 314 TCWG (FEA) Ashiya AB Japan 22Jun54 – 316 TCWG (FEA) Ashiya AB Japan 54-55 – 528 ADF GP (ADC) Presque Isle 20Apr55; assigned 82 Fighter SQ – 517 ADF GP (ADC) Burlington 09Jun55 – BMC (AMC) Birmingham 04Aug55 – 14 Fighter GP (ADC) Burlington 05Dec55; to 2750 AB WG (AMC) WPAFB 28Jul56 mnt; depl 4756 ADF GP (ADC) Moody 05Aug56 – 32 ADH DV (ADC) Hancock 13Sep56 – 4624 AB SQ (ADC) Hancock 10May58 – 14 CLM SQ (ADC) Burlington 04Jun58 – 446 TCWG (AFR) Ellington 23Sep58; 924 TCG (446 TCWG) (AFR) Ellington 31May63; 926 TCG (446 TCWG) (AFR) NAS New Orleans 20May65 – Wfu to MASDC (LOG) Davis-Monthan 21Aug65 for storage – REC: 21Jun66 – Final Disposition: Scrapped.

10502 / 51-2544 / C-119C-22-FA

Avble: 18Jul51; Acc: 08Aug51; Del: 23Aug51 to USAF – 580 ARC WG (MATS) Mt. Home 23Aug51; to Wheelus AB Libya date unk; to 85 AD WG (AFE) Erding AB W. Germany 17Aug53 mnt; to 7559 MND GP (AFE) RAF Burtonwood England 15Sep53 mnt; to 7100 SUT WG (AFE) Wiesbaden AB W. Germany 29Mar54 – 96 Bomb WG (SAC) Altus 04Nov55; to 7373 MAI GP (AFE) Chateauroux AB France 07Nov55 mnt; to 3131 MAI GP (AMC) Chateauroux AB France 01Jan56 mnt – 1400 FDM SQ (MATS) Keflavik AS Iceland 16Mar56 mnt – 2252 RFC (CNR) Clinton 03May57 – 302 TCWG (AFR) Clinton 19Mar59; 907 TCG (302 TCWG) (AFR) Clinton 11Feb63 – 4500 AB WG (TAC) Langley 29Jan64 lost due to ground accident – DFI: 30Jan64 – Final Disposition: Accident.

10503 / 51-2545 / C-119C-22-FA

Avble: 21Jul51; Acc: 13Aug51; Del: 20Aug51 to USAF – 435 TCWG (TAC) Miami 20Aug51 – 403 TCWG (FEA) Ashiya AB Japan 21Jul52 – 314 TCWG (FEA) Ashiya AB Japan 04Nov52; loan to French Armée de l'Air Indochina 31Mar54-24Jul54, French markings applied – 6332 AB WG (FEA) Kadena AB Japan 54 – 18 Fighter-Bomber WG (FEA) Kadena AB Japan 04May55 – to 6486 AB WG (FEA) Hickham 29Jul55 – SMAAR (AMC) McClellan 31Oct55 – BMC (AMC) Birmingham 10Jan56 – 824 AB GP (SAC) Carswell 27Mar56; to 806 AB GP (SAC) Lake Charles 31Jan57; to 3415 TTA WG (TC) Lowry 13Jul57 mnt – 2589 RFC (CNR) Dobbins 19Nov57 – 2559 RFC (CNR) Andrews 23Jul58 – 2255 RFC (CNR) Bradley 04Oct58 – Hayes (AMC) Birmingham 24Feb59 – 337 TCSq (514 TCWG) (AFR) Bradley 11Apr59; 905 TCG (514 TCWG) (AFR) Bradley 18Jan63 – Wfu to MASDC (LOG) Davis-Monthan 09Oct65 for storage – DFI: 06May68 – Final Disposition: Scrapped.

10504 / 51-2546 / C-119C-22-FA

Avble: 21Jul51; Acc: 15Aug51; Del: 05Sep51 to USAF – 435 TCWG (TAC) Miami 03Sep51; depl Grenier New Hampshire 03Jan52 – 403 TCWG (FEA) Ashiya AB Japan 15Aug52 – 483 TCWG (FEA) Ashiya AB Japan 01Jan53 – 314 TCWG (FEA) Ashiya AB Japan 09Mar54; loan to French Armée de l'Air Indochina 11Mar54, French markings applied. Made an emergency landing due to engine trouble on an airstrip at Dien Bien Phu N. Vietnam 11Mar54. The parked C-119 was then shelled and destroyed that evening

Section Four—Service Histories

by Communist forces from their hill positions. Listed as officially transferred out of USAF 09Mar54 – DFI: 11May54 – Final Disposition: Combat Loss.

10505 / 51-2547 / C-119C-22-FA

Avble: 02Jul51; Acc: 16Aug51; Del: 05Sep51 to USAF – 435 TCWG (TAC) Miami 03Sep51 – 456 TCWG (TAC) Miami IAP 01Dec52 – 483 TCWG (FEA) Ashiya AB Japan 16Feb53; depl Bofu Japan 24May53; depl Clark AB The Philippines 14Apr54 – 314 TCWG (FEA) Ashiya AB Japan 12Jul54 – 316 TCWG (FEA) Ashiya AB Japan 54 – 519 ADF GP (ADC) Suffolk 02Mar55; to 2750 AB WG (AMC) WPAFB 30Jul55 mnt – 52 Fighter GP (ADC) Suffolk 18Aug55; to 1611 FDM SQ (MATS) McGuire 16Nov55 mnt – BMC (AMC) Birmingham 19Jun56 – 52 CLM SQ (ADC) Suffolk 13Feb57; to 810 AB GP (SAC) Biggs 26Oct58 mnt – 69 TCSq (433 TCWG) (AFR) NAS Dallas 30Dec59; 923 TCG (433 TCWG) (AFR) NAS Dallas 17Jan63; to Carswell 02Mar63 – 907 TCG (302 TCWG) (AFR) Clinton 18Sep65 – Wfu to MASDC (LOG) Davis-Monthan 17Nov65 for storage – DFI: 06May68 – Final Disposition: Scrapped.

10506 / 51-2548 / C-119C-22-FA

Avble: 26Jul51; Acc: 18Aug51; Del: 05Sep51 to USAF – 435 TCWG (TAC) Miami 03Sep51; depl Grenier New Hampshire 16Jan52 – 403 TCWG (FEA) Ashiya AB Japan 13Aug52 – SMAAR (AMC) McClellan 03Oct52 mnt – 483 TCWG (FEA) Ashiya AB Japan 06Jan54; depl Clark AB The Philippines 14Apr54 – 314 TCWG (FEA) Ashiya AB Japan 13Jul54 – 316 TCWG (FEA) Ashiya AB Japan 54 – BMC (AMC) Birmingham 54-55 – 93 Bomb WG (SAC) Castle 17Mar55; to 4900 AB GP (ARD) Kirtland 08Aug55 mnt – 9 AB GP (SAC) Mt. Home 10Aug56; to 2750 AB WG (AMC) WPAFB 14Apr57 mnt; depl 9 Bomb WG (SAC) Mt. Home 19Apr57 – 2577 RFC (CNR) Brooks 29Jun58 – 433 TCWG (AFR) Brooks 19Sep58; to 839 AB GP (TAC) Sewart 20Feb59 mnt; to Kelly 31Aug60; 922 TCG (433 TCWG) (AFR) Kelly 17Jan63; 921 TCG (433 TCWG) (AFR) Kelly 10Oct65. Wfu REC / DFI: 06Jun66 – Final Disposition: Scrapped.

10507 / 51-2549 / C-119C-22-FA

Avble: 27Jul51; Acc: 27Aug51; Del: 05Sep51 to USAF – 435 TCWG (TAC) Miami 03Sep51; depls Grenier New Hampshire 23Feb52 & 05Nov52 – 403 TCWG (FEA) Ashiya AB Japan 31Jul52 – 314 TCWG (FEA) Ashiya AB Japan 20Aug52 – OKLAR (AMC) Tinker 31May54 mnt – MIDAR (AMC) Olmsted 03Jun54 mnt – **RECORDS INDECIPHERABLE** – 509 Bomb WG (SAC) Walker date unk – 812 AB GP (SAC) Walker 25Jul56 – 2252 RFC (CNR) Clinton 25Apr58 – 302 TCWG (AFR) Clinton 19Mar59 – 706 TCSq (446 TCWG) (AFR) NAS New Orleans 20Jun62. Departed Alvin Challendar Field New Orleans on a training flight to Florida 10Mar62 with 6 crew onboard. Engine problems then developed that necessitated a return to base but they were unable to make the runway and were all forced to bail out. The abandoned aircraft came down on a residential house killing 4 civilian occupants – DFI: 21Jun62 – Final Disposition: Accident.

10508 / 51-2550 / C-119C-22-FA

Avble: 28Jul51; Acc: 27Aug51; Del: 05Sep51 to USAF – 435 TCWG (TAC) Miami 03Sep51; depl Grenier New Hampshire 13Jan52 – 403 TCWG (FEA) Ashiya AB Japan 52 – 314 TCWG (FEA) Ashiya AB Japan 13Oct52; loan to French Armée de l'Air Indochina 03May–30Jul53, French markings applied callsign: F-OYXD; to 24 MND GP (FEA) Clark AB The Philippines 30Jul53 mnt; depl 483 TCWG (FEA) Ashiya AB Japan 14Mar54; depl (483 TCWG) Clark AB The Philippines 15Apr54; depl Clark AB The Philippines 21Jul54 –

316 TCWG (FEA) Ashiya AB Japan 54-55 – 438 Fighter SQ (534 ADF GP) (ADC) Kinross 02Mar55 – 3450 TTN WG (TC) Francis E. Warren 04Aug55 – 56 Fighter GP (ADC) O'Hare 28Oct55 – BMC (AMC) Birmingham 19Dec55 – 327 Fighter GP (ADC) Truax 10Mar56; to 3345 TTA WG (TC) Chanute 27Jul56 mnt; to 810 AB GP (SAC) Biggs 02Dec56 mnt; to 329 CLM SQ (ADC) Stewart Feb57 – 327 CLM SQ (ADC) Truax 08Nov57 – 2584 RFC (CNR) Memphis 08Feb58 – 2578 RFC (CNR) Ellington 12Sep58 – 446 TCWG (AFR) Ellington 20Sep58; 357 TCSq (446 TCWG) (AFR) NAS New Orleans 12Apr59; 706 TCSq (446 TCWG) (AFR) NAS New Orleans 08May61; 926 TCG (446 TCWG) (AFR) NAS New Orleans 17Jan63 – Wfu to MASDC (LOG) Davis-Monthan 21Aug65 for storage – REC: 21Jun66 – Final Disposition: Scrapped.

10509 / 51-2551 / C-119C-22-FA

Avble: 03Aug51; Acc: 29Aug51; Del: 05Sep51 to USAF – 435 TCWG (TAC) Miami 06Sep51; depl 314 TCWG (TAC) Sewart 12Oct51; depl Grenier New Hampshire 13Jan52 – 403 TCWG (FEA) Ashiya AB Japan 13Aug52, assigned 63 TCSq. Flew into a mountain 20 miles East of Seoul AB (K-16) S. Korea 14Nov52 while returning servicemen back to Korea after rest in Japan. The 7 crew and 37 passengers killed were: Capt William C. Moskosky, Sr., Capt Robert L. Schenck, 2Lt John C. Mortensen (pilot), Cwo Alfred H. Auger, A1C Leroy J. Beer, A2C Marvin L. Gainey, MSgt Arthur J. Holland, SSgt Ray W. Mansholt, SSgt Thaddeus L. Smith, Jr., Sgt Francis C. Berger, Sgt Albert W. Dzinwkowski, Sgt Robert W. Irwin, Sgt Richard L. Menninger, Sgt Donald R. Northrup, Sgt Arthur J. Scowcraft, Sgt Ray A. Shepherd, Sgt John C. Stauch, Sr., Sgt Harry N. Tsuruoka, Cpl Antonio Calaustro, Cpl William E. Clark, Cpl Patrick E. Connolly, Cpl Donald D. Drinnen, Cpl James A. Flory, Cpl Frank Gfroerer, Cpl Thomas E. Giglio, Cpl John W. Hanlon, Cpl John H. Williams, Pvt Robert J. Beachy, Pvt Gilberto A. Berrios, Pvt C. Concepcion-Esquilio, Pvt Delbert Coulam, Pvt Lloyd O. Fogt, Pvt Robert H. Koehler, Pvt Leon Letts, Jr., Pvt Bronson J. Mastne, Pvt Erling P. Miller, Pvt Billy G. Mouney, Pvt Byron H. Pittman, Pvt Lovell E. Prater, Pvt Juan Rivera-Gonzales, Pvt (1C) Juan Rivera-Gonzales, Pvt Raul Rosalez, Pvt Eugene R. Serra and Pvt Freelan Shrewsbury – DFI: 10Dec52 – Final Disposition: Accident.

10510 / 51-2552 / C-119C-22-FA

Avble: 09Aug51; Acc: 30Aug51; Del: 11Sep51 to USAF – 435 TCWG (TAC) Miami 10Sep51; depl Grenier New Hampshire 18Jan52 – 403 TCWG (FEA) Ashiya AB Japan 13Aug52 – 314 TCWG (FEA) Ashiya AB Japan 53; depl Tachikawa AB Japan 23Jun53; loan to French Armée de l'Air Indochina 17Apr54-31May54, French markings applied. Crashed after an engine failure 1.7 miles E Daegu AB (K-2) S. Korea 20Oct54, at least 1 crew member killed. To 6480 AD WG (FEA) Tachikawa AB Japan for salvage – DFI: 25Oct54 – Final Disposition: Accident.

10511 / 51-2553 / C-119C-22-FA

Avble: 15Aug51; Acc: 31Aug51; Del: 11Sep51 to USAF – 435 TCWG (TAC) Miami 10Sep51; depl Grenier New Hampshire 05Feb52 – 403 TCWG (FEA) Ashiya AB Japan 08Aug52 – 314 TCWG (FEA) Ashiya AB Japan 20Aug52; depl Bofu Japan 18May53; depl 483 TCWG (FEA) Clark AB The Philippines 16Apr54 – 316 TCWG (FEA) Ashiya AB Japan 54 – BMC (AMC) Birmingham 54-55 – 807 AB GP (SAC) March 20Feb55 – 40 Bomb WG (SAC) Smoky Hill 23Sep55 – 802 AB GP (SAC) Smoky Hill 01Feb57; to 817 AB GP (SAC) Portsmouth IAP 10Aug57 – 2578 RFC (CNR) Ellington 30Apr58 – 446 TCWG (AFR)

Ellington 20Nov58; to 3800 AB WG (AU) Maxwell 22Jan61mnt; 925 TCG (446 TCWG) (AFR) Ellington 31May63 – Wfu to MASDC (LOG) Davis-Monthan 18Sep65 for storage – REC: 21Jun66 – Final Disposition: Scrapped.

10512 / 51-2554 / C-119C-22-FA

Avble: 21Aug51; Acc: 12Sep51; Del: 27Oct51 to USAF – 3904 CMP WG (SAC) Camp Carson 26Oct51; depl Peterson Dec51; multiple assignments OGDAR (AMC) Hill from 04Feb52; to Stead 22Jan53; SMAAR (AMC) McClellan 15Feb53 – BMC (AMC) Birmingham 25Sep53; to WRAAR (AMC) Robins 20Nov53 – 42 Bomb WG (SAC) Limestone 19Dec53; to 3310 TTN WG (TC) Scott 27Aug54 mnt – 42 AB GP (SAC) Loring 19Feb55 – 91 SM WG (SAC) Lockbourne 15May55; to 3635 CCT WG (TC) Stead 08Nov56 mnt – 801 AB GP (SAC) Lockbourne 13Dec57; to 4082 AB GP (SAC) Goose Bay Canada 17May57; to 28 AB GP (SAC) Ellsworth 09Mar58 – 2578 RFC (CNR) Ellington 04Jul58 – 446 TCWG (AFR) Ellington 20Sep58; to 1607 TSH WG (MATS) Dover 59 mnt; to 823 COS GP (SAC) Homestead 22Jan61 mnt; 925 TCG (446 TCWG) (AFR) Ellington 31May63 – Wfu to MASDC (LOG) Davis-Monthan 18Sep65 for storage – REC: 21Jun66 – Final Disposition: Scrapped.

10513 / 51-2555 / C-119C-22-FA

Avble: 27Aug51; Acc: 22Sep51; Del: 10Oct51 to USAF – 581 ARC WG (MATS) Mt. Home 10Oct51; to 3200 PTS WG & 3203 MSU GP (APG) Eglin 29Nov51; to Clark AB The Philippines 10Jul52; depl 403 TCWG (FEA) Ashiya AB Japan 09Dec52; depl 483 TCWG (FEA) Ashiya AB Japan 01Jan53 – BMC (AMC) Birmingham 02Nov53 – 483 TCWG (FEA) Ashiya AB Japan 25Mar54; depl 314 TCWG (FEA) Ashiya AB Japan 20Apr54 – 316 TCWG (FEA) Ashiya AB Japan Nov54 – 527 ADF GP (ADC) Wurtsmith 24May55 – 575 ADF GP (ADC) Selfridge 11Aug55 – 412 Fighter GP (ADC) Wurtsmith 01Sep55; to 2750 AB WG (AMC) WPAFB 08Mar57 mnt – 412 CLM SQ (ADC) Wurtsmith 08Sep57 – 1 MSU GP (ADC) Selfridge 23Dec57 – 446 TCWG (AFR) Ellington 26Dec58; 357 TCSq (446 TCWG) (AFR) NAS New Orleans 10Jun59; 706 TCSq (446 TCWG) (AFR) NAS New Orleans 08May61; 926 TCG (446 TCWG) (AFR) NAS New Orleans 17Jan63 – Wfu to 2704 ASD GP (LOG) Davis-Monthan 02Dec64 for storage – REC: 03May65 – Final Disposition: Scrapped.

10514 / 51-2556 / C-119C-22-FA

Avble: 31Aug51; Acc: 18Sep51; Del: 10Oct51 to USAF – 581 ARC WG (MATS) Mt. Home 10Oct51; depl 316 TCG (TAC) Sewart 16Oct51; to Clark AB The Philippines 18Jul52; depl 35 Fighter WG (FEA) Johnson 02Oct52; depl 403 TCWG (FEA) Ashiya AB Japan 17Nov52; depl 483 TCWG (FEA) Ashiya AB Japan 01Jan53 – 6200 MSU GP (FEA) Clark AB The Philippines 04Mar54 for rec/salvage – 6424 AD WG (FEA) Clark AB The Philippines 10Jun54 for salvage due to abnormal deterioration in use – REC: 29Jul54 – Final Disposition: Scrapped.

10515 / 51-2557 / C-119C-23-FA

Avble: 08Sep51; Acc: 21Sep51; Del: 05Oct51 to USAF – 435 TCWG (TAC) Miami 05Oct51; depl Grenier New Hampshire 05Jan52; **RECORDS INDECIPHERABLE** – 403 TCWG (FEA) Ashiya AB Japan 13Aug52 – 483 TCWG (FEA) Ashiya AB Japan 01Jan53; depl 314 TCWG (FEA) Ashiya AB Japan 17Apr54; loan to French Armée de l'Air Indochina 17Apr54-03Jun54, French markings applied; depl (314 TCWG) Clark AB The Philippines 03Sep54 – 316 TCWG (FEA) Ashiya AB Japan Nov54 – 502 ADF GP (ADC) Youngstown 28Mar55; to 4750 ADF WG (ADC) Yuma 27Jun55 – BMC (AMC) Birmingham 27Jul55 – 79 Fighter GP (ADC) Youngstown 04Nov55 – 1 MSU GP (ADC) Selfridge 01Oct57; to 2750

AB WG (AMC) WPAFB 01Apr58 mnt – 4750 ADF WG (ADC) Vincent 07Jan59 – 446 TCWG (AFR) Ellington 29May59; 924 TCG (446 TCWG) (AFR) Ellington 31May63 – Wfu to MASDC (LOG) Davis-Monthan 26Aug65 for storage – REC: 21Jun66 – Final Disposition: Scrapped.

10516 / 51-2558 / C-119C-23-FA

Avble: 04Sep51; Acc: 22Sep51; Del: 05Oct51 to USAF – 435 TCWG (TAC) Miami 05Oct51; depl Grenier New Hampshire 13Jan52 – 403 TCWG (FEA) Ashiya AB Japan 13Aug52 – 314 TCWG (FEA) Ashiya AB Japan 21Oct52; depl Bofu Japan 18May53; loan to French Armée de l'Air Indochina 14Mar54-01Aug54, French markings applied – 483 TCWG (FEA) Ashiya AB Japan Nov54 – 316 TCWG (FEA) Ashiya AB Japan Nov54 – 54 Fighter SQ (29 ADV) (ADC) Ellsworth 26Apr55 – 4676 ADF GP (ADC) Grandview 03May55 – BMC (AMC) Birmingham 11Jul55 – 328 Fighter GP (ADC) Grandview 02Nov55 – 328 CLM SQ (ADC) Richards 01Aug57 – 2233 RFC (CNR) Mitchel 21Oct57; to 2500 AB WG (CNC) Mitchel 10Mar58 mnt – 514 TCWG (AFR) Mitchel 19Dec58; 337 TCSq (514 TCWG) (AFR) Bradley 20Mar61; 905 TCG (514 TCWG) (AFR) Bradley 31May63; 903 TCG (514 TCWG) (AFR) McGuire 06Nov65 – Wfu to MASDC (LOG) Davis-Monthan 06Apr66 for storage – DFI: 06May68 – Final Disposition: Scrapped.

10517 / 51-2559 / C-119C-23-FA

Avble: 08Sep51; Acc: 21Sep51; Del: 05Oct51 to USAF – 435 TCWG (TAC) Miami 05Oct51; depl Grenier New Hampshire 13Jan52 – 403 TCWG (FEA) Ashiya AB Japan 22Oct52 – 314 TCWG (FEA) Ashiya AB Japan 22Oct52; loan to French Armée de l'Air Indochina 54, French markings applied; depl 483 TCWG (FEA) Clark AB The Philippines 16Apr54 – BMC (AMC) Birmingham Oct54 – 814 AB GP (SAC) Fairchild 20Feb55 – 303 Bomb WG (SAC) Davis-Monthan 23Mar55 – 43 Bomb WG (SAC) Davis-Monthan 01Jul56 – SBNAR (AMC) Norton 06May57 – 5 AB GP (SAC) Travis 22Jul57 – 2596 RFC (CNR) NAS Dallas 13Nov57 – 2577 RFC DT (CNR) NAS Dallas 01Dec57 – 69 TCSq (433 TCWG) (AFR) NAS Dallas 20Sep58; to 3500 PTN WG (ATC) Reese 22Jul59 mnt; 923 TCG (433 TCWG) (AFR) NAS Dallas 17Jan63; to Carswell 02Mar63 – 907 TCG (302 TCWG) (AFR) Clinton 18Sep65 – Wfu to MASDC (LOG) Davis-Monthan 04Dec65 for storage – DFI: 06May68 – Final Disposition: Scrapped.

10518 / 51-2560 / C-119C-23-FA

Avble: 10Sep51; Acc: 22Sep51; Del: 05Oct51 to USAF – 435 TCWG (TAC) Miami 05Oct51, assigned 76 TCSq; depl Grenier AFB New Hampshire 13Jan52. Had a minor structural failure over Jacksonville Florida 27May52. Depl Elmendorf 26Oct52. Crashed into Mt. Silverthorne about 133 miles NNW of departure base Elmendorf AFB while heading to Fairbanks Alaska 07Nov1952 as part of Exercise *Warm Wind*. A mislabeled navigation chart put the aircraft off course and at night in icy conditions the aircraft unknowingly struck terrain. 19 fatalities consisted of the 5 crew: Capt Glenn H. Wall (pilot), 1Lt Frank Mates (co-pilot), 2Lt Enoch G. Crowe (navigator), SSgt Isham C. Pope (radio operator), A1C Gene G. Wood (flight engineer) and 14 USAF & U.S. Army passengers: Capt Walter Radmeister, 1Lt Nobel E. Williams, 2Lt Thomas B. Keen, Jr., MSgt Donald L. Ehnat, MSgt Fred B. McGee, MSgt William H. Moore, Cpl Gail W. Daugherty, Cpl William J. Newton, Cpl Judith R. Statler, Pvt Daniel W. Blasi, Pvt Leo C. Block, Pvt Alphonse S. Grzelka, Pvt Raymond L. Housler and Pvt John W. Salmon. The impact zone was located several days later but a recovery could not be attempted due to avalanche conditions. The wreckage was located decades later in Aug 2016 about 4 miles down an adjoining glacier flow – REC: 09Nov52 – DFI: 06Feb53 – Final Disposition: Accident.

10519 / 51-2561 / C-119C-23-FA

Avble: 12Sep51; Acc: 26Sep51; Del: 05Oct51 to USAF – 435 TCWG (TAC) Miami 05Oct51; depl Grenier New Hampshire 13Jan52 – 403 TCWG (FEA) Ashiya AB Japan 08Aug52 – 483 TCWG (FEA) Ashiya AB Japan 01Jan53; loan to French Armée de l'Air Indochina 06Apr54-23May54, French markings applied – 314 TCWG (FEA) Clark AB The Philippines (depl) 01Jul54; to Ashiya AB Japan 29Aug54 – 316 TCWG (FEA) Ashiya AB Japan Nov54 – 521 ADF GP (ADC) Sioux City 28Feb55 – bailment Fairchild (AMC) Hagerstown 08Aug55 – 53 Fighter GP (ADC) Sioux City 14Sep55 – 328 Fighter GP (ADC) Grandview 16Oct56 – BMC (AMC) Birmingham 26Jun57 – 328 CLM SQ (ADC) Richards 05Nov57 – 2596 RFC (CNR) NAS Dallas 13Nov57 – 2577 RFC DT (CNR) NAS Dallas 01Dec57 – 69 TCSq (433 TCWG) (AFR) NAS Dallas 20Sep58; depl 357 TCSq (446 TCG) (AFR) Alvin Callender Field 10Nov58; 923 TCG (433 TCWG) (AFR) NAS Dallas 17Jan63; to Carswell 02Mar63 – 907 TCG (302 TCWG) (AFR) Clinton 18Sep65 – Wfu to MASDC (LOG) Davis-Monthan 14Jan66 for storage – REC: 21Jun66 – Final Disposition: Scrapped.

10520 / 51-2562 / C-119C-23-FA

Avble: 14Sep51; Acc: 26Sep51; Del: 05Oct51 to USAF – 435 TCWG (TAC) Miami 05Oct51; depl Grenier New Hampshire 13Jan52 – 403 TCWG (FEA) Ashiya AB Japan 13Aug52 – 314 TCWG (FEA) Ashiya AB Japan 53; depl Tachikawa AB Japan 23Jun53; loan to French Armée de l'Air Indochina 05Apr54-07May54, French markings applied – BMC (AMC) Birmingham 13Jul54 – 483 TCWG (FEA) Ashiya AB Japan 54 – 5 SRH WG (SAC) Travis 01Mar55 – 43 Bomb WG (SAC) Davis-Monthan 14Mar55 – 28 AB WG (SAC) Ellsworth 16Jul56 – 5 AB GP (SAC) Travis 21May57 – 2578 RFC (CNR) Ellington 15Jun58 – 446 TCWG (AFR) Ellington 20Sep58; 357 TCSq (446 TCWG) (AFR) Alvin Callender Field 25Sep58; to NAS New Orleans 01Dec58; 706 TCSq (446 TCWG) (AFR) NAS New Orleans 08May61 – 926 TCG (446 TCWG) (AFR) NAS New Orleans 17Jan63 – Wfu to MASDC (LOG) Davis-Monthan 25Sep65 for storage – REC: 21Jun66 – Final Disposition: Scrapped.

10521 / 51-2563 / C-119C-23-FA

Avble: 17Sep51; Acc: 28Sep51; Del: 05Oct51 to USAF – 435 TCWG (TAC) Miami 05Oct51; depl Grenier New Hampshire 13Jan52 – 403 TCWG (FEA) Ashiya AB Japan 31Jul52 – 483 TCWG (FEA) Ashiya AB Japan 01Jan53; depl Bofu Japan 08Nov53; loan to French Armée de l'Air Indochina 06-29Apr54, French markings applied – 444 Fighter SQ (35 ADH DV) (ADC) Charleston 12Apr55; to 463 TCWG (TAC) Ardmore 10Nov55 mnt – BMC (AMC) Birmingham 02Dec55 – 35 ADH DV (ADC) Dobbins 09Mar56 – 32 ADH DV (ADC) Dobbins 15Nov58 – 446 TCWG (AFR) Ellington 15Feb59; to 819 COS GP (SAC) Dyess 05Jan60 mnt; 925 TCG (446 TCWG) (AFR) Ellington 31May63 – Wfu to MASDC (LOG) Davis-Monthan 18Sep65 for storage – REC: 21Jun66 – Final Disposition: Scrapped.

10522 / 51-2564 / C-119C-23-FA

Avble: 18Sep51; Acc: 28Sep51; Del: 05Oct51 to USAF – 435 TCWG (TAC) Miami 05Oct51; depl Grenier New Hampshire 13Jan52 – 456 TCWG (TAC) Miami 01Dec52 – 483 TCWG (FEA) Ashiya AB Japan 16Feb53; to 19 Bomb WG (FEA) Andersen AFB Guam 53; depl Bofu Japan 24May53; depl 314 TCWG (FEA) Ashiya AB Japan 16Apr54; loan to French Armée de l'Air Indochina 17Apr54- 14Jun54, French markings applied; depl (314 TCWG) Clark AB

The Philippines 21Jul54 – 316 TCWG (FEA) Ashiya AB Japan Nov54 – 528 ADF GP (ADC) Presque Isle 02Mar55 – 517 ADF GP (ADC) Burlington 12Apr55 – BMC (AMC) Birmingham 12Jun55 – 14 Fighter GP (ADC) Burlington 07Sep55; depl 475 Fighter GP (ADC) Minneapolis-St. Paul IAP 29Dec55 – 14 CLM SQ (ADC) Burlington 19Aug57 – 2577 RFC (CNR) Brooks 18Apr58; to 2750 AB WG (AMC) WPAFB 19Apr58 mnt – 433 TCWG (AFR) Brooks 18Sep58; to 807 COS GP (SAC) Little Rock 12Jun59 mnt; to Kelly 31Aug60; 922 TCG (433 TCWG) (AFR) Kelly 31May63; 921 TCG (433 TCWG) (AFR) Kelly 10Oct65 – Wfu to MASDC (LOG) Davis-Monthan 14Jan66 for storage – DFI: 06May68 – Final Disposition: Scrapped.

10523 / 51-2565 / C-119C-23-FA

Avble: 18Sep51; Acc: 28Sep51; Del: 05Oct51 to USAF – 435 TCWG (TAC) Miami 05Oct51; depl Grenier New Hampshire 03Jan52 – 403 TCWG (FEA) Ashiya AB Japan 52 – 483 TCWG (FEA) Ashiya AB Japan 01Jan53; depl Clark AB The Philippines 14Apr54 – 314 TCWG (FEA) Ashiya AB Japan 03Jul54; depl Clark AB The Philippines 09Aug54 – 316 TCWG (FEA) Ashiya AB Japan 07Mar55 – 539 Fighter SQ (4709 ADF WG) (ADC) Stewart 08Jan56 – 519 ADF GP (ADC) Suffolk 12Jul56 – bailment Fairchild (AMC) Hagerstown 02Aug56 – 52 Fighter GP (ADC) Suffolk 07Sep56 – 2750 AB WG (AMC) WPAFB 05Jan57 mnt – 52 CLM SQ (ADC) Suffolk 21Feb57; depl 355 Fighter GP (ADC) McGhee Tyson 22Apr57 – MIDAR (AMC) Olmsted 12Feb58 – 757 TCSq (459 TCWG) (AFR) Youngstown 08Sep60 – 302 TCWG (AFR) Clinton 10Nov60; 906 TCG (302 TCWG) (AFR) Clinton 11Feb63; 907 TCG (302 TCWG) (AFR) Clinton 29Aug63 – Wfu to MASDC (LOG) Davis-Monthan 06Nov65 for storage – DFI: 06May68 – Final Disposition: Scrapped.

10524 / 51-2566 / C-119C-23-FA

Avble: 18Sep51; Acc: 28Sep51; Del: 05Oct51 to USAF – 435 TCWG (TAC) Miami 05Oct51; depl Grenier New Hampshire 03Jan52; depl Elmendorf 24Oct52 – 456 TCWG (TAC) Miami 02Dec52 – 582 ARC WG (MATS) Great Falls 22May53; to RAF Molesworth England 29Jan54 – 301 Bomb WG (SAC) Barksdale 22May55 – 308 Bomb WG (SAC) Hunter 17Dec55; depls 2 Bomb WG (SAC) Hunter 22Dec55 & 04Oct56 – 804 AB GP (SAC) Hunter 05Aug57; to 2750 AB WG (AMC) WPAFB 15Oct57 mnt – 2577 RFC (CNR) Brooks 25Apr58 – 433 TCWG (AFR) Brooks 18Sep58; to Kelly 31Aug60; 922 TCG (433 TCWG) (AFR) Kelly 31May63; 921 TCG (433 TCWG) (AFR) Kelly 10Oct65 – Wfu 07Feb66 to museum / school status. Loan to CAF Harlingen Texas 1966–1986. Displayed from 1986 at the Museum of Aviation Robins AFB Georgia – Final Disposition: Extant.

10525 / 51-2567 / C-119C-23-FA

Avble: 22Sep51; Acc: 19Oct51; Del: 06Nov51 to USAF – 581 ARC (MATS) Mt. Home 06Nov51; depl 316 TCG (TAC) Sewart 51; to Clark AB The Philippines Jul52; depl 403 TCWG (FEA) Ashiya AB Japan 24Nov52; depls 483 TCWG (FEA) Ashiya AB Japan 01Jan53 & date unk – BMC (AMC) Birmingham 54 – 5 SRH WG (SAC) Travis 02Mar55; to 4900 AB GP (ARD) Kirtland 07Aug55 mnt – bailment Fairchild (AMC) Hagerstown 18Sep55 – 5 Bomb WG (SAC) Travis 15Oct55 – 5 AB GP (SAC) Travis 28Sep56; to 479 Fighter WG (TAC) George 06May57 mnt – 2577 RFC (CNR) Brooks 22Apr58 – 4510 CCT WG (TAC) Luke 21Jul58 mnt – 433 TCWG (AFR) Brooks 12Sep58; to 1 MSU GP (ADC) Selfridge 09Jan59 mnt; to Kelly 31Aug60; 922 TCG (433 TCWG) (AFR) Kelly 31May63; 921 TCG (433 TCWG) (AFR) Kelly 10Oct65 – Wfu 16Mar66 to museum / school status. Presently displayed at the USAF History & Traditions Museum Lackland AFB Texas – Final Disposition: Extant.

Section Four—Service Histories

10526 / 51-2568 / C-119C-23-FA

Avble: 25Sep51; Acc: 25Oct51; Del: 06Nov51 to USAF – 435 TCWG (TAC) Miami 08Nov51; depl Grenier New Hampshire 13Jan52 – 456 TCWG (TAC) Miami IAP 01Dec52 – 64 TCWG (TAC) Greenville 19Jun53 – 1739 FRY SQ (MATS) Amarillo 13Oct53 – BMC (AMC) Birmingham 02Feb54 – 3 AMO SQ (MATS) Tinker 21Jul54; to 3320 TTN WG (TC) Amarillo 27Feb56 mnt; depl 1800 ACS WG (MATS) Tinker 24Mar56; to 3345 TTA WG (TC) Chanute 14Jun57 mnt – 2578 RFC (CNR) Ellington 15Aug58 – 446 TCWG (AFR) Ellington 20Sep58; 924 TCG (446 TCWG) (AFR) Ellington 31May63 – 906 TCG (302 TCWG) (AFR) Clinton 04Sep65; 907 TCG (302 TCWG) (AFR) Clinton 05Oct65 – Wfu to MASDC (LOG) Davis-Monthan 29Dec65 for storage – DFI: 06May68 – Final Disposition: Scrapped.

10527 / 51-2569 / C-119C-23-FA

Avble: 28Sep51; Acc: 12Oct51; Del: 06Nov51 to USAF – 435 TCWG (TAC) Miami 08Nov51; depl Grenier New Hampshire 03Jan52 – 403 TCWG (FEA) Ashiya AB Japan 13Aug52 – 314 TCWG (FEA) Ashiya AB Japan 52; depl Bofu Japan 18May53 – BMC (AMC) Birmingham 30Nov53 – 483 TCWG (FEA) Ashiya AB Japan 14Sep54 – MIDAR (AMC) Olmsted 31May55 – 82 Fighter GP (ADC) New Castle 07Feb56 – 82 CLM SQ (ADC) New Castle 08Jul57 – 1 MSU GP (ADC) Selfridge 25Nov57 – Hayes (AMC) Birmingham Nov58 – 446 TCWG (AFR) Ellington 29Jan59; to 4756 ADF WG (ADC) Tyndall 12Feb59 mnt; to 2578 AB GP (CNC) Ellington 24Aug60 mnt. Wfu reasons unk – REC: 19Jan61 – Final Disposition: Scrapped.

10528 / 51-2570 / C-119C-23-FA

Avble: 01Oct51; Acc: 26Oct51; Del: 06Nov51 to USAF – 435 TCWG (TAC) Miami 08Nov51; depl Grenier New Hampshire 13Jan52; depl Elmendorf AFB Alaska 24Oct52. Disappeared without trace 15Nov52 while on a flight from NAS Kodiak to Elmendorf AFB Alaska as part of Exercise *Warm Wind*, 5 crew and 15 passengers onboard, pilot: Russell G. Peek – REC: 16Nov52 – DFI: 01Sep54 – Final Disposition: Accident.

10529 / 51-2571 / C-119C-23-FA

Avble: 01Oct51; Acc: 29Oct51; Del: 08Nov51 to USAF – 435 TCWG (TAC) Miami 08Nov51; depl Grenier New Hampshire 20Jan52 – 403 TCWG (FEA) Ashiya AB Japan 13Aug52 – 483 TCWG (FEA) Ashiya AB Japan 01Jan53; depl Bofu Japan 24May53; loan to French Armée de l'Air Indochina 54, French markings applied; depl Clark AB The Philippines 15Apr54; depl 314 TCWG (FEA) Ashiya AB Japan 28Jun54; depl (314 TCWG) Clark AB The Philippines 28Jul54 – 316 TCWG (FEA) Ashiya AB Japan 54 – BMC (AMC) Birmingham Nov54 – 55 SM WG (SAC) Forbes 21Apr55 – 310 Bomb WG (SAC) Smoky Hill 16May55; to 818 AB GP (SAC) Lincoln 19Jun56 – 802 AB GP (SAC) Smoky Hill 01Feb57; to 56 Fighter GP (ADC) O'Hare 11Mar57 mnt; to 340 AB GP (SAC) Whiteman 28Mar58 – 2233 RFC (CNR) Mitchel 14Jul58; to 4238 AB GP (SAC) Barksdale 23Aug58 mnt; to 825 COS GP (SAC) Little Rock 17Oct58 mnt – 839 AB GP (TAC) Sewart Nov58 mnt – 514 TCWG (AFR) Mitchel 11Feb59; 337 TCSq (514 TCWG) (AFR) Bradley 14Mar61; 905 TCG (514 TCWG) (AFR) Bradley 31May63; 903 TCG (514 TCWG) (AFR) McGuire 16Jul65 – Wfu to MASDC (LOG) Davis-Monthan 25Jun66 for storage – DFI: 06May68 – Final Disposition: Scrapped.

10530 / 51-2572 / C-119C-23-FA

Avble: 03Oct51; Acc: 08Nov51; Del: 08Nov51 to USAF – 435 TCWG (TAC) Miami 08Nov51; depl Grenier New Hampshire 02Jan52 – 403 TCWG (FEA) Ashiya AB Japan Aug52 – 483 TCWG (FEA) Ashiya AB Japan 01Jan53; depl Bofu Japan 17Jan53; loan to French

Armée de l'Air Indochina 07May54-25Jun54, French markings applied – 316 TCWG (FEA) Ashiya AB Japan 54 – 567 ADF GP (ADC) McChord 22May55 – 566 ADF GP (ADC) Hamilton 06Jun55 – BMC (AMC) Birmingham 29Jun55 – 78 Fighter GP (ADC) Hamilton 07Sep55; to 2584 RFC (CNR) Memphis 09Nov55; to 3595 CCT WG (TC) Nellis 14Mar56 mnt – 2584 RFC (CNR) Memphis 05Mar58 – 2584 AB SQ (CNC) Memphis 19Dec58 – 758 TCSq (459 TCWG) (AFR) Greater Pittsburgh 17Mar59; 911 TCG (459 TCWG) (AFR) Greater Pittsburgh 17Jan63 – 926 TCG (446 TCWG) (AFR) NAS New Orleans 20Feb63 – Wfu to MASDC (LOG) Davis-Monthan 25Sep65 for storage – REC: 21Jun66 – Final Disposition: Scrapped.

10531 / 51-2573 / C-119C-23-FA

Avble: 04Oct51; Acc: 30Oct51; Del: 08Nov51 to USAF – 435 TCWG (TAC) Miami 08Nov51; depl Grenier New Hampshire 02Jan52; to 6603 AB GP (NEA) Goose Bay Canada 52 mnt – 403 TCWG (FEA) Ashiya AB Japan 02Aug52 – 314 TCWG (FEA) Ashiya AB Japan 21Aug52; depl Tachikawa AB Japan 23Jun53; loan to French Armée de l'Air Indochina 54, French markings applied; depl 483 TCWG (FEA) Clark AB The Philippines 15Apr54; to (483 TCWG) Ashiya AB Japan 26Jun54 – 316 TCWG (FEA) Ashiya AB Japan 54 – BMC (AMC) Birmingham 54 – 814 AB GP (SAC) Fairchild 04Apr55 – 90 SM WG (SAC) Forbes 25Apr55 – 310 Bomb WG (SAC) Smoky Hill 14May56; to 815 AB GP (SAC) Forbes 06Sep56 – 802 AB GP (SAC) Smoky Hill 01Feb57 – 2252 RFC (CNR) Clinton 20Dec57; to 2750 AB WG (AMC) WPAFB 24Dec57 mnt – 302 TCWG (AFR) Clinton 19Mar59; 906 TCG (302 TCWG) (AFR) Clinton 11Feb63; 907 TCG (302 TCWG) (AFR) Clinton 31Jan64 – Wfu 15Dec65 to museum / school status. No record can be found of 51-2573 at any museum, likely bku for school instructional use – Final Disposition: Scrapped.

10532 / 51-2574 / C-119C-23-FA

Avble: 04Oct51; Acc: 30Oct51; Del: 08Nov51 to USAF – 435 TCWG (TAC) Miami 08Nov51; depl Grenier New Hampshire 13Jan52 – 403 TCWG (FEA) Ashiya AB Japan 01Aug52 – 483 TCWG (FEA) Ashiya AB Japan 01Jan53; depl Bofu Japan 24May53; depl Clark AB The Philippines 16Apr54 – 314 TCWG (FEA) Ashiya AB Japan 30Jun54 – 316 TCWG (FEA) Ashiya AB Japan 54 – 564 ADF GP (ADC) Otis 27Mar55 – 810 AB GP (SAC) Biggs 05Aug55 mnt – 33 Fighter GP (ADC) Otis 23Aug55; depl 4750 ADF WG (ADC) Youngstown 24Aug55; to Yuma 28Aug55; to 3800 AB WG (AU) Maxwell 15Sep56 mnt – 4900 AB GP (ARD) Kirtland 23Sep56 mnt – 33 MSU GP (ADC) Otis 18Oct56 – 551 MSU GP (ADC) Otis 18Aug57 – 551 AEW WG (ADC) Otis 01Sep57 – 2578 RFC (CNR) Ellington 23Apr58; 2578 RFC DT (CNR) Alvin Callender Field 58 – 357 TCSq (446 TCG) (AFR) Alvin Callender Field 20Sep58; to NAS New Orleans 01Dec58; 706 TCSq (446 TCWG) (AFR) NAS New Orleans 08May61; 926 TCG (446 TCWG) (AFR) NAS New Orleans 17Jan63 – 2850 AB GP (LOG) Brookley 13May65 – MOAAR (LOG) Brookley 22May65 – DFI: 27May65 – Final Disposition: Scrapped.

10533 / 51-2575 / C-119C-23-FA

Avble: 08Oct51; Acc: 30Oct51; Del: 08Nov51 to USAF – 435 TCWG (TAC) Miami 08Nov51; depl Grenier New Hampshire 13Jan52; to 6530 AB WG (ARD) Griffiss 12Feb52 mnt; to ROD Center (ARD) Griffiss 01Aug52 mnt – 456 TCWG (TAC) Miami IAP 01Dec52 – 483 TCWG (FEA) Ashiya AB Japan 16Feb53; depl 314 TCWG (FEA) Ashiya AB Japan 53; depl (314 TCWG) Tachikawa AB Japan 23Jun53; loan to French Armée de l'Air Indochina 17Apr54-25Jun54, French markings applied – 316 TCWG (FEA) Ashiya AB Japan 54 – 567 ADF GP (ADC) McChord 02May55 – BMC (AMC) Birmingham 10Aug55 – 325 Fighter GP (ADC) McChord 03Nov55 – 325 MSU GP (ADC) McChord 30Apr58 – 446 TCWG

(AFR) Ellington 29Sep58; 924 TCG (446 TCWG) (AFR) Ellington 31May63 – Wfu to REC 930 TCG (434 TCWG) (AFR) Bakalar 01Sep65, noted derelict at Bakalar 1969 – Final Disposition: Scrapped.

10534 / 51-2576 / C-119C-23-FA

Avble: 12Oct51; Acc: 31Oct51; Del: 08Nov51 to USAF – 435 TCWG (TAC) Miami 08Nov51; depl Grenier New Hampshire 14Feb52 – 403 TCWG (FEA) Ashiya AB Japan 52 – 483 TCWG (FEA) Ashiya AB Japan 01Jan53; depl 314 TCWG (FEA) Ashiya AB Japan 14Jan54; depl Clark AB The Philippines 14Apr54; depl (314 TCWG) Clark AB The Philippines 23Jul54; loan to French Armée de l'Air Indochina 26-27Jul54, French markings applied – 316 TCWG (FEA) Ashiya AB Japan 54 – 567 ADF GP (ADC) McChord 17May55 – 325 Fighter GP (ADC) McChord 18Aug55 – 325 MSU GP (ADC) McChord 30Apr58 – ODGAR (AMC) Hill 03Jun58 mnt at Wendover – 446 TCWG (AFR) Ellington 01Dec58; 925 TCG (446 TCWG) (AFR) Ellington 31May63 – Wfu to MASDC (LOG) Davis-Monthan 18Sep65 for storage – REC: 21Jun66 – Final Disposition: Scrapped.

10535 / 51-2577 / C-119C-23-FA

Avble: 15Oct51; Acc: 29Oct51; Del: 08Nov51 to USAF – 435 TCWG (TAC) Miami 08Nov51; depl Grenier New Hampshire 16Jan52 – 403 TCWG (FEA) Ashiya AB Japan 52 – 314 TCWG (FEA) Ashiya AB Japan 52; depl 483 TCWG (FEA) Clark AB The Philippines 13Apr54 – BMC (AMC) Birmingham 54 – 5 AB GP (SAC) Travis 54-55 – 807 AB GP (SAC) March 01Mar55 mnt – 303 Bomb WG (SAC) Davis- Monthan 31Mar55; to 4900 AB GP (ARD) Kirtland 14Sep55 mnt – 43 Bomb WG (SAC) Davis-Monthan 01Jul56 – 5 AB GP (SAC) Travis 04Oct56 – 2577 RFC (CNR) Brooks 22Mar58 – 433 TCWG (AFR) Brooks 18Sep58; 69 TCSq (433 TCWG) (AFR) NAS Dallas 01Mar59; to Kelly 31Aug60; 922 TCG (433 TCWG) (AFR) Kelly 31May63; 921 TCG (433 TCWG) (AFR) Kelly 10Oct65. Lost to operator outside USAF 25Feb66 – Final Disposition: Unknown.

10536 / 51-2578 / C-119C-23-FA

Avble: 16Oct51; Acc: 31Oct51; Del: 08Nov51 to USAF – 435 TCWG (TAC) Miami 08Nov51; to WRDCN (ARD) WPAFB Dec51; to 2750 AB WG (AMC) WPAFB 12Feb52 mnt – 403 TCWG (FEA) Ashiya AB Japan 52 – 483 TCWG (FEA) Ashiya AB Japan 01Jan53; depl Bofu Japan 17Nov53; loan to French Armée de l'Air Indochina 17Apr54-03Jun54, French markings applied – 581 ASL GP (MATS) Kadena AB Japan 54; depl 8 Fighter-Bomber WG (FEA) Itazuke AB Japan 14Aug56 – 1 Fighter GP (ADC) Selfridge 06Sep56 – 1 MSU GP (ADC) Selfridge 18Oct56; to 3800 AB WG (AU) Maxwell 20May57 mnt; to 3635 FTA WG (ATC) Stead 13Aug58 mnt – 446 TCWG (AFR) Ellington 26Dec58; to 839 AB GP (TAC) Sewart 01Sep60 mnt; 924 TCG (446 TCWG) (AFR) Ellington 31May63 – Wfu to MASDC (LOG) Davis-Monthan 24Aug65 for storage – REC: 21Jun66 – Final Disposition: Scrapped.

10537 / 51-2579 / C-119C-23-FA

Avble: 18Oct51; Acc: 31Oct51; Del: 08Nov51 to USAF – 435 TCWG (TAC) Miami 08Nov51; depl Grenier New Hampshire 17Jan52; depl Elmendorf Alaska 24Oct52 – 456 TCWG (TAC) Miami 01Dec52 – 64 TCWG (TAC) Donaldson 02Jul53 – 1739 FRY SQ (MATS) Amarillo 17Sep53 – 1 IM SQ (MATS) Tinker 25Mar54 – 1881 IM SQ (MATS) Tinker 16Mar55; to 3310 TTA WG (TC) Scott 27Apr56 mnt – 1800 ACS WG (MATS) Tinker 28Oct57 – COAAR (MATS) Tinker 01Nov57 – 2596 RFC (CNR) NAS Dallas 20Nov57 – 2578 RFC (CNR) Ellington Nov57; depl 2577 RFC DT (CNR) NAS Dallas 01Dec57 –

446 TCWG (AFR) Ellington 25Sep58; 357 TCSq (446 TCG) (AFR) Alvin Callender Field 25Sep58; to NAS New Orleans 01Dec58; 706 TCSq (446 TCWG) (AFR) NAS New Orleans 08May61; 926 TCG (446 TCWG) (AFR) NAS New Orleans 17Jan63 – Wfu to MASDC (LOG) Davis-Monthan 23Oct65 for storage – DFI: 06May68 – Final Disposition: Scrapped.

10538 / 51-2580 / C-119C-23-FA

Avble: 22Oct51; Acc: 21Nov51; Del: 30Nov51 to USAF – 581 ARC WG (MATS) Mt. Home 30Nov51; to Clark AB The Philippines 10Jul52; depl 403 TCWG (FEA) Ashiya AB Japan 01Nov52; depl 483 TCWG (FEA) Ashiya AB Japan 01Jan53; to Kadena AB Japan 54 – 92 Bomb WG (SAC) Fairchild 20Feb55 – 814 AB GP (SAC) Fairchild 25Jul56 – 92 AB GP (SAC) Fairchild 04Sep56; to 812 AB GP (SAC) Fairchild 08Dec56 – 2577 RFC (CNR) Brooks 01Apr58 – 1611 TSM WG (MATS) McGuire 21Jul58 – 433 TCWG (AFR) Brooks 31Jan59; to 2584 AB WG (CNC) Memphis 04Feb59 mnt; to Kelly 31Aug60; to 306 COS GP (SAC) MacDill 15Nov60 mnt; 922 TCG (433 TCWG) (AFR) Kelly 17Jan63; 921 TCG (433 TCWG) (AFR) Kelly 10Oct65 – 903 TCG (514 TCWG) (AFR) McGuire 16Feb66 – Wfu to MASDC (LOG) Davis-Monthan 13May66 for storage – DFI: 06May68 – Final Disposition: Scrapped.

10539 / 51-2581 / C-119C-23-FA

Avble: 25Oct51; Acc: 21Nov51; Del: 29Nov51 to USAF – 435 TCWG (TAC) Miami 29Nov51; depl Grenier New Hampshire 15Jan52; depl Elmendorf 24Oct52 – 456 TCWG (TAC) Miami 01Dec52 – 483 TCWG (FEA) Ashiya AB Japan 17Feb53; loan to French Armée de l'Air Indochina 03May53- 24Jul53, French markings applied callsign: F-OYXE; to 24 MND GP (FEA) Clark AB The Philippines 24Jul53 mnt; depl 314 TCWG (FEA) Ashiya AB Japan 16Apr54; loan to French Armée de l'Air Indochina 16-26Apr54, French markings applied; depl (314 TCWG) Clark AB The Philippines 29Aug54 – 316 TCWG (FEA) Ashiya AB Japan 54 – BMC (AMC) Birmingham 23Feb55 – 71 Fighter SQ (500 ADF GP) (ADC) Greater Pittsburgh 06Jun55 – bailment Fairchild (AMC) Hagerstown 04Aug55 – 54 Fighter GP (ADC) Greater Pittsburgh 20Sep55 – 551 AEW WG (ADC) Otis 03Oct57 – 2578 RFC (CNR) Ellington 06May58; 2578 RFC DT (CNR) Alvin Callender Field 02Oct58 – 357 TCSq (446 TCG) (AFR) Alvin Callender Field 01Aug58; to NAS New Orleans 01Dec58; 706 TCSq (446 TCWG) (AFR) NAS New Orleans 08May61; 926 TCG (446 TCWG) (AFR) NAS New Orleans 17Jan63 – Wfu to MASDC (LOG) Davis-Monthan 21Aug65 for storage – REC: 21Jun66 – Final Disposition: Scrapped.

10540 / 51-2582 / C-119C-23-FA

Avble: 30Oct51; Acc: 21Nov51; Del: 21Nov51 to USAF – bailment Kaiser-Frazer (AMC) Willow Run 21Nov51 – WRDCN (ARD) WPAFB 16May52; cvtd **EC-119C** 09Sep52; to 5064 CMT SQ (AAC) Ladd 24Oct52 for test duties; cvtd **C-119C** 07Feb53 – 580 ASL GP (MATS) Wheelus AB Libya 05Dec53 – 9 Bomb WG (SAC) Mt. Home 13Jul55; to 803 AB GP (SAC) Davis-Monthan 20Apr56 – 479 Fighter GP (TAC) George 13Jul56 mnt – 9 AB GP (SAC) Mt. Home 25Sep56 – 2252 RFC (CNR) Clinton 02Apr58 – 302 TCWG (AFR) Clinton 19Mar59; 906 TCG (302 TCWG) (AFR) Clinton 11Feb63; 907 TCG (302 TCWG) (AFR) Clinton 15Nov65 – Wfu to MASDC (LOG) Davis-Monthan 11Dec65 for storage – DFI: 06May68 – Final Disposition: Scrapped.

10541 / 51-2583 / C-119C-23-FA

Avble: 01Nov51; Acc: 30Nov51; Del: 05Jan52 to USAF – 435 TCWG (TAC) Miami 04Jan52; depl Grenier New Hampshire 13Jan52 – 403 TCWG (FEA) Ashiya AB Japan 21Jul52 –

483 TCWG (FEA) Ashiya AB Japan 01Jan53; to 19 Bomb WG (FEA) Anderson 14Apr53; loan to French Armée de l'Air Indochina 03May–08Jul53, French markings applied callsign: F-OYXF; to 24 MND GP (FEA) Clark AB The Philippines 08Jul53 mnt; depl Clark AB The Philippines 14Apr54; depl 314 TCWG (FEA) Ashiya AB Japan 01Jul54; depl (314 TCWG) Clark AB The Philippines 31Jul54 – 316 TCWG (FEA) Ashiya AB Japan Nov54 – BMC (AMC) Birmingham Nov54 – 93 Bomb WG (SAC) Castle 25Apr55; to 803 AB GP (SAC) Davis-Monthan 07Jun55; to 62 TCWG (TAC) Larson 10Oct55 mnt; to 93 AB WG (SAC) Castle 28Nov55; to 6515 MAI GP (ARD) Edwards 06Dec55 mnt – 812 AB GP (SAC) Walker 12Jul56; to 96 AB GP (SAC) Altus 26Nov56; to 1405 AB WG (MATS) Scott 21Jan58 – 2584 RFC (CNR) Memphis 07May58 – 446 TCWG (AFR) Ellington 12Oct58; 357 TCSq (446 TCWG) (AFR) NAS New Orleans 20Jun59; 706 TCSq (446 TCWG) (AFR) NAS New Orleans 08May61; 926 TCG (446 TCWG) (AFR) NAS New Orleans 17Jan63 – Wfu to MASDC (LOG) Davis-Monthan 28Aug65 for storage – REC: 21Jun66 – Final Disposition: Scrapped.

10542 / 51-2584 / C-119C-23-FA

Avble: 06Nov51; Acc: 30Nov51; Del: 05Jan52 to USAF – 435 TCWG (TAC) Miami 04Jan52; depl Grenier New Hampshire 03Jan52; 403 TCWG (FEA) Ashiya AB Japan 21Jul52 – 483 TCWG (FEA) Ashiya AB Japan 01Jan53; depl Bofu Japan 24May53; depl Clark AB The Philippines 14Apr54; depl 314 TCWG (FEA) Ashiya AB Japan 21Jul54 – 316 TCWG (FEA) Ashiya AB Japan Nov 54 – BMC (AMC) Birmingham 54-55 – 376 Bomb WG (SAC) Barksdale 02Jun55; to 807 AB GP (SAC) March 04Jul55; to 805 AB GP (SAC) Barksdale 55 – 806 AB GP (SAC) Lake Charles 30Sep55 – 506 Fighter WG (SAC) Tinker 26Apr56 – 44 Bomb WG (SAC) Lake Charles 20Aug56 – 384 Bomb WG (SAC) Little Rock 14Nov56 – 814 AB GP (SAC) Westover 14Mar57 – 825 AB GP (SAC) Little Rock 13Apr57; to 805 AB GP (SAC) Barksdale 09May57; to 384 Bomb WG (SAC) Little Rock 27May57 – 2577 RFC DT (CNR) NAS Dallas 03Dec57 – 69 TCSq (433 TCWG) (AFR) NAS Dallas 20Sep58; 923 TCG (433 TCWG) (AFR) NAS Dallas 17Jan63; to Carswell 02Mar63 – 906 TCG (302 TCWG) (AFR) Clinton 18Sep65; 907 TCG (302 TCWG) (AFR) Clinton 05Oct65 – Wfu to MASDC (LOG) Davis-Monthan 29Dec65 for storage – DFI: 06May68 – Final Disposition: Scrapped.

10543 / BuNo. 126574 / R4Q-1

Avble: 07Nov51; Acc: 30Nov51; Del: 11Dec51 to USMC – NATC NAS Patuxent River 10Dec51 for tests – O&R MCAS Cherry Point 22Jan52 mnt – VMR-252 MCAS Cherry Point 21Mar52 – MTG-10 MCAS El Toro 25Jun53 – VMR-253 MCAF Itami Japan 24Nov53; to MCAF Iwakuni Japan 31Jan55 – Wfu to NAF Litchfield Park 17Jul59 for storage – DFI: 13May60 – Final Disposition: Scrapped.

10544 / BuNo. 126575 / R4Q-1

Avble: 08Nov51; Acc: 30Nov51; Del: 11Dec51 to USMC – O&R MCAS Cherry Point 10Dec51 mnt – VMR-252 MCAS Cherry Point 21Mar52 – AES-12 MCAS Quantico 12May53 – MTG-10 MCAS El Toro 29Oct53 – HEDRON-25 MCAS El Toro 01Mar54 – O&R NAS Corpus Christi 25Mar55 mnt – VMR-253 MCAF Iwakuni Japan 28Dec55 – Wfu to NAF Litchfield Park 17Mar59 for storage – DFI: 13May60 – Final Disposition: Scrapped.

10545 / BuNo. 126576 / R4Q-1

Avble: 13Nov51; Acc: 30Nov51; Del: 11Dec51 to USMC – O&R MCAS Cherry Point 11Dec51 mnt – VMR-252 MCAS Cherry Point 19Mar52 – MTG-10 MCAS El Toro 15Jun53 – HEDRON-25 MCAS El Toro 13Mar54 – H&MS MAW-1 Pohang AB (K-3) S. Korea 16Sep54 –

VMR-253 MCAF Iwakuni Japan 31Aug55 – Wfu to NAF Litchfield Park 18Dec58 for storage – DFI: 13May60 – Final Disposition: Scrapped.

10546 / BuNo. 126577 / R4Q-1

Avble: 13Nov51; Acc: 30Nov51; Del: 13Dec51 to USMC – O&R MCAS Cherry Point 13Dec51 mnt – VMR-252 MCAS Cherry Point 21Mar52 – MTG-10 MCAS El Toro 18May53 – VMR-253 MCAF Itami Japan 19Dec53; to MCAF Iwakuni Japan 10May55 – Wfu to NAF Litchfield Park 18Jun59 for storage – DFI: 13May60 – Final Disposition: Scrapped.

10547 / BuNo. 126578 / R4Q-1

Avble: 13Nov51; Acc: 30Nov51; Del: 13Dec51 to USMC – O&R MCAS Cherry Point 13Dec51 mnt – AirFMLANT MCAS Cherry Point 21Apr52 – VMR-252 MCAS Cherry Point 05Jun52 – HEDRON-2 MCAS Cherry Point 29May53 – MARS-27 MCAS Cherry Point 02Sep53 – HAMRON-14 MCAS Edenton 28Oct53 – H&MS-25 MCAS El Toro 17Aug54 – VMR-253 MCAF Iwakuni Japan 10Mar55 – Wfu to NAF Litchfield Park 02Feb59 for storage – DFI: 13May60 – Final Disposition: Scrapped.

10548 / BuNo. 126579 / R4Q-1

Avble: 16Nov51; Acc: 30Nov51; Del: 13Dec51 to USMC – O&R MCAS Cherry Point 13Dec51 mnt – VMR-252 MCAS Cherry Point 18Mar52 – MTG-10 MCAS El Toro 19Jun53 – VMR-253 MCAF Itami Japan 22Dec53. While on a routine flight from Kunsan AB (K-8) S. Korea to MCAF Itami Japan on 04Jun54 at about 22.5 miles off the coast of S. Korea the left propeller failed producing an imbalance so violent that it shook the left engine off its mountings. A bail out order was issued and 10 of the 11 onboard parachuted into the sea. Aircraft Commander Maj Gene M. Badgley stayed onboard but was killed when the aircraft impacted the water. 6 of the 10 that parachuted were later rescued with 3 of the 4 lost being: Capt Rupert Nelson, SSgt William McBride and civilian contractor Donald W. Eastridge – Final Disposition: Accident.

10549 / BuNo. 126580 / R4Q-1

Avble: 20Nov51; Acc: 30Nov51; Del: 04Jan52 to USMC – O&R MCAS Cherry Point 07Jan52 mnt – VMR-252 MCAS Cherry Point 21Mar52 – MTG-10 MCAS El Toro 08Jun53 – HEDRON-25 MCAS El Toro 13Mar54 – H&MS-25 MCAS El Toro 09Apr54 – VMR-253 MCAF Itami Japan 08Nov54; to MCAF Iwakuni Japan 25Jan55 – Wfu to NAF Litchfield Park 02Feb59 for storage – DFI: 13May60 – Film appearance *The Flight of the Phoenix* (1965). Wings, nacelles and boom sections used for a ground based C-82 mock-up, parts acquired and readied through Allied Aircraft, Inc. Tucson Arizona early 1965; bku after filming – Final Disposition: Scrapped.

10550 / BuNo. 126581 / R4Q-1

Avble: 19Nov51; Acc: 30Nov51; Del: 04Jan52 to USMC – O&R MCAS Cherry Point 07Jan52 mnt – VMR-252 MCAS Cherry Point 19Mar52 – MTG-10 MCAS El Toro 10Jun53 – VMR-253 MCAS El Toro 13Jul53; depl MTG-10 MCAS El Toro 05Aug53; to MCAF Itami Japan 27Nov53; to MCAF Iwakuni Japan 05Feb55; to O&R NAS Corpus Christi 16Feb55 mnt; depl H&MS-25 MCAS El Toro 13May55 – Wfu to NAF Litchfield Park 26May59 for storage – DFI: 13May60 – Final Disposition: Scrapped.

10551 / BuNo. 126582 / R4Q-1

Avble: 23Nov51; Acc: 30Nov51; Del: 19Dec51 to USMC – VMR-253 MCAS El Toro 22Dec51 – MTG-10 MCAS El Toro 19Aug53 – O&R MCAS Cherry Point 21Sep53 mnt – NATC NAS Patuxent River 05Feb54 for tests – O&R NAS Corpus Christi & NAS San Diego 31Aug54 mnt – VMR-253 MCAF Iwakuni Japan 12Mar57 – Wfu to NAF Litchfield Park 10Jul59 for storage – DFI: 13May60 – Final Disposition: Scrapped.

10552 / BuNo. 128723 / R4Q-1

Avble: 27Nov51; Acc: 21Dec51; Del: 14Feb52 to USMC – VMR-253 MCAS El Toro 13Feb52; to MCAF Itami Japan 20Oct53 – O&R NAS San Diego & MCAS Cherry Point 04Jun54 mnt – 5 Naval District NAS Norfolk 01Oct56 – O&R NAS Corpus Christi 23Jul58 mnt – VMR-253 MCAF Iwakuni Japan 15Jan59 – O&R NAS Corpus Christi 13Feb59 mnt – Wfu to NAF Litchfield Park 15May59 for storage – DFI: 13May60 – Final Disposition: Scrapped.

10553 / BuNo. 128724 / R4Q-1

Avble: 29Nov51; Acc: 21Dec51; Del: 14Feb52 to USMC – VMR-253 MCAS El Toro 14Feb52; to MCAF Itami Japan 07Dec53; to MCAF Iwakuni Japan 17Jan55 – O&R NAS Corpus Christi 01Feb55 mnt – H&MS-25 MCAS El Toro 23May55 – MARS-37 MCAS El Toro 16Sep55 – VMR-253 MCAF Iwakuni Japan 29Feb56 – Wfu to NAF Litchfield Park 23Jun59 for storage – DFI: 13May60 – Final Disposition: Scrapped.

10554 / BuNo. 128725 / R4Q-1

Avble: 30Nov51; Acc: 21Dec51; Del: 14Feb52 to USMC – VMR-253 MCAS El Toro 13Feb52; to MCAF Itami Japan 02Dec53; to MCAF Iwakuni Japan 17Jan55; to O&R NAS Corpus Christi 01Feb55 mnt – Wfu to NAF Litchfield Park 12Jun59 for storage – DFI: 13May60 – Final Disposition: Scrapped.

10555 / BuNo. 128726 / R4Q-1

Avble: 03Dec51; Acc: 21Dec51; Del: 14Feb52 to USMC – VMR-253 MCAS El Toro 14Feb52; to MCAF Itami Japan 30Nov53 – O&R NAS San Diego & MCAS Cherry Point 07Dec53 mnt – H&MS-25 MCAS El Toro 31Jan55 – VMR-253 MCAF Iwakuni Japan 09Mar55. Listed as a crash 16May56, nfd – Final Disposition: Accident.

10556 / BuNo. 128727 / R4Q-1

Avble: 05Dec51; Acc: 26Dec51; Del: 14Feb52 to USMC – VMR-253 MCAS El Toro 14Feb52; to MCAF Itami Japan 07Dec53 – O&R NAS San Diego & MCAS Cherry Point 21Dec53 mnt – 11 Naval District NAS North Island 07May58 – Wfu to NAF Litchfield Park 12Mar59 for storage – DFI: 13May60 – Final Disposition: Scrapped.

10557 / BuNo. 128728 / R4Q-1

Avble: 07Dec51; Acc: 26Dec51; Del: 14Feb52 to USMC – VMR-253 MCAS El Toro 14Feb52; to MCAF Itami Japan 01Sep53; to MCAF Iwakuni Japan 25May55 – Wfu to NAF Litchfield Park 10Jul59 for storage – DFI: 13May60 – Final Disposition: Scrapped.

10558 / BuNo. 128729 / R4Q-1

Avble: 07Dec51; Acc: 27Dec51; Del: 14Feb52 to USMC – VMR-253 MCAS El Toro 14Feb52; to MCAF Itami Japan 08Nov53 – O&R NAS San Diego & MCAS Cherry Point

10Dec53 mnt – 12 Naval District NAS Alameda 02Jul58 – Wfu to NAF Litchfield Park 20Mar59 for storage – DFI: 13May60 – Final Disposition: Scrapped.

10559 / BuNo. 128730 / R4Q-1

Avble: 11Dec51; Acc: 27Dec51; Del: 14Feb52 to USMC – VMR-253 MCAS El Toro 15Feb52; to MCAF Itami Japan 04Dec53 – O&R MCAS Cherry Point & NAS San Diego 09Dec53 mnt – H&MS-25 MCAS El Toro 04Feb55 – VMR-253 MCAF Iwakuni Japan 23Mar55 – Wfu to NAF Litchfield Park 30Jun59 for storage – DFI: 13May60 – Final Disposition: Scrapped.

10560 / BuNo. 128731 / R4Q-1

Avble: 10Dec51; Acc: 27Dec51; Del: 19Feb52 to USMC – VMR-253 MCAS El Toro 19Feb52 – O&R MCAS Cherry Point & NAS San Diego 11Aug52 mnt – VMR-253 Itami Japan 06Dec54; to MCAF Iwakuni Japan 13Aug55 – Wfu to NAF Litchfield Park 19Dec58 for storage – DFI: 13May60 – Final Disposition: Scrapped.

10561 / BuNo. 128732 / R4Q-1

Avble: 13Dec51; Acc: 27Dec51; Del: 19Feb52 to USMC – VMR-253 MCAS El Toro 19Feb52; to MCAF Itami Japan 07Dec53 – O&R NAS San Diego & MCAS Cherry Point 19Oct54 mnt – Wfu to NAF Litchfield Park 29Nov59 for storage – DFI: 13May60 – Final Disposition: Scrapped.

10562 / BuNo. 128733 / R4Q-1

Avble: 17Dec51; Acc: 29Dec51; Del: 19Feb52 to USMC – VMR-253 MCAS El Toro 19Feb52; to MCAF Itami Japan 14Sep53; to MCAF Iwakuni Japan 01Jan55 – Wfu to NAF Litchfield Park 04May59 for storage – DFI: 13May60 – Final Disposition: Scrapped.

10563 / BuNo. 128734 / R4Q-1

Avble: 17Dec51; Acc: 29Dec51; Del: 19Feb52 to USMC – VMR-253 MCAS El Toro 19Feb52; to MCAF Itami Japan 07Dec53 – O&R NAS San Diego & MCAS Cherry Point 13Sep54 mnt – Wfu to NAF Litchfield Park 19Jan59 for storage – DFI: 13May60 – Final Disposition: Scrapped.

10564 / BuNo. 128735 / R4Q-1

Avble: 17Dec51; Acc: 29Dec51; Del: 19Feb52 to USMC – VMR-253 MCAS El Toro 19Feb52; to MCAF Itami Japan 07Dec53 – O&R NAS San Diego & MCAS Cherry Point 15Dec53 mnt – NATC NAS Patuxent River 12Jan57 for tests – O&R MCAS Cherry Point 24Apr57 mnt – 6 Naval District NAS Jacksonville 23Jan58 – Wfu to NAF Litchfield Park 27Oct59 for storage – DFI: 13May60 – Final Disposition: Scrapped.

10565 / BuNo. 128736 / R4Q-1

Avble: 18Dec51; Acc: 29Dec51; Del: 19Feb52 to USMC – VMR-253 MCAS El Toro 19Feb52; to MCAF Itami Japan 31Oct53 – O&R NAS San Diego & MCAS Cherry Point 15Sep54 mnt – SO&ES MCAS Cherry Point 08Sep58 – Wfu to NAF Litchfield Park 25Mar59 for storage – DFI: 13May60 – Final Disposition: Scrapped.

10566 / BuNo. 128737 / R4Q-1

Avble: 26Dec51; Acc: 28Jan52; Del: 18Mar52 to USMC – VMR-252 MCAS Cherry Point 02Apr52 – MTG-10 MCAS El Toro 18May53; depl HAMRON-15 MCAS El Toro

22May53 – HEDRON-25 MCAS El Toro 12Jan54 – O&R NAS Corpus Christi 11Jan55 mnt – H&MS-25 MCAS El Toro 11Mar55 – VMR-253 MCAF Iwakuni Japan 10Apr55 – Wfu to NAF Litchfield Park 02Feb59 for storage – DFI: 13May60 – Final Disposition: Scrapped.

10567 / BuNo. 128738 / R4Q-1

Avble: 29Dec51; Acc: 28Jan52; Del: 18Mar52 to USMC – VMR-252 MCAS Cherry Point 03Apr52 – MTG-10 MCAS El Toro 08Jun53 – VMR-253 Itami Japan 24Nov53; to MCAF Iwakuni Japan 23Feb55; to O&R NAS Corpus Christi 03Mar55 mnt; to MCAF Iwakuni Japan 24Sep55. Made an emergency belly-landing at MCAF Iwakuni 03May58 after a landing gear malfunction, repaired – Wfu to NAF Litchfield Park 17Mar59 for storage – DFI: 13May60 – Final Disposition: Scrapped.

10568 / BuNo. 128739 / R4Q-1

Avble: 29Dec51; Acc: 28Jan52; Del: 18Mar52 to USMC – AirFMFLANT MCAS Cherry Point 04Apr52 – VMR-252 MCAS Cherry Point 23Jun52 – MTG-10 MCAS El Toro 28May53; depl HAMRON-15 MCAS El Toro 16Jul53 – H&MS-25 MCAS El Toro 13Mar54 – O&R NAS Corpus Christi 09Mar55 mnt – NABTC NAS Pensacola 12Oct55 – O&R NAS Corpus Christi 13Nov57 mnt – NAATC NAS Corpus Christi 01Jul58 – Wfu to NAF Litchfield Park 31May59 for storage – DFI: 13May60 – Final Disposition: Scrapped.

10569 / BuNo. 128740 / R4Q-1

Avble: 08Jan52; Acc: 29Jan52; Del: 20Mar52 to USMC – AirFMFLANT MCAS Cherry Point 04Apr52 – VMR-252 MCAS Cherry Point 23Jun52 – MTG-10 MCAS El Toro 19Jun53 – VMR-253 MCAS El Toro 11Aug53; to MCAF Itami Japan 21Oct53; to MCAF Iwakuni Japan 25May55 – Wfu to NAF Litchfield Park 17Mar59 for storage – DFI: 13May60 – Final Disposition: Scrapped.

10570 / BuNo. 128741 / R4Q-1

Avble: 05Jan52; Acc: 29Jan52; Del: 20Mar52 to USMC – AirFMFLANT MCAS Cherry Point 08Apr52 – VMR-252 MCAS Cherry Point 05Jun52 – MTG-10 MCAS El Toro 08Jun53 – VMR-253 MCAF Itami Japan 28Oct53; to MCAF Iwakuni Japan 01Jan55. Mid-air collision with Douglas AD-6 (BuNo. 135350) while on approach to land at Okinawa-Naha AFB Japan 07Mar58, both aircraft had been flying in formation from The Philippines. It was an instrument approach with the AD-6 having comms or nav problems so the pilot elected to fly wing on the R4Q-1 for the approach. It is believed the cause was mainly due to the failure of the crew on both aircraft to maintain a safe flying distance and that the Skyraider pilot may have lost control of his aircraft due to the turbulent weather conditions. Both aircraft fell into the ocean some miles from their destination killing all 7 crew and 19 passengers on the R4Q-1 and the pilot in the Skyraider. 8 of the fatalities are known to be: 1Lt Thomas A. Annexstad, 1Lt Lowell E. Hendrick, Jr., SSgt Lyle D. DeYoung, SSgt Thomas F. Palmer, Jr., SSgt Royce Taliaferro, TSgt John F. Maher, TSgt Ernst A. Sohn and Sgt Eugene W. Christy. All bodies and the wreckage were later recovered in a valiant salvage operation – DFI: 15Mar58 – Final Disposition: Accident.

10571 / BuNo. 128742 / R4Q-1

Avble: 09Jan52; Acc: 30Jan52; Del: 20Mar52 to USMC – AirFMFLANT MCAS Cherry Point 04Apr52 – VMR-252 MCAS Cherry Point 05Jun52; to MCAF New River Jacksonville 15Jul53 – O&R MCAS Cherry Point 21Oct54 mnt – Transport Pool NAS Corpus Christi

02Jun55 – NAATC NAS Corpus Christi 31Aug55 – O&R NAS Corpus Christi 24Apr56 mnt – Wfu to NAF Litchfield Park 09May56 for storage – O&R NAS Corpus Christi 19Aug58 mnt – Wfu to NAF Litchfield Park 06Mar59 for storage – DFI: 13May60 – Final Disposition: Scrapped.

10572 / BuNo. 128743 / R4Q-1

Avble: 10Jan52; Acc: 31Jan52; Del: 20Mar52 to USMC – AirFMFLANT MCAS Cherry Point 08Apr52 – VMR-252 MCAS Cherry Point 23Jun52 – HAMRON-26 MCAS Cherry Point 08Jun53 – HAMRON-14 MCAS Edenton 01Oct53 – H&MS-25 MCAS El Toro 17Aug54; to O&R NAS Corpus Christi 05Jan55 mnt – VMR-253 MCAF Iwakuni Japan 01Jul55 – Wfu to NAF Litchfield Park 26May59 for storage – DFI: 13May60 – Final Disposition: Scrapped.

10573 / BuNo. 128744 / R4Q-1

Avble: 14Jan52; Acc: 31Jan52; Del: 16Feb52 diverted to The Martin Aircraft Co. Baltimore Maryland under AMC (USAF) / BAR (USN) 17Feb52 for project – O&R MCAS Cherry Point 18Jun52mnt – VMR-253 MCAS El Toro 20Mar53; to MCAF Itami Japan 25Nov53; to MCAF Iwakuni Japan 25May55. Listed as damaged beyond repair in an incident at MCAF Iwakuni Japan 11May56, nfd – DFI: 22May56 – Final Disposition: Accident.

10574 / 51-2585 / YC-119H-FA Skyvan

Avble: 29May52; Acc: 31Oct52; Del: 28Nov52 to USAF. Experimental long-span heavyweight prototype. WRDCN (ARD) WPAFB 28Nov52 for test duties, redes **EYC-119H** 14May53 – Bailment Fairchild (AMC) Hagerstown 28May53 for test duties – OKLAR (AMC) Tinker 19Dec53 – Wfu to 3750 TTN WG (ATC) Sheppard 22Sep54 redes **YC-119H**, assigned to ground instruction – Final Disposition: Scrapped.

10575 / 51-2586 / C-119F-FA / (YC-119F-FA)

Avble: 07Jan52; Acc: 29Feb52; Del: 20Jun52 to USAF. Unofficially designated as **YC-119F** prototype. Bailment Fairchild (AMC) Hagerstown early 52 for test duties – 6510 AB WG (ARD) Edwards 14Jul54 – 6511 PDT GP (ARD) NAS El Centro 19Jul54; to Fairchild (AMC) Hagerstown 05Jun56, cvtd **C-119G** 29Jun56 – BMC (AMC) Birmingham cvtd **JC-119G** 23Aug56 – 6511 MAI GP (ARD) NAS El Centro 30Oct56 – AFT Center (ARD) NAS El Centro (depl) cvtd **C-119G** 01Apr57; to Edwards 01Jul57; depl NAS El Centro 19Sep57; TV appearance *Steve Canyon* (1958) episode 6 *Operation Survival* filmed at El Centro 9-11Sep58; to Edwards Jul59; cvtd **JC-119G** 59; cvtd **C-119G** 25Jan60 – 452 TCWG (AFR) March 11Jan61; 942 TCG (452 TCWG) (AFR) March 11Feb63 – 904 TCG (514 TCWG) (AFR) Stewart 04Jun65; 903 TCG (514 TCWG) (AFR) McGuire 30Jul67 – 928 TAL GP (403 TCWG) (AFR) O'Hare 08May68 – MAP Taiwan 10Apr69 – Final Disposition: Unknown.

10576 / 51-2587 / C-119C-25-FA / (C-119CF)

Avble: 21Jan52; Acc: 31Jan52; Del: 27Mar52 to USAF – 443 TCWG (TAC) Greenville 27Mar52; depl Ladd Alaska 29Oct52 – 465 TCWG (TAC) Donaldson 25Aug53 – 465 TCWG (AFE) Neubiberg AB W. Germany 02Dec53; to Toul-Rosières AB France 16Apr54; depls Landstuhl W. Germany 01Mar55 & 04Apr55; to Évreux-Fauville AB France 18May55; depl unit unk (AFE) Rhein-Main AB W. Germany 56; to 7100 SUT WG (AFE) Wiesbaden AB W. Germany 28Apr57 – 317 TCWG (AFE) Évreux-Fauville AB France 08Jul57 – 2578 RFC (CNR) Ellington 18Feb58; 2578 RFC DT (CNR) Alvin Callender Field 58 – 357 TCSq (446 TCG) (AFR) Alvin Callender Field 20Sep58; to NAS New Orleans 01Dec58; to 2750

AB WG (AMC) WPAFB 02Apr59; to 53 MAT SQ (ADC) Sioux City MAP 09Nov59; 706 TCSq (446 TCWG) (AFR) NAS New Orleans 08May61; 926 TCG (446 TCWG) (AFR) NAS New Orleans 17Jan63 – Wfu to MASDC (LOG) Davis-Monthan 11Dec65 for storage – DFI: 21Jan69 – Final Disposition: Scrapped.

10577 / 51-2588 / C-119C-25-FA / (C-119CF)

Avble: 23Jan52; Acc: 31Jan52; Del: 27Mar52 to USAF – 443 TCWG (TAC) Greenville 27Mar52; depl Ladd Alaska 29Oct52 – 465 TCWG (TAC) Greenville 01Feb53; depl Thule Greenland 06Jul53 – 465 TCWG (AFE) Neubiberg AB W. Germany 02Dec53; to Toul-Rosières AB France 16Apr54; to Évreux- Fauville AB France 18May55 – 317 TCWG (AFE) Évreux-Fauville AB France 08Jul57 – 2252 RFC (CNR) Clinton 08Oct57 – 302 TCWG (AFR) Clinton 19Mar59; 906 TCG (302 TCWG) (AFR) Clinton 11Feb63; 907 TCG (302 TCWG) (AFR) Clinton 11Dec65 – Wfu to MASDC (LOG) Davis-Monthan 05Aug66 for storage – DFI: 18Jun69 – Final Disposition: Scrapped.

10578 / 51-2589 / C-119C-25-FA / (C-119CF)

Avble: 28Jan52; Acc: 31Jan52; Del: 07Feb52 to USAF – 443 TCWG (TAC) Greenville 07Feb52; depl Ladd Alaska 29Oct52 – 465 TCWG (TAC) Greenville 01Feb53; depl Thule Greenland 06Jul53; to 6603 AB GP (NEA) Goose Bay Canada 09Jul53 mnt – 465 TCWG (AFE) Neubiberg AB W. Germany 02Dec53; to Toul-Rosières AB France 16Apr54; to Évreux-Fauville AB France 18May55; depl 7330 FTN WG (AFE) Fürstenfeldbruck AB W. Germany 18Aug55 – 317 TCWG (AFE) Évreux-Fauville AB France 08Jul57 – 3131 MAI GP (AMC) Chateauroux AB France 20Dec57 mnt; depl 2589 RFC (CNR) Dobbins 19Apr58 – 2584 RFC (CNR) Memphis 24Sep58 – Hayes (AMC) Birmingham 19Dec58 – 2584 AB SQ (CNC) Memphis 18Mar59 – 337 TCSq (514 TCWG) (AFR) Bradley IAP 25Apr59; 905 TCG (514 TCWG) (AFR) Bradley IAP 17Jan63 – Wfu to MASDC (LOG) Davis-Monthan 26Feb66 for storage – DFI: 21Jan69 – Final Disposition: Scrapped.

10579 / 51-2590 / C-119C-25-FA / (C-119CF)

Avble: 30Jan52; Acc: 18Feb52; Del: 29Feb52 to USAF – 443 TCWG (TAC) Greenville 29Feb52; depl Ladd Alaska 29Oct52. Made a forced landing due to a catastrophic engine failure 17Nov52. While enroute from Edmonton Canada to Denver Colorado at 9,000 ft the no. 1 propeller began to give trouble, it subsequently failed and punctured the fuselage. Pilot: William E. Shutles, 8 onboard were killed and 8 others injured after the aircraft made a forced landing in a field 12 miles NW Billings Montana while trying to make an emergency landing at Billings Airport. The vibrations had shaken the no.1 engine off its mounts sending it falling to the ground near a local highway prior to the forced landing – 1701 AB GP (MATS) Great Falls 17Nov52 for parts salvage – REC: 26Jan53 – DFI: 25Feb53 – Final Disposition: Accident.

10580 / 51-2591 / C-119C-25-FA / (C-119CF)

Avble: 31Jan52; Acc: 21Feb52; Del: 29Feb52 to USAF – 443 TCWG (TAC) Greenville 29Feb52; depl Ladd Alaska 31Oct52 – 465 TCWG (TAC) Greenville 01Feb53 – 465 TCWG (AFE) Neubiberg AB W. Germany 05Dec53; to Toul-Rosières AB France 16Apr54; to Évreux-Fauville AB France 22May55 – 317 TCWG (AFE) Évreux-Fauville AB France 08Jul57 – 2584 RFC (CNR) Memphis 15Feb58 – 2252 RFC (CNR) Clinton 26Jul58 – 302 TCWG (AFR) Clinton 19Mar59; 906 TCG (302 TCWG) (AFR) Clinton 11Feb63 – Wfu to MASDC (LOG) Davis-Monthan 04Dec65 for storage – DFI: 21Jan69 – Final Disposition: Scrapped.

10581 / 51-2592 / C-119C-25-FA / (C-119CF)

Avble: 01Feb52; Acc: 25Feb52; Del: 29Feb52 to USAF – 443 TCWG (TAC) Greenville 29Feb52; depl Ladd Alaska 29Oct52 – 465 TCWG (TAC) Greenville 01Feb53; depl Thule Greenland 17Aug53 – 465 TCWG (AFE) Neubiberg AB W. Germany 05Dec53; to Toul-Rosières AB France 16Apr54; to Évreux- Fauville AB France 18May55 – 317 TCWG (AFE) Évreux-Fauville AB France 08Jul57; to 3131 MAI GP (AMC) Chateauroux AB France 03Aug57 mnt – 2584 RFC (CNR) Memphis 08Feb58 – 3131 MAI GP (AMC) Chateauroux AB France 16Mar58 mnt – 4440 ADE GP (TAC) Langley 15Apr59 – 302 TCWG (AFR) Clinton 24Apr59; 906 TCG (302 TCWG) (AFR) Clinton 11Feb63 – Wfu to MASDC (LOG) Davis- Monthan 04Dec65 for storage – DFI: 21Jan69 – Final Disposition: Scrapped.

10582 / 51-2593 / C-119C-25-FA / (C-119CF)

Avble: 06Feb52; Acc: 25Feb52; Del: 29Feb52 to USAF – 443 TCWG (TAC) Greenville 29Feb52; depl Ladd Alaska 29Oct52 – 465 TCWG (TAC) Greenville 01Feb53; depl Thule Greenland 01Jul53; to 6612 AB GP (NEA) Thule AB Greenland 13Aug53 mnt – 465 TCWG (AFE) Neubiberg AB W. Germany 05Dec53; to Toul-Rosières AB France 16Apr54; to Évreux-Fauville AB France 18May55 – 317 TCWG (AFE) Évreux-Fauville AB France 08Jul57 – 2577 RFC (CNR) Brooks 07Mar58 – 4081 AB GP (SAC) Ernest Harmon Canada 11Mar58 mnt – 433 TCWG (AFR) Brooks 18Nov58; to Kelly 31Aug60; 922 TCG (433 TCWG) (AFR) Kelly 10Jul63; 921 TCG (433 TCWG) (AFR) Kelly 10Oct65 – Wfu to MASDC (LOG) Davis-Monthan 28Jan66 for storage – DFI: 21Jan69 – Final Disposition: Scrapped.

10583 / 51-2594 / C-119C-25-FA / (C-119CF)

Avble: 12Feb52; Acc: 29Feb52; Del: 06Mar52 to USAF – 443 TCWG (TAC) Greenville 08Mar52; depl Ladd Alaska 29Oct52 – 465 TCWG (TAC) Greenville 01Feb53; depl Thule Greenland 06Jul53; to 6603 AB GP (NEA) Goose Bay Canada 07Jul53 mnt – 465 TCWG (AFE) Neubiberg AB W. Germany 05Dec53; to 6603 AB GP (NEA) Goose Bay Canada 10Dec53 mnt; to Toul-Rosières AB France 16Apr54; to Évreux-Fauville AB France 18May55; to 3150 MAI GP (AMC) Nouasseur AB Morocco 03Dec57 – 317 TCWG (AFE) Évreux-Fauville AB France 08Jul57 – 2578 RFC (CNR) Ellington 06Feb58 – 446 TCWG (AFR) Ellington 20Sep58; to 839 AB GP (TAC) Sewart 24Sep58 mnt; to 95 COS GP (SAC) Biggs 15Apr59 mnt; 924 TCG (446 TCWG) (AFR) Ellington 31May63; 926 TCG (446 TCWG) (AFR) NAS New Orleans 15Aug65 – Wfu to MASDC (LOG) Davis-Monthan 18Dec65 for storage – DFI: 21Jan69 – Final Disposition: Scrapped.

10584 / 51-2595 / C-119C-25-FA / (C-119CF)

Avble: 12Feb52; Acc: 29Feb52; Del: 06Mar52 to USAF – 443 TCWG (TAC) Greenville 06Mar52; depl Ladd Alaska 29Oct52 – 465 TCWG (TAC) Greenville 01Feb53; depl Thule Greenland 21Sep53; to 6603 AB GP (NEA) Goose Bay Canada 24Sep53 mnt – 465 TCWG (AFE) Neubiberg AB W. Germany 08Dec53; to Toul-Rosières AB France 16Apr54; to Évreux-Fauville AB France 18May55 – 317 TCWG (AFE) Évreux-Fauville AB France 08Jul57 – 2233 RFC (CNR) Mitchel 10Feb58; to 4737 AB WG (ADC) Pepperell AFB Canada 24Feb58 mnt; to MIDAR (AMC) Olmsted 23May58; to 4440 ADE GP (TAC) Langley 10Dec58 – 514 TCWG (AFR) Mitchel 19Dec58; to 1608 TSH WG (MATS) Charleston 09Apr59 mnt; assigned 337 TCSq Bradley 24Mar61; 905 TCG (514 TCWG) (AFR) Bradley 17Jan63 – 906 TCG (302 TCWG) (AFR) Clinton 10Oct65 – Wfu to MASDC (LOG) Davis-Monthan 04Dec65 for storage – DFI: 21Jan69 – Final Disposition: Scrapped.

10585 / 51-2596 / C-119C-25-FA / (C-119CF)

Avble: 13Feb52; Acc: 29Feb52; Del: 06Mar52 to USAF – 443 TCWG (TAC) Greenville 06Mar52; depl Ladd Alaska 29Oct52 – 465 TCWG (TAC) Greenville 01Feb53; depl Thule Greenland 24Jul53 – 465 TCWG (AFE) Neubiberg AB W. Germany 08Dec53; to Toul-Rosières AB France 16Apr54; depls Landstuhl W. Germany 09Aug54, 07Sep54 & 15Mar55; to Évreux-Fauville AB France 18May55; to 7207 AB SQ (AFE) Aviano AB Italy 15Aug56; to 7602 SUT WG (AFE) Seville Spain 27Jan57 – 317 TCWG (AFE) Évreux-Fauville AB France 08Jul57 – 2577 RFC (CNR) Brooks 26Mar58 – 433 TCWG (AFR) Brooks 18Sep58; to 464 TCWG (TAC) Pope 22Aug60; to Kelly 31Aug60; 922 TCG (433 TCWG) (AFR) Kelly 31May63; 921 TCG (433 TCWG) (AFR) Kelly 10Oct65 – Wfu to MASDC (LOG) Davis-Monthan 25Feb66 for storage – DFI: 21Jan69 – Final Disposition: Scrapped.

10586 / 51-2597 / C-119C-25-FA / (C-119CF)

Avble: 14Feb52; Acc: 29Feb52; Del: 06Mar52 to USAF – 443 TCWG (TAC) Greenville 06Mar52; depl Ladd Alaska 29Oct52 – 465 TCWG (TAC) Greenville 01Feb53 – 465 TCWG (AFE) Neubiberg AB W. Germany 08Dec53; to Toul-Rosières AB France 16Apr54; to 7373 MAI GP (AFE) Chateauroux AB France 26Apr55; to 3131 MAI GP (AMC) Chateauroux AB France 23Apr56; to Évreux-Fauville AB France 23Apr56; depl 48 Fighter-Bomber WG (AFE) Chaumont 01Nov56 – 2252 RFC (CNR) Clinton 25Dec56 – 302 TCWG (AFR) Clinton 19Mar59; 906 TCG (302 TCWG) (AFR) Clinton 11Feb63; 907 TCG (302 TCWG) (AFR) Clinton 30Nov65 – 910 TCG (459 TCWG) (AFR) Youngstown 23Jul66 – Wfu to MASDC (LOG) Davis-Monthan 01Oct66 for storage – DFI: 18Jun69 – Final Disposition: Scrapped.

10587 / 51-2598 / C-119C-25-FA / (C-119CF)

Avble: 16Feb52; Acc: 29Feb52; Del: 06Mar52 to USAF – 443 TCWG (TAC) Greenville 06Mar52; depl Ladd Alaska 29Oct52 – 465 TCWG (TAC) Greenville 01Feb53 – 465 TCWG (AFE) Neubiberg AB W. Germany 08Dec53; to Toul-Rosières AB France 16Apr54; to 7373 MAI GP (AFE) Chateauroux AB France date unk; to Évreux-Fauville AB France 27May55; depl 317 TCWG (AFE) Neubiberg AB W. Germany 07Sep55; to 7312 AB GP (AFE) Rhein-Main AB W. Germany 19Jan56; to 7272 AB WG (AFE) Wheelus AB Libya 13Jan57 – 317 TCWG (AFE) Évreux-Fauville AB France 08Jul57 – 2584 RFC (CNR) Memphis 02Mar58 – 2259 RFC DT (CNR) Youngstown 09Jul58 – Hayes (AMC) Birmingham 22Oct58 – 757 TCSq (459 TCG) (AFR) Youngstown 30Dec58; 910 TCG (459 TCWG) (AFR) Youngstown 17Jan63 – Wfu to MASDC (LOG) Davis-Monthan 14May66 for storage – DFI: 21Jan69 – Final Disposition: Scrapped.

10588 / 51-2599 / C-119C-25-FA / (C-119CF)

Avble: 20Feb52; Acc: 29Feb52; Del: 07Mar52 to USAF – 443 TCWG (TAC) Greenville 07Mar52 – 465 TCWG (TAC) Greenville 01Feb53; depl Thule Greenland 13Jul53 – 465 TCWG (AFE) Neubiberg AB W. Germany 12Dec53; to Toul-Rosières AB France 16Apr54; to Évreux-Fauville AB France 18May55; to 7109 SUT WG (AFE) Wiesbaden AB W. Germany 20Apr56 – 2259 RFC (CNR) Andrews 20Feb57; depl Clinton Ohio 03Aug57; to 3600 CCT WG (TC) Luke 14Mar58 mnt; to 1001 AB WG (HQC) Andrews 23May58 mnt – 459 TCWG (AFR) Andrews 04Nov58; to 839 AB GP (TAC) Sewart 23Oct60; 909 TCG (459 TCWG) (AFR) Andrews 01Feb63 – 905 TCG (514 TCWG) (AFR) Bradley date unk – Wfu to MASDC (LOG) Davis-Monthan 20Jun66 for storage – DFI: 21Jan69 – Final Disposition: Scrapped.

10589 / 51-2600 / C-119C-25-FA / (C-119CF)

Avble: 25Feb52; Acc: 20Mar52; Del: 26Mar52 to USAF – 443 TCWG (TAC) Greenville 26Mar52; depl Ladd Alaska 29Oct52 – 465 TCWG (TAC) Greenville 01Feb53; to 6603 AB GP (NEA) Goose Bay Canada 12Aug53; depl Thule Greenland 17Aug53 – 465 TCWG (AFE) Neubiberg AB W. Germany 12Dec53; to Toul-Rosières AB France 16Apr54; to Évreux-Fauville AB France 18May55; to 7373 MAI GP (AFE) Chateauroux AB France 19Sep55 mnt; to 3131 MAI GP (AMC) Chateauroux AB France 01Jan56; to 3150 MAI GP (AMC) Nouasseur AB Morocco 10Dec56 – 317 TCWG (AFE) Évreux-Fauville AB France 08Jul57 – 2589 RFC (CNR) Dobbins 29Oct57 – 2584 RFC (CNR) Memphis 05Nov58 – 2584 AB SQ (CNC) Memphis 19Dec58 – 514 TCWG (AFR) Mitchel 08Jun59; assigned 337 TCSq Bradley 23Mar61; 905 TCG (514 TCWG) (AFR) Bradley 17Jan63 – Wfu to MASDC (LOG) Davis-Monthan 26Feb66 for storage – DFI: 21Jan69 – Final Disposition: Scrapped.

10590 / 51-2601 / C-119C-25-FA / (C-119CF)

Avble: 27Feb52; Acc: 21Mar52; Del: 27Mar52 to USAF – 443 TCWG (TAC) Greenville 27Mar52; depl Ladd Alaska 29Oct52 – 465 TCWG (TAC) Greenville 01Feb53 – 465 TCWG (AFE) Neubiberg AB W. Germany 12Dec53; to Toul-Rosières AB France 16Apr54; to Évreux-Fauville AB France 18May55; **RECORDS INDECIPHERABLE** → to Dreux AB France date unk ← **RECORDS INDECIPHERABLE** – 317 TCWG (AFE) Évreux-Fauville AB France 08Jul57 – 2589 RFC (CNR) Dobbins 29Oct57 – 2578 RFC (CNR) Ellington 17Dec57 – 446 TCWG (AFR) Ellington 20Sep58; assigned 357 TCSq NAS New Orleans 12Apr59 – 187 AML SQ (WY-ANG) Cheyenne 05May61 – 129 ACO GP (ANG) Hayward 06Sep63 – Wfu to MASDC (LOG) Davis-Monthan 22Jan69 for storage – DFI: 03Feb69 – Final Disposition: Scrapped.

10591 / 51-2602 / C-119C-25-FA / (C-119CF)

Avble: 28Feb52; Acc: 21Mar52; Del: 27Mar52 to USAF – 443 TCWG (TAC) Greenville 27Mar52; depl Ladd Alaska 29Oct52 – 465 TCWG (TAC) Greenville 01Feb53 – 465 TCWG (AFE) Neubiberg AB W. Germany 14Dec53; to Toul-Rosières AB France 16Apr54; to Évreux-Fauville AB France 18May55. Right engine failed on take-off at Évreux France 03Jul57. As the pilots setup to make an emergency landing back at the air base the unattended, and unfrictioned, throttle levers vibrated back to idle power casing a loss of airspeed too low to recover from. The aircraft came down heavy collapsing the wing structure with the wreckage sliding off the end of the runway, no fatalities and no injuries to the crew – Final Disposition: Accident.

10592 / 51-2603 / C-119C-25-FA / (C-119CF)

Avble: 27Feb52; Acc: 21Mar52; Del: 27Mar52 to USAF – 316 TCG (TAC) Sewart 27Mar52 – 435 TCWG (TAC) Miami Jun52; depl Elmendorf 24Oct52 – 456 TCWG (TAC) Miami IAP 01Dec52 – 60 TCWG (AFE) Rhein-Main AB W. Germany 12Jan53; depl RAF Sculthorpe England 23Mar54 – 44 Bomb WG (SAC) Lake Charles 54; to 60 TCWG (AFE) Rhein-Main AB W. Germany 10Mar55 mnt; to 3635 CCT WG (TC) Stead 19Oct55 mnt – 806 AB GP (SAC) Lake Charles 05Aug57 – 2578 RFC (CNR) Ellington 16May58 – 446 TCWG (AFR) Ellington 20Sep58; 924 TCG (446 TCWG) (AFR) Ellington 31May63; 926 TCG (446 TCWG) (AFR) NAS New Orleans 25Aug65 – Wfu to MASDC (LOG) Davis-Monthan 18Dec65 for storage – DFI: 21Jan69 – Final Disposition: Scrapped.

10593 / 51-2604 / C-119C-25-FA / (C-119CF)

   Avble: 03Mar52; Acc: 26Mar52; Del: 31Mar52 to USAF – 316 TCG (TAC) Sewart 31Mar52 – 435 TCWG (TAC) Miami Jun52; depl Elmendorf 24Oct52 – 456 TCWG (TAC) Miami IAP 01Dec52 – 64 TCWG (TAC) Donaldson 09Jul53 – 317 TCWG (AFE) Neubiberg AB W. Germany 13Nov53 – 60 TCWG (AFE) Rhein-Main AB W. Germany 13Mar55 – 4676 ADF GP (ADC) Grandview 06Jun55 – 328 Fighter GP (ADC) Grandview 18Aug55 – 328 CLM SQ (ADC) Richards 01Aug57 – 2577 RFC DT (CNR) NAS Dallas 01Dec57 – 69 TCSq (433 TCWG) (AFR) NAS Dallas 20Sep58 – 923 TCG (433 TCWG) (AFR) NAS Dallas 17Jan63; to Carswell 02Mar63 – 909 TCG (459 TCWG) (AFR) Andrews 17Sep65 – Wfu to MASDC (LOG) Davis-Monthan 17Jul66 for storage – DFI: 18Jun69 – Final Disposition: Scrapped.

10594 / 51-2605 / C-119C-25-FA / (C-119CF)

   Avble: 05Mar52; Acc: 26Mar52; Del: 31Mar52 to USAF – 316 TCG (TAC) Sewart 31Mar52 – 435 TCWG (TAC) Miami Oct52; depl Elmendorf 24Oct52 – 456 TCWG (TAC) Miami IAP 01Dec52 – 60 TCWG (AFE) Rhein-Main AB W. Germany 12Jan53; multiple depls RAF Sculthorpe England from Nov53 – 580 ASL GP (MATS) Wheelus AB Libya 24Mar55 – 7272 AB WG (AFE) Wheelus AB Libya 15Sep56 – 2252 RFC (CNR) Clinton 14Jan57 – 2259 RFC (CNR) Andrews 03Aug57 – 459 TCWG (AFR) Andrews 04Nov58 – 187 AML SQ (WY-ANG) Cheyenne 31May61 – 129 ACO GP (ANG) Hayward 20Sep63 – Wfu to MASDC (LOG) Davis-Monthan 22Jan69 for storage – DFI: 03Feb69 – Aero Union Corp. Chico California 1969, spares airframe – Hemet Valley Flying Service Co. Hemet California CofR **N9966F** 15Oct76, spares airframe, bku 1980s – Final Disposition: Scrapped.

10595 / 51-2606 / C-119C-25-FA / (C-119CF)

   Avble: 07Mar52; Acc: 27Mar52; Del: 01Apr52 to USAF – 316 TCG (TAC) Sewart 01Apr52 – 435 TCWG (TAC) Miami 04Jun52; depl Elmendorf 30Oct52 – 456 TCWG (TAC) Miami IAP 01Dec52 – 64 TCWG (TAC) Greenville 25Jun53 – 465 TCWG (TAC) Donaldson Aug53 – 465 TCWG (AFE) Neubiberg AB W. Germany 12Dec53; to Toul-Rosières AB France 16Apr54; to Évreux-Fauville AB France 18May55 – 317 TCWG (AFE) Évreux-Fauville AB France 08Jul57 – 2259 RFC DT (CNR) Youngstown 21Dec57 – 757 TCSq (459 TCG) (AFR) Youngstown 19Dec58; 910 TCG (459 TCWG) (AFR) Youngstown 17Jan63 – Wfu to MASDC (LOG) Davis-Monthan 26Nov66 for storage – DFI: 21Jan69 – Final Disposition: Scrapped.

10596 / 51-2607 / C-119C-25-FA / (C-119CF)

   Avble: 07Mar52; Acc: 28Mar52; Del: 01Apr52 to USAF – 316 TCG (TAC) Sewart 01Apr52 – 435 TCWG (TAC) Miami Jul52; depl Elmendorf 24Oct52 – 456 TCWG (TAC) Miami IAP 01Dec52 – 60 TCWG (AFE) Rhein-Main AB W. Germany 04Feb53 – 68 Bomb WG (SAC) Lake Charles 12Feb55; to 3800 AB WG (AU) Maxwell 25Jun56 mnt; to 3345 TTA WG (TC) Chanute 20Aug56 mnt; to 4900 AB GP (ARD) Kirtland 10Oct56 mnt; to 824 AB GP (SAC) Carswell 22May57 – 806 AB GP (SAC) Lake Charles 02Aug57 – 2596 RFC (CNR) NAS Dallas 15Oct57 – 2577 RFC DT (CNR) NAS Dallas 01Dec57 – 69 TCSq (433 TCWG) (AFR) NAS Dallas 20Sep58; 923 TCG (433 TCWG) (AFR) NAS Dallas 17Jan63; to Carswell 02Mar63; 921 TCG (433 TCWG) (AFR) Kelly 24Sep65 – Wfu to MASDC (LOG) Davis-Monthan 28Jan66 for storage – DFI: 21Jan69 – Final Disposition: Scrapped.

10597 / 51-2608 / C-119C-25-FA / (C-119CF)

   Avble: 11Mar52; Acc: 27Mar52; Del: 01Apr52 to USAF – 316 TCG (TAC) Sewart 01Apr52 – 435 TCWG (TAC) Miami 52; depl Elmendorf 24Oct52 – 456 TCWG (TAC)

Miami IAP 01Dec52 – 60 TCWG (AFE) Rhein-Main AB W. Germany 15Jan53; depl RAF Sculthorpe England 10Nov53 – 580 ASL GP (MATS) Wheelus AB Libya 24Mar55; depl 317 TCWG (AFE) Neubiberg AB W. Germany 13Feb56 – 7272 AB WG (AFE) Wheelus AB Libya 15Sep56 – 2252 RFC (CNR) Clinton 03Nov56; to 3415 TTA WG (TC) Lowry 14Sep57 mnt – 302 TCWG (AFR) Clinton 19Mar59; to 1608 TSH WG (MATS) Charleston 08May60 mnt – 187 AML SQ (WY-ANG) Cheyenne 24Apr61 – 129 ACO GP (ANG) Hayward 06Sep63 – Wfu to MASDC (LOG) Davis-Monthan 27Mar69 for storage – DFI: 09Apr69 – Aero Union Corp. Chico California 1969, spares airframe – Hemet Valley Flying Service Co. Hemet California CofR **N9959F** 15Oct76, spares airframe, bku 1980s – Reg N9959F canx 01April1 – Final Disposition: Scrapped.

10598 / 51-2609 / C-119C-25-FA / (C-119CF)

Avble: 12Mar52; Acc: 28Mar52; Del: 01Apr52 to USAF – 316 TCG (TAC) Sewart 01Apr52 – 435 TCWG (TAC) Miami Jul52; depl Elmendorf 24Oct52 – 456 TCWG (TAC) Miami IAP 01Dec52 – 1739 FRY SQ (MATS) Amarillo 13Dec52 – 2370 AB SQ (CNC) Long Beach 26Dec52 mnt – 60 TCWG (AFE) Rhein- Main AB W. Germany 07May53; to 80 AB SQ (MATS) Dover 02Jul53 mnt; to 1607 MSU SQ (MATS) Dover 01Aug53 mnt; depl 317 TCWG (AFE) Neubiberg AB W. Germany 04Jan54 – 566 ADF GP (ADC) Hamilton 31May55 – Bailment Fairchild (AMC) Hagerstown 08Aug55 – 78 Fighter GP (ADC) Hamilton 06Dec55; depl 4750 ADF WG (ADC) Yuma 07Jul56; to 1100 OP GP (HQC) Bolling 18Jan57 mnt – 78 MSU GP (ADC) Hamilton 30Apr58 – 4061 AB GP (SAC) Malmstrom 10May58 mnt. Wfu to REC: 13Aug58, reasons unk – Final Disposition: Scrapped.

10599 / 51-2610 / C-119C-25-FA / (C-119CF)

Avble: 14Mar52; Acc: 28Mar52; Del: 01Apr52 to USAF – 316 TCG (TAC) Sewart 01Apr52 – 435 TCWG (TAC) Miami 16Jul52; depl Elmendorf 24Oct52 – 456 TCWG (TAC) Miami IAP 01Dec52 – 60 TCWG (AFE) Rhein-Main AB W. Germany 14Jan53; depl RAF Sculthorpe England 29Dec53 – 68 Bomb WG (SAC) Lake Charles date unk; to 1631 MAT SQ (MATS) Prestwick Scotland 24Feb55 mnt – 806 AB GP (SAC) Lake Charles 02Aug57 – 2589 RFC (CNR) Dobbins 19Dec57 – 446 TCWG (AFR) Ellington 04Nov58; to 445 TCA DT (AFR) Memphis 05May60 mnt – 187 AML SQ (WY-ANG) Cheyenne 13Apr61 – 129 ACO GP (ANG) Hayward 11Oct63 – Wfu to MASDC (LOG) Davis-Monthan 27Mar69 for storage – DFI: 09Apr69 – Aero Union Corp. Chico California 1969, spares airframe – Hemet Valley Flying Service Co. Hemet California CofR **N9961F** 15Oct76, spares airframe, bku 1980s – Final Disposition: Scrapped.

10600 / 51-2611 / C-119C-25-FA / (C-119CF)

Avble: 18Mar52; Acc: 28Mar52; Del: 07Apr52 to USAF – 316 TCG (TAC) Sewart 04Apr52 – 435 TCWG (TAC) Miami 16Jul52; depl Elmendorf 24Oct52 – 456 TCWG (TAC) Miami 01Dec52 – 60 TCWG (AFE) Rhein-Main AB W. Germany 15Jan53; to 6602 AB GP (NEA) Ernest Harmon Canada 27Jan53 mnt; loan to RAE Farnborough England 23Apr53–18Sep56; to 7559 MAI GP (AFE) RAF Burtonwood England 07Aug53 mnt – 465 TCWG (AFE) Évreux-Fauville AB France 18Sep56 – 317 TCWG (AFE) Évreux-Fauville AB France 08Jul57 – 2578 RFC (CNR) Ellington 25Feb58 – 2589 RFC (CNR) Dobbins 02Spr58 – 2584 RFC (CNR) Memphis 05Oct58 – 514 TCWG (AFR) Mitchel 13Jun59; assigned 337 TCSq Bradley 20Mar61; 905 TCG (514 TCWG) (AFR) Bradley 17Jan63. Crashed 08Jan66 while enroute from Bradley Field Connecticut to Binghamton Airport New York. An uncontained engine failure led to a loss of control. 1 crewman managed to bail out at 2,000 ft but the pilot and co-pilot could not and were killed on impact when the aircraft crashed into a lakefront house near Scranton Pennsylvania. A young boy in the house at the time was also killed – 4603 AB GP (ADC) Stewart 10Jan66 for SAL – DFI: 11Jan66 – Final Disposition: Accident.

10601 / 51-2612 / C-119C-25-FA / (C-119CF)

Avble: 19Mar52; Acc: 21Apr52; Del: 26Apr52 to USAF – 316 TCG (TAC) Sewart 25Apr52 – 435 TCWG (TAC) Miami 24Oct52; depl Elmendorf 24Oct52 – 456 TCWG (TAC) Miami 01Dec52 – 60 TCWG (AFE) Rhein-Main AB W. Germany 28Jan53; depl RAF Sculthorpe England 10Sep54; to Dreux AB France 12Sep55; to 7559 MAI GP (AFE) RAF Burtonwood England 07Oct55 mnt; to 3110 MAI GP (AMC) RAF Burtonwood England 01Jan56 mnt – 580 ASL GP (MATS) Wheelus AB Libya 09Jul56; to 7272 AB WG (AFE) Wheelus AB Libya 15Sep56 – 465 TCWG (AFE) Évreux-Fauville AB France 01Oct56 – 2259 RFC (CNR) Andrews 03Mar57 – 459 TCWG (AFR) Andrews 04Nov58; to KTC (ATC) Keesler 26Jun60 mnt; 909 TCG (459 TCWG) (AFR) Andrews 01Feb63 – 905 TCG (514 TCWG) (AFR) Bradley date unk – Wfu to MASDC (LOG) Davis-Monthan 20Jun66 for storage – DFI: 21Jan69 – Final Disposition: Scrapped.

10602 / 51-2613 / C-119C-25-FA / (C-119CF)

Avble: 21Mar52; Acc: 21Apr52; Del: 26Apr52 to USAF – 316 TCG (TAC) Sewart 25Apr52 – 435 TCWG (TAC) Miami 07Jun52; depl Elmendorf 24Oct52 – 456 TCWG (TAC) Miami 01Dec52 – 64 TCWG (TAC) Greenville 19Jun53 – 317 TCWG (AFE) Neubiberg AB W. Germany 03Nov53 – 60 TCWG (AFE) Rhein-Main AB W. Germany 03Dec53; depl RAF Sculthorpe England 01Apr54 – 1603 TSP WG (AFE) Wheelus AB Libya 30Apr55 – 580 ASL GP (MATS) Wheelus AB Libya 04May55; depl 7330 FTN WG (AFE) Fürstenfeldbruck AB, West Germany 09Jun55 – SABBE (AMC) Brussels Belgium 12Sep56 mnt – 465 TCWG (AFE) Évreux-Fauville AB France 01Feb57 – 2259 RFC (CNR) Andrews 12Feb57 – 459 TCWG (AFR) Andrews 04Nov58; to AOT Center (ATC) Amarillo 03Mar59 mnt; 909 TCG (459 TCWG) (AFR) Andrews 01Feb63 – Wfu to MASDC (LOG) Davis-Monthan 04Apr66 for storage – DFI: 21Jan69 – Final Disposition: Scrapped.

10603 / 51-2614 / C-119C-25-FA / (C-119CF)

Avble: 24Mar52; Acc: 21Apr52; Del: 26Apr52 to USAF – 316 TCG (TAC) Sewart 25Apr52 – 435 TCWG (TAC) Miami 04Jun52; depl Elmendorf 24Oct52 – 456 TCWG (TAC) Miami 01Dec52 – 60 TCWG (AFE) Rhein-Main AB W. Germany 17Jan53; to 73 AD WG (AFE) Chateauroux AB France Mar53; depls RAF Sculthorpe England 05Jan54 & 06May54; to 1604 MAT SQ (MATS) Kindley 12Feb55 mnt; to Dreux AB France 13Sep55; to 7310 AB GP (AFE) Rhein-Main AB W. Germany 15Oct55; to 86 Fighter WG (AFE) Landstuhl AB W. Germany 18Feb56 – 580 ASL GP (MATS) Wheelus AB Libya 12Aug56 – 7272 AB WG (AFE) Wheelus AB Libya 15Sep56 – 2252 RFC (CNR) Clinton 22Dec56 – 2259 RFC (CNR) Andrews 03Aug57; to 144 CLM SQ (CAL-ANG) Fresno 10Aug58 mnt – 459 TCWG (AFR) Andrews 04Nov58; 909 TCG (459 TCWG) (AFR) Andrews 01Feb63 – Wfu to MASDC (LOG) Davis-Monthan 04Apr66 for storage – DFI: 21Jan69 – Final Disposition: Scrapped.

10604 / 51-2615 / C-119C-25-FA / (C-119CF)

Avble: 25Mar52; Acc: 21Apr52; Del: 26Apr52 to USAF – 316 TCG (TAC) Sewart 25Apr52 – 435 TCWG (TAC) Miami 22Aug52; depl Elmendorf 24Oct52 – 456 TCWG (TAC) Miami 01Dec52 – 443 TCWG (TAC) Greenville 22Jan53 – 465 TCG (TAC) Greenville 01Feb53; depl Thule Greenland 07Sep53 – 465 TCWG (AFE) Neubiberg AB W. Germany

02Dec53; to Toul-Rosières AB France 16Apr54; to 7559 MAI GP (AFE) RAF Burtonwood England 21Apr55 mnt; to Évreux-Fauville AB France 08Jun55; to 3131 MAI GP (AMC) Chateauroux AB France 18Feb56 mnt – 2252 RFC (CNR) Clinton 29Jan57 – 302 TCWG (AFR) Clinton 19Mar59; to 839 AB GP (TAC) Sewart 05Feb60 mnt; 906 TCG (302 TCWG) (AFR) Clinton 11Feb63; 907 TCG (302 TCWG) (AFR) Clinton 30Nov65 – Wfu to MASDC (LOG) Davis-Monthan 03Jun66 for storage – DFI: 18Jun69 – Final Disposition: Scrapped.

10605 / 51-2616 / C-119C-25-FA / (C-119CF)

Avble: 27Mar52; Acc: 21Apr52; Del: 26Apr52 to USAF – 316 TCG (TAC) Sewart 25Apr52 – 435 TCWG (TAC) Miami 12Oct52; depl Elmendorf 24Oct52 – 456 TCWG (TAC) Miami 01Dec52 – 443 TCWG (TAC) Greenville 20Jan53 – 465 TCG (TAC) Greenville 01Feb53; depl 64 TCWG (TAC) Donaldson 03Sep53 – 97 Bomb WG (SAC) Biggs 20Sep53 mnt – SAAAR (AMC) Kelly 04Jan54 for project duties; to MIDAR (AMC) Olmsted 07May54 mnt – 1607 FDM SQ (MATS) Dover 54-55 mnt – MIDAR (AMC) Olmsted 02May55 – 407 AB GP (SAC) Great Falls 19Sep55 – 407 Fighter WG (SAC) Malmstrom 24Oct55 – 92 AB GP (SAC) Fairchild 04Sep56; to 814 AB GP (SAC) Fairchild 12Jul56; to 2750 AB WG (AMC) WPAFB 28Jul57 mnt – 2233 RFC (CNR) Mitchel 13Feb58 – 514 TCWG (AFR) Mitchel 19Dec58; to 464 TCWG (TAC) Pope 17Dec59 mnt; to 2585 AB SQ (CNC) Miami 10May60 mnt; depl 435 TCWG (AFR) Miami 14Jun60; 337 TCSq Bradley 25Mar61; 905 TCG (514 TCWG) (AFR) Bradley 17Jan63; depl 903 TCG (514 TCWG) (AFR) McGuire date unk – Wfu 11Mar66 to museum / school status. Displayed on loan at Bradley Air Museum Windsor Locks Connecticut. Destroyed by a tornado 03Oct79 that also wiped out many of the other exhibits – Final Disposition: Destroyed.

10606 / 51-2617 / C-119C-25-FA / (C-119CF)

Rsvd YC-119E prototype canx. Avble: 28Mar52; Acc: 19Apr52; Del: 26Apr52 to USAF – 316 TCG (TAC) Sewart 25Apr52 – 435 TCWG (TAC) Miami 22Jun52; depl Elmendorf 24Oct52 – 456 TCWG (TAC) Miami 01Dec52; depl 64 TCWG (TAC) Greenville 19Jun53 – 465 TCWG (TAC) Greenville 53; to 6603 AB GP (NEA) Goose Bay Canada 26Dec53 mnt – 465 TCWG (AFE) Toul-Rosières AB France 14Dec53; to Évreux-Fauville AB France 14Jul55; to 7373 MAI GP (AFE) Chateauroux AB France 21Dec55 mnt; to 3131 MAI GP (AMC) Chateauroux AB France 01Jan56 mnt – 317 TCWG (AFE) Évreux-Fauville AB France 08Jul57; to 3131 MAI GP (AMC) Chateauroux AB France 25Jul57; depl 7330 FTN WG (AFE) Fürstenfeldbruck AB, W. Germany 25Nov57 – 2578 RFC (CNR) Ellington 25Feb58 – 446 TCWG (AFR) Ellington 20Sep58; to LTC (ATC) Lowry 24Sep59 mnt; 925 TCG (446 TCWG) (AFR) Ellington 31May63 – 926 TCG (446 TCWG) (AFR) NAS New Orleans 12Sep65 – Wfu to MASDC (LOG) Davis-Monthan 11Dec65 for storage – DFI: 21Jan69 – Final Disposition: Scrapped.

10607 / 51-2618 / C-119C-26-FA / (C-119CF)

Avble: 31Mar52; Acc: 23Apr52; Del: 29Apr52 to USAF – 316 TCG (TAC) Sewart 28Apr52 – 435 TCWG (TAC) Miami 14Jun52; depl Elmendorf 24Oct52 – 456 TCWG (TAC) Miami 01Dec52 – 443 TCG (TAC) Greenville 28Jan53 – 465 TCG (TAC) Greenville 01Feb53 – 465 TCG (AFE) Wiesbaden AB W. Germany 14Dec53; to Toul-Rosières AB France 04May54; to Évreux-Fauville AB France 30May55; depl 317 TCWG (AFE) Neubiberg AB W. Germany 28Jun55; to 7310 AB GP (AFE) Rhein-Main AB W. Germany Feb56 – 317 TCWG (AFE) Évreux-Fauville AB France 08Jul57 – 2584 RFC (CNR) Memphis 10Mar58 – 2259 RFC DT (CNR) Youngstown 24Jul58 – 757 TCSq (459 TCG) (AFR)

Youngstown 19Dec58; depl 187 AML SQ (WY-ANG) Cheyenne 12Apr61 – 187 AML SQ (ANG) Cheyenne 18Jan62-129 ACO GP (ANG) Hayward 07Nov63 – Wfu to MASDC (LOG) Davis-Monthan 08Apr69 for storage – DFI: 09Apr69 – Aero Union Corp. Chico California 1969, spares airframe – Hemet Valley Flying Service Co. Hemet California CofR **N9960F** 15Oct76, spares airframe, bku 1980s – Final Disposition: Scrapped.

10608 / 51-2619 / C-119C-26-FA / (C-119CF)

Avble: 02Apr52; Acc: 23Apr52; Del: 29Apr52 to USAF – 316 TCG (TAC) Sewart 25Apr52 – 435 TCWG (TAC) Miami Jun52; depl Elmendorf 24Oct52 – 456 TCWG (TAC) Miami 01Dec52; to 1739 FRY SQ (MATS) Amarillo 12Jan53 – 60 TCWG (AFE) Rhein-Main AB W. Germany 16Jan53; depls RAF Sculthorpe England 03Jan53 & 11May54 – 91 SM WG (SAC) Lockbourne 13Feb55 – 801 AB GP (SAC) Lockbourne 01Feb57 – 2259 RFC DT (CNR) Greater Pittsburgh 02Apr58; to Youngstown 09Apr58 – 1405 AB WG (MATS) Scott 16Sep58 mnt – 757 TCSq (459 TCG) (AFR) Youngstown 11Jan59; 910 TCG (459 TCWG) (AFR) Youngstown 17Jan63 – Wfu to MASDC (LOG) Davis- Monthan 22Oct66 for storage – DFI: 18Jun69 – Final Disposition: Scrapped.

10609 / 51-2620 / C-119C-26-FA / (C-119CF)

Avble: 02Apr52; Acc: 23Apr52; Del: 29Apr52 to USAF – 316 TCG (TAC) Sewart 25Apr52 – 435 TCWG (TAC) Miami 14Aug52; depl Elmendorf 24Oct52 – 456 TCWG (TAC) Miami 01Dec52 – 60 TCWG (AFE) Rhein-Main AB W. Germany 27Jan53; depls RAF Sculthorpe England 28Apr54 & 08Sep54 – 1603 TSP WG (AFE) Wheelus AB Libya 30Apr55 – 580 ASL GP (MATS) Wheelus AB Libya 04May55; depl 317 TCWG (AFE) Neubiberg AB W. Germany 25Jul55; to 7100 SUT WG (AFE) Wiesbaden AB W. Germany 13Feb56 – 7272 AB WG (AFE) Wheelus AB Libya 15Sep56 – 2252 RFC (CNR) Clinton 06May57 – 302 TCWG (AFR) Clinton 19Mar59 – 757 TCSq (459 TCWG) (AFR) Youngstown 14Jan60; 910 TCG (459 TCWG) (AFR) Youngstown 17Jan63 – Wfu to MASDC (LOG) Davis-Monthan 30Jul66 for storage – DFI: 21Jan69 – Final Disposition: Scrapped.

10610 / 51-2621 / C-119C-26-FA / (C-119CF)

Avble: 04Apr52; Acc: 24Apr52; Del: 29Apr52 to USAF – 316 TCG (TAC) Sewart 25Apr52 – 435 TCWG (TAC) Miami 22Aug52; depl Elmendorf 24Oct52 – 456 TCWG (TAC) Miami 01Dec52 – 64 TCWG (TAC) Greenville 09Jul53 – 317 TCWG (AFE) Neubiberg AB W. Germany 20Nov53. Crashed 27Nov53 while on its delivery flight to the 317 TCWG 2.5 miles NE of Paris-Orly Airport France while on final approach to land. Pilot: Samuel J. Salem and five other crew members were killed. Witnesses on the ground stated the aircraft was having engine trouble but the USAF investigation found the clamshell doors broke away inflight smashing through the horizontal stabilizer causing a fatal structural failure and loss of control that witnesses described as an "explosion." One crewman unsuccessfully attempted a bailout at 700 ft, his body was found 40 yards from the wreck site in a group of trees, his parachute only partially open. The wreckage came down and burned in a Paris neighborhood but there were no civilian casualties – DFI: 05May54 – Final Disposition: Accident.

10611 / 51-2622 / C-119C-26-FA / (C-119CF)

Avble: 07Apr52; Acc: 26Apr52; Del: 03May52 to USAF – 316 TCG (TAC) Sewart 02May52 – 435 TCWG (TAC) Miami 14Aug52; depl Elmendorf 24Oct52 – 456 TCWG (TAC) Miami 01Dec52 – 64 TCWG (TAC) Donaldson 09Jul53 – 465 TCWG (TAC)

Donaldson 53 – 465 TCWG (AFE) Wiesbaden AB W. Germany 26Dec53; to Toul-Rosières AB France 04May54 **RECORDS INDECIPHERABLE** depls Landstuhl AB W. Germany 03Mar55 & 30May55; to Évreux-Fauville AB France 30May55; depl 48 Fighter-Bomber WG (AFE) Chaumont AB France 10Jan57 – 317 TCWG (AFE) Évreux-Fauville AB France 08Jul57 – 2259 RFC DT (CNR) Youngstown 20Dec57 – 757 TCSq (459 TCG) (AFR) Youngstown 20Nov58; 910 TCG (459 TCWG) (AFR) Youngstown 17Jan63 – Wfu to MASDC (LOG) Davis-Monthan 21Jan67 for storage – DFI: 18Jun69 – Final Disposition: Scrapped.

10612 / 51-2623 / C-119C-26-FA / (C-119CF)

Avble: 08Apr52; Acc: 26Apr52; Del: 03May52 to USAF – 316 TCG (TAC) Sewart 02May52 – 435 TCWG (TAC) Miami 12Aug52 – 456 TCWG (TAC) Miami 01Dec52 – 64 TCWG (TAC) Donaldson 09Jul53 – 465 TCWG (TAC) Donaldson 53 – 465 TCWG (AFE) Wiesbaden AB W. Germany 14Dec53; to 6603 AB GP (NEA) Goose Bay Canada 23Dec53 mnt; depls Landstuhl AB W. Germany 27Mar54; 11May54 & 30Jul54; to Toul-Rosières AB France 04May54; to Évreux-Fauville AB France 30May55; depl 406 Fighter WG (AFE) RAF Manston England 10Aug55; to 7100 SUT WG (AFE) Wiesbaden AB W. Germany 19Jan56 – 3131 MAI GP (AMC) Chateauroux AB France 24Dec56 – 317 TCWG (AFE) Évreux-Fauville AB France 03Dec57 – 2584 RFC (CNR) Memphis 30Apr58 – 2252 RFC (CNR) Clinton 30Jul58 – 302 TCWG (AFR) Clinton 19Mar59 – 337 TCSq (514 TCWG) (AFR) Bradley 16Dec59; 905 TCG (514 TCWG) (AFR) Bradley 17Jan63 – Wfu to MASDC (LOG) Davis-Monthan 02Feb66 for storage – DFI: 21Jan69 – Final Disposition: Scrapped.

10613 / 51-2624 / C-119C-26-FA / (C-119CF)

Avble: 09Apr52; Acc: 28Apr52; Del: 03May52 to USAF – 316 TCG (TAC) Sewart 02May52 – 435 TCWG (TAC) Miami 52; depl Elmendorf 24Oct52 – 456 TCWG (TAC) Miami 01Dec52 – 64 TCWG (TAC) Donaldson 02Jul53 – 317 TCWG (AFE) Neubiberg AB W. Germany 18Nov53 – 60 TCWG (AFE) Rhein-Main AB W. Germany 27Nov53 – 91 SM WG (SAC) Lockbourne 01Jul55 – 801 AB GP (SAC) Lockbourne 01Feb57 – 2577 RFC (CNR) Brooks 06Jun58 – 433 TCWG (AFR) Brooks 18Sep58. Crashed in a snow storm during a ground controlled approach to Bunker Hill AFB Indiana 03Jan59, all 3 crew were killed – Final Disposition: Accident.

10614 / 51-2625 / C-119C-26-FA / (C-119CF)

Avble: 10Apr52; Acc: 28Apr52; Del: 03May52 to USAF – 316 TCG (TAC) Sewart 02May52 – 435 TCWG (TAC) Miami 14Aug52; depl Elmendorf 24Oct52 – 456 TCWG (TAC) Miami 01Dec52 – 60 TCWG (AFE) Rhein-Main AB W. Germany 08Nov53; depl RAF Sculthorpe England 07Dec53 & 09Mar54 – BMC (AMC) Birmingham 54-55 – 566 ADF GP (ADC) Hamilton 13Apr55 – 78 Fighter GP (ADC) Hamilton 18Aug55; depl 337 Fighter GP (ADC) Portland 25Jan56; to 3320 TTA WG (TC) Amarillo 31Jul56 mnt; to 187 Fighter SQ (WY-ANG) Cheyenne 15Mar58 mnt – 78 MSU GP (ADC) Hamilton 20Apr58 – 2578 RFC (CNR) Ellington 11Jul58; DT to Alvin Callender Field 01Aug58 – 357 TCSq (446 TCG) (AFR) Alvin Callender Field 20Sep58; to NAS New Orleans 19Nov58; 706 TCSq (446 TCWG) (AFR) NAS New Orleans 08May61; 926 TCG (446 TCWG) (AFR) NAS New Orleans 17Jan63 – Wfu to MASDC (LOG) Davis-Monthan 11Dec65 for storage – DFI: 21Jan69 – Final Disposition: Scrapped.

10615 / 51-2626 / C-119C-26-FA / (C-119CF)

    Avble: 11Apr52; Acc: 08May52; Del: 14May52 to USAF – 443 TCWG (TAC) Greenville 14May52; depl Ladd Alaska 29Oct52 – 465 TCWG (TAC) Greenville 01Feb53; depl Thule Greenland 10Jul53 – 465 TCWG (AFE) Wiesbaden AB W. Germany 01Dec53; to Toul-Rosières AB France 04May54; to Évreux-Fauville AB France 30May55; bailment SABBE (AMC) Brussels Belgium 02Nov55 – 317 TCWG (AFE) Évreux-Fauville AB France 08Jul57 – 2252 RFC (CNR) Clinton 13Oct57; to 1001 AB WG (HQC) Andrews 08May58 – 302 TCWG (AFR) Clinton 19Mar59; 906 TCG (302 TCWG) (AFR) Clinton 11Feb63; 907 TCG (302 TCWG) (AFR) Clinton 11Dec65 – Wfu to MASDC (LOG) Davis-Monthan 18Apr66 for storage – DFI: 21Jan69 – Final Disposition: Scrapped.

10616 / 51-2627 / C-119C-26-FA / (C-119CF)

    Avble: 14Apr52; Acc: 08May52; Del: 14May52 to USAF – 443 TCWG (TAC) Greenville 14May52; depl Ladd Alaska 29Oct52 – 465 TCWG (TAC) Greenville 01Feb53; depl Thule Greenland 11Jun53 – 465 TCWG (AFE) Wiesbaden AB W. Germany 01Dec53; to Toul-Rosières AB France 04May54; to Évreux-Fauville AB France 20Jun55; depl Landstuhl AB W. Germany 26Jul56 – 317 TCWG (AFE) Évreux-Fauville AB France 08Jul57 – 2577 RFC (CNR) Brooks 26Mar58 – 433 TCWG (AFR) Brooks 18Sep58; to 464 TCWG (TAC) Pope 15Oct59 mnt; to Kelly 31Aug60; 922 TCG (433 TCWG) (AFR) Kelly 31May63; 921 TCG (433 TCWG) (AFR) Kelly 10Oct65 – Wfu to MASDC (LOG) Davis-Monthan 12Mar66 for storage – DFI: 21Jan69 – Final Disposition: Scrapped.

10617 / 51-2628 / C-119C-26-FA / (C-119CF)

    Avble: 15Apr52; Acc: 08May52; Del: 14May52 to USAF – 443 TCWG (TAC) Greenville 14May52; depl Ladd Alaska 29Oct52 – 465 TCWG (TAC) Greenville 01Feb53; depl 479 Fighter-Bomber WG (TAC) George 03May53 – 465 TCWG (AFE) Wiesbaden AB W. Germany 01Dec53; to Toul-Rosières AB France 04May54; to Évreux-Fauville AB France 30May55; to 7272 AB WG (AFE) Wheelus AB Libya 13Jan56 – 317 TCWG (AFE) Évreux-Fauville AB France 08Jul57 – 2584 RFC (CNR) Memphis 08Mar58 – 1605 AB WG (MATS) Lajes Field, Azores 30Mar58 mnt – 2577 RFC (CNR) Brooks 11Jun58 – 433 TCWG (AFR) Brooks 18Sep58; to 2596 AB SQ (CNC) NAS Dallas 20May59 mnt; to 839 AB GP (TAC) Sewart 11May60; to Kelly 31Aug60; 922 TCG (433 TCWG) (AFR) Kelly 31May63; 921 TCG (433 TCWG) (AFR) Kelly 10Oct65 – Wfu to MASDC (LOG) Davis-Monthan 12Mar66 for storage – DFI: 21Jan69 – Final Disposition: Scrapped.

10618 / 51-2629 / C-119C-26-FA / (C-119CF)

    Avble: 16Apr52; Acc: 19May52; Del: 22May52 to USAF – 443 TCWG (TAC) Greenville 22May52 **RECORDS INDECIPHERABLE** → 64 TCWG (TAC) Greenville 53; depl 316 TCG (TAC) Burlington Feb-Mar53; 465 TCWG (TAC) Donaldson 53; depls Thule Greenland 53 ← **RECORDS INDECIPHERABLE** 465 TCWG (AFE) Wiesbaden AB W. Germany 03Dec53; depl Landstuhl AB W. Germany 20Apr54; to Toul-Rosières AB France 11May54; to Évreux-Fauville AB France 30May55; to 7272 AB WG (AFE) Wheelus AB Libya 08Mar57; 317 TCWG (AFE) Évreux-Fauville AB France 08Jul57; depl 36 Fighter WG (AFE) Hahn AB W. Germany 05Aug57 – 2577 RFC DT (CNR) NAS Dallas 12Dec57 – 69 TCSq (433 TCWG) (AFR) NAS Dallas 20Sep58; 923 TCG (433 TCWG) (AFR) Carswell 02Mar63; 921 TCG (433 TCWG) (AFR) Kelly 23Sep65 – Wfu to MASDC (LOG) Davis-Monthan 11Mar66 for storage – DFI: 21Jan69 – Final Disposition: Scrapped.

10619 / 51-2630 / C-119C-26-FA / (C-119CF)

Avble: 18Apr52; Acc: 19May52; Del: 22May52 to USAF – 443 TCWG (TAC) Greenville 22May52; depl Ladd Alaska 29Oct52 – 5001 CMP WG (AAC) Ladd 18Dec52 mnt – 465 TCWG (TAC) Greenville 18Feb53; depl Thule Greenland 15Jun53; to 6603 AB GP (NEA) Goose Bay Canada 17Jun53 mnt – 465 TCWG (AFE) Wiesbaden AB W. Germany 03Dec53; to Toul-Rosières AB France 04May54; to Évreux-Fauville AB France 30May55; to 7207 AB SQ (AFE) Aviano AB Italy 27Jul56; 317 TCWG (AFE) Évreux-Fauville AB France 08Jul57 – 2589 RFC (CNR) Dobbins 21Dec57 – 2259 RFC (CNR) Andrews 25Jun58 – 459 TCWG (AFR) Andrews 04Nov58; 909 TCG (459 TCWG) (AFR) Andrews 01Feb63 – Wfu to 2704 ASD GP (LOG) Davis-Monthan 16Nov64 for storage – REC: 03May65 – Final Disposition: Scrapped.

10620 / 51-2631 / C-119C-26-FA / (C-119CF)

Avble: 21Apr52; Acc: 19May52; Del: 22May52 to USAF – 443 TCWG (TAC) Greenville 22May52; depl Ladd Alaska 29Oct52 – 465 TCWG (TAC) Greenville 01Feb53; depl Thule Greenland 15Jun53 – 465 TCWG (AFE) Wiesbaden AB W. Germany 03Dec53; to Toul-Rosières AB France 04May54; to Évreux-Fauville AB France 30May55; depl Landstuhl AB W. Germany 05Jul55; to 7312 AB GP (AFE) Rhein-Main AB W. Germany 03May56; depl 317 TCWG (AFE) Neubiberg AB W. Germany 26Mar57 – 317 TCWG (AFE) Évreux-Fauville AB France 08Jul57; to 7413 SUT GP (AFE) Bordeaux AB France 12Aug57 – 2259 RFC DT (CNR) Youngstown 19Dec57; to 2585 AB SQ (CNC) Miami 26Oct58 mnt – 804 COS GP (SAC) Hunter 25Nov58 mnt – 757 TCSq (459 TCG) (AFR) Youngstown 11Jan59; 910 TCG (459 TCWG) (AFR) Youngstown 17Jan63 – Wfu to MASDC (LOG) Davis-Monthan 26Nov66 for storage – DFI: 18Jun69 – Final Disposition: Scrapped.

10621 / 51-2632 / C-119C-26-FA / (C-119CF)

Avble: 22Apr52; Acc: 19May52; Del: 22May52 to USAF – 443 TCWG (TAC) Greenville 22May52; depl Ladd Alaska 29Oct52 – 465 TCWG (TAC) Greenville 01Feb53; depl Thule Greenland 15Jun53 – 465 TCWG (AFE) Wiesbaden AB W. Germany 03Dec53; to Toul-Rosières AB France 04May54 **RECORDS INDECIPHERABLE** → to Évreux-Fauville AB France 30May55 – 2259 RFC (CNR) Andrews Jun57 ← **RECORDS INDECIPHERABLE** 909 TCG (459 TCWG) (AFR) Andrews 01Feb63 – 910 TCG (459 TCWG) (AFR) Youngstown 01Oct64 – 916 TCG (435 TCWG) (AFR) Carswell 24Jul65 – 906 TCG (302TCWG) (AFR) Clinton 18Sep65 – 907 TCG (302 TCWG) (AFR) Clinton 30Nov65 – Wfu to MASDC (LOG) Davis-Monthan 03Jun66 for storage – DFI: 21Jan69 – Final Disposition: Scrapped.

10622 / 51-2633 / C-119C-26-FA / (C-119CF)

Avble: 23Apr52; Acc: 19May52; Del: 22May52 to USAF – 443 TCWG (TAC) Greenville 22May52; depl Ladd Alaska 29Oct52 – 465 TCWG (TAC) Greenville 01Feb53 – 465 TCWG (AFE) Wiesbaden AB W. Germany 08Dec53; to 7373 AD GP (AFE) Chateauroux AB France 08Feb54; to Toul-Rosières AB France 04May54; to Évreux-Fauville AB France 30May55 **RECORDS INDECIPHERABLE** → to 7310 AB GP (AFE) Rhein-Main AB W. Germany 28Oct56 – 317 TCWG (AFE) Évreux-Fauville AB, France 08Jul57 – 2578 RFC (CNR) Ellington Feb58 – 446 TCWG (AFR) Ellington date unk ← **RECORDS INDECIPHERABLE** 925 TCG (446 TCWG) (AFR) Ellington 63; 926 TCG (446 TCWG) (AFR) NAS New Orleans 30Sep65 – Wfu to MASDC (LOG) Davis-Monthan 11Dec65 for storage – DFI: 21Jan69 – Final Disposition: Scrapped.

10623 / 51-2634 / C-119C-26-FA / (C-119CF)

Avble: 24Apr52; Acc: 28May52; Del: 03Jun52 to USAF – 443 TCWG (TAC) Greenville 03Jun52; depl Ladd Alaska 29Oct52 – 465 TCWG (TAC) Greenville 01Feb53; to 807 AB GP (SAC) March 28Feb53 mnt – 465 TCWG (AFE) Wiesbaden AB W. Germany 08Dec53; to 7280 MAI GP (AFE) Nouasseur AB Morocco 24Mar54; to Toul-Rosières AB France 04May54; depl Landstuhl AB W. Germany 10Jun54; to 7373 MAI GP (AFE) Chateauroux AB France 29Jul54 mnt; to Évreux-Fauville AB France 30May55 – 317 TCWG (AFE) Évreux-Fauville AB France 08Jul57 **RECORDS INDECIPHERABLE** → 2578 RFC (CNR) Ellington May58 – 925 TCG (446 TCWG) (AFR) Ellington 63 ← **RECORDS INDECIPHERABLE** 926 TCG (446 TCWG) (AFR) NAS New Orleans 12Sep65 – Wfu to MASDC (LOG) Davis-Monthan 28Dec65 for storage – DFI: 06May68 – Final Disposition: Scrapped.

10624 / 51-2635 / C-119C-26-FA / (C-119CF)

Avble: 25Apr52; Acc: 28May52; Del: 03Jun52 to USAF – 443 TCWG (TAC) Donaldson 03Jun52; depl Ladd Alaska 29Oct52 – 465 TCG (TAC) Donaldson 01Feb53; assigned 781 TCSq. Damaged beyond repair after landing short of the runway at Donaldson AFB South Carolina 09Feb53. Pilot: Joe M. McLelland Jr. and crew survived – REC: 16Feb53 – Final Disposition: Accident.

10625 / 51-2636 / C-119C-26-FA / (C-119CF)

Avble: 28Apr52; Acc: 28May52; Del: 03Jun52 to USAF – 443 TCWG (TAC) Greenville 03Jun52; depl Ladd Alaska 29Oct52 – 465 TCWG (TAC) Greenville 01Feb53 – 465 TCWG (AFE) Wiesbaden AB W. Germany 08Dec53; to Toul-Rosières AB France 04May54; to 7100 SUT WG (AFE) Wiesbaden AB W. Germany 04May54; depl Landstuhl AB W. Germany Nov53; to Évreux-Fauville AB France 30May55; to 7310 AB GP (AFE) Rhein-Main AB W. Germany 28Dec56 – 317 TCWG (AFE) Évreux-Fauville AB France 08Jul57 – **RECORDS MISSING** – 2577 RFC (CNR) Brooks Apr58 – 433 TCWG (AFR) Brooks 16Mar59; to Kelly 31Aug60; **RECORDS MISSING** – 921 TCG (433 TCWG) (AFR) Kelly 63 – Wfu to MASDC (LOG) Davis-Monthan 13Mar66 for storage – DFI: 21Jan69 – Final Disposition: Scrapped.

10626 / 51-2637 / C-119C-26-FA / (C-119CF)

Avble: 29Apr52; Acc: 28May52; Del: 03Jun52 to USAF – 443 TCWG (TAC) Greenville 03Jun52 – 465 TCWG (TAC) Greenville 01Feb53; depl Thule Greenland 03Aug53 – 465 TCWG (AFE) Wiesbaden AB W. Germany 08Dec53; to Toul-Rosières AB France 04May54; to 7373 MAI GP (AFE) Chateauroux AB France 12May54; depls Landstuhl AB W. Germany 10Jul54 & 28Sep54 **RECORDS INDECIPHERABLE** → to Évreux-Fauville AB France 30May55 – 317 TCWG (AFE) Évreux-Fauville AB France 08Jul57 – 2584 RFC (CNR) Memphis Feb58 – 757 TCSq (459 TCWG) (AFR) Youngstown date unk ← **RECORDS INDECIPHERABLE** 910 TCG (459 TCWG) (AFR) Youngstown 63 – Wfu to MASDC (LOG) Davis-Monthan 14May66 for storage – DFI: 21Jan69 – Final Disposition: Scrapped.

10627 / 51-2638 / C-119C-26-FA / (C-119CF)

Avble: 30Apr52; Acc: 28May52; Del: 03Jun52 to USAF – 443 TCWG (TAC) Greenville 03Jun52; depl Ladd Alaska 29Oct52 – 465 TCWG (TAC) Greenville 01Feb53; depl Thule Greenland 03Aug53 – 465 TCWG (AFE) Wiesbaden AB W. Germany 26Dec53; to Toul-Rosières AB France 04May54; to Évreux- Fauville AB France 30May55 **RECORDS INDECIPHERABLE** → 317 TCWG (AFE) Évreux-Fauville AB France 08Jul57 – 2584 RFC (CNR) Memphis Jan58 – 337 TCSq (514 TCWG) (AFR) Bradley date unk ← **RECORDS**

INDECIPHERABLE 905 TCG (514 TCWG) (AFR) Bradley 17Jan63 – 907 TCG (302 TCWG) (AFR) Clinton 07Nov65 – Wfu to MASDC (LOG) Davis-Monthan 03Jun66 for storage – DFI: 21Jan69 – Final Disposition: Scrapped.

10628 / 51-2639 / C-119C-26-FA / (C-119CF)

Avble: 02May52; Acc: 29May52; Del: 04Jun52 to USAF – 443 TCWG (TAC) Greenville 04Jun52; depl Ladd Alaska 29Oct52 – 465 TCWG (TAC) Greenville 01Feb53; depl Thule Greenland 22Jul53 – 465 TCWG (AFE) Wiesbaden AB W. Germany 26Dec53; to Toul-Rosières AB France 04May54; to Évreux-Fauville AB France 30May55 – 317 TCWG (AFE) Évreux-Fauville AB France 08Jul57 – **RECORDS MISSING** – 2577 RFC (CNR) Brooks Mar58 – 433 TCWG (AFR) Brooks date unk. Crashed into a hill 3.9 miles NE Hopkinsville-Fort Campbell Kentucky 15Jan61, 5 crew killed – 839 AB GP (TAC) Sewart 15Jan61 for SAL – DFI: 26Jan61 – Final Disposition: Accident.

10629 / 51-2640 / C-119C-26-FA / (C-119CF)

Avble: 05May52; Acc: 29May52; Del: 04Jun52 to USAF – 443 TCWG (TAC) Greenville 04Jun52; depl Ladd Alaska 29Oct52 – 465 TCWG (TAC) Greenville 01Feb53; depl Thule Greenland 11Jun53; to 6603 AB GP (NEA) Goose Bay Canada 24Jul53 mnt – 465 TCWG (AFE) Wiesbaden AB W. Germany 26Dec53; to Toul-Rosières AB France 04May54; to 7100 SUT WG (AFE) Wiesbaden AB W. Germany 29Jun54; to Évreux-Fauville AB France 30May55; to 3110 MAI GP (AMC) RAF Burtonwood England 14Nov56 mnt; to 7272 AB WG (AFE) Wheelus AB Libya 04Dec56 – 317 TCWG (AFE) Évreux-Fauville AB France 08Jul57 – 2589 RFC (CNR) Dobbins Dec57 – 2259 RFC (CNR) Andrews 1958–2255 RFC (CNR) Bradley 16Oct58 – 337 TCSq (514 TCWG) (AFR) Bradley 19Mar59; 905 TCG (514 TCWG) (AFR) Bradley 17Jan63 – 4411 CCT GP (TAC) Shaw 22Aug63. Listed as accident 07Oct63, nfd – Final Disposition: Accident.

10630 / 51-2641 / C-119C-26-FA / (C-119CF)

Avble: 07May52; Acc: 29May52; Del: 04Jun52 to USAF – 443 TCWG (TAC) Greenville 04Jun52; depl Ladd Alaska 29Oct52 – 465 TCWG (TAC) Greenville 01Feb53; depl Thule Greenland 11Jun53; to 6612 AB GP (NEA) Thule AB Greenland 20Jul53 mnt – 465 TCWG (AFE) Wiesbaden AB W. Germany 01Dec53; to Toul-Rosières AB France 04May54; to Évreux-Fauville AB France 30May55 **RECORDS INDECIPHERABLE** → 317 TCWG (AFE) Évreux-Fauville AB France 08Jul57 – 2578 RFC (CNR) Ellington Jan58; to DT Alvin Callender Field date unk – 357 TCSq (446 TCWG) (AFR) NAS New Orleans date unk ← **RECORDS INDECIPHERABLE** 706 TCSq (446 TCWG) (AFR) NAS New Orleans 08May61; 926 TCG (446 TCWG) (AFR) NAS New Orleans 07Feb63 – Wfu to MASDC (LOG) Davis-Monthan 11Dec65 for storage – DFI: 21Jan69 – Final Disposition: Scrapped.

10631 / 51-2642 / C-119C-26-FA / (C-119CF)

Avble: 07May52; Acc: 29May52; Del: 04Jun52 to USAF – 443 TCWG (TAC) Greenville 04Jun52 – 465 TCWG (TAC) Greenville 01Feb53; depl Thule Greenland 15Jun53 – 465 TCWG (AFE) Toul-Rosières AB France 26Dec53; depl Landstuhl AB W. Germany Oct54; **RECORDS MISSING**; to Évreux-Fauville AB France 30May55; to 3131 MAI GP (AMC) Chateauroux AB France 04Jun57 mnt **RECORDS INDECIPHERABLE** → 317 TCWG (AFE) Évreux-Fauville AB France 08Jul57 – 2584 RFC (CNR) Memphis Feb58 – 757 TCSq (459 TCWG) (AFR) Youngstown date unk ← **RECORDS INDECIPHERABLE**

910 TCG (459 TCWG) (AFR) Youngstown 17Jan63 – Wfu to MASDC (LOG) Davis-Monthan 30Jul66 for storage – DFI: 21Jan69 – Final Disposition: Scrapped.

10632 / 51-2643 / C-119C-26-FA / (C-119CF)

Avble: 08May52; Acc: 04Jun52; Del: 09Jun52 to USAF – 443 TCWG (TAC) Greenville 09Jun52 – 465 TCWG (TAC) Greenville 01Feb53; depl Thule Greenland 15Jun53 – 465 TCWG (AFE) Toul-Rosières AB France 26Dec53; depl Landstuhl AB W. Germany Oct54; to Évreux-Fauville AB France 30May55; **RECORDS MISSING**; to 7373 MAI GP (AFE) Chateauroux AB France 22Nov56 mnt – 317 TCWG (AFE) Évreux-Fauville AB France 08Jul57 – 2577 RFC (CNR) Brooks Mar58 – **RECORDS MISSING** – 433 TCWG (AFR) Brooks date unk; to Kelly 31Aug60; **RECORDS INDECIPHERABLE**; 921 TCG (433 TCWG) (AFR) Kelly 63; to 3800 AU WG (AU) Maxwell 30Nov65 mnt – Wfu to MASDC (LOG) Davis-Monthan 14Jan66 for storage – DFI: 21Jan69 – Final Disposition: Scrapped.

10633 / 51-2644 / C-119C-26-FA / (C-119CF)

Avble: 09May52; Acc: 14Jun52; Del: 19Jun52 to USAF – 443 TCWG (TAC) Greenville 19Jun52; depl Ladd Alaska 29Oct52 – 465 TCWG (TAC) Greenville 01Feb53; depl Thule Greenland 23Jul53 – 465 TCWG (AFE) Toul-Rosières AB France 26Dec53; to Évreux-Fauville AB France 23May55 **RECORDS INDECIPHERABLE** → 317 TCWG (AFE) Évreux-Fauville AB France 08Jul57 – 2578 RFC (CNR) Ellington Jan58; to 1607 TSH WG (MATS) Dover 58 mnt; DT to Alvin Callender Field 58 – 706 TCSq (446 TCWG) (AFR) NAS New Orleans 08May61; 926 TCG (446 TCWG) (AFR) NAS New Orleans 17Jan63 ← **RECORDS INDECIPHERABLE** Wfu to MASDC (LOG) Davis-Monthan 18Dec65 for storage – DFI: 21Jan69 – Final Disposition: Scrapped.

10634 / 51-2645 / C-119C-26-FA / (C-119CF)

Avble: 13May52; Acc: 14Jun52; Del: 19Jun52 to USAF – 443 TCWG (TAC) Greenville 19Jun52; depl Ladd Alaska 29Oct52 – 465 TCWG (TAC) Greenville 01Feb53; depl Thule Greenland 23Jul53 – 465 TCWG (AFE) Toul-Rosières AB France 26Dec53; depl 60 TCWG (AFE) Rhein-Main AB W. Germany 18Sep54; to Évreux-Fauville AB France Dec54. Had an inflight engine failure 9.4 miles SE Moulins France 18Apr55 while on a resupply mission, crew bailed out with one injured – Final Disposition: Accident.

10635 / 51-2646 / C-119C-26-FA / (C-119CF)

Avble: 14May52; Acc: 26Jun52; Del: 07Jul52 to USAF – 443 TCWG (TAC) Greenville 07Jul52; depl Ladd Alaska 29Oct52 – 465 TCWG (TAC) Greenville 01Feb53 – 465 TCWG (AFE) Toul-Rosières AB France 05Dec53; depl 86 Fighter-Bomber WG (AFE) Landstuhl AB W. Germany 16Jan54; depl 317 TCWG (AFE) Neubiberg AB W. Germany 13Oct53; to Évreux-Fauville AB France Dec54 **RECORDS INDECIPHERABLE** → to 3131 MAI GP (AMC) Chateauroux AB France date unk mnt – 317 TCWG (AFE) Évreux-Fauville AB France 08Jul57 – 2259 RFC (CNR) Andrews Nov57 – 459 TCWG (AFR) Andrews date unk ← **RECORDS INDECIPHERABLE** 909 TCG (459 TCWG) (AFR) Andrews 63 – Wfu to MASDC (LOG) Davis-Monthan 21Aug66 for storage – DFI: 21Jan69 – Final Disposition: Scrapped.

10636 / 51-2647 / C-119C-26-FA / (C-119CF)

Avble: 15May52; Acc: 16Jun52; Del: 28Jun52 to USAF – 443 TCWG (TAC) Greenville 27Jun52; depl Ladd Alaska 29Oct52 – 465 TCWG (TAC) Greenville 01Feb53; depl Thule Greenland 15Jun53 – 465 TCWG (AFE) Toul-Rosières AB France 05Dec53; to 7100 HQS

WG (AFE) Wiesbaden AB W. Germany 01Feb54; to Évreux-Fauville AB France Dec54 **RECORDS INDECIPHERABLE** → to 3110 MAI GP (AMC) RAF Burtonwood England 05Nov56 mnt – 317 TCWG (AFE) Évreux-Fauville AB France 08Jul57 – 2584 RFC (CNR) Memphis Mar58 – 302 TCWG (AFR) Clinton date unk ← **RECORDS INDECIPHERABLE** 906 TCG (302 TCWG) (AFR) Clinton 63; 907 TCG (302 TCWG) (AFR) Clinton 11Dec65 – Wfu to MASDC (LOG) Davis-Monthan 05Aug66 for storage – DFI: 18Jun69 – Final Disposition: Scrapped.

10637 / 51-2648 / C-119C-26-FA / (C-119CF)

Avble: 16May52; Acc: 16Jun52; Del: 28Jun52 to USAF – 443 TCWG (TAC) Greenville 27Jun52; depl Ladd Alaska 29Oct52 – 465 TCWG (TAC) Greenville 01Feb53; depl Thule Greenland 15Jun53; to 6603 AB GP (NEA) Goose Bay Canada 02Aug53 mnt – 465 TCWG (AFE) Toul-Rosières AB France 05Dec53; to 86 Fighter-Bomber WG (AFE) Landstuhl AB W. Germany 19Dec53 mnt; to Évreux-Fauville AB France Jan55 **RECORDS INDECIPHERABLE** Crash-landed due to fuel starvation at Ghisonaccia- Alzitone Field Corsica (France) 18Apr55, 4 crew onboard suffered minor injuries – DFI: 17Aug55 – Final Disposition: Accident.

10638 / 51-2649 / C-119C-26-FA / (C-119CF)

Avble: 19May52; Acc: 17Jun52; Del: 30Jun52 to USAF – 443 TCWG (TAC) Greenville 30Jun52; depl Ladd Alaska 29Oct52 – 465 TCWG (TAC) Greenville 01Feb53; to 2500 AB WG (CNC) Mitchel 15Jun53 mnt – 465 TCWG (AFE) Toul-Rosières AB France 05Dec53; to Évreux-Fauville AB France Dec54 – **RECORDS MISSING** – 317 TCWG (AFE) Évreux-Fauville AB France 08Jul57 – 2259 RFC DT (CNR) Youngstown Apr58 – **RECORDS MISSING** – 757 TCSq (459 TCWG) (AFR) Youngstown date unk; 910 TCG (459 TCWG) (AFR) Youngstown 17Jan63 – Wfu to MASDC (LOG) Davis-Monthan 30Jul66 for storage – DFI: 21Jan69 – Final Disposition: Scrapped.

10639 / 51-2650 / C-119C-26-FA / (C-119CF)

Avble: 21May52; Acc: 17Jun52; Del: 30Jun52 to USAF – 443 TCWG (TAC) Greenville 30Jun52; depl Ladd Alaska 29Oct52 – 465 TCWG (TAC) Greenville 01Feb53; depl Thule Greenland 15Jun53 – 465 TCWG (AFE) Toul-Rosières AB France 09Dec53; depl Landstuhl AB W. Germany 31Aug54; to Évreux-Fauville AB France Dec54 **RECORDS INDECIPHERABLE** → 317 TCWG (AFE) Évreux-Fauville AB France 08Jul57 – 2284 RFC (CNR) Memphis Mar58 ← **RECORDS INDECIPHERABLE** 757 TCSq (459 TCWG) (AFR) Youngstown date unk; 910 TCG (459 TCWG) (AFR) Youngstown 17Jan63 – Wfu to MASDC (LOG) Davis-Monthan 14May66 for storage – DFI: 21Jan69 – Final Disposition: Scrapped.

10640 / 51-2651 / C-119C-26-FA / (C-119CF)

Avble: 22May52; Acc: 25Jun52; Del: 30Jun52 to USAF – 443 TCWG (TAC) Greenville 30Jun52; depl Ladd Alaska 29Oct52 – 465 TCWG (TAC) Greenville 01Feb53; depl Thule Greenland 27Jul53 – 465 TCWG (AFE) Toul-Rosières AB France 09Dec53; to Évreux-Fauville AB France Dec54 **RECORDS INDECIPHERABLE** → 317 TCWG (AFE) Évreux-Fauville AB France 08Jul57 – 2578 RFC (CNR) Ellington Feb58 ← **RECORDS INDECIPHERABLE** 706 TCSq (446 TCWG) (AFR) NAS New Orleans 08May61; 926 TCG (446 TCWG) (AFR) NAS New Orleans 63 – Wfu to MASDC (LOG) Davis-Monthan 19Nov65 for storage – DFI: 21Jan69 – Final Disposition: Scrapped.

10641 / 51-2652 / C-119C-26-FA / (C-119CF)

Avble: 23May52; Acc: 25Jun52; Del: 30Jun52 to USAF – 443 TCWG (TAC) Greenville 30Jun52; depl Ladd Alaska 29Oct52 – 465 TCWG (TAC) Greenville 01Feb53; depl Thule Greenland 24Aug53 – 465 TCWG (AFE) Toul-Rosières AB France 09Dec53; depl Landstuhl AB W. Germany Oct54; to Évreux- Fauville AB France Dec54 **RECORDS INDECIPHERABLE** → 317 TCWG (AFE) Évreux-Fauville AB France 08Jul57 – 2589 RFC (CNR) Dobbins Dec57 ← **RECORDS INDECIPHERABLE** 446 TCWG (AFR) Ellington date unk; 925 TCG (446 TCWG) (AFR) Ellington 63 – 910 TCG (459 TCWG) (AFR) Youngstown 21Sep65 – Wfu to MASDC (LOG) Davis-Monthan 30Jul66 for storage – DFI: 21Jan69 – Final Disposition: Scrapped.

10642 / 51-2653 / C-119C-26-FA / (C-119CF)

Avble: 28May52; Acc: 25Jun52; Del: 30Jun52 to USAF – 443 TCWG (TAC) Greenville 30Jun52; depl Ladd Alaska 29Oct52 – 465 TCWG (TAC) Greenville 01Feb53 – 465 TCWG (AFE) Toul-Rosières AB France 09Dec53; to Évreux-Fauville AB France Dec54 **RECORDS INDECIPHERABLE** → to 1603 TSP WG (AFE) Wheelus AB Libya 20Jun55 ← **RECORDS INDECIPHERABLE** Landed 200 ft short of the runway at Chateauroux AB France 12Feb56 during a night landing, no fatalities – 3131 MAI GP (AMC) Chateauroux AB France 29Feb56 for salvage – DFI: 18Apr56 – Final Disposition: Accident.

10643 / 51-2654 / C-119C-26-FA / (C-119CF)

Avble: 28May52; Acc: 25Jun52; Del: 30Jun52 to USAF – 443 TCWG (TAC) Greenville 30Jun52; depl Ladd Alaska 29Oct52 – 465 TCWG (TAC) Greenville 01Feb53 – 465 TCWG (AFE) Toul-Rosières AB France 13Dec53; to 7373 AD WG (AFE) Chateauroux AB France 11Feb54; to Évreux-Fauville AB France Dec54 **RECORDS INDECIPHERABLE** → 317 TCWG (AFE) Évreux-Fauville AB France 08Jul57 – 2577 RFC (CNR) Brooks Mar58 ← **RECORDS INDECIPHERABLE** 433 TCWG (AFR) Brooks date unk; to Kelly 31Aug60; 921 TCG (433 TCWG) (AFR) Kelly 31May63 – Wfu to MASDC (LOG) Davis- Monthan 25Feb66 for storage – DFI: 21Jan69 – Final Disposition: Scrapped.

10644 / 51-2655 / C-119C-26-FA / (C-119CF)

Avble: 28May52; Acc: 25Jun52; Del: 30Jun52 to USAF – 443 TCWG (TAC) Greenville 30Jun52; depl Ladd Alaska 29Oct52 – 465 TCWG (TAC) Greenville 01Feb53; depl Thule Greenland 15Jun53 – 465 TCWG (AFE) Toul-Rosières AB France 12Dec53; to Évreux-Fauville AB France Dec54; **RECORDS MISSING** – 317 TCWG (AFE) Évreux-Fauville AB France 08Jul57 – 2577 RFC DT (CNR) NAS Dallas 02Dec57 – **RECORDS MISSING** – 69 TCSq (433 TCWG) (AFR) NAS Dallas date unk; 923 TCG (433 TCWG) (AFR) NAS Dallas 17Jan63; to Carswell 02Mar63 – 907 TCG (302 TCWG) (AFR) Clinton 18Sep65; assigned 906 TCG Clinton 05Oct65-30Nov65 – Wfu to MASDC (LOG) Davis-Monthan 05Aug66 for storage – DFI: 18Jun69 – Final Disposition: Scrapped.

10645 / 51-2656 / C-119C-26-FA / (C-119CF)

Avble: 30May52; Acc: 26Jun52; Del: 02Jul52 to USAF – 443 TCWG (TAC) Greenville 02Jul52; depl Ladd Alaska 29Oct52 – 465 TCWG (TAC) Greenville 01Feb53 – 465 TCWG (AFE) Toul-Rosières AB France 14Dec53; to Évreux-Fauville AB France Dec54 – **RECORDS MISSING** – 317 TCWG (AFE) Évreux-Fauville AB France 08Jul57 – 2596 RFC (CNR) NAS Dallas 29Oct57 – **RECORDS MISSING** – 69 TCSq (433 TCWG) (AFR) NAS Dallas date unk; 923 TCG (433 TCWG) (AFR) NAS Dallas 17Jan63; to Carswell 02Mar63 –

246 Section Four—Service Histories

906 TCG (302 TCWG) (AFR) Clinton 18Sep65; 907 TCG (302 TCWG) (AFR) Clinton 30Nov65 – Wfu to MASDC (LOG) Davis-Monthan 18Apr66 for storage – DFI: 21Jan69 – Final Disposition: Scrapped.

10646 / 51-2657 / C-119C-26-FA / (C-119CF)

Avble: 02Jun52; Acc: 26Jun52; Del: 02Jul52 to USAF – 443 TCWG (TAC) Greenville 02Jul52; depl Ladd Alaska 29Oct52 – 465 TCWG (TAC) Greenville 01Feb53 – 465 TCWG (AFE) Toul-Rosières AB France 12Dec53; to Évreux-Fauville AB France Dec54; **RECORDS MISSING** – 317 TCWG (AFE) Évreux- Fauville AB France 08Jul57 – 2589 RFC (CNR) Dobbins 11Dec57; 1605 FDM SQ (MATS) Lajes Field, Azores 19Dec57 mnt – **RECORDS MISSING** – 337 TCSq (514 TCWG) Bradley date unk; to 2253 AB GP (CNC) Greater Pittsburgh 01Jun59 mnt; **RECORDS INDECIPHERABLE**; 905 TCG (514 TCWG) (AFR) Bradley 63 – Wfu to MASDC (LOG) Davis-Monthan 17Feb66 for storage – DFI: 21Jan69 – Final Disposition: Scrapped.

10647 / 51-2658 / C-119C-26-FA / (C-119CF)

Avble: 04Jun52; Acc: 26Jun52; Del: 09Jul52 to USAF – 435 TCWG (TAC) Miami 09Jul52; depl Elmendorf 24Oct52 – 456 TCWG (TAC) Miami IAP 01Dec52 – 64 TCWG (TAC) Greenville 19Jun53 – 317 TCWG (AFE) Neubiberg AB W. Germany 01Nov53 **RECORDS INDECIPHERABLE** → 60 TCWG (AFE) Rhein-Main AB W. Germany date unk – 521 ADF GP (ADC) Sioux City 27Apr55; to 1400 FDM SQ (MATS) Keflavik AS Iceland 27Apr55 mnt – 53 Fighter GP (ASDC) Sioux City 18Aug56 ← **RECORDS INDECIPHERABLE** 923 TCG (433 TCWG) (AFR) Carswell 63; 921 TCG (433 TCWG) (AFR) Kelly 23Sep65 – Wfu to MASDC (LOG) Davis-Monthan 14Jan66 for storage – DFI: 21Jan69 – Final Disposition: Scrapped.

10648 / 51-2659 / C-119C-26-FA / (C-119CF)

Avble: 05Jun52; Acc: 26Jun52; Del: 09Jul52 to USAF – 435 TCWG (TAC) Miami 09Jul52; depl Elmendorf 01Nov52 – 456 TCWG (TAC) Miami 01Dec52 – 64 TCWG (TAC) Greenville 19Jun53 – 317 TCWG (AFE) Neubiberg AB W. Germany 04Nov53 – 60 TCWG (AFE) Rhein-Main AB W. Germany 28Jan54 – 501 ADF GP (ADC) O'Hare 11May55 – SBNAR (AMC) Norton 12Aug55 mnt – 56 Fighter GP (ADC) O'Hare 19Nov55 – 56 CLM SQ (ADC) O'Hare 08Sep57; to 328 CLM SQ (ADC) Richards 10Jul58 – 446 TCWG (AFR) Ellington 24Jan59; to 2578 AB GP (CNC) Ellington 24Aug60 mnt – REC: 19Jan61 – Final Disposition: Scrapped.

10649 / 51-2660 / C-119C-26-FA / (C-119CF)

Avble: 06Jun52; Acc: 26Jun52; Del: 09Jul52 to USAF – 435 TCWG (TAC) Miami IAP 09Jul52; depl Elmendorf 24Oct52 – 456 TCWG (TAC) Miami 01Dec52 – 64 TCWG (TAC) Greenville 19Jun53 – 317 TCWG (AFE) Neubiberg AB W. Germany 02Nov53 **RECORDS INDECIPHERABLE** → 60 TCWG (AFE) Rhein-Main AB W. Germany date unk – 33 Fighter GP (ADC) Otis 30Aug55 – 33 MSU GP (ADC) Otis 26Dec56; to 801 AB GP (SAC) Lockbourne 04Apr57 – 706 TCSq (446 TCWG) (AFR) NAS New Orleans 08May61 ← **RECORDS INDECIPHERABLE** 926 TCG (446 TCWG) (AFR) NAS New Orleans 63 – Wfu to MASDC (LOG) Davis-Monthan 19Nov65 for storage – DFI: 21Jan69 – Final Disposition: Scrapped.

10650 / 51-2661 / C-119C-26-FA / (C-119CF)

Avble: 09Jun52; Acc: 26Jun52; Del: 09Jul52 to USAF – 435 TCWG (TAC) Miami 09Jul52; depl Elmendorf 24Oct52 – 456 TCWG (TAC) Miami 01Dec52 – 64 TCWG (TAC)

Greenville 09Jul53 – 465 TCWG (TAC) Donaldson Aug53 – 465 TCWG (AFE) Toul-Rosières AB France 12Dec53; to Évreux-Fauville AB France Dec54 – **RECORDS MISSING** – 2259 RFC (CNR) Andrews Apr57 – 459 TCWG (AFR) Andrews date unk; 909 TCG (459 TCWG) (AFR) Andrews 63 – Wfu to MASDC (LOG) Davis-Monthan 04Apr66 for storage – DFI: 21Jan69 – Final Disposition: Scrapped.

10651 / 51-2662 / C-119F-26-FA

Avble: 12Jun52; Acc: 11Jul52; Del: 17Jul52 to USAF – 516 TCWG (TAC) Memphis 17Jul52 – 463 TCWG (TAC) Memphis 16Jan53; to Ardmore 17Aug53; depl Charleston 25Apr54; to Fairchild (AMC) Hagerstown 54, cvtd **C-119G** 15Feb55; depl unk base 07Apr55 – 2466 RFC (CNR) Bakalar 08Aug57 – 434 TCWG (AFR) Bakalar 19Apr59 – 403 TCWG (AFR) Selfridge date unk; 927 TCG (403 TCWG) (AFR) Selfridge 63 – 930 SOP GP (434 TCWG) (AFR) Bakalar 01Jul69; to Grissom 12Jan70 – Wfu to MASDC (LOG) Davis-Monthan 04Sep70 for storage, PCN: **CJ357** – DFI: 16Sep70 – Kolar, Inc. Tucson Arizona 12Feb76 for scrap – Final Disposition: Scrapped.

10652 / 51-2663 / C-119F-26-FA

Avble: 20Jun52; Acc: 17Jul52; Del: 22Jul52 to USAF – 316 TCG (TAC) Sewart 22Jul52; depl Burlington 06Feb53; depl Pope 19Apr54; depl Sumpter Alabama 24Aug54 – 314 TCWG (TAC) Sewart 54; depl Elmendorf Jan55; depl Ladd Alaska 01Jun55 – 463 TCWG (TAC) Sewart (depl) 19Aug55 – 513 TFW GP (TAC) Sewart 08Nov55 **RECORDS INDECIPHERABLE** → cvtd **C-119G** date unk – 403 TCWG (AFR) Selfridge date unk ← **RECORDS INDECIPHERABLE** 927 TCG (403 TCWG) (AFR) Selfridge 10Feb63 – 930 SOP GP (434 TCWG) (AFR) Bakalar 01Jul69; to Grissom 12Jan70 – Wfu to MASDC (LOG) Davis- Monthan 04Sep70 for storage, PCN: **CJ358** – DFI: 16Sep70 – Kolar, Inc. Tucson Arizona 05Feb76 for scrap Final Disposition: Scrapped.

10653 / 51-2664 / C-119F-26-FA

Avble: 16Jun52; Acc: 17Jul52; Del: 22Jul52 to USAF – 316 TCG (TAC) Sewart 22Jul52; depl Burlington 06Feb53; depl Sumpter Alabama 24Aug54 – 314 TCWG (TAC) Sewart 54; depl Elmendorf Jan55 – 463 TCWG (TAC) Sewart 10Aug55 – 513 TFW GP (TAC) Sewart 08Nov55 **RECORDS INDECIPHERABLE** → cvtd **C-119G** date unk – 464 TCWG (TAC) Pope 01Sep57 – **RECORDS MISSING** – 512 TCWG (AFR) NAS Willow Grove date unk – 77 TCSq (435 TCWG) (AFR) Donaldson 23Mar59 – 403 TCWG (AFR) Selfridge date unk; 927 TCG (403 TCWG) (AFR) Selfridge 63 ← **RECORDS INDECIPHERABLE** 930 SOP GP (434 TCWG) (AFR) Bakalar 01Jul69; to Grissom 12Jan70 – Wfu to MASDC (LOG) Davis-Monthan 04Sep70 for storage, PCN: **CJ359** – DFI: 16Sep70 – Kolar, Inc. Tucson Arizona 10Feb76 for scrap – Final Disposition: Scrapped.

10654 / 51-2665 / C-119F-26-FA

Avble: 17Jun52; Acc: 17Jul52; Del: 22Jul52 to USAF – 316 TCG (TAC) Sewart 22Jul52; depl Burlington 06Feb53; depl Pope 20Apr54; depl Sumpter Alabama 24Aug54 – 314 TCWG (TAC) Sewart 54; depl Elmendorf Jan55. Crashed after a loss of control due to mechanical failure during a training flight 1.9 miles E Lapine Alabama 01Apr55. Pilot: Kenneth Rasmussen and 5 other crew bailed out with the aircraft coming down in a wooded area resulting in a forest fire – Final Disposition: Accident.

10655 / 51-2666 / C-119F-26-FA

Avble: 17Jun52; Acc: 17Jul52; Del: 22Jul52 to USAF – 316 TCG (TAC) Sewart 22Jul52; depl Burlington 06Feb53; depl Sumpter Alabama 24Aug54 – 314 TCWG (TAC) Sewart 54; depl Elmendorf Jan55; bailment Fairchild (AMC) Hagerstown 17Oct55, cvtd **C-119G** 23Nov55 **RECORDS INDECIPHERABLE** → 76 TCSq (435 TCWG) (AFR) Homestead date unk ← **RECORDS INDECIPHERABLE** 915 TCG (435 TCWG) (AFR) Homestead 17Jan63 – 906 TCG (302 TCWG) (AFR) Clinton 15Dec65 – Wfu to MASDC (LOG) Davis-Monthan 18Feb69 for storage, PCN: **CJ288** – DFI: 09Dec74 – Final Disposition: Scrapped.

10656 / 51-2667 / C-119F-26-FA

Avble: 18Jun52; Acc: 23Jul52; Del: 26Jul52 to USAF – 316 TCG (TAC) Sewart 25Jul52; depl Burlington 06Feb53; depl Pope 20Apr54 – 314 TCWG (TAC) Sewart 54 – **RECORDS MISSING** – cvtd **C-119G** date unk – 2466 RFC (CNR) Bakalar 18Aug57 – 2469 RFC (CNR) Scott 01Sep57, to DT Scott 01Dec57 – **RECORDS MISSING** – 73 TCSq (434 TCWG) (AFR) Scott 18Apr59 **RECORDS INDECIPHERABLE** 932 TCG (434 TCWG) (AFR) Scott 63 – 910 TCG (459 TCWG) (AFR) Youngstown 05Nov66 – 72 Bomb WG (SAC) Ramey 14Dec68. Crashed into Pico Del Este Mountain 23 miles ESE San Juan Puerto Rico 14Dec68. The C-119 departed NAS Roosevelt Roads Puerto Rico destined for Homestead AFB Florida but after developing a radio fault the crew decided to return. The aircraft impacted the mountain at 3,400 ft while on approach killing all 8 crew onboard, the wreckage was found the next day – DFI: 15Dec68 – Final Disposition: Accident.

10657 / 51-2668 / C-119F-FA

Avble: 20Jun52; Acc: 23Jul52; Del: 26Jul52 to USAF – 316 TCG (TAC) Sewart 25Jul52; depl Burlington 06Feb53; depl Pope 20Apr54; depl Sumpter Alabama 24Aug54 – 314 TCWG (TAC) Sewart 54; **RECORDS INDECIPHERABLE** → cvtd **C-119G** date unk – 452 TCWG (AFR) March date unk ← **RECORDS INDECIPHERABLE** 944 TCG (452 TCWG) (AFR) March 63 – 933 TAL GP (440 TCWG) (AFR) Gen Mitchell 27Jan68 – Wfu to MASDC (LOG) Davis-Monthan 28Nov70 for storage, PCN: **CJ369** – DFI: 21May71 – Kolar, Inc. Tucson Arizona 27Feb76 for scrap – Final Disposition: Scrapped.

10658 / 51-2669 / C-119F-FA

Avble: 23Jun52; Acc: 23Jul52; Del: 26Jul52 to USAF – 316 TCG (TAC) Sewart 25Jul52; depl Burlington 06Feb53; depl Pope 20Apr54; depl Sumpter Alabama 24Aug54 – 314 TCWG (TAC) Sewart 54; depl Elmendorf Jan55; bailment Fairchild (AMC) Hagerstown 18Aug55, cvtd **C-119G** 22Dec55 **RECORDS INDECIPHERABLE** → 435 TCWG (AFR) Homestead date unk; assigned 76 TCSq date unk ← **RECORDS INDECIPHERABLE** 915 TCG (435 TCWG) (AFR) Homestead 63 – 926 TCG (446 TCWG) (AFR) NAS New Orleans 08Nov65 – 906 TAL GP (AFR) Clinton 10Jan70 – MAP Taiwan 13Mar70, RoCAF s/n: **3208** – Final Disposition: Unknown.

10659 / 51-2670 / C-119F-FA

Avble: 15Jul52; Acc: 23Jul52; Del: 26Jul52 to USAF – 316 TCG (TAC) Sewart 25Jul52; depl Burlington 06Feb53; depl Pope 20Apr54; depl Sumpter Alabama 24Aug54 – 314 TCWG (TAC) Sewart 54; depl Elmendorf Jan55; bailment Fairchild (AMC) Hagerstown 28Nov55, cvtd **C-119G** 04Jan56 **RECORDS INDECIPHERABLE** → 2472 RFC (CNR) Richards 28Jun57 – 442 TCWG (AFR) Richards 02Mar59 – 76 TCSq (435 TCWG) (AFR) Homestead date unk; 915 TCG (435 TCWG) (AFR) Homestead 63 ← **RECORDS**

Section Four—Service Histories 249

INDECIPHERABLE 906 TCG (302 TCWG) (AFR) Clinton 26Nov65 – 910 TAL GP (459 TCWG) (AFR) Youngstown 06Nov67 – Wfu to MASDC (LOG) Davis-Monthan 06Dec69 for storage, PCN: **CJ336** – DFI: 19Jan70 – Kolar, Inc. Tucson Arizona 23Feb76 for scrap – Final Disposition: Scrapped.

10660 / 51-2671 / C-119F-FA

Avble: 15Jul52; Acc: 23Jul52; Del: 26Jul52 to USAF – 316 TCG (TAC) Sewart 25Jul52; depl Pope 20Apr54; depl Sumpter Alabama 24Aug54 – 314 TCWG (TAC) Sewart 54; depl Elmendorf Jan55 **RECORDS INDECIPHERABLE** → cvtd **C-119G** date unk – 442 TCWG (AFR) Richards date unk – 73 TCSq (434 TCWG) (AFR) Scott date unk; 932 TCG (434 TCWG) (AFR) Scott 63 ← **RECORDS INDECIPHERABLE** 928 TCG (403 TCWG) (AFR) O'Hare 19Jan67 – MAP Taiwan 03Dec69 – Final Disposition: Unknown.

10661 / 51-2672 / C-119F-FA

Avble: 26Jun52; Acc: 23Jul52; Del: 26Jul52 to USAF – 316 TCG (TAC) Sewart 25Jul52; depl Burlington 06Feb53; depl Pope 20Apr54; depl Sumpter Alabama 24Aug54 – 314 TCWG (TAC) Sewart 54; depl Elmendorf Jan55; bailment Fairchild (AMC) Hagerstown 21Nov55, cvtd **C-119G** 22Dec55 **RECORDS INDECIPHERABLE** → 73 TCSq (434 TCWG) (AFR) Scott date unk; 932 TCG (434 TCWG) (AFR) Scott 63 ← **RECORDS INDECIPHERABLE** 910 TCG (459 TCWG) (AFR) Youngstown 05Nov66 – Wfu to MASDC (LOG) Davis-Monthan 13Sep69 for storage, PCN: **CJ309** – DFI: 31Mar76 – Final Disposition: Scrapped.

10662 / 51-2673 / C-119F-FA

Avble: 27Jun52; Acc: 28Jul52; Del: 31Jul52 to USAF – 316 TCG (TAC) Sewart 31Jul52; depl Burlington 06Feb53; depl Pope 19Apr54; depl Sumpter Alabama 24Aug54 – 314 TCWG (TAC) Sewart 54; depl Elmendorf Jan55; bailment Fairchild (AMC) Hagerstown 24Oct55, cvtd **C-119G** 06Dec55 – **RECORDS MISSING** – 2466 RFC (CNR) Bakalar 18Aug57 – 2469 RFC (CNR) Scott 01Sep57, to DT Scott 01Dec57 – **RECORDS MISSING** – 65 TCSq (403 TCWG) (AFR) Davis date unk **RECORDS INDECIPHERABLE** 929 TCG (403 TCWG) (AFR) Davis 63 – 922 TCG (433 TCWG) (AFR) Kelly 17Oct65; to 464 TAL WG (TAC) Pope 21Aug67 – Wfu to MASDC (LOG) Davis-Monthan 07Nov68 for storage, PCN: **CJ272** – DFI: 26Dec73 – Final Disposition: Scrapped.

10663 / 51-2674 / C-119F-FA

Avble: 30Jun52; Acc: 28Jul52; Del: 31Jul52 to USAF – 316 TCG (TAC) Sewart 31Jul52; depl Burlington 06Feb53; depl Pope 19Apr54; depl Sumpter Alabama 24Aug54; depl Elmendorf 01Sep54 – 314 TCWG (TAC) Sewart 15Nov54; depl Elmendorf Jan55; bailment BMC (AMC) Birmingham 03Oct55, cvtd **C-119G** 29Nov55 – **RECORDS MISSING** – 403 TCWG (AFR) Selfridge date unk; 927 TCG (403 TCWG) (AFR) Selfridge 10Feb63 – 930 SOP GP (434 TCWG) (AFR) Bakalar 02Jul69 – to Grissom 12Jan70 – Wfu to MASDC (LOG) Davis-Monthan 23Jul70 for storage, PCN: **CJ349** – DFI: 25Aug70 – Kolar, Inc. Tucson Arizona 05Feb76 for scrap – Final Disposition: Scrapped.

10664 / 51-2675 / C-119F-FA

Avble: 30Jun52; Acc: 28Jul52; Del: 31Jul52 to USAF – 316 TCG (TAC) Sewart 31Jul52; depl Burlington 06Feb53; depl Pope 20Apr54; depl Sumpter Alabama 24Aug54 – 314 TCWG (TAC) Sewart Nov54; depl Elmendorf Jan55; bailment Fairchild (AMC) Hagerstown 12Dec55, cvtd **C-119G** 56 – **RECORDS MISSING** – 2466 RFC (CNR) Bakalar

24Jul57 – 2469 RFC (CNR) Scott 01Sep57, to DT Scott 01Dec57 – **RECORDS MISSING** – 403 TCWG (AFR) Selfridge date unk; 927 TCG (403 TCWG) (AFR) Selfridge 10Feb63 – 930 SOP GP (434 TCWG) (AFR) Bakalar 02Jul69; to Grissom 12Jan70; brief depl 922 TAL GP (433 TCWG) (AFR) Kelly 11Sep70 – Wfu 13Oct70 to museum / school status. Displayed on loan at the Pate Museum of Transportation Fort Worth Texas. Disassembled and moved in 2012 to the civilian owned U.S. Veterans Museum Grandbury Texas. It currently appears the museum is closed with the C-119 stored in parts – Final Disposition: Extant.

10665 / 51-2676 / C-119F-FA

Avble: 02Jul52; Acc: 28Jul52; Del: 31Jul52 to USAF – 316 TCG (TAC) Sewart 31Jul52; depl Burlington 06Feb53; depl Sumpter Alabama 24Aug54 – 314 TCWG (TAC) Sewart Nov54; depl Elmendorf Jan55; bailment Fairchild (AMC) Hagerstown 31Oct55, cvtd **C-119G** 06Dec55 – **RECORDS MISSING** – 732 TCSq (94 TCWG) (AFR) Grenier date unk – **RECORDS MISSING** – 902 TCG (94 TCWG) (AFR) Grenier 63 – 915 TCG (435 TCWG) (AFR) Homestead date unk – 926 TCG (446 TCWG) (AFR) NAS New Orleans 12Nov65 – 915 TCG (435 TCWG) (AFR) Homestead 15Jan66 – 926 TCG (446 TCWG) (AFR) NAS New Orleans 02Mar66 – Wfu to MASDC (LOG) Davis-Monthan 30Nov69 for storage, PCN: **CJ326** – DFI: 19Jan70 – Kolar, Inc. Tucson Arizona 05Feb76 for scrap – Final Disposition: Scrapped.

10666 / 51-2677 / C-119F-FA

Avble: 03Jul52; Acc: 28Jul52; Del: 31Jul52 to USAF – 316 TCG (TAC) Sewart 31Jul52; depl Burlington 06Feb53; depl Pope 20Apr54; depl Sumpter Alabama 24Aug54 – 314 TCWG (TAC) Sewart Nov54; depl Elmendorf Jan55; depl Ladd Alaska 01Aug55; bailment BMC (AMC) Birmingham 19Oct55, cvtd **C-119G** 12Jan56 **RECORDS INDECIPHERABLE** → 357 TCSq (302 TCWG) (AFR) Bates 08May61; 908 TCG (302 TCWG) (AFR) Bates 11Feb63 – 908 TCG (435 TCWG) (AFR) Bates 18Mar63 ← **RECORDS INDECIPHERABLE** to Brookley 30Oct64 – MAP Taiwan 03Feb69 – Final Disposition: Unknown.

10667 / 51-2678 / C-119F-FA

Avble: 07Jul52; Acc: 28Jul52; Del: 31Jul52 to USAF – 316 TCG (TAC) Sewart 31Jul52; depl Burlington 06Feb53; depl Pope 20Apr54; depl Sumpter Alabama 24Aug54 – 314 TCWG (TAC) Sewart Nov54; depl Elmendorf Jan55 **RECORDS INDECIPHERABLE** → cvtd **C-119G** date unk – 73 TCSq (434 TCWG) (AFR) Scott date unk ← **RECORDS INDECIPHERABLE** 932 TCG (434 TCWG) (AFR) Scott 63 – 914 TAL GP (512 TCWG) (AFR) Niagara Falls IAP 26Jan67 – 910 TAL GP (459 TCWG) (AFR) Youngstown 06Nov67 – Wfu to MASDC (LOG) Davis-Monthan 30Nov69 for storage, PCN: **CJ328** – DFI: 19Jan70 – Kolar, Inc. Tucson Arizona 12Feb76 for scrap – Final Disposition: Scrapped.

10668 / 51-2679 / C-119F-FA

Avble: 08Jul52; Acc: 28Jul52; Del: 31Jul52 to USAF – 316 TCG (TAC) Sewart 31Jul52; assigned 37 TCSq; depl 314 TCWG (TAC) Sewart 25Nov52; depl Burlington 07Feb53. Had a catastrophic engine fire after takeoff at Pope AFB North Carolina 30Mar54. The aircraft tried to line up for an emergency landing but veered off coming down skirting an officers' building at Fort Bragg then slid across a parade ground before hitting a mess hall creating a well involved fire. 4 of the C-119 crew died in the crash plus 2 personnel that were in the mess hall. 2 of the crew and 2 soldiers onboard the C-119 survived with injuries along with 5 others in the mess hall. The 6 killed were: 1Lt Albert W. Parks (pilot), A1C Rudolph Valentino Short, Cpl Robert Dervan, Cpl Donald F. Greenlee, Cpl Osman S. Palmer and Pvt

James A. Macre. The 9 injured were: Capt Charles L. Shirley, 1Lt Raymond Fitzsimmons (co-pilot), Cwo William Angeloff, Sgt Henry C. Clay, A1C Eugene R. Snyder, Pvt Ralph E. Salisbury, Pvt William Cook, Pvt Edward Ellison and Edward R. Ross. The fire burned for over three hours due to the quantity of aviation fuel plus gas tanks situated in the mess hall – REC: 27May54 – Final Disposition: Accident.

10669 / 51-2680 / C-119F-FA

Avble: 09Jul52; Acc: 31Jul52; Del: 06Aug52 to USAF – 316 TCG (TAC) Sewart 06Aug52; depl 64 TCWG (TAC) Greenville 12Sep52; depl Burlington 07Feb53; depl Sumpter Alabama 24Aug54 – 314 TCWG (TAC) Sewart Nov54 **RECORDS INDECIPHERABLE** → 463 TCWG (TAC) Sewart (depl) 11Aug55; depl 513 TFW GP (TAC) Sewart 08Nov55 – cvtd **C-119G** date unk – 440 TCWG (AFR) Gen Mitchel date unk; 933 TCG (440 TCWG) (AFR) Gen Mitchel 11Feb63 ← **RECORDS INDECIPHERABLE** Disappeared without trace about 177 miles from destination Grand Turk Is. In The Bahamas 05Jun65. The C-119 departed Homestead AFB Florida late in the day with 5 crew, 4 mechanics and 1 passenger assigned a mission to fly a spare engine to a stranded C-119 on Grand Turk Is. The last radio report was off Crooked Is. With no indication anything was amiss but nothing was heard from the aircraft again, an extensive search was undertaken without success. A wheel chock labelled "680" and other debris was found on 18Jul65 along a beach on Acklins Is. (near Crooked Is.). The 10 personnel onboard were 5 aircrew: Maj Louis A. Giuntoli (pilot), Capt Richard J. Bassett (navigator), 1Lt Lawrence F. Gares (co-pilot), MSgt Milton E. Adams (flight engineer) A1C Thomas P. Nugent (loadmaster) and 5 passengers: Raoul P. Benedict (mnt crew), Duane W. Brooks (mnt crew), Frank Ellison (mnt crew), Norman J. Mimier (mnt crew) and John W. Lazenry (airman, passenger). The mysterious nature of this loss has led to the incident being cited by many conspiracy theorists as one of the more sensational disappearances in the so-called "Bermuda Triangle" – Final Disposition: Accident.

10670 / 51-2681 / C-119F-FA

Avble: 10Jul52; Acc: 31Jul52; Del: 06Aug52 to USAF – 316 TCG (TAC) Sewart 06Aug52; depl Burlington 07Feb53; depl Pope 20Apr54; depl Sumpter Alabama 24Aug54 – 314 TCWG (TAC) Sewart Nov54; depl Elmendorf Jan55 **RECORDS INDECIPHERABLE** → 463 TCWG (TAC) Sewart 10Aug55 – 513 TFW GP (TAC) Sewart (depl) 08Nov55 – cvtd **C-119G** date unk – 2472 RFC (CNR) Richards 58 – 452 TCWG (AFR) March date unk ← **RECORDS INDECIPHERABLE** 942 TCG (452 TCWG) (AFR) March 11Jul63 – 924 TCG (446 TCWG) (AFR) Ellington 21Aug65 – 908 TCG (435 TCWG) (AFR) Brookley 25May68 – 926 TAL GP (446 TCWG) (AFR) NAS New Orleans 26Feb69 – 913 TAL GP (512 TCWG) (AFR) NAS Willow Grove 01Feb70 – 906 SOP GP (302 TCWG) (AFR) Clinton 04Nov70 – MAP Taiwan 27Nov70 – Final Disposition: Unknown.

10671 / 51-2682 / C-119F-FA

Avble: 11Jul52; Acc: 12Aug52; Del: 16Aug52 to USAF – 316 TCG (TAC) Sewart 16Aug52; depl Burlington 07Feb53; depl Pope 20Apr54 – 314 TCWG (TAC) Sewart Nov54; depl Elmendorf Jan55 **RECORDS INDECIPHERABLE** → 463 TCWG (TAC) Sewart (depl) 12Aug55 – 513 TFW GP (TAC) Sewart 08Nov55 – cvtd **C-119G** date unk – 434 TCWG (AFR) Bakalar date unk; 931 TCG (434 TCWG) (AFR) Bakalar 63 ← **RECORDS INDECIPHERABLE** Wfu to MASDC (LOG) Davis-Monthan 07Nov68 for storage, PCN: **CJ271** – DFI: 19Oct76 – Final Disposition: Scrapped.

10672 / 51-2683 / C-119F-FA

Avble: 14Jul52; Acc: 12Aug52; Del: 16Aug52 to USAF – 316 TCG (TAC) Sewart 16Aug52; depl Burlington 07Feb53; depl Pope 20Apr54; depl Sumpter Alabama 24Aug54 – 314 TCWG (TAC) Sewart Nov54; depl Elmendorf Jan55 **RECORDS INDECIPHERABLE** → 463 TCWG (TAC) Sewart (depl) 11Aug55; depl 513 TFW GP (TAC) Sewart 08Nov55 – cvtd **C-119G** date unk – 2472 RFC (CNR) Richards 13Jul58 – 442 TCWG (AFR) Richards 02Mar59 – **RECORDS MISSING** – 944 TCG (452 TCWG) (AFR) March 11Feb63 ← **RECORDS INDECIPHERABLE** 942 TCG (452 TCWG) (AFR) March 31Mar65 – 925 TCG (446 TCWG) (AFR) Ellington 11Sep65 – 927 TAL GP (403 TCWG) (AFR) Selfridge 27Jan68 – MAP Taiwan 06Mar69 – Final Disposition: Unknown.

10673 / 51-2684 / C-119F-FA

Avble: 15Jul52; Acc: 25Aug52; Del: 29Aug52 to USAF – 316 TCG (TAC) Sewart 29Aug52; depl Burlington 06Feb53; depl Pope 19Apr54; depl Sumpter Alabama 24Aug54 – 314 TCWG (TAC) Sewart Nov54; depl Elmendorf Jan55 **RECORDS INDECIPHERABLE** → 463 TCWG (TAC) Sewart (depl) 08Oct55; depl 513 TFW GP (TAC) Sewart 08Nov55 – cvtd **C-119G** date unk – 2466 RFC (CNR) Bakalar 05Jan57; depl 2472 RFC (CNR) Richards 16Jun57 – 434 TCWG (AFR) Bakalar 19Apr59 – 349 TCWG (AFR) Portland 28Oct62; assigned 313 TCSq 29Oct62; 939 TCG (349 TCWG) (AFR) Portland 63 ← **RECORDS INDECIPHERABLE** 943 TAL GP (452 TCWG) (AFR) March 27Jun68 – 931 TAL GP (434 TCWG) (AFR) Bakalar 10Apr69; 930 SOP GP (434 TCWG) (AFR) Bakalar 30Jun69; to Grissom 12Jan70 – Wfu to MASDC (LOG) Davis-Monthan 11Sep70 for storage, PCN: **CJ362** – DFI: 14Oct70 – Kolar Inc. Tucson Arizona 05Feb76 for scrap – Final Disposition: Scrapped.

10674 / 51-2685 / C-119F-FA

Avble: 16Jul52; Acc: 25Aug52; Del: 29Aug52 to USAF – 316 TCG (TAC) Sewart 28Aug52; depl Burlington 06Feb53; depl Sumpter Alabama 24Aug54 – 314 TCWG (TAC) Sewart Nov54; depl Elmendorf Jan55 **RECORDS INDECIPHERABLE** → 463 TCWG (TAC) Sewart (depl) 12Aug55; depl 513 TFW GP (TAC) Sewart 08Nov55 – cvtd **C-119G** date unk – 2466 RFC (CNR) Bakalar date unk – 434 TCWG (AFR) Bakalar date unk ← **RECORDS INDECIPHERABLE** 931 TCG (434 TCWG) (AFR) Bakalar 63; 930 SOP GP (434 TCWG) (AFR) Bakalar 30Jun69; to Grissom 12Jan70 – Wfu to MASDC (LOG) Davis-Monthan 28Jul70 for storage, PCN: **CJ351** – DFI: 25Aug70 – Kolar Inc. Tucson Arizona 05Feb76 for scrap – Final Disposition: Scrapped.

10675 / 51-2686 / C-119F-FA

Avble: 17Jul52; Acc: 26Aug52; Del: 29Aug52 to USAF – 316 TCG (TAC) Sewart 28Aug52; depl Burlington 07Feb53; depl Pope 20Apr54; depl Elmendorf 01Jul54; depl Sumpter Alabama 24Aug54; depl Elmendorf 01Sep54 – 314 TCWG (TAC) Sewart Nov54; depl Elmendorf Jan55 **RECORDS INDECIPHERABLE** → depl 463 TCWG (TAC) Sewart (depl) 11Aug55 – cvtd **C-119G** date unk – 2472 RFC (CNR) Richards 16Jul57 – 442 TCWG (AFR) Richards 02Mar59 – 452 TCWG (AFR) March date unk; 943 TCG (452 TCWG) (AFR) March 63 ← **RECORDS INDECIPHERABLE** 942 TCG (452 TCWG) (AFR) March 31Mar65 – 924 TCG (446 TCWG) (AFR) Ellington 12Aug65 – 910 TAL GP (459 TCWG) (AFR) Youngstown 20Jan68 – Wfu to MASDC (LOG) Davis-Monthan 07Nov68 for storage, PCN: **CJ270** – DFI: 09Dec74 – Final Disposition: Scrapped.

10676 / 51-2687 / C-119F-FA

Avble: 18Jul52; Acc: 26Aug52; Del: 27Aug52 to Canada, RCAF s/n: **22101** – BOC: 08Sep52 435 SQ Edmonton; dropped from USAF inventory 10Sep52 – 436 SQ Dorval 07Dec54; cvtd **C-119G** 11Jan57 – 435 SQ Namao 14Nov58 – 436 SQ Uplands 29Jul64 – Wfu into storage 1005 TSD Saskatoon 15Jul65 – SOC: 09May67 to Crown Assets Disposal Corp. – BofS Aircraft International Associates Dallas Texas 03May67; reg **N15505** 04Oct68 – BofS Hawkins & Powers Aviation, Inc. Greybull Wyoming 16Dec75; used as a spares source – Don F. Pratt Memorial Museum Fort Campbell Kentucky Sep95, restored for display marked as: "563." Possibly used parts salvaged from ex-Norway wreck msn: 10684, s/n: 12695 / BW-E as dataplates found inside confirm – Reg N15505 can 07Aug13 – Final Disposition: Extant.

10677 / 51-2688 / C-119F-FA

Avble: 28Jul52; Acc: 26Aug52; Del: 27Aug52 to Canada, RCAF s/n: **22102** – BOC: 08Sep52 435 SQ Edmonton; dropped from USAF inventory 10Sep52 – 436 SQ Dorval 07Dec54; cvtd **C-119G** 11Jan57 – Wfu into storage 1005 TSD Saskatoon 01Mar65 – SOC: 02Mar67 to Crown Assets Disposal Corp. – Listed in RCAF records as going to Aircraft International Associates but no further history is available, likely bku for spares – Final Disposition: Scrapped.

10678 / 51-2689 / C-119F-FA

Avble: 21Jul52; Acc: 26Aug52; Del: 27Aug52 to Canada, RCAF s/n: **22103** – BOC: 09Sep52 CJATC Rivers; dropped from USAF inventory 10Sep52 – 436 SQ Dorval 07Dec54; cvtd **C-119G** 12Dec55 – 435 SQ Namao 27Nov56 – 436 SQ Downsview 21Oct60 – Wfu into storage 1005 TSD Saskatoon 22Apr64 – SOC: 06Jun67 to Crown Assets Disposal Corp. – BofS Aircraft International Associates Dallas Texas 02Jun67; reg **N8092** 19Aug70 – BofS Hawkins & Powers Aviation, Inc. Greybull Wyoming 16Dec75, used as a spares airframe – To U.S. Forest Service 1991, ferried to National Warplane Museum Geneseo New York for static display – Reg N8092 canx 24Apr13 – Final Disposition: Extant.

10679 / 51-2690 / C-119F-20-FA

Avble: 22Jul52; Acc: 26Aug52; Del: 28Sep52 to USAF – 7373 AB WG (AFE) Chateauroux AB France 24Oct52 for training Belgian crews. DFI: 27Nov53 – MDAP Belgium, BAF s/n: **CP-17**; BOC: 27Nov53 15 WG / 20 SQ Melsbroek AB Belgium, callsign: **OT-CAQ** – 15 WG / 40 SQ Apr54. SOC: 12Sep55 – Back to USAF/MAAG – MAP to Spain 07Dec55 but not accepted – Back to USAF and stored – Purchased by Belgium 20May59; cvtd **C-119G** 1959 by SABENA; BOC: 01Jun61 15 WG / 40 SQ Melsbroek AB Belgium. Right wing-tip hit hangar while parking 31May72 – Wfu: 02Jul73 and stored at Koksijde AB Belgium; SOC: 22Oct75 – to civil company International Engine Parts, scrapped 1977-78 – Final Disposition: Scrapped.

10680 / 51-2691 / C-119F-20-FA

Avble: 23Jul52; Acc: 26Aug52; Del: 29Sep52 to USAF – 7373 AB WG (AFE) Chateauxroux AB France 24Oct52 for training Belgian crews; depl 60 RCWG (AFE) Rhein-Main AB W. Germany 09Jan53. DFI: 24Dec53 – MDAP Belgium, BAF s/n: **CP-18**; BOC: 24Dec53 15 WG / 20 SQ Melsbroek AB Belgium, callsign: **OT-CAR** – 15 WG / 40 SQ Apr54; SOC: 12Sep55 – Back to USAF/MAAG – MAP Spain 07Dec55 but not accepted – Back to USAF and stored – Purchased by Belgium 20May59; cvtd **C-119G** 1959 by SABENA; BOC: 12Jul60 15 WG / 40 SQ Melsbroek AB Belgium. Heavy shimmy on landing 11Oct61, two right u/c doors torn off. 15 WG / 20 SQ 04Jun62 – Wfu: 09Jul73 and stored at Koksijde AB Belgium; SOC: 22Oct75 – To civil company International Engine Parts, scrapped 1977-78 – Final Disposition: Scrapped.

10681 / 51-2692 / C-119F-20-FA

Avble: 24Jul52; Acc: 26Aug52; Del: 17Sep52 to MDAP Belgium, BAF s/n: **CP-1**; dropped from USAF inventory 18Sep52 – BOC: 25Sep52 15 WG / 20 SQ Melsbroek AB Belgium, callsign: **OT-CAA** – SOC: 12Sep55, back to USAF/MAAG; cvtd **C-119G** 1956 by SABENA – MAP Norway, BOC: 18Jul56 335 SQ Gardermoen Norway, RNAF s/n: **12692** / **BW-C**, named: *Cappy* – SOC: 22May69 back to USAF – Wfu to MASDC (LOG) Davis-Monthan 10Jul69 for storage, PCN: **CJ305** – DFI: 26Dec73 – Kolar, Inc. Tucson Arizona 24Feb76 for scrap – Final Disposition: Scrapped.

10682 / 51-2693 / C-119F-20-FA

Avble: 25Jul52; Acc: 26Aug52; Del: 16Sep52 to MDAP Belgium, BAF s/n: **CP-2**; dropped from USAF inventory 17Sep52 – BOC: 24Sep52 15 WG / 20 SQ Melsbroek AB Belgium, callsign: **OT-CAB** – SOC: 12Sep55 back to USAF/MAAG; cvtd **C-119G** 1956 by SABENA – MAP Norway, BOC: 26Jun56 335 SQ Gardermoen Norway, RNAF s/n: **12693** / **BW-B**, named: *Bamse* – SOC: 07May69 back to USAF – Wfu to MASDC (LOG) Davis-Monthan 07May69 for storage, PCN: **CJ303** – DFI: 26Dec73 – Kolar, Inc. Tucson Arizona 23Jan76 for scrap – Final Disposition: Scrapped.

10683 / 51-2694 / C-119F-20-FA

Avble: 28Jul52; Acc: 27Aug52; Del: 17Sep52 to MDAP Belgium, BAF s/n: **CP-3**; dropped from USAF inventory 18Sep52 – BOC: 24Sep52 15 WG / 20 SQ Melsbroek AB Belgium, callsign: **OT-CAC** – SOC: 12Sep55 back to USAF/MAAG – MAP Spain 07Dec55 but not accepted – Back to USAF and stored. Purchased by Belgium 20May59, re-assigned BAF s/n: **CP-14**, callsign: **OT-CAN**; cvtd **C-119G** 1959 by SABENA; BOC: 19Sep60 15 WG / 40 SQ Melsbroek AB Belgium. Had a possible left engine fire after take-off at Melsbroek 31Aug67. Pilot made a belly landing with the aircraft sliding 2950ft before stopping. The 5 crew were uninjured but 1 of the 3 passengers suffered minor injuries. The aircraft was repaired and returned to service 11Mar68. While being moved by ground crew at Melsbroek on 10Nov71, the nose section suffered serious damage after colliding with a hangar door. Deemed as damaged beyond economic repair; SOC: 08Feb72 – Considered as a gate guardian but instead scrapped Dec74 at Melsbroek after a period of storage – Final Disposition: Accident.

10684 / 51-2695 / C-119F-20-FA

Avble: 28Jul52; Acc: 27Aug52; Del: 18Sep52 to MDAP Belgium, BAF s/n: **CP-4**; dropped from USAF inventory 22Sep52 – BOC: 25Sep52 15 WG / 20 SQ Melsbroek AB Belgium, callsign: **OT-CAD** – SOC: 12Sep55 back to USAF/MAAG; cvtd **C-119G** 1956 by SABENA – MAP Norway, BOC: 31Jul56 335 SQ Gardermoen Norway, RNAF s/n: **12695** / **BW-E**, named: *Elmer*. While on approach to Roros Airport Norway 06Dec68, the pilot misjudged the runway threshold. The left main u/c struck the ground 43ft short of the runway causing it to collapse sending the aircraft skidding off the runway. The aircraft was deemed as damaged beyond repair and later sold for scrap to Brodrene London of Oslo Norway – SOC: Dec68 – Final Disposition: Accident.

10685 / 51-2696 / C-119F-20-FA

Avble: 30Jul52; Acc: 29Aug52; Del: 19Sep52 to MDAP Belgium, BAF s/n: **CP-5**; dropped from USAF inventory 22Sep52 – BOC: 25Sep52 15 WG / 20 SQ Melsbroek AB Belgium, callsign: **OT-CAE** – SOC: 12Sep55 back to USAF/MAAG – MAP Spain 07Dec55 but not accepted – Back to USAF and stored. Purchased by Belgium 20May59, re-assigned BAF s/n: **CP-11**, callsign: **OT-CAK**; cvtd **C-119G** 1959 by SABENA; BOC: 13Feb61 15 WG / 40 SQ Melsbroek AB Belgium – Wfu: 09Jul73 and stored at Koksijde AB Belgium; SOC: 22Oct75 – To civil company International Engine Parts. Bku to provide parts for s/n: CP-9 (Bata Intl.), CP-10 (Melsbroek Barracks) and CP-46 (Brussels Air Museum) – Final Disposition: Scrapped.

10686 / 51-2697 / C-119F-20-FA

Avble: 31Jul52; Acc: 28Aug52; Del: 23Sep52 to MDAP Belgium, BAF s/n: **CP-6**; dropped from USAF inventory 24Sep52 – BOC: 25Sep52 15 WG / 20 SQ Melsbroek AB Belgium, callsign: **OT-CAF**. Had a minor accident 02Feb53 details unk – SOC: 12Sep55 back to USAF/MAAG; cvtd **C-119G** 1956 by SABENA – MAP Norway, BOC: 24Jun56 335 SQ Gardermoen Norway, RNAF s/n: **12697** / **BW-A**, named: *Anton* – SOC: 14Jul69 back to USAF – Wfu to MASDC (LOG) Davis-Monthan 15Jul69 for storage, PCN: **CJ307** – DFI: 09Dec74 – Southwestern Alloys, Inc. Tucson Arizona 09Jul76 for scrap – Final Disposition: Scrapped.

10687 / 51-2698 / C-119F-20-FA

Avble: 31Jul52; Acc: 28Aug52; Del: 24Sep52 to MDAP Belgium, BAF s/n: **CP-7**; dropped from USAF inventory 26Sep52 – BOC: 25Sep52 15 WG / 20 SQ Melsbroek AB Belgium, callsign: **OT-CAG**. Had a minor accident 24Dec52 details unk – SOC: 12Sep55 back to USAF/MAAG; cvtd **C-119G** 1956 by SABENA – MAP Norway, BOC: 08Aug56 335 SQ Gardermoen Norway, RNAF s/n: **12698** / **BW-F**, named: *Fiinbeck* – SOC: 22May69 back to USAF – Wfu to MASDC (LOG) Davis-Monthan 04May69 for storage, PCN: **CJ302** – DFI: 09Dec74 – Southwestern Alloys, Inc. Tucson Arizona 30Jun76 for scrap – Final Disposition: Scrapped.

10688 / 51-2699 / C-119F-20-FA

Avble: 04Aug52; Acc: 29Aug52; Del: 30Sep52 to MDAP Belgium, BAF s/n: **CP-8**; dropped from USAF inventory 01Oct52 – BOC: 25Sep52 15 WG / 20 SQ Melsbroek AB Belgium, callsign: **OT-CAH**. Had a minor mid-air collision with s/n: CP-14 07May55 – SOC: 12Sep55 back to USAF/MAAG; cvtd **C-119G** 1956 by SABENA – MAP Norway, BOC: 02Aug56 335 SQ Gardermoen Norway, RNAF s/n: **12699** / **BW-D**, named: *Donald* – SOC: 22May69 back to USAF – Wfu to MASDC (LOG) Davis-Monthan 27May69 for storage, PCN: **CJ304** – DFI: 26Dec73 – Kolar, Inc. Tucson Arizona 14Jan76 for scrap – Final Disposition: Scrapped.

10689 / 51-2700 / C-119F-20-FA

Avble: 09Aug52; Acc: 28Aug52; Del: 01Oct52 to MDAP Belgium, BAF s/n: **CP-9**; dropped from USAF inventory 02Oct52 – BOC: 28Oct52 15 WG / 20 SQ Melsbroek AB Belgium, callsign: **OT-CAI** – 15 WG / 40 SQ Apr54 – SOC: 12Sep55 back to USAF/MAAG – MAP Spain 07Dec55 but not accepted – Back to USAF and stored – Purchased by Belgium 20May59; cvtd **C-119G** 1959 by SABENA; BOC: 21May59 15 WG / 40 SQ Melsbroek AB Belgium. Minor bird strike during landing 17Mar64. Minor ground collision during taxi with s/n: CP-43 25May67 – Wfu: 03Nov72 and stored at Koksijde AB Belgium; SOC: 22Oct75 – to civil company International Engine Parts – Bata International Airlines Equatorial Guinea Africa reg **3C-ABA**, flown to Manston Airport UK 09Apr81 in preparation for use in cargo operations in Africa but the aircraft remained stored at

Manston – Aces High Ltd. Surrey UK reg **G-BLSW** 28Dec84. Made airworthy and test flown by Aces High 05Apr85 – BofS Consolidated Aviation Enterprises, Inc. East Middlebury Vermont 19Feb85 reg **N2700** – BofS John P. Downey (d/b/a Consolidated Aviation Enterprises) East Middlebury Vermont 10Jun85, second BofS to John P. Downey from Aces High Ltd. 02Oct85. G-BLSW reg canx 04Oct85. Aircraft flown to North Weald Airfield UK 25Jul87 and stored. N2700 reg canx 18Apr91. Bku 94, nose section noted at North Weald up to 2007 when it was acquired by The Wings Museum Surrey UK. Subsequently acquired by a private owner, current status unk – Final Disposition: Extant.

10690 / 51-2701 / C-119F-20-FA

Avble: 13Aug52; Acc: 29Aug52; Del: 10Sep52 to MDAP Belgium, BAF s/n: **CP-10**; dropped from USAF inventory 12Sep52 – BOC: 28Oct52 15 WG / 20 SQ Melsbroek AB Belgium, callsign: **OT-CAJ** – 15 WG / 40 SQ Apr54 – SOC: 12Sep55 back to USAF/MAAG – MAP Spain 07Dec55 but not accepted – Back to USAF and stored – Purchased by Belgium 20May59; cvtd **C-119G** 1959 by SABENA; BOC: 13May60 15 WG / 40 SQ Melsbroek AB Belgium. Minor bird strike incident 11Mar68. Struck turbulence 23Jun70, passenger broke leg. After take-off 11Sep71 the right propeller went into overspeed causing a loss in altitude, the aircraft suffered belly and ventral fin damage when it skimmed through some tree tops. Flew into a flock of birds 09May72, minor damage to left wing – Wfu: 02Jul72 and stored at Koksijde AB Belgium; SOC: 22Oct75 – To civil company International Engine Parts – Christian Decot Neuville 25May90, transported by road for use as a clubhouse at Neuville Field – Transported to Melsbroek AB Groenveld Barracks Aug94-Feb95, restored for display Aug96 using boom sections of s/n: CP-11. Recognized by the government of the Flanders Region 07Feb12 as an item of "cultural heritage" – Final Disposition: Extant.

10691 / 51-2702 / C-119F-20-FA

Avble: 20Aug52; Acc: 23Sep52; Del: 06Oct52 to MDAP Belgium, BAF s/n: **CP-11**; dropped from USAF inventory 07Oct52 – BOC: 28Oct52 15 WG / 20 SQ Melsbroek AB Belgium, callsign: **OT-CAK** – 15 WG / 40 SQ Apr54 – SOC: 12Sep55 back to USAF/MAAG; cvtd **C-119G** 1956 by SABENA – MAP Norway, BOC: 31Aug56 335 SQ Gardermoen Norway, RNAF s/n: **12702** / **BW-H**, named: *Hiawatha* – SOC: Jun69 back to USAF – Wfu to MASDC (LOG) Davis-Monthan 18Jul69 for storage, PCN: **CJ308** – DFI: 26Dec73 – Kolar, Inc. Tucson Arizona 20Jan76 for scrap – Final Disposition: Scrapped.

10692 / 51-2703 / C-119F-20-FA

Avble: 21Aug52; Acc: 17Sep52; Del: 07Oct52 to MDAP Belgium, BAF s/n: **CP-12**; dropped from USAF inventory 09Oct52 – BOC: 28Oct52 15 WG / 20 SQ Melsbroek AB Belgium, callsign: **OT-CAL** – 15 WG / 40 SQ Apr54 – SOC: 12Sep55 back to USAF/MAAG – MAP Spain 07Dec55 but not accepted – Back to USAF and stored – Purchased by Belgium 20May59; cvtd **C-119G** 1959 by SABENA; BOC: 28Dec60 15 WG / 20 SQ Melsbroek AB Belgium. 15 WG / 40 SQ 03May61 – Wfu: 02Jul73 and stored at Koksijde AB Belgium; SOC: 22Oct75 – To civil company International Engine Parts, scrapped 1977-78 – Final Disposition: Scrapped.

10693 / 51-2704 / C-119F-20-FA

Avble: 22Aug52; Acc: 17Sep52; Del: 07Oct52 to MDAP Belgium, BAF s/n: **CP-13**; dropped from USAF inventory 14Oct52 – BOC: 28Oct52 15 WG / 20 SQ Melsbroek AB Belgium, callsign: **OT-CAM** – 15 WG / 40 SQ Apr54 – SOC: 12Sep55 back to USAF/MAAG – MAP Spain 07Dec55 but not accepted – Back to USAF and stored – Purchased by Belgium

20May59; cvtd **C-119G** 1959 by SABENA; BOC: 17Aug60 15 WG / 40 SQ Melsbroek AB Belgium. Minor incident 26Sep61 where nose wheel exerted a heavy skimmy on landing. Minor accident 29Mar63 with wing-tip damaged during taxi – Wfu: 02Jul73 and stored at Koksijde AB Belgium; SOC: 22Oct75 – To civil company International Engine Parts, scrapped 1977-78 – Final Disposition: Scrapped.

10694 / 51-2705 / C-119F-20-FA

Avble: 25Aug52; Acc: 17Sep52; Del: 09Oct52 to MDAP Belgium, BAF s/n: **CP-14**; dropped from USAF inventory 14Oct52 – BOC: 28Oct52 15 WG / 20 SQ Melsbroek AB Belgium, callsign: **OT-CAN** – 15 WG / 40 SQ Apr54. Had a minor mid-air collision with s/n: CP-8 07May55 – SOC: 12Sep55 back to USAF/MAAG; cvtd **C-119G** 1956 by SABENA – MAP Norway, BOC: 07Sep56 335 SQ Gardermoen Norway, RNAF s/n: **12705** / **BW-G**, named: *Goofey* – SOC: Jun69 back to USAF – Wfu to MASDC (LOG) Davis-Monthan 10Jul69 for storage, PCN: **CJ306** – DFI: 09Dec74 – Southwestern Alloys, Inc. Tucson Arizona 13Jul76 for scrap – Final Disposition: Scrapped.

10695 / 51-2706 / C-119F-20-FA

Avble: 26Aug52; Acc: 17Sep52; Del: 13Oct52 to MDAP Belgium,, BAF s/n: **CP-15**; dropped from USAF inventory 15Oct52 – BOC: 28Oct52 15 WG / 20 SQ Melsbroek AB Belgium, callsign: **OT-CAO**. Minor incident collided with s/n: CP-20 17Sep54 while being towed at Melsbroek – SOC: 12Sep55 back to USAF/MAAG – MAP Spain 07Dec55 but not accepted – Back to USAF and stored – Purchased by Belgium 20May59; cvtd **C-119G** 1959 by SABENA; BOC: 28Apr60 15 WG / 40 SQ Melsbroek AB Belgium. Lost trailing antenna near Paris 15May62 – Wfu: 01Jun72 and stored at Koksijde AB Belgium; SOC: 22Oct75 – To civil company International Engine Parts, scrapped 1977-78 – Final Disposition: Scrapped.

10696 / 51-2707 / C-119F-FA

Avble: 27Aug52; Acc: 19Sep52; Del: 14Oct52 to MDAP Belgium, BAF s/n: **CP-16**; dropped from USAF inventory 16Oct52 – BOC: 28Oct52 15 WG / 20 SQ Melsbroek AB Belgium, callsign: **OT-CAP** – 15 WG / 40 SQ Apr54 – SOC: 12Sep55 back to USAF/MAAG – MAP Spain 07Dec55 but not accepted – Back to USAF and stored – Purchased by Belgium 20May59; cvtd **C-119G** 1959 by SABENA; BOC: 26Apr61 15 WG / 40 SQ Melsbroek AB Belgium – Wfu: 02Jul73 and stored at Koksijde AB Belgium; SOC: 22Oct75 – To civil company International Engine Parts, scrapped 1977-78 – Final Disposition: Scrapped.

10697 / 51-2708 / C-119F-FA

Avble: 28Aug52; Acc: 19Sep52; Del: 24Sep52 to USAF – 316 TCG (TAC) Sewart 24Sep52; depl Burlington 07Feb53; depl Pope 20Apr54; depl Sumpter Alabama 24Aug54 – 314 TCWG (TAC) Sewart 18Nov54; depl Elmendorf 11Jan55; depl Ladd Alaska 01Jun55; depl 463 TCWG (TAC) Sewart (depl) 13Aug55; bailment BMC (AMC) Birmingham 18Aug55, cvtd **C-119G** 19Dec55 **RECORDS INDECIPHERABLE** → 73 TCSq (434 TCWG) (AFR) Scott date unk; 932 TCG (434 TCWG) (AFR) Scott 63 ← **RECORDS INDECIPHERABLE** 933 TCG (440 TCWG) (AFR) Gen Mitchel IAP 28Jan67 – 930 TCG (434 TCWG) (AFR) Bakalar 04Apr67; depl 71 ACO SQ (1 ACO WG) (TAC) Lockbourne 17Jun68 – Wfu to MASDC (LOG) Davis-Monthan 04Nov68 for storage, PCN: **CJ266** – DFI: 09Dec74 – Final Disposition: Scrapped.

10698 / 51-2709 / C-119F-FA

Avble: 29Aug52; Acc: 19Sep52; Del: 24Sep52 to USAF – 316 TCG (TAC) Sewart 24Sep52; depl Burlington 07Feb53; depl Elmendorf 01Jul54; depl Sumpter Alabama 24Aug54; depl Elmendorf 01Sep54 – 314 TCWG (TAC) Sewart 18Nov54; depl Elmendorf 11Jan55; depl Ladd Alaska 01Jun55; depl 463 TCWG (TAC) Sewart (depl) 11Aug55; to 6606 AB WG (NEA) Goose Bay Canada 06Oct55 mnt **RECORDS INDECIPHERABLE** → cvtd **C-119G** date unk – 434 TCWG (AFR) Bakalar date unk ← **RECORDS INDECIPHERABLE** 931 TCG (434 TCWG) (AFR) Bakalar 63 – MAP Taiwan 06Mar69, RoCAF s/n: **3125** – Final Disposition: Unknown.

10699 / 51-2710 / C-119F-FA

Avble: 02Sep52; Acc: 22Sep52; Del: 28Sep52 to USAF – 316 TCG (TAC) Sewart 28Sep52; depl Burlington 07Feb53; depl Sumpter Alabama 24Aug54 – 314 TCWG (TAC) Sewart 18Nov54; depl Elmendorf 11Jan55; depl Ladd Alaska 01Jun55; depl 463 TCWG (TAC) Sewart (depl) 12Aug55; to 6606 AB WG (NEA) Goose Bay Canada 17Oct55 mnt **RECORDS INDECIPHERABLE** → cvtd **C-119G** date unk – 2466 RFC (CNR) Bakalar 18Aug57 – 2469 RFC (CNR) Scott 01Sep57; to DT Scott 01Dec57 – 73 TCSq (434 TCWG) (AFR) Scott date unk; 932 TCG (434 TCWG) (AFR) Scott 63 ← **RECORDS INDECIPHERABLE** 910 TCG (459 TCWG) (AFR) Youngstown 08Oct66 – MAP Taiwan 02Dec69, RoCAF s/n: **3174** – Final Disposition: Unknown.

10700 / 51-2711 / C-119F-FA

Avble: 02Sep52; Acc: 22Sep52; Del: 28Sep52 to USAF – 316 TCG (TAC) Sewart 26Sep52; depl Burlington 06Feb53; depl Pope 20Apr54; depl Sumpter Alabama 24Aug54 – 314 TCWG (TAC) Sewart 15Nov54; depl Elmendorf 11Jan55 **RECORDS INDECIPHERABLE** → depl 463 TCWG (TAC) Sewart (depl) 55 – 2472 RFC (CNR) Richards 03Jun57; cvtd **C-119G** 22May57; depl 2466 RFC (CNR) Bakalar 17Jun57 – 65 TCSq (403 TCWG) (AFR) Davis date unk ← **RECORDS INDECIPHERABLE** 929 TCG (403 TCWG) (AFR) Davis Field 63 – 922 TCG (433 TCWG) (AFR) Kelly 17Oct65 – MAP Taiwan 08Apr69, RoCAF s/n: **3162** – Final Disposition: Unknown.

10701 / 51-2712 / C-119F-FA

Avble: 03Sep52; Acc: 22Sep52; Del: 28Sep52 to USAF – 516 TCWG (TAC) Memphis 29Sep52 – 463 TCWG (TAC) Memphis 16Jan53; to Ardmore 17Aug53; depl Charleston 23Apr54 **RECORDS INDECIPHERABLE** → cvtd **C-119G** date unk – 2466 RFC (CNR) Bakalar 17Jun57 – 440 TCWG (AFR) Gen Mitchel IAP date unk ← **RECORDS INDECIPHERABLE** 933 TCG (440 TCWG) (AFR) Gen Mitchel 63 – Wfu to MASDC (LOG) Davis-Monthan 18Feb71 for storage, PCN: **CJ404** – DFI: 31Mar76 – Final Disposition: Scrapped.

10702 / 51-2713 / C-119F-FA

Avble: 04Sep52; Acc: 24Sep52; Del: 29Sep52 to USAF – 516 TCWG (TAC) Memphis 29Sep52 – 463 TCWG (TAC) Memphis 16Jan53; to Ardmore 17Aug53; depl Charleston 23Apr54; depl 405 Fighter- Bomber WG (TAC) Langley 06May54; depl Gray Texas 16May54; depl Laurinburg-Maxton Airport North Carolina 02Aug55; bailment Fairchild (AMC) Hagerstown 07Nov55, cvtd **C-119G** 06Dec55 **RECORDS INDECIPHERABLE** 942 TCG (452 TCWG) (AFR) March 11Feb63 – 944 TCG (452 TCWG) (AFR) March 31Mar65 – 939 TAL GP (349 TCWG) (AFR) Portland 16Jan68; brief depl 944 TAL GP (452 TCWG) (AFR) March 24Jan68 – 943 TAL GP (452 TCWG) (AFR) March 20Jan69 – Wfu to

MASDC (LOG) Davis-Monthan 11Feb69 for storage, PCN: **CJ284** – DFI: 31Mar76 – Final Disposition: Scrapped.

10703 / 51-2714 / C-119F-FA

Avble: 05Sep52; Acc: 26Sep52; Del: 01Oct52 to USAF – 516 TCWG (TAC) Memphis 01Oct52 – 463 TCWG (TAC) Memphis 16Jan53; to Ardmore 17Aug53; depl Charleston 23Apr54; depl base unk 17Apr55 – 366 Fighter-Bomber WG (TAC) Alexandria 09Jun55 **RECORDS INDECIPHERABLE** → cvtd **C-119G** date unk – 452 TCWG (AFR) March date unk – 76 TCSq (435 TCWG) (AFR) Homestead 31Jul62; 915 TCG (435 TCWG) (AFR) Homestead 63 ← **RECORDS INDECIPHERABLE** 906 TCG (302 TCWG) (AFR) Clinton 20Nov65 – Wfu to MASDC (LOG) Davis-Monthan 12Feb69 for storage, PCN: **CJ287** – DFI: 31Mar76 – Final Disposition: Scrapped.

10704 / 51-2715 / C-119F-FA

Avble: 08Sep52; Acc: 26Sep52; Del: 01Oct52 to USAF – 516 TCWG (TAC) Memphis 01Oct52 – 463 TCWG (TAC) Memphis 16Jan53; to Ardmore 17Aug53; depl Charleston 21Apr54; depl base unk 17Apr55 – 4002 AB SQ (SAC) Campbell 27Dec55 mnt **RECORDS INDECIPHERABLE** → cvtd **C-119G** date unk – 732 TCSq (94 TCWG) (AFR) Grenier date unk; 902 TCG (94 TCWG) (AFR) Grenier 11Feb63 ← **RECORDS INDECIPHERABLE** 906 TCG (302 TCWG) (AFR) Clinton 19Nov65 – MAP Taiwan 04Jan70 – Final Disposition: Unknown.

10705 / 51-2716 / C-119F-FA

Avble: 09Sep52; Acc: 26Sep52; Del: 01Oct52 to USAF – 516 TCWG (TAC) Memphis 01Oct52 – 463 TCWG (TAC) Memphis 16Jan53; to Ardmore 17Aug53; depl Charleston 21Apr54 **RECORDS INDECIPHERABLE** → cvtd **C-119G** date unk – 313 TCSq (349 TCWG) (AFR) Portland date unk; 939 TCG (349 TCWG) (AFR) Portland 63 ← **RECORDS INDECIPHERABLE** 908 TAL GP (435 TCWG) (AFR) Brookley, to MAP Taiwan 03Feb69, RoCAF s/n: **3112** – Final Disposition: Unknown.

10706 / 51-8016 / C-119F-FA

Avble: 15Dec52; Acc: 30Dec52; Del: 06Jan53 to USAF **RECORDS INDECIPHERABLE** → 514 TCWG (TAC) Mitchel 06Jan53 – 313 TCG (TAC) Mitchel 53; to Sewart 02Oct53; depl Sumpter Alabama 19Jul54 ← **RECORDS INDECIPHERABLE** bailment Fairchild (AMC) Hagerstown 24Jan55, cvtd **C-119G** 17Feb55 – 314 TCWG (TAC) Sewart 09May55 – 2472 RFC (CNR) Richards 10Jul57 – 442 TCWG (AFR) Richards 02Mar59 – 452 TCWG (AFR) March 14Apr61; 944 TCG (452 TCWG) (AFR) March 11Feb63; 942 TCG (452 TCWG) (AFR) March 31Mar65 – 924 TCG (446 TCWG) (AFR) Ellington 12Aug65 – 906 TAL GP (302 TCWG) (AFR) Clinton 20Jan68 – MAP Taiwan 26Feb70, RoCAF s/n: **3202** – Presently displayed in Yuanzhiluxiuxian Park Nanhua District Taiwan – Final Disposition: Extant.

10707 / 51-7968 / C-119F-FA

Avble: 11Sep52; Acc: 30Sep52; Del: 12Nov52 to USAF – Bailment Fairchild (AMC) Hagerstown 12Nov52 for experimental duties; cvtd **KC-119F** 14Aug53 – WRDCN (ARD) WPAFB 14Aug53 for test duties – 6520 TES WG (ARD) Hanscom 07Jun54; cvtd **C-119F** – CAR (ARD) Hanscom 01Jan55 – 456 TCWG (TAC) Charleston 09Mar55; to Shiroi AB Japan 10Nov55; to Ardmore 05May56 – BMC (AMC) Birmingham 19Oct56; cvtd **C-119J** 24Jul57 – 2465 RFC (CNR) Minneapolis-St. Paul 29Aug57 – 2473 RFC DT (CNR) Minneapolis-St. Paul 01Dec57 – 96 TCSq (440 TCWG) (AFR) Minneapolis-St. Paul 19Dec58 –

140 AML SQ (PA-ANG) Olmsted 23Mar61 – 96 TCSq (440 TCWG) (AFR) Minneapolis-St. Paul 28Nov62 – Wfu to 2704 ASD GP (LOG) Davis-Monthan 16Jan63 for storage – Fairchild (LOG) St. Augustine 09Nov63 – MAP Italy 28Jan64 as a spares airframe – Final Disposition: Scrapped.

10708 / 51-7969 / C-119F-FA

Avble: 12Sep52; Acc: 30Sep52; Del: 06Oct52 to USAF – 1300 AB WG (MATS) Great Falls 06Oct52 – 313 TCG (TAC) Mitchel 29Aug53; depl 465 TCWG (TAC) Mitchel Jan53; to Sewart 01Oct53 – 316 TCG (TAC) Sewart 23Apr54; depl Pope 23Apr54; depl Sumpter Alabama Aug54 – 314 TCWG (TAC) Sewart 15Nov54; depl Elmendorf 11Jan55; depl Ladd Alaska 01Aug55; depl 463 TCWG (TAC) Sewart (depl) 02Sep55; bailment Fairchild (AMC) Hagerstown 12Dec55, cvtd **C-119G** 21Jan56 – 2472 RFC (CNR) Richards 02Jul57; depl 2466 RFC (CNR) Bakalar 07Jul57; to 5 AB GP (SAC) Travis 20Mar58 mnt – Hayes (AMC) Birmingham 18Feb59 mnt – 442 TCWG (AFR) Richards 08Apr59 – 452 TCWG (AFR) March 27Apr61; 943 TCG (452 TCWG) (AFR) March 11Feb63 – 4603 AB GP (ADC) Stewart 24Jan69 – Wfu to MASDC (LOG) Davis-Monthan 10Nov69 for storage, PCN: **CJ315** – DFI: 19Jan70 – Final Disposition: Scrapped.

10709 / 51-7970 / C-119F-FA

Avble: 13Sep52; Acc: 30Sep52; Del: 07Oct52 to USAF – 1300 AB WG (MATS) Great Falls 06Oct52 **RECORDS INDECIPHERABLE** 313 TCG (TAC) Sewart 06Jan53 – 316 TCG (TAC) Sewart 06Mar54; depl Pope 20Apr54; depl Sumpter Alabama 4Aug54 – 314 TCWG (TAC) Sewart 15Nov54; depl Elmendorf 11Jan55; depl 463 TCWG (TAC) Sewart (depl) 13Aug55; to BMC (AMC) Birmingham 21Feb56, cvtd **C-119G** 12Apr56 – 463 TCWG (TAC) Ardmore 24Apr56; depl North Aux AF 05May56 – 2469 RFC (CNR) Scott 14Jun57; depl 2466 RFC (CNR) Bakalar 18Aug57 – 2466 RFC DT (CNR) Scott 01Dec57 – 73 TCSq (434 TCWG) (AFR) Scott 18Apr59; 932 TCG (434 TCWG) (AFR) Scott 11Feb63 – 910 TCG (459 TCWG) (AFR) Youngstown 05Nov66 – MAP Taiwan 02Dec69 – Final Disposition: Unknown.

10710 / 51-7971 / C-119F-FA

Avble: 15Sep52; Acc: 30Sep52; Del: 06Oct52 to USAF – 1300 AB WG (MATS) Great Falls 06Oct52 – 313 TCG (TAC) Mitchel 22Aug53; depl 465 TCWG (TAC) Mitchel Jan53; to Sewart 04Feb53; depl Pope 20Apr54 – 314 TCWG (TAC) Sewart 54; depl 463 TCWG (TAC) Sewart (depl) 08Jun55 & 13Aug55 – bailment Fairchild (AMC) Hagerstown 27Feb56, cvtd **C-119G** 28Mar56 – 463 TCWG (TAC) Ardmore 05Apr56 – 2585 RFC (CNR) Miami 14May57 – 435 TCWG (AFR) Miami 19Dec58; assigned 78 TCSq Bates 17Nov59; to 2589 AB GP (CNC) Dobbins 27Jan60 mnt – 357 TCSq (302 TCWG) Bates 08May61; 908 TCG (302 TCWG) (AFR) Bates 11Feb63 – 908 TCG (435 TCWG) (AFR) Bates 18Mar63; to Brookley 30Oct64 – MAP Taiwan 03Feb69, RoCAF s/n: **3158** – Presently displayed at the China University of Science and Technology Taipei Taiwan – Final Disposition: Extant.

10711 / 51-7972 / C-119F-FA

Avble: 16Sep52; Acc: 30Sep52; Del: 13Oct52 to USAF – 1300 AB WG (MATS) Mt. Home 13Oct52; to Great Falls date unk – SAAAR (AMC) Kelly 18Sep53 – 582 ARC GP (MATS) Great Falls 04Feb54 – 313 TCG (TAC) Sewart 08Feb54; depl Sumpter Alabama 07Aug54; bailment Fairchild (AMC) Hagerstown 12Dec54, cvtd **C-119G** 11Jan55; depl 463 TCWG (TAC) Sewart (depl) 08Jun55 – **RECORDS MISSING** – 314 TCWG (TAC) Sewart 55; depl Dreux AB France 21Oct55 – 2585 RFC (CNR) Miami IAP 03May57 **RECORDS**

INDECIPHERABLE 435 TCWG (AFR) Miami 19Dec58; to Homestead 25Jul60; assigned 76 TCSq Homestead 01Oct61; 915 TCG (435 TCWG) (AFR) Homestead 17Jan63 – 926 TCG (446 TCWG) (AFR) NAS New Orleans 28Nov65 – Wfu to MASDC (LOG) Davis-Monthan 30Nov69 for storage, PCN: **CJ327** – DFI: 19Jan70 – Kolar, Inc. Tucson Arizona 18Feb76 for scrap – Final Disposition: Scrapped.

10712 / 51-7973 / C-119F-FA

Avble: 17Sep52; Acc: 30Sep52; Del: 03Oct52 to USAF **RECORDS INDECIPHERABLE** 463 TCWG (TAC) Memphis 16Jan53; to Ardmore 17Aug53; to 5700 AB GP (CAC) Albrook 03Mar54 mnt; depl Charleston 21Apr54; depl Évreux-Fauville AB, France 55; bailment Fairchild (AMC) Hagerstown 07Feb56, cvtd **C-119G** 22Mar56; depl North Aux AF 02Apr56 – 464 TCWG (TAC) Pope 14Oct56 – 2346 RFC DT (CNR) Portland 31Jul58 – 313 TCSq (349 TCWG) (AFR) Portland 20Dec58; 939 TCG (349 TCWG) (AFR) Portland 19Aug63 – 926 TCG (446 TCWG) (AFR) NAS New Orleans 24May68 – 906 TAL GP (302 TCWG) (AFR) Clinton 10Jan70 – MAP Taiwan 15Mar70, RoCAF s/n: **3210** – Final Disposition: Unknown.

10713 / 51-7974 / C-119F-FA

Avble: 18Sep52; Acc: 30Sep52; Del: 03Oct52 to USAF **RECORDS INDECIPHERABLE** → 463 TCWG (TAC) Memphis 16Jan53; to Ardmore 17Aug53; depl Gray Texas 07May54; bailment Fairchild (AMC) Hagerstown 03Jan55, cvtd **C-119G** 31Jan55 ← **RECORDS INDECIPHERABLE** depl Laurinburg-Maxton Airport North Carolina 02Aug55 – 464 TCWG (TAC) Pope 29Sep56 – Hayes (AMC) Birmingham 29Jul57 – 2472 RFC (CNR) Richards 10Oct57 – 442 TCWG (AFR) Richards 02Mar59; to 4347 CCT WG (SAC) McConnell 26Sep60 mnt – 452 TCWG (AFR) March 27Apr61; 943 TCG (452 TCWG) (AFR) March 11Feb63; 942 TCG (452 TCWG) (AFR) March 10May65; 943 TCG (452 TCWG) (AFR) March 01Feb66 – 928 TAL GP (403 TCWG) (AFR) O'Hare 22Apr69 – 943 TCG (452 TCWG) (AFR) March 28Apr69 – 928 TAL GP (403 TCWG) (AFR) O'Hare 08May69 – MAP Taiwan 03Dec69 – Final Disposition: Unknown.

10714 / 51-7975 / C-119F-FA

Avble: 19Sep52; Acc: 21Oct52; Del: 25Oct52 to USAF **RECORDS INDECIPHERABLE** → 516 TCWG (TAC) Memphis 24Oct52 – 463 TCWG (TAC) Memphis 16Jan53; to Ardmore 17Aug53 ← **RECORDS INDECIPHERABLE** depl 63 TCWG (TAC) Donaldson 10Mar55; depl North Aux AF 17Apr55; bailment Fairchild (AMC) Hagerstown 09Jan56, cvtd **C-119G** 14Feb56; to MIDAR (AMC) Olmsted 22Feb56 mnt; depl North Aux AF 05May56 – 2235 RFC (CNR) Grenier 21Mar57 – 328 TCSq (512 TCWG) (AFR) Niagara Falls 25Feb59 – 732 TCSQ (94 TCWG) (AFR) Grenier 21Mar59; 902 TCG (94 TCWG) (AFR) Grenier 11Feb63 – 926 TCG (446 TCWG) (AFR) NAS New Orleans 15Nov65 – Wfu to MASDC (LOG) Davis-Monthan 23Nov69 for storage, PCN: **CJ320** – DFI: 19Jan70 – Kolar, Inc. Tucson Arizona 04Feb76 for scrap – Final Disposition: Scrapped.

10715 / 51-7976 / C-119F-FA

Avble: 22Sep52; Acc: 21Oct52; Del: 25Oct52 to USAF **RECORDS INDECIPHERABLE** → 516 TCWG (TAC) Memphis 24Oct52 – 463 TCWG (TAC) Memphis 16Jan53; to Ardmore 17Aug53; to 4415 AB GP (TAC) Pope 19Feb54; depl Charleston 21Apr54 ← **RECORDS INDECIPHERABLE** bailment Fairchild (AMC) Hagerstown 31Jan56, cvtd **C-119G** 01Mar56; depl North Aux AF 05May56 – 2466 RFC (CNR) Bakalar 06Mar57 – 434 TCWG (AFR) Bakalar 19Apr59; 930 TCG (434 TCWG) (AFR) Bakalar 11Feb63; depl 71

ACO SQ (1 ACO WG) (TAC) Lockbourne 20Jun68 – Wfu to MASDC (LOG) Davis-Monthan 04Nov68 for storage, PCN: **CJ267** – DFI: 09Dec74 – Final Disposition: Scrapped.

10716 / 51-7977 / C-119F-FA

Avble: 22Sep52; Acc: 21Oct52; Del: 25Oct52 to USAF **RECORDS INDECIPHERABLE** → 516 TCWG (TAC) Memphis 24Oct52 – 463 TCWG (TAC) Memphis 16Jan53; to Ardmore 17Aug53; depl Charleston 21Apr54; depl Gray Texas 16May54 ← **RECORDS INDECIPHERABLE** bailment Fairchild (AMC) Hagerstown 31Jan55, cvtd **C-119G** 28Feb55; depls North Aux AF 07Apr55 & 05May56 – BMC (AMC) Birmingham 03Dec56 – 2466 RFC (CNR) Bakalar 27Feb57 – 434 TCWG (AFR) Bakalar 19Apr59; 931 TCG (434 TCWG) (AFR) Bakalar 11Feb63 – MAP Taiwan 06Mar69 – Final Disposition: Unknown.

10717 / 51-7978 / C-119F-FA

Avble: 24Sep52; Acc: 21Oct52; Del: 25Oct52 to USAF **RECORDS INDECIPHERABLE** → 516 TCWG (TAC) Memphis 24Oct52 – 463 TCWG (TAC) Memphis 16Jan53; to Ardmore 17Aug53; depl Gray Texas 07May54 ← **RECORDS INDECIPHERABLE – RECORDS MISSING** – depl Évreux-Fauville AB France 55; bailment Fairchild (AMC) Hagerstown 20Mar56, cvtd **C-119G** 06Apr56 – BMC (AMC) Birmingham 11Dec56 – 2466 RFC (CNR) Bakalar 11Mar57; depl 2242 RFC (CNR) Selfridge 27Sep57; to 837 AB GP (TAC) Shaw 23May58 mnt – 434 TCWG (AFR) Bakalar 19Apr59; 931 TCG (434 TCWG) (AFR) Bakalar 11Feb63 – MAP Taiwan 06Mar69, RoCAF s/n: **3123** – Final Disposition: Unknown.

10718 / 51-7979 / C-119F-FA

Avble: 24Sep52; Acc: 21Oct52; Del: 25Oct52 to USAF **RECORDS INDECIPHERABLE** → 516 TCWG (TAC) Memphis 24Oct52 – 463 TCWG (TAC) Memphis 16Jan53; to Ardmore 17Aug53; to 2750 AB WG (AMC) WPAFB 19Aug53 mnt ← **RECORDS INDECIPHERABLE** depl Évreux-Fauville AB France 55; bailment Fairchild (AMC) Hagerstown 07Nov55, cvtd **C-119G** 15Dec55; depl North? 02Apr56 – BMC (AMC) Birmingham 01Nov56 – 17 BTA WG (TAC) Eglin 07Feb57; to 2347 RFC (CNR) Long Beach 05Mar57 mnt – 2347 RFC (CNR) Long Beach 25Jun58 – 452 TCWG (AFR) Long Beach 13Jun59; to March 01Oct60; 944 TCG (452 TCWG) (AFR) March 11Feb63 – 934 TAL GP (440 TCWG) (AFR) Minneapolis-St. Paul 25Jan68 – MAP Taiwan 08Apr69 – Final Disposition: Unknown.

10719 / 51-7980 / C-119F-FA

Avble: 25Sep52; Acc: 21Oct52; Del: 25Oct52 to USAF **RECORDS INDECIPHERABLE** → 516 TCWG (TAC) Memphis 24Oct52 – 463 TCWG (TAC) Memphis 16Jan53; to Ardmore 17Aug53; depl Charleston 23Apr54; depl Gray Texas 07May54 ← **RECORDS INDECIPHERABLE** depl North Aux AF 17Apr55; bailment Fairchild (AMC) Hagerstown 20Mar56, cvtd **C-119G** 10Apr56; depl North Aux AF 22May56 – 2472 RFC (CNR) Richards 03Jun57; depl 2466 RFC (CNR) Bakalar 17Jun57 – 442 TCWG (AFR) Richards 02Mar59; depl 96 TCSq (440 TCWG) (AFR) Minneapolis-St. Paul 22Jan61 – 452 TCWG (AFR) March 04May61; 942 TCG (452 TCWG) (AFR) March 31May63 – 925 TCG (446 TCWG) (AFR) Ellington 02Sep65; 924 TAL GP (446 TCWG) (AFR) Ellington 14Jan68 – 908 TAL GP (435 TCWG) (AFR) Brookley 28May68 – MAP Taiwan 03Feb69 – Final Disposition: Unknown.

10720 / 51-7981 / C-119F-FA

Avble: 26Sep52; Acc: 21Oct52; Del: 25Oct52 to USAF **RECORDS INDECIPHERABLE** → 516 TCWG (TAC) Memphis 24Oct52 – 463 TCWG (TAC) Memphis 16Jan53; to

Ardmore 17Aug53 ← **RECORDS INDECIPHERABLE** depl Laurinburg-Maxton Airport North Carolina 03Aug55; bailment Fairchild (AMC) Hagerstown 17Jan56, cvtd **C-119G** 15Feb56; depl 464 TCWG (TAC) Pope 23Feb56 – 2235 RFC (CNR) Grenier 01May57 – 2230 RFC (CNR) NAS New York 26Jul57 – Hayes (AMC) Birmingham 06Sep57 – 2234 RFC (CNR) Hanscom 04Dec57 – 94 TCWG (AFR) Hanscom 19Mar59; 901 TCG (94 TCWG) (AFR) Hanscom 31May63 – 907 TCG (302 TCWG) (AFR) Clinton 16Jul66 – MAP Taiwan 05Jan70, RoCAF s/n: **3195** – Final Disposition: Unknown.

10721 / 51-7982 / C-119F-FA

Avble: 29Sep52; Acc: 21Oct52; Del: 25Oct52 to USAF **RECORDS INDECIPHERABLE** → 516 TCWG (TAC) Memphis 24Oct52 – 463 TCWG (TAC) Memphis 16Jan53; to Ardmore 17Aug53 ← **RECORDS INDECIPHERABLE** depl Laurinburg-Maxton Airport North Carolina 02Aug55; bailment Fairchild (AMC) Hagerstown 03Jan56, cvtd **C-119G** 01Feb56 – BMC (AMC) Birmingham 20Dec56 – 2235 RFC (CNR) Grenier 27Mar57 – Hayes (AMC) Birmingham 06Feb59 – 732 TCSq (94 TCWG) (AFR) Grenier 25Mar59; 902 TCG (94 TCWG) (AFR) Grenier 11Feb63 – 926 TCG (446 TCWG) (AFR) NAS New Orleans 12Nov65 – Wfu to MASDC (LOG) Davis-Monthan 23Nov69 for storage, PCN: **CJ323** – DFI: 19Jan70 – Kolar, Inc. Tucson Arizona 19Feb76 for scrap – Final Disposition: Scrapped.

10722 / 51-7983 / C-119F-FA

Avble: 30Sep52; Acc: 22Oct52; Del: 27Oct52 to USAF **RECORDS INDECIPHERABLE** → 516 TCWG (TAC) Memphis 27Oct52 – 463 TCWG (TAC) Memphis 16Jan53; to Ardmore 17Aug53 ← **RECORDS INDECIPHERABLE** depl North Aux AF 07Apr55; bailment Fairchild (AMC) Hagerstown 19Dec55, cvtd **C-119G** 25Jan56; depl North Aux AF 18May56 – 2466 RFC (CNR) Bakalar 29Jan57; depl 2469 RFC (CNR) Scott 03Jun57; depl 2465 RFC (CNR) Minneapolis-St. Paul 06Aug57 – 434 TCWG (AFR) Bakalar 19Apr59; 930 TCG (434 TCWG) (AFR) Bakalar 31May63; depl 71 ACO SQ (1 ACO WG) (TAC) Lockbourne 20Jun68 – 908 TAL GP (435 TCWG) (AFR) Brookley 17Oct68 – 922 TAL GP (433 TCWG) (AFR) Kelly 27Feb69 – Fairchild (LOG) Hagerstown 08Apr69 – MAP S. Vietnam 11Aug69, VNAF s/n: **517983**, 413 TSQ (53 TWG) Tan Son Nhut AB – 405 Fighter WG (PAF) Clark AB The Philippines 04Jan73 for disposal – DFI: 05Jun73 – Final Disposition: Scrapped.

10723 / 51-7984 / C-119F-FA

Avble: 02Oct52; Acc: 22Oct52; Del: 27Oct52 to USAF **RECORDS INDECIPHERABLE** → 516 TCWG (TAC) Memphis 27Oct52 – 463 TCWG (TAC) Memphis 16Jan53; to Ardmore 17Aug53; depl Charleston 21Apr54; depl Gray Texas 06May54 ← **RECORDS INDECIPHERABLE** bailment Fairchild (AMC) Hagerstown 19Dec55, cvtd **C-119G** 25Jan56; depl North Aux AF 02Apr56 – BMC (AMC) Birmingham 13Nov56 – 354 Fighter WG (TAC) Myrtle Beach 25Jan57 – 2242 RFC (CNR) Slefridge 19Nov57 **RECORDS INDECIPHERABLE** 403 TCWG (AFR) Selfridge date unk; 927 TCG (403 TCWG) (AFR) Selfridge 11Feb63 – 934 TAL GP (440 TCWG) (AFR) Minneapolis-St. Paul 19Apr69 – 913 TAL GP (512 TCWG) (AFR) NAS Willow Grove 07Feb70 – 906 SOP GP (302 TCWG) (AFR) Clinton 04Nov70 – MAP Taiwan 04Jan71, RoCAF s/n: **3220** – Final Disposition: Unknown.

10724 / 51-7985 / C-119F-FA

Avble: 08Oct52; Acc: 22Oct52; Del: 27Oct52 to USAF **RECORDS INDECIPHERABLE** → 516 TCWG (TAC) Memphis 27Oct52 – 463 TCWG (TAC) Memphis 16Jan53; to Ardmore 17Aug53; depl 5001 AB GP (AAC) Ladd date unk; depl 5039 MNT GP (AAC)

264 Section Four—Service Histories

Elmendorf date unk ← **RECORDS INDECIPHERABLE** depl Évreux-Fauville AB France 55; bailment Fairchild (AMC) Hagerstown 20Mar56, cvtd **C-119G** 04Apr56; depl North Aux AF 06Apr56 – BMC (AMC) Birmingham 02Nov56 – 450 Fighter WG (TAC) Foster 10Jan57 **RECORDS INDECIPHERABLE** 4520 CCT WG (TAC) Nellis date unk – 512 TCWG (AFR) NAS Willow Grove 10Jul59; 913 TCG (512 TCWG) (AFR) NAS Willow Grove 11Feb63 – MAP Taiwan 21Apr69, RoCAF s/n: **3160** – Presently displayed in Xihujunji Park, Xihu, Changhua County Taiwan – Final Disposition: Extant.

10725 / 51-7986 / C-119F-FA

Avble: 03Oct52; Acc: 22Oct52; Del: 27Oct52 to USAF **RECORDS INDECIPHERABLE** → 516 TCWG (TAC) Memphis 27Oct52 – 463 TCWG (TAC) Memphis 16Jan53; to Ardmore 17Aug53; depl 63 TCWG (TAC) Donaldson 09Jan54 ← **RECORDS INDECIPHERABLE** bailment Fairchild (AMC) Hagerstown 31Jan55, cvtd **C-119G** 11Mar55; depl Laurinburg-Maxton Airport North Carolina 02Aug55 – 2472 RFC (CNR) Richards 03Jun57; depl 2466 RFC (CNR) Bakalar 17Jun57 – 442 TCWG (AFR) Richards 02Mar59 – 452 TCWG (AFR) March 04May61; 942 TCG (452 TCWG) (AFR) March 11Feb63 – 904 TCG (514 TCWG) (AFR) Sewart 04Jun65; 903 TCG (514 TCWG) (AFR) McGuire 03May66; depl Myrtle Beach 16Jul67 – 933 TAL (440 TCWG) (AFR) Gen Mitchell IAP 08Aug68 – Wfu to MASDC (LOG) Davis-Monthan 04Nov68 for storage, PCN: **CJ263** – DFI: 09Dec74 – Final Disposition: Scrapped.

10726 / 51-7987 / C-119F-FA

Avble: 06Oct52; Acc: 22Oct52; Del: 27Oct52 to USAF – 516 TCWG (TAC) Memphis 27Oct52; to 4415 AB GP (TAC) Pope 15Dec52 – 463 TCWG (TAC) Memphis 16Jan53; to Ardmore 17Aug53; depl Charleston 23Apr54; bailment Fairchild (AMC) Hagerstown 10Jan55, cvtd **C-119G** 08Feb55; depl 366 Fighter-Bomber WG (TAC) England 24Sep55; to 1608 FDM SQ (MATS) Charleston 11Sep56 – BMC (AMC) Birmingham 27Nov56 – 461 BTA WG (TAC) Blytheville 11Feb57; depl 363 TR WG (TAC) Shaw 09Mar57 – 2347 RFC (CNR) Long Beach 09Feb58; to 3320 TTA WG (TC) Amarillo 10Feb58 mnt – 452 TCWG (AFR) Long Beach 13Jun59; to March 01Oct60; 943 TCG (452 TCWG) (AFR) March 11Feb63; 942 TCG (452 TCWG) (AFR) March 31Mar65 – 924 TCG (446 TCWG) (AFR) Ellington 21Aug65 – 910 TAL (459 TCWG) (AFR) Youngstown 20Jan68 – Wfu to MASDC (LOG) Davis- Monthan 13Sep69 for storage, PCN: **CJ310** – DFI: 31Mar76 – Final Disposition: Scrapped.

10727 / 51-7988 / C-119F-FA

Avble: 07Oct52; Acc: 23Oct52; Del: 29Oct52 to USAF – 516 TCWG (TAC) Memphis 29Oct52 – 463 TCWG (TAC) Memphis 16Jan53; to Ardmore 17Aug53; depl Charleston 23Apr54; bailment Fairchild (AMC) Hagerstown 25Jan55, cvtd **C-119G** 23Feb55 – 2466 RFC (CNR) Bakalar 08Jan57 – 434 TCWG (AFR) Bakalar 19Apr59; 930 TCG (434 TCWG) (AFR) Bakalar 11Feb63; depl 71 ACO SQ (1 ACO WG) (TAC) Lockbourne 20Jun68 – 910 TAL (459 TCWG) (AFR) Youngstown 10Sep68 – Wfu to MASDC (LOG) Davis-Monthan 06Dec69 for storage, PCN: **CJ334** – DFI: 19Jan70 – Kolar, Inc. Tucson Arizona 18Feb76 for scrap-Final Disposition: Scrapped.

10728 / 51-7989 / C-119F-FA

Avble: 08Oct52; Acc: 23Oct52; Del: 29Oct52 to USAF – 516 TCWG (TAC) Memphis 29Oct52 – 463 TCWG (TAC) Memphis 16Jan53; to Ardmore 17Aug53; depl Charleston 23Apr54; depl 479 Fighter WG (TAC) George 04Jun54; depl Évreux-Fauville AB France 55; bailment BMC (AMC) Birmingham 10Oct55, cvtd **C-119G** 30Dec55; depl 314 TCWG

Section Four—Service Histories

(TAC) Sewart 21Feb56; depl North Aux AF 02Apr56 – 2466 RFC (CNR) Bakalar 28Jun57 – 434 TCWG (AFR) Bakalar 19Apr59; 931 TCG (434 TCWG) (AFR) Bakalar 11Feb63; to 19 Bomb WG (SAC) Homestead 03Dec65 mnt – MAP Taiwan 06Mar69, RoCAF s/n: **3119** – Final Disposition: Unknown.

10729 / 51-7990 / C-119F-FA

Avble: 09Oct52; Acc: 27Oct52; Del: 30Oct52 to USAF – 516 TCWG (TAC) Memphis 30Oct52 – 463 TCWG (TAC) Memphis 16Jan53; to Ardmore 17Aug53; depls North Aux AF 07Apr55 & 05May56; bailment Fairchild (AMC) Hagerstown 14Nov55, cvtd **C-119G** 15Dec55 – 2235 RFC (CNR) Grenier 23Mar57 **RECORDS INDECIPHERABLE** 512 TCWG (AFR) NAS Willow Grove 04Mar59 – 732 TCSq (94 TCWG) (AFR) Grenier 15Apr59; 902 TCG (94 TCWG) (AFR) Grenier 11Feb63 – 926 TCG (446 TCWG) (AFR) NAS New Orleans 12Nov65 – 906 TAL GP (302 TCWG) (AFR) Clinton 10Jan70 – MAP Taiwan 03Mar70, RoCAF s/n: **3206** – Final Disposition: Unknown.

10730 / 51-7991 / C-119F-FA

Avble: 10Oct52; Acc: 23Oct52; Del: 29Oct52 to USAF **RECORDS INDECIPHERABLE** → 516 TCWG (TAC) Memphis 29Oct52 – 463 TCWG (TAC) Memphis 16Jan53; to Ardmore 17Aug53 ← **RECORDS INDECIPHERABLE** bailment Fairchild (AMC) Hagerstown 25Jan55, cvtd **C-119G** 24Feb55; depls North Aux AF 07Apr55 & 05May56 – BMC (AMC) Birmingham 16Nov56 – 2466 RFC (CNR) Bakalar 22Feb57; depl DT Dress Memorial AP 05Aug57 – 434 TCWG (AFR) Bakalar 19Apr59; 931 TCG (434 TCWG) (AFR) Bakalar 11Feb63; 930 SOP GP (434 TCWG) (AFR) Bakalar 30Jun69; to Grissom 12Jan70 – Wfu to MASDC (LOG) Davis-Monthan 18Sep70 for storage, PCN: **CJ363** – DFI: 21May71 – Kolar, Inc. Tucson Arizona 27Feb76 for scrap – Final Disposition: Scrapped.

10731 / 51-7992 / C-119F-FA

Avble: 12Oct52; Acc: 27Oct52; Del: 30Oct52 to USAF **RECORDS INDECIPHERABLE** → 516 TCWG (TAC) Memphis 30Oct52 – 463 TCWG (TAC) Memphis 16Jan53; to Ardmore 17Aug53; depl Charleston 23Apr54; depl Gray Texas 07May54 ← **RECORDS INDECIPHERABLE** bailment Fairchild (AMC) Hagerstown 16Dec54, cvtd **C-119G** 18Jan55; depls North Aux AF 07Apr55 & 05May56; to 3310 TTA WG (TC) Scott 13Apr56 mnt – BMC (AMC) Birmingham 21Nov56 – 366 Fighter-Bomber WG (TAC) England 02Feb57 – 834 AB GP (TAC) England 25Sep57 – 506 Fighter-Bomber WG (TAC) Tinker 02Apr58 **RECORDS INDECIPHERABLE** → 4520 CCT WG (TAC) Nellis date unk – 77 TCSq (435 TCWG) (AFR) Donaldson 29Jun59; to 4600 AB WG (ADC) Peterson 17Sep59 mnt ← **RECORDS INDECIPHERABLE** 65 TCSq (403 TCWG) (AFR) Davis 03Jan61; 929 TCG (403 TCWG) (AFR) Davis 11Feb63 – 922 TCG (433 TCWG) (AFR) Kelly 17Oct65 – MAP Taiwan 08Apr69 – Final Disposition: Unknown.

10732 / 51-7993 / C-119F-FA

Avble: 13Oct52; Acc: 27Oct52; Del: 30Oct52 to USAF – 516 TCWG (TAC) Memphis 30Oct52 – 463 TCWG (TAC) Memphis 16Jan53; to Ardmore 17Aug53. Crashed near Lothian (1 mile SSW Mt. Zion) Maryland 19Mar54 while enroute from Bolling AFB Washington D.C. to Mitchel Field New York. Pilot: Gene T. Coppedge, 5 other crew and 12 passengers killed on impact after the aircraft lost control in poor night conditions where the crew were flying under VFR when they should have been under IFR flight rules – DFI: 20Jun54 – Final Disposition: Accident.

10733 / 51-7994 / C-119F-FA

Avble: 14Oct52; Acc: 30Oct52; Del: 04Nov52 to USAF **RECORDS INDECIPHERABLE** → 516 TCWG (TAC) Memphis 52 – 463 TCWG (TAC) Memphis 16Jan53; to Ardmore 17Aug53; depl Charleston 23Apr54; depl Gray Texas May54 ← **RECORDS INDECIPHERABLE** depls North Aux AF 17Apr55 & 05May56; bailment Fairchild (AMC) Hagerstown 20Dec55, cvtd **C-119G** 27Jan56 – 2235 RFC (CNR) Grenier 20Apr57; to 464 TCWG (TAC) Pope 03Feb58 mnt **RECORDS INDECIPHERABLE** 328 TCSq (512 TCWG) (AFR) Niagara Falls 28Feb59 – 732 TCSq (94 TCWG) (AFR) Grenier 21Mar59; 902 TCG (94 TCWG) (AFR) Grenier 11Feb63. Crashed on approach to Barnes MAP Westfield Massachusetts 08May63, crew fatalities or injuries unk – Final Disposition: Accident.

10734 / 51-7995 / C-119F-FA

Avble: 15Oct52; Acc: 30Oct52; Del: 04Nov52 to USAF – 516 TCWG (TAC) Memphis 04Nov52 – 463 TCWG (TAC) Memphis 16Jan53; to Ardmore 17Aug53; depl Gray Texas 06May54; bailment Fairchild (AMC) Hagerstown 03Jan55, cvtd **C-119G** 28Jan55. Landed short of the runway at Hagerstown Airport Maryland 03Mar55. The C-119 bounced back into the air but slammed down again collapsing the right u/c gear, the wreck skidded off the runway at speed and came to a stop only 40 ft from a line of brand new C-119G Boxcars. The 13 crew onboard evacuated safely with the aircraft being engulfed by fire several minutes later – Final Disposition: Accident.

10735 / (RCAF) 22104 / C-119F-FA

Avble: 17Oct52; Acc: 31Oct52; Del: 05Nov52 to Canada, RCAF s/n: **22104** – BOC: 12Nov52 CJATC Rivers; dropped from USAF inventory 20Nov52; cvtd **C-119G** 17Aug56 – 435 SQ Namao 13Dec56 – 436 SQ Downsview 02Oct57 – Wfu into storage 1005 TSD Saskatoon 07May64 – SOC: 06Jun67 to Crown Assets Disposal Corp. – Aircraft International Associates Dallas Texas 02Jun67; reg **N3558** 06Oct71. Listed as "destroyed / scrapped in Saskatoon" 04Feb75 – Reg N3558 canx 08May75 – Final Disposition: Scrapped.

10736 / (RCAF) 22105 / C-119F-FA

Avble: 17Oct52; Acc: 31Oct52; Del: 05Nov52 to Canada, RCAF s/n: **22105** – BOC: 12Nov52 435 SQ Edmonton; dropped from USAF inventory 20Nov52; cvtd **C-119G** 30Jan56 – 436 SQ Downsview 03Aug56 – 435 SQ Namao 08Jan59 – Wfu into storage 1005 TSD Saskatoon 25Jun65 – SOC: 01Nov66 to Crown Assets Disposal Corp. – BofS Aircraft International Associates Dallas Texas 26Oct66; reg **N15506** 04Oct68 – BofS Hawkins & Powers Aviation, Inc. Greybull Wyoming 16Dec75; used as a spares source – USAF Museum WPAFB Ohio 04Nov88. Static display at General Mitchell ANGB Milwaukee Wisconsin marked as: "51-2680" to honor a 440 TCWG C-119 lost in 1965 – Reg N15506 canx 09Dec88 – Dismantled and trucked Jul06 to Niagara Falls ARS New York where it is presently displayed – Final Disposition: Extant.

10737 / (RCAF) 22106 / C-119F-FA

Avble: 20Oct52; Acc: 25Nov52; Del: 03Dec52 to Canada, RCAF s/n: **22106**; dropped from USAF inventory 04Dec52 – BOC: 06Dec52 435 SQ Edmonton; cvtd **C-119G** 26Apr56 – 436 SQ Downsview 16Sep60 – Wfu into storage 1005 TSD Saskatoon 26Jun64 – SOC: 22Aug67 to Crown Assets Disposal Corp. – BofS Aircraft International Associates Dallas Texas 01Aug67; reg **N3003** 16Sep71 – BofS Hawkins & Powers Aviation, Inc. Greybull Wyoming 16Dec75; cvtd **C-119G-3E** 1980 for cargo and agricultural work. Ag sprayer fleet

No. **03** – BofS Heims Seafoods, Inc. Myrtle Point Oregon 18Jun80 – BofS Hawkins & Powers Aviation, Inc. Greybull Wyoming 23Jun82; lease to Northern Pacific Transport, Inc. Anchorage Alaska 11May83 – BofS D&G, Inc. Greybull Wyoming 01Jan93 – BofS The Pride Capital Group, LLC Deerfield Illinois 07Sep05 for auction – BofS B&G Industries, LLC Greybull Wyoming 24Aug06. Reg N3003 canx 13Aug13, renewed 12Mar14 – Atterbury-Bakalar Air Museum Columbus Indiana Sep19 for static display, dismantled and moved by road Nov20 to Columbus MAP and reassembled Jun21 for public display – Final Disposition: Extant.

10738 / (RCAF) 22107 / C-119F-FA

Avble: 22Oct52; Acc: 25Nov52; Del: 03Dec52 to Canada, RCAF s/n: **22107**; dropped from USAF inventory 04Dec52 – BOC: 06Dec52 CE&PE Namao – CJATC Rivers 13Nov56, cvtd **C-119G** 02Jan57 – 435 SQ Namao 24Nov61 – Wfu into storage 1005 TSD Saskatoon 25May64 – SOC: 22Aug67 to Crown Assets Disposal Corp. – BofS Aircraft International Associates Dallas Texas 01Aug67; ferried from Canada to Love Field Texas Nov67; reg **N966S** 04Oct68; BofS Hawkins & Powers Aviation, Inc. Greybull Wyoming 16Dec75; not Jet-Pak cvtd – USAF Museum WPAFB Dayton Ohio circa 1988. Displayed at Hill Aerospace Museum Hill AFB Utah marked as: "52-2107" – Reg N966S canx 20Mar13 – Final Disposition: Extant.

10739 / 51-7996 / C-119F-FA

Avble: 22Oct52; Acc: 21Nov52; Del: 04Dec52 to USAF **RECORDS INDECIPHERABLE** → 516 TCWG (TAC) Memphis 04Dec52 – 463 TCWG (TAC) Memphis 16Jan53; to Ardmore 17Aug53; depl Charleston 23Apr54; depl Gray Texas 07May54 ← **RECORDS INDECIPHERABLE** depl Évreux-Fauville AB France 55; bailment Fairchild (AMC) Hagerstown 28Dec55, cvtd **C-119G** 01Feb56; depl North Aux AF 02Apr56 – BMC (AMC) Birmingham 21Nov56 – 2466 RFC (CNR) Bakalar 20Feb57 – 434 TCWG (AFR) Bakalar 19Apr59; 930 TCG (434 TCWG) (AFR) Bakalar 11Feb63; depl 71 ACO SQ (1 ACO WG) (TAC) Lockbourne 21Jun68 – 907 TAL GP (302 TCWG) (AFR) Clinton 25Sep68 – MAP Taiwan 04Jan70, RoCAF s/n: **3199** – Presently displayed at the War Memorial of Korea Seoul S. Korea – Final Disposition: Extant.

10740 / 51-7997 / C-119F-FA

Avble: 24Oct52; Acc: 21Nov52; Del: 03Dec52 to USAF **RECORDS INDECIPHERABLE** → 516 TCWG (TAC) Memphis 03Dec52 – 463 TCWG (TAC) Memphis 16Jan53; to Ardmore 17Aug53 ← **RECORDS INDECIPHERABLE** depl Évreux-Fauville AB France 55; bailment Fairchild (AMC) Hagerstown 13Jan56, cvtd **C-119G** 07Feb56; depl North Aux AF 02Apr56 – 2466 RFC (CNR) Bakalar 17Dec56; to 3345 TTA WG (TC) Chanute 24Jun57 mnt; to 839 AB GP (TAC) Sewart 13Jan59 mnt – 434 TCWG (AFR) Bakalar 19Apr59; to 63 TCWG (MATS) Donaldson 26Aug60 mnt; 930 TCG (434 TCWG) (AFR) Bakalar 11Feb63; depl 71 ACO SQ (1 ACO WG) (TAC) Lockbourne 21Jun68 – 933 TAL GP (440 TCWG) (AFR) Gen Mitchell 21Oct68 – Wfu to MASDC (LOG) Davis-Monthan 28Nov70 for storage, PCN: **CJ366** – DFI: 21May71 – Kolar, Inc. Tucson Arizona 02Mar76 for scrap – Final Disposition: Scrapped.

10741 / 51-7998 / C-119F-FA

Avble: 24Oct52; Acc: 21Nov52; Del: 03Dec52 to USAF **RECORDS INDECIPHERABLE** → 516 TCWG (TAC) Memphis 03Dec52 – 463 TCWG (TAC) Memphis 16Jan53; to Ardmore 17Aug53; depl Charleston 21Apr54 ← **RECORDS INDECIPHERABLE** to 317

TCWG (AFE) Neubiberg AB W. Germany 22Dec54 mnt; depl Évreux-Fauville AB France 55; bailment Fairchild (AMC) Hagerstown 27Feb56, cvtd **C-119G** 23Mar56; depl North Aux AF 02Apr56 – BMC (AMC) Birmingham 04Dec56 – 2466 RFC (CNR) Bakalar 25Feb57 – 2472 RFC DT (CNR) Tinker 03May58; depl Davis Oklahoma 20Jul58 – 305 TCSq (442 TCWG) (AFR) Tinker 19Mar59 – 434 TCWG (AFR) Bakalar 20Jan61; 931 TCG (434 TCWG) (AFR) Bakalar 11Feb63; 930 SOP GP (434 TCWG) (AFR) Bakalar 30Jun69; to Grissom 12Jan70 – Wfu to MASDC (LOG) Davis-Monthan 09Sep70 for storage, PCN: **CJ360** – DFI: 16Sep70 – Kolar, Inc. Tucson Arizona 12Feb76 for scrap – Final Disposition: Scrapped.

10742 / 51-7999 / C-119F-FA

Avble: 23Oct52; Acc: 21Nov52; Del: 03Dec52 to USAF **RECORDS INDECIPHERABLE** → 516 TCWG (TAC) Memphis 03Dec52 – 463 TCWG (TAC) Memphis 16Jan53; to Ardmore 17Aug53; depl Charleston 23Apr54 ← **RECORDS INDECIPHERABLE** depl Évreux-Fauville AB France 55; bailment BMC (AMC) Birmingham 10Aug55, cvtd **C-119G** 29Nov55; depl North Aux AF 02Apr56 – 2466 RFC (CNR) Bakalar 15Mar57; to 3310 TTA WG (TC) Scott 19Apr57 mnt; depl 2465 RFC (CNR) Minneapolis-St. Paul 12Aug57 – 2473 RFC (CBR) Gen Mitchell 06Mar58 – 440 TCWG (AFR) Gen Mitchell 20Nov58; 933 TCG (440 TCWG) (AFR) Gen Mitchell 11Feb63 – Wfu to MASDC (LOG) Davis- Monthan 28Nov70 for storage, PCN: **CJ373** – DFI: 21May71 – Kolar, Inc. Tucson Arizona 01Mar76 for scrap – Final Disposition: Scrapped.

10743 / 51-8000 / C-119F-FA

Avble: 27Oct52; Acc: 21Nov52; Del: 04Dec52 to USAF – 516 TCWG (TAC) Memphis 04Dec52 – 463 TCWG (TAC) Memphis 16Jan53; to Ardmore 17Aug53; depl Charleston 21Apr54; depl Laurinburg- Maxton Airport North Carolina 03Aug55; bailment Fairchild (AMC) Hagerstown 07Feb56, cvtd **C-119G** 01Mar56 – BMC (AMC) Birmingham 04Jan57 – 2235 RFC (CNR) Grenier 15Apr57; to RDP (AMC) Griffiss 05May57 mnt – Hayes (AMC) Birmingham 10Mar59 – 732 TCSq (94 TCWG) (AFR) Grenier 08Apr59; 902 TCG (94 TCWG) (AFR) Grenier 11Feb63 – 906 TCG (302 TCWG) (AFR) Clinton 19Nov65 – MAP Taiwan 02Dec69, RoCAF s/n: **3181** – Destroyed in a ground fire at Pingtung AB Taiwan 01Jun96, no fatalities, notable as being the last C-119 accident – Final Disposition: Accident.

10744 / 51-8001 / C-119F-FA

Avble: 28Oct52; Acc: 25Nov52; Del: 03Dec52 to USAF **RECORDS INDECIPHERABLE** → 516 TCWG (TAC) Memphis 03Dec52 – 463 TCWG (TAC) Memphis 16Jan53; to Ardmore 17Aug53; depl Charleston 21Apr54 ← **RECORDS INDECIPHERABLE** depl Laurinburg-Maxton Airport North Carolina 02Aug55; bailment Fairchild (AMC) Hagerstown 05Dec55, cvtd **C-119G** 05Jan56 – 2585 RFC (CNR) Miami 26May57; to DT Donaldson 25Feb58 – 77 TCSq (435 TCWG) (AFR) Donaldson 19Dec58 – 512 TCWG (AFR) NAS Willow Grove 10Jan61; 912 TCG (512 TCWG) (AFR) NAS Willow Grove 31May63 – 910 TAL GP (459 TCWG) (AFR) Youngstown 19Aug68 – Wfu to MASDC (LOG) Davis-Monthan 13Sep69 for storage, PCN: **CJ312** – DFI: 31Mar76 – Final Disposition: Scrapped.

10745 / 51-8002 / C-119F-FA

Avble: 28Oct52; Acc: 26Nov52; Del: 03Dec52 to USAF **RECORDS INDECIPHERABLE** → 516 TCWG (TAC) Memphis 03Dec52 – 463 TCWG (TAC) Memphis 16Jan53; to Ardmore 17Aug53; depl Gray Texas 07May54; depl 479 Fighter WG (TAC) George 21Jul54 ← **RECORDS INDECIPHERABLE** to 3625 CCT WG (TC) Tyndall 01Feb56 mnt;

bailment Fairchild (AMC) Hagerstown 15Mar56, cvtd **C-119G** 03Apr56; depl North Aux AF 05May56 – 2235 RFC (CNR) Grenier 26May57 – 2230 RFC (CNR) NAS New York 26Jul57 – 2234 RFC (CNR) Hanscom 13Oct57 – 94 TCWG (AFR) Hanscom 19Mar59; 901 TCG (94 TCWG) (AFR) Hanscom 31May63 – 907 TCG (302 TCWG) (AFR) Clinton 16Jul66 – MAP Taiwan 03Jan70 – Final Disposition: Unknown.

10746 / 51-8003 / C-119F-FA

Avble: 31Oct52; Acc: 26Nov52; Del: 03Dec52 to USAF **RECORDS INDECIPHERABLE** → 516 TCWG (TAC) Memphis 03Dec52 – 463 TCWG (TAC) Memphis 16Jan53; to Ardmore 17Aug53; depl Charleston 27Apr54; depl Gray Texas 06May54 ← **RECORDS INDECIPHERABLE** depl Laurinburg- Maxton Airport North Carolina 02Aug55; bailment Fairchild (AMC) Hagerstown 07Feb56, cvtd **C-119G** 01Mar56 – BMC (AMC) Birmingham 03Jan57 – 2235 RFC (CNR) Grenier 12Apr57 – 732 TCSq (94 TCWG) (AFR) Grenier 19Mar59; 902 TCG (94 TCWG) (AFR) Grenier 11Feb63 – 926 TCG (446 TCWG) (AFR) NAS New Orleans 15Nov65 – Wfu to MASDC (LOG) Davis-Monthan 23Nov69 for storage, PCN: **CJ321** – DFI: 19Jan70 – Kolar, Inc. Tucson Arizona 04Feb76 for scrap – Final Disposition: Scrapped.

10747 / 51-8004 / C-119F-FA

Avble: 30Oct52; Acc: 25Nov52; Del: 03Dec52 to USAF – 516 TCWG (TAC) Memphis 03Dec52 – 463 TCWG (TAC) Memphis 16Jan53; to Ardmore 17Aug53; depl Charleston 21Apr54; depl Laurinburg- Maxton Airport North Carolina 02Aug55; bailment BMC (AMC) Birmingham 07Nov55, cvtd **C-119G** 23Jan56 – 2585 RFC (CNR) Miami 27Jul57; to DT Donaldson 24Apr58 – 77 TCSq (435 TCWG) (AFR) Donaldson 19Dec58 – 94 TCWG (AFR) Hanscom 09Jan61; 901 TCG (94 TCWG) (AFR) Hanscom 31May63; to 4080 STR WG (SAC) Hanscom 14Oct63 – 907 TCG (302 TCWG) (AFR) Clinton 23Jul66 – MAP Taiwan 27Jan70, RoCAF s/n: **3186** – Final Disposition: Unknown.

10748 / 51-8005 / C-119F-FA

Avble: 31Oct52; Acc: 21Nov52; Del: 04Dec52 to USAF **RECORDS INDECIPHERABLE** → 516 TCWG (TAC) Memphis 04Dec52 – 463 TCWG (TAC) Memphis 16Jan53; to Ardmore 17Aug53; depl Charleston 24Apr54 ← **RECORDS INDECIPHERABLE** depl 314 TCWG (TAC) Sewart 09Nov54; depl Laurinburg-Maxton Airport North Carolina 02Aug55; bailment Fairchild (AMC) Hagerstown 14Nov55, cvtd **C-119G** 16Dec55 – BMC (AMC) Birmingham 07Nov56 – 83 Fighter WG (TAC) Seymour 24Jan57; depl 464 TCWG (TAC) Pope 26Feb57 – 4 Fighter WG (TAC) Seymour 08Dec57 – 2242 RFC (CNR) Selfridge 31Jan58 – 403 TCWG (AFR) Selfridge 13Nov58; 927 TCG (403 TCWG) (AFR) Selfridge 10Feb63; 928 TAL GP (403 TCWG) (AFR) O'Hare 19May69; 927 TCG (403 TCWG) (AFR) Selfridge 23May69; 928 TAL GP (403 TCWG) (AFR) O'Hare 05Jun69 – 907 TAL GP (302 TCWG) (AFR) Clinton 12Dec69 – MAP Taiwan 26Feb70 – Final Disposition: Unknown.

10749 / 51-8006 / C-119F-FA

Avble: 03Nov52; Acc: 22Nov52; Del: 03Dec52 to USAF **RECORDS INDECIPHERABLE** → 516 TCWG (TAC) Memphis 03Dec52 – 463 TCWG (TAC) Memphis 16Jan53; to Ardmore 17Aug53; depl Charleston 21Apr54 ← **RECORDS INDECIPHERABLE** depl Évreux-Fauville AB France 55; bailment Fairchild (AMC) Hagerstown 03Jan56, cvtd **C-119G** 27Jan56; depl North Aux AF 02Apr56 – BMC (AMC) Birmingham 07Dec56 – 2235 RFC (CNR) Grenier 12Mar57 – 732 TCSq (94 TCWG) (AFR) Grenier 19Mar59; 902 TCG (94 TCWG) (AFR)

Grenier 11Feb63 – 926 TCG (446 TCWG) (AFR) NAS New Orleans 30Nov65 – 913 TAL GP (512 TCWG) (AFR) NAS Willow Grove 01Feb70 – Wfu to MASDC (LOG) Davis-Monthan 07Feb71 for storage, PCN: **CJ399** – DFI: 31Mar76 – Final Disposition: Scrapped.

10750 / 51-8007 / C-119F-FA

Avble: 03Nov52; Acc: 22Nov52; Del: 04Dec52 to USAF **RECORDS INDECIPHERABLE** → 516 TCWG (TAC) Memphis 04Dec52 – 463 TCWG (TAC) Memphis 16Jan53; to Ardmore 17Aug53; depl Charleston 23Apr54; depl Gray Texas 07May54 ← **RECORDS INDECIPHERABLE** depl Laurinburg- Maxton Airport North Carolina 03Aug55; bailment Fairchild (AMC) Hagerstown 15Mar56, cvtd **C-119G** 02Apr56 – 2235 RFC (CNR) Grenier 04Apr57 – Hayes (AMC) Birmingham 05Jan59 – 732 TCSq (94 TCWG) (AFR) Grenier 31Mar59; 902 TCG (94 TCWG) (AFR) Grenier 11Feb63 – 906 TCG (302 TCWG) (AFR) Clinton 19Nov65 – MAP Taiwan 01Dec69, RoCAF s/n: **3167** – Final Disposition: Unknown.

10751 / 51-8008 / C-119F-FA

Avble: 03Nov52; Acc: 22Nov52; Del: 03Dec52 to USAF **RECORDS INDECIPHERABLE** → 516 TCWG (TAC) Memphis 03Dec52 – 463 TCWG (TAC) Memphis 16Jan53; to Ardmore 17Aug53; depl Charleston 23Apr54 ← **RECORDS INDECIPHERABLE** depl Évreux-Fauville AB France 55; bailment Fairchild (AMC) Hagerstown 01Feb56, cvtd **C-119G** 27Feb56; dep North Aux AF 02Apr56 – BMC (AMC) Birmingham 21Dec56 – 2235 RFC (CNR) Grenier 29Mar57; depl 328 TCSq (512 TCWG) (AFR) Niagara Falls 25Feb59 – 732 TCSq (94 TCWG) (AFR) Grenier 19Mar59; 902 TCG (94 TCWG) (AFR) Grenier 11Feb63 – 906 TCG (302 TCWG) (AFR) Clinton 19Nov65 – MAP Taiwan 03Jan70 – Final Disposition: Unknown.

10752 / 51-8009 / C-119F-FA

Avble: 05Nov52; Acc: 25Nov52; Del: 03Dec52 to USAF **RECORDS INDECIPHERABLE** → 516 TCWG (TAC) Memphis 03Dec52 – 463 TCWG (TAC) Memphis 16Jan53; to Ardmore 17Aug53; depl Charleston 23Apr54 ← **RECORDS INDECIPHERABLE** bailment Fairchild (AMC) Hagerstown 31Jan55, cvtd **C-119G** 28Feb55 – 2466 RFC (CNR) Bakalar 08Jan57; to 4756 ADF WG (ADC) Tyndall 22Nov57 mnt; to 56 CLM SQ (ADC) O'Hare IAP 28Feb59 mnt; to PGC Center (ARD) Eglin 16Mar59 mnt – 434 TCWG (AFR) Bakalar 19Apr59; 930 TCG (434 TCWG) (AFR) Bakalar 11Feb63; depl 71 ACO SQ (1 ACO WG) (TAC) Lockbourne 20Jun68 – 914 TAL GP (512 TCWG) (AFR) Niagara Falls 30Aug68 – Wfu to MASDC (LOG) Davis-Monthan 01Feb71 for storage, PCN: **CJ394** – DFI: 31Mar76 – Final Disposition: Scrapped.

10753 / 51-8010 / C-119F-FA

Avble: 06Nov52; Acc: 25Nov52; Del: 03Dec52 to USAF **RECORDS INDECIPHERABLE** → 516 TCWG (TAC) Memphis 03Dec52 – 463 TCWG (TAC) Memphis 16Jan53; to Ardmore 17Aug53; depl Charleston 23Apr54; depl Gray Texas 07May54 ← **RECORDS INDECIPHERABLE** depls North Aux AF 07Apr55 & 06May56; bailment Fairchild (AMC) Hagerstown 27Feb56, cvtd **C-119G** 15Mar56 – 2466 RFC (CNR) Bakalar 05Jan57 – 434 TCWG (AFR) Bakalar 19Apr59; to LTC (ATC) Lowry 23Jun59 mnt; 930 TCG (434 TCWG) (AFR) Bakalar 11Feb63; depl 71 ACO SQ (1 ACO WG) (TAC) Lockbourne 20Jun68 – 926 TAL GP (446 TCWG) (AFR) NAS New Orleans 03Oct68 – Wfu to MASDC (LOG) Davis-Monthan 09Nov69 for storage, PCN: **CJ313** – DFI: 19Jan70 – Final Disposition: Scrapped.

Section Four—Service Histories

10754 / 51-8011 / C-119F-FA

Avble: 07Nov52; Acc: 25Nov52; Del: 03Dec52 to USAF **RECORDS INDECIPHERABLE** → 516 TCWG (TAC) Memphis 03Dec52 – 463 TCWG (TAC) Memphis 16Jan53; to Ardmore 17Aug53 ← **RECORDS INDECIPHERABLE** depl Évreux-Fauville AB France 55; bailment BMC (AMC) Birmingham 10Oct55, cvtd **C-119G** 19Dec55; depl North Aux AF 02Apr56 – 2469 RFC (CNR) Scott 20Jun57; depl 2466 RFC (CNR) Bakalar 18Aug57 – 2466 RFC DT (CNR) Scott 01Dec57 – 2347 RFC (CNR) Long Beach 08Jun58 – 452 TCWG (AFR) Long Beach 13Jun59; to March 01Oct60; 944 TCG (452 TCWG) (AFR) March 11Feb63; 942 TCG (452 TCWG) (AFR) March 31Mar65 – 924 TCG (446 TCWG) (AFR) Ellington 02Aug65; 925 TCG (446 TCWG) (AFR) Ellington 11Sep65 – 907 TAL GP (302 TCWG) (AFR) Clinton 20Jan68 – MAP Taiwan 03Jan70 – Final Disposition: Unknown.

10755 / 51-8012 / C-119F-FA

Avble: 24Nov52; Acc: 26Nov52; Del: 03Dec52 to USAF **RECORDS INDECIPHERABLE** → 516 TCWG (TAC) Memphis 03Dec52 – 463 TCWG (TAC) Memphis 16Jan53; to Ardmore 17Aug53; depl Charleston 21Apr54 ← **RECORDS INDECIPHERABLE** bailment Fairchild (AMC) Hagerstown 16Jan56, cvtd **C-119G** 09Feb56 – 464 TCWG (TAC) Pope 31Oct56 – Hayes (AMC) Birmingham 28Aug57 – 2473 RFC (CNR) Gen Mitchell 25Nov57 – 440 TCWG (AFR) Gen Mitchell 19Dec58; to 354 Fighter WG (TAC) Myrtle Beach 15Nov59 mnt; 933 TCG (440 TCWG) (AFR) Gen Mitchell 11Feb63 – Wfu to MASDC (LOG) Davis-Monthan 28Nov70 for storage, PCN: **CJ367** – DFI: 21May71 – Kolar, Inc. Tucson Arizona 27Jan76 for scrap-Final Disposition: Scrapped.

10756 / 51-8013 / C-119F-FA

Avble: 13Nov52; Acc: 26Nov52; Del: 03Dec52 to USAF **RECORDS INDECIPHERABLE** → 316 TCG (TAC) Sewart 03Dec52; depl Burlington date unk; depl Pope 20Apr54 – 314 TCWG (TAC) Sewart 15Nov54; depl Elmendorf 11Jan55 ← **RECORDS INDECIPHERABLE** bailment Fairchild (AMC) Hagerstown 28Nov55, cvtd **C-119G** 04Jan56 – 2471 RFC (CNR) O'Hare 16May57; depl 2466 RFC (CNR) Bakalar 21Jul57 – 2473 RFC DT (CNR) O'Hare 01Dec57 – 2242 RFC DT (CNR) O'Hare 01Apr58 – Hayes (AMC) Birmingham 21Nov58 – 64 TCSq (403 TCWG) (AFR) O'Hare 11Feb59; 928 TCG (403 TCWG) (AFR) O'Hare 11Feb63 – MAP Taiwan 01Dec69 – Final Disposition: Unknown.

10757 / 51-8014 / C-119F-FA

Avble: 13Nov52; Acc: 26Nov52; Del: 03Dec52 to USAF **RECORDS INDECIPHERABLE** → 316 TCG (TAC) Sewart 03Dec52; depl Burlington date unk; depl Pope 20Apr54 – 314 TCWG (TAC) Sewart 15Nov54; depl Elmendorf 11Jan55 ← **RECORDS INDECIPHERABLE** depl Big Delta Alaska 28Nov55 – bailment Fairchild (AMC) Hagerstown 20Feb56, cvtd **C-119G** 28Mar56 – 463 TCWG (TAC) Ardmore 04Apr56 – 2466 RFC (CNR) Bakalar 15Mar57 – 2473 RFC (CNR) Gen Mitchell 07Apr58 – Hayes (AMC) Birmingham 29Oct58 – 440 TCWG (AFR) Gen Mitchell 29Dec58; 933 TCG (440 TCWG) (AFR) Gen Mitchell 11Feb63 – Wfu to MASDC (LOG) Davis-Monthan 12Feb71 for storage, PCN: **CJ402** – DFI: 26Dec73 – Final Disposition: Scrapped.

10758 / 51-8015 / C-119F-FA

Avble: 13Nov52; Acc: 26Nov52; Del: 03Dec52 to USAF **RECORDS INDECIPHERABLE** → 316 TCG (TAC) Sewart 03Dec52; depl Burlington date unk; depl Pope 20Apr54 – 314 TCWG (TAC) Sewart 15Nov54; depl Elmendorf 11Jan55 ← **RECORDS**

INDECIPHERABLE bailment Fairchild (AMC) Hagerstown 28Nov55, cvtd **C-119G** 28Dec55 – BMC (AMC) Birmingham 22Jan57 – 2469 RFC (CNR) Scott 24Apr57 – 2466 RFC DT (CNR) Scott 01Dec57 – 73 TCSq (434 TCWG) (AFR) Scott 18Apr59; 932 TCG (434 TCWG) (AFR) Scott 11Feb63 – 908 TCG (435 TCWG) (AFR) Brookley 14Jan67 – 913 TAL GP (512 TCWG) (AFR) NAS Willow Grove 25Feb69 – 907 TAL GP (302 TCWG) (AFR) Clinton 20Dec69 – MAP Taiwan 09Jan70 – Final Disposition: Unknown.

10759 / 51-2717 / C-119F-FA

Avble: 14Nov52; Acc: 26Nov52; Del: 03Dec52 to USAF – 316 TCG (TAC) Sewart 03Dec52; depl Burlington 05Feb53; depl Pope 20Apr54; depl Sumpter Alabama 24Aug54 – 314 TCWG (TAC) Sewart 15Nov54 **RECORDS INDECIPHERABLE** → 3345 TTA WG (TC) Chanute 26Dec57, grounded awaiting disposition orders ← **RECORDS INDECIPHERABLE** Final Disposition: Unknown.

10760 / 51-8017 / C-119F-FA

Avble: 17Nov52; Acc: 17Dec52; Del: 22Dec52 to USAF **RECORDS INDECIPHERABLE** → 316 TCG (TAC) Sewart 22Dec52; depl Burlington date unk; depl Pope 20Apr54; depl Elmendorf 01Jul54; depl Sumpter Alabama 24Aug54 – 314 TCWG (TAC) Sewart 15Nov54; depl Elmendorf 11Jan55 ← **RECORDS INDECIPHERABLE** depl Ladd Alaska 01Jun55; bailment BMC (AMC) Birmingham 19Oct55, cvtd **C-119G** 02Jan56 – 2472 RFC (CNR) Richards 15Jun57; depl DT Tinker 25Jan58 – 442 TCWG (AFR) Richards 02Mar59 – 452 TCWG (AFR) March 14Apr61; 944 TCG (452 TCWG) (AFR) March 11Feb63; 942 TCG (452 TCWG) (AFR) March 31Mar65 – 925 TCG (446 TCWG) (AFR) Ellington 11Sep65; 924 TAL GP (446 TCWG) (AFR) Ellington 14Jan68 – 922 TAL GP (433 TCWG) (AFR) Kelly 09May68 – MAP Taiwan 08Apr69, RoCAF s/n: **3147** – Final Disposition: Unknown.

10761 / 51-8018 / C-119F-FA

Avble: 17Nov52; Acc: 17Dec52; Del: 22Dec52 to USAF **RECORDS INDECIPHERABLE** → 316 TCG (TAC) Sewart 22Dec52; depl Burlington date unk; depl Pope 20Apr54; depl Sumpter Alabama 24Aug54 – 314 TCWG (TAC) Sewart 15Nov54; depl Elmendorf 11Jan55 ← **RECORDS INDECIPHERABLE** bailment Fairchild (AMC) Hagerstown 28Nov55, cvtd **C-119G** 19Jan56 – BMC (AMC) Birmingham 15Jan57 – 2469 RFC (CNR) Scott 18Apr57; depl 2466 RFC (CNR) Bakalar 18Aug57 – 2466 RFC DT (CNR) Scott 01Dec57; to 1405 AB WG (MATS) Scott 03Dec57 mnt; depl 2472 RFC (CNR) Richards 15Mar58 – 65 TCSq (403 TCWG) (AFR) Davis 20Sep58; 929 TCG (403 TCWG) (AFR) Davis 11Feb63 – 922 TCG (433 TCWG) (AFR) Kelly 17Oct65 – 906 SOP GP (302 TCWG) (AFR) Clinton 06Nov70 – MAP Taiwan 27Nov70 – Final Disposition: Unknown.

10762 / 51-8019 / C-119F-FA

Avble: 18Nov52; Acc: 17Dec52; Del: 22Dec52 to USAF **RECORDS INDECIPHERABLE** → 316 TCG (TAC) Sewart 22Dec52; depl Burlington date unk; depl Pope 20Apr54; depl Sumpter Alabama 24Aug54 – 314 TCWG (TAC) Sewart 15Nov54; depl Elmendorf 11Jan55 ← **RECORDS INDECIPHERABLE** bailment Fairchild (AMC) Hagerstown 17Oct55, cvtd **C-119G** 26Nov55; depl Big Delta Alaska 03Jan56 – BMC (AMC) Birmingham 01Feb57 – 2469 RFC (CNR) Scott 27Apr57; depls 2466 RFC (CNR) Bakalar 16Jun57 & 18Aug57 – 2466 RFC DT (CNR) Scott 01Dec57 – 73 TCSq (434 TCWG) (AFR) Scott 18Apr59; 932 TCG (434 TCWG) (AFR) Scott 11Feb63. Crashed during a check flight at Scott AFB Illinois 17Apr66. While in flight the no. 2 propeller failed to fully feather so an

Section Four—Service Histories

emergency landing was requested and the no. 2 engine shut down. The first approach was high so a go-round was initiated. The second approach was also too high and too fast so another go-round was initiated. At this point it appears the descent rate was not recovered before the aircraft struck the runway at speed and slid off some way across a neighboring field. The 3 crew survived but the entire front section of the C-119 was destroyed – REC: 22Apr66 – Final Disposition: Accident.

10763 / 51-8020 / C-119F-FA

Avble: 19Nov52; Acc: 17Dec52; Del: 22Dec52 to USAF **RECORDS INDECIPHERABLE** → 316 TCG (TAC) Sewart 22Dec52; depl Burlington date unk; depl Pope 20Apr54; depl Sumpter Alabama 24Aug54 – 314 TCWG (TAC) Sewart 15Nov54; depl Elmendorf 11Jan55 ← **RECORDS INDECIPHERABLE** bailment Fairchild (AMC) Hagerstown 24Oct55, cvtd **C-119G** 13Dec55 – BMC (AMC) Birmingham 08Feb57 – 2235 RFC (CNR) Grenier 30Apr57; depl 328 TCSq (512 TCWG) (AFR) Niagara Falls 07Mar59 – 732 TCSq (94 TCWG) (AFR) Grenier 19Mar59; 902 TCG (94 TCWG) (AFR) Grenier 11Feb63 – 906 TCG (302 TCWG) (AFR) Clinton 19Nov65 – MAP Taiwan 01Dec69 – Final Disposition: Unknown.

10764 / 51-8021 / C-119F-FA

Avble: 19Nov52; Acc: 18Dec52; Del: 29Dec52 to USAF **RECORDS INDECIPHERABLE** → 316 TCG (TAC) Sewart 29Dec52; depl Burlington date unk; depl Pope 20Apr54; depl Sumpter Alabama 24Aug54 – 314 TCWG (TAC) Sewart 15Nov54; depl Elmendorf 11Jan55 ← **RECORDS INDECIPHERABLE** bailment BMC (AMC) Birmingham 15Aug55, cvtd **C-119G** 01Dec55 – 2471 RFC (CNR) O'Hare IAP 06Jun57; depl 2466 RFC (CNR) Bakalar 16Jul57 – 2473 RFC DT (CNR) O'Hare 01Dec57. Had an engine failure at O'Hare IAP Illinois 04Feb58 during take-off sending the aircraft off the runway into a snow bank with such force the forward fuselage broke off. The 4 crew onboard were not injured – Assigned 56 CLM SQ (ADC) O'Hare IAP 04Feb58 for REC – 2242 RFC DT (CNR) O'Hare 18Mar58 for REC – REC: 20Mar58 – Final Disposition: Accident.

10765 / 51-8022 / C-119F-FA

Avble: 20Nov52; Acc: 18Dec52; Del: 29Dec52 to USAF **RECORDS INDECIPHERABLE** → 316 TCG (TAC) Sewart 29Dec52; depl Burlington date unk; depl Pope 20Apr54; depl Sumpter Alabama 24Aug54 – 314 TCWG (TAC) Sewart 15Nov54; depl Elmendorf 11Jan55 ← **RECORDS INDECIPHERABLE** bailment BMC (AMC) Birmingham 03Oct55, cvtd **C-119G** 01Dec55 – 2466 RFC (CNR) Bakalar 30Jul57; depl DT Dress Memorial AP 05Aug57 – 434 TCWG (AFR) Bakalar 19Apr59; 931 TCG (434 TCWG) (AFR) Bakalar 11Feb63; 930 SOP GP (434 TCWG) (AFR) Bakalar 30Jun69; to Grissom 12Jan70 – Wfu to MASDC (LOG) Davis-Monthan 27Jul70 for storage, PCN: **CJ348** – DFI: 25Aug70 – Kolar, Inc. Tucson Arizona 04Feb76 for scrap – Final Disposition: Scrapped.

10766 / 51-8023 / C-119F-FA

Avble: 21Nov52; Acc: 23Dec52; Del: 05Jan53 to USAF **RECORDS INDECIPHERABLE** → 316 TCG (TAC) Sewart 05Jan53; depl Burlington date unk; depl Pope 20Apr54; depl Sumpter Alabama 24Aug54 – 314 TCWG (TAC) Sewart 15Nov54; depl Elmendorf 11Jan55 ← **RECORDS INDECIPHERABLE** bailment Fairchild (AMC) Hagerstown 10Oct55, cvtd **C-119G** 26Nov55 – BMC (AMC) Birmingham 11Feb57 – 2585 RFC (CNR) Miami 07May57; to DT Donaldson 23Mar58 – 77 TCSq (435 TCWG) (AFR) Donaldson 19Dec58; to 14 AF HQ (CNC) Robins 06Mar59 mnt – 732 TCSq (94 TCWG) (AFR) Grenier 06Jan61; 902 TCG

(94 TCWG) (AFR) Grenier 31May63 – 906 TCG (302 TCWG) (AFR) Clinton 19Nov65 – Wfu to MASDC (LOG) Davis-Monthan 31Aug70 for storage, PCN: **CJ352** – DFI: 16Sep70 – Kolar, Inc. Tucson Arizona 04Feb76 for scrap – Final Disposition: Scrapped.

10767 / 51-8024 / C-119F-FA

Avble: 22Nov52; Acc: 23Dec52; Del: 05Jan53 to USAF **RECORDS INDECIPHERABLE** → 316 TCG (TAC) Sewart 05Jan53; depl Burlington date unk; depl Elmendorf 01Jul54; depl Sumpter Alabama 24Aug54 – 314 TCWG (TAC) Sewart 15Nov54; depl Elmendorf 11Jan55 ← **RECORDS INDECIPHERABLE** bailment BMC (AMC) Birmingham 13Dec55, cvtd **C-119G** 21Feb56 – 2466 RFC (CNR) Bakalar 17Jul57 – 2472 RFC DT (CNR) Tinker 09May58 – 305 TCSq (442 TCWG) (AFR) Tinker 19Mar59 – 434 TCWG (AFR) Bakalar 20Jan61; 930 TCG (434 TCWG) (AFR) Bakalar 11Feb63; depl 71 ACO SQ (1 ACO WG) (TAC) Lockbourne 20Jun68 – 4413 CCT SQ (1 ACO WG) (TAC) Lockbourne 07Oct68 – 1 SOP WG (TAC) Lockbourne 17Feb69 – 4410 CCT WG (TAC) Lockbourne 14Jul69 – Wfu 20Mar70 to museum / school status. Displayed at SAC & Aerospace Museum Offutt AFB Nebraska. Relocated 1997 to new SAC museum at Ashland Nebraska, fully restored – Final Disposition: Extant.

10768 / 51-8025 / C-119F-FA

Avble: 17Dec52; Acc: 23Dec52; Del: 05Jan53 to USAF **RECORDS INDECIPHERABLE** → 316 TCG (TAC) Sewart 05Jan53; depl Burlington date unk; depl 4415 AB GP (TAC) Pope 13Jul53; depl Pope 20Apr54; depl Sumpter Alabama 24Aug54 – 314 TCWG (TAC) Sewart 15Nov54; depl Elmendorf 11Jan55 ← **RECORDS INDECIPHERABLE** bailment Fairchild (AMC) Hagerstown 31Oct55, cvtd **C-119G** 06Dec55 – BMC (AMC) Birmingham 14Feb57 – 2585 RFC (CNR) Miami 08May57; depl 2589 RFC DT (CNR) Donaldson 25Feb58; to 1611 TSM WG (MATS) McGuire 06Mar58 mnt; to 2585 RFC DT (CNR) Donaldson 25Apr58 – 77 TCSq (435 TCWG) (AFR) Donaldson 19Dec58; depl 512 TCWG (AFR) NAS Willow Grove 02Mar59 – 96 TCSq (440 TCWG) (AFR) Minneapolis-St. Paul 05Jan61; 934 TCG (440 TCWG) (AFR) Minneapolis-St. Paul 11Feb63; 933 TAL GP (440 TCWG) (AFR) Gen Mitchell 30Jan70 – 928 TAL GP (403 TCWG) (AFR) O'Hare 28Mar70 – 906 SOP GP (302 TCWG) (AFR) Clinton 27Oct70 – MAP Taiwan 27Nov70, RoCAF s/n: **3213** – Final Disposition: Unknown.

10769 / 51-8026 / C-119F-FA

Avble: 25Nov52; Acc: 23Dec52; Del: 05Jan53 to USAF **RECORDS INDECIPHERABLE** → 316 TCG (TAC) Sewart 05Jan53; depl Burlington date unk; depl Pope 20Apr54; depl Sumpter Alabama 24Aug54 – 314 TCWG (TAC) Sewart 15Nov54; depl Elmendorf 11Jan55 ← **RECORDS INDECIPHERABLE** bailment Fairchild (AMC) Hagerstown 21Nov55, cvtd **C-119G** 22Dec55. Hit the side of a mountain 6.9 miles N Newburg Pennsylvania 26Oct56 while on a flight from Sewart AFB to Olmsted AFB in bad weather. 3 crew and 1 passenger killed on impact were: 1Lt Robert S. Hantsch (pilot), 2Lt Walter B. Gordon, Jr. (co-pilot), TSgt Marvin W. Seigler (engineer) and 1Lt Gracye E. Young (nurse 4457 USAF Hospital) – DFI: 06Nov56 – Final Disposition: Accident.

10770 / 51-8027 / C-119F-FA

Avble: 26Nov52; Acc: 23Dec52; Del: 05Jan53 to USAF **RECORDS INDECIPHERABLE** → 316 TCG (TAC) Sewart 05Jan53; depl Burlington date unk; depl Pope 19Apr54; depl Sumpter Alabama 24Aug54 – 314 TCWG (TAC) Sewart 15Nov54; depl Elmendorf 11Jan55 ← **RECORDS INDECIPHERABLE** bailment Fairchild (AMC) Hagerstown

Section Four—Service Histories

05Dec55, cvtd **C-119G** 06Jan56 – BMC (AMC) Birmingham 20Feb57 – 2471 RFC (CNR) O'Hare 16May57; depl 2466 RFC (CNR) Bakalar 21Jul57 – 2473 RFC DT (CNR) O'Hare 01Dec57 – 2242 RFC DT (CNR) O'Hare 01Apr58 – 64 TCSq (403 TCWG) (AFR) O'Hare 20Dec58; to 818 COS GP (SAC) Lincoln 25Oct59 mnt; 928 TCG (403 TCWG) (AFR) O'Hare 11Feb63 – MAP Taiwan 03Dec69 – Final Disposition: Unknown.

10771 / 51-8028 / C-119F-FA

Avble: 26Nov52; Acc: 24Dec52; Del: 31Dec52 to USAF **RECORDS INDECIPHERABLE** → 514 TCWG (TAC) Mitchel 31Dec52 – 313 TCG (TAC) Mitchel 01Feb53; to Sewart 02Oct53; depl Pope 20Apr54; depl Sumpter Alabama 19Jul54; bailment Fairchild (AMC) Hagerstown 06Dec54, cvtd **C-119G** 27Dec54; depl Elmendorf 12Jan55 ← **RECORDS INDECIPHERABLE** 314 TCWG (TAC) Sewart 15May55 – 2466 RFC (CNR) Bakalar 11Aug57; depl DT Dress Memorial AP 29Aug57; to 837 AB GP (TAC) Shaw 23May58 mnt – 434 TCWG (AFR) Bakalar 19Apr59; 930 TCG (434 TCWG) (AFR) Bakalar 11Feb63 – 2578 AB SQ (CNC) Ellington 16Oct66 mnt. Listed as having accident 02Mar67, nfd – Final Disposition: Accident.

10772 / 51-8029 / C-119F-FA

Avble: 01Dec52; Acc: 24Dec52; Del: 31Dec52 to USAF **RECORDS INDECIPHERABLE** → 514 TCWG (TAC) Mitchel 31Dec52 – 313 TCG (TAC) Mitchel 01Feb53; to Sewart 02Oct53; depl Sumpter Alabama 19Jul54; bailment Fairchild (AMC) Hagerstown 06Dec54, cvtd **C-119G** 30Dec54; depl Elmendorf 12Jan55 ← **RECORDS INDECIPHERABLE** 314 TCWG (TAC) Sewart 11May55 – 2472 RFC (CNR) Richards 11Jul57 – 442 TCWG (AFR) Richards 02Mar59 – 452 TCWG (AFR) March 04May61; 943 TCG (452 TCWG) (AFR) March 11Feb63 – Wfu to MASDC (LOG) Davis-Monthan 11Feb69 for storage, PCN: **CJ282** – DFI: 26Dec73 – Final Disposition: Scrapped.

10773 / (RCAF) 22108 / C-119F-FA

Avble: 01Dec52; Acc: 24Dec52; Del: 06Jan53 to Canada, RCAF s/n: **22108**; dropped from USAF inventory 06Jan53 – BOC: 12Jan53 435 SQ Edmonton – 436 SQ Dorval 30Sep54; cvtd **C-119G** 12Dec55 – 408 SQ Rivers 20Apr64 – Wfu into storage 1005 TSD Saskatoon 04Aug64 – SOC: 13Sep67 to Crown Assets Disposal Corp. – BofS Aircraft International Associates Dallas Texas 07Sep67; reg **N5215R** 30Sep71 – BofS Hawkins & Powers Aviation, Inc. Greybull Wyoming 16Dec75; used as a spares source – To U.S. Forest Service for display at the Museum of Flight & Aerial Firefighting Greybull Wyoming marked as: "06" – Reg N5215R canx 19Aug13 – Final Disposition: Extant.

10774 / (RCAF) 22109 / C-119F-FA

Avble: 03Dec52; Acc: 24Dec52; Del: 06Jan53 to Canada, RCAF s/n: **22109**; dropped from USAF inventory 06Jan53 – BOC: 12Jan53 435 SQ Edmonton; cvtd **C-119G** 14Jan57 – 436 SQ Downsview 23Feb60 – Wfu into storage 1005 TSD Saskatoon 24Jun65 – SOC: 01Dec66 to Crown Assets Disposal Corp. – BofS Aircraft International Associates Dallas Texas 30Nov66; CofR **N15504** 07Oct68 – Government of Morocco 23Dec68, RMAF s/n: **CN-AMQ**, 1 AT SQ; reg N15504 canx 03Jan69 – Final Disposition: Unknown.

10775 / (RCAF) 22110 / C-119F-FA

Avble: 03Dec52; Acc: 24Dec52; Del: 06Jan53 to Canada, RCAF s/n: **22110**; dropped from USAF inventory 06Jan53 – BOC: 12Jan53 435 SQ Edmonton – 436 SQ Dorval

05Jan56 – 114 Com Flt Capodichino Italy 03Jun57; cvtd **C-119G** 11Dec57-435 SQ Namao 16Jan58 – 436 SQ Downsview 25Jan61 – Wfu into storage 1005 TSD Saskatoon 13Apr64 – SOC: 13Sep67 to Crown Assets Disposal Corp. – BofS Aircraft International Associates Dallas Texas 07Sep67; reg **N15509** 04Oct68 – BofS Hawkins & Powers Aviation, Inc. Greybull Wyoming 16Dec75 – BofS Sergio A. Tomassoni d/b/a Sergio Aviation Buckeye Arizona 30Dec75; cvtd **C-119G-3E** 1976, Tanker **36** – BofS William A. Grantham & Sergio A. Tomassoni Buckeye Arizona 01Jul77; cvtd to 3-bladed propellers 14Apr79 – BofS T&G Aviation, Inc. Chandler Arizona 03Jun80; film appearance *Wrong Is Right* (1982) filmed Aug-Dec80 at Mojave Airport California; bulk fuselage tank fitted 24Jun83. Skewed off a snow-covered runway during take-off from Tobin Creek Mine Airstrip Alaska 21Apr84. 2 crew and 2 passengers survived but the aircraft was deemed as damaged beyond repair. The wreck remains on-site to this day. Nearby is another wreck, C-46 N92853 – Reg N15509 canx 21Oct85 – Final Disposition: Accident.

10776 / (RCAF) 22111 / C-119F-FA

Avble: 04Dec52; Acc: 29Dec52; Del: 07Jan53 to Canada, RCAF s/n: **22111**; dropped from USAF inventory 07Jan53 – BOC: 12Jan53 435 SQ Edmonton – 436 SQ Dorval 08Feb54; cvtd **C-119G** 13Nov56 – Wfu into storage 1005 TSD Saskatoon 24Jun65 – SOC: 06Jan67 to Crown Assets Disposal Corp. – BofS Aircraft International Associates Dallas Texas 04Jan67; reg **N8093** 19Aug70 – BofS Hawkins & Powers Aviation, Inc. Greybull Wyoming 16Dec75; cvtd **C-119G-3E** 23Apr79 with chemical tank for firefighting, Tanker **132**; cargo deck floor reinforcement 05Jun79 – BofS Salamatof Seafoods, Inc. Kenai Alaska 05Jun79 – BofS Hawkins & Powers Aviation, Inc. Greybull Wyoming 08Jun79 – BofS Heims Seafoods, Inc. Myrtle Point Oregon 01Jun80 – BofS Hawkins & Powers Aviation, Inc. Greybull Wyoming 16Jun80; circa 1986 became Fire Tanker **140**; wfu 1987; film appearance in Steven Spielberg's *Always* (1989) performing fire retardant drops – BofS D.A. Powers Greybull Wyoming 24Oct91 – BofS Hawkins & Powers Aviation, Inc. Greybull Wyoming 03Dec91 – BofS D&G, Inc. Greybull Wyoming 01Jan93; still in flying service 1997 – BofS The Pride Capital Group LLC Deerfield Illinois 07Sep05 for auction – BofS Hagerstown Aviation Museum Hagerstown Maryland 10Oct06. Restored to flying condition for ferry flight from Greybull to Hagerstown for static museum display, CofA 12Nov08 – Reg N8093 canx 13May13 – Final Disposition: Extant.

10777 / 51-17365 / C-119G-1-FA

Avble: 04Nov52; Acc: 31Dec52; Del: 08May53 to MDAP Italy, AMI s/n: **MM51-17365** – Final Disposition: Scrapped.

10778 / 51-17366 / C-119G-1-FA

Avble: 05Dec52; Acc: 31Dec52; Del: 12May53 to MDAP Italy, AMI s/n: **MM51-17366** – Final Disposition: Scrapped.

10779 / 51-17367 / C-119G-1-FA

Avble: 08Dec52; Acc: 31Dec52; Del: 02Jun53 to MDAP Italy, AMI s/n: **MM51-17367** – Final Disposition: Scrapped.

10780 / 52-6000 / C-119G-5-FA

Avble: 09Dec52; Acc: 31Dec52; Del: 19May53 to MDAP Italy, AMI s/n: **MM52-6000** – Final Disposition: Scrapped.

Section Four—Service Histories

10781 / 51-8233 / C-119C-70-FA / (C-119CF)

Avble: 10Dec52; Acc: 29Dec52; Del: 04Feb53 to USAF – 60 TCWG (AFE) Rhein-Main AB W. Germany 04Feb53; depls RAF Sculthorpe England May53 & 08Jun54 – SABBE (AMC) Brussels Belgium 18May55 – 465 TCWG (AFE) Évreux-Fauville AB France 16Aug55 – 2252 RFC (CNR) Clinton 28Dec56 – 2259 RFC (CNR) Andrews 18Jul57; depl 2233 RFC (CNR) Mitchel 23Sep57; to 3800 AB WG (AU) Maxwell 12Mar58 mnt; to 1001 AB WG (HQC) Andrews 23May58 mnt – AEMCO (AMC) Oakland 28Jun58 mnt – 459 TCWG (AFR) Andrews 09Jan59; briefly assigned 758 TCSq Greater Pittsburgh 01Aug60; 909 TCG (459 TCWG) (AFR) Andrews 01Feb63 – Wfu to MASDC (LOG) Davis-Monthan 13Jun66 for storage – DFI: 21Jan69 – Final Disposition: Scrapped.

10782 / 51-8234 / C-119C-70-FA / (C-119CF)

Avble: 11Dec52; Acc: 29Dec52; Del: 04Feb53 to USAF – 60 TCWG (AFE) Rhein-Main AB W. Germany 04Feb53; depls RAF Sculthorpe England 19Jan54 & 26Jun54 – SABBE (AMC) Brussels Belgium 17Apr55 – 465 TCWG (AFE) Évreux-Fauville AB France 02Jul55. Suffered a hydraulic failure after landing at Algiers Algeria 31Aug56 which caused the aircraft to skew off the runway into a ditch with no fatalities – 3131 MAI GP (AMC) Chateauroux AB France 19Oct56 for assessment and parts salvage, hull left in Algeria – REC: 13Dec56 – Final Disposition: Accident.

10783 / 51-8235 / C-119C-70-FA / (C-119CF)

Avble: 11Dec52; Acc: 29Dec52; Del: 04Feb53 to USAF – 60 TCWG (AFE) Rhein-Main AB W. Germany 04Feb53. Had a catastrophic mid-air collision at 5,000ft 10 miles NE Mannheim W. Germany 15May53. A group of F-84E fighters from the 22 Fighter-Bomber SQ flew through the formation of 18 C-119s. One F-84E, 51-628, hit C-119CF 51-8235, which then broke up hitting another C-119CF, 51-8241, resulting in the loss of both aircraft. The pilot of the F-84E managed to eject and parachute to the ground but suffered burns. All 6 crew onboard 51-8235 were killed along with 2 on 51-8241 but the other 4 crew managed to bail out of the doomed aircraft, although several of these had suffered varying degrees of injury. Also damaged but not downed in the collision were C-119CF 51-8249 and 51-8259, the latter having its hull punctured by chunks of the F-84 including pieces of the radio set! – REC: 15May55 – DFI: 24Jun54 – Final Disposition: Accident.

10784 / 51-8236 / C-119C-70-FA / (C-119CF)

Avble: 12Dec52; Acc: 29Dec52; Del: 04Feb53 to USAF – 60 TCWG (AFE) Rhein-Main AB W. Germany 04Feb53; depls RAF Sculthorpe England 22Jun54 & 28Jul54 – 47 BLJ WG (AFE) RAF Sculthorpe England 03May55 – 433 TCWG (AFR) Brooks 07Oct58; to Kelly 31Aug60 – 187 AML SQ (WY-ANG) Cheyenne 06May61 – 909 TCG (459 TCWG) (AFR) Andrews 29May63; briefly 910 TCG (459 TCWG) (AFR) Youngstown date unk – Wfu to MASDC (LOG) Davis-Monthan 16Sep66 for storage – DFI: 18Jun69 – Final Disposition: Scrapped.

10785 / 51-8237 / C-119C-70-FA / (C-119CF)

Avble: 16Dec52; Acc: 08Jan53; Del: 05Feb53 to USAF – 60 TCWG (AFE) Rhein-Main AB W. Germany 05Feb53; depl RAF Sculthorpe England 28Sep54; to Dreux AB France 13Sep55 – 580 ASL GP (MATS) Wheelus AB Libya 08May56 – 7272 AB WG (AFE) Wheelus AB Libya 15Sep56 – 465 TCWG (AFE) Évreux-Fauville AB France 20Oct56 – 2259 RFC (CNR) Andrews 20Feb57; depl 2233 RFC (CNR) Mitchel 23Sep57 – 459 TCWG (AFR)

Andrews 04Nov58; 909 TCG (459 TCWG) (AFR) Andrews 01Feb63 – Wfu to MASDC (LOG) Davis-Monthan 21Aug66 for storage – DFI: 21Jan69 – Final Disposition: Scrapped.

10786 / 51-8238 / C-119C-70-FA / (C-119CF)

Avble: 17Dec52; Acc: 19Jan53; Del: 05Feb53 to USAF – 60 TCWG (AFE) Rhein-Main AB W. Germany 05Feb53; depls RAF Sculthorpe England 09Jan54, 06Jul54 & 16Oct54; depl 317 TCWG (AFE) Neubiberg AB W. Germany 27Feb55 – SABBE (AMC) Brussels Belgium 22May55 – 465 TCWG (AFE) Évreux-Fauville AB France 21Jul55 – 2252 RFC (CNR) Clinton 05Feb57; depl 2233 RFC (CNR) Mitchel 23Sep57 – 2234 RFC DT (CNR) Youngstown 14Oct57 – 2259 RFC DT (CNR) Youngstown 01Dec57 – AEMCO (AMC) Oakland 26Jun58 – **RECORDS MISSING 1958–1964** – 910 TCG (459 TCWG) (AFR) Youngstown date unk – Wfu to MASDC (LOG) Davis-Monthan 22Oct66 for storage – DFI: 18Jun69 – Final Disposition: Scrapped.

10787 / 51-8239 / C-119C-70-FA / (C-119CF)

Avble: 17Dec52; Acc: 19Jan53; Del: 06Feb53 to USAF – 60 TCWG (AFE) Rhein-Main AB W. Germany 06Feb53; depls RAF Sculthorpe England 23Feb53 & 01Jun54; depl 317 TCWG (AFE) Neubiberg AB W. Germany 30Nov54 – **RECORDS MISSING 1955–1964** – 906 TCG (302 TCWG) (AFR) Clinton 63; 907 TCG (302 TCWG) (AFR) Clinton 30Nov65 – 910 TCG (459 TCWG) (AFR) Youngstown 23Jul66 – Wfu to MASDC (LOG) Davis-Monthan 22Oct66 for storage – DFI: 18Jun69 – Final Disposition: Scrapped.

10788 / 51-8240 / C-119C-70-FA / (C-119CF)

Avble: 18Dec52; Acc: 19Jan53; Del: 06Feb53 to USAF – 60 TCWG (AFE) Rhein-Main AB W. Germany 06Feb53; depls RAF Sculthorpe England 20Feb54 & 16Sep54 – **RECORDS MISSING 1955–1964** – 909 TCG (459 TCWG) (AFR) Andrews date unk – Wfu to MASDC (LOG) Davis-Monthan 20Jun66 for storage – DFI: 21Jan69 – Final Disposition: Scrapped.

10789 / 51-8241 / C-119C-70-FA / (C-119CF)

Avble: 19Dec52; Acc: 21Jan53; Del: 09Feb53 to USAF – 60 TCWG (AFE) Rhein-Main AB W. Germany 09Feb53. Had a catastrophic mid-air collision at 5,000ft 10 miles NE Mannheim W. Germany 15May53. A group of F-84E fighters from the 22 Fighter-Bomber SQ flew through the formation of 18 C-119s. One F-84E, 51-628, hit C-119CF 51-8235, which then broke up hitting another C-119CF, 51-8241, resulting in the loss of both aircraft. The pilot of the F-84E managed to eject and parachute to the ground but suffered burns. All 6 crew onboard 51-8235 were killed along with 2 on 51-8241 but the other 4 crew managed to bail out of the doomed aircraft, although several of these had suffered varying degrees of injury. Also damaged but not downed in the collision were C-119CF 51-8249 and 51-8259, the latter having its hull punctured by chunks of the F-84 including pieces of the radio set! – REC: 15May55 – DFI: 24Jun54 – Final Disposition: Accident.

10790 / 51-8242 / C-119C-70-FA / (C-119CF)

Avble: 20Dec52; Acc: 21Jan53; Del: 09Feb53 to USAF – 60 TCWG (AFE) Rhein-Main AB W. Germany 09Feb53; depl RAF Sculthorpe England 23Nov54 – **RECORDS MISSING 1955–1964** – 909 TCG (459 TCWG) (AFR) Andrews date unk – Wfu to MASDC (LOG) Davis-Monthan 20Jun66 for storage – DFI: 21Jan69 – Final Disposition: Scrapped.

Section Four—Service Histories

10791 / 51-8243 / C-119C-70-FA / (C-119CF)

Avble: 22Dec52; Acc: 23Jan53; Del: 19Feb53 to USAF – 60 TCWG (AFE) Rhein-Main AB W. Germany 19Feb53; depls RAF Sculthorpe England 53 & 30Jan54; depl 317 TCWG (AFE) Neubiberg AB W. Germany 08Nov54 – **RECORDS MISSING 1955–1964** – 909 TCG (459 TCWG) (AFR) Andrews date unk – Wfu to MASDC (LOG) Davis-Monthan 16Jul66 for storage – DFI: 18Jun69 – Final Disposition: Scrapped.

10792 / 51-8244 / C-119C-70-FA / (C-119CF)

Avble: 23Dec52; Acc: 23Jan53; Del: 19Feb53 to USAF – 60 TCWG (AFE) Rhein-Main AB W. Germany 19Feb53; to 7373 AB GP (AFE) Chateauroux AB France 02Dec53; depl RAF Sculthorpe England 16Feb54 – 47 BLJ WG (AFE) RAF Sculthorpe England 22Mar55; to 3150 MAI GP (AMC) Nouasseur AB, Morocco 21Jan56 mnt – 2578 RFC (CNR) Ellington 19May58 – 446 TCWG (AFR) Ellington 20Sep58 – 4238 COS GP (SAC) Barksdale 16Dec59. Wfu and sold by commercial sale to unknown civil entity 21Jun60, nfd – Final Disposition: Unknown.

10793 / 51-8245 / C-119C-70-FA / (C-119CF)

Avble: 24Dec52; Acc: 23Jan53; Del: 19Feb53 to USAF – 60 TCWG (AFE) Rhein-Main AB W. Germany 19Feb53; depl RAF Sculthorpe England 13Apr54 – **RECORDS MISSING 1955–1964** – 909 TCG (459 TCWG) (AFR) Andrews date unk – Wfu to MASDC (LOG) Davis-Monthan 04Apr66 for storage – DFI: 21Jan69 – Final Disposition: Scrapped.

10794 / 51-8246 / C-119C-70-FA / (C-119CF)

Avble: 24Dec52; Acc: 23Jan53; Del: 19Feb53 to USAF – 60 TCWG (AFE) Rhein-Main AB W. Germany 19Feb53; to 7373 AD WG (AFE) Chateauroux AB France 02Feb54; depls RAF Sculthorpe England 13Apr54 & 17Aug54 – **RECORDS MISSING 1955–1964** – 926 TCG (446 TCWG) (AFR) NAS New Orleans date unk – Wfu to MASDC (LOG) Davis-Monthan 18Dec65 for storage – DFI: 21Jan69 – Final Disposition: Scrapped.

10795 / 51-8247 / C-119C-70-FA / (C-119CF)

Avble: 29Dec52; Acc: 23Jan53; Del: 20Feb53 to USAF – 60 TCWG (AFE) Rhein-Main AB W. Germany 20Feb53; to 7373 AD WG (AFE) Chateauroux AB France 01Nov53; depls RAF Sculthorpe England 20Oct54 & 26Jan55 – **RECORDS MISSING 1955–1964** – 906 TCG (302 TCWG) (AFR) Clinton 63; 907 TCG (302 TCWG) (AFR) Clinton 11Dec65 – 910 TCG (459 TCWG) (AFR) Youngstown 16Apr66 – Wfu to MASDC (LOG) Davis-Monthan 01Oct66 for storage – DFI: 18Jun69 – Final Disposition: Scrapped.

10796 / 51-8248 / C-119C-70-FA / (C-119CF)

Avble: 30Dec52; Acc: 26Jan53; Del: 20Feb53 to USAF – 60 TCWG (AFE) Rhein-Main AB W. Germany 20Feb53; depl RAF Sculthorpe England 22Dec53 – **RECORDS MISSING 1955–1964** – 905 TCG (514 TCWG) (AFR) Bradley date unk – 906 TCG (302 TCWG) (AFR) Clinton 10Oct65; 907 TCG (302 TCWG) (AFR) Clinton 11Dec65 – Wfu to MASDC (LOG) Davis-Monthan 18Apr66 for storage – DFI: 21Jan69 – Final Disposition: Scrapped.

10797 / 51-8249 / C-119C-70-FA / (C-119CF)

Avble: 31Dec52; Acc: 26Jan53; Del: 19Feb53 to USAF – MIDAR (AMC) Olmsted 19Feb53 for project – 60 TCWG (AFE) Rhein-Main AB W. Germany 06Mar53. Damaged 15May53 in mid-air collision incident 10 miles NE Mannheim W. Germany. Depls

RAF Sculthorpe England 27Apr54 & 17Aug54 – **RECORDS MISSING 1955–1964** – 910 TCG (459 TCWG) (AFR) Youngstown date unk – Wfu to MASDC (LOG) Davis-Monthan 22Oct66 for storage – DFI: 21Jan69 – Final Disposition: Scrapped.

10798 / 51-8250 / C-119C-70-FA / (C-119CF)

Avble: 31Dec52; Acc: 26Jan53; Del: 19Feb53 to USAF – MIDAR (AMC) Olmsted 19Feb53 for project – 60 TCWG (AFE) Rhein-Main AB W. Germany 06Mar53; depls RAF Sculthorpe England 09Feb54 & 05Nov54 – **RECORDS MISSING 1955–1964** – 924 TCG (446 TCWG) (AFR) Ellington date unk; 926 TCG (446 TCWG) (AFR) NAS New Orleans 23Aug65 – Wfu to MASDC (LOG) Davis-Monthan 11Dec65 for storage – DFI: 21Jan69 – Final Disposition: Scrapped.

10799 / 51-8251 / C-119C-70-FA / (C-119CF)

Avble: 31Dec52; Acc: 26Jan53; Del: 19Feb53 to USAF – MIDAR (AMC) Olmsted 19Feb53 for project – 60 TCWG (AFE) Rhein-Main AB W. Germany 05Mar53; depl RAF Sculthorpe England 16Mar54; to 7373 MAI GP (AFE) Chateauroux AB France 19Jul54 mnt – **RECORDS MISSING 1955–1964** – 906 TCG (302 TCWG) (AFR) Clinton date unk; 907 TCG (302 TCWG) (AFR) Clinton 30Nov65 – 910 TCG (459 TCWG) (AFR) Youngstown 23Jul66 – Wfu to MASDC (LOG) Davis-Monthan 26Nov66 for storage – DFI: 18Jun69 – Final Disposition: Scrapped.

10800 / 51-8252 / C-119C-70-FA / (C-119CF)

Avble: 07Jan53; Acc: 29Jan53; Del: 19Feb53 to USAF – MIDAR (AMC) Olmsted 19Feb53 for project – 60 TCWG (AFE) Rhein-Main AB W. Germany; depls RAF Sculthorpe England 20May54, 14Sep54 & 16Oct54; to 3920 AB GP (SAC) RAF Brize Norton England 11Oct54 mnt – photographic evidence shows assignment to 47 BLJ WG (AFE) RAF Sculthorpe England date unk – **RECORDS MISSING 1955–1964** – 905 TCG (514 TCWG) (AFR) Bradley date unk – Wfu to MASDC (LOG) Davis-Monthan 12Feb66 for storage – DFI: 21Jan69 – Final Disposition: Scrapped.

10801 / 51-8253 / C-119C-70-FA / (C-119CF)

Avble: 07Jan53; Acc: 26Jan53; Del: 19Feb53 to USAF – MIDAR (AMC) Olmsted 19Feb53 for project – OKLAR (AMC) Tinker 18Mar53 mnt – 60 TCWG (AFE) Rhein-Main AB W. Germany 23Mar53; depls RAF Sculthorpe England May53 & 07Jan55 – 47 BLJ WG (AFE) RAF Sculthorpe England 05Feb55 – **RECORDS MISSING 1955–1964** – 924 TCG (446 TCWG) (AFR) Ellington date unk; 926 TCG (446 TCWG) (AFR) NAS New Orleans 02Sep64 – Wfu to MASDC (LOG) Davis-Monthan 18Dec65 for storage – DFI: 21Jan69 – Final Disposition: Scrapped.

10802 / 51-8254 / C-119C-70-FA / (C-119CF)

Avble: 09Jan53; Acc: 29Jan53; Del: 19Feb53 to USAF – MIDAR (AMC) Olmsted 19Feb53 for project – 60 TCWG (AFE) Rhein-Main AB W. Germany 05Mar53. Loan to RAE Farnborough England 53 for pilot training; to SABBE (AMC) Brussels Belgium 01Nov54 – **RECORDS MISSING 1955–1964** – 909 TCG (459 TCWG) (AFR) Andrews date unk – Wfu to MASDC (LOG) Davis-Monthan 04Mar66 for storage – DFI: 21Jan69 – Final Disposition: Scrapped.

Section Four—Service Histories

10803 / 51-8255 / C-119C-70-FA / (C-119CF)

Avble: 09Jan53; Acc: 29Jan53; Del: 19Feb53 to USAF – MIDAR (AMC) Olmsted 19Feb53 for project – OKLAR (AMC) Tinker 18Mar53 mnt – 60 TCWG (AFE) Rhein-Main AB W. Germany 21Mar53; depl RAF Sculthorpe England 03Nov53 – SABBE (AMC) Brussels Belgium 05Jan55 – **RECORDS MISSING 1955–1964** – 909 TCG (459 TCWG) (AFR) Andrews date unk – Wfu to MASDC (LOG) Davis-Monthan 21Aug66 for storage – DFI: 21Jan69 – Final Disposition: Scrapped.

10804 / 51-8256 / C-119C-70-FA / (C-119CF)

Avble: 12Jan53; Acc: 28Jan53; Del: 06Feb53 to USAF – 60 TCWG (AFE) Rhein-Main AB W. Germany 06Feb53; to 7373 MAI GP (AFE) Chateauroux AB France 02Jul54; depl RAF Sculthorpe England 05Oct53 – **RECORDS MISSING 1955–1964** – 129 ACO GP (ANG) Hayward 64 – Wfu to MASDC (LOG) Davis-Monthan 02Feb69 for storage, PCN: **CJ278** – DFI: 03Feb69 – Final Disposition: Scrapped.

10805 / 51-8257 / C-119C-70-FA / (C-119CF)

Avble: 12Jan53; Acc: 28Jan53; Del: 19Feb53 to USAF – MIDAR (AMC) Olmsted 19Feb53 for project – 60 TCWG (AFE) Rhein-Main AB W. Germany 16Mar53; depls RAF Sculthorpe England 02Feb54 & 18May54 – **RECORDS MISSING 1955–1964** – 909 TCG (459 TCWG) (AFR) Andrews date unk – Wfu to MASDC (LOG) Davis-Monthan 16Jul66 for storage – DFI: 21Jan69 – Final Disposition: Scrapped.

10806 / 51-8258 / C-119C-70-FA / (C-119CF)

Avble: 12Jan53; Acc: 28Jan53; Del: 19Feb53 to USAF – MIDAR (AMC) Olmsted 19Feb53 for project – 60 TCWG (AFE) Rhein-Main AB W. Germany 14Mar53; depls RAF Sculthorpe England 53 & 21Dec54 – 47 BLJ WG (AFE) RAF Sculthorpe England 05Feb55 – **RECORDS MISSING 1955–1964** – 926 TCG (446 TCWG) (AFR) NAS New Orleans date unk – Wfu to MASDC (LOG) Davis-Monthan 18Dec65 for storage – DFI: 06May68 – Final Disposition: Scrapped.

10807 / 51-8259 / C-119C-70-FA / (C-119CF)

Avble: 14Jan53; Acc: 30Jan53; Del: 19Feb53 to USAF – MIDAR (AMC) Olmsted 19Feb53 for project – 60 TCWG (AFE) Rhein-Main AB W. Germany 16Mar53. Damaged 15May53 in mid-air collision incident 10 miles NE Mannheim W. Germany. Depl RAF Sculthorpe England 15Jun54 – **RECORDS MISSING 1955–1964** – 129 ACO GP (ANG) Hayward date unk – Wfu to MASDC (LOG) Davis- Monthan 06Feb69 for storage – DFI: 14Feb69 – Final Disposition: Scrapped.

10808 / 51-8260 / C-119C-70-FA / (C-119CF)

Avble: 14Jan53; Acc: 30Jan53; Del: 04Mar53 to USAF – 60 TCWG (AFE) Rhein-Main AB W. Germany 04Mar53; depls RAF Sculthorpe England 02Mar54 & 10Aug54 – **RECORDS MISSING 1955–1964** – 923 TCG (433 TCWG) (AFR) Carswell date unk – 906 TCG (302 TCWG) (AFR) Clinton 18Sep65; 907 TCG (302 TCWG) (AFR) Clinton 30Nov65 – 910 TCG (459 TCWG) (AFR) Youngstown 16Apr66 – Wfu to MASDC (LOG) Davis-Monthan 28Jan67 for storage – DFI: 21Jan69 – Final Disposition: Scrapped.

10809 / 51-8261 / C-119C-70-FA / (C-119CF)

Avble: 15Jan53; Acc: 10Feb53; Del: 04Mar53 to USAF – 60 TCWG (AFE) Rhein-Main AB W. Germany 04Mar53; to 80 AD WG (AFE) Nouasseur AB Morocco 27Mar53; depls RAF Sculthorpe England 02Feb54 & 04Aug54 – **RECORDS MISSING 1955–1964** – 905 TCG (514 TCWG) (AFR) Bradley date unk – 906 TCG (302 TCWG) (AFR) Clinton 07Nov65; 907 TCG (302 TCWG) (AFR) Clinton Nov65 – Wfu to MASDC (LOG) Davis-Monthan 17Nov65 for storage – DFI: 21Jan69 – Final Disposition: Scrapped.

10810 / 51-8262 / C-119C-70-FA / (C-119CF)

Avble: 16Jan53; Acc: 29Jan53; Del: 21Feb53 to USAF – 60 TCWG (AFE) Rhein-Main AB W. Germany 21Feb53; depl RAF Sculthorpe England 29Jan54 – 465 TCWG (AFE) Toul-Rosières AB France 05Jan55; depl 60 TCWG (AFE) Rhein-Main AB W. Germany 13Feb55; to Évreux-Fauville AB France 02Jun55; to 3131 MAI GP (AMC) Chateauroux AB France 19Nov56 mnt – 317 TCWG (AFE) Évreux-Fauville AB France 08Jul57 – 2259 RFC (CNR) Andrews 03Jan58; to 1001 AB WG (HQC) Andrews 01Mar58 mnt – Hayes (AMC) Birmingham 07Oct58 mnt – 459 TCWG (AFR) Andrews 09Jan59; to KTC (ATC) Keesler 26Jun60 mnt; to Fairchild (LOG) St. Augustine 10Jan63 mnt; 909 TCG (459 TCWG) (AFR) Andrews 18Feb63 – REC: 13Nov63 – Final Disposition: Scrapped.

10811 / 51-8263 / C-119C-70-FA / (C-119CF)

Avble: 17Jan53; Acc: 31Jan53; Del: 17Mar53 to USAF – 60 TCWG (AFE) Rhein-Main AB W. Germany 17Mar53; depl RAF Sculthorpe England 27Jan54 – **RECORDS MISSING 1955–1964** – 129 ACO GP (ANG) Hayward date unk – Wfu to MASDC (LOG) Davis-Monthan 06Feb69 for storage – DFI: 14Feb69 – Aero Union Corp. Chico California 1969, spares airframe – Hemet Valley Flying Service Co. Hemet California CofR **N9956F** 15Oct76, spares airframe, bku 1980s – Reg N9956F canx 01Apr11 – Final Disposition: Scrapped.

10812 / 51-8264 / C-119C-70-FA / (C-119CF)

Avble: 17Jan53; Acc: 20Feb53; Del: 17Mar53 to USAF – 60 TCWG (AFE) Rhein-Main AB W. Germany 17Mar53; depls RAF Sculthorpe England 53, 06Mar54 & 28Sep54; depl 317 TCWG (AFE) Neubiberg AB W. Germany 18Nov54 – **RECORDS MISSING 1955–1964** – 910 TCG (459 TCWG) (AFR) Youngstown date unk – Wfu to MASDC (LOG) Davis-Monthan 21Jan67 for storage – DFI: 18Jun69 – Final Disposition: Scrapped.

10813 / 51-8265 / C-119C-70-FA / (C-119CF)

Avble: 20Jan53; Acc: 20Feb53; Del: 17Mar53 to USAF – 60 TCWG (AFE) Rhein-Main AB W. Germany 17Mar53; **RECORDS INDECIPHERABLE 1953–1955** – **RECORDS MISSING 1955–1964** – 910 TCG (459 TCWG) (AFR) Youngstown date unk – Wfu to MASDC (LOG) Davis-Monthan 26Nov66 for storage – DFI: 18Jun69 – Final Disposition: Scrapped.

10814 / 51-8266 / C-119C-70-FA / (C-119CF)

Avble: 23Jan53; Acc: 25Feb53; Del: 17Mar53 to USAF – 60 TCWG (AFE) Rhein-Main AB W. Germany 17Mar53; depls RAF Sculthorpe England 27Oct54 & 21Dec54 – 47 BLJ WG (AFE) RAF Sculthorpe England 08Feb55 – **RECORDS MISSING 1955–1964** – 924 TCG (446 TCWG) (AFR) Ellington date unk; 926 TCG (446 TCWG) (AFR) NAS New Orleans 15Aug65 – Wfu to MASDC (LOG) Davis-Monthan 19Nov65 for storage – DFI: 21Jan69 – Final Disposition: Scrapped.

Section Four—Service Histories

10815 / 51-8267 / C-119C-70-FA / (C-119CF)

Avble: 23Jan53; Acc: 25Feb53; Del: 17Mar53 to USAF – 60 TCWG (AFE) Rhein-Main AB W. Germany 17Mar53; depls RAF Sculthorpe England May53 & 14Jul54 – **RECORDS MISSING 1955–1964** – 905 TCG (514 TCWG) (AFR) Bradley date unk – 906 TCG (302 TCWG) (AFR) Clinton 07Nov65; 907 TCG (302 TCWG) (AFR) Clinton 07Nov65 – 910 TCG (459 TCWG) (AFR) Youngstown 16Apr66 – Wfu to MASDC (LOG) Davis-Monthan 22Oct66 for storage – DFI: 18Jun69 – Final Disposition: Scrapped.

10816 / 51-8268 / C-119C-70-FA / (C-119CF)

Avble: 23Jan53; Acc: 25Feb53; Del: 21Mar53 to USAF – 60 TCWG (AFE) Rhein-Main AB W. Germany 20Mar53; depls RAF Sculthorpe England May53 & 12Aug54; to 7373 AD WG (AFE) Chateauroux AB France 05Feb54 – **RECORDS MISSING 1955–1964** – 910 TCG (459 TCWG) (AFR) Youngstown date unk – Wfu to MASDC (LOG) Davis-Monthan 21Jan67 for storage – DFI: 18Jun69 – Final Disposition: Scrapped.

10817 / 51-8269 / C-119C-70-FA / (C-119CF)

Avble: 21Jan53; Acc: 25Feb53; Del: 21Mar53 to USAF – 60 TCWG (AFE) Rhein-Main AB W. Germany 20Mar53; **RECORDS MISSING 1953–1964** – 910 TCG (459 TCWG) (AFR) Youngstown date unk – Wfu to MASDC (LOG) Davis-Monthan 01Oct66 for storage – DFI: 06May68 – Final Disposition: Scrapped.

10818 / 51-8270 / C-119C-70-FA / (C-119CF)

Avble: 26Jan53; Acc: 25Feb53; Del: 21Mar53 to USAF – 60 TCWG (AFE) Rhein-Main AB W. Germany 20Mar53; depls RAF Sculthorpe England May53, 08Sep54 & 01Dec54 – **RECORDS MISSING 1955–1964** – 905 TCG (514 TCWG) (AFR) Bradley date unk – Wfu to MASDC (LOG) Davis-Monthan 29Jun66 for storage – DFI: 21Jan69 – Final Disposition: Scrapped.

10819 / 51-8271 / C-119C-70-FA / (C-119CF)

Avble: 26Jan53; Acc: 25Feb53; Del: 21Mar53 to USAF – 60 TCWG (AFE) Rhein-Main AB W. Germany 20Mar53 – **RECORDS MISSING 1955–1964** – 906 TCG (302 TCWG) (AFR) Clinton date unk; 907 TCG (302 TCWG) (AFR) Clinton 30Nov65 – 910 TCG (459 TCWG) (AFR) Youngstown 14Jun66 – Wfu to MASDC (LOG) Davis-Monthan 21Jan67 for storage – DFI: 18Jun69 – Final Disposition: Scrapped.

10820 / 51-8272 / C-119C-70-FA / (C-119CF)

Avble: 23Jan53; Acc: 25Feb53; Del: 21Mar53 to USAF – 60 TCWG (AFE) Rhein-Main AB W. Germany 20Mar53; to 59 AD WG (AFE) RAF Burtonwood England 18Aug53 mnt; to 7559 MND GP (AFE) RAF Burtonwood England 15Sep53 mnt; depls RAF Sculthorpe England 25May54 & 09Dec54 – 47 BLJ WG (AFE) RAF Sculthorpe England 05Feb55 – **RECORDS MISSING 1955–1964** – 924 TCG (446 TCWG) (AFR) Ellington date unk; 926 TCG (446 TCWG) (AFR) NAS New Orleans 25Aug65 – Wfu to MASDC (LOG) Davis-Monthan 18Dec65 for storage – DFI: 21Jan69 – Final Disposition: Scrapped.

10821 / 51-8273 / C-119C-70-FA / (C-119CF)

Avble: 27Jan53; Acc: 25Feb53; Del: 04May53 to USAF – bailment Fairchild (AMC) Hagerstown 04May53 – 60 TCWG (AFE) Rhein-Main AB W. Germany 15May53; depl RAF Sculthorpe England 12Jan54. Lost a propeller near Stint-Truiden Belgium 20Jun54 followed by the engine tearing itself off the wing. The 4 crew and 3 passengers bailed out – REC: 20Jun54 – DFI: 01Aug54 – Final Disposition: Accident.

10822 / 51-8030 / C-119F-FA

Avble: 29Jan53; Acc: 25Feb53; Del: 05Mar53 to USAF **RECORDS INDECIPHERABLE** → unit unk 05Mar53 – 456 TCWG (TAC) Charleston 15Aug53; depl 63 TCWG (TAC) Donaldson 12May54; depl Langley Virginia 01Oct54 ← **RECORDS INDECIPHERABLE** to Shiroi AB Japan 10Nov55; to Ardmore 05May56 – OGDAR (AMC) Hill 10Sep56 – bailment BMC (AMC) Birmingham 18Oct56, cvtd **C-119G** 17Mar57 – CAR Center (ARD) Hanscom 11Apr57, cvtd **C-119J** 30Jul57, cvtd **JC-119J** 15Jul58; to SMAAR (AMC) McClellan 01Feb59 mnt; to Fairchild (AMC) St. Augustine 01Feb60 mnt – AED DV (ARD) WPAFB 01Apr60 – 3245 AB WG (SYS) Hanscom 31Jul62 – Wfu to 2704 ASD GP (LOG) Davis- Monthan 25Mar64 for storage – REC: 27Jul64 – Final Disposition: Scrapped.

10823 / (RCAF) 22112 / C-119F-FA

Avble: 29Jan53; Acc: 26Feb53; Del: 10Mar53 to Canada, RCAF s/n: **22112**; dropped from USAF inventory 10Mar53 – BOC: 18Mar53 4 OTU Trenton; cvtd **C-119G** and ECM prototype 27Apr56 – 104 Com Flt St. Hubert 25Oct56 – Wfu into storage 1005 TSD Saskatoon 14Sep65 – SOC: 05Apr67 to Crown Assets Disposal Corp. – BofS Aircraft International Associates Dallas Texas 31Mar67; ferried from Canada to Dallas Texas Nov67; CofR **N964S** 29Feb68 – Government of Morocco 23Dec68, RMAF s/n: **CN-AMP**, 1 AT SQ; reg N964S canx 27Feb69 – Final Disposition: Unknown.

10824 / (RCAF) 22113 / C-119F-FA

Avble: 30Jan53; Acc: 26Feb53; Del: 10Mar53 to Canada, RCAF s/n: **22113**; dropped from USAF inventory 10Mar53 – BOC: 18Mar53 4 OTU Trenton; cvtd **C-119G** with ECM package 22Oct56 – 104 Com Flt St. Hubert 22May56; had a minor accident 21Sep56 at St. Hubert and repaired – Wfu into storage 1005 TSD Saskatoon 14Sep65 – SOC: 05Apr67 to Crown Assets Disposal Corp. – BofS Aircraft International Associates Dallas Texas 31Mar67; reg **N3935** 06Oct71 – BofS Hawkins & Powers Aviation, Inc. Greybull Wyoming 26Jan72; cvtd **C-119G-3E** 24May72 with chemical tank for aerial fire-fighting, Tanker **139**; wfu and stored 1987; film appearance *Stop! Or My Mom Will Shoot* (1992); stored at Greybull falsely marked as: "N5216R / 136" which actually belongs to msn: 10956 / (RCAF) 22131 – Donated to Museum of Flight & Aerial Firefighting Greybull Wyoming since 2006-Reg N3935 canx 21Jun12 – Final Disposition: Extant.

10825 / (RCAF) 22114 / C-119F-FA

Avble: 02Feb53; Acc: 26Feb53; Del: 10Mar53 to Canada, RCAF s/n: **22114**; dropped from USAF inventory 10Mar53 – BOC: 18Mar53 4 OTU Trenton – CJATC Rivers 07Sep55; cvtd **C-119G** 25Apr57 – 436 SQ Downsview 08Mar62 – Wfu into storage 1005 TSD Saskatoon 24Jun65 – SOC: 01Feb67 to Crown Assets Disposal Corp. – BofS Aircraft International Associates Dallas Texas 31Jan67; reg **N15502** 04Oct68 – BofS Hawkins & Powers Aviation, Inc. Greybull Wyoming 16Dec75; used as a spares source – USAF Museum WPAFB Ohio 24Oct88. Displayed at McClellan AFB Museum, California, marked as: "22114" – Reg N15502 canx 09Dec88 – Final Disposition: Extant.

10826 / 52-6001 / C-119G-5-FA

Avble: 03Feb53; Acc: 27Feb52; Del: 11Mar53 – bailment Aeroproducts Division (AMC) Vandalia & Fairchild (AMC) Hagerstown 11Mar53 – MDAP Italy 27Jun53, AMI s/n: **MM52-6001** – Final Disposition: Scrapped.

10827 / 52-6002 / C-119G-5-FA

Avble: 05Feb53; Acc: 27Feb53; Del: 11Mar53 – bailment Aeroproducts Division (AMC) Vandalia & Fairchild (AMC) Hagerstown 11Mar53 – MDAP Italy 16Jun53, AMI s/n: **MM52-6002** – Final Disposition: Scrapped.

10828 / 52-6003 / C-119G-5-FA

Avble: 06Feb53; Acc: 27Feb53; Del: 29May53 to MDAP Italy, AMI s/n: **MM52-6003** – Final Disposition: Scrapped.

10829 / BuNo. 131662 / R4Q-2

Avble: 10Feb53; Acc: 26Feb53; Del: 04Mar53 to USMC – VMR-153 MCAS Cherry Point 12Mar53 – VMR-252 MCAS Cherry Point 13May59 – VMR-353 MCAS Cherry Point 15Sep61; redes **C-119F** 18Sep62 – Wfu to NAS Litchfield Park 02Apr63 for storage – Tf MASDC (LOG) Davis-Monthan 17Aug65, PCN: **4C002** – DFI: 08May69 – Kolar, Inc. Tucson Arizona 06Feb76 for scrap – Final Disposition: Scrapped.

10830 / BuNo. 131663 / R4Q-2

Avble: 10Feb53; Acc: 27Feb53; Del: 04Mar53 to USMC – VMR-153 MCAS Cherry Point 12Mar53. Suffered a left engine failure during climb out after taking off from NAS Whiting Field Florida 17Jul53. Fully loaded and fueled the aircraft was unable to maintain altitude and control and came down on a farm about 1.5 miles NW of the departure airfield near the town of Milton destroying 2 parked cars and a barn upon impact. 46 crew and passengers were onboard with 40 killed in the crash and rescuers finding 6 survivors but 1 died enroute to hospital and 3 more in the days that followed. The only survivors were crewman Cpl Jerome P. Tuttle and passenger Jay B. Weidler, Jr. Crew: Capt Charles E. Graff (pilot, died later), Capt Grady L. Yoder (co-pilot), MSgt David L. Sabel, TSgt Jerome L. Farley, Sgt Ned J. Lyons and Cpl Jerome P. Tuttle (navigator). Passengers: Edward L. Bailey, Jr., Eldred D. Bates, Jr., William R. Biles, Frank M. Caldwell, Edward R. Clayton, George W. Coyle, Jr., Raymond A. Daniel, Robert K.M. Dickson, Wallis C. Elston, Emil E. Fahrenkamp, Charles S. Heddleson, John P. Hughes, James J. Kingen, Roy V. Lulow, Jr., Thomas F. Maggard (died later), Billy E. Mills, James L. Munkres, Ted G. Phillips, George H. Prentiss, Jr., James P. Railbourn, Robert E. Rhyne, Robert E. Richardson, John B. Rushing, Richard W. Schleiff, George F. Schwaebe, Dale E. Scott (died later), David R. Smith, Gordon H. Smith, Lee W. Smith, Lloyd M. Smith, James C. Stafford, Jr., Kenneth R. Starr, Darrell E. Stricklin, Jerald R. Russell, Dennis M. Sheets, Elwood A. Tracy, Jay B. Weidler, Jr., Bowden W. Wilson, Jr., William W. Wohn and Allen L. Wright. The 40 passengers, some teenagers, were all midshipmen from various universities undertaking active-duty training as part of their summer break with the Naval Reserve Officer Training Corps. (NROTC). The tragic loss of life notably disrupted the NROTC program in the coming years. It was found the flaps were retracted at the time of crash indicating they were never deployed which would have severely hampered the aircraft's ability to climb – DFI: 08May69 – Final Disposition: Accident.

10831 / BuNo. 131664 / R4Q-2

Avble: 10Feb53; Acc: 27Feb53; Del: 04Mar53 to USMC – VMR-153 MCAS Cherry Point 13Mar53 – VMR-252 MCAS Cherry Point 17Jun58 – VMR-253 MCAF Iwakuni Japan 02Jun59 – VMR-234 (MARTD) NAS Minneapolis 24Dec61; redes **C-119F** 18Sep62 – VMR-222 (MARTD) NAS Grosse Ile 22Aug63 – **RECORDS MISSING 1963–1966** – VMR-216 (MARTD) NAS Seattle 24Jun66; depl VMR-222 (MARTD) Grosse Ile 11Jul69; to NAS Whidbey Island 28Apr70 – Wfu to MASDC (LOG) Davis-Monthan 05Aug72, PCN: **4C018** – DFI: 29Jul74 – BofS Dross Metals, Inc. Tucson Arizona 28Jul81 (departed MASDC 11Mar81) reg **N131DM** ARAp 02Dec86. Bku, fuselage used as a workshop at DMI's yard until at least 2013 – Reg N131DM canx 22Dec14 – Final Disposition: Unknown (Presumed Scrapped).

10832 / BuNo. 131665 / R4Q-2

Avble: 12Feb53; Acc: 27Feb53; Del: 04Mar53 to USMC – VMR-153 MCAS Cherry Point 12Mar53; depl NATC NAS Patuxent River 13Apr54 for tests; depl H&MS-14 MCAS Edenton 22Sep54; to VMR-153 Detachment NAS Port Lyautey Morocco 05Jan57 – VMR-252 MCAS Cherry Point 25Apr58; to VMR-252 Detachment NAS Port Lyautey Morocco 13Oct60 – VR-24 NAS Port Lyautey Morocco 22Jan61 – Wfu to NAF Litchfield Park 26Jan62 for storage; redes **C-119F** 18Sep62 – DFI: 24Feb64 – Final Disposition: Scrapped.

10833 / BuNo. 131666 / R4Q-2

Avble: 12Feb53; Acc: 27Feb53; Del: 06Mar53 to USMC – VMR-153 MCAS Cherry Point 30Mar53 – VMR-353 MCAS Miami 19Nov57 – VMR-253 MCAF Iwakuni Japan 19May59 – Wfu to NAF Litchfield Park 01Dec61 for storage; redes **C-119F** 18Sep62 – DFI: 24Feb64 – Final Disposition: Scrapped.

10834 / BuNo. 131667 / R4Q-2

Avble: 21Feb53; Acc: 27Feb53; Del: 06Mar53 to USMC – VMR-153 MCAS Cherry Point 12Mar53 – VMR-252 MCAS Cherry Point 01Apr59; to VMR-252 Detachment NAS Port Lyautey Morocco 22Nov59 – Wfu to NAF Litchfield Park 08Jun61 for storage; redes **C-119F** 18Sep62 – DFI: 24Feb64 – Final Disposition: Scrapped.

10835 / BuNo. 131668 / R4Q-2

Avble: 13Feb53; Acc: 27Feb53; Del: 06Mar53 to USMC – VMR-153 MCAS Cherry Point 12Mar53 – VMR-252 MCAS Cherry Point 24Apr59; to VMR-252 Detachment NAS Port Lyautey Morocco 28Sep60 – VR-24 NAS Port Lyautey Morocco 20Dec60 – Wfu to NAF Litchfield Park 17Feb62 for storage; redes **C-119F** 18Sep62 – DFI: 24Feb64 – Final Disposition: Scrapped.

10836 / BuNo. 131669 / R4Q-2

Avble: 17Feb53; Acc: 27Feb53; Del: 07Mar53 to USMC – VMR-153 MCAS Cherry Point 13Mar53 – VMR-353 MCAS Miami 28Sep57 – VMR-252 MCAS Cherry Point 12Apr60 – VMR-353 MCAS Cherry Point 01Sep61; redes **C-119F** 18Sep62 – VMR-234 (MARTD) NAS Minneapolis 20Sep62 – VMR-222 (MARTD) NAS Grosse Ile 25Feb65 – Fairchild St. Augustine 31May67 mnt – VMR-234 (MARTD) NAS Twin Cities 30Nov67 – Hayes (NPRO) Dothan 09Oct69 mnt – VMR-216 (MARTD) NAS Seattle 25Mar70; to NAS Whidbey Island 28Apr70 – VMR-234 (MARTD) NAS Glenview 10Aug72 – Wfu to MASDC (LOG) Davis-Monthan 14Sep73 for storage, PCN: **4C023** – DFI: 29Jul74 – BofS Dross Metals, Inc.

Tucson Arizona 20Oct80 (departed MASDC 05May81) reg **N13626** ARAp 12Oct81; used for crew training and filmmaking near Jean Nevada Oct81 – BofS Chandalar Development Associates Bellevue Washington 24Dec81; used for crew training near Tucson Arizona Jan82; radio, nav and instrument upgrade 21Feb82; used in crew training near Boeing Field Washington Mar82 – BofS Supra International, Inc. Fairbanks Alaska 27Jul82. Sank through ice while taxiing on the River Kagoak Alaska 08May83, the crew escaped but the aircraft was damaged beyond repair – N13626 de-reg 17Feb84 – Final Disposition: Accident.

10837 / BuNo. 131670 / R4Q-2

Avble: 17Feb53; Acc: 27Feb53; Del: 07Mar53 to USMC – VMR-153 MCAS Cherry Point 13Mar53 – VMR-253 MCAF Iwakuni Japan 07May59 – VMGR-152 MCAF Iwakuni Japan 11Mar62 – VMR-216 (MARTD) NAS Seattle 12Apr62; redes **C-119F** 18Sep62 – VMR-222 (MARTD) NAS Grosse Ile 07Mar64; to NARF Cherry Point 02Sep66 mnt – Hayes (NPRO) Dothan 11Jun69 mnt – VMR-216 (MARTD) NAS Seattle 13Jan70; to NAS Whidbey Island 28Apr70; depl VMR-234 (MARTD) NAS Glenview 02Aug70 – VMR-234 (MARTD) NAS Glenview 09Aug72 – Wfu to MASDC (LOG) Davis-Monthan 01Jun73 for storage, PCN: **4C020** – DFI: 02Jun73 – Dross Metals, Inc. Tucson Arizona 14Jul81 – Final Disposition: Scrapped.

10838 / BuNo. 131671 / R4Q-2

Avble: 17Feb53; Acc: 19Mar53; Del: 26Mar53 to USMC – VMR-153 MCAS Cherry Point 30Mar53 – VMR-253 MCAF Iwakuni Japan 07May59 – Wfu to NAF Litchfield Park 01Dec61 for storage; redes **C-119F** 18Sep62 – DFI: 24Feb64 – Final Disposition: Scrapped.

10839 / BuNo. 131672 / R4Q-2

Avble: 17Feb53; Acc: 18Mar53; Del: 26Mar53 to USMC – VMR-153 MCAS Cherry Point 30Mar53 – VMR-353 MCAS Miami 07Oct57 – VMR-252 MCAS Cherry Point 19Jan60; to VMR-252 Detachment Port Lyautey Morocco 28Sep60 – Wfu to NAF Litchfield Park 14Sep61 for storage; redes **C-119F** 18Sep62 – DFI: 24Feb64 – Final Disposition: Scrapped.

10840 / BuNo. 131673 / R4Q-2

Avble: 19Feb53; Acc: 20Mar53; Del: 26Mar53 to USMC – VMR-153 MCAS Cherry Point 30Mar53 – VMR-252 MCAS Cherry Point 04Apr58 – VMR-353 MCAS Cherry Point 01Sep61 – VMR-216 (MARTD) NAS Seattle 20Apr62; redes **C-119F** 18Sep62 – VMR-222 (MARTD) NAS Grosse Ile 09May63; to NAS Glenview date unk – Hayes (NPRO) Dothan 02Dec68 mnt – VMR-216 (MARTD) NAS Seattle 23Jul69; to NAS Whidbey Island 28Apr70 – Wfu to MASDC (LOG) Davis-Monthan 03Aug72 for storage, PCN: **4C016** – DFI: 29Jul74 – BofS Dross Metals, Inc. Tucson Arizona 20Oct80 (departed MASDC 01Oct80) reg **N1394N** ARAp 23Feb82 – BofS Elling Halvorson, Inc. Redmond Washington 15Apr82; radio, nav and instrument upgrade 20Apr82; cvtd to **C-119F Jet-Pak** (J34-WE) 12Jun82 – Delta Associates, Inc. Anchorage Alaska 15Aug85 – Delta Leasing Anchorage Alaska 10Oct87; leased to Stebbins & Ambler Air Transport which is owned by Delta Leasing – BofS John S. Reffett & Terrence E. Luther Anchorage Alaska 21Mar89. Made a forced landing at Port Lions Kodiak Island Alaska in Jul89 due to engine trouble. The owners couldn't, at the time, afford the repair costs so N1394N remained in storage at the local airstrip for the next 13 and a half years. Periodic trips were made by Reffett and a group of dedicated volunteers to gradually replace the faulty right engine and make good other parts of the C-119. The infamous departure from Port Lions took place on 04Nov02 with the lumbering C-119 only just making it into the air with a smokey left engine and a gradual climb out with gear kept lowered. Onboard were 3 crew, Capt Roger Bartels, co-pilot Alex Roesch and flight engineer / owner John Reffett – BofS John S. Reffett Eagle River Alaska 10Jul15. Currently stored at Palmer Municipal Airport Palmer Alaska – Final Disposition: Extant.

10841 / BuNo. 131674 / R4Q-2

Avble: 18Feb53; Acc: 20Mar53; Del: 26Mar53 to USMC – VMR-153 MCAS Cherry Point 30Mar53 – VMR-353 MCAS Miami 12Oct57 – VMR-252 MCAS Cherry Point 18Dec59 – Wfu to NAF Litchfield Park 14Sep61 for storage; redes **C-119F** 18Sep62 – DFI: 24Feb64 – Final Disposition: Scrapped.

10842 / BuNo. 131675 / R4Q-2

Avble: 19Feb53; Acc: 23Mar53; Del: 26Mar53 to USMC – VMR-153 MCAS Cherry Point 30Mar53 – VMR-252 MCAS Cherry Point 20Apr59 – VMR-353 MCAS Cherry Point 01Sep61 – Wfu to NAF Litchfield Park 26Feb62 for storage; redes **C-119F** 18Sep62 – DFI: 28Feb65 – Tf MASDC (LOG) Davis-Monthan 17Aug65 for storage, PCN: **4C001** – DFI: 08May69 – Kolar, Inc. Tucson Arizona 06Feb76 for scrap – Final Disposition: Scrapped.

10843 / BuNo. 131676 / R4Q-2

Avble: 20Feb53; Acc: 23Mar53; Del: 28Mar53 to USMC – VMR-153 MCAS Cherry Point 30Mar53 – VMR-353 MCAS Miami 26Jan58; to MCAS Cherry Point 31Aug59 – Wfu to NAF Litchfield Park 26Feb62 for storage; redes **C-119F** 18Sep62 – DFI: 28Feb65 – Tf MASDC (LOG) Davis-Monthan 25Aug65 for storage, PCN: **4C003** – DFI: 08May69 – Fate unk, presumed scr – Final Disposition: Scrapped.

10844 / BuNo. 131677 / R4Q-2

Avble: 19Feb53; Acc: 23Mar53; Del: 28Mar53 to USMC – VMR-252 MCAS Cherry Point 01Apr53; VMR-252 Detachment Port Lyautey Morocco 07Jan57 – VMR-353 MCAS Miami 08Mar58 – VMR-253 MCAF Iwakuni Japan 06Jun59 – VMR-234 (MARTD) NAS Minneapolis 19Dec61; redes **C-119F** 18Sep62 – VMR-216 (MARTD) NAS Seattle 25May65 – Fairchild St. Augustine 02Oct67 mnt – VMR-234 (MARTD) NAS Twin Cities 17Apr68; to NAS Glenview 12Feb70 – Wfu to MASDC (LOG) Davis-Monthan 30Jul75 for storage, PCN: **4C030** – DFI: 25Nov75 – BofS Southwestern Alloys Corp. Tucson Arizona 21Jul80 (departed MASDC 25Jun80) – BofS D.M.I. Aviation Tucson Arizona 02Apr84 reg **N49543**; recovered control surfaces, major avionics upgrade 03Jun84 – BofS Marine Lumber, Inc. Nantucket Massachusetts 08Jun84, re-reg **N175ML** 12Jun84, used to fly builders and materials to Nantucket Is. During the building boom at the time – BofS Mid–Atlantic Air Museum Reading Pennsylvania 16Dec94; currently on static display – Final Disposition: Extant.

10845 / BuNo. 131678 / R4Q-2

Avble: 21Feb53; Acc: 24Mar53; Del: 28Mar53 to USMC – VMR-252 MCAS Cherry Point 01Apr53; depl VMR-153 MCAS Cherry Point 05Jun58 – VMR-353 MCAS Cherry Point 31Aug61; redes **C-119F** 18Sep62 – Wfu to NAF Litchfield Park 28Feb63 for storage – DFI: 28Feb65 – Tf MASDC (LOG) Davis-Monthan 27Aug65 for storage, PCN: **4C004** – DFI: 08May69 – Kolar, Inc. Tucson Arizona 06Feb76 for scrap – Final Disposition: Scrapped.

10846 / BuNo. 131679 / R4Q-2

Avble: 25Feb53; Acc: 24Mar53; Del: 28Mar53 to USMC – VMR-252 MCAS Cherry Point 01Apr53; depl VMR-153 MCAS Cherry Point 10Sep56; to VMR-252 Detachment NAS Port Lyautey Morocco 21Dec56; depl VMR-153 MCAS Cherry Point 04Apr58 – VMR-353 MCAS Cherry Point 15Sep61 – VMR-216 (MARTD) NAS Seattle 09Sep62; redes **C-119F** 18Sep62 – VMR-234 (MARTD) NAS Twin Cities 26Jan65; depl NAS Glenview 12Feb70 – Wfu to MASDC (LOG) Davis-Monthan 11Apr74 for storage, PCN: **4C024** – DFI: 12Apr74 – Don F. Pratt Memorial Museum Fort Campbell Kentucky for static display 25Sep81 – Final Disposition: Extant.

10847 / BuNo. 131680 / R4Q-2

Avble: 27Feb53; Acc: 24Mar53; Del: 28Mar53 to USMC – VMR-252 MCAS Cherry Point 01Apr53 – VMR-353 MCAS Cherry Point 01Sep61 – Wfu to NAF Litchfield Park 15Apr62 for storage; redes **C-119F** 18Sep62 – DFI: 28Feb65 – Tf MASDC (LOG) Davis-Monthan 27Sep65 for storage, PCN: **4C005** – DFI: 08May69 – Fate unk, presumed scr – Final Disposition: Scrapped.

10848 / BuNo. 131681 / R4Q-2

Avble: 26Feb53; Acc: 25Mar53; Del: 01Apr53 to USMC – VMR-252 MCAS Cherry Point 07Apr53. Accident listed at MCAS Cherry Point 15Apr55, cause and fatalities unk – DFI: 03May55 – Final Disposition: Accident.

10849 / BuNo. 131682 / R4Q-2

Avble: 02Mar53; Acc: 25Mar53; Del: 30Mar53 to USMC – VMR-252 MCAS Cherry Point 07Apr53 – VMR-353 MCAS Miami 10Sep57 – VMR-252 MCAS Cherry Point 21Jan60 – Wfu to NAF Litchfield Park 15Sep61 for storage; redes **C-119F** 18Sep62 – DFI: 24Feb64 – Final Disposition: Scrapped.

10850 / BuNo. 131683 / R4Q-2

Avble: 03Mar53; Acc: 26Mar53; Del: 30Mar53 to USMC – VMR-252 MCAS Cherry Point 07Apr53 – VMR-153 MCAS Cherry Point 30Apr57 – VMR-353 MCAS Miami 05Dec58; to MCAS Cherry Point 31Aug59; depl VMR-252 MCAS Cherry Point 17May61; redes **C-119F** 18Sep62 – MARS-27 MCAS Cherry Point 31Mar63 – Wfu to NAF Litchfield Park 02Apr63 for storage – DFI: 28Feb65 – Tf MASDC (LOG) Davis-Monthan 08Nov65 for storage, PCN: **4C007** – DFI: 08May69 – Fate unk, presumed scr – Final Disposition: Scrapped.

10851 / BuNo. 131684 / R4Q-2

Avble: 03Mar53; Acc: 26Mar53; Del: 30Mar53 to USMC – VMR-252 MCAS Cherry Point 07Apr53 – Wfu to NAF Litchfield Park 07Sep61 for storage; redes **C-119F** 18Sep62 – DFI: 24Feb64 – Final Disposition: Scrapped.

10852 / BuNo. 131685 / R4Q-2

Avble: 04Mar53; Acc: 26Mar53; Del: 01Apr53 to USMC – VMR-252 MCAS Cherry Point 07Apr53; to VMR-252 Detachment NAS Port Lyautey Morocco 28Nov56 – VMR-153 MCAS Cherry Point 09Jan59 – VMR-352 MCAS El Toro 14May59 – Wfu to NAF Litchfield Park 20Mar61 for storage; redes **C-119F** 18Sep62 – DFI: 24Feb64 – Final Disposition: Scrapped.

10853 / BuNo. 131686 / R4Q-2

Avble: 05Mar53; Acc: 26Mar53; Del: 28Mar53 to USMC – VMR-252 MCAS Cherry Point 01Apr53 – VMR-353 MCAS Miami 09Oct57; to MCAS Cherry Point 31Aug59 – Wfu to NAF Litchfield Park 01Jul61 for storage; redes **C-119F** 18Sep62 – DFI: 24Feb64 – Final Disposition: Scrapped.

10854 / BuNo. 131687 / R4Q-2

Avble: 06Mar53; Acc: 27Mar53; Del: 28Mar53 to USMC – VMR-252 MCAS Cherry Point 01Apr53 – VMR-153 MCAS Cherry Point 04Apr58 – VMR-353 MCAS Miami 15May59; to MCAS Cherry Point 31Aug59 – VMR-252 MCAS Cherry Point 26Aug60 – VMR-353 MCAS Cherry Point 15Sep61 – Wfu to NAF Litchfield Park 08Jun62 for storage; redes **C-119F** 18Sep62 – DFI: 26Feb64 – Tf MASDC (LOG) Davis-Monthan 02Nov65 for storage, PCN: **4C006** – DFI: 08May69 – Fate unk, presumed scr – Final Disposition: Scrapped.

10855 / BuNo. 131688 / R4Q-2

Avble: 06Mar53; Acc: 27Mar53; Del: 28Mar53 to USMC – VMR-252 MCAS Cherry Point 01Apr53 – VMR-153 MCAS Cherry Point 03May57 – VMR-252 MCAS Cherry Point 03Mar59 – VMR-352 MCAS El Toro 15May59 – Wfu to NAF Litchfield Park 03Apr61 for storage; redes **C-119F** 18Sep62 – DFI: 24Feb64 – Tf MASDC (LOG) Davis-Monthan 21Dec65 for storage, PCN: **4C010** – Fairchild St. Augustine 08May67 mnt – VMR-216 (MARTD) NAS Seattle 03Nov67; to NAS Whidbey Island 28Apr70; depl VMR-234 (MARTD) NAS Glenview 02Aug70 – VMR-234 (MARTD) NAS Glenview 28Nov72 – Wfu to MASDC (LOG) Davis-Monthan 27Aug73 for storage, PCN: **4C022** – DFI: 09May75 – BofS The City of Pueblo Colorado 28Feb77 (departed MASDC 25Apr77); CofR **N99574** 15Jun80. Donated to the Pueblo Weisbrod Aircraft Museum Pueblo Colorado for static display in its original USMC livery – Reg N99574 canx 08Apr11 – Final Disposition: Extant.

10856 / BuNo. 131689 / R4Q-2

Avble: 07Mar53; Acc: 28Mar53; Del: 28Mar53 to USMC – VMR-252 MCAS Cherry Point 01Apr53 – VMR-153 MCAS Cherry Point 28Feb58 – VMR-353 MCAS Miami 25Feb59; to MCAS Cherry Point 31Aug59 – VMR-222 (MARTD) NAS Grosse Ile 29Dec61; redes **C-119F** 18Sep62 – VMR-216 (MARTD) NAS Seattle 02Jan64 – VMR-234 (MARTD) NAS Twin Cities 20May66; to NAS Glenview 25Feb70 – Wfu to MASDC (LOG) Davis-Monthan 25Aug72 for storage, PCN: **4C019** – DFI: 26Aug72 – Kolar, Inc. Tucson Arizona 14Jan76 for scrap – Final Disposition: Scrapped.

10857 / 51-8031 / C-119F-FA

Avble: 18Feb53; Acc: 30Mar53; Del: 08Apr53 to USAF **RECORDS INDECIPHERABLE** → assignment unk Apr53 – 6614 TSP SQ (NEA) Ernest Harmon Canada 01Jul53 ← **RECORDS INDECIPHERABLE** bailment Fairchild (AMC) Hagerstown 05Jan55, cvtd **C-119G** 07Feb55; to 6606 AB WG (NEA) Goose Bay Canada 21Mar56 – 4087 TSM SQ (SAC) Ernest Harmon Canada 01Apr57 – 2472 RFC DT (CNR) Davis Field 28Mar58; to 463 TCWG (TAC) Ardmore 18May58 mnt; to 125 Fighter SQ (OK-ANG) Tulsa MAP 01Jun58 mnt – Hayes (AMC) Birmingham 20Aug58 – 65 TCSq (403 TCWG) (AFR) Davis Field 06Nov58; 929 TCG (403 TCWG) (AFR) Davis Field 11Feb63 – 922 TCG (433 TCWG) (AFR) Kelly 17Oct65 – 906 SOP GP (302 TCWG) (AFR) Clinton 06Nov70 – MAP Taiwan 27Nov70, RoCAF s/n: **3212** – Final Disposition: Unknown.

Section Four—Service Histories

10858 / 51-8032 / C-119F-FA

Avble: 11Mar53; Acc: 31Mar53; Del: 17Apr53 to USAF **RECORDS INDECIPHERABLE** → assignment unk Apr53 – 6614 TSP SQ (NEA) Ernest Harmon Canada 01Jul53; to 6600 AB GP (NEA) Pepperrell Canada 29Sep53; to 6603 AB GP (NEA) Goose Bay Canada 16Feb54 mnt ← **RECORDS INDECIPHERABLE** bailment Fairchild (AMC) Hagerstown 12Feb55, cvtd **C-119G** 10Mar55; to 6606 AB WG (NEA) Goose Bay Canada 20May55; to 6604 AB WG (NEA) Pepperrell Canada 01Jul55; to HIRAICTGP? (NEA) Sondrestrom AB Greenland 03May56; to 6606 AB WG (NEA) Goose Bay Canada 19Jul56. Listed as crashed 25Oct56, nfd – DFI: 01Jun57 – Final Disposition: Accident.

10859 / (RCAF) 22115 / C-119F-FA

Avble: 11Mar53; Acc: 31Mar53; Del: 14Apr53 to Canada, RCAF s/n: **22115**; dropped from USAF inventory 14Apr53 – BOC: 24Apr53 25 AMB Lincoln Park – CJATC Rivers 22Feb55; cvtd **C-119G** 02Jan57 – 435 SQ Namao 21Oct59; depl to 408 SQ Rivers 22Apr64 – Wfu into storage 1005 TSD Saskatoon 15Jul65 – SOC: 09May67 to Crown Assets Disposal Corp. – BofS Aircraft International Associates Dallas Texas 03May67; reg **N8682** 28Aug70 – BofS Hawkins & Powers Aviation, Inc. Greybull Wyoming 15Nov71; cvtd as the **C-119G-3E** prototype Jan72 with chemical tank for aerial firefighting, Tanker **138**; Aero-Union radio installation 29May72; main wing repair 08Sep72 after wing strike – BofS Salamatof Seafoods, Inc. Kenai Alaska 05Jun79 – BofS Hawkins & Powers Aviation, Inc. Greybull Wyoming 08Jun79. Suffered a catastrophic failure of the no. 2 engine resulting in a fire during cargo drops to ground crews fighting forest fires in the Bettles Alaska region on 27Jun81. The engine fire spread the length of the starboard boom while the crew jettisoned the cargo, many of the electrical systems on the aircraft began to fail and altitude could not be maintained. 4 smoke-jumpers and the co-pilot bailed out (at 200 ft) and pilot Ed Dugan belly-landed the C-119 on a sand bar along the Kayokuk River 25 miles south of Bettles Airstrip settlement. All 6 crew were rescued within hours by helicopter, the wreck remains at the site to this day – Reg N8682 canx 05Apr91 – Final Disposition: Accident.

10860 / (RCAF) 22116 / C-119F-FA

Avble: 12Mar53; Acc: 31Mar53; Del: 14Apr53 to Canada, RCAF s/n: **22116**; dropped from USAF inventory 14Apr53 – BOC: 24Apr53 25 AMB Lincoln Park – 436 SQ Dorval 18Feb54; cvtd **C-119G** 29Apr57 – 435 SQ Namao 14Dec62 – 408 SQ Rivers 06Jan65 – Wfu into storage 1005 TSD Saskatoon 25May65 – SOC: 07Jul67 to Crown Assets Disposal Corp. – BofS Aircraft International Associates Dallas Texas 05Jul67; reg **N5217R** 30Sep71 – BofS Hawkins & Powers Aviation, Inc. Greybull Wyoming 16Dec75, used as a spares source – To USAF Museum Program for static display at National Infantry Museum Fort Benning Georgia circa 1992 marked as: "22116" – Reg N5217R canx 20Aug13 – Final Disposition: Extant.

10861 / (RCAF) 22117 / C-119F-FA

Avble: 12Mar53; Acc: 31Mar53; Del: 15Apr53 to Canada, RCAF s/n: **22117**; dropped from USAF inventory 14Apr53 – BOC: 24Apr53 25 AMB Lincoln Park – CE&PE Rockcliffe 15Jun54; cvtd **C-119G** 07Jun56 – 435 SQ Namao 30Oct58 – 408 SQ Rivers 22Apr64 – Wfu into storage 1005 TSD Saskatoon 31May65 – SOC: 07Jul67 to Crown Assets Disposal Corp. – BofS Aircraft International Associates Dallas Texas 05Jul67; CofR **N15507** 07Oct68 – Government of Morocco 23Dec68, RMAF s/n: **CN-AMR**, 1 AT SQ; reg N15507 canx 03Jan69 – Final Disposition: Unknown.

10862 / 52-6004 / C-119G-5-FA

    Avble: 13Mar53; Acc: 31Mar53; Del: 20Jun53 to MDAP Italy, AMI s/n: **MM52-6004** – Final Disposition: Scrapped.

10863 / 52-6005 / C-119G-10-FA

    Avble: 16Mar53; Acc: 31Mar53; Del: 29Jun53 to MDAP Italy, AMI s/n: **MM52-6005**. Damaged beyond repair after making a belly landing at Pisa-Arturo dell'Oro Air Base Italy 30Mar63, no fatalities – Final Disposition: Accident.

10864 / 52-6006 / C-119G-10-FA

    Avble: 17Mar53; Acc: 31Mar53; Del: 26Jun53 to MDAP Italy, AMI s/n: **MM52-6006** – Final Disposition: Scrapped.

10865 / 52-6007 / C-119G-10-FA

    Avble: 17Mar53; Acc: 31Mar53; Del: 29Jun53 to MDAP Italy, AMI s/n: **MM52-6007**; diverted MIDAR (AMC) Olmsted 09Jul53-03Aug53 for pilot training – Final Disposition: Scrapped.

10866 / 52-6008 / C-119G-10-FA

    Avble: 18Mar53; Acc: 15Apr53; Del: 29Jun53 to MDAP Italy, AMI s/n: **MM52-6008**; diverted MIDAR (AMC) Olmsted 14Jul53-15Sep53 for pilot training – Final Disposition: Scrapped.

10867 / 52-6009 / C-119G-10-FA

    Avble: 19Mar53; Acc: 15Apr53; Del: 10Jul53 to MDAP Italy, AMI s/n: **MM52-6009**; diverted MIDAR (AMC) Olmsted 23Jul53-28Jul53 for pilot training – Final Disposition: Scrapped.

10868 / 52-6010 / C-119G-10-FA

    Avble: 19Mar53; Acc: 15Apr53; Del: 09Jul53 to MDAP Italy, AMI s/n: **MM52-6010** – Final Disposition: Scrapped.

10869 / 52-6011 / C-119G-10-FA

    Avble: 20Mar53; Acc: 15Apr53; Del: 16Jul53 to MDAP Italy, AMI s/n: **MM52-6011**. Crashed due to failure of the no. 1 engine after take-off from Luluabourg Airport The Congo 15Feb61. 3 crew killed were: Capt Sergio Celli (pilot), Lt Dario Giorgi (pilot) and Italo Quadri. 6 passengers survived with injuries included some Pakistani soldiers on a UNO mission – Final Disposition: Accident.

10870 / (RCAF) 22118 / C-119F-FA

    Avble: 23Mar53; Acc: 07Apr53; Del: 15Apr53 to Canada, RCAF s/n: **22118**; dropped from USAF inventory 15Apr53 – BOC: 24Apr53 25 AMB Lincoln Park – 4 OTU Trenton 09Dec55; cvtd **C-119G** 02Apr57 – 114 Com Flt Capodichino Italy 13Sep57 – 435 SQ Namao 11Dec59 – 436 SQ Downsview 16Oct63 – Wfu into storage 1005 TSD Saskatoon 17Sep64 – SOC: 13Sep67 to Crown Assets Disposal Corp. – BofS Aircraft International Associates Dallas Texas 07Sep67; reg **N3559** 06Oct71 – BofS Hawkins & Powers Aviation, Inc. Greybull Wyoming 03Mar75; control surfaces reconditioned 07Feb75; cvtd

Section Four—Service Histories 293

C-119G-3E 19Feb75 with chemical tank for firefighting, Tanker **137**; wfu and stored 1987 – To USAF Museum Program and acquired by Dover AFB Museum (Air Mobility Command Museum) Dover AFB Delaware Oct91; ferried to the museum from Greybull. Restored 2004–2005 and displayed marked as: "0-12881" – Reg N3559 canx 18Sep12 – Final Disposition: Extant.

10871 / (RCAF) 22119 / C-119F-FA

Avble: 23Mar53; Acc: 07Apr53; Del: 15Apr53 to Canada, RCAF s/n: **22119**; dropped from USAF inventory 15Apr53 – BOC: 11May53 6 RD Trenton – 436 SQ Dorval 28Jun54; cvtd **C-119G** 01Aug56 – 435 SQ Namao 14Nov56 – 436 SQ Downsview 20Sep60; depl to 408 SQ Rivers 29Jul64 – Wfu into storage 1005 TSD Saskatoon 17Mar65 – SOC: 05Apr67 to Crown Assets Disposal Corp. – Listed in RCAF records as going to Aircraft International Associates but no further history is available, likely bku for spares – Final Disposition: Scrapped.

10872 / (RCAF) 22120 / C-119F-FA

Avble: 24Mar53; Acc: 07Apr53; Del: 15Apr53 to Canada, RCAF s/n: **22120**; dropped from USAF inventory 15Apr53 – BOC: 11May53 6 RD Trenton – CE&PE Rockcliffe 26Mar54; cvtd **C-119G** 23Apr58 – 435 SQ Namao 25May64 – Wfu into storage 1005 TSD Saskatoon 06May65 – SOC: 07Jul67 to Crown Assets Disposal Corp. – BofS Aircraft International Associates Dallas Texas 05Jul67; reg **N961S** 06Nov67; ferried from Canada to Love Field Texas Nov67 – BofS Hawkins & Powers Aviation, Inc. Greybull Wyoming 16Dec75; cvtd **C-119G-3E** Jan80, RCAF livery removed. Does not appear to have entered forestry service as a tanker, no tanker number assigned. Listed as "donated" to U.S. Forest Service prior to Apr94, no further history is known – Reg N961S canx 20Mar13 – Final Disposition: Unknown.

10873 / 51-8033 / C-119F-FA

Avble: 25Mar53; Acc: 07Apr53; Del: 17Apr53 to USAF **RECORDS INDECIPHERABLE** → assignment unk Apr53 – 6614 TSP SQ (NEA) Ernest Harmon Canada 01Jul53; to 6606 AB WG (NEA) Goose Bay Canada 25Aug54, 13Dec54 & 13Mar55 ← **RECORDS INDECIPHERABLE** bailment Fairchild (AMC) Hagerstown 10Oct55, cvtd **C-119G** 06Dec55; to 2500 AB WG (CNC) Mitchel 06Jul56 mnt; to 6605 AB WG (NEA) Ernest Harmon Canada 01Feb57 – 4087 TSM SQ (SAC) Ernest Harmon Canada 01Apr57 – 2237 RFC (CNR) New Castle 04Mar58; to NAS Willow Grove 21Jul58 – Hayes (AMC) Birmingham 27Oct58 – 512 TCWG (AFR) NAS Willow Grove 29Dec58; 912 TCG (512 TCWG) (AFR) NAS Willow Grove 11Feb63 – 910 TAL GP (459 TCWG) (AFR) Youngstown 07May68 – Wfu to MASDC (LOG) Davis-Monthan 13Sep69 for storage, PCN: **CJ311** – DFI: 26Dec73 – Final Disposition: Scrapped.

10874 / 51-8034 / C-119F-FA

Avble: 26Mar53; Acc: 15Apr53; Del: 23Apr53 to USAF **RECORDS INDECIPHERABLE** → unit unk Apr53 – 6614 TSP SQ (NEA) Ernest Harmon Canada 01Jul53; to 6612 AB GP (NEA) Thule AB Greenland Aug53 ← **RECORDS INDECIPHERABLE** to 6606 AB WG (NEA) Goose Bay Canada 28Mar55 – BMC (AC) Birmingham 25May55 – 3499 MBT WG (TC) Chanute 27Jul55; bailment Fairchild (AMC) Hagerstown 16Nov55, cvtd **C-119G** 15Dec55 – 3345 TTA WG (TC) Chanute 21Mar56; to AFM Center (ARD) Patrick 27Apr56 mnt. Lost an engine near Columbia South Carolina 04Mar57, 12 crew and passengers bailed out safely – Final Disposition: Accident.

10875 / BuNo. 131690 / R4Q-2

Avble: 30Mar53; Acc: 22Apr53; Del: 27Apr53 to USMC – VMR-353 MCAS Miami 14May53 – VMR-252 MCAS Cherry Point 04Apr58 – VMR-253 MCAF Iwakuni 01Jun59 – VMR-234 (MARTD) NAS Minneapolis 02Dec61; redes **C-119F** 18Sep62 – VMR-216 (MARTD) NAS Seattle 14Dec62 – VMR-222 (MARTD) NAS Grosse Ile 22Oct63 – VMR-234 (MARTD) NAS Twin Cities 09Apr65 – Fairchild St. Augustine 14Jun67 mnt – VMR-216 (MARTD) NAS Seattle 02Dec67; to NAS Whidbey Island 28Apr70 – Hayes (NPRO) Dothan 25May70 mnt – VMR-234 (MARTD) NAS Glenview 01Oct70 – Wfu to MASDC (LOG) Davis-Monthan 13May74 for storage, PCN: **4C025** – DFI: 09May75 – BofS Dross Metals, Inc. Tucson Arizona 21Jul80 (departed MASDC 16Jun80) – BofS Pacific International Foods, Inc. Kenai Alaska 22Sep80, reg N8501W, paperwork in error actually purchased C-119 msn: 10880 – BofS Tobacco Road Farms, Inc. Ronda North Carolina 09Apr81 reg **N4234S**. The aircraft has no known FAA airworthiness documents but is known to have been flown to Colombia late 1981-82. Noted impounded at El Dorado Intl. Airport Bogota Colombia 24Nov82 by Colombian authorities for reasons unknown marked with false reg "N4234C." Likely scrapped along with other impounded aircraft by 1997 – Reg N4234S canx 07Jun13 – Final Disposition: Scrapped.

10876 / BuNo. 131691 / R4Q-2

Avble: 30Mar53; Acc: 17Apr53; Del: 23Apr53 to USMC – VMR-252 MCAS Cherry Point 29Apr53 – VMR-353 MCAS Miami 07Mar58; to MCAS Cherry Point 31Aug59 – VMR-252 MCAS Cherry Point 07Jul60; to VMR-252 Detachment NAS Port Lyautey Morocco 13Nov60 – VR-24 NAS Port Lyautey Morocco 22Jan61 – Wfu to NAF Litchfield Park 12Jun62 for storage; redes **C-119F** 18Sep62 – SO&ES MCAS Cherry Point 09Jan63 – **RECORDS MISSING 1963-1965** – VMR-234 (MARTD) NAS Twin Cities 27Dec65 – Fairchild St. Augustine 30Nov67 mnt – VMR-216 (MARTD) NAS Seattle 11May68; to NAS Whidbey Island 28Apr70; depl VMR-234 (MARTD) NAS Glenview 02Aug70 – VMR-234 (MARTD) NAS Glenview 14Aug72 – Wfu to MASDC (LOG) Davis-Monthan 24Apr75 for storage, PCN: **4C027** – DFI: 25Nov75 – Southwestern Alloys Corp. Tucson Arizona 25Jun80 for scrap – 20th Century Fox Film Corp. 2003, film appearance *Flight of the Phoenix* (2004). Fuselage used for interior filming. Wings, nacelles and boom sections used for static mock-up airplane, bku after filming – Final Disposition: Scrapped.

10877 / BuNo. 131692 / R4Q-2

Avble: 30Mar53; Acc: 17Apr53; Del: 23Apr53 to USMC – VMR-252 MCAS Cherry Point 29Apr53 – VMR-353 MCAS Miami 28Sep57; to MCAS Cherry Point 31Oct59 – Wfu to NAF Litchfield Park 01Jul61 for storage; redes **C-119F** 18Sep62 – DFI: 24Feb64 – Final Disposition: Scrapped.

10878 / BuNo. 131693 / R4Q-2

Avble: 01Apr53; Acc: 20Apr53; Del: 23Apr53 to USMC – VMR-353 MCAS Miami 10May53 – VMR-253 MCAF Iwakuni Japan 01Jun59 – Wfu to NAF Litchfield Park 01Dec61 for storage; redes **C-119F** 18Sep62 – DFI: 24Feb64 – Final Disposition: Scrapped.

10879 / BuNo. 131694 / R4Q-2

Avble: 07Apr53; Acc: 20Apr53; Del: 23Apr53 to USMC – VMR-353 MCAS Miami 05May53 – VMR-252 MCAS Cherry Point 13Jun58; to VMR-252 Detachment NAS Port Lyautey Morocco 06Feb60 – VMR-353 MCAS Cherry Point 12Dec60; redes **C-119F**

18Sep62 – Wfu to NAF Litchfield Park 14Mar63 for storage – DFI: 28Feb65 – Tf MASDC (LOG) Davis-Monthan 20Dec65 for storage, PCN: **4C009** – DFI: 08May69 – Fate unk, presumed scr – Final Disposition: Scrapped.

10880 / BuNo. 131695 / R4Q-2

Avble: 07Apr53; Acc: 20Apr53; Del: 23Apr53 to USMC – VMR-353 MCAS Miami 05May53 – VMR-153 MCAS Cherry Point 11Jan57 – VMR-352 MCAS El Toro 24May59 – VMR-252 MCAS Cherry Point 08May61 – VMR-353 MCAS Cherry Point 31Aug61; redes **C-119F** 18Sep62 – Wfu to NAF Litchfield Park 09Mar63 for storage – **RECORDS MISSING 1963–1967** – VMR-216 (MARTD) NAS Seattle 17Apr67; to NAS Whidbey Island 28Apr70; depl VMR-234 (MARTD) NAS Glenview 02Aug70 – VMR-234 (MARTD) NAS Glenview 13Aug72 – Wfu to MASDC (LOG) Davis-Monthan 22Aug73 for storage, PCN: **4C021** – DFI: 09May75 – BofS Dross Metals, Inc. Tucson Arizona 21Jul80 (departed MASDC 16Jun80) – BofS Pacific International Foods, Inc. Kenai Alaska 08Apr81 reg **N8501W**; onboard equipment removed to lighten airframe 11Apr81; control surfaces recovered 12Apr81; major radio, nav, flight instrument upgrade 27Apr81 – Seattle-First National Bank Olympia Washington 15Sep81 through repossession from Pacific Intl. Foods – BofS Elling Halvorson, Inc. Redmond Washington 17Jun82 – BofS Delta Associates, Inc. Anchorage Alaska 22Aug85 – BofS Delta Leasing Anchorage Alaska 10Oct87 – BofS John S. Reffett Eagle River Alaska 30Apr98. Currently stored at Palmer Municipal Airport Palmer Alaska – Final Disposition: Extant.

10881 / BuNo. 131696 / R4Q-2

Avble: 03Apr53; Acc: 21Apr53; Del: 27Apr53 to USMC – VMR-353 MCAS Miami 21May53 – Wfu to NAF Litchfield Park 08Mar56 for storage – VMR-252 MCAS Cherry Point 18Jan57 – VMR-153 MCAS Cherry Point 11Nov58 – VMR-352 MCAS El Toro 11May59 – Wfu to NAF Litchfield Park 20Mar61 for storage; redes **C-119F** 18Sep62 – DFI: 24Feb64 – Final Disposition: Scrapped.

10882 / BuNo. 131697 / R4Q-2

Avble: 03Apr53; Acc: 21Apr53; Del: 27Apr53 to USMC – VMR-353 MCAS Miami 01May53 – VMR-252 MCAS Cherry Point 20Nov56. Crashed 06Mar59 while on a night approach in a severe rainstorm 5 miles from MCAS Cherry Point. 9 crew and passengers were onboard, 8 were killed including: Maj William E. Zane, MSgt Irving J. Tompkins, SSgt Jack C. Sillman, GSgt Walter R. Archambault, GSgt Willis L. Jones, Andrew F. Franzoni, Jr., and 1 (Sgt Ralph J. Mauro) survived with injuries – Final Disposition: Accident.

10883 / BuNo. 131698 / R4Q-2

Avble: 06Apr53; Acc: 21Apr53; Del: 27Apr53 to USMC – VMR-353 MCAS Miami 09May53 – VMR-252 MCAS Cherry Point 29Nov56 – VMR-353 MCAS Cherry Point 01Sep61; redes **C-119F** 18Sep62 – Wfu to NAF Litchfield Park 14Mar63 for storage – DFI: 28Feb65 – Tf MASDC (LOG) Davis-Monthan 09Dec65 for storage, PCN: **4C008** – DFI: 08May69 – Fate unk, presumed scr – Final Disposition: Scrapped.

10884 / BuNo. 131699 / R4Q-2

Avble: 07Apr53; Acc: 21Apr53; Del: 27Apr53 to USMC – VMR-353 MCAS Miami 01May53 – VMR-252 MCAS Cherry Point 22Mar58 – VMR-353 MCAS Cherry Point 16Jun60 – VMR-222 (MARTD) NAS Grosse Ile 26Jun62; redes **C-119F** 18Sep62 – VMR-216

(MARTD) NAS Seattle 31May64; to NAS Whidbey Island 28Apr70 – Wfu to MASDC (LOG) Davis-Monthan 03Aug72 for storage, PCN: **4C017** – DFI: 29Jul74 – Dross Metals, Inc. Tucson Arizona 01Oct80 for scrap – Final Disposition: Scrapped.

10885 / BuNo. 131700 / R4Q-2

Avble: 08Apr53; Acc: 23Apr53; Del: 06May53 to USMC – VMR-353 MCAS Miami 22May53 – NATC NAS Patuxent River 11Sep56 for tests – VMR-252 MCAS Cherry Point 28Nov56 – VMR-153 MCAS Cherry Point 10Jul58 – VMR-253 MCAF Iawkuni Japan 11May59 – VMR-234 (MARTD) NAS Minneapolis 19Dec61; redes **C-119F** 18Sep62 – VMR-216 (MARTD) NAS Seattle 06Jul63 – **RECORDS MISSING 1963–1966** – VMR-234 (MARTD) NAS Twin Cities 08Mar66 – Fairchild St. Augustine 18Oct67 mnt – VMR-222 (MARTD) NAS Grosse Ile 24Mar68 – VMR-234 (MARTD) NAS Twin Cities 27Jul69; to NAS Glenview 11Feb70 – Wfu to MASDC (LOG) Davis-Monthan 03Jul75 for storage, PCN: **4C029** – DFI: 25Nov75 – BofS Dross Metals, Inc. Tucson Arizona 20Oct80 (departed MASDC 01Oct80) reg **N3267U** Aug82, company renamed D.M.I. Aviation, Inc. 01Aug82. Major overhaul undertaken in late 1982 after a deposit was made by ABBAS Intl. To purchase the aircraft – BofS ABBAS International, Inc. Dallas Texas 07Jun83 – BofS D.M.I. Aviation, Inc. Tucson Arizona 05Jun87 – BofS Comutair Gering Nebraska 06Jul87, CofR 04Aug87; noted at London Stansted Airport Great Britain in early 1988 in Comutair livery and markings. Republic of Cyprus reg **5B-CFG** assigned but ntu, circumstances and dates unk. Operated in Kenya under lease to the United Nations / World Food Program air-dropping relief supplies into Sudan. Abandoned at Jomo Kenyatta Intl. Airport around 1996 after making an illegal landing in Uganda while lost in bad weather – 20th Century Fox Film Corp. 2003, film appearance *Flight of the Phoenix* (2004). Transported to Namibia Africa for use as the main crash site set. Bku after filming – Reg N3267U canx 14May13 – Final Disposition: Scrapped.

10886 / BuNo. 131701 / R4Q-2

Avble: 08Apr53; Acc: 23Apr53; Del: 06May53 to USMC – VMR-353 MCAS Miami 21May53 – Wfu to NAF Litchfield Park 10Apr56 for storage – VMR-153 MCAS Cherry Point 05Dec56 – VMR-252 MCAS Cherry Point 11Sep58 – to VMR-252 Detachment NAS Port Lyautey Morocco 22Nov59 – VMR-353 MCAS Cherry Point 20Mar61; redes **C-119F** 18Sep62 – Wfu to NAF Litchfield Park 21Mar63 for storage – DFI: 28Feb65 – Tf MASDC (LOG) Davis-Monthan 13Jan66 for storage, PCN: **4C012** – DFI: 08May69 – Fate unk, presumed scr – Final Disposition: Scrapped.

10887 / BuNo. 131702 / R4Q-2

Avble: 09Apr53; Acc: 30Apr53; Del: 06May53 to USMC – VMR-353 MCAS Miami 21May53; depl VMR-252 Detachment NAS Port Lyautey Morocco 31Aug56; to MCAS Cherry Point 31Aug59 – VMR-252 MCAS Cherry Point 14Mar60; to VMR-252 Detachment NAS Port Lyautey Morocco 13Aug60 – VMR-353 MCAS Cherry Point 31Aug61 – Wfu to NAF Litchfield Park 29Mar62 for storage; redes **C-119F** 18Sep62 – DFI: 28Feb65 – Tf MASDC (LOG) Davis-Monthan 29Dec65 for storage, PCN: **4C011** – DFI: 08May69 – Fate unk, presumed scr – Final Disposition: Scrapped.

10888 / BuNo. 131703 / R4Q-2

Avble: 10Apr53; Acc: 28Apr53; Del: 06May53 to USMC – VMR-353 MCAS Miami 18May53. Departed MCAS Miami 04Feb56 for a routine training flight when an engine fire occurred during climb-out while passing over Fort Lauderdale Florida. All 4 crew bailed out over the ocean but 2 drowned after landing: Capt Peter Looney and TSgt William O. Packard, Jr. – Final Disposition: Accident.

10889 / BuNo. 131704 / R4Q-2

Avble: 13Apr53; Acc: 28Apr53; Del: 11May53 to USMC – VMR-353 MCAS Miami 22May53 – VMR-252 MCAS Cherry Point 07Dec57 – VMR-353 MCAS Cherry Point 15Sep61 – VMR-216 (MARTD) NAS Seattle 06Aug62; redes **C-119F** 18Sep62 – VMR-234 (MARTD) NAS Twin Cities 01Sep64 – Wfu to MASDC (LOG) Davis-Monthan 24Feb67 for storage, PCN: **4C013** – DFI: 08May69 – Fate unk, presumed scr – Final Disposition: Scrapped.

10890 / BuNo. 131705 / R4Q-2

Avble: 13Apr53; Acc: 28Apr53; Del: 11May53 to USMC – VMR-353 MCAS Miami 23May53 – Wfu to NAF Litchfield Park 10Apr56 for storage – VMR-252 MCAS Cherry Point 07Jun57 – VMR-353 MCAS Cherry Point 05Nov59 – Wfu to NAF Litchfield Park 01Jul61 for storage; redes **C-119F** 18Sep62 – DFI: 24Feb64 – Final Disposition: Scrapped.

10891 / BuNo. 131706 / R4Q-2

Avble: 15Apr53; Acc: 30Apr53; Del: 11May53 to USMC – HAMRON-24 MCAS Cherry Point 25May53 – VMR-153 MCAS Cherry Point 16Mar55 – VMR-353 MCAS Miami 07Jul56 – VMR-153 MCAS Cherry Point 11Jun58 – VMR-252 MCAS Cherry Point 13May59 – VMR-353 MCAS Cherry Point 15Sep61 – SO&ES MCAS Cherry Point 10May62; redes **C-119F** 18Sep62 – VMR-234 (MARTD) NAS Minneapolis 24May63 – VMR-222 (MARTD) NAS Grosse Ile 22Feb66 – VMR-234 (MARTD) NAS Twin Cities 27Jul69; to NAS Glenview 17Feb70 – Wfu to MASDC (LOG) Davis-Monthan 19Jun75 for storage, PCN: **4C028** – DFI: 25Nov75 – Southwestern Alloys Corp. Tucson Arizona 25Jun80 for scrap – 20th Century Fox Film Corp. 2003, film appearance *Flight of the Phoenix* (2004). Wings, right nacelle and boom sections used for motorized mock-up airplane with working radial engine, bku after filming. Fuselage survives stored in Tucson – Final Disposition: Extant.

10892 / BuNo. 131707 / R4Q-2

Avble: 15Apr53; Acc: 28Apr53; Del: 11May53 to USMC – HAMRON-24 MCAS Cherry Point 22May53 – VMR-252 MCAS Cherry Point 01Mar55 – Wfu to NAF Litchfield Park 06Mar56 for storage – VMR-153 MCAS Cherry Point 29Dec56 – VMR-252 MCAS Cherry Point 21Apr59 – VMR-353 MCAS Cherry Point 11Jul61; redes **C-119F** 18Sep62 – Wfu to NAF Litchfield Park 16Mar63 for storage – DFI: 28Feb65 – Final Disposition: Scrapped.

10893 / BuNo. 131708 / R4Q-2

Avble: 16Apr53; Acc: 30Apr53; Del: 11May53 to USMC – VMR-353 MCAS Miami 22May53 – MARS/MWSG-37 MCAS Miami 01Nov54 – VMR-252 MCAS Cherry Point 09Feb55 – Wfu to NAF Litchfield Park 06Mar56 for storage – VMR-353 MCAS Miami 26Mar57 – VMR-153 MCAS Cherry Point 19Aug58 – VMR-252 MCAS Cherry Point 15May59 – VMR-253 MCAF Iwakuni Japan 19May59 – VMR-222 (MARTD) NAS Grosse Ile 18Dec61; redes **C-119F** 18Sep62 – VMR-216 (MARTD) NAS Seattle 28Aug65 – Fairchild St. Augustine 03Nov67 mnt – VMR-222 (MARTD) NAS Grosse Ile 04May68 – VMR-234 (MARTD) NAS Twin Cities 18May68; to NAS Glenview 17Feb70. Minor accident 28Feb71 when u/c failed. Made a safe belly-landing at NAS Glenview Illinois, no fatalities aircraft repaired – Wfu to MASDC (LOG) Davis-Monthan 19Mar75 for storage, PCN: **4C026** – DFI: 09May75 – BofS Dross Metals, Inc. Tucson Arizona 20Oct80 (departed MASDC

02Oct80) – BofS Hawkins & Powers Aviation, Inc. Greybull Wyoming 01Jun88 reg **N7051U** Dec88 – BofS Marine Corps Air-Ground Museum Quantico Virginia 25Nov89 for static display at MCAS El Toro California. Relocated for display to Flying Leatherneck Aviation Museum MCAS Miramar California from 2002 – Reg N7051U canx 10Jul13 – Final Disposition: Extant.

10894 / BuNo. 131709 / R4Q-2

Avble: 17Apr53; Acc: 13May53; Del: 18May53 to USMC – HAMRON-14 NAS Roosevelt Roads Puerto Rico 22May53; to MCAS Edenton 05Oct53 – MARS-27 MCAS Cherry Point 13Oct53 – HEDRON/MWSG-27 MCAS Cherry Point 11Feb54 – VMR-153 MCAS Cherry Point 29Nov54 – Wfu to NAF Litchfield Park 06Mar56 for storage – VMR-353 MCAS Miami 27Jul57; to MCAS Cherry Point 31Aug59 – Wfu to NAF Litchfield Park 11Sep61 for storage; redes **C-119F** 18Sep62 – DFI: 24Feb64 – Final Disposition: Scrapped.

10895 / BuNo. 131710 / R4Q-2

Avble: 20Apr53; Acc: 13May53; Del: 18May53 to USMC – HAMRON-14 NAS Roosevelt Roads 26May53 – HAMRON-26 MCAS Cherry Point 14Nov53 – H&MS-26 MCAF New River 12Aug54 – VMR-153 MCAS Cherry Point 29Jan55; to VMR-153 Detachment NAS Port Lyautey Morocco 22Nov56 – VMR-252 MCAS Cherry Point 15May59 – VMR-353 MCAS Cherry Point 03May60 – Wfu to NAF Litchfield Park 22Jan62 for storage; redes **C-119F** 18Sep62 – DFI: 24Feb64 – Final Disposition: Scrapped.

10896 / BuNo. 131711 / R4Q-2

Avble: 20Apr53; Acc: 13May53; Del: 18May53 to USMC – HAMRON-11 MCAS Edenton 25May53 – VMR-252 MCAS Cherry Point 05Aug53 – MTG-20 MCAS Cherry Point 15Oct53 – HEDRON/ MWSG-27 MCAS Cherry Point 01Apr54 – H&MS-32 MCAS Cherry Point 02Jun54 – VMR-252 MCAS Cherry Point 11May55 – VMR-353 MCAS Cherry Point 31Aug61; redes **C-119F** 18Sep62 – Wfu to NAF Litchfield Park 21Mar63 for storage – DFI: 24Feb64 – Final Disposition: Scrapped.

10897 / BuNo. 131712 / R4Q-2

Avble: 20Apr53; Acc: 13May53; Del: 18May53 to USMC – HAMRON-11 MCAS Edenton 27May53 – VMR-252 MCAS Cherry Point 04Aug53 – MTG-20 MCAS Cherry Point 28Sep53 – HEDRON/ MWSG-27 MCAS Cherry Point 01Apr54 – MARS/MWSG-27 MCAS Cherry Point 01Feb55 – VMR-252 MCAS Cherry Point 09Feb55 – VMR-153 MCAS Cherry Point 08Sep58 – VMR-352 MCAS El Toro 14Sep59 – VMGR-252 MCAS Cherry Point 08Mar61 – VMR-353 MCAS Cherry Point 21Jul61; redes **C-119F** 18Sep62 – Wfu to NAF Litchfield Park 16Mar63 for storage – DFI: 24Feb64 – Final Disposition: Scrapped.

10898 / BuNo. 131713 / R4Q-2

Avble: 21Apr53; Acc: 15May53; Del: 29May53 to USMC – HAMRON-31 MCAS Miami 08Jun53; to NAS Roosevelt Roads 01Feb54 – H&MS-31 MCAS Miami 06May54 – H&MS-32 MCAS Cherry Point 18May55 – VMR-353 MCAS Miami 28Jul55; to MCAS Cherry Point 31Aug59 – VMR-252 MCAS Cherry Point 03Oct60 – VR-24 NAS Port Lyautey Morocco 12Mar61 – Wfu to NAF Litchfield Park 13Jun62 for storage; redes **C-119F** 18Sep62 – DFI: 24Feb64 – Final Disposition: Scrapped.

10899 / BuNo. 131714 / R4Q-2

Avble: 22Apr53; Acc: 18May53; Del: 29May53 to USMC – HEDRON/MTG-20 MCAS Cherry Point 10Jun53 – VMR-153 MCAS Cherry Point 19Jul53 – VMR-252 MCAS Cherry Point 08May58 – VMR-353 MCAS Cherry Point 01Sep61 – VC-1 NAS Barber's Point 30Nov61; redes **C-119F** 18Sep62 – Wfu to NAF Litchfield Park 31Jan63 for storage – DFI: 28Feb65 – Final Disposition: Scrapped.

10900 / BuNo. 131715 / R4Q-2

Avble: 23Apr53; Acc: 18May53; Del: 29May53 to USMC – HAMRON-31 MCAS Miami 09Jun53 – VMR-353 MCAS Miami 10Dec55 – VMR-252 MCAS Cherry Point 06Mar58 – VMR-353 MCAS Cherry Point 15Sep61; redes **C-119F** 18Sep62 – VMR-234 (MARTD) NAS Minneapolis 20Sep62 – VMR-216 (MARTD) NAS Seattle 22Oct64 – NARF Cherry Point 22Oct66 mnt – VMR-234 (MARTD) NAS Twin Cities 04Feb67 – Hayes (NPRO) Dothan 20May69 mnt – VMR-216 NAS Seattle 02Nov69; to NAS Whidbey Island 28Apr70 – Wfu to MASDC (LOG) Davis-Monthan 02Aug72 for storage, PCN: **4C015** – DFI: 29Jul74 – Dross Metals, Inc. Tucson Arizona 03Oct80 for scrap, noted bku 2011 – Final Disposition: Scrapped.

10901 / BuNo. 131716 / R4Q-2

Avble: 24Apr53; Acc: 21May53; Del: 29May53 to USMC – HAMRON-32 MCAS Miami 08Jun53 – H&MS-32 MCAS Cherry Point 10Feb54 – VMR-153 MCAS Cherry Point 06May54 – VMR-353 MCAS Miami 11Sep56. While on a local training flight 20Sep56 experienced propeller control problems. The crew shut down both engines, feathered the props and made a graceful belly-landing at MCAS Miami but slid into a canal off the runway damaging the aircraft beyond repair, no fatalities – DFI: 11Oct56 – Final Disposition: Accident.

10902 / BuNo. 131717 / R4Q-2

Avble: 27Apr53; Acc: 22May53; Del: 29May53 to USMC – HAMRON-32 MCAS Miami 12Jun53 – H&MS-32 MCAS Cherry Point 05Apr54 – VMR-153 MCAS Cherry Point 25May55 – VMR-252 MCAS Cherry Point 10Sep56 – VMR-353 MCAS Miami 19Jun58; to MCAS Cherry Point 31Aug59; depl VMR-252 MCAS Cherry Point 23Nov60 – VMR-222 (MARTD) NAS Grosse Ile 21Jun62 – VMR-234 (MARTD) NAS Minneapolis 05Aug62; redes **C-119F** 18Sep62 – VMR-216 (MARTD) NAS Seattle 01Apr66; to NAS Whidbey Island 28Apr70; depl VMR-234 (MARTD) NAS Glenview 02Aug70 Wfu to MASDC (LOG) Davis-Monthan 01Aug72 for storage, PCN: **4C014** – DFI: 02Aug72 – Kolar, Inc. Tucson Arizona 14Jan76 for scrap – Final Disposition: Scrapped.

10903 / BuNo. 131718 / R4Q-2

Avble: 27Apr53; Acc: 21May53; Del: 10Jun53 to USMC – HEDRON/MTG-20 MCAS Cherry Point 19Jun53; depl AirFMFLANT MCAS Cherry Point 05Aug53 – HEDRON/MWSG-37 MCAS Miami 22Apr54 – VMR-353 MCAS Miami 01Nov54 – VMR-253 MCAF Iwakuni Japan 20May59 – VMR-222 (MARTD) NAS Grosse Ile 10Dec61; redes **C-119F** 18Sep62 – NARF Cherry Point 30Aug63 – VMR-234 (MARTD) NAS Twin Cities 30Nov63 – NARF Cherry Point 17May66 mnt – VMR-222 (MARTD) NAS Grosse Ile 12Aug66 – Hayes (NPRO) Dothan 29Aug69 mnt – Wfu & DFI: 06Jan70 – Final Disposition: Scrapped.

10904 / BuNo. 131719 / R4Q-2

Avble: 29Apr53; Acc: 25May53; Del: 10Jun53 to USMC – HEDRON/MTG-20 MCAS Cherry Point 19Jun53; depl AirFMFLANT MCAS Cherry Point 05Aug53 – HEDRON/

MWSG-27 MCAS Cherry Point 01Apr54 – MARS/MWSG-27 MCAS Cherry Point 24Nov54 – VMR-252 MCAS Cherry Point 22Jan55; to VMR-252 Detachment NAS Port Lyautey Morocco 12Aug60 – VMR-353 MCAS Cherry Point 01Mar61; redes **C-119F** 18Sep62 – Wfu to NAF Litchfield Park 09Mar63 for storage – DFI: 28Feb65 – Final Disposition: Scrapped.

10905 / (RCAF) 22121 / C-119F-FA

Avble: 29Apr53; Acc: 15May53; Del: 25May53 to Canada, RCAF s/n: **22121**; dropped from USAF inventory 25May53 – BOC: 16Jun53 6 RD Trenton – 435 SQ Edmonton 29Apr54; cvtd **C-119G** 11Jan57 – 436 SQ Downsview 11Mar57 – 435 SQ Namao 23Jul63 – Wfu into storage 1005 TSD Saskatoon 06May65 – SOC: 07Jul67 to Crown Assets Disposal Corp. – BofS Aircraft International Associates Dallas Texas 05Jul67; ferried from Canada to Dallas Texas Nov67; CofR **N962S** 29Feb68 – Government of Morocco 23Dec68, RMAF s/n: **CN-AMM**, 1 AT SQ; reg N962S canx 27Feb69 – Final Disposition: Unknown.

10906 / (RCAF) 22122 / C-119F-FA

Avble: 29Apr53; Acc: 15May53; Del: 25May53 to Canada, RCAF s/n: **22122**; dropped from USAF inventory 25May53 – BOC: 16Jun53 6 RD Trenton – 435 SQ Namao 12Dec55; cvtd **C-119G** with ECM package 12Dec55 – 104 Com Flt St. Hubert 13Dec56 – Wfu into storage 1005 TSD Saskatoon 14Sep65 – SOC: 09May67 to Crown Assets Disposal Corp. – BofS Aircraft International Associates Dallas Texas 03May67; reg **N8091** 19Aug70 – BofS Hawkins & Powers Aviation, Inc. Greybull Wyoming 16Dec75; used as a spares airframe – BofS USAF Museum WPAFB Dayton Ohio. Aircraft restored to flying condition for ferry flight from Greybull to March AFB arriving 29Sep88, CofA 12Aug88. On static display at March Field Air Museum March AFB California marked as: "0-22122" – N8091 de-reg 30Jan89 – Final Disposition: Extant.

10907 / (RCAF) 22123 / C-119F-FA

Avble: 30Apr53; Acc: 25May53; Del: 03Jun53 to Canada, RCAF s/n: **22123**; dropped from USAF inventory 03Jun53 – BOC: 16Jun53 6 RD Trenton – 436 SQ Dorval 25Nov54; cvtd **C-119G** 01Apr57 – 435 SQ Namao 23Nov60 – 436 SQ Downsview 07May64 – Wfu into storage 1005 TSD Saskatoon 06May65 – SOC: 22Aug67 to Crown Assets Disposal Corp. – BofS Aircraft International Associates Dallas Texas 10Nov66; reg **N8832** 30Sep71; used in aerialtanking role from Nov71 – BofS Hawkins & Powers Aviation, Inc. Greybull Wyoming 16Dec75; cvtd **C-119G-3E** 04Mar76 with chemical tank, radio upgrade, Tanker **134**; minor incident requiring skin around nose wheel to be repaired 11May77. Made a hard landing at Battle Mountain Municipal Airport Nevada 17Aug85 that resulted in the starboard boom snapping off and so damaging much of the rear tail area. It seems the aircraft was entirely repairable but given the C-119s days were numbered, it appears H&P elected to remove the props and other reusable equipment then simply leave the aircraft where it was. The Battle Mountain Air Museum apparently used parts from N8832 for their static display C-119 N5216R (10956). What remained was bku and scrapped some years later – N8832 de-reg 08Apr91 – Final Disposition: Accident.

10908 / (RCAF) 22124 / C-119F-FA

Avble: 01May53; Acc: 25May53; Del: 03Jun53 to Canada, RCAF s/n: **22124**; dropped from USAF inventory 03Jun53 – BOC: 16Jun53 6 RD Trenton – 435 SQ Edmonton 10Jul53 – 436 SQ Dorval 02Aug55; cvtd **C-119G** 02Aug55. Destroyed in a hangar fire at RCAF Dorval, Quebec 19Mar56, no fatalities – SOC: 27Apr56 – Final Disposition: Accident.

Section Four—Service Histories 301

10909 / 52-6012 / C-119G-10-FA

Avble: 05May53; Acc: 28May53; Del: 20Jun53 to MDAP Italy, AMI s/n: **MM52-6012** – Final Disposition: Scrapped.

10910 / 52-6013 / C-119G-10-FA

Avble: 05May53; Acc: 29May53; Del: 10Jun53 to MDAP Italy, AMI s/n: **MM52-6013** – Final Disposition: Scrapped.

10911 / 52-6014 / C-119G-10-FA

Avble: 06May53; Acc: 29May53; Del: 13Jun53 to MDAP Italy, AMI s/n: **MM52-6014**. Forced to make an emergency landing due to engine problems near the village of Kalungwi Tanzania 17Nov61. Landing on rough ground one u/c leg collapsed causing the aircraft to violently yaw smashing the flightdeck. 4 of the 7 crew were killed and the 3 passengers survived – Final Disposition: Accident.

10912 / 52-6015 / C-119G-10-FA

Avble: 06May53; Acc: 29May53; Del: 15Jun53 to MDAP Italy, AMI s/n: **MM52-6015** – Final Disposition: Scrapped.

10913 / 51-8035 / C-119F-FA

Avble: 07May53; Acc: 28May53; Del: 03Jun53 to USAF – 456 TCWG (TAC) Miami 03Jun53; to Charleston 15Aug53; to Shiroi AB Japan 10Nov55; to Ardmore 05May56 – OGDAR (AMC) Hill 10Sep56 – BMC (AMC) Birmingham 22Oct56, cvtd **C-119G** 02Apr57 – CAR Center (ARD) Hanscom Field 03May57, cvtd **C-119J** 30Jul57, cvtd **JC-119J** 15Jul58 – AED DV (ARD) Hanscom Field cvtd **C-119J** 01Apr60 – Fairchild (AMC) S. Augustine 05May60 – 452 TCWG (AFR) Long Beach 20Jun60; to March 01Oct60 – 167 Fighter SQ (WV-ANG) Martinsburg 25Mar61 – 140 AML SQ (PA-ANG) Olmsted 26Apr61 – 452 TCWG (AFR) March 28Sep62 – Wfu to 2704 ASD GP (LOG) Davis-Monthan 29Nov62 for storage – Fairchild (LOG) St. Augustine 21Nov63 – MAP Italy 10Feb64 as a spares airframe – Final Disposition: Scrapped.

10914 / 51-8036 / C-119F-FA

Avble: 08May53; Acc: 28May53; Del: 03Jun53 to USAF – 456 TCWG (TAC) Miami 03Jun53; to Charleston 15Aug53; to Shiroi AB Japan 10Nov55; to Ardmore 05May56 – OGDAR (AMC) Hill 10Sep56 – BMC (AMC) Birmingham 12Oct56, cvtd **C-119G** 29Dec56 – 2585 RFC (CNR) Miami 31Jan57; depl Dobbins 13Apr57; cvtd **C-119J** 26Jul57 – 102 AML SQ (NY-ANG) New York ANGB 23Sep58 – Wfu to 2704 ASD GP (LOG) Davis-Monthan 09Mar64 for storage – Fairchild (LOG) St. Augustine 17Jan67 – 78 Fighter WG (ADC) Hamilton 04Mar67. Lost to ground accident 08Aug68 when the no. 2 engine caught fire after mnt, the entire right boom section burned and collapsed – DFI: 05Dec68 – Final Disposition: Accident.

10915 / 51-8037 / C-119F-FA

Avble: 09May53; Acc: 26May53; Del: 02Jun53 to USAF – 456 TCWG (TAC) Miami 02Jun53; to Charleston 15Aug53; depl Langley Virginia 01Oct54; to Shiroi AB Japan 10Nov55; to Ardmore 05May56 – OGDAR (AMC) Hill 10Sep56 – BMC (AMC) Birmingham 12Oct56, cvtd **C-119J** 13Aug57 – 2472 RFC (CNR) Richards 14Sep57 – 2465

RFC (CNR) Minneapolis-St. Paul 24Sep57 – 2473 RFC DT (CNR) Minneapolis-St. Paul 01Dec57; to 2465 AB GP (CNC) Minneapolis-St. Paul 27Feb58 mnt – Fairchild (AMC) St. Augustine 03Aug58 mnt – 6593 Test SQ (ARD) Edwards 19Sep58; to Hickham 01Dec58 – 4440 ADE GP (TAC) Langley 04Dec61 – 434 TCWG (AFR) Bakalar 13Dec61; 931 TCG (434 TCWG) (AFR) Bakalar 11Feb63 – Wfu 13Nov63 to museum / school status. Ferried to the National Museum of the USAF WPAFB Dayton Ohio for storage. Placed on public display 04Aug71 – Final Disposition: Extant.

10916 / 51-8038 / C-119F-FA

Avble: 11May53; Acc: 26May53; Del: 02Jun53 to USAF – 456 TCWG (TAC) Miami 02Jun53; to Charleston 15Aug53; depl Langley Virginia 01Oct54; to Shiroi AB Japan 10Nov55; to Ardmore 05May56 – OGDAR (AMC) Hill 10Sep56 – BMC (AMC) Birmingham 12Oct56, cvtd **C-119J** 22Aug57 – 2230 RFC (CNR) NAS New York 13Sep57 – 2234 RFC (CNR) Hanscom Field 05Oct57 – Fairchild (AMC) St. Augustine 30Jun58 – 6593 Test SQ (ARD) Edwards 04Sep58; to Hickham 01Dec58 – 4440 ADE GP (TAC) Langley 04Dec61 – 434 TCWG (AFR) Bakalar 13Dec61; 931 TCG (434 TCWG) (AFR) Bakalar 11Feb63 – Wfu to 2704 ASD GP (LOG) Davis-Monthan 27Dec63 for storage – Fairchild (LOG) St. Augustine 03Aug66 – 4603 AB GP (ADC) Stewart 22Dec66 – 4676 AB GP (ADC) Richards 17Nov69 – 4650 COS SQ (ADC) Richards 30Jun70 – Wfu to MASDC (LOG) Davis-Monthan 13Jul72 for storage, PCN: **CJ433** – DFI: 09Dec74 – Final Disposition: Scrapped.

10917 / 51-8039 / C-119F-FA

Avble: 12May53; Acc: 29May53; Del: 09Jun53 to USAF – 456 TCWG (TAC) Miami 09Jun53; to Charleston 15Aug53; to Shiroi AB Japan 10Nov55; to 325 Fighter GP (ADC) McChord 16Nov55 mnt; to Ardmore 05May56 – OGDAR (AMC) Hill 10Sep56 – BMC (AMC) Birmingham 22Oct56, cvtd **C-119G** 20Jul57 – 2256 RFC (CNR) Niagara Falls 16Aug57 – 2242 RFC DT (CNR) Niagara Falls cvtd **C-119J** 01Dec57 – 2237 RFC DT (CNR) Niagara Falls 14Jan58 – Fairchild (AMC) St. Augustine 30Jun58 mnt – 6593 Test SQ (ARD) Edwards 21Aug58; to Hickham 01Dec58 – 73 TCSq (434 TCWG) (AFR) Scott 28Jan62; 932 (434 TCWG) (AFR) Scott 11Feb63 – Wfu to 2704 ASD GP (LOG) Davis-Monthan 26Dec63 for storage – Fairchild (LOG) St. Augustine 04Aug66 – 78 Fighter WG (ADC) Hamilton 08Dec66 – 4676 AB GP (ADC) Richards 01Dec69 – 4650 COS SQ (ADC) Richards 30Jun70; to CRC DT (SYS) NAS Dallas 02Mar71; to AFT Center (SYS) Edwards 27Apr71; to 92 STA WG (SAC) Fairchild 11Feb72 mnt – Wfu to MASDC (LOG) Davis-Monthan 29Jun72 for storage, PCN: **CJ427** – DFI: 21Mar73 – Final Disposition: Scrapped.

10918 / 51-8040 / C-119F-FA

Avble: 12May53; Acc: 28May53; Del: 03Jun53 to USAF – 456 TCWG (TAC) Miami 03Jun53; to Charleston 15Aug53; depl Langley Virginia 01Oct54; to Shiroi AB Japan 10Nov55; to Ardmore 05May56 – OGDAR (AMC) Hill 10Sep56 – BMC (AMC) Birmingham 18Oct56, cvtd **C-119G** 07Mar57 – WRDCN (ARD) WPAFB 08Apr57, cvtd **JC-119G** 09Apr57, cvtd **JC-119J** 25Jul57; to SAAAR (AMC) Kelly 14Oct57 mnt – Bell Aircraft Corp. (AMC) Niagara Falls 18May58, cvtd **C-119J** 14May59 test duties – 102 AML SQ (NY-ANG) New York ANGB 03Jun59 – Wfu to 2704 ASD GP (LOG) Davis-Monthan 15Feb63 for storage – Fairchild (LOG) St. Augustine 26Nov63 – MAP Italy 10Feb64 as a spares airframe – Final Disposition: Scrapped.

Section Four—Service Histories

10919 / 51-8041 / C-119F-FA

Avble: 16May53; Acc: 28May53; Del: 03Jun53 to USAF – 456 TCWG (TAC) Miami 03Jun53; to Charleston 15Aug53; depl Langley Virginia 01Oct54; to Shiroi AB Japan 10Nov55; to Ardmore 05May56 – **RECORDS MISSING** – cvtd **C-119G** date unk – 2235 RFC (CNR) Grenier date unk; cvtd **C-119J** 03Aug57; to 1604 AB WG (MATS) Kindley 14Jun58 mnt – Fairchild (AMC) St. Augustine 07Nov58 – AFT Center (ARD) Edwards 10Dec58; bailment ARD CM (ARD) Andrews 10Dec58 – 434 TCWG (AFR) Bakalar 18Sep61; 931 TCG (434 TCWG) (AFR) Bakalar 11Feb63 – Wfu to 2704 ASD GP (LOG) Davis-Monthan 27Dec63 for storage – REC: 27Jul64 – Final Disposition: Scrapped.

10920 / 51-8042 / C-119F-FA

Avble: 13May53; Acc: 29May53; Del: 09Jun53 to USAF – 456 TCWG (TAC) Miami 09Jun53; to Charleston 15Aug53; depl Langley Virginia 01Oct54; to Shiroi AB Japan 10Nov55; to Ardmore 05May56 – OGDAR (AMC) Hill 10Sep56 – BMC (AMC) Birmingham 12Oct56, cvtd **C-119J** 29Aug57 – 2471 RFC (CNR) O'Hare 10Sep57 – 2473 RFC DT (CNR) O'Hare 01Dec57 – 2242 RFC DT (CNR) O'Hare 01Apr58 – Fairchild (AMC) St. Augustine Jun58 – 6593 Test SQ (ARD) Edwards 29Aug58; to Hickham 01Dec59 – 4440 ADE GP (TAC) Langley 05Dec61 – 434 TCWG (AFR) Bakalar 10Dec61; 930 TCG (434 TCWG) (AFR) Bakalar 11Feb63 – Wfu to 2704 ASD GP (LOG) Davis-Monthan 27Dec63 for storage – REC: 27Jul64 – Final Disposition: Scrapped.

10921 / 51-8043 / C-119F-FA

Avble: 14May53; Acc: 29May53; Del: 09Jun53 to USAF – 456 TCWG (TAC) Miami 09Jun53; to Charleston 15Aug53; depl 64 TCWG (TAC) Donaldson 17Sep53; to Shiroi AB Japan 10Nov55; to Ardmore 05May56 – OGDAR (AMC) Hill 10Sep56 – BMC (AMC) Birmingham 24Oct56, cvtd **C-119J** 05Aug57 – 2469 RFC (CNR) Scott 05Sep57 – 2466 RFC DT (CNR) Scott 01Dec57 – Fairchild (AMC) St. Augustine Jun58 – 6593 Test SQ (ARD) Edwards 27Aug58; to Hickham 01Dec58 – 4440 ADE GP (TAC) Langley 23Nov61 – 434 TCWG (AFR) Bakalar 27Nov61; 930 TCG (434 TCWG) (AFR) Bakalar 11Feb63 – Wfu to 2704 ASD GP (LOG) Davis-Monthan 27Dec63 for storage – Fairchild (LOG) St. Augustine 04Aug66 – 4603 AB GP (ADC) Stewart 13Dec66 – 4676 AB GP (ADC) Richards 17Nov69 – 4650 COS SQ (ADC) Richards 30Jun70 – Wfu to MASDC (LOG) Davis-Monthan 29Jun72, PCN: **CJ428** – DFI: 21Mar73 – Final Disposition: Scrapped.

10922 / 51-8044 / C-119F-FA

Avble: 15May53; Acc: 09Jun53; Del: 12Jun53 to USAF – 313 TCWG (TAC) Mitchel 12Jun53; to Sewart 02Oct53 – 6614 TSP SQ (NEA) Ernest Harmon Canada 18Feb54; to 6611 AB GP (NEA) Narsarsuaq AB Greenland 20Jul54; to 6621 AB SQ (NEA) Sondrestrom AB Greenland 24Jul54; to 6606 AB GP (NEA) Goose Bay Canada 15Sep54 & 21Dec54; bailment Fairchild (AMC) Hagerstown 20Jan55, cvtd **C-119G** 15Feb55 – 3499 MBT WG (TC) Chanute 05May55 – 2585 RFC (CNR) Miami IAP 17May57; to DT Donaldson 18Apr58 – Hayes (AMC) Birmingham 21Nov58 – 77 TCSq (435 TCWG) (AFR) Donaldson 06Mar59; 78 TCSq (435 TCWG) (AFR) Bates Field 07Nov60 – 357 TCSq (302 TCWG) (AFR) Bates Field 08May61; 908 TCG (302 TCWG) (AFR) Bates Field 11Feb63 – 908 TCG (435 TCWG) (AFR) Bates Field 18Mar63; to Brookley 30Oct64 – 926 TAL GP (446 TCWG) (AFR) NAS New Orleans 20Feb69 – MAP Taiwan 08Apr69 – Final Disposition: Unknown.

10923 / 51-8045 / C-119F-FA

Avble: 16May53; Acc: 09Jun53; Del: 23Jun53 to USAF – 456 TCWG (TAC) Miami 23Jun53; to Charleston 29Aug53; to Shiroi AB Japan 10Nov55; to Ardmore 05May56 – OGDAR (AMC) Hill 10Sep56 – BMC (AMC) Birmingham 19Oct56, cvtd **C-119J** 05Aug57 – 2465 RFC (CNR) Minneapolis-St. Paul 19Aug57 – 2473 RFC DT (CNR) Minneapolis-St. Paul 01Dec57 – 2347 RFC (CNR) Long Beach 10Jun58 – Fairchild (AMC) St. Augustine 02Aug58 – 6593 Test SQ (ARD) Edwards 18Sep58; to Hickham 01Dec58 – 4440 ADE GP (TAC) Langley 25Jul61 – 434 TCWG (AFR) Bakalar 28Jul61; 930 TCG (434 TCWG) (AFR) Bakalar 11Feb63 – Wfu to 2704 ASD GP (LOG) Davis-Monthan 27Dec63 for storage – Fairchild (LOG) St. Augustine 01Sep66 – 78 Fighter WG (ADC) Hamilton 31May67 – 4676 AB GP (ADC) Richards 23Nov69 – 4650 COS SQ (ADC) Richards 30Jun70; to 2849 AB GP (LOG) Hill 22Feb72 mnt – Wfu to MASDC (LOG) Davis-Monthan 07Jul72 for storage, PCN: **CJ429** – DFI: 21Mar73 – Final Disposition: Scrapped.

10924 / 51-8046 / C-119F-FA

Avble: 26May53; Acc: 09Jun53; Del: 23Jun53 to USAF – 456 TCWG (TAC) Miami 23Jun53; to Charleston 15Aug53; to Shiroi AB Japan 10Nov55; to Ardmore 05May56 – OGDAR (AMC) Hill 10Sep56 – BMC (AMC) Birmingham 18Oct56, cvtd **C-119G** 19Feb57 – WRDCE (ARD) WPAFB 08Mar57; cvtd **JC-119G** 12Mar57; cvtd **JC-119J** 25Jul57; cvtd **C-119J** 13Jul59 – 512 TCWG (AFR) NAS Willow Grove 04Aug59 – 78 TCSq (435 TCWG) (AFR) Bates Field 06Apr60; 435 TCWG (AFR) Homestead 02Dec60 – 156 AML SQ (NC-ANG) Douglas 21Mar61 – 183 AML SQ (MS-ANG) Hawkins 27May61 – 76 TCSq (435 TCWG) (AFR) Homestead 15Aug62 – Wfu to 2704 ASD GP (LOG) Davis-Monthan 29Nov62 for storage – Fairchild (LOG) St. Augustine 12Nov63 – MAP Italy 28Jan64, AMI s/n: **MM51-8046**. Ground accident 14Dec64, no fatalities. Assigned as a ground instructional airframe 1965 then used as a spares source up to 1978 – Final Disposition: Accident.

10925 / 51-8047 / C-119F-FA

Avble: 21May53; Acc: 10Jun53; Del: 13Jun53 to USAF – 313 TCWG (TAC) Mitchel 12Jun53; to Sewart 02Oct53; depl Pope 20Apr54; depl Sumpter Alabama 19Jul54; bailment Fairchild (AMC) Hagerstown 06Dec54, cvtd **C-119G** 03Jan55; depl Elmendorf 12Jan55 – 463 TCWG (TAC) Sewart (depl) 08Jun55 – 314 TCWG (TAC) Sewart 13Aug55; depl Dreux AB France 21Oct55 – 2471 RFC (CNR) O'Hare IAP 07Jun57; depl 2466 RFC (CNR) Bakalar 21Jul57 – Hayes (AMC) Birmingham 27Sep57 – 2473 RFC DT (CNR) O'Hare IAP 26Dec57 – 2242 RFC DT (CNR) O'Hare IAP 01Apr58 – 64 TCSq (403 TCWG) (AFR) O'Hare IAP 20Dec58; 928 TCG (403 TCWG) (AFR) O'Hare IAP 11Feb63 – MAP Taiwan 10Apr69 – Final Disposition: Unknown.

10926 / 51-8048 / C-119F-FA

Avble: 21May53; Acc: 10Jun53; Del: 24Jun53 to USAF – 313 TCWG (TAC) Mitchel 24Jun53; to Sewart 02Oct53 – 6614 TSP SQ (NEA) Ernest Harmon Canada 18Feb54; to 463 TCWG (TAC) Ardmore 21Jun54 mnt; to 6604 AB WG (NEA) Pepperrell Canada 20Jul54; to 6606 AB WG (NEA) Goose Bay Canada 02Nov54 & 31Mar55 – 3499 MBT WG (TC) Chanute 20Jun55; bailment Fairchild (AMC) Hagerstown 01Dec55, cvtd **C-119G** 30Dec55; to 3345 TTA WG (TC) Chanute 19Mar57 – 2472 RFC (CNR) Richards 31May57; depl 2466 RFC (CNR) Bakalar 17Jun57 – Hayes (AMC) Birmingham 01Jan59 – 442 TCWG (AFR) Richards 31Mar59; to 819 COS GP (SAC) Dyess 20Nov60 mnt – 452 TCWG (AFR) March 27Apr61; 943 TCG (452 TCWG) (AFR) March 11Feb63; 942 TCG (452 TCWG) (AFR)

March 31Mar65 – 924 TCG (446 TCWG) (AFR) Ellington 02Aug65 – 908 TAL GP (435 TCWG) (AFR) Brookley 25May68 – 913 TAL GP (512 TCWG) (AFR) NAS Willow Grove 25Feb69 – 907 TAL GP (302 TCWG) (AFR) Clinton 20Dec69 – MAP Taiwan 25Feb70, RoCAF s/n: **3209** – Final Disposition: Unknown.

10927 / 51-8049 / C-119F-FA

Avble: 21May53; Acc: 10Jun53; Del: 15Jun53 to USAF – 456 TCWG (TAC) Miami 15Jun53; to Charleston 15Aug53; depl Langley Virginia 01Oct54; to Shiroi AB Japan 10Nov55; to Ardmore 05May56 – OGDAR (AMC) Hill 10Sep56 – BMC (AMC) Birmingham 05Nov56, cvtd **C-119G** 10Jul57 – 2471 RFC (CNR) O'Hare IAP cvtd **C-119J** 03Aug57 – 2473 RFC DT (CNR) O'Hare IAP 01Dec57; to 2465 AB GP (CNC) Minneapolis-St. Paul 15Feb58 mnt – 2242 RFC DT (CNR) O'Hare IAP 01Apr58 – Fairchild (AMC) St. Augustine 18Jul58 – 6593 Test SQ (ARD) Edwards 11Sep58; to Hickham 01Dec58 – 73 TCSq (434 TCWG) (AFR) Scott 27Jan62; 434 TCWG (AFR) Scott 11Mar63; 932 TCG (434 TCWG) (AFR) Scott 02Jun63 – Wfu to 2704 ASD GP (LOG) Davis-Monthan 26Dec63 for storage – Fairchild (LOG) St. Augustine 20Sep66 – 78 Fighter WG (ADC) Hamilton 30Dec66 – 4676 AB GP (ADC) Richards 08Dec69 – 4650 COS SQ (ADC) Richards 30Jun70 – Wfu to MASDC (LOG) Davis-Monthan 20Jun72 for storage, PCN: **CJ425** – DFI: 21Mar73 – Final Disposition: Scrapped.

10928 / 51-8050 / C-119F-FA

Avble: 21May53; Acc: 10Jun53; Del: 15Jun53 to USAF – 456 TCWG (TAC) Miami 15Jun53; to Charleston 15Aug53; to Shiroi AB Japan 10Nov55; to Ardmore 05May56 – OGDAR (AMC) Hill 10Sep56 – BMC (AMC) Birmingham 15Oct56, cvtd **C-119J** 04Sep57 – 2466 RFC (CNR) Bakalar 06Sep57; to 328 CLM SQ (ADC) Richards 18Apr58 mnt – Fairchild (AMC) St. Augustine 17Jul58 – 6593 Test SQ (ARD) Edwards 08Sep58; depl AFT Center (ARD) Edwards 01Dec58 tests; to Hickham 09Apr59 – 4440 ADE GP (TAC) Langley 26Jul61 – 434 TCWG (AFR) Bakalar 28Jul61; 73 TCSq (434 TCWG) (AFR) Scott 12Sep61; 932 TCG (434 TCWG) (AFR) Scott 11Feb63 – Wfu to 2704 ASD GP (LOG) Davis-Monthan 26Dec63 for storage – Fairchild (LOG) St. Augustine 04Aug66 – 78 Fighter WG (ADC) Hamilton 27Nov66 – Wfu to MASDC (LOG) Davis-Monthan 12Nov69 for storage, PCN: **CJ316** – DFI: 20Jan70 – Final Disposition: Scrapped.

10929 / 51-8051 / C-119F-FA

Avble: 25May53; Acc: 10Jun53; Del: 15Jun53 to USAF – 456 TCWG (TAC) Miami 15Jun53; to Charleston 15Aug53; to Shiroi AB Japan 10Nov55; to Ardmore 05May56 – OGDAR (AMC) Hill 10Sep56 – BMC (AMC) Birmingham 16Oct56, cvtd **C-119G** 24Mar57 – AFT Center (ARD) Edwards 10Apr57; to OGDAR (AMC) Hill 13Jun57 mnt; cvtd **C-119J** 26Jul57 – 2242 RFC (CNR) Selfridge 18Nov57 – 403 TCWG (AFR) Selfridge 13Nov58. No. 1 engine caught fire just after takeoff from Selfridge AFB Michigan 07Mar59. The aircraft made a heavy landing in a field 5 miles South of the base that sheared off the u/c and broke the fuselage into four pieces. The 5 crew escaped with minor injuries – Final Disposition: Accident.

10930 / 51-8052 / C-119F-FA

Avble: 25May53; Acc: 10Jun53; Del: 15Jun53 to USAF – 456 TCWG (TAC) Miami 15Jun53; to Charleston 15Aug53; to Shiroi AB Japan 10Nov55; to Ardmore 05May56 – OGDAR (AMC) Hill 10Sep56 – BMC (AMC) Birmingham 16Oct56, cvtd **C-119J** 21Aug57 – 2230 RFC (CNR) NAS New York 06Sep57 – 2234 RFC (CNR) Hanscom Field 09Oct57 –

2235 RFC (CNR) Grenier 31Jan58 – 732 TCSq (94 TCWG) (AFR) Grenier 19Mar59 – 78 TCSq (435 TCWG) (AFR) Bates Field 07Apr60; 77 TCSq (435 TCWG) (AFR) Donaldson 08Nov60 – 145 CLM SQ (NC-ANG) Douglas 31Jan61 – 156 AML SQ (NC-ANG) Douglas 01Feb61 – 183 AML SQ (MS-ANG) Hawkins 05Apr61 – 94 TCWG (AFR) Hanscom 29Oct63 – Wfu to 2704 ASD GP (LOG) Davis-Monthan 09Mar64 for storage – REC: 27Jul64 – Final Disposition: Scrapped.

10931 / 51-8053 / C-119G-1-FA

Avble: 13Mar53; Acc: 01Apr53; Del: 28May53 to USAF – 6510 AB WG (ARD) Edwards 27May53 for experimental duties – 464 TCWG (TAC) Lawson 15Mar54; to Pope 07Sep54; 1604 MAT SQ (MATS) Kindley 12Mar55 mnt; depl North Aux AF 17Jun55; to WRDCN (ARD) WPAFB 27Jul55 mnt; depl Aberdeen 05May56 – Hayes (AMC) Birmingham 28Aug57 – 2473 RFC (CNR) Gen Mitchell 21Nov57 – 440 TCWG (AFR) Gen Mitchell 10Dec58; 933 TCG (440 TCWG) (AFR) Gen Mitchell 11Feb63 – Wfu to MASDC (LOG) Davis-Monthan 28Nov70 for storage, PCN: **CJ365** – DFI: 21May71 – Kolar, Inc. Tucson Arizona 24Feb76 for scrap – Final Disposition: Scrapped.

10932 / 51-8054 / C-119G-1-FA

Avble: 26May53; Acc: 16Jun53; Del: 19Jun53 to USAF – 64 TCWG (TAC) Greenville 19Jun53 – 464 TCWG (TAC) Lawson 03Aug54; to Pope 01Sep54; depl North Aux AF 06Jun55; depl Aberdeen 05May56 – 2466 RFC (CNR) Bakalar 23Oct58 – 434 TCWG (AFR) Bakalar 19Apr59; 931 TCG (434 TCWG) (AFR) Bakalar 11Feb63; 930 SOP GP (434 TCWG) (AFR) Bakalar 30Jun69; to Grissom 12Jan70 – Wfu to MASDC (LOG) Davis-Monthan 23Jul70 for storage, PCN: **CJ350** – DFI: 25Aug70 – Kolar, Inc. Tucson Arizona 04Feb76 for scrap-Final Disposition: Scrapped.

10933 / 51-8055 / C-119G-1-FA

Avble: 27May53; Acc: 26Jun53; Del: 01Jul53 to USAF – 64 TCWG (TAC) Donaldson 01Jul53; depl Seymour 27Apr54 – 483 TCWG (FEA) Ashiya AB Japan 18Aug54; depl 314 TCWG (FEA) Ashiya AB Japan 17Sep54; depl 316 TCWG (FEA) Ashiya AB Japan 15Nov54; bailment CAT (FEA) Taiwan 22Nov54 – 2585 RFC (CNR) Miami 01May57; to DT Donaldson 17Apr58; to 836 AB GP (TAC) Langley 21Sep58 mnt – 77 TCSq (435 TCWG) (AFR) Donaldson 19Dec58 – 440 TCWG (AFR) Gen Mitchell 16Feb61; 933 TCG (440 TCWG) (AFR) Gen Mitchell 11Feb63 – Wfu to MASDC (LOG) Davis-Monthan 28Nov70 for storage, PCN: **CJ372** – DFI: 21May71 – Kolar, Inc. Tucson Arizona 24Feb76 for scrap – Final Disposition: Scrapped.

10934 / 51-8056 / C-119G-1-FA

Avble: 28May53; Acc: 22Jun53; Del: 24Jun53 to USAF – 64 TCWG (TAC) Greenville 24Jun53; depl Seymour 23Apr54 – 483 TCWG (FEA) Ashiya AB Japan 23Aug54; depl 314 TCWG (FEA) Ashiya AB Japan 23Sep54; depl 316 TCWG (FEA) Ashiya AB Japan 15Nov54; bailment CAT (FEA) Taiwan 20Jan55 – 2465 RFC (CNR) Minneapolis-St. Paul 20May57 – 2473 RFC DT (CNR) Minneapolis-St. Paul 01Dec57 – 96 TCSq (440 TCWG) (AFR) Minneapolis-St. Paul 19Dec59; 934 TCG (440 TCWG) (AFR) Minneapolis-St. Paul 31May63 – 906 TAL GP (302 TCWG) (AFR) Clinton 12Jan70 – MAP Taiwan 11Mar70 – Final Disposition: Unknown.

10935 / 51-8057 / C-119G-1-FA

Avble: 28May53; Acc: 25Jun53; Del: 29Jun53 to USAF – 64 TCWG (TAC) Greenville 29Jun53; depl Seymour 26Apr54 – 464 TCWG (TAC) Lawson 06Aug54; to Pope 01Sep54; depl Aberdeen 05May56 – Hayes (AMC) Birmingham 26Sep58 – 2347 RFC (CNR) Long Beach 30Dec58 – 452 TCWG (AFR) Long Beach 13Jun59; to 4756 ADF WG (ADC) Tyndall 24Aug59 mnt; to March 01Oct60; 942 TCG (452 TCWG) (AFR) March 11Feb63 – 904 TCG (514 TCWG) (AFR) Stewart 04Jun65; 903 TCG (514 TCWG) (AFR) McGuire 11May66; depl Myrtle Beach 16Jul67 – 928 TAL GP (403 TCWG) (AFR) O'Hare 08Aug68 – MAP Taiwan 10Apr69, RoCAF s/n: **3157** – Final Disposition: Unknown.

10936 / 51-8058 / C-119G-1-FA

Avble: 29May53; Acc: 23Jun53; Del: 29Jun53 to USAF – 64 TCWG (TAC) Greenville 29Jun53; depl Seymour 26Apr54 – 464 TCWG (TAC) Lawson 06Aug54; to Pope 01Sep54; depl Aberdeen 05May56 – Hayes (AMC) Birmingham 26Sep58 – 97 TCSq (349 TCWG) (AFR) Paine 30Dec58; 941 TCG (349 TCWG) (AFR) Paine 31May63 – 925 TCG (446 TCWG) (AFR) Ellington 02Oct65; 924 TAL GP (446 TCWG) (AFR) Ellington 14Jan68 – 908 TAL GP (435 TCWG) (AFR) Brookley 25May68 – 926 TAL GP (446 TCWG) (AFR) NAS New Orleans 12Feb69 – MAP Taiwan 08Apr69, RoCAF s/n: **3151** – Final Disposition: Unknown.

10937 / 51-8059 / C-119G-1-FA

Avble: 29May53; Acc: 19Jun53; Del: 24Jun53 to USAF – 64 TCWG (TAC) Greenville 24Jun53; depl Seymour 23Apr54 – 464 TCWG (TAC) Lawson 08Aug54; to Pope 01Sep54; depl Aberdeen 05May56 – 2346 RFC DT (CNR) Portland 31Jul58 – Hayes (AMC) Birmingham 08Dec58 – 313 TCSq (349 TCWG) (AFR) Portland 04Mar59; 939 TCG (349 TCWG) (AFR) Portland 16Jul63 – 926 TAL GP (446 TCWG) (AFR) NAS New Orleans 24May68 – Wfu to MASDC (LOG) Davis-Monthan 30Nov69 for storage, PCN: **CJ329** – DFI: 20Jan70 – Kolar, Inc. Tucson Arizona 03Feb76 for scrap – Final Disposition: Scrapped.

10938 / 51-8060 / C-119G-1-FA

Avble: 01Jun53; Acc: 30Jun53; Del: 06Jul53 to USAF – 64 TCWG (TAC) Donaldson 06Jul53; depl Seymour 23Apr54 – 313 TCG (TAC) Sumpter (depl) 07Aug54; to Sewart date unk; depl Elmendorf 12Jan55 – BMC (AMC) Birmingham 06Apr55 – 314 TCWG (TAC) Sewart 23Jun55 – 464 TCWG (TAC) Sewart (depl) 10Aug55; to Pope 07Sep55; depl Aberdeen 05May56 – Hayes (AMC) Birmingham 15Dec58 – 2347 RFC (CNR) Long Beach 27Mar59 – 314 TCSq (349 TCWG) (AFR) McClellan 02Apr59; 940 TCG (349 TCWG) (AFR) McClellan 31May63; listed as cvtd **C-119C** 31Jan64-18Aug64 for project – 904 TCG (514 TCWG) (AFR) Stewart 02Apr65; 903 TCG (514 TCWG) (AFR) McGuire 20May66 – 927 TAL GP (403 TCWG) (AFR) Selfridge 05Aug68 – MAP Taiwan 06Mar69 – Final Disposition: Unknown.

10939 / 51-8061 / C-119G-1-FA

Avble: 02Jun53; Acc: 26Jun53; Del: 01Jul53 to USAF – 64 TCWG (TAC) Donaldson 01Jul53; depl Seymour 23Apr54 – 313 TCG (TAC) Sumpter (depl) 04Aug54; to Sewart 19Aug54; depl Elmendorf 12Jan55; to 3902 AB WG (SAC) Offutt 17Jan55 mnt – BMC (AMC) Birmingham 04Apr55 – 464 TCWG (TAC) Sewart (depl) 08Jun55; to Pope 07Sep55; depl Aberdeen 05May56 – 2472 RFC (CNR) Richards 07Jun57; depl 2466 RFC (CNR) Bakalar 17Jun57 – 442 TCWG (AFR) Richards 02Mar59 – 328 TCSq (512 TCWG) (AFR) Niagara Falls 23Apr61; 914 TCG (512 TCWG) (AFR) Niagara Falls 11Feb63 – Wfu to MASDC (LOG) Davis-Monthan 30Nov70 for storage, PCN: **CJ374** – DFI: 21May71 but reinstated 12Apr72 to storage status – DFI: 26Dec73 – Kolar Inc. Tucson Arizona 24Feb76 – Kolar, Inc. Tucson Arizona 24Feb76 for scrap – Final Disposition: Scrapped.

10940 / 51-8062 / C-119G-1-FA

Avble: 02Jun53; Acc: 25Jun53; Del: 29Jun53 to USAF – 64 TCWG (TAC) Greenville 29Jun53; depl Seymour 23Apr54 – Fairchild (AMC) Hagerstown 09Aug54 – 483 TCWG (FEA) Ashiya AB Japan 25Aug54; depl 314 TCWG (FEA) Ashiya AB Japan 11Sep54; depl 316 TCWG (FEA) Ashiya AB Japan 15Nov54; bailment CAT (FEA) Taiwan 21Jan55; depl (316 TCWG) Andersen 05May55 – 2237 RFC (CNR) New Castle 04Apr57; to NAS Willow Grove 21Jul58 – 512 TCWG (AFR) NAS Willow Grove 01Dec58; 912 TCG (512 TCWG) (AFR) NAS Willow Grove 11Feb63 – 934 TAL GP (440 TCWG) (AFR) Minneapolis-St. Paul 15Aug68 – Wfu to MASDC (LOG) Davis-Monthan 20Feb69 for storage, PCN: **CJ289** – DFI: 31Mar76 – Final Disposition: Scrapped.

10941 / 51-8063 / C-119G-1-FA

Avble: 02Jun53; Acc: 24Jun53; Del: 29Jun53 to USAF – 64 TCWG (TAC) Greenville 29Jun53; depl Seymour 23Apr54 – 313 TCG (TAC) Sumpter (depl) 04Aug54; to Sewart 19Aug54; depl Elmendorf 12Jan55 – BMC (AMC) Birmingham 05Apr55 – 314 TCWG (TAC) Sewart 13Jun55 – 464 TCWG (TAC) Sewart (depl) 09Aug55; to Pope 07Sep55; depl Aberdeen 05May56 – 2472 RFC (CNR) Richards 06Jun57; depl 2466 RFC (CNR) Bakalar 17Jun57; to DT Davis 02Mar58; to DT Tinker 09May58 – 305 TCSq (442 TCWG) (AFR) Tinker 19Mar59 – 434 TCWG (AFR) Bakalar 21Feb61; 930 TCG (434 TCWG) (AFR) Bakalar 11Feb63; depl 71 ACO SQ (1 ACO WG) (TAC) Lockbourne 20Jun68 – 910 TAL GP (459 TCWG) (AFR) Youngstown 13Oct68 – Wfu to MASDC (LOG) Davis-Monthan 03Dec69 for storage, PCN: **CJ333** – DFI: 20Jan70 – Kolar, Inc. Tucson Arizona 17Feb76 for scrap – Final Disposition: Scrapped.

10942 / (RCAF) 22125 / C-119F-FA

Avble: 03Jun53; Acc: 19Jun53; Del: 29Jun53 to Canada, RCAF s/n: **22125**; dropped from USAF inventory 29Jun53 – BOC: 10Jul53 435 SQ Edmonton; cvtd **C-119G** 21Nov56. Suffered an engine failure and subsequently lost control on landing at RCAF Namao, Alberta 13Jan58, 13 crew and passengers survived. The wreckage was retained for spares – SOC: 20Jan58 – Final Disposition: Accident.

10943 / (RCAF) 22126 / C-119F-FA

Avble: 04Jun53; Acc: 22Jun53; Del: 29Jun53 to Canada, RCAF s/n: **22126**; dropped from USAF inventory 29Jun53 – BOC: 10Jul53 435 SQ Edmonton; cvtd **C-119G** 11Jan57 – 436 SQ Downsview 11Mar57 – Wfu into storage 1005 TSD Saskatoon 30Apr65 – SOC: 13Sep67 to Crown Assets Disposal Corp. – BofS Aircraft International Associates Dallas Texas 07Sep67; ferried from Canada to Dallas Texas Nov67; CofR **N965S** 29Feb68 – Government of Morocco 23Dec68, RMAF s/n: **CN-AMN**, 1 AT SQ; reg N965S canx 27Feb69. Displayed as a gate guardian at Kenitra AB Kenitra Morocco – Final Disposition: Extant.

10944 / (RCAF) 22127 / C-119F-FA

Avble: 05Jun53; Acc: 25Jun53; Del: 06Jul53 to Canada, RCAF s/n: **22127**; dropped from USAF inventory 06Jul53 – BOC: 10Jul53 435 SQ Edmonton; cvtd **C-119G** 13Dec56 – 436 SQ Downsview 24Apr59. While landing at Thule Greenland on 11Aug60, the aircraft skidded off the runway resulting in damage beyond repair, the 5 crew onboard were uninjured – SOC: 20Sep60 – Final Disposition: Accident.

10945 / (RCAF) 22128 / C-119F-FA

Avble: 08Jun53; Acc: 26Jun53; Del: 06Jul53 to Canada, RCAF s/n: **22128**; dropped from USAF inventory 06Jul53 – BOC: 10Jul53 435 SQ Edmonton; cvtd **C-119G** 14Nov56 – 436 SQ Downsview 22May58. Crash-landed due to fuel starvation 50 miles SW Moderna Italy 13Dec58 enroute from Athens to Pisa. The aircraft was carrying the remains of an RCAF DHC-3 Otter that had itself crashed in Gaza on 19Sep58. The 7 onboard survived but 3 were injured – SOC: 04Mar59 – Final Disposition: Accident.

10946 / 52-6016 / C-119G-10-FA

Avble: 10Jun53; Acc: 30Jun53; Del: 23Jul53; to MIDAR (AMC) Olmsted 22Jul53 mnt; to MDAP Italy 08Jan54, AMI s/n: **MM52-6016** – Final Disposition: Scrapped.

10947 / 52-6017 / C-119G-10-FA

Avble: 30Jun53; Acc: 30Jun53; Del: 06Jul53 to MDAP Italy, AMI s/n: **MM52-6017**; diverted MIDAR (AMC) Olmsted 10Aug53-12Aug53 for pilot training – Final Disposition: Scrapped.

10948 / 52-6018 / C-119G-10-FA

Avble: 10Jun53; Acc: 30Jun53; Del: 08Jul53 to MDAP Italy, AMI s/n: **MM52-6018**. Crashed during climb out from Rivolto Air Base Italy due to a no. 1 engine failure 25Apr70. 17 were killed consisting of 7 crew and 10 passengers, 2 passengers survived – Final Disposition: Accident.

10949 / 52-6019 / C-119G-15-FA

Avble: 12Jun53; Acc: 30Jun53; Del: 28Jul53 to MDAP Italy, AMI s/n: **MM52-6019**; diverted MIDAR (AMC) Olmsted 31Jul53-07Aug53 for pilot training – Final Disposition: Scrapped.

10950 / 52-6020 / C-119G-15-FA

Avble: 12Jun53; Acc: 06Jul53; Del: 23Jul53 to MDAP Italy, AMI s/n: **MM52-6020**; diverted MIDAR (AMC) Olmsted 05Aug53-13Aug53 for pilot training. Preserved at Rivolto Air Base Udine Italy – Final Disposition: Extant.

10951 / 52-6021 / C-119G-15-FA

Avble: 15Jun53; Acc: 29Jul53; Del: 06Aug53 to MDAP Belgium, BAF s/n: **CP-23**; BOC: 09Sep53 15 WG / 20 SQ Melsbroek AB Belgium, callsign: **OT-CBC**. Hit tail of parked s/n: CP-31 during a taxi run 01Feb61. Catastrophic mid-air collision with C-119G s/n: CP-25 (11082) over Cambron-St. Vincent Belgium 12Dec61, both aircraft came down at Montignies-lez-Lens. 5 crew killed on CP-23 were: Andre Bolle, Gilbert Deneef, Jules Gerard (pilot), David Vandemeersche and Albert Vermeerbergen – Final Disposition: Accident.

10952 / 52-6022 / C-119G-15-FA

Avble: 16Jun53; Acc: 07Jul53; Del: 21Jul53 to MDAP Belgium, BAF s/n: **CP-21**; BOC: 04Aug53 15 WG / 40 SQ Melsbroek AB Belgium, callsign: **OT-CBA**, later to 15 WG / 20 SQ. Minor accident 29Dec54. Suffered a lightning strike 04Apr62. Lost antenna counterweight in-flight 13Jan69 – Wfu 02May72 and stored at Koksijde AB Belgium; SOC: Jul74 – Sold for scrap to J. Trappeniers 04Oct74 – Fuselage sold to a private civil owner, Mr. Yvens of Sint-Niklaas of Belgium. Transported by road to his backyard for conversion into a garage 1976. Current status unk – Final Disposition: Extant.

10953 / 52-6023 / C-119G-15-FA

Avble: 16Jun53; Acc: 10Jul53; Del: 21Jul53 to MDAP Belgium, BAF s/n: **CP-22**; BOC: 08Aug53 15 WG / 40 SQ Melsbroek AB Belgium, callsign: **OT-CBB**, later to 15 WG / 20 SQ. Several minor accidents 29Dec54 and 01Mar69 – Wfu 02Oct72 and stored at Koksijde AB Belgium; SOC: Jul74 – Sold for scrap to J. Trappeniers 04Oct74, scrapped 1977-78 – Final Disposition: Scrapped.

10954 / (RCAF) 22129 / C-119F-FA

Avble: 17Jun53; Acc: 22Jul53; Del: 28Jul53 to Canada, RCAF s/n: **22129**; dropped from USAF inventory 28Jul53 – BOC: 31Jul53 435 SQ Edmonton; cvtd **C-119G** 13Dec56 – 436 SQ Downsview 15Dec58 – 435 SQ Namao 19Apr60 – 436 SQ Downsview 26Apr63 – Wfu into storage 1005 TSD Saskatoon 14May65 – SOC: 22Aug67 to Crown Assets Disposal Corp. – BofS Aircraft International Associates Dallas Texas 01Aug67; ferried from Canada to Dallas Texas Nov67; CofR **N963S** 29Feb68-Government of Morocco 23Dec68, RMAF s/n: **CN-AMO**, 1 AT SQ; reg N963S canx 27Feb69 – Final Disposition: Unknown.

10955 / (RCAF) 22130 / C-119F-FA

Avble: 19Jun53; Acc: 22Jul53; Del: 28Jul53 to Canada, RCAF s/n: **22130**; dropped from USAF inventory 28Jul53 – BOC: 31Jul53 CJATC Rivers – 4 OTU Trenton 16Feb56; cvtd **C-119G** 01Apr57 – 435 SQ Namao 26Feb58 – 436 SQ Downsview 19Dec63 – Wfu into storage 1005 TSD Saskatoon 15Jul65 – SOC: 06Jun67 to Crown Assets Disposal Corp. – BofS Aircraft International Associates Dallas Texas 02Jun67; reg **N15501** 04Oct68 – BofS Hawkins & Powers Aviation, Inc. Greybull Wyoming 16Dec75; noted 1979 stored without engines; cvtd **C-119G-3E** 1981 for aerial firefighting, cargo and agricultural work. Fire Tanker **138**, ag sprayer fleet No. **01**. Installation of aerial spraying system 14Apr87-14Oct87, fitted with 3-bladed props. Placed into storage from 1988, noted stored at Tucson Arizona Apr88. Restored to airworthy condition 2002–2003 for film appearance in *Flight of the Phoenix* (2004), ferried to Namibia 08Dec03 for motion picture filming with false reg "H0180-H," returned to the U.S. 01Jun04 – BofS The Pride Capital Group LLC Deerfield Illinois 07Sep05 for auction – BofS Hans O. Lauridsen Phoenix Arizona 23Jan07. Maintained in fully restored condition at the Lauridsen Aviation Museum, Buckeye Arizona as probably the best preserved C-119 remaining in the world. – Final Disposition: Extant.

10956 / (RCAF) 22131 / C-119F-FA

Avble: 22Jun53; Acc: 22Jul53; Del: 30Jul53 to Canada, RCAF s/n: **22131**; dropped from USAF inventory 30Jul53 – BOC: 17Aug53 6 RD Trenton – 435 SQ Namao 09Nov55; cvtd **C-119G** 13Dec56 – 436 SQ Downsview 13Mar59 – Wfu into storage 1005 TSD Saskatoon 06May65 – SOC: 13Sep67 to Crown Assets Disposal Corp. – BofS Aircraft International Associates Dallas Texas 1967; reg **N5216R** 30Sep71 – BofS Hawkins & Powers Aviation, Inc. Greybull Wyoming 28Jun74; cvtd **C-119G-3E** 19Dec74 with chemical tank for aerial firefighting, Tanker **136**; control surfaces recovered 26Feb75; wfu and stored 1987 – To U.S. Forest Service for display at Battle Mountain Air Museum Nevada circa 1991, noted at Battle Mountain Airport Jun92. Dorsal Jet-Pak from C-119G N3559 (10870) has been fitted and so incorrectly displayed as Tanker "137." Aircraft became derelict in subsequent years – Reg N5216R revoked 24Feb04, canx 11Sep12 – Rolling Boxcar, Inc. Alaska Feb19, dismantled for restoration Jun19 – Final Disposition: Extant.

10957 / (RCAF) 22132 / C-119F-FA

Avble: 22Jun53; Acc: 22Jul53; Del: 30Jul53 to Canada, RCAF s/n: **22132**; dropped from USAF inventory 30Jul53 – BOC: 17Aug53 6 RD Trenton – 435 SQ Edmonton 17Mar54 – 436 SQ Dorval 24Feb56; cvtd **C-119G** 11Jan57; depl to 408 SQ Rivers 22Apr64 – Wfu into storage 1005 TSD Saskatoon 25May65 – SOC: 02Mar67 to Crown Assets Disposal Corp. – BofS Aircraft International Associates Dallas Texas 28Feb67; reg **N3560** 06Oct71 – BofS Hawkins & Powers Aviation, Inc. Greybull Wyoming 13Dec72; cvtd **C-119G-3E** 03Jan73 with chemical tank for aerial firefighting, Tanker **140**. Crashed due to a runaway no. 2 engine propeller while on a test flight out of Greybull Wyoming 10Jun78. The aircraft belly crash-landed 5 miles NE of the airfield killing the 3 crew but the 1 passenger onboard survived – Reg N3560 canx 20Nov89 – Final Disposition: Accident.

10958 / 51-8064 / C-119G-1-FA

Avble: 23Jun53; Acc: 17Jul53; Del: 22Jul53 to USAF **RECORDS INDECIPHERABLE** → 464 TCWG (TAC) Pope date unk ← **RECORDS INDECIPHERABLE** depl North Aux AF 06Jun55; depl Aberdeen 05May56; to 5700 AB GP (CAC) Albrook 14Aug56 mnt – Hayes (AMC) Birmingham 26Sep58 – 97 TCSq (349 TCWG) (AFR) Paine 29Dec58 – Fairchild (LOG) St. Augustine 03Apr62 – MAP Brazil 02Jul62, FAB s/n: **2306**, No. 2/1 Grupo de Transporte de Tropa (2 SQ / 1 GTT) Aerea dos Afonsos AB Rio de Janeiro – Wfu 11Nov74 – Final Disposition: Scrapped.

10959 / 51-8065 / C-119G-1-FA

Avble: 23Jun53; Acc: 17Jul53; Del: 22Jul53 to USAF – 64 TCWG (TAC) Donaldson 22Jul53 – Fairchild (AMC) Hagerstown 09Aug54 – 483 TCWG (FEA) Ashiya AB Japan 26Sep54; depl 316 TCWG (FEA) Ashiya AB Japan 17Apr55; bailment CAT (FEA) Taiwan 27Sep55 – 2237 RFC (CNR) New Castle 26Apr57; to NAS Willow Grove 21Jul58 – 512 TCWG (AFR) NAS Willow Grove 01Dec58; to 4500 AB WG (TAC) Langley 15Jan61 mnt – Fairchild (LOG) St. Augustine 20Jun62 – MAP Brazil 24Jan62, FAB s/n: **2300**, No. 2/1 Grupo de Transporte de Tropa (2 SQ /1 GTT) Aerea dos Afonsos AB Rio de Janeiro – Wfu 11Nov74 – Final Disposition: Scrapped.

10960 / 51-8066 / C-119G-1-FA

Avble: 24Jun53; Acc: 17Jul53; Del: 22Jul53 to USAF – 64 TCWG (TAC) Donaldson 22Jul53 – Fairchild (AMC) Hagerstown 02Aug54 – 483 TCWG (FEA) Ashiya AB Japan 11Aug54; depl 314 TCWG (FEA) Ashiya AB Japan 08Nov54; depl 316 TCWG (FEA) Ashiya AB Japan 15Nov54; bailment CAT (FEA) Taiwan 27Feb55; to 18 Fighter-Bomber WG (FEA) Kadena AB Japan 13May55 – 2256 RFC (CNR) Niagara Falls 27Apr57; depl 2235 RFC (CNR) Grenier 14May57 – 2242 RFC DT (CNR) Niagara Falls 01Dec57 – 2237 RFC DT (CNR) Niagara Falls 01Apr58 – 328 TCSq (512 TCWG) (AFR) Niagara Falls 19Dec58 – Fairchild (LOG) St. Augustine 12Jun61 – MAP Brazil 24Jan62, FAB s/n: **2301**, No. 2/1 Grupo de Transporte de Tropa (2 SQ / 1 GTT) Aerea dos Afonsos AB Rio de Janeiro – Wfu 11Nov74 – Final Disposition: Scrapped.

10961 / 51-8067 / C-119G-1-FA

Avble: 26Jun53; Acc: 17Jul53; Del: 22Jul53 to USAF – 64 TCWG (TAC) Donaldson 22Jul53; depl Seymour 23Apr54 – 313 TCG (TAC) Sumpter (depl) 05Aug54; to Sewart 19Aug54; to WRDCN (ARD) WPAFB 27Sep54 mnt; to Elmendorf 12Jan55 – 463 TCWG (TAC) Sewart (depl) 08Jun55 – 314 TCWG (TAC) Sewart 11Aug55; depl Dreux AB France 21Oct55 – 2585 RFC (CNR) Miami 10Oct57; to DT Donaldson 26Apr58 – 77 TCSq (435 TCWG) (AFR) Donaldson 19Dec58 – 328 TCSq (512 TCWG) (AFR) Niagara Falls 09Jan61 – Fairchild (LOG) St. Augustine 12Jun61 – MAP Brazil 24Jan62, FAB s/n: **2302**, No. 2/1 Grupo de Transporte de Tropa (2 SQ / 1 GTT) Aerea dos Afonsos AB Rio de Janeiro. Made a forced landing in mountainous terrain in the Sao Francisco de Paula region of Brazil while on a training flight out of the Rio de Janeiro base 01Apr62. 5 Brazilian crew and 2 USAF crew were onboard with one of these sustaining serious injuries – Final Disposition: Accident.

10962 / 51-8068 / C-119G-1-FA

Avble: 26Jun53; Acc: 22Jul53; Del: 27Jul53 to USAF – 313 TCG (TAC) Mitchel 27Jul53; depl 465 TCWG (TAC) Mitchel 01Oct53; to Sewart 02Oct53; depl Sumpter Alabama 19Jul54 – Fairchild (AMC) Hagerstown 21Aug54 – 483 TCWG (FEA) Ashiya AB Japan 24Sep54; depl 316 TCWG (FEA) Ashiya AB Japan 24Apr55; bailment CAT (FEA) Taiwan 19May55 – 2237 RFC (CNR) New Castle 03May57; to NAS Willow Grove 21Jul58 – Hayes (AMC) Birmingham 27Oct58 – 512 TCWG (AFR) NAS Willow Grove 10Dec58 – Fairchild (LOG) St. Augustine 20Jun61 – MAP Brazil 24Jan62, FAB s/n: **2303**, No. 2/1 Grupo de Transporte de Tropa (2 SQ / 1 GTT) Aerea dos Afonsos AB Rio de Janeiro – Wfu 11Nov74 – Final Disposition: Scrapped.

10963 / 51-8069 / C-119G-1-FA

Avble: 29Jun53; Acc: 24Jul53; Del: 27Jul53 to USAF – 313 TCG (TAC) Mitchel 27Jul53; depl 465 TCWG (TAC) Mitchel 01Oct53; to Sewart 02Oct53; to Pope 20Apr54; depl Sumpter Alabama 19Jul54; depl Elmendorf 12Jan55 – 464 TCWG (TAC) Sewart 08Jun55; to Pope 07Sep55; depl Aberdeen 05May56 – 2346 RFC DT (CNR) Portland 04Aug58 – 337 CLM SQ (ADC) Portland 27Sep58 mnt – Hayes (AMC) Birmingham 13Jan59 mnt – 512 TCWG (AFR) NAS Willow Grove 29May59; to 464 TCWG (AFR) Pope 04Nov59 mnt; 912 TCG (512 TCWG) (AFR) NAS Willow Grove 11Feb63; 913 TAL GP (512 TCWG) (AFR) NAS Willow Grove 24Jul68 – Wfu to MASDC (LOG) Davis-Monthan 14Dec69 for storage, PCN: **CJ339** – DFI: 20Jan70 – Kolar, Inc. Tucson Arizona 03Feb76 for scrap – Final Disposition: Scrapped.

10964 / 51-8070 / C-119G-1-FA

Avble: 29Jun53; Acc: 22Jul53; Del: 03Aug53 to USAF – 313 TCG (TAC) Mitchel 03Aug53; depl 465 TCWG (TAC) Mitchel Oct53; to Sewart 02Oct53; depl Pope 20Apr54; depl Sumpter Alabama 19Jul54 – Fairchild (AMC) Hagerstown 11Aug54 – 483 TCWG (FEA) Ashiya AB Japan 10Sep54; depl 314 TCWG (FEA) Ashiya AB Japan 04Oct54; depl 316 TCWG (FEA) Ashiya AB Japan 15Nov54; bailment CAT (FEA) Taiwan 13May55 – 2237 RFC (CNR) New Castle 25Apr57; to NAS Willow Grove 21Jul58 – 512 TCWG (AFR) NAS Willow Grove 01Dec58; 912 TCG (512 TCWG) (AFR) NAS Willow Grove 11Feb63 – 927 TAL GP (403 TCWG) (AFR) Selfridge 19Aug68 – MAP Taiwan 06Mar69, RoCAF s/n: **3120** – Presently displayed at Pingtung AB Taiwan – Final Disposition: Extant.

10965 / 51-8071 / C-119G-1-FA

Avble: 29Jun53; Acc: 28Jul53; Del: 03Aug53 to USAF – 313 TCG (TAC) Mitchel 03Aug53; depl 465 TCWG (TAC) Mitchel Oct53; to Sewart 02Oct53; depl Pope 20Apr54; depl Sumpter Alabama 19Jul54; depl Elmendorf 12Jan55; to 6606 AB WG (NEA) Goose Bay Canada 16Mar55 mnt – BMC (AMC) Birmingham 27May55 – 464 TCWG (TAC) Sewart 05Aug55; to Pope 07Sep55; depl Aberdeen 05May56 – Hayes (AMC) Birmingham 10Dec58 – 2347 RFC (CNR) Long Beach 13Mar59 – 452 TCWG (AFR) Long Beach 13Jun59; to March 01Oct60; 944 TCG (452 TCWG) (AFR) March 11Feb63; 942 TCG (452 TCWG) (AFR) March 31Mar65 – 924 TCG (446 TCWG) (AFR) Ellington 12Aug66 – 906 TAL GP (302 TCWG) (AFR) Clinton 20Jan68 – MAP Taiwan 03Jan70, RoCAF s/n: **3183** – Presently displayed in Junji Park, Changhua City Taiwan – Final Disposition: Extant.

10966 / 51-8072 / C-119G-1-FA

Avble: 01Jul53; Acc: 24Jul53; Del: 31Jul53 to USAF – 313 TCG (TAC) Mitchel 31Jul53; depl 465 TCWG (TAC) Mitchel Oct53; to Sewart 02Oct53; depl Pope 20Apr54; depl Sumpter Alabama 19Jul54; depl Elmendorf 12Jan55 – 464 TCWG (TAC) Sewart 08Jun55; to Pope 07Sep55; depl Aberdeen 05May56 – Hayes (AMC) Birmingham 07Nov58 – 2347 RFC (CNR) Long Beach 20Feb59 – 452 TCWG (AFR) Long Beach 13Jun59; to March 01Oct60; 943 TCG (452 TCWG) (AFR) March 11Feb63 – 931 TAL GP (434 TCWG) (AFR) Bakalar 22Apr69; 930 SOP GP (434 TCWG) (AFR) Bakalar 30Jun69 – Wfu to MASDC (LOG) Davis-Monthan 03Dec69 for storage, PCN: **CJ332** – DFI: 20Jan70 – Kolar, Inc. Tucson Arizona 30Jan76 for scrap – Final Disposition: Scrapped.

10967 / 51-8073 / C-119G-1-FA

Avble: 02Jul53; Acc: 24Jul53; Del: 31Jul53 to USAF – 313 TCG (TAC) Mitchel 31Jul53; depl 465 TCWG (TAC) Mitchel Oct53; to Sewart 02Oct53; depl Pope 20Apr54; depl Sumpter Alabama 19Jul54; depl Elmendorf 12Jan55 – 314 TCWG (TAC) Sewart 09May55 – 2472 RFC (CNR) Richards 24Jul57; depl 2466 RFC (CNR) Bakalar 30Jul57; depl 2585 RFC (CNR) Miami 27Aug57 – 442 TCWG (AFR) Richards 02Mar59 – 357 TCSq (302 TCWG) (AFR) Bates 23May61; 908 TCG (302 TCWG) (AFR) Bates 11Feb63 – 908 TCG (435 TCWG) (AFR) Bates 18Mar63; to Brookley 30Oct64 – 926 TAL GP (446 TCWG) (AFR) NAS New Orleans 26Feb69 – 913 TAL GP (512 TCWG) (AFR) NAS Willow Grove 01Feb70 – Wfu to MASDC (LOG) Davis-Monthan 28Feb71 for storage, PCN: **CJ407** – DFI: 31Mar76 – Final Disposition: Scrapped.

10968 / 51-8074 / C-119G-1-FA

Avble: 03Jul53; Acc: 30Jul53; Del: 04Aug53 to USAF – 64 TCWG (TAC) Donaldson 04Aug53 – Fairchild (AMC) Hagerstown 12Aug54 – 483 TCWG (FEA) Ashiya AB Japan 16Sep54; depl 316 TCWG (FEA) Ashiya AB Japan 11Dec54; bailment CAT (AMC) Taiwan 21Jan55 – 2465 RFC (CNR) Minneapolis-St. Paul 14Apr57; depls 2466 RFC (CNR) Bakalar 07Jul57 & 04Aug57 – 2473 RFC DT (CNR) Minneapolis-St. Paul 01Dec57 – Hayes (AMC) Birmingham 31Dec58 – 96 TCSq (440 TCWG) (AFR) Minneapolis-St. Paul 02Apr59; depl 434 TCWG (AFR) Bakalar 25Jan60; to Gen Mitchell 14Jun60 – Fairchild (LOG) St. Augustine 29Jun61 – MAP Brazil 24Jan62, FAB s/n: **2304**, No. 2/1 Grupo de Transporte de Tropa (2 SQ / 1 GTT) Aerea dos Afonsos AB Rio de Janeiro – Wfu 11Nov74. Displayed from 1975 outdoors at the 8 Parachutist Field Artillery Group building Rio de Janeiro – Final Disposition: Extant.

10969 / 51-8075 / C-119G-1-FA

Avble: 06Jul53; Acc: 24Jul53; Del: 31Jul53 to USAF – 313 TCG (TAC) Mitchel 31Jul53; depl 465 TCWG (TAC) Mitchel Oct53; to Sewart 02Oct53; depl Pope 20Apr54; depl Sumpter Alabama 19Jul54; depl Elmendorf 12Jan55 – 463 TCWG (TAC) Sewart (depl) 08Jun55 – 314 TCWG (TAC) Sewart 12Aug55; depl Dreux AB France 21Oct55 – 2585 RFC (CNR) Miami 08Aug57; to DT Donaldson 24Jun58 – 77 TCSq (435 TCWG) (AFR) Donaldson 19Dec58 – 64 TCSq (403 TCWG) (AFR) O'Hare 11Jan61 – Fairchild (LOG) St. Augustine 09Apr62 – 357 TCSq (302 TCWG) (AFR) Bates 21Aug62 – MAP Brazil 22Aug62, FAB s/n: **2308**, No. 2/1 Grupo de Transporte de Tropa (2 SQ / 1 GTT) Aerea dos Afonsos AB Rio de Janeiro – Wfu 11Nov74 – Final Disposition: Scrapped.

10970 / 51-8076 / C-119G-1-FA

Avble: 06Jul53; Acc: 28Jul53; Del: 03Aug53 to USAF – 313 TCG (TAC) Mitchel 03Aug53; depl 465 TCWG (TAC) Mitchel Oct53; to Sewart 02Oct53; depl Sumpter Alabama 19Jul54 – 483 TCWG (FEA) Ashiya AB Japan 54; depl 316 TCWG (FEA) Ashiya AB Japan 25Jun55; depl Tachikawa 04Dec56 – 2235 RFC (CNR) Grenier 12Jun57 – 2230 RFC (CNR) NAS New York 26Jul57 – 2466 RFC (CNR) Bakalar 05Oct57 – 2242 RFC (CNR) Selfridge 19Oct57 – 2473 RFC (CNR) Gen Mitchell 12Dec57; to 2465 AB GP (CNC) Minneapolis-St. Paul 15Feb58 – 440 TCWG (AFR) Gen Mitchell 19Dec58 – Fairchild (LOG) St. Augustine 29Jun62 – MAP Brazil 18Apr62, FAB s/n: **2305**, to No. 2/1 Grupo de Transporte de Tropa (2 SQ / 1 GTT) Aerea dos Afonsos AB Rio de Janeiro – Wfu 11Nov74. Preserved from 1975 for display at Museu Aeroespacial (MUSAL), Campos dos Afonos AB, Rio de Janeiro – Final Disposition: Extant.

10971 / 51-8077 / C-119G-1-FA

Avble: 06Jul53; Acc: 28Jul53; Del: 03Aug53 to USAF – 313 TCG (TAC) Mitchel 03Aug53; depl 465 TCWG (TAC) Mitchel Oct53; to Sewart 02Oct53; depl Pope 21Apr54; depl Sumpter Alabama 19Jul54; depl Elmendorf 12Jan55 – 463 TCWG (TAC) Sewart 08Jun55 – 314 TCWG (TAC) Sewart (depl) 12Aug55; depl Dreux AB France 21Oct55 – 2472 RFC (CNR) Richards 23Jun57; to DT Davis 03Sep58 – 65 TCSq (403 TCWG) (AFR) Davis 18Sep58 – Fairchild (LOG) St. Augustine 09Apr62 – **RECORDS MISSING** – MAP Brazil 1962, FAB s/n: **2309**, No. 2/1 Grupo de Transporte de Tropa (2 SQ / 1 GTT) Aerea dos Afonsos AB Rio de Janeiro – Wfu 11Nov74 – Final Disposition: Scrapped.

10972 / 51-8078 / C-119G-1-FA

Avble: 06Jul53; Acc: 30Jul53; Del: 05Aug53 to USAF – 64 TCWG (TAC) Donaldson 05Aug53 – 313 TCG (TAC) Sumpter (depl) 30Jul54; to Sewart 19Aug54; depl Elmendorf 16Jan55 – BMC (AMC) Birmingham 11Apr55 – 463 TCWG (TAC) Sewart (depl) 22Jun55 – 314 TCWG (TAC) Sewart 13Aug55; depl Dreux AB France 21Oct55 – 2472 RFC (CNR) Richards 07Sep57 – 2465 RFC (CNR) Minneapolis-St. Paul 14Sep57 – 2473 RFC DT (CNR) Minneapolis-St. Paul 01Dec57 – 96 TCSq (440 TCWG) (AFR) Minneapolis-St. Paul 19Dec59; 934 TCG (440 TCWG) (AFR) Minneapolis-St. Paul 10Mar63 – Wfu to MASDC (LOG) Davis-Monthan 01Dec69 for storage, PCN: **CJ330** – DFI: 20Jan70 – Kolar, Inc. Tucson Arizona 18Feb76 for scrap – Final Disposition: Scrapped.

10973 / 51-8079 / C-119G-1-FA

Avble: 07Jul53; Acc: 29Jul53; Del: 03Aug53 to USAF – 313 TCG (TAC) Mitchel 03Aug53; depl 465 TCWG (TAC) Mitchel Oct53; to Sewart 02Oct53; depls Sumpter

Alabama 19Jul54 & 02Aug54; depl 314 TCWG (TAC) Sewart 26Jul54; depl Elmendorf 12Jan55; to 28 SRH WG (SAC) Ellsworth 19Jan55 mnt – BMC (AMC) Birmingham 03May55 – 463 TCWG (TAC) Sewart (depl) 01Jul55 – 314 TCWG (TAC) Sewart 12Aug55; depl Dreux AB France 21Oct55 – 2466 RFC (CNR) Bakalar 10Aug57; to PGC Center (ARD/PGC) Eglin 18Oct58 mnt – 434 TCWG (AFR) Bakalar 19Apr59; 931 TCG (434 TCWG) (AFR) Bakalar 11Feb63 – MAP Taiwan 06Mar69, RoCAF s/n: **3144** – Presently at Kao Yuan Institute of Technology Kaohsiung City Taiwan – Final Disposition: Extant.

10974 / 51-8080 / C-119G-1-FA

Avble: 08Jul53; Acc: 29Jul53; Del: 03Aug53 to USAF – 313 TCG (TAC) Mitchel 03Aug53; depl 465 TCWG (TAC) Mitchel Oct53; to Sewart 02Oct53; depl Pope 20Apr54; depl Sumpter Alabama 19Jul54; depl Elmendorf 12Jan55 – 464 TCWG (TAC) Sewart (depl) 08Jun55; to Pope 07Sep55; depl Aberdeen 05May56 – 2347 RFC (CNR) Long Beach 19Dec58 – 452 TCWG (AFR) Long Beach 13Jun59; to March 01Oct60 – Fairchild (LOG) St. Augustine 31Jul62 – **RECORDS MISSING** – MAP Brazil 1962, FAB s/n: **2310**, No. 2/1 Grupo de Transporte de Tropa (2 SQ / 1 GTT) Aerea dos Afonsos AB Rio de Janeiro – Wfu 11Nov74 – Final Disposition: Scrapped.

10975 / 51-8081 / C-119G-1-FA

Avble: 09Jul53; Acc: 30Jul53; Del: 04Aug53 to USAF – 64 TCWG (TAC) Donaldson 04Aug53; depl Seymour 23Apr54 – 802 AB GP (SAC) Smokey Hill 02Jul54 mnt – 313 TCG (TAC) Sewart 21Aug54 – 314 TCWG (TAC) Sewart 08Jun55 – 464 TCWG (TAC) Sewart 10Aug55; to Pope 08Sep55; depl Aberdeen 05May56 – Hayes (AMC) Birmingham 26Sep58 – 2347 RFC (CNR) Long Beach 21Mar59 – 452 TCWG (AFR) Long Beach 13Jun59; to March 01Oct60; 943 TCG (452 TCWG) (AFR) March 11Feb63 – Wfu to MASDC (LOG) Davis-Monthan 12Feb69 for storage, PCN: **CJ286** – DFI: 31Mar76 – Final Disposition: Scrapped.

10976 / 51-8082 / C-119G-1-FA

Avble: 09Jul53; Acc: 30Jul53; Del: 04Aug53 to USAF – 64 TCWG (TAC) Donaldson 04Aug53 – 4501 SUT SQ (TAC) Donaldson 06Aug54; depl 464 TCWG (TAC) Pope 12Apr56; to 4405 OP SQ (TAC) Langley 03Jan57 – Hayes (AMC) Birmingham 03Sep57 – 2473 RFC (CNR) Gen Mitchell 23Nov57 – 440 TCWG (AFR) Gen Mitchell 19Dec58; to 801 COS GP (SAC) Lockbourne 10Jan59 mnt; 933 TCG (440 TCWG) (AFR) Gen Mitchell 31May63 – Wfu to MASDC (LOG) Davis-Monthan 28Nov70 for storage, PCN: **CJ368** – DFI: 21May71 – Kolar, Inc. Tucson Arizona 27Jan76 for scrap – Final Disposition: Scrapped.

10977 / 51-8083 / C-119G-1-FA

Avble: 13Jul53; Acc: 31Jul53; Del: 07Aug53 to USAF **RECORDS INDECIPHERABLE** → 64 TCWG (TAC) Donaldson 07Aug53 – 464 TCWG (TAC) Lawson 04Aug54 ← **RECORDS INDECIPHERABLE** to Pope 22Nov54; depl North Aux AF 03Jun55; depl Aberdeen 05May56 – 2585 RFC DT (CNR) Donaldson 24Oct58 – 77 TCSq (435 TCWG) (AFR) Donaldson 19Dec58; 78 TCSq (435 TCWG) (AFR) Bates 07Nov60 – 357 TCSq (302 TCWG) (AFR) Bates 08May61; 908 TCG (302 TCWG) (AFR) Bates 11Feb63 – 908 TCG (435 TCWG) (AFR) Bates 18Mar63; to Brookley 30Oct64 – 3960 STR WG (SAC) Andersen AFB Guam 19Feb69 – MAP Taiwan 20Feb69, RoCAF s/n: **3116** – Final Disposition: Unknown.

10978 / 51-8084 / C-119G-1-FA

Avble: 13Jul53; Acc: 07Aug53; Del: 20Aug53 to USAF – 313 TCG (TAC) Mitchel 20Aug53; to 4415 AB GP (TAC) Pope 07Sep53; depl Pope 20Apr54; to Sewart 30Apr54; depl Elmendorf 12Jan55 – 314 TCWG (TAC) Sewart 08Jun55 – 464 TCWG (TAC) Sewart 11Aug55; to Pope 07Sep55; depl Aberdeen 05May56 – Hayes (AMC) Birmingham 18Nov58 – 2347 RFC (CNR) Long Beach 10Feb59 – 452 TCWG (AFR) Long Beach 13Jun59; to March 01Oct60 – Fairchild (LOG) St. Augustine 04Apr62 – MAP Brazil 03Aug62, FAB s/n: **2307**, to No. 2/1 Grupo de Transporte de Tropa (2 SQ / 1 GTT) Aerea dos Afonsos AB Rio de Janeiro. Crashed into the Marechal Hermes area of Rio de Janeiro 27Jun74 during a training flight, 2 of the 4 crew were killed – Final Disposition: Accident.

10979 / 51-8085 / C-119G-1-FA

Avble: 14Jul53; Acc: 07Aug53; Del: 12Aug53 to USAF **RECORDS INDECIPHERABLE** → 64 TCWG (TAC) Donaldson 12Aug53 – Fairchild (AMC) Hagerstown 12Aug54 – 2750 AB WG (AMC) WPAFB 08Sep54 – 483 TCWG (FEA) Ashiya AB Japan 22Sep54; depl 316 TCWG (FEA) Ashiya AB Japan 01Dec54; assigned 37 TCSq ← **RECORDS INDECIPHERABLE** Crashed after an engine failure during take-off from Matsushima AB Japan 16Apr55. Pilot: Lowell S. Smith plus 4 crew and 12 passengers onboard, 1 sustained serious injuries and may have died but this is unconfirmed – Final Disposition: Accident.

10980 / 51-8086 / C-119G-1-FA

Avble: 15Jul53; Acc: 10Aug53; Del: 13Aug53 to USAF – 313 TCWG (TAC) Mitchel 13Aug53; depl 465 TCWG (TAC) Mitchel Oct53; to Sewart 02Oct53, assigned 48 TCSq. Crashed in a wooded area for reasons unknown 1.5 miles S Newton Falls Ohio 09Dec53. 2 crew killed but a third bailed out at 200 ft and survived – Final Disposition: Accident.

10981 / 51-8087 / C-119G-1-FA

Avble: 15Jul53; Acc: 14Aug53; Del: 19Aug53 to USAF – 313 TCG (TAC) Mitchel 19Aug53; depl 465 TCWG (TAC) Mitchel Oct53; to Sewart 02Oct53; depl Pope 20Apr54; depl Sumpter Alabama 19Jul54 – Fairchild (AMC) Hagerstown 14Aug54 – 483 TCWG (FEA) Ashiya AB Japan 16Aug54; depl 314 TCWG (FEA) Ashiya AB Japan 12Nov54; depl 316 TCWG (FEA) Ashiya AB Japan 15Nov54; bailment CAT (FEA) Taiwan 24Jun55; depl Akddoae? 25Aug55 – 2471 RFC (CNR) O'Hare 17May57; to 6486 AB WG (FEA) Hickham 28May57 mnt – 2473 RFC DT (CNR) O'Hare 01Dec57 – 2242 RFC DT (CNR) O'Hare 01Apr58 – Hayes (AMC) Birmingham 30Oct58 – 2346 RFC DT (CNR) McClellan 14Nov58 – Hayes (AMC) Birmingham 10Jan59 – 64 TCSq (403 TCWG) (AFR) O'Hare 05Feb59; to CTC (ATC) Chanute 24May59 mnt – 435 TCWG (AFR) Homestead 07Jul61; assigned 76 TCSq Homestead 01Oct61; 915 TCG (435 TCWG) (AFR) Homestead 17Jan63 – 926 TCG (446 TCWG) (AFR) NAS New Orleans 10Nov65 – 913 TAL GP (512 TCWG) (AFR) NAS Willow Grove 11Feb70 – Wfu to MASDC (LOG) Davis-Monthan 07Feb71 for storage, PCN: **CJ398** – DFI: 31Mar76 – Final Disposition: Scrapped.

10982 / 51-8088 / C-119G-1-FA

Avble: 16Jul53; Acc: 14Aug53; Del: 19Aug53 to USAF – 313 TCG (TAC) Mitchel 19Aug53; depl 465 TCWG (TAC) Mitchel Oct53; to Sewart 02Oct53; depl Sumpter Alabama 19Jul54; depl Elmendorf 12Jan55; to 1600 FDM SQ (MATS) Westover 23Mar55 mnt; to 6606 AB WG (NEA) Goose Bay Canada 12Apr55 mnt – BMC (AMC) Birmingham 03May55 – 464 TCWG (TAC) Sewart 11Jul55; to Pope 07Sep55; depl Aberdeen 05May56 – Hayes

(AMC) Birmingham 04Dec58; to MOBAR (AMC) Brookley 06Feb59 – 2347 RFC (CNR) Long Beach 12Mar59 – 452 TCWG (AFR) Long Beach 13Jun59; to March 01Oct60; 942 TCG (452 TCWG) (AFR) March 11Feb63 – 925 TCG (446 TCWG) (AFR) Ellington 01Sep65 – 910 TAL GP (459 TCWG) (AFR) Youngstown 20Jan68 – Wfu to MASDC (LOG) Davis-Monthan 10Feb69 for storage, PCN: **CJ285** – DFI: 09Dec74 – Final Disposition: Scrapped.

10983 / 51-8089 / C-119G-1-FA

Avble: 16Jul53; Acc: 10Aug53; Del: 13Aug53 to USAF – 313 TCG (TAC) Mitchel 13Aug53; depl 465 TCWG (TAC) Mitchel Oct53; to Sewart 02Oct53; depl Pope 20Apr54; depl Sumpter Alabama 19Jul54; depl Elmendorf 12Jan55 – 314 TCWG (TAC) Sewart 08Jun55 – 464 TCWG (TAC) Sewart 10Aug55; to Pope 07Sep55; depl Aberdeen 05May56 – 2346 RFC DT (CNR) Portland 29Jul58; to 1611 TSM WG (MATS) McGuire 15Nov58 mnt – 313 TCSq (349 TCWG) (AFR) Portland 23Dec58; 939 TCG (349 TCWG) (AFR) Portland 63 – 922 TAL GP (433 TCWG) (AFR) Kelly 13Jun68 – Wfu to MASDC (LOG) Davis-Monthan 03Mar71 for storage, PCN: **CJ410** – DFI: 21May71 – Final Disposition: Scrapped.

10984 / 51-8090 / C-119G-1-FA

Avble: 17Jul53; Acc: 10Aug53; Del: 13Aug53 to USAF – 313 TCG (TAC) Mitchel 13Aug53; to 1401 AB WG (MATS) Andrews 31Aug53; depl 465 TCWG (TAC) Mitchel Oct53; to Sewart 02Oct53; depl Pope 20Apr54; depl Sumpter Alabama 19Jul54; depl Elmendorf 12Jan55 – BMC (AMC) Birmingham 03May55 – 314 TCWG (TAC) Sewart 08Jul55 – 464 TCWG (TAC) Sewart 09Aug55; to Pope 07Sep55; depl Aberdeen 05May56; to 2750 AB WG (AMC) WPAFB 02Sep56 mnt – Hayes (AMC) Birmingham 07Nov58 – 2347 RFC (CNR) Long Beach 06Feb59 – 452 TCWG (AFR) Long Beach 13Jun59; to March 01Oct60; 943 TCG (452 TCWG) (AFR) March 11Feb63. Force landed in a cotton field near Seale Alabama 11May64 after an engine failure. 2 fatalities among the 4 crew and 43 student skydivers onboard; to 401 Fighter WG (TAC) England 11May64 for SAL – DFI: 12May64 – Final Disposition: Accident.

10985 / 51-8091 / C-119G-1-FA

Avble: 20Jul53; Acc: 14Aug53; Del: 19Aug53 to USAF – 313 TCG (TAC) Mitchel 19Aug53; depl 465 TCWG (TAC) Mitchel Oct53; to Sewart 02Oct53; depl Pope 20Apr54; depl Sumpter Alabama 19Jul54; depl Elmendorf 12Jan55 – BMC (AMC) Birmingham 27May55 – 463 TCWG (TAC) Sewart 29Jul55 – 314 TCWG (TAC) Sewart 12Aug55; depl Dreux AB France 21Oct55 – 2472 RFC (CNR) Richards 25Jul57; to DT Tinker 13Mar58; briefly DT Davis 27Mar58 – 305 TCSq (442 TCWG) (AFR) Tinker 19Mar59 – 434 TCWG (AFR) Bakalar 20Jan61; 931 TCG (434 TCWG) (AFR) Bakalar 11Feb63 – MAP Taiwan 06Mar69, RoCAF s/n: **3126** – Final Disposition: Unknown.

10986 / 51-8092 / C-119G-1-FA

Avble: 20Jul53; Acc: 14Aug53; Del: 28Aug53 to USAF – 313 TCG (TAC) Mitchel 29Aug53; depl 465 TCWG (TAC) Mitchel Oct53; to Sewart 02Oct53; depl Pope 20Apr54; depl Sumpter Alabama 19Jul54; depl Elmendorf 12Jan55 – 314 TCWG (TAC) Sewart 08Jun55 – 464 TCWG (TAC) Sewart 11Aug55; to Pope 07Sep55; depl Aberdeen 05May56; to 72 Bomb WG (SAC) Ramey Puerto Rico 58 mnt – 2347 RFC (CNR) Long Beach 16Dec58 – 452 TCWG (AFR) Long Beach 13Jun59; to March 01Oct60 – Fairchild (AMC) St. Augustine 16Apr62 – MAP Brazil 12Oct62 (USAF records list as lost to ground accident), FAB s/n: **2311**, No. 2/1 Grupo de Transporte de Tropa (2 SQ / 1 GTT) Aerea dos Afonsos AB Rio de Janeiro – Wfu 11Nov74 – Final Disposition: Scrapped.

10987 / 51-8093 / C-119G-1-FA

Avble: 21Jul53; Acc: 14Aug53; Del: 19Aug53 to USAF – 313 TCG (TAC) Mitchel 19Aug53; depl 465 TCWG (TAC) Mitchel Oct53; to Sewart 02Oct53; depl Pope 20Apr54; depl Sumpter Alabama 19Jul54; depl Elmendorf 12Jan55 – 314 TCWG (TAC) Sewart 08Jun55; depl Dreux AB France 21Oct55; to 3150 MAI GP (AMC) Nouasseur AB Morocco 21Jan56 mnt – 2472 RFC (CNR) Richards 08Jul57 – 442 TCWG (AFR) Richards 02Mar59 – 328 TCSq (512 TCWG) (AFR) Niagara Falls 23Apr61; 914 TCG (512 TCWG) (AFR) Niagara Falls 11Feb63 – Wfu to MASDC (LOG) Davis-Monthan 09Feb71 for storage, PCN: **CJ400** – DFI: 09Dec74 – Final Disposition: Scrapped.

10988 / 51-8094 / C-119G-1-FA

Avble: 22Jul53; Acc: 19Aug53; Del: 24Aug53 to USAF – 64 TCWG (TAC) Donaldson 24Aug53; depl Seymour 23Apr54 – Fairchild (AMC) Hagerstown 10Aug54 – 483 TCWG (FEA) Ashiya AB Japan 22Sep54; depl 316 TCG (FEA) Ashiya AB Japan 27Jul55 – 2256 RFC (CNR) Niagara Falls 03Apr57 – 2237 RFC (CNR) New Castle 21Apr57; to NAS Willow Grove 21Jul58 – 512 TCWG (AFR) NAS Willow Grove 01Dec58; 913 TCG (512 TCWG) (AFR) NAS Willow Grove 11Feb63 – MAP Taiwan 21Apr69 – Final Disposition: Unknown.

10989 / 51-8095 / C-119G-1-FA

Avble: 23Jul53; Acc: 19Aug53; Del: 24Aug53 to USAF – 64 TCWG (TAC) Donaldson 24Aug53; depl Seymour 24Apr54 – 313 TCG (TAC) Sumpter (depl) 05Aug54; to Sewart 54; depl Elmendorf 12Jan55 – 314 TCG (TAC) Sewart 55, assigned 18 TCSq. Crashed and burned near Cullman Alabama 10Mar55 after the 11 personnel bailed out due to an engine failure. Pilot: James L. Morrill, Jr., and others parachuted to the ground safely – Final Disposition: Accident.

10990 / 51-8096 / C-119G-1-FA

Avble: 23Jul53; Acc: 19Aug53; Del: 24Aug53 to USAF – 64 TCWG (TAC) Donaldson 24Aug53; depl Seymour 23Apr54 – Fairchild (AMC) Hagerstown 06Aug54 – 483 TCWG (FEA) Ashiya AB Japan 23Aug54; to 1500 MAI SQ (MATS) Hickham 19Sep54 mnt; depl 314 TCWG (FEA) Ashiya AB Japan 17Oct54; depl 316 TCWG (FEA) Ashiya AB Japan 15Nov54; bailment CAT (AMC) Taiwan 23Aug55 – 2256 RFC (CNR) Niagara Falls 30Mar57; depl 2235 RFC (CNR) Grenier 18Apr57 – 2242 RFC DT (CNR) Niagara Falls 01Dec57 – 2237 RFC DT (CNR) Niagara Falls 19Mar58 – 328 TCSq (512 TCWG) (AFR) Niagara Falls 19Dec58; 914 TCG (512 TCWG) (AFR) Niagara Falls 11Feb63 – Wfu to MASDC (LOG) Davis-Monthan 13Nov68 for storage, PCN: **CJ273** – DFI: 26Dec74 – Final Disposition: Scrapped.

10991 / 51-8097 / C-119G-1-FA

Avble: 24Jul53; Acc: 20Aug53; Del: 26Aug53 to USAF – 64 TCWG (TAC) Donaldson 26Aug53 – 464 TCWG (TAC) Lawson 06Aug54; to Pope 13Nov54; depl Aberdeen 05May56 – 2242 RFC (CNR) Selfridge 26Sep58 – 403 TCWG (AFR) Selfridge 13Nov58; to CTC (ATC) Chanute 17Mar59 mnt; 927 TCG (403 TCWG) (AFR) Selfridge 28Feb63 – 913 TAL GP (512 TCWG) (AFR) NAS Willow Grove 08Jul69 – Wfu to MASDC (LOG) Davis-Monthan 03Jan70 for storage, PCN: **CJ346** – DFI: 20Jan70 – Kolar, Inc. Tucson Arizona 24Feb76 for scrap – Final Disposition: Scrapped.

10992 / (RCAF) 22133 / C-119F-FA

Avble: 07Aug53; Acc: 20Aug53; Del: 26Aug53 to Canada, RCAF s/n: **22133**; dropped from USAF inventory 26Aug53 – BOC: 03Sep53 436 SQ Dorval; cvtd **C-119G** 12Dec55 –

435 SQ Namao 12Nov63 – Wfu into storage 1005 TSD Saskatoon 25Jun65 – SOC: 02Mar67 to Crown Assets Disposal Corp. – Dual BofS Frank Shelley Electronics, Inc. Los Angeles California & Aircraft International Associates Dallas Texas both 28Feb67; reg **N383S** 04May67; cvtd **C-119G STOLmaster** prototype May67, engineering and initial flight tests provided by Steward-Davis, Inc. – BofS Aircraft International Associates Omaha Nebraska 10Nov67 – BofS Lee-Argyle Corp. Miami Florida 17Jun69 – BofS Carl W. Renstrom Omaha Nebraska 24Nov69 – BofS Ren-Aire Aviation, Inc. Omaha Nebraska 24Feb70 – BofS Hawkins & Powers Aviation, Inc. Greybull Wyoming 16Dec75; cvtd **C-119G-3E** 04Mar76 with chemical tank for firefighting, Tanker **133**. Crashed during fire-bombing operations in Banning Pass, California 08Jun79 after a wing separated from the main fuselage. Pilot Denny L. Connor (44yrs) and co-pilot Richard M. Ray (50yrs) were killed – Reg N383S canx 05May87 – Final Disposition: Accident.

10993 / (RCAF) 22134 / C-119F-FA

Avble: 28Jul53; Acc: 20Aug53; Del: 26Aug53 to Canada, RCAF s/n: **22134**; dropped from USAF inventory 26Aug53 – BOC: 06Oct53 4 OTU Trenton – 435 SQ Namao 12Dec55; cvtd **C-119G** 11Jan57 – 436 SQ Downsview 14Apr58 – 435 SQ Namao 29Jul64 – Wfu into storage 1005 TSD Saskatoon 15Jul65 – SOC: 06Jun67 to Crown Assets Disposal Corp. – BofS Aircraft International Associates Dallas Texas 02Jun67; reg **N15508** 04Oct68 – BofS Hawkins & Powers Aviation, Inc. Greybull Wyoming 16Dec75; not Jet-Pak cvtd; chemical and spray system installed 27May77 – BofS Conair Aviation, Ltd. Abbotsford Canada 24Oct88, sale likely not completed – USAF Museum WPAFB Ohio 28Oct88. Displayed at Travis AFB California marked as: "53-22134" – Reg N15508 canx 23Aug05 – Final Disposition: Extant.

10994 / (RCAF) 22135 / C-119F-FA

Avble: 28Jul53; Acc: 27Aug53; Del: 27Aug53 to Canada, RCAF s/n: **22135**; dropped from USAF inventory 27Aug53 – BOC: 03Sep53 436 SQ Dorval; cvtd **C-119G** 13Nov56 – 435 SQ Namao 14Feb61 – 436 SQ Downsview 17Apr62 – 435 SQ Namao 15May62 – Wfu into storage 1005 TSD Saskatoon 17Aug64 – SOC: 13Sep67 to Crown Assets Disposal Corp. – BofS Aircraft International Associates Dallas Texas 07Sep67; reg **N8094** 19Aug70 – BofS Hawkins & Powers Aviation, Inc. Greybull Wyoming 16Dec75; used as spares airframe – BofS D&G, Inc. Greybull Wyoming 01Jan93 – BofS The Pride Capital Group LLC Deerfield Illinois 07Sep05 for auction – BofS Harold Sheppard, Jr., d/b/a Sheppard Trucking Riverton Wyoming 24Aug06. Presently the aircraft remains derelict at Greybull Wyoming along with several other C-119 and piston-engined aircraft – Final Disposition: Extant.

10995 / 52-6024 / C-119G-15-FA

Avble: 30Jul53; Acc: 27Aug53; Del: 03Sep53 to MDAP Italy, AMI s/n: **MM52-6024**; diverted MIDAR (AMC) Olmsted 03Sep53-12Jan54 for pilot training – Final Disposition: Scrapped.

10996 / 52-6025 / C-119G-15-FA

Avble: 30Jul53; Acc: 24Aug53; Del: 02Sep53 to MDAP Italy, AMI s/n: **MM52-6025**; diverted MIDAR (AMC) Olmsted 53-22Jan54 mnt – Final Disposition: Scrapped.

10997 / 52-6026 / C-119G-15-FA

Avble: 31Jul53; Acc: 24Aug53; Del: 31Aug53 to MDAP Belgium, BAF s/n: **CP-27**; BOC: 18Jan54 15 WG / 40 SQ Melsbroek AB Belgium, callsign: **OT-CBG**, later to 15 WG / 20 SQ.

Minor technical incident 08May56 – Wfu 02Jul73 and stored at Koksijde AB Belgium; SOC: Jul74 – Sold for scrap to J. Trappeniers 04Oct74, scrapped 1977-78 – Final Disposition: Scrapped.

10998 / 52-6027 / C-119G-15-FA

Avble: 03Aug53; Acc: 27Aug53; Del: 04Sep53 to MDAP Belgium, BAF s/n: **CP-30**; diverted MIDAR (AMC) Olmsted 04Sep53-13Jan54 for pilot training; BOC: 27Jan54 15 WG / 40 SQ Melsbroek AB Belgium, callsign: **OT-CBJ**, later to 15 WG / 20 SQ. Minor ground incidents 16Dec57 and 01Feb61. Minor in-flight incident 17Nov67 – Wfu 31Jan72 and stored at Koksijde AB Belgium; SOC: Jul74 – Sold for scrap to J. Trappeniers 04Oct74, scrapped 1977-78 – Final Disposition: Scrapped.

10999 / 52-5840 / C-119G-1-FA

Avble: 01Aug53; Acc: 28Aug53; Del: 02Sep53 to USAF – 464 TCWG (TAC) Lawson 02Sep53; to Pope 13Sep54; depl North Aux AF 06Jun55; depl Aberdeen 05May56 – 2466 RFC (CNR) Bakalar 18Jun57; depl 2469 RFC (CNR) Scott 31Jul57 – 2347 RFC (CNR) Long Beach 18Jun58; to 1607 TSH WG (MATS) Dover 13Oct58 mnt – 452 TCWG (AFR) Long Beach 13Jun59; to March 01Oct60; 944 TCG (452 TCWG) (AFR) March 63 – 933 TAL GP (440 TCWG) (AFR) Gen Mitchell 27Jan68 – Wfu to MASDC (LOG) Davis-Monthan 28Nov70 for storage, PCN: **CJ370** – DFI: 21May71 – Kolar, Inc. Tucson Arizona 01Mar76 for scrap – Final Disposition: Scrapped.

11000 / 52-5841 / C-119G-1-FA

Avble: 05Aug53; Acc: 28Aug53; Del: 02Sep53 to USAF – 464 TCWG (TAC) Lawson 02Sep53; depl Seymour 12Apr54; to Pope 07Sep54; depl North Aux AF 15Jun55; depl Aberdeen 05May56 – Hayes (AMC) Birmingham 05Jul57 – 2465 RFC (CNR) Minneapolis-St. Paul 19Oct57 – 2473 RFC DT (CNR) Minneapolis-St. Paul 01Dec57 – 96 TCSq (440 TCWG) (AFR) Minneapolis-St. Paul 19Dec59; 934 TCG (440 TCWG) (AFR) Minneapolis-St. Paul 11Feb63 – Orchid (LOG) St. Augustine 03Jun63 – MAP India 16Jul63 (USAF list as ground accident) – Final Disposition: Unknown.

11001 / 52-5842 / C-119G-1-FA

Avble: 06Aug53; Acc: 28Aug53; Del: 02Sep53 to USAF – 464 TCWG (TAC) Lawson 02Sep53; depl Seymour 12Apr54 – 483 TCWG (FEA) Ashiya AB Japan 03Sep54; depls 316 TCWG (FEA) Ashiya AB Japan 01May55 & 01Jul55 – 2256 RFC (CNR) Niagara Falls 28Apr57 – 2242 RFC DT (CNR) Niagara Falls 01Dec57 – 328 TCSq (512 TCWG) (AFR) Niagara Falls 19Dec58; 914 TCG (512 TCWG) (AFR) Niagara Falls 11Feb63 – Fairchild (LOG) St. Augustine 63 – MAP India 06Jun63 – Final Disposition: Unknown.

11002 / 52-5843 / C-119G-1-FA

Avble: 10Aug53; Acc: 31Aug53; Del: 03Sep53 to USAF – 313 TCWG (TAC) Mitchel 03Sep53; depl 465 TCWG (TAC) Mitchel Oct53; to Sewart 04Dec53; depl Pope 27Apr54; depl Sumpter Alabama 16Aug54 – 483 TCWG (FEA) Ashiya AB Japan 20Sep54; depl 316 TCG (FEA) Ashiya AB Japan Oct54; bailment CAT (AMC) Taiwan 15Dec55; depl Albrook Panama CZ 57 – 2237 RFC (CNR) New Castle 02May57; to NAS Willow Grove 21Jul58 – 512 TCWG (AFR) NAS Willow Grove 01Dec58; 913 TCG (512 TCWG) (AFR) NAS Willow Grove 11Feb63 – Fairchild (LOG) St. Augustine 04May63 – MAP India 06Jun63 – Final Disposition: Unknown.

11003 / 52-5844 / C-119G-1-FA

Avble: 11Aug53; Acc: 31Aug53; Del: 05Sep53 to USAF – 313 TCWG (TAC) Mitchel 04Sep53; depl 4415 AB GP (TAC) Pope 29Sep53; to Sewart 05Nov53; depl Sumpter Alabama 19Jul54 – 314 TCWG (TAC) Sewart 55; depl England Louisiana 05Nov56 – BMC (AMC) Birmingham 05Mar57 – 2471 RFC (CNR) O'Hare 24May57; to 3345 TTA WG (TC) Chanute 18Jul57 mnt – 2473 RFC DT (CNR) O'Hare 01Dec57 **RECORDS INDECIPHERABLE** 2242 RFC DT (CNR) O'Hare 58 – 64 TCSq (403 TCWG) (AFR) O'Hare 20Dec58; 928 TCG (403 TCWG) (AFR) O'Hare 31May63 – 907 TAL GP (302 TCWG) (AFR) Clinton 12Dec69 – MAP Taiwan 27Jan70, RoCAF s/n: **3184** – Presently displayed at Junshi Park Jiji City Taiwan – Final Disposition: Extant.

11004 / 52-5845 / C-119G-1-FA

Avble: 13Aug53; Acc: 31Aug53; Del: 03Sep53 to USAF – 64 TCWG (TAC) Donaldson 03Sep53; depl Seymour 23Apr54 – 313 TCWG (TAC) Sumpter (depl) 04Aug54; to Sewart 19Aug54; depl Elmendorf 19Feb55 – 314 TCWG (TAC) Sewart 55; depl England Louisiana 05Nov56 – BMC (AMC) Birmingham 01Mar57 – 2471 RFC (CNR) O'Hare 21May57; depl 2466 RFC (CNR) Bakalar 21Jul57; cvtd **C-119J** 22Jul57 – 2473 RFC DT (CNR) O'Hare 01Dec57 **RECORDS INDECIPHERABLE** 2242 RFC DT (CNR) O'Hare date unk – 102 AML SQ (NY-ANG) New York ANGB 23Sep58 **RECORDS INDECIPHERABLE** Wfu to 2704 ASD GP (LOG) Davis-Monthan 11May63 for storage – Fairchild (LOG) St. Augustine 16Dec66 – 4603 AB GP (ADC) Stewart 05Feb67 – 4676 AB GP (ADC) Richards 26Nov69 – 4650 COS SQ (ADC) Richards 30Jun70 – Wfu to MASDC (LOG) Davis-Monthan 14Jul72 for storage, PCN: **CJ435** – DFI: 09Dec74 – Final Disposition: Scrapped.

11005 / 52-5846 / C-119G-1-FA

Avble: 17Aug53; Acc: 31Aug53; Del: 03Sep53 to USAF – 64 TCWG (TAC) Donaldson 03Sep53; depl Seymour 23Apr54 – Fairchild (AMC) Hagerstown 05Aug54 – 483 TCWG (FEA) Ashiya AB Japan 13Sep54; depl 314 TCWG (FEA) Ashiya AB Japan Nov54; depl 316 TCWG (FEA) Ashiya AB Japan Nov54; bailment CAT (AMC) Taiwan 19Oct55 – 2465 RFC (CNR) Minneapolis-St. Paul 25Apr57; depl 2466 RFC (CNR) Bakalar 07Aug57 – 2473 RFC DT (CNR) Minneapolis-St. Paul 01Dec57; depl 2465 RFC (CNR) Minneapolis-St. Paul 24May58 – Hayes (AMC) Birmingham 19Nov58 – 96 TCSq (440 TCWG) (AFR) Minneapolis-St. Paul 12Feb59; 934 TCG (440 TCWG) (AFR) Minneapolis-St. Paul 11Feb63; depl 4413 CCT SQ (1 ACO WG) (TAC) Lockbourne 10May68 – Fairchild (LOG) St. Augustine 09Dec68 – 930 SOP GP (434 TCWG) (AFR) Bakalar 18Nov69; to Grissom 12Jan70 – 907 SOP GP (302 TCWG) (AFR) Clinton 22Jul70; to Lockbourne 24Jul71 – Wfu 13Sep71 to museum / school status. Loan to the Michigan Military Museum Saginaw Michigan 1971–1975 – BofS Jack R. Munson Muncie Indiana 14Apr75 – BofS Hawkins & Powers Aviation, Inc. Greybull Wyoming 20May75; reg **N48076** 23May75; cvtd **C-119G-3E** 11Feb76 with chemical tank, radio upgrade for aerial firefighting, Tanker No. **135**. Suffered a catastrophic structural failure while on a fire-bombing run in Shasta-Trinity National Forest California 16Sep87. The crew made a successful retardant drop but were reporting trouble maintaining airspeed. Soon after the right main wing outside of the right boom separated from the airframe followed by both booms. The main fuselage with both engines and most of the left wing spiraled into the ground killing pilot Bill Berg (48yrs), co-pilot Charles Peterson (28yrs) and mechanic Stephen Harrell (26yrs). A resulting fire was suppressed by ground workers. This crash resulted in the C-119 Flying Boxcar being withdrawn from use as a fire-bomber by the U.S. Forest Service – N48076 de-reg 20Nov89 – Final Disposition: Accident.

11006 / 52-5847 / C-119G-1-FA

Avble: 20Aug53; Acc: 09Sep53; Del: 15Sep53 to USAF – **RECORDS MISSING 1953–1955** – 483 TCWG (FEA) Ashiya AB Japan date unk; to 21 TCSq (483 TCWG) (FEA) Tachikawa AB Japan 16Oct56 – 2256 RFC (CNR) Niagara Falls 01Apr57 – 2242 RFC DT (CNR) Niagara Falls 01Dec57 – 2237 RFC DT (CNR) Niagara Falls 58 – 328 TCSq (512 TCWG) (AFR) Niagara Falls 19Dec58; 914 TCG (512 TCWG) (AFR) Niagara Falls 11Feb63 – Wfu to MASDC (LOG) Davis-Monthan 30Nov70 for storage, PCN: **CJ376** – DFI: 21May71 – Kolar, Inc. Tucson Arizona 01Mar76 for scrap – Final Disposition: Scrapped.

11007 / 52-5848 / C-119G-1-FA

Avble: 20Aug53; Acc: 09Sep53; Del: 15Sep53 to USAF – 64 TCWG (TAC) Donaldson 15Sep53; depl Seymour 23Apr54 – 464 TCWG (TAC) Lawson 30Jul54; to Pope 01Sep54; depl Aberdeen 05May56 – 2346 RFC DT (CNR) McClellan 08Sep58; to 1501 TSH WG (MATS) Travis 20Oct58 mnt – 314 TCSq (349 TCWG) (AFR) McClellan 23Dec58; 940 TCG (349 TCWG) (AFR) McClellan 10Feb63 – 924 TCG (446 TCWG) (AFR) Ellington 18Aug65; 925 TAL GP (446 TCWG) (AFR) Ellington 20Nov67 – 906 TAL GP (302 TCWG) (AFR) Clinton 20Jan68 – MAP Taiwan 03Jan70, RoCAF s/n: **3191** – Final Disposition: Unknown.

11008 / 52-5849 / C-119G-1-FA

Avble: 21Aug53; Acc: 11Sep53; Del: 17Oct53 to USAF – 64 TCWG (TAC) Donaldson 17Oct53; depl Seymour Apr54 – 4501 HQS SQ (TAC) Donaldson 05Aug54 – 464 TCWG (TAC) Pope 13Apr55; depl North Aux AF 03Jun55 – BMC (AMC) Birmingham Jul56 – 363 TR WG (TAC) Shaw 23Oct56; cvtd **C-119J** 02Aug57 – Hayes (AMC) Birmingham 17Sep57 – 2242 RFC (CNR) Selfridge 04Dec57 – 102 AML SQ (NY-ANG) New York ANGB 24Sep58 – Wfu to 2704 ASD GP (LOG) Davis-Monthan 14Jan63 for storage – MAP Italy 20Jan64, AMI s/n: **MM52-5849** – Final Disposition: Scrapped.

11009 / 52-5850 / C-119G-1-FA

Avble: 24Aug53; Acc: 09Sep53; Del: 15Sep53 to USAF – 64 TCWG (TAC) Donaldson 15Sep53; depl Seymour 23Apr54 – 313 TCWG (TAC) Sumpter (depl) 06Aug54; to Sewart 19Aug54; depl 464 TCWG (TAC) Pope 13Sep54; depl Elmendorf 12Jan55 – 463 TCWG (TAC) Sewart 08Jun55 – 314 TCWG (TAC) Sewart 10Aug56; depl England Louisiana 05Nov56 – BMC (AMC) Birmingham 14Mar57 – 2469 RFC (CNR) Scott 07Jun57; depls 2466 RFC (CNR) Bakalar 07Jul57 & 18Aug57 – 2466 RFC DT (CNR) Scott 01Dec57 – 73 TCSq (434 TCWG) (AFR) Scott 18Apr59; 932 TCG (434 TCWG) (AFR) Scott 11Feb63 – 934 TCG (440 TCWG) (AFR) Minneapolis-St. Paul 27Jan67; 933 TAL GP (440 TCWG) (AFR) Gen Mitchell 30Jan70 – 928 TAL GP (403 TCWG) (AFR) O'Hare 25Mar70 – Wfu to 305 ARH WG (SAC) Grissom 19Jan71 to museum / school status. Presently displayed at Grissom Air Museum Grissom AFB Indiana – Final Disposition: Extant.

11010 / 52-5851 / C-119G-1-FA

Avble: 25Aug53; Acc: 09Sep53; Del: 15Sep53 to USAF – 64 TCWG (TAC) Donaldson 15Sep53 – MIDAR (AMC) Olmsted 06Jul54 mnt – 464 TCWG (TAC) Pope 22Sep54 – BMC (AMC) Birmingham 15Oct56 – 323 Fighter WG (TAC) Bunker Hill 03Jan57 – 2230 RFC (CNR) NAS New York 24Jul57; cvtd **C-119J** 01Aug57 – 2234 RFC (CNR) Hanscom 09Oct57 – 2237 RFC (CNR) New Castle Mar58; to NAS Willow Grove 21Jul58; depl 2234 RFC (CNR) Hanscom 10Aug58 – 512 TCWG (AFR) NAS Willow Grove 01Dec58 **RECORDS INDECIPHERABLE** 78 TCSq (435 TCWG) (AFR) Bates date unk – 187 AML SQ

(WY-ANG) Cheyenne 06Mar61 – 147 AML SQ (PA-ANG) Greater Pittsburgh 11May61 – Wfu to 2704 ASD GP (LOG) Davis-Monthan 19Nov62 for storage – Fairchild (LOG) St. Augustine 05Dec63 – MAP Italy 27Feb64 (USAF list as ground accident), AMI s/n: **MM52-5851** – Final Disposition: Scrapped.

11011 / 52-5852 / C-119G-1-FA

Avble: 26Aug53; Acc: 09Sep53; Del: 15Sep53 to USAF – 64 TCWG (TAC) Donaldson 15Sep53 – 464 TCWG (TAC) Lawson 07Jul54 – Fairchild (AMC) Hagerstown 10Aug54 – 483 TCWG (FEA) Ashiya AB Japan 07Sep54; depl 316 TCWG (FEA) Ashiya AB Japan Dec54; bailment CAT (AMC) Taiwan 27Nov55 – 2235 RFC (CNR) Grenier 21May57; to 6486 AB GP (FEA) Hickham 26May57 mnt – Hayes (AMC) Birmingham 29Dec58 – 732 TCSq (94 TCWG) (AFR) Grenier 08Apr59; 902 TCG (94 TCWG) (AFR) Grenier 11Feb63 – Fairchild (AMC) St. Augustine 06May63 – MAP India 26Jun63 – Final Disposition: Unknown.

11012 / 52-5853 / C-119G-1-FA

Avble: 26Aug53; Acc: 10Sep53; Del: 16Sep53 to USAF – 464 TCWG (TAC) Lawson 16Sep53. Destroyed on the ground when a tornado struck Lawson AFB 13Mar54, 5 other C-119s were also destroyed – DFI: 14Apr54 – Final Disposition: Destroyed.

11013 / 52-5854 / C-119G-1-FA

Avble: 27Aug53; Acc: 10Sep53; Del: 16Sep53 to USAF – 464 TCWG (TAC) Lawson 16Sep53. Destroyed on the ground when a tornado struck Lawson AFB 13Mar54, 5 other C-119s were also destroyed – DFI: 14Apr54 – Final Disposition: Destroyed.

11014 / 52-5855 / C-119G-1-FA

Avble: 28Aug53; Acc: 14Sep54; Del: 17Sep54 to USAF – 464 TCWG (TAC) Lawson 17Sep53; depl Seymour 16Apr54; to Pope 01Sep54; depl Aberdeen 05May56 – 2242 RFC DT (CNR) O'Hare 27Sep58 – 64 TCSq (403 TCWG) (AFR) O'Hare 20Dec58 **RECORDS INDECIPHERABLE** 4440 ADE GP (TAC) Langley 27Jun60 – MAP India 10Jul60 – Final Disposition: Unknown.

11015 / 52-5856 / C-119G-1-FA

Avble: 31Aug53; Acc: 18Sep53; Del: 17Oct53 to USAF – 64 TCWG (TAC) Donaldson 17Oct53; depl Seymour 12Apr54 – 464 TCWG (TAC) Lawson 30Jul54; to Pope 07Sep54; depl North Aux AF 06Jun55; depl Aberdeen 05May56; depl 2234 RFC (CNR) Hanscom 25Nov57 – 2585 RFC DT (CNR) Donaldson 10Oct58 – 77 TCSq (435 TCWG) (AFR) Donaldson 19Dec58 – 73 TCSq (434 TCWG) (AFR) Scott 10Jan61; 932 TCG (434 TCWG) (AFR) Scott 11Feb63 – 922 TCG (433 TCWG) (AFR) Kelly 30Jan67 – 906 SOP GP (302 TCWG) (AFR) Clinton 06Nov70 – MAP Taiwan 04Jan71 – Final Disposition: Unknown.

11016 / 52-5857 / C-119G-1-FA

Avble: 01Sep53; Acc: 15Sep53; Del: 21Oct53 to USAF – 64 TCWG (TAC) Donaldson 21Oct53; depl Seymour 12Apr54 – Fairchild (AMC) Hagerstown 12Aug54 – 483 TCWG (FEA) Ashiya AB Japan 09Sep54; depl 314 TCWG (FEA) Ashiya AB Japan 05Oct54; depls 316 TCWG (FEA) Ashiya AB Japan Nov54 & 04May55; bailment CAT (FEA) Taiwan 27Jun55; depl Akddoae? 07Sep56 – 2237 RFC (CNR) New Castle 03Apr57; to NAS Willow Grove 21Jul58 – 512 TCWG (AFR) NAS Willow Grove 01Dec58; 913 TCG (512 TCWG) (AFR) NAS Willow Grove 11Feb63 – Fairchild (LOG) St. Augustine 04May63 – MAP India 19Jun63 – Final Disposition: Unknown.

11017 / 52-5858 / C-119G-1-FA

Avble: 02Sep53; Acc: 16Sep53; Del: 17Oct53 to USAF – **RECORDS MISSING 1953–1955** – 464 TCWG (TAC) Pope date unk; to WRDCN (ARD) WPAFB 25Jul55 mnt; depl Aberdeen 05May56; to PGC Center (ARD) Eglin 58 – Hayes (AMC) Birmingham 04Dec58 – 2234 RFC (RFC) Hanscom 16Mar59 – 94 TCWG (AFR) Hanscom 19Mar59 – Fairchild (LOG) St. Augustine 12Dec62 – MAP India 25Jun63 – Final Disposition: Unknown.

11018 / 52-5859 / C-119G-1-FA

Avble: 02Sep53; Acc: 15Sep53; Del: 21Oct53 to USAF – 64 TCWG (TAC) Donaldson 21Oct53; depl Seymour 12Apr54 – 313 TCWG (TAC) Sumpter (depl) 06Aug54 – 464 TCWG (TAC) Pope 17Sep54. Crashed after take-off from Fort Bragg North Carolina 06Oct54 due to a catastrophic engine fire. 2 of the crew were killed – 1 Lt W.L. Wyatt (pilot) and 1 Lt F.N. Fulton (co-pilot); the other 3 crew and 7 army passengers survived with injuries. The C-119 came down on a barracks construction site injuring 2 workers – DFI: 08Nov54 – Final Disposition: Accident.

11019 / 52-5860 / C-119G-1-FA

Avble: 04Sep53; Acc: 21Sep53; Del: 17Oct53 to USAF – 64 TCWG (TAC) Donaldson 17Oct53; depl Seymour 12Apr54 – 313 TCWG (TAC) Sumpter (depl) 06Aug54; to Sewart 19Aug54; depl Elmendorf 12Jan55 – 463 TCWG (TAC) Sewart 08Jun55 – 314 TCWG (TAC) Sewart 06Dec55; depl England Louisiana 05Nov56 – BMC (AMC) Birmingham 18Mar57 – 2471 RFC (CNR) O'Hare 06Jun57; depl 2466 RFC (CNR) Bakalar 21Jul57 – 2473 RFC DT (CNR) O'Hare 01Dec57 – 328 CLM SQ (ADC) Richards 14Feb58 mnt – 2242 RFC DT (CNR) O'Hare 11Apr58 – Hayes (AMC) Birmingham 19Dec58 – 64 TCSq (403 TCWG) (AFR) O'Hare 06Mar59; briefly to Selfridge 01Apr60; 928 TCG (403 TCWG) (AFR) O'Hare 11Feb63 – Fairchild (LOG) St. Augustine 17May63 – MAP India 01Jul63 – Final Disposition: Unknown.

11020 / 52-5861 / C-119G-1-FA

Avble: 04Sep53; Acc: 18Sep53; Del: 17Oct53 to USAF – 64 TCWG (TAC) Donaldson 17Oct53; depl Seymour 12Apr54 – Fairchild (AMC) Hagerstown 16Aug54 – 483 TCWG (FEA) Ashiya AB Japan 17Sep54; depl 316 TCWG (FEA) Ashiya AB Japan 27Feb55 – 2471 RFC (CNR) O'Hare 16May57; depl 2466 RFC (CNR) Bakalar 21Jul57 – 2473 RFC DT (CNR) O'Hare 01Dec57 **RECORDS INDECIPHERABLE** 2242 RFC DT (CNR) O'Hare 58 – 64 TCSq (403 TCWG) (AFR) O'Hare 20Dec58; 928 TCG (403 TCWG) (AFR) O'Hare 11Feb63 – 907 TAL GP (302 TCWG) (AFR) Clinton 12Dec69 – MAP Taiwan 28Jan70 – Final Disposition: Unknown.

11021 / 52-5862 / C-119G-1-FA

Avble: 08Sep53; Acc: 23Sep53; Del: 17Oct53 to USAF – 64 TCWG (TAC) Donaldson 17Oct53; depl Seymour 10Apr54 – Fairchild (AMC) Hagerstown 10Aug54 – 483 TCWG (FEA) Ashiya AB Japan 20Sep54; depl 316 TCWG (FEA) Ashiya AB Japan 01Jan55; bailment CAT (AMC) Taiwan 19Mar55 **RECORDS INDECIPHERABLE** 2256 RFC (CNR) Niagara Falls 26Apr57; depl 2230 RFC (CNR) NAS New York 25May57 – 2242 RFC DT (CNR) Niagara Falls 01Dec57 **RECORDS INDECIPHERABLE** 328 TCSq (512 TCWG) (AFR) Niagara Falls 19Dec58; 914 TCG (512 TCWG) (AFR) Niagara Falls 11Feb63 – Wfu to MASDC (LOG) Davis-Monthan 23Feb71 for storage, PCN: **CJ406** – DFI: 09Dec74 – Final Disposition: Scrapped.

11022 / 52-5863 / C-119G-1-FA

Avble: 09Sep53; Acc: 22Sep53; Del: 17Oct53 to USAF – 64 TCWG (TAC) Donaldson 17Oct53; depl Seymour 12Apr54 – 313 TCWG (TAC) Sumpter (depl) 05Aug54; to Sewart 19Aug54 – BMC (AMC) Birmingham 03May55 – 464 TCWG (TAC) Sewart 27Jun55 – 314 TCWG (TAC) Sewart 09Aug56; depl England Louisiana 05Nov56 – 2472 RFC (CNR) Richards 04Jul57 – 442 TCWG (AFR) Richards 02Mar59 – 512 TCWG (AFR) NAS Willow Grove 27Apr61; 912 TCG (512 TCWG) (AFR) NAS Willow Grove 11Feb63 – 931 TAL GP (434 TCWG) (AFR) Bakalar 01Jul68; 930 TAL GP (434 TCWG) (AFR) Bakalar 30Jun69; to Grissom 12Jan70 – 906 TAL GP (302 TCWG) (AFR) Clinton 24Jul70 – Wfu to MASDC (LOG) Davis-Monthan 22Sep71 for storage, PCN: **CJ412** – Fairchild (LOG) Crestview cvtd **C-119K** 28Oct71 – 907 TAL GP (302 TCWG) (AFR) Lockbourne 05May72 – MAP Jordan 23Sep72, RJAF s/n: **115**, nfd – Final Disposition: Unknown.

11023 / 52-5864 / C-119G-1-FA

Avble: 10Sep53; Acc: 23Sep53; Del: 21Oct53 to USAF – 64 TCWG (TAC) Donaldson 21Oct53; depl Seymour 10Apr54 – 313 TCWG (TAC) Sumpter (depl) 05Aug54; to Sewart 19Aug54 – 314 TCWG (TAC) Sewart 08Jun55; depl England Louisiana 05Nov56 – BMC (AMC) Birmingham 20Mar57 – 2469 RFC (CNR) Scott 12Jun57; depl 2466 RFC (CNR) Bakalar 28Jul57 – 2466 RFC DT (CNR) Scott 01Dec57 – 73 TCSq (434 TCWG) (AFR) Scott 18Apr59; 932 TCG (434 TCWG) (AFR) Scott 11Feb63 – 927 TCG (403 TCWG) (AFR) Selfridge 31May67 – Fairchild (LOG) St. Augustine 09Sep68, cvtd **AC-119K Stinger** 17Sep68 – 1 SOP WG (TAC) Lockbourne 26May69 – 4410 CCT WG (TAC) Lockbourne 14Jul69 – 14 SOP WG (PAF) Phan Rang AB S. Vietnam 27Dec69; assigned variously Phu Cat, Da Nang, Tan Son Nhut S. Vietnam & Nakhon Phanom, Udorn Thailand from 30Jan70; to 313 ADH DV (PAF) Kadena AB Japan 18Oct70; to 366 Fighter WG (PAF) Da Nang S. Vietnam 03Jun71 – 56 SOP WG (PAF) Da Nang AB S. Vietnam 24Aug71; multiple assignments to Nakhon Phanom Thailand from 03Sep71 – MAP S. Vietnam 10Nov72, VNAF s/n: **525864**, 821 ASQ (53 TWG) Tan Son Nhut AB – Final Disposition: Unknown.

11024 / 52-5865 / C-119G-1-FA

Avble: 11Sep53; Acc: 23Sep53; Del: 01Oct53 to USAF – **RECORDS MISSING 1953–1955** – bailment General Motors Oct53 – bailment Aeroproducts Division (AMC) Vandalia date unk – bailment Bellanca Aircraft Corp. (AMC) New Castle date unk – 464 TCWG (TAC) Pope 01Nov55; depl Aberdeen 05May56 – 512 TCWG (AFR) NAS Willow Grove 13Dec58 – 4440 ADE GP (TAC) Langley 27Jun60 – MAP India 10Jul60 – Final Disposition: Unknown.

11025 / 52-5866 / C-119G-1-FA

Avble: 11Sep53; Acc: 30Sep53; Del: 23Oct53 to USAF – 64 TCWG (TAC) Donaldson 23Oct53; depl Seymour 12Apr54 – 464 TCWG (TAC) Lawson 04Aug54; to Pope 01Sep54 – 17 BTA WG (TAC) Eglin 02Nov56; cvtd **C-119J** 01Aug57 **RECORDS INDECIPHERABLE** → 150 AML SQ (NJ-ANG) Newark Jun58 ← **RECORDS INDECIPHERABLE** Wfu to 2704 ASD GP (LOG) Davis-Monthan 17Jan63 for storage – Fairchild (LOG) St. Augustine 17Dec63 – MAP Italy 27Feb64, AMI s/n: **MM52-5866** – Final Disposition: Scrapped.

11026 / 52-5867 / C-119G-1-FA

Avble: 14Sep53; Acc: 30Sep53; Del: 23Oct53 to USAF – 64 TCWG (TAC) Donaldson 23Oct53; depl Seymour 12Apr54 – 313 TCWG (TAC) Sumpter (depl) 05Aug54; to Sewart 19Aug54; depl Elmendorf 12Jan55 – 464 TCWG (TAC) Sewart 08Jun55 – 314 TCWG (TAC)

Sewart 11Aug56; depl England Louisiana 05Nov56 – BMC (AMC) Birmingham 01Apr57 – 2472 RFC (CNR) Richards 25Jun57 – 442 TCWG (AFR) Richards 02Mar59 – 512 TCWG (AFR) NAS Willow Grove 18May61 – Fairchild (LOG) St. Augustine 06Jun63 – MAP India 07Jun63 – Final Disposition: Unknown.

11027 / 52-5868 / C-119G-1-FA

Avble: 14Sep53; Acc: 30Sep53; Del: 21Oct53 to USAF – 64 TCWG (TAC) Donaldson 21Oct53; depl Seymour 12Apr54 – 464 TCWG (TAC) Lawson 06Aug54 – Fairchild (AMC) Hagerstown 16Aug54 – 483 TCWG (FEA) Ashiya AB Japan 23Sep54; depl 316 TCWG (FEA) Ashiya AB Japan 01Aug55 – 2465 RFC (CNR) Minneapolis-St. Paul 26Apr57 – 2466 RFC (CNR) Bakalar 04Aug57 – 2473 RFC DT (CNR) Minneapolis-St. Paul 01Dec57 – 2472 RFC (CNR) Richards 05Mar58; 2472 RFC DT (CNR) Davis 14Mar58 – 65 TCSq (403 TCWG) (AFR) Davis 18Sep58; 929 TCG (403 TCWG) (AFR) Davis 11Feb63 – 922 TCG (433 TCWG) (AFR) Kelly 18Oct65 – Wfu to MASDC (LOG) Davis-Monthan 26Jan71 for storage, PCN: **CJ388** – DFI: 09Dec74 – Final Disposition: Scrapped.

11028 / 52-5869 / C-119G-1-FA

Avble: 15Sep53; Acc: 30Sep53; Del: 21Oct53 to USAF – 64 TCWG (TAC) Donaldson 21Oct53; depl Seymour 24Apr54 – 464 TCWG (TAC) Lawson 06Aug54; to Pope 07Sep54; depl North Aux AF 06Jun55; depl Aberdeen 05May56 – Hayes (AMC) Birmingham 04Nov58 – 314 TCSq (349 TCWG) (AFR) McClellan 13Jan59; 940 TCG (349 TCWG) (AFR) McClellan 10Feb63 – 924 TCG (446 TCWG) (AFR) Ellington 18Aug65 – 910 TAL GP (459 TWG) (AFR) Youngstown 20Jan68 – 930 SOP GP (434 TCWG) (AFR) Bakalar 18Oct69; to Grissom 12Jan70 – 906 TAL GP (302 TCWG) (AFR) Clinton 22Jul70 – MAP Taiwan 28Nov70, RoCAF s/n: **3214** – Final Disposition: Unknown.

11029 / 52-6028 / C-119G-15-FA

Avble: 16Sep53; Acc: 30Sep53; Del: 29Dec53 to MDAP Belgium, BAF s/n: **CP-26**; BOC: 11Jan54 15 WG / 40 SQ Melsbroek AB Belgium, callsign: **OT-CBF**, later to 15 WG / 20 SQ. Near in-flight collision with F-84 03Jan55. Minor ground incident 03Jan57. Smoking right engine due to oil leak 23Aug57 – Wfu 01Sep72 and stored at Koksijde AB Belgium; SOC: Jul74 – Sold for scrap to J. Trappeniers 04Oct74, scrapped 1977-78 – Final Disposition: Scrapped.

11030 / 52-6029 / C-119G-15-FA

Avble: 17Sep53; Acc: 30Sep53; Del: 29Dec53 to MDAP Italy, AMI s/n: **MM52-6029** – Stored and displayed at the Friulano Aero Club Campoformido Italy after being flown in 06Dec78, airframe has deteriorated over subsequent years – Dismantled and transported to the Museo Storico dell'Aeronauctica Militare (Italian Air Force Museum) Vigna di Valle Italy Nov04 for restoration – Final Disposition: Extant.

11031 / 52-6030 / C-119G-15-FA

Avble: 18Sep53; Acc: 30Sep53; Del: 29Dec53 to MDAP Italy, AMI s/n: **MM52-6030**. Damaged beyond repair following a hard landing at Pisa-Arturo dell'Oro Air Base Italy 24Jan79, no fatalities – Final Disposition: Accident.

11032 / 52-6031 / C-119G-15-FA

Avble: 18Sep53; Acc: 30Sep53; Del: 29Dec53 to MDAP Italy, AMI s/n: **MM52-6031**; cvtd **VC-119G** 1960; cvtd **EC-119G** 1975 – Final Disposition: Scrapped.

11033 / 52-6032 / C-119G-15-FA

Avble: 21Sep53; Acc: 23Oct53; Del: 23Jan54 to MDAP Belgium, BAF s/n: **CP-33**; BOC: 14Feb54 15 WG / 40 SQ Melsbroek AB Belgium, callsign: **OT-CBM**, later to 15 WG / 20 SQ – Wfu 18Nov71 and stored at Koksijde AB Belgium; SOC: Jul74 – Sold for scrap to J. Trappeniers 04Oct74, scrapped 1977-78 – Final Disposition: Scrapped.

11034 / 52-6033 / C-119G-15-FA

Avble: 22Sep53; Acc: 28Oct53; Del: 23Jan54 to MDAP Belgium, BAF s/n: **CP-20**; BOC: 23Mar54 15 WG / 20 SQ Melsbroek AB Belgium, callsign: **OT-CAT**. Minor collision with s/n: CP-15 17Sep54 while being towed at Melsbroek. Nose damaged by towing truck towbar 01Jan56. 5 passengers injured after flying through severe turbulence 15Oct66 – Wfu 13Dec71 and stored at Koksijde AB Belgium; SOC: Jul74 – Sold for scrap to J. Trappeniers 04Oct74, scrapped 1977-78 – Final Disposition: Scrapped.

11035 / 52-6034 / C-119G-15-FA

Avble: 23Sep53; Acc: 27Oct53; Del: 26Jan54 to MDAP Belgium, BAF s/n: **CP-19**; BOC: 26Mar54 15 WG / 20 SQ Melsbroek AB Belgium, callsign: **OT-CAS**. Hit by a truck while on a taxi apron 03Aug55 requiring replacement of clamshell doors. Right wing-tip damaged 23May56 when it hit a power pole while taxiing. Minor ground collision with DC-3 s/n: K-21 16Oct56 while aircraft under tow. Flew into a hill during night operations 22Oct65 near Hofgeisnar, West Germany totally destroying the airframe. Pilot: Capt Luc Mommer, 4 other crew and 3 paratroops killed. Final Disposition: Accident.

11036 / 52-6035 / C-119G-15-FA

Avble: 23Sep53; Acc: 27Oct53; Del: 26Jan54 to MDAP Belgium, BAF s/n: **CP-31**; BOC: 09Feb54 15 WG / 40 SQ Melsbroek AB Belgium, callsign: **OT-CBK**, later to 15 WG / 20 SQ. Tail hit by taxiing C-119G s/n: CP-31 01Feb61 – Wfu 29Dec72 and stored at Koksijde AB Belgium; SOC: Jul74 – Sold for scrap to J. Trappeniers 04Oct74, scrapped 1977-78 – Final Disposition: Scrapped.

11037 / 52-5870 / C-119G-1-FA

Avble: 24Sep53; Acc: 28Oct53; Del: 02Nov53 to USAF – 464 TCWG (TAC) Lawson 02Oct53; depl Seymour 12Apr54; to Pope 07Sep54; depl North Aux AF 06Jun55; depl Aberdeen 05May56 – 2472 RFC (CNR) Richards 18Oct58 – 442 TCWG (AFR) Richards 02Mar59 – 64 TCSq (403 TCWG) (AFR) O'Hare 21Apr61; 928 TCG (403 TCWG) (AFR) O'Hare 11Feb63 – 907 TAL GP (302 TCWG) (AFR) Clinton 12Dec69 – MAP Taiwan 27Jan70, RoCAF s/n: **3187** – Final Disposition: Unknown.

11038 / 52-5871 / C-119G-1-FA

Avble: 25Sep53; Acc: 23Oct53; Del: 29Oct53 to USAF – 464 TCWG (TAC) Lawson 29Oct53; depl Seymour 22Apr54 – Fairchild (AMC) Hagerstown 11Aug54 – 483 TCWG (FEA) Ashiya AB Japan 03Sep54; depl 314 TCWG (FEA) Ashiya AB Japan 04Oct54; depl 316 TCWG (FEA) Ashiya AB Japan Nov54; bailment CAT (AMC) Taiwan 04Dec55 – SMAAR (AMC) McClellan 09Jun57 mnt – 2237 RFC (CNR) New Castle 03Jun57; to NAS Willow Grove 21Jul58 – 512 TCWG (AFR) NAS Willow Grove 01Dec58; 912 TCG (512 TCWG) (AFR) NAS Willow Grove 11Feb63; 913 TAL GP (512 TCWG) (AFR) NAS Willow Grove 06May68 – Wfu to MASDC (LOG) Davis-Monthan 14Dec69 for storage, PCN: **CJ343** – DFI: 20Jan70 – Kolar, Inc. Tucson Arizona 17Feb76 for scrap – Final Disposition: Scrapped.

11039 / 52-5872 / C-119G-1-FA

Avble: 25Sep53; Acc: 30Oct53; Del: 04Nov53 to USAF – 464 TCWG (TAC) Lawson 04Nov53; depl Seymour 12Apr54 – Fairchild (AMC) Hagerstown 10Aug54 – 483 TCWG (FEA) Ashiya AB Japan 31Aug54; depl 316 TCWG (FEA) Ashiya AB Japan Nov54 – 2256 RFC (CNR) Niagara Falls 07Apr57; depl 2235 RFC (CNR) Grenier 25Apr57 – 2242 RFC DT (CNR) Niagara Falls 01Dec57 – 2237 RFC DT (CNR) Niagara Falls 14Jan58 – 328 TCSq (512 TCWG) (AFR) Niagara Falls 19Dec58; 914 TCG (512 TCWG) (AFR) Niagara Falls 11Feb63 – Wfu to MASDC (LOG) Davis-Monthan 07Dec70 for storage, PCN: **CJ380** – DFI: 21May71 – Kolar, Inc. Tucson Arizona 05Feb76 for scrap – Final Disposition: Scrapped.

11040 / 52-5873 / C-119G-1-FA

Avble: 28Sep53; Acc: 28Oct53; Del: 02Nov53 to USAF – 464 TCWG (TAC) Lawson 02Nov53; depl Seymour 12Apr54 – Fairchild (AMC) Hagerstown 13Aug54 – 483 TCWG (FEA) Ashiya AB Japan 13Sep54; depl 314 TCWG (FEA) Ashiya AB Japan Oct54; depl 316 TCWG (FEA) Ashiya AB Japan 18Nov54 – 2235 RFC (CNR) Grenier 22May57 – 2237 RFC DT (CNR) Niagara Falls 20Apr58 – Hayes (AMC) Birmingham 30Oct58 – 328 TCSq (512 TCWG) (AFR) Niagara Falls 02Feb59; 914 TCG (512 TCWG) (AFR) Niagara Falls 11Feb63 – Wfu to MASDC (LOG) Davis-Monthan 07Dec70 for storage, PCN: **CJ379** – DFI: 21May71 – Kolar, Inc. Tucson Arizona 03Mar76 for scrap – Final Disposition: Scrapped.

11041 / 52-5874 / C-119G-1-FA

Avble: 29Sep53; Acc: 29Oct53; Del: 03Nov53 to USAF – 464 TCWG (TAC) Lawson 03Nov53 – Fairchild (AMC) Hagerstown 13Aug54 – 483 TCWG (FEA) Ashiya AB Japan 13Sep54; depl 316 TCWG (FEA) Ashiya AB Japan Feb55; depl 8 Fighter WG (FEA) Itazuke AB Japan 27Feb57 – 2465 RFC (CNR) Minneapolis-St. Paul 16May57; to 6486 AB WG (FEA) Hickham 19May57 mnt – 2466 RFC (CNR) Bakalar 04Aug57 – 2473 RFC DT (CNR) Minneapolis-St. Paul 01Dec57 – 96 TCSq (440 TCWG) (AFR) Minneapolis-St. Paul 19Dec59; 934 TCG (440 TCWG) (AFR) Minneapolis-St. Paul 11Feb63; 933 TAL GP (440 TCWG) (AFR) Gen Mitchell 30Jan70 – 913 TAL GP (512 TCWG) (AFR) NAS Willow Grove 25Mar70 – Wfu to MASDC (LOG) Davis-Monthan 06Apr71 for storage, PCN: **CJ411** – DFI: 21May71 – Final Disposition: Scrapped.

11042 / 52-5875 / C-119G-1-FA

Avble: 30Sep53; Acc: 28Oct53; Del: 02Nov53 to USAF – 464 TCWG (TAC) Lawson 02Nov53; depl Seymour 22Apr54; to Pope 13Sep54; depl North Aux AF 06Jun55; depl Aberdeen 05May56 – 363 TR WG (TAC) Shaw 23Oct56; cvtd **C-119J** 02Aug57 **RECORDS INDECIPHERABLE** 145 AML SQ (OH-ANG) Akron-Canton 58; to Clinton 09Jul62 – Fairchild (LOG) St. Augustine 14Dec63 – MAP Italy 21Feb64 as a spares airframe – Final Disposition: Scrapped.

11043 / 52-5876 / C-119G-1-FA

Avble: 01Oct53; Acc: 29Oct53; Del: 05Nov53 to USAF – 64 TCWG (TAC) Donaldson 05Nov53; depl Seymour 12Apr54 – Fairchild (AMC) Hagerstown 12Aug54 – 483 TCWG (FEA) Ashiya AB Japan 07Sep54; depl 314 TCWG (FEA) Ashiya AB Japan Oct54; depl 316 TCWG (FEA) Ashiya AB Japan Nov54; bailment CAT (AMC) Taiwan 17Aug55 – 2465 RFC (CNR) Minneapolis-St. Paul 29Apr57; depl 2472 RFC (CNR) Richards 10May57; depl 2466 RFC (CNR) Bakalar 23Jun57 – 2466 RFC (CNR) Bakalar 04Aug57 – 2473 RFC DT (CNR) Minneapolis-St. Paul 01Dec57; to 1611 TSM WG (MATS) McGuire 15Jul58 mnt – 96 TCSq

(440 TCWG) (AFR) Minneapolis-St. Paul 19Dec58; 934 TCG (440 TCWG) (AFR) Minneapolis-St. Paul 11Feb63 – 913 TAL GP (512 TCWG) (AFR) NAS Willow Grove 07Feb70 – Wfu to MASDC (LOG) Davis-Monthan 08Jul70 for storage, PCN: **CJ347** – DFI: 28Jul70 – Kolar, Inc. Tucson Arizona 20Feb76 for scrap – Final Disposition: Scrapped.

11044 / 52-5877 / C-119G-1-FA

Avble: 02Oct53; Acc: 29Oct53; Del: 03Nov53 to USAF – 64 TCWG (TAC) Donaldson 03Nov53; depl Seymour 10Apr54 – 313 TCWG (TAC) Sumpter (depl) 03Aug54; to Sewart 19Aug54 – 314 TCWG (TAC) Sewart 30Apr55 – 2471 RFC (CNR) O'Hare 28May57; depl 2466 RFC (CNR) Bakalar 21Jul57; cvtd **C-119J** 22Jul57 – 2473 RFC DT (CNR) O'Hare 01Dec57 – 2242 RFC DT (CNR) O'Hare Jun58 – 2472 RFC DT (CNR) Davis 30Jun58 – 65 TCSq (403 TCWG) (AFR) Davis 18Sep58 – 167 Fighter SQ (WV-ANG) Martinsburg 11Mar61 – 150 AML SQ (NJ-ANG) Newark 03May61 – Wfu to 2704 ASD GP (LOG) Davis-Monthan 09Mar64 for storage – 4603 AB GP (ADC) Stewart 16Mar67 – 4676 AB GP (ADC) Richards 21Nov69 – 4650 COS SQ (ADC) Richards 30Jun70; to 4500 AB WG (TAC) Langley 26Feb71 mnt – Wfu to MASDC (LOG) Davis-Monthan 14Jul72 for storage, PCN: **CJ436** – DFI: 09Dec74 – Final Disposition: Scrapped.

11045 / 52-5878 / C-119G-1-FA

Avble: 02Oct53; Acc: 29Oct53; Del: 03Nov53 to USAF – 64 TCWG (TAC) Donaldson 03Nov53; depl Seymour 12Apr54 – 464 TCWG (TAC) Lawson 04Aug54; to Pope 03Nov54; depl North Aux AF 06Jun55; depl Aberdeen 05May56 – 2585 RFC DT (CNR) Donaldson 04Oct58 – 77 TCSq (435 TCWG) (AFR) Donaldson 19Dec58; to 2585 AB SQ (CNC) Miami 26May59 mnt **RECORDS INDECIPHERABLE** 78 TCSq (435 TCWG) (AFR) Bates date unk – 357 TCSq (302 TCWG) (AFR) Bates 08May61 – 908 TCG (435 TCWG) (AFR) Bates 18Mar63 – Fairchild (LOG) St. Augustine 29Apr63 – MAP India 28May63 – Final Disposition: Unknown.

11046 / 52-5879 / C-119G-1-FA

Avble: 05Oct53; Acc: 30Oct53; Del: 04Nov53 to USAF – 464 TCWG (TAC) Lawson 04Nov53. Destroyed on the ground when a tornado struck Lawson AFB 13Mar54, 5 other C-119s were also destroyed – DFI: 14Apr54 – Final Disposition: Destroyed.

11047 / 52-5880 / C-119G-1-FA

Avble: 05Oct53; Acc: 30Oct53; Del: 04Nov53 to USAF – 464 TCWG (TAC) Lawson 04Nov53; depl Seymour 12Apr54; to Pope 07Sep54; to 1604 MAT SQ (MATS) Kindley 12Mar55 mnt; depl Aberdeen 05May56; to 3600 CCT WG (ATC) Luke 09Jun58 mnt – Hayes (AMC) Birmingham 04Nov58 mnt – 2347 RFC(CNR) Long Beach 13Feb59 – 452 TCWG (AFR) Long Beach 13Jun59; to March 01Oct60; 944 TCG (452 TCWG) (AFR) March 11Feb63 – 931 TAL GP (434 TCWG) (AFR) Bakalar 24Jan68; 930 TAL GP (434 TCWG) (AFR) Bakalar 30Jun69 – 907 SOP GP (302 TCWG) (AFR) Clinton 22Jul70 – Wfu to MASDC (LOG) Davis-Monthan 25Sep71 for storage – Fairchild (LOG) Crestview cvtd **C-119K** 25Oct71 – 907 TAL GP (302 TCWG) (AFR) Lockbourne 05May72 – MAP Jordan 23Sep72, RJAF s/n: **116**, nfd – Final Disposition: Unknown.

11048 / 52-5881 / C-119G-1-FA

Avble: 06Oct53; Acc: 30Oct53; Del: 04Nov53 to USAF – 464 TCWG (TAC) Lawson 04Nov53; depl Seymour 12Apr54 – Fairchild (AMC) Hagerstown 13Aug54 – 483 TCWG (FEA) Ashiya AB Japan 13Sep54; depls 316 TCWG (FEA) Ashiya AB Japan Oct54 & Nov54.

Suffered an engine failure over Kagoshima Japan 30Nov55. Unable to feather the propeller the 5 crew bailed out and the empty C-119 came down in a rice field – DFI: 02Feb56 – Final Disposition: Accident.

11049 / 52-5882 / C-119G-1-FA

Avble: 06Oct53; Acc: 30Oct53; Del: 04Nov53 to USAF – 464 TCWG (TAC) Lawson 04Nov53. Destroyed on the ground when a tornado struck Lawson AFB 13Mar54, 5 other C-119s were also destroyed – DFI: 14Apr54 – Final Disposition: Destroyed.

11050 / 52-5883 / C-119G-1-FA

Avble: 08Oct53; Acc: 30Oct53; Del: 05Nov53 to USAF – 464 TCWG (TAC) Lawson 05Nov53 – Fairchild (AMC) Hagerstown 11Aug54 – 483 TCWG (FEA) Ashiya AB Japan 22Sep54; depl 316 TCWG (FEA) Ashiya AB Japan 30Mar55 – 2471 RFC (CNR) O'Hare 31May57; depl 2466 RFC (CNR) Bakalar 21Jul57 – 2473 RFC DT (CNR) O'Hare 01Dec57 – 2242 RFC DT (CNR) O'Hare 01Apr58 – 64 TCSq (403 TCWG) (AFR) O'Hare 20Dec58; 928 TCG (403 TCWG) (AFR) O'Hare 11Feb63 – Wfu to MASDC (LOG) Davis-Monthan 11Dec69 for storage, PCN: **CJ337** – DFI: 20Jan70 – Kolar, Inc. Tucson Arizona 24Feb76 for scrap – Final Disposition: Scrapped.

11051 / 52-5884 / C-119G-1-FA

Avble: 08Oct53; Acc: 30Oct53; Del: 04Nov53 to USAF – 464 TCWG (TAC) Lawson 04Nov53; depl Seymour 12Apr54; to Pope 01Sep54; depl Aberdeen 05May56 – 461 BTA WG (TAC) Blytheville 14Nov56; cvtd **C-119J** 05Aug57 – Hayes (AMC) Birmingham 09Oct57 – 2347 RFC (CNR) Long Beach 58 – 452 TCWG (AFR) Long Beach 13Jun59; to March 01Oct60 – 150 AML SQ (NJ-ANG) Newark 24Mar61 – Wfu to 2704 ASD GP (LOG) Davis-Monthan 09Jan63 for storage – Fairchild (LOG) St. Augustine 20Dec63 – MAP Italy 03Mar64, AMI s/n: **MM52-5884**; cvtd **EC-119J** 1969 – Final Disposition: Scrapped.

11052 / 52-5885 / C-119G-1-FA

Avble: 08Oct53; Acc: 30Oct53; Del: 04Nov53 to USAF – 464 TCWG (TAC) Lawson 04Nov53; depl Seymour 12Apr54 – Fairchild (AMC) Hagerstown 11Aug54 – 483 TCWG (FEA) Ashiya AB Japan 02Sep54; depl 316 TCWG (FEA) Ashiya AB Japan 11Dec54; bailment CAT (AMC) Taiwan 07Feb56. Crashed into the sea for reasons unknown 20Jul56 while on a local flight out of Ashiya AB Japan with the 5 crew onboard all killed – Final Disposition: Accident.

11053 / 52-5886 / C-119G-1-FA

Avble: 12Oct53; Acc: 30Oct53; Del: 05Nov53 to USAF – 464 TCWG (TAC) Lawson 05Nov53; depl Seymour 22Apr54 – Fairchild (AMC) Hagerstown 13Aug54 – 483 TCWG (FEA) Ashiya AB Japan 13Sep54; depl 314 TCWG (FEA) Ashiya AB Japan 09Nov54; depl 316 TCWG (FEA) Ashiya AB Japan 16Dec54. A propeller malfunction resulted in a severe loss of control 18 miles NW Ashiya AB Japan 01Mar55. Of the 7 crew onboard, 5 managed to bail out of the spiraling aircraft, with one of these dying from injuries after being picked up by a Japanese fishing boat – DFI: 29Apr55 – Final Disposition: Accident.

11054 / 52-5887 / C-119G-1-FA

Avble: 12Oct53; Acc: 30Oct53; Del: 05Nov53 to USAF – 464 TCWG (TAC) Lawson 05Nov53; depl Seymour 12Apr54 – Fairchild (AMC) Hagerstown 13Aug54 – 483 TCWG (FEA) Ashiya AB Japan 23Sep54; depl 316 TCWG (FEA) Ashiya AB Japan 13Apr55 – 2585

Section Four—Service Histories

RFC (CNR) Miami 17May57 – OKLAR (AMC) Tinker 30Nov58 – 435 TCWG (AFR) Miami 12Jan59; to unk unit 59; to Homestead 25Jul60; assigned 76 TCSq Homestead 01Oct61; 915 TCG (435 TCWG) (AFR) Homestead 17Jan63 – Fairchild (LOG) St. Augustine 29Apr63 – MAP India 28May63 – Final Disposition: Unknown.

11055 / 52-5888 / C-119G-1-FA

Avble: 13Oct53; Acc: 30Oct53; Del: 05Nov53 to USAF – 464 TCWG (TAC) Lawson 05Nov53; to Pope 18Sep54 – **RECORDS MISSING 1954–1956** – depl Aberdeen 05May56 – 512 TCWG (AFR) NAS Willow Grove 13Dec58; 912 TCG (512 TCWG) (AFR) NAS Willow Grove 11Feb63 – Fairchild (LOG) St. Augustine 04May63 – MAP India 19Jun63 – Final Disposition: Unknown.

11056 / 52-5889 / C-119G-1-FA

Avble: 17Oct53; Acc: 19Nov53; Del: 25Nov53 to USAF – 464 TCWG (TAC) Lawson 25Nov53 – Fairchild (AMC) Hagerstown 10Aug54 – SMAAR (AMC) McClellan 01Jan55 – 483 TCWG (FEA) Ashiya AB Japan 06Aug55; depl 316 TCWG (FEA) Akddoae? 24Aug55 – 2465 RFC (CNR) Minneapolis-St. Paul 06May57 – 2466 RFC (CNR) Bakalar 04Aug57 – 2473 RFC DT (CNR) Minneapolis-St. Paul 01Dec57 – 96 TCSq (440 TCWG) (AFR) Minneapolis-St. Paul 19Dec59; 934 TCG (440 TCWG) (AFR) Minneapolis-St. Paul 11Feb63 – Fairchild (LOG) St. Augustine 30Sep68, cvtd **AC-119K Stinger** 07Oct68 – 1 SOP WG (TAC) Lockbourne 14May69 – 4410 CCT WG (TAC) Lockbourne 14Jul69 – 14 SOP WG (PAF) Phan Rang AB S. Vietnam 23Oct69; assigned variously Phu Cat, Da Nang, Tan Son Nhut S. Vietnam & Nakhon Phanom Thailand from 29Dec69; to 313 ADH DV (PAF) Kadena AB Japan 24Sep70; to 460 TR WG (PAF) Tan Son Nhut AB S. Vietnam 13Dec70; to 366 Fighter WG (PAF) Da Nang AB S. Vietnam 05Jun71 – 56 SOP WG (PAF) Nakhon Phanom AB Thailand 23Aug71; assigned variously Phan Rang, Da Nang S. Vietnam from 10Sep71 – MAP S. Vietnam 10Nov72, VNAF s/n: **525889**, 821 ASQ (53 TWG) Tan Son Nhut AB. Shot down by a communist SA-7 missile that exploded under the left engine nacelle while circling over Tan Son Nhut AB S. Vietnam in the early hours of 29Apr75. The wing completely failed a short time later sending the stricken aircraft spiraling into the ground. 1Lt Trang Van Thanh (pilot); MSgt Phan Quoc Tuan (flight engineer) plus up to 4 other crew were killed on impact. Another victim was Sgt Bui-Minh-Tan (gunner) who, in the panic, jumped without a parachute and fell to his death. MSgt Nguyen Van Chin (gunner) barely survived when he bailed out, his parachute opening fully just before hitting the ground – Final Disposition: Combat Loss.

11057 / 52-5890 / C-119G-1-FA

Avble: 16Oct53; Acc: 19Nov53; Del: 25Nov53 to USAF – 464 TCWG (TAC) Lawson 25Nov53; depl Seymour 12Apr54; to Pope Oct54; depl Aberdeen 05May56 – 2346 RFC (CNR) Hamilton 11Sep58; to DT Paine 22Sep58; to 3415 TTA WG (ATC) Lowry 20Oct58 mnt – Hayes (AMC) Birmingham 17Dec58 – 97 TCSq (349 TCWG) (AFR) Paine 18Mar59; to unk unit 59; 941 TCG (349 TCWG) (AFR) Paine 63 – 924 TCG (446 TCWG) (AFR) Ellington 18Aug65 – 908 TAL GP (435 TCWG) (AFR) Brookley 12Jan68 – Wfu to MASDC (LOG) Davis-Monthan 25Feb69 for storage – DFI: 26Dec73 – Final Disposition: Scrapped.

11058 / 52-5891 / C-119G-1-FA

Avble: 17Oct53; Acc: 19Nov53; Del: 25Nov53 to USAF – 464 TCWG (TAC) Lawson 25Nov53; to Seymour 12Apr54 – Fairchild (AMC) Hagerstown 12Aug54 – 483 TCWG

(FEA) Ashiya AB Japan 23Sep54. While on a ferry flight to Japan, enroute over the Pacific Ocean, the aircraft suffered a runaway propeller 654 miles NE Hickham AFB Hawaii 09Nov55. Unable to feather the prop the 5 crew were forced to bail out, 4 were later rescued with 1 remaining missing – DFI: 04Jan56 – Final Disposition: Accident.

11059 / 52-5892 / C-119G-1-FA

Avble: 20Oct53; Acc: 19Nov53; Del: 25Nov53 to USAF – 464 TCWG (TAC) Lawson 25Nov53; to Seymour 12Apr54 – Fairchild (AMC) Hagerstown 12Aug54 – 483 TCWG (FEA) Ashiya AB Japan 20Sep54; depl 316 TCWG (FEA) Ashiya AB Japan 03Feb55 – 2465 RFC (CNR) Minneapolis-St. Paul 12Apr57 – 2466 RFC (CNR) Bakalar 04Aug57 – 2473 RFC DT (CNR) Minneapolis-St. Paul 01Dec57; to 2465 AB GP (CNC) Minneapolis-St. Paul 06Mar58 mnt – 65 TCSq (403 TCWG) (AFR) Davis 23Sep58; 929 TCG (403 TCWG) (AFR) Davis 11Feb63 – 922 TCG (433 TCWG) (AFR) Kelly 18Oct65 – Fairchild (LOG) St. Augustine 13May68, cvtd **AC-119G Shadow** 01Jul68 – 71 SOP SQ (1 ACO WG) (TAC) Lockbourne 07Sep68 – Fairchild (LOG) St. Augustine 17Dec68 – 14 SOP WG (PAF) Nha Trang AB S. Vietnam 21Jan69; assigned variously Phan Rang, Tan Son Nhut, Tuy Hoa, Phu Cat, Da Nang S. Vietnam from 14Mar69; to 313 ADH DV (PAF) Kadena AB Japan 12May71 – MAP S. Vietnam 20Aug71, VNAF s/n: **525892**, 819 ASQ (53 TWG) Tan Son Nhut AB – Final Disposition: Unknown.

11060 / 52-5893 / C-119G-1-FA

Avble: 21Oct53; Acc: 20Nov53; Del: 25Nov53 to USAF – 464 TCWG (TAC) Lawson 25Nov53. Destroyed on the ground when a tornado struck Lawson AFB 13Mar54, 5 other C-119s were also destroyed – DFI: 14Apr54 – Final Disposition: Destroyed.

11061 / 52-5894 / C-119G-1-FA

Avble: 21Oct53; Acc: 24Nov53; Del: 25Nov53 to USAF – 464 TCWG (TAC) Lawson 25Nov53, assigned 777 TCSq. On 26Feb54 the aircraft was assigned a training flight out of Lawson AFB Georgia that would involve multiple landings and take-offs at Lawson, Maxwell AFB, Alabama and the local Muscogee Airport adjacent Lawson. The 4 crew were: 1Lt Jack C. Jenkins (25yrs) (pilot), 2Lt John Peachey (co-pilot), A2C Franklin D. Levy (student flight engineer) and A2C David A. Probus (flight engineer). After take-off the C-119G immediately departed the assigned area of operation and proceeded on a 320 mile course NW to Huntingdon Tennessee, the pilot's hometown. Upon arrival over Huntingdon local witnesses stated they saw the C-119 make a high-speed low pass over the town, it then made a steep turn for another run, this time at a speed in excess of 250 mph (it was found post-crash that the throttle levers were pushed all the way forward). As the aircraft made a second pass, the pilot initiated a sharp pull-up and the right wing was then seen to buckle and breakaway, the fuel load spilling and igniting mid-air. In quick succession the aircraft began to roll with the left wing breaking off followed by the booms and rear clamshell doors. The main fuselage and other pieces still attached impacted a field just outside the town boundary killing all onboard. Large sections of the aircraft that had broken away came down on the town damaging cars and property. Two men working in a field sustained third degree burns from the burning gasoline that rained down on them. The investigation concluded a reckless, foolhardy violation of air rules and flying on Lt Jenkins' part that led to a catastrophic airframe structural failure due to aerodynamic loads exceeding design limits, basically the pilot flew the wings off the aircraft performing aerobatic "buzz" maneuvers. It was later learnt that Lt Jenkins, a Korean War Veteran, had committed the same violation over Huntingdon some weeks prior on 09Feb54 with a different crew onboard. This incident had been reported in a local newspaper but somehow, it appears, was not related by any party to senior USAF personnel in the form of a complaint – DFI: 14Apr54 – Final Disposition: Accident.

11062 / 52-5895 / C-119G-1-FA

Avble: 23Oct53; Acc: 24Nov53; Del: 25Nov53 to USAF – 464 TCWG (TAC) Lawson 25Nov53; depls Seymour 12Apr54 & 25Apr54; to Pope 07Sep54; depl North Aux AF 06Jun55; depl Aberdeen 05May56 – BMC (AMC) Birmingham 13Sep56 – 354 Fighter WG (TAC) Myrtle Beach 28Nov56; cvtd **C-119J** 29Jul57; to 3345 TTA WG (TC) Chanute 21Jan57 mnt – 2235 RFC (CNR) Grenier 25Oct57 – 2585 RFC DT (CNR) Donaldson 15Jun58 – 77 TCSq (435 TCWG) (AFR) Donaldson 19Dec58 **RECORDS INDECIPHERABLE** 78 RCSq (435 TCWG) (AFR) Bates date unk; to 3800 AB WG (AU) Maxwell 06Aug60 mnt – 153 CLM SQ (WY-ANG) Cheyenne 24Mar61 – 187 AML SQ (WY-ANG) Cheyenne 01Apr61 – 147 AML SQ (PA-ANG) Greater Pittsburgh 11May61 – Wfu to 2704 ASD GP (LOG) Davis-Monthan 09Mar64 for storage – Fairchild (LOG) St. Augustine 08Dec66 – 4603 AB GP (ADC) Stewart 31Jan67 – 4676 AB GP (ADC) Richards 18Nov69 – 4650 COS SQ (ADC) Richards 30Jun70 – Wfu to MASDC (LOG) Davis-Monthan 13Jul72 for storage, PCN: **CJ434** – DFI: 21Mar73 – Final Disposition: Scrapped.

11063 / 52-5896 / C-119G-1-FA

Avble: 23Oct53; Acc: 20Nov53; Del: 25Nov53 to USAF – 464 TCWG (TAC) Lawson 25Nov53; depl Seymour 12Apr54; to Pope 07Sep54; depl North Aux AF 06Jun55 – BMC (AMC) Birmingham 19Sep56 – 83 Fighter WG (TAC) Seymour 26Nov56; cvtd **C-119J** 01Aug57 – 4 Fighter WG (TAC) Seymour 08Dec57 – 2472 RFC DT (CNR) Tinker 20Mar58; depl Davis 20Jul58 – 305 TCSq (442 TCWG) (AFR) Tinker 19Mar59 **RECORDS INDECIPHERABLE** 153 CLM SQ (WY-ANG) Cheyenne 24Feb61 – 187 AML SQ (WY-ANG) Cheyenne 01Apr61 – 147 AML SQ (PA-ANG) Greater Pittsburgh 11May61 – 732 TCSq (94 TCWG) (AFR) Grenier 16Aug62 – Wfu to 2704 ASD GP (LOG) Davis-Monthan 09Jan63 for storage – Fairchild (LOG) St. Augustine 63 – MAP Italy 19Feb64, AMI s/n: **MM52-5896**; cvtd **EC-119J** 1973 – Final Disposition: Scrapped.

11064 / 52-5897 / C-119G-1-FA

Avble: 27Oct53; Acc: 20Nov53; Del: 25Nov53 to USAF – 464 TCWG (TAC) Lawson 25Nov53; depl Seymour 12Apr54; to Pope 13Sep54; depl North Aux AF 06Jun55; depl Aberdeen 05May56 – 323 Fighter WG (TAC) Bunker Hill 06Oct56 – MOBAR (AMC) Brookley 12Jun57 – 2466 RFC (CNR) Bakalar cvtd **C-119J** 03Nov57 – Hayes (AMC) Birmingham 05Feb59 – 434 TCWG (AFR) Bakalar 04May59; to unk unit (CNC) 59 – 167 TFG SQ (WV-ANG) Martinsburg 09Mar61 – 147 AML SQ (PA-ANG) Greater Pittsburgh 07Apr61 – Wfu to 2704 ASD GP (LOG) Davis-Monthan 20Nov63 for storage – Fairchild (LOG) St. Augustine 63 – MAP Italy 21Feb64, AMI s/n: **MM52-5897** – Final Disposition: Scrapped.

11065 / 52-5898 / C-119G-1-FA

Avble: 28Oct53; Acc: 24Nov53; Del: 25Nov53 to USAF – 464 TCWG (TAC) Lawson 25Nov53; depl Seymour 12Apr54; to Pope 01Sep54; depl Aberdeen 05May56 – Hayes (AMC) Birmingham 26Sep58 – 2347 RFC (CNR) Long Beach 19Dec58; to DT Hill 26Feb59 – 733 TCSq (452 TCWG) (AFR) Hill 03Apr59; 452 TCWG (AFR) March 20Dec62; 944 TCG (452 TCWG) (AFR) March 11Feb63; 943 TCG (452 TCWG) (AFR) March 07Jan65 –

Fairchild (LOG) St. Augustine 25Feb68 – 4413 CCT SQ (1 ACO WG) (TAC) Lockbourne 03Jul68; cvtd **AC-119G Shadow** 11Jul68 – Fairchild (LOG) St. Augustine 22Jan69 – 1 SOP WG (TAC) Lockbourne 25Feb69 – 4410 CCT WG (TAC) Lockbourne 14Jul69 – 14 SOP WG (PAF) Phan Rang AB S. Vietnam 27Dec69; assigned variously Tan Son Nhut, Phu Cat, Da Nang S. Vietnam from 14Feb70; to 13 ADH DV (PAF) Taipei Taiwan 10Apr71 – MAP S. Vietnam 24Aug71, VNAF s/n: **525898**, 819 ASQ (53 TWG) Tan Son Nhut AB – Final Disposition: Unknown.

11066 / 52-5899 / C-119G-1-FA

Avble: 30Oct53; Acc: 24Nov53; Del: 25Nov53 to USAF – **RECORDS MISSING 1953–1955** – 464 TCWG (TAC) Pope date unk; depl Rhein-Main AB W. Germany 27Mar55; depl Aberdeen 05May56 – 1001 AB WG (HQC) Andrews 21Oct57 mnt – Hayes (AMC) Birmingham 20Aug58 – 512 TCWG (AFR) NAS Willow Grove 15Jan59; 912 TCG (512 TCWG) (AFR) NAS Willow Grove 11Feb63; 913 TCG (512 TCWG) (AFR) NAS Willow Grove 22Apr64 – Wfu to MASDC (LOG) Davis-Monthan 14Dec69 for storage – Fairchild (LOG) St. Augustine 02Mar70 – MAP Ethiopia 12Sep70, ETAF s/n: **912** – Wfu 86. Noted 90s stored at Debre Zeyit AB Ethiopia – Final Disposition: Scrapped.

11067 / 52-5900 / C-119G-1-FA

Avble: 30Oct53; Acc: 25Nov53; Del: 25Nov53 to USAF – 464 TCWG (TAC) Lawson 25Nov53; depl Seymour 12Apr54 – Fairchild (AMC) Hagerstown 10Aug54 – 483 TCWG (FEA) Ashiya AB Japan 23Sep54; depl 316 TCWG (FEA) Ashiya AB Japan 26Jun55 – 2465 RFC (CNR) Minneapolis-St. Paul 05Apr57 – 2466 RFC (CNR) Bakalar 04Aug57 – 2473 RFC DT (CNR) Minneapolis-St. Paul 01Dec57 – 2347 RFC (CNR) Long Beach Jun58 – 452 TCWG (AFR) Long Beach 13Jun59; to March 01Oct60; 943 TCG (452 TCWG) (AFR) March 11Feb63 – 931 TAL GP (434 TCWG) (AFR) Bakalar 17Apr69; 930 SOP GP (434 TCWG) (AFR) Bakalar 30Jun69; to Grissom 12Jan70 – 907 SOP GP (302 TCWG) (AFR) Clinton 24Jul70; to Lockbourne 24Jul71 – Wfu to MASDC (LOG) Davis-Monthan 23Sep71 for storage, PCN: **CJ413** – DFI: 09Dec74 – Final Disposition: Scrapped.

11068 / 52-5901 / C-119G-1-FA

Avble: 30Oct53; Acc: 24Nov53; Del: 25Nov53 to USAF – 464 TCWG (TAC) Lawson 25Nov53; depl Seymour 12Apr54; to Pope 01Sep54; to WRDCN (ARD) WPAFB 20Jul55 mnt; depl Aberdeen 05May56 – 2242 RFC DT (CNR) O'Hare 01Oct58 – 64 TCSq (403 TCWG) (AFR) O'Hare 20Dec58; 928 TCG (403 TCWG) (AFR) O'Hare 11Feb63 – Wfu to MASDC (LOG) Davis-Monthan 11Dec69 for storage, PCN: **CJ338** – DFI: 20Jan70 – Kolar, Inc. Tucson Arizona 20Feb76 for scrap – Final Disposition: Scrapped.

11069 / 52-5902 / C-119G-1-FA

Avble: 02Nov53; Acc: 25Nov53; Del: 02Dec53 to USAF – 464 TCWG (TAC) Lawson 02Dec53; depl Seymour 22Apr54 – Fairchild (AMC) Hagerstown 11Aug54 – 483 TCWG (FEA) Ashiya AB Japan 20Sep54; depl 316 TCWG (FEA) Ashiya AB Japan 11Dec54 – 2465 RFC (CNR) Minneapolis-St. Paul 29Apr57; depl 2472 RFC (CNR) Richards 10May57 – 2466 RFC (CNR) Bakalar 04Aug57 – 2473 RFC DT (CNR) Minneapolis-St. Paul 01Dec57 – 96 TCSq (440 TCWG) (AFR) Minneapolis-St. Paul 19Dec59 **RECORDS INDECIPHERABLE** 934 TCG (440 TCWG) (AFR) Minneapolis-St. Paul 11Feb63 – Fairchild (LOG) St. Augustine 03Jun63 – MAP India 16Jul63 – Final Disposition: Unknown.

11070 / 52-5903 / C-119G-1-FA

Avble: 02Nov53; Acc: 25Nov53; Del: 02Dec53 to USAF – 464 TCWG (TAC) Lawson 02Dec53; depl Seymour 12Apr54; to Pope 01Sep54; to 5700 AB GP (CAC) Albrook 22Mar56 mnt; depl Aberdeen 05May56 – 17 BTA WG (TAC) Eglin 07Nov56; cvtd **C-119J** 01Aug57 – 2347 RFC (CNR) Long Beach Jun58 – 452 TCWG (AFR) Long Beach 13Jun59; to March 01Oct60 – 112 CML SQ (PA-ANG) Greater Pittsburgh 20Mar61 – 147 AML SQ (PA-ANG) Greater Pittsburgh 01Apr61 – Wfu 2704 ASD GP (LOG) Davis-Monthan 13Nov63 for storage – Fairchild (LOG) St. Augustine 15Dec66 – 78 Fighter WG (ADC) Hamilton 03Feb67 – 4676 AB GP (ADC) Richards 08Dec69 – 4650 COS SQ (ADC) Richards 30Jun70 – Wfu to MASDC (LOG) Davis-Monthan 07Jul72 for storage, PCN: **CJ430** – DFI: 21Mar73 – Final Disposition: Scrapped.

11071 / 52-5904 / C-119G-1-FA

Avble: 03Nov53; Acc: 25Nov53; Del: 02Dec53 to USAF – 464 TCWG (TAC) Lawson 02Dec53, assigned 777 TCSq; depl Seymour 12Apr54. Crashed due to structural failure 20 miles S Goldsboro North Carolina 22Apr54 during a deployment to Seymour-Johnson Field. 2 of the 4 crew managed to escape and parachute to safety before the C-119 struck the ground and burned, these were Capt Irvin B. Wilson (navigator) and 2Lt Eldon Olsen (co-pilot). The stricken aircraft brought down power lines and damaged a local civilian house when it plummeted from the sky – Final Disposition: Accident.

11072 / 52-5905 / C-119G-1-FA

Avble: 04Nov53; Acc: 25Nov53; Del: 02Dec53 to USAF – 464 TCWG (TAC) Lawson 02Dec53 – Fairchild (AMC) Hagerstown 11Aug54 – 483 TCWG (FEA) Ashiya AB Japan 27Sep54; depl 316 TCWG (FEA) Ashiya AB Japan 27Jul55 – 2471 RFC (CNR) O'Hare 22May57 – 2473 RFC DT (CNR) O'Hare 01Dec57 – 2242 RFC DT (CNR) O'Hare 20Nov58 – 64 TCSq (403 TCWG) (AFR) O'Hare 20Dec58; to 328 CLM SQ (ADC) Richards 27Jun60 mnt; 928 TCG (403 TCWG) (AFR) O'Hare 11Feb63 – Fairchild (LOG) St. Augustine 08Jun68, cvtd **AC-119G Shadow** 01Jul68 – 71 SOP SQ (1 ACO WG) (TAC) Lockbourne 25Sep68 – Fairchild (LOG) St. Augustine 04Nov68 – 14 SOP WG (PAF) Nha Trang AB S. Vietnam 10Dec68; assigned variously Phan Rang, Tan Son Nhut, Tuy Hoa, Phu Cat, Da Nang S. Vietnam from 10Mar69; to 313 ADH DV (PAF) Kadena AB Japan 23Apr71 – MAP S. Vietnam 31Aug71, VNAF s/n: **525905**, 819 ASQ (53 TWG) Tan Son Nhut AB – Final Disposition: Unknown.

11073 / 52-5906 / C-119G-1-FA

Avble: 05Nov53; Acc: 25Nov53; Del: 02Dec53 to USAF – 464 TCWG (TAC) Lawson 02Dec53; depl Seymour 12Apr54; to Pope 01Sep54; depl Aberdeen 05May56 – 363 TR WG (TAC) Shaw 08Oct56; cvtd **C-119J** 02Aug57 – Hayes (AMC) Birmingham 15Oct57 – 2347 RFC (CNR) Long Beach 58 – 452 TCWG (AFR) Long Beach 13Jun59 **RECORDS INDECIPHERABLE** to March 01Oct60 – 150 AML SQ (NJ-ANG) Newark 24Mar61 – Wfu to 2704 ASD GP (LOG) Davis-Monthan 17Nov62 for storage – Fairchild (LOG) St. Augustine 16Dec66 – 78 Fighter WG (ADC) Hamilton 31May67 – 4676 AB GP (ADC) Richards 08Dec69 – 4650 COS SQ (ADC) Richards 30Jun70; depl 317 TAL WG (TAC) Lockbourne 08Jul70; to 4600 CLM SQ (ADC) Peterson 26Mar71 mnt – Wfu to MASDC (LOG) Davis-Monthan 11Jul72 for storage – DFI: 21Mar73 – Final Disposition: Scrapped.

11074 / 52-5907 / C-119G-1-FA

Avble: 06Nov53; Acc: 30Nov53; Del: 04Dec53 to USAF – 464 TCWG (TAC) Lawson 04Dec53; depl Seymour 12Apr54; to Pope 01Sep54; to WRDCN (ARD) WPAFB 27Jul55 mnt; to 3201 AB WG (APG) Eglin 13SDec55 mnt; depl Aberdeen 05May56; depl 63 TCWG (TAC) Donaldson Oct56 – 2242 RFC DT (CNR) O'Hare 27Sep58 – 64 TCSq (403 TCWG) (AFR) O'Hare 20Dec58; to 4750 ADF WG (ADC) Vincent 04Jan59 mnt; 928 TCG (403 TCWG) (AFR) O'Hare 11Feb63 – Fairchild (LOG) St. Augustine 14Mar68, cvtd **AC-119G Shadow** 01Jul68 – 71 SOP SQ (1 ACO WG) (TAC) Lockbourne 07Aug68 – Fairchild (LOG) St. Augustine 22Nov68 – 14 SOP WG (PAF) Nha Trang AB S. Vietnam 20Dec68; assigned variously Phan Rang, Tuy Hoa, Tan Son Nhut S. Vietnam from 15Feb69. Crashed after take-off from Tan Son Nhut AB S. Vietnam 11Oct69 after an engine failed and caught fire. Fully laden with fuel and ammunition the C-119 struggled to maintain height and came down outside the air base. 5 of the 8 crew were killed plus a photographer from the 600 Photographic SQ, a Vietnamese interpreter and a civilian on the ground: Lt Col Bernard R. Knapic (pilot), Maj Moses L. Alves (navigator), Maj Jerome J. Rice (navigator), Capt John H.V. Hathaway (co-pilot), SSgt Abraham L. Moore (flight engineer) and SSgt Ellsworth S. Bradford (USAF photographer). 3 other unknown crew survived and appear to all be part of the gunship technical crew who would have been seated in the rear section of the aircraft – Final Disposition: Accident.

11075 / 52-6036 / C-119G-15-FA

Avble: 09Nov53; Acc: 30Nov53; Del: 04Dec53 to MDAP Italy, AMI s/n: **MM52-6036**. While on a low-level flight off the coast near Pisa-Arturo dell'Oro Air Base Italy 20Apr64, a wing touched the water causing the aircraft to lose controlled flight, bounce and break up. All 6 onboard were killed – Final Disposition: Accident.

11076 / 52-6037 / C-119G-15-FA

Avble: 10Nov53; Acc: 30Nov53; Del: 04Dec53 to MDAP Italy, AMI s/n: **MM52-6037**. Written-off while making an emergency landing at Kwamouth The Congo 02Feb61, no fatalities – Final Disposition: Accident.

11077 / 52-6038 / C-119G-15-FA

Avble: 12Nov53; Acc: 30Nov53; Del: 04Dec53 to MDAP Belgium, BAF s/n: **CP-24**; BOC: 17Dec53 15 WG / 40 SQ Melsbroek AB Belgium, callsign: **OT-CBD**, later to 15 WG / 20 SQ. Multiple damage to airframe from ground incidents 10Nov55, 21Jun58 and 08Jan68 – Wfu 22Dec71 and stored at Koksijde AB Belgium; SOC: Jul74 – Sold for scrap to J. Trappeniers 04Oct74, scrapped 1977-78 – Final Disposition: Scrapped.

11078 / 52-6039 / C-119G-15-FA

Avble: 12Nov53; Acc: 30Nov53; Del: 14Dec53 to MDAP Belgium, BAF s/n: **CP-28**; BOC: 18Jan54 15 WG / 40 SQ Melsbroek AB Belgium, callsign: **OT-CBH**, later to 15 WG / 20 SQ. Minor incidents 03May56 and 01Mar60 – Wfu 09Jul73 and stored at Koksijde AB Belgium; SOC: Jul74 – Sold for scrap to J. Trappeniers 04Oct74, scrapped 1977-78 – Final Disposition: Scrapped.

11079 / 52-6040 / C-119G-15-FA

Avble: 12Nov53; Acc: 04Dec53; Del: 29Dec53 to MDAP Italy, AMI s/n: **MM52-6040** – Final Disposition: Scrapped.

11080 / 52-6041 / C-119G-15-FA

Avble: 16Nov53; Acc: 08Dec53; Del: 21Dec53 to MDAP Italy, AMI s/n: **MM52-6041**. An engine and u/c failure occurred during take-off from Chateauroux Airport France 22 Oct 64, apparently the engine "exploded" which caused the u/c leg to subsequently collapse. The aircraft was deemed as damaged beyond repair, no fatalities – Final Disposition: Accident.

11081 / 52-6042 / C-119G-30-FA

Avble: 17Nov53; Acc: 08Dec53; Del: 04Jan54 to MDAP Italy, AMI s/n: **MM52-6042** – Final Disposition: Scrapped.

11082 / 52-6043 / C-119G-30-FA

Avble: 17Nov53; Acc: 08Dec53; Del: 15Dec53 to MDAP Belgium, BAF s/n: **CP-25**; BOC: 22Dec53 15 WG / 40 SQ Melsbroek AB Belgium, callsign: **OT-CBE**, later to 15 WG / 20 SQ. Wing-tip damaged in minor ground collision with s/n: CP-35 05Nov57. Catastrophic mid-air collision with C-119G s/n: CP-23 (10951) over Cambron-St. Vincent Belgium 12Dec61, both aircraft came down at Montignies-lez-Lens. 7 crew killed on CP-25 were: Camille Bauvois, Jose DeCaigny (pilot), Romeo DeCot, Florimont DeLroeux, Raoul DePoorter, Alfred Donckier, Gilbert Francois and Yvan Passchiersens – Final Disposition: Accident.

11083 / 52-6044 / C-119G-30-FA

Avble: 18Nov53; Acc: 14Dec53; Del: 29Dec53 to MDAP Belgium, BAF s/n: **CP-36**; BOC: 18Feb54 15 WG / 40 SQ Melsbroek AB Belgium, callsign: **OT-CBP**, later to 15 WG / 20 SQ. Take-off accident 11May59 at N'Djili Airport in The Congo where the aircraft appears to have stalled after lift-off then settled back onto the runway with u/c retracted. Much belly damage but the aircraft was repaired. Crashed into a hill due to a starboard engine failure while fully loaded 46 miles N of Goma in The Congo 19Jul60. Pilot: Jean Van Gompel and 4 other crew with 36 paratroops onboard. 33 were killed in the crash with the 8 survivors deciding to hike to safety but 4 of them perished along the way – Final Disposition: Accident.

11084 / 52-6045 / C-119G-30-FA

Avble: 19Nov53; Acc: 21Dec53; Del: 15Jan54 to MDAP Belgium, BAF s/n: **CP-32**; BOC: 14Feb54 15 WG / 40 SQ Melsbroek AB Belgium, callsign: **OT-CBL**, later to 15 WG / 20 SQ. Minor incident at Tours AB 25Apr60 – Wfu 10Feb72 and stored at Koksijde AB Belgium; SOC: Jul74 – Sold for scrap to J. Trappeniers 04Oct74, scrapped 1977-78 – Final Disposition: Scrapped.

11085 / 52-6046 / C-119G-30-FA

Avble: 20Nov53; Acc: 21Dec53; Del: 06Jan54 to MDAP Belgium, BAF s/n: **CP-39**; BOC: 22Feb54 15 WG / 40 SQ Melsbroek AB Belgium, callsign: **OT-CBS**, later to 15 WG / 20 SQ – Wfu 01Jun72 and stored at Koksijde AB Belgium; SOC: Jul74 – Sold for scrap to J. Trappeniers 04Oct74, scrapped 1977-78 – Final Disposition: Scrapped.

11086 / 52-6047 / C-119G-30-FA

Avble: 23Nov53; Acc: 21Dec53; Del: 12Jan54 to MDAP Belgium, BAF s/n: **CP-29**; BOC: 19Jan54 15 WG / 40 SQ Melsbroek AB Belgium, callsign: **OT-CBI**, later to 15 WG / 20 SQ. Minor incident at Melsbroek 11May57. Lost hydraulic pressure while taxiing at Zaventem Airport 16Dec65 – Wfu 03Jul72 and stored at Koksijde AB Belgium – Ethiopia 29Sep72, ETAF s/n: **918** – Wfu 86. Noted 90s stored at Debre Zeyit AB Ethiopia – Final Disposition: Scrapped.

11087 / 52-5908 / C-119G-1-FA

Avble: 24Nov53; Acc: 21Dec53; Del: 31Dec53 to USAF – 464 TCWG (TAC) Lawson 31Dec53. Destroyed on the ground when a tornado struck Lawson AFB 13Mar54, 5 other C-119s were also destroyed – DFI: 14Apr54 – Final Disposition: Destroyed.

11088 / 52-5909 / C-119G-1-FA

Avble: 25Nov53; Acc: 21Dec53; Del: 31Dec53 to USAF – 464 TCWG (TAC) Lawson 31Dec53; depl Seymour 12Apr54; to Pope 07Sep54; depl North Aux AF 06Jun55; depl Aberdeen 05May56 – 2466 RFC DT (CNR) Scott 25Nov58 – 73 TCSq (434 TCWG) (AFR) Scott 18Apr59; to 2585 AB SQ (CNC) Miami 21Jun59 mnt – 4440 ADE GP (TAC) Langley 28Jun60 – MAP India 10Jul60 – Final Disposition: Unknown.

11089 / 52-5910 / C-119G-1-FA

Avble: 25Nov53; Acc: 21Dec53; Del: 31Dec53 to USAF – 464 TCWG (TAC) Lawson 31Dec53; to Pope 13Sep54; to 6511 AB GP (NEA) Narsarsuaq AB Greenland date unk mnt; depl Aberdeen 05May56 – 2585 RFC DT (CNR) Donaldson 03Oct58 – 77 TCSq (435 TCWG) (AFR) Donaldson 19Dec58; to 2585 AB SQ (CNC) Miami 21Apr59 mnt – 349 TCWG (AFR) Hamilton 23Jan61; 938 TCG (349 TCWG) (AFR) Hamilton 10Feb63 – 944 TCG (452 TCWG) (AFR) March 22Jun66; 943 TAL GP (452 TCWG) (AFR) March 30Jun67 – Fairchild (LOG) St. Augustine 17May68, cvtd **AC-119K Stinger** 01Jul68 – 4413 CCT SQ (1 ACO WG) (TAC) Lockbourne 02Dec68 – 1 SOP WG (TAC) Lockbourne 17Feb69 – Fairchild (LOG) St. Augustine 09Jun69 – 4410 CCT WG (TAC) Lockbourne 15Jul69 – 1 SOP WG (TAC) Eglin 01Jul71; to Hurlburt 30Oct72 – MAP S. Vietnam 06Nov72, VNAF s/n: **525910**, 821 ASQ (53 TWG) Tan Son Nhut AB – Final Disposition: Unknown.

11090 / 52-5911 / C-119G-1-FA

Avble: 30Nov53; Acc: 21Dec53; Del: 31Dec53 to USAF – 464 TCWG (TAC) Lawson 31Dec53; depl Seymour 12Apr54; to Pope 13Sep54; depl Rhein-Main AB W. Germany 19Oct54; depl Aberdeen 05May56 – 2473 RFC DT (CNR) Minneapolis-St. Paul 01Oct58 – 96 TCSq (440 TCWG) (AFR) Minneapolis-St. Paul 19Dec59; 934 TCG (440 TCWG) (AFR) Minneapolis-St. Paul 11Feb63 – Fairchild (LOG) St. Augustine 31May68, cvtd **AC-119K Stinger** 01Jul68 – 4413 CCT SQ (1 ACO WG) (TAC) Lockbourne 06Dec68 – 1 SOP WG (TAC) Lockbourne 17Feb69 – Fairchild (LOG) St. Augustine 09Jun69 – 4410 CCT WG (TAC) Lockbourne 13Oct69 – 1 SOP WG (TAC) Eglin 07Jul71; to Hurlburt 31Oct72 – MAP S. Vietnam 06Nov72, VNAF s/n: **525911**, 821 ASQ (53 TWG) Tan Son Nhut AB – Final Disposition: Unknown.

11091 / 52-5912 / C-119G-1-FA

Avble: 01Dec53; Acc: 22Dec53; Del: 31Dec53 to USAF – 464 TCWG (TAC) Lawson 31Dec53; depl Seymour 12Apr54; to Pope 07Sep54; depl North Aux AF 06Jun55; to WRDCN (ARD) WPAFB 23Jun55 test support; depl Aberdeen 05May56; to 5700 AB GP (CAC) Albrook AFS Panama CZ 17Jul58 mnt – 2466 RFC (CNR) Bakalar 12Oct58 – 434 TCWG (AFR) Bakalar 19Apr59 **RECORDS INDECIPHERABLE** 4440 ADE GP (TAC) Langley 14Jun60 – MAP India 26Jun60 – Final Disposition: Unknown.

11092 / 52-5913 / C-119G-1-FA

Avble: 02Dec53; Acc: 22Dec53; Del: 31Dec53 to USAF – 464 TCWG (TAC) Lawson 31Dec53; depl Seymour 12Apr54; to Pope 01Sep54 – **RECORDS MISSING 1954–1956** –

depl Aberdeen 05May56 – 2237 RFC DT (CNR) Niagara Falls 13Dec58 – 328 TCSq (512 TCWG) (AFR) Niagara Falls 19Dec58; 914 TCG (512 TCWG) (AFR) Niagara Falls 11Feb63 – Wfu to MASDC (LOG) Davis-Monthan 30Nov70 for storage, PCN: **CJ375** – DFI: 21May71 – Kolar, Inc. Tucson Arizona 09Mar76 for scrap – Final Disposition: Scrapped.

11093 / 52-5914 / C-119G-1-FA

Avble: 03Dec53; Acc: 22Dec53; Del: 31Dec53 to USAF – 464 TCWG (TAC) Lawson 31Dec53; depl Seymour 12Apr54; to Pope 13Sep54 – **RECORDS MISSING 1954–1955** – depl Rhein-Main AB W. Germany date unk; depl Aberdeen 05May56 – 2473 RFC DT (CNR) Minneapolis-St. Paul 23Dec58 – 96 TCSq (440 TCWG) (AFR) Minneapolis-St. Paul 23Dec59; 934 TCG (440 TCWG) (AFR) Minneapolis-St. Paul 11Feb63; 933 TAL GP (440 TCWG) (AFR) Gen Mitchell 30Jan70 – 913 TAL GP (512 TCWG) (AFR) NAS Willow Grove 18Apr70 – Wfu to MASDC (LOG) Davis-Monthan 28Feb71 for storage, PCN: **CJ409** – DFI: 21May71 – Final Disposition: Scrapped.

11094 / 52-5915 / C-119G-1-FA

Avble: 03Dec53; Acc: 23Dec53; Del: 31Dec53 to USAF – 464 TCWG (TAC) Lawson 31Dec53; to Pope 13Sep54; depl Rhein-Main AB W. Germany 19Oct54; to 6511 AB GP (NEA) Narsarsuaq AB Greenland 18Nov55 mnt; depl Aberdeen 05May56; to 1405 AB WG (MATS) Scott 24Jun58 mnt – 2237 RFC DT (CNR) Niagara Falls 26Nov58 – 328 TCSq (512 TCWG) (AFR) Niagara Falls 19Dec58; 914 TCG (512 TCWG) (AFR) Niagara Falls 11Feb63 – Wfu to MASDC (LOG) Davis-Monthan 01Feb71 for storage, PCN: **CJ395** – DFI: 31Mar76 – Final Disposition: Scrapped.

11095 / 52-5916 / C-119G-1-FA

Avble: 04Dec53; Acc: 28Dec53; Del: 06Jan54 to USAF – 464 TCWG (TAC) Lawson 06Jan54; depl Seymour 12Apr54; to Pope 13Sep54; depl Rhein-Main AB W. Germany 19Oct54; depl Aberdeen 05May56 – 2472 RFC (CNR) Richards 25Oct58 – 442 TCWG (AFR) Richards 02Mar59 – 733 TCSq (452 TCWG) (AFR) Hill 08Apr61; 945 TCG (452 TCWG) (AFR) Hill 17Jan63 – MAP India 17Jul63 – Final Disposition: Unknown.

11096 / 52-5917 / C-119G-1-FA

Avble: 07Dec53; Acc: 29Dec53; Del: 06Jan54 to USAF – 64 TCWG (TAC) Donaldson 06Jan54; depl Seymour 23Apr54 – Fairchild (AMC) Hagerstown 10Aug54 – 483 TCWG (FEA) Ashiya AB Japan 01Oct54; depl 316 TCWG (FEA) Ashiya AB Japan 14Jun55 – 2237 RFC (CNR) New Castle 14Jun57; to NAS Willow Grove 21Jul58 – 512 TCWG (AFR) NAS Willow Grove 01Dec58; 912 TCG (512 TCWG) (AFR) NAS Willow Grove 11Feb63 – 934 TAL GP (440 TCWG) (AFR) Minneapolis-St. Paul 15Aug68 – Wfu to MASDC (LOG) Davis-Monthan 17Mar69 for storage, PCN: **CJ293** – DFI: 09Dec74 – Final Disposition: Scrapped.

11097 / 52-5918 / C-119G-1-FA

Avble: 08Dec53; Acc: 28Dec53; Del: 06Jan54 to USAF – 64 TCWG (TAC) Donaldson 06Jan54; to 363 TR WG (TAC) Shaw 04Feb54; depl Seymour 23Apr54 – 313 TCWG (TAC) Sumpter (depl) 03Aug54; to Sewart 19Aug54; to Elmendorf 12Jan55 – 464 TCWG (TAC) Sewart 08Jun55 – 314 TCWG (TAC) Sewart 09Aug55; to 5001 MAI SQ (AAC) Ladd 12Dec55 mnt; depl Big Delta Alaska 28Nov55 – BMC (AMC) Birmingham 09Apr57 – 2472 RFC (CNR) Richards 01Jul57; depl 2466 RFC (CNR) Bakalar 07Jul57; to 2578 AB GP (CNC) Ellington 01Sep58 mnt – 442 TCWG (AFR) Richards 02Mar59 – 733 TCSq (452 TCWG) (AFR) Hill 08Apr61;

945 TCG (452 TCWG) (AFR) Hill 17Jan63 – 907 TCG (302 TCWG) (AFR) Clinton 03Jun66; to Lockbourne 24Jul71 – Wfu to MASDC (LOG) Davis-Monthan 25Sep71 for storage – Fairchild (LOG) Crestview 28Oct71, cvtd **C-119K** 15May72 – 907 TAL GP (302 TCWG) (AFR) Lockbourne 26May72 – MAP Jordan 21Oct72, RJAF s/n: **117**, nfd – Final Disposition: Unknown.

11098 / 52-5919 / C-119G-1-FA

Avble: 09Dec53; Acc: 29Dec53; Del: 06Jan54 to USAF – 64 TCWG (TAC) Donaldson 06Jan54; depl Seymour 23Apr54 – Fairchild (AMC) Hagerstown 11Aug54 – 483 TCWG (FEA) Ashiya AB Japan 20Sep54; depl 316 TCWG (FEA) Ashiya AB Japan 18Feb55 – 2235 RFC (CNR) Grenier 16May57 – 732 TCSq (94 TCWG) (AFR) Grenier 19Mar59 – 357 TCSq (302 TCWG) (AFR) Bates 08May61 – 908 TCG (435 TCWG) (AFR) Bates 18Mar63; to Brookley 30Oct64. After taking-off from Jacksonville NAS Florida 16Jul66 the no. 1 engine caught fire which quickly spread. 4 crew and 30 paratroopers bailed out with the aircraft coming down in a swampy pasture. The flight crew were awarded the Distinguished Flying Cross (DFC) medal for the quick evacuation of a full aircraft – DFI: 02Aug66 – Final Disposition: Accident.

11099 / 52-5920 / C-119G-1-FA

Avble: 10Dec53; Acc: 29Dec53; Del: 06Jan54 to USAF – 64 TCWG (TAC) Donaldson 06Jan54 – Fairchild (AMC) Hagerstown 10Aug54 – 483 TCWG (FEA) Ashiya AB Japan 16Sep54; depl 316 TCG (FEA) Ashiya AB Japan 13Nov54; bailment CAT (AMC) Taiwan 20Oct55 – 2235 RFC (CNR) Grenier 13May57 – 2230 RFC (CNR) NAS New York 26Jul57 – 2466 RFC (CNR) Bakalar 28Oct57 – 2242 RFC (CNR) Selfridge 31Oct57 – 403 TCWG (AFR) Selfridge 13Nov58; 927 TCG (403 TCWG) (AFR) Selfridge 10Feb63 – MAP India 01Jul63 – Final Disposition: Unknown.

11100 / 52-5921 / C-119G-1-FA

Avble: 10Dec53; Acc: 29Dec53; Del: 18Jan54 to USAF – 64 TCWG (TAC) Donaldson 18Jan54; to Seymour 10Apr54 – Fairchild (AMC) Hagerstown 05Aug54 – 483 TCWG (FEA) Ashiya AB Japan 20Sep54; depl 314 TCWG (FEA) Ashiya AB Japan 11Nov54; depl 316 TCWG (FEA) Ashiya AB Japan 15Nov54. Forced to ditch in the ocean after engine failure 5.5 miles W Ashiya AB Japan 06Dec54. Pilot: Joseph H. Obendorf and crew were rescued and the aircraft sank – Final Disposition: Accident.

11101 / 53-4637 / C-119G-36-FA

Avble: 14Dec53; Acc & Del: 28Jan54 to MDAP India, IAF s/n: **IK441** – Final Disposition: Unknown.

11102 / 53-4638 / C-119G-36-FA

Avble: 14Dec53; Acc: 27Jan54; Del: 28Jan54 to MDAP India, IAF s/n: **IK442** – Final Disposition: Unknown.

11103 / 52-5924 / C-119G-1-FA

Avble: 16Dec53; Acc: 21Jan54; Del: 26Jan54 to USAF – 313 TCWG (TAC) Sewart 26Jan54; depl Sumpter 07Aug54; depl Elmendorf 12Jan55 – 314 TCWG (TAC) Sewart 08Jun55; depl England Louisiana 05Nov56; to 6606 AB WG (NEA) Goose Bay Canada 22Nov56 mnt – BMC (AMC) Birmingham 16Apr57 – 2472 RFC (CNR) Richards 11Jul57 – 442 TCWG (AFR) Richards 02Mar59 – 314 TCSq (349 TCWG) (AFR) McClellan 15Apr61;

941 TCG (349 TCWG) (AFR) Paine 63 – 925 TCG (446 TCWG) (AFR) Ellington 11Sep65 – 922 TAL GP (433 TCWG) (AFR) Kelly 16Jan68 – 906 SOP GP (302 TCWG) (AFR) Clinton 06Nov70 – MAP Taiwan 27Nov70 RoCAF s/n: **3217** – Final Disposition: Unknown.

11104 / 52-5925 / C-119G-1-FA

Avble: 16Dec53; Acc: 18Jan54; Del: 22Jan54 to USAF – 483 TCWG (FEA) Ashiya AB Japan 22Jan54; to 21 TCSq (483 TCWG) (FEA) Tachikawa AB Japan 18Sep56 – 2256 RFC (CNR) Niagara Falls 27Mar57; depl 2235 RFC (CNR) Grenier 10Apr57 – 2242 RFC DT (CNR) Niagara Falls 01Dec57 – 2237 RFC DT (CNR) Niagara Falls 01Apr58 – 328 TCSq (512 TCWG) (AFR) Niagara Falls 19Dec58; 914 TCG (512 TCWG) (AFR) Niagara Falls 11Feb63 – Fairchild (LOG) St. Augustine 01May68, cvtd **AC-119G Shadow** 01Jul68 – 71 SOP SQ (1 ACO WG) (TAC) Lockbourne 29Aug68 – Fairchild (LOG) St. Augustine 06Dec68 – 14 SOP WG (PAF) Nha Trang AB S. Vietnam 07Jan69; assigned variously Phan Rang, Tan Son Nhut, Tuy Hoa, Phu Cat S. Vietnam from 18Feb69 – MAP S. Vietnam 27Aug71, VNAF s/n: **525925**, 819 ASQ (53 TWG) Tan Son Nhut AB – Final Disposition: Unknown.

11105 / 52-5926 / C-119G-1-FA

Avble: 17Dec53; Acc: 29Jan54; Del: 09Feb54 to USAF – 483 TCWG (FEA) Ashiya AB Japan 09Feb54; bailment CAT (AMC) Taiwan 29Oct56 – 2585 RFC (CNR) Miami 21May57 – 435 TCWG (AFR) Miami 19Dec58; to Homestead 25Jul60; assigned 76 TCSq Homestead 01Oct61; 915 TCG (435 TCWG) (AFR) Homestead 17Jan63 – 926 TCG (446 TCWG) (AFR) NAS New Orleans 13Nov65 – Fairchild (LOG) St. Augustine 24Jun68, cvtd **AC-119K Stinger** 01Jul68 – 1 SOP WG (TAC) Lockbourne 28May69 – 4410 CCT WG (TAC) Lockbourne 14Jul69 – 14 SOP WG (PAF) Phan Rang AB S. Vietnam 18Nov69; assigned variously Phu Cat, Da Nang S. Vietnam & Udorn, Nakhon Phanom Thailand from 29Dec69; to 313 ADH DV (PAF) Kadena AB Japan 07Aug70 – China Airlines (LOG) Taipei Taiwan 22Jul71 mnt – 56 SOP WG (PAF) Nakhon Phanom AB Thailand 05 Oct71; assigned variously Da Nang S. Vietnam from 30Apr72 – MAP S. Vietnam 10Nov72, VNAF s/n: **525926**, 821 ASQ (53 TWG) Tan Son Nhut AB – Final Disposition: Unknown.

11106 / 52-5927 / C-119G-1-FA

Avble: 18Dec53; Acc: 27Jan54; Del: 04Feb54 to USAF – 483 TCWG (FEA) Ashiya AB Japan 04Feb54; bailment CAT (FEA) Taiwan 29Jun55 – 2346 RFC (CNR) Hamilton 31Jul58; to 6486 AB WG (PAF) Hickham 06Aug58 mnt – 349 TCWG (AFR) Hamilton 23Dec58; 938 TCG (349 TCWG) (AFR) Hamilton 10Feb63 – 908 TAL GP (435 TCWG) (AFR) Brookley 01Jul67 – Fairchild (LOG) St. Augustine 20Mar68, cvtd **AC-119G Shadow** 01Jul68 – 71 SOP SQ (1 ACO WG) (TAC) Lockbourne 07Aug68 – Fairchild (LOG) St. Augustine 19Nov68 – 14 SOP WG (PAF) Nha Trang AB S. Vietnam 18Dec68; assigned variously Phan Rang, Tuy Hoa, Tan Son Nhut, Phu Cat, Da Nang S. Vietnam from 08Mar69 – MAP S. Vietnam 25Aug71, VNAF s/n: **525927**, 819 ASQ (53 TWG) Tan Son Nhut AB – Final Disposition: Unknown.

11107 / 52-5928 / C-119G-1-FA

Avble: 21Dec53; Acc: 27Jan54; Del: 04Feb54 to USAF – 483 TCWG (FEA) Ashiya AB Japan 04Feb54; bailment CAT (AMC) Taiwan 17Mar56; to 21 TCSq (483 TCWG) (FEA) & Tachikawa AB Japan 18Sep56 – Wfu to 868 TM SQ (PAF) Tainan Taiwan 02Oct58 – MAP Taiwan 01Jul59 – Final Disposition: Unknown.

11108 / 52-5929 / C-119G-1-FA

Avble: 22Dec53; Acc: 27Jan54; Del: 09Feb54 to USAF – 483 TCWG (FEA) Ashiya AB Japan 09Feb54; to 1500 MAI SQ (MATS) Hickham 24Feb54 mnt; depl Tachikawa AB Japan 18Sep56 – 2347 RFC (CNR) Long Beach 04Mar58 – 452 TCWG (AFR) Long Beach 13Jun59; to March 01Oct60; 942 TCG (452 TCWG) (AFR) March 11Feb63; 944 TCG (452 TCWG) (AFR) March 31Mar65 – 931 TAL GP (434 TCWG) (AFR) Bakalar 24Jan68; 930 TAL GP (434 TCWG) (AFR) Bakalar 25Apr68; depl 71 ACO SQ (1 ACO WG) (TAC) Lockbourne 20Jun68 – 927 TAL GP (403 TCWG) (AFR) Selfridge 09Oct68 – MAP Taiwan 06Mar69 – Final Disposition: Unknown.

11109 / 52-5930 / C-119G-1-FA

Avble: 22Dec53; Acc: 27Jan54; Del: 04Feb54 to USAF – 483 TCWG (FEA) Ashiya AB Japan 04Feb54; bailment CAT (AMC) Taiwan 19Feb56. Crashed in mountainous terrain 2.5 miles S Yawata Japan 22Apr57, all 4 crew onboard were killed on impact. A Sikorsky H-19 rescue helicopter sent to search for the missing aircraft also crashed 2 miles W of the crash site, the 3 crew were injured – Final Disposition: Accident.

11110 / 52-5931 / C-119G-1-FA

Avble: 23Dec53; Acc: 27Jan54; Del: 04Feb54 to USAF – 483 TCWG (FEA) Ashiya AB Japan 04Feb54 **RECORDS INDECIPHERABLE** → depl Tachikawa AB Japan 58 – 2237 RFC DT (CNR) Niagara Falls Jun58 ← **RECORDS INDECIPHERABLE** 328 TCSq (512 TCWG) (AFR) Niagara Falls 19Dec58; 914 TCG (512 TCWG) (AFR) Niagara Falls 11Feb63 – Wfu to MASDC (LOG) Davis-Monthan 23Feb71 for storage, PCN: **CJ405** – DFI: 31Mar76 – Final Disposition: Scrapped.

11111 / 52-5932 / C-119G-1-FA

Avble: 28Dec53; Acc: 28Jan54; Del: 03Feb54 to USAF – 313 TCWG (TAC) Sewart 03Feb54; depl Pope 20Apr54; depl Sumpter 19Jul54 – Fairchild (AMC) Hagerstown 20Mar54 – 483 TCWG (FEA) Ashiya AB Japan 12Oct54; depl 316 TCG (FEA) Ashiya AB Japan 01Jul55 – 2237 RFC (CNR) New Castle 06May57; to 6486 AB WG (FEA) Hickham 16May57 mnt; to NAS Willow Grove 21Jul58 – 512 TCWG (AFR) NAS Willow Grove 01Dec58; 913 TCG (512 TCWG) (AFR) NAS Willow Grove 11Feb63 – Wfu to MASDC (LOG) Davis-Monthan 14Dec69 for storage – Fairchild (LOG) St. Augustine 05Mar70 – MAP Ethiopia 24Nov70, ETAF s/n: **913**; cvtd **C-119K**, nfd – Final Disposition: Unknown.

11112 / 52-5933 / C-119G-1-FA

Avble: 29Dec53; Acc: 28Jan54; Del: 02Feb54 to USAF – 313 TCWG (TAC) Sewart 02Feb54; depl Pope 20Apr54; depl Sumpter 19Jul54 – Fairchild (AMC) Hagerstown 26Mar54 – 483 TCWG (FEA) Ashiya AB Japan 27Sep54; depl 316 TCG (FEA) Ashiya AB Japan 16Feb55 – 2256 RFC (CNR) Niagara Falls 27Mar57; depl 2235 RFC (CNR) Grenier 14Apr57; depl 2230 RFC (CNR) NAS New York 08Jun57 – 2242 RFC DT (CNR) Niagara Falls 01Dec57 – 328 TCSq (512 TCWG) (AFR) Niagara Falls 19Dec58; 914 TCG (512 TCWG) (AFR) Niagara Falls 01Jun63 – Wfu to MASDC (LOG) Davis-Monthan 07Dec70 for storage, PCN: **CJ378** – DFI: 21May71 – Kolar, Inc. Tucson Arizona 08Mar76 for scrap – Final Disposition: Scrapped.

11113 / 52-5934 / C-119G-1-FA

Avble: 30Dec53; Acc: 28Jan54; Del: 02Feb54 to USAF – 313 TCWG (TAC) Sewart 02Feb54; depl Pope 20Apr54; depl Sumpter 19Jul54 – Fairchild (AMC) Hagerstown

20Mar54 – 483 TCWG (FEA) Ashiya AB Japan 02Oct54; depl 316 TCG (FEA) Ashiya AB Japan 03Apr55 – 2585 RFC (CNR) Miami 25Jun57 – 435 TCWG (AFR) Miami 19Dec58; to Homestead 25Jul60; assigned 78 TCSq Bates 29Nov60 – 357 TCSq (302 TCWG) (AFR) Bates 08May61 – 908 TCG (435 TCWG) (AFR) Bates 18Mar63; to Brookley 30Oct64 – Wfu to MASDC (LOG) Davis-Monthan 25Feb69 for storage – DFI: 26Dec74 – Final Disposition: Scrapped.

11114 / 52-5935 / C-119G-1-FA

Avble: 01Jan54; Acc: 28Jan54; Del: 02Feb54 to USAF – 313 TCWG (TAC) Sewart 02Feb54; depl Pope 19Apr54; depl Sumpter 19Jul54 – Fairchild (AMC) Hagerstown 09Aug54 – 483 TCWG (FEA) Ashiya AB Japan Sep54; depl 314 TCWG (FEA) Ashiya AB Japan 09Nov54; depl 316 TCG (FEA) Ashiya AB Japan 13Nov54; bailment CAT (AMC) Taiwan 01Mar56 – 2237 RFC (CNR) New Castle 25Apr57; to unit unk (ARD) Hanscom Jan58; to NAS Willow Grove 21Jul58 – 512 TCWG (AFR) NAS Willow Grove 01Dec58; to unit unk 59; 912 TCG (512 TCWG) (AFR) NAS Willow Grove 11Feb63; 913 TAL GP (512 TCWG) (AFR) NAS Willow Grove 19Aug68 – Fairchild (LOG) St. Augustine 23Sep68, cvtd **AC-119K Stinger** 02Oct68 – 1 SOP WG (TAC) Lockbourne 30Apr69 – 4410 CCT WG (TAC) Lockbourne 14Jul69 – 14 SOP WG (PAF) Phan Rang AB S. Vietnam 18Nov69; assigned variously Da Nang, Phu Cat S. Vietnam & Udorn Thailand from 07Dec69. Suffered a runaway propeller on the no. 1 engine after taking off from Da Nang AB S. Vietnam on the night of 06Jun70. While attempting a return to base the aircraft became uncontrollable and was abandoned over the sea east of Da Nang. 9 of the 10 crew were able to be rescued except for TSgt Clyde D. Alloway who is listed as lost at sea. The abandoned AC-119K caused some concern for a period when it flew on toward Chinese airspace near Hainan Is., however it soon impacted the South China Sea and sank – Final Disposition: Accident.

11115 / 52-5936 / C-119G-1-FA

Avble: 04Jan54; Acc: 28Jan54; Del: 02Feb54 to USAF – 313 TCWG (TAC) Sewart 02Feb54; depl Pope 20Apr54; depl Sumpter 19Jul54 – Fairchild (AMC) Hagerstown 26Aug54 – 483 TCWG (FEA) Ashiya AB Japan Sep54; depl 316 TCG (FEA) Ashiya AB Japan 01May55 – 2585 RFC (CNR) Miami 07Jun57 – 435 TCWG (AFR) Miami 19Dec58 **RECORDS INDECIPHERABLE** to 445 TCA DT (CNC) Memphis 26May60 mnt; to Homestead 25Jul60; assigned 76 TCSq Homestead 01Oct61; 915 TCG (435 TCWG) (AFR) Homestead 17Jan63 – 906 TCG (302 TCWG) (AFR) Clinton 07Dec65; to Lockbourne 24Jul71 – Fairchild (LOG) Crestview 27Dec71, cvtd **C-119K** 26Jun72 – 907 TAL GP (302 TCWG) (AFR) Lockbourne 18Jul72 – MAP Jordan 21Oct72, RJAF s/n: **118** (ntu). Crashed into a mountain near Devès Pass near Aulan France in poor weather after departing Wiesbaden AB W. Germany on its delivery flight to Amman Jordan 27Oct72. It appears the aircraft deviated from its intended flight route likely due to the strong winds at the time. 5 ANG delivery crew were killed: Lt Col Dale R. Anderson (co-pilot), Maj Francis T. Durkin (pilot), Maj Marion R. Mackentroth (navigator), TSgt William A. Barbor (mechanic) and TSgt William L. Champion – Final Disposition: Accident.

11116 / 52-6048 / C-119G-30-FA

Avble: 04Jan54; Acc: 29Jan54; Del: 11Feb54 to MDAP Italy, AMI s/n: **MM52-6048** – Final Disposition: Scrapped.

11117 / 52-6049 / C-119G-30-FA

Avble: 06Jan54; Acc: 28Jan54; Del: 08Feb54 to MDAP Italy, AMI s/n: **MM52-6049** – Final Disposition: Scrapped.

11118 / 52-6050 / C-119G-30-FA

Avble: 07Jan54; Acc: 29Jan54; Del: 08Feb54 to MDAP Belgium, BAF s/n: **CP-34**; BOC: 15Feb54 15 WG / 40 SQ Melsbroek AB Belgium, callsign: **OT-CBN**, later to 15 WG / 20 SQ. Minor incident at Melsbroek 05Nov57. Ground collision with RAF Meteor s/n: WD752 02Sep59 – Wfu 09Jul73 and stored at Koksijde AB Belgium; SOC: Jul74 – Sold for scrap to J. Trappeniers 04Oct74, scrapped 1977-78 – Final Disposition: Scrapped.

11119 / 52-6051 / C-119G-30-FA

Avble: 08Jan54; Acc: 29Jan54; Del: 08Feb54 to MDAP Belgium, BAF s/n: **CP-38**; BOC: 21Feb54 15 WG / 40 SQ Melsbroek AB Belgium, callsign: **OT-CBR**, later to 15 WG / 20 SQ. Violent nose wheel shimmy damaged steering system while landing at Melsbroek 10Oct55 – Wfu 01Dec72 and stored at Koksijde AB Belgium; SOC: Jul74 – Sold for scrap to J. Trappeniers 04Oct74, scrapped 1977-78 – Final Disposition: Scrapped.

11120 / 52-6052 / C-119G-30-FA

Avble: 09Jan54; Acc: 29Jan54; Del: 10Feb54 to MDAP Belgium, BAF s/n: **CP-35**; BOC: 17Feb54 15 WG / 40 SQ Melsbroek AB Belgium, callsign: **OT-CBO**, later to 15 WG / 20 SQ. Hard landing causing damage to the nose gear 19Jul55. Fuselage ripped open during the loading of a Harvard aircraft 07Jul56. Minor incidents 09Jul57, 05Nov57. Hit tree during taxiing 04Jan61. Minor incidents 19Oct65, 19Nov65, 18Mar70 and 08Dec70. Cargo drop mishap 06Jul72 with packages falling well off target onto a factory and a garden – Wfu 02Jul73 and stored at Koksijde AB Belgium; SOC: Jul74 – Sold for scrap to J. Trappeniers 04Oct74, scrapped 1977-78 – Final Disposition: Scrapped.

11121 / 52-6053 / C-119G-30-FA

Avble: 11Jan54; Acc: 05Feb54; Del: 19Feb54 to MDAP Italy, AMI s/n: **MM52-6053** – Final Disposition: Scrapped.

11122 / 52-6054 / C-119G-30-FA

Avble: 12Jan54; Acc: 12Feb54; Del: 19Feb54 to MDAP Italy, AMI s/n: **MM52-6054** – Final Disposition: Scrapped.

11123 / 52-6055 / C-119G-35-FA

Avble: 13Jan54; Acc: 05Feb54; Del: 10Feb54 to MDAP Belgium, BAF s/n: **CP-37**; BOC: 18Feb54 15 WG / 40 SQ Melsbroek AB Belgium, callsign: **OT-CBQ**, later to 15 WG / 20 SQ. Minor accidents 04May55 and 20Nov67 – Wfu 03Jul72 and stored at Koksijde AB Belgium – MAP Ethiopia 29Sep72, ETAF s/n: **919** – Wfu 86. Noted 90s stored at Debre Zeyit AB Ethiopia – Final Disposition: Scrapped.

11124 / 52-5937 / C-119G-1-FA

Avble: 14Jan54; Acc: 05Feb54; Del: 19Mar54 to USAF – 483 TCWG (FEA) Ashiya AB Japan 19Mar54; to 21 TCSq (483 TCWG) Tachikawa AB Japan 28May57 – Wfu to 868 TM SQ (PAF) Tainan Taiwan 02Oct58 – MAP Taiwan 01Jul59, RoCAF s/n: **3137** – Final Disposition: Unknown.

Section Four—Service Histories 345

11125 / 52-5938 / C-119G-1-FA

Avble: 15Jan54; Acc: 12Feb54; Del: 19Mar54 to USAF – **RECORDS MISSING 1954-55** – 483 TCWG (FEA) Ashiya AB Japan 19Mar54; bailment CAT (AMC) Taiwan 30Dec55 **RECORDS INDECIPHERABLE** 2346 RFC (CNR) Hamilton 58 – Hayes (AMC) Birmingham 16Dec58 – 349 TCWG (AFR) Hamilton 07Mar59; 938 TCG (349 TCWG) (AFR) Hamilton 10Feb63 – 944 TCG (452 TCWG) (AFR) March 02Mar66 – 930 TAL GP (434 TCWG) (AFR) Bakalar 24Jan68 – Fairchild (LOG) St. Augustine 26Mar68, cvtd **AC-119G Shadow** 01Jul68 – 71 SOP WG (1 ACO WG) (TAC) Lockbourne 07Aug68 – Fairchild (LOG) St. Augustine 09Dec68 – 14 SOP WG (PAF) Nha Trang AB S. Vietnam 15Jan69; assigned variously Phan Rang, Tan Son Nhut, Tuy Hoa, Phu Cat, Da Nang S. Vietnam from 27Feb69 – MAP S. Vietnam 04Sep71, VNAF s/n: **525938**, 819 ASQ (53 TWG) Tan Son Nhut AB – Final Disposition: Unknown.

11126 / 52-5939 / C-119G-1-FA

Avble: 16Jan54; Acc: 10Feb54; Del: 19Mar54 to USAF – 483 TCWG (FEA) Ashiya AB Japan 19Mar54; to 1500 MAI SQ (MATS) Hickham 24Aug54 mnt; bailment CAT (AMC) Taiwan 01Mar56; to 6200 AB WG (FEA) Clark AB The Philippines 26Dec56 – 2346 RFC (CNR) Hamilton 01Aug58 – 349 TCWG (AFR) Hamilton 23Dec58 **RECORDS INDECIPHERABLE** 938 TCG (349 TCWG) (AFR) Hamilton 10Feb63 – 907 TCG (302 TCWG) (AFR) Clinton 18Apr66 – Wfu to MASDC (LOG) Davis-Monthan 25Sep71 for storage, PCN: **CJ414** – DFI: 26Dec73 – Final Disposition: Scrapped.

11127 / 52-5940 / C-119G-1-FA

Avble: 18Jan54; Acc: 12Feb54; Del: 19Mar54 to USAF – 483 TCWG (FEA) Ashiya AB Japan 19Mar54; bailment CAT (AMC) Taiwan 07Jan56 – 2472 RFC DT (CNR) Tinker Mar58 – 305 TCSq (442 TCWG) (AFR) Tinker 19Mar59 – 65 TCSq (403 TCWG) (AFR) Davis 27Feb61; 929 TCG (403 TCWG) (AFR) Davis 11Feb63 – 922 TCG (433 TCWG) (AFR) Kelly 18Oct65 – Fairchild (LOG) St. Augustine 01Jul68, cvtd **AC-119K Stinger** 26Jul68 – 1 SOP WG (TAC) Lockbourne 02Jun69 – 4410 CCT WG (TAC) Lockbourne 14Jul69 – 14 SOP WG (PAF) Phan Rang AB S. Vietnam 26Dec69; assigned variously Phu Cat, Da Nang S. Vietnam & Udorn, Nakhon Phanom Thailand from 30Jan70; depl 460 TR WG (PAF) Tan Son Nhut S. Vietnam 13Dec70 – 56 SOP WG (PAF) Da Nang AB S. Vietnam 24Aug71; assigned variously Nakhon Phanom Thailand from 21Sep71; to 432 TR WG (PAF) Udorn AB Thailand 25May72 – MAP S. Vietnam 10Nov72, VNAF s/n: **525940**, 821 ASQ (53 TWG) Tan Son Nhut AB – Final Disposition: Unknown.

11128 / 52-5941 / C-119G-1-FA

Avble: 21Jan54; Acc: 12Feb54; Del: 19Mar54 to USAF – 483 TCWG (FEA) Ashiya AB Japan 19Mar54 – 2347 RFC (CNR) Long Beach 24May58 – 452 TCWG (AFR) Long Beach 13Jun59; to March 01Oct60; 943 TCG (452 TCWG) (AFR) March 11Feb63 – Fairchild (LOG) St. Augustine 05Jun63 – MAP India 17Jul63 – Final Disposition: Unknown.

11129 / 52-5942 / C-119G-1-FA

Avble: 21Jan54; Acc: 18Feb54; Del: 19Mar54 to USAF – 483 TCWG (FEA) Ashiya AB Japan 19Mar54 – 2347 RFC (CNR) Long Beach 01Jul58 – 2346 RFC DT (CNR) McClellan 23Jul58 – 314 TCSq (349 TCWG) (AFR) McClellan 23Dec58; 940 TCG (349 TCWG) (AFR) McClellan 10Feb63 – 904 TCG (514 TCWG) (AFR) Stewart 13May65; 903 TCG (514 TCWG) (AFR) McGuire 31May67; depl Myrtle Beach South Carolina 17Jul67 – Fairchild

(LOG) St. Augustine 04Apr68, cvtd **AC-119G Shadow** 01Jul68 – 71 SOP SQ (1 ACO WG) (TAC) Lockbourne 07Aug68 – Fairchild (LOG) St. Augustine 26Nov68 – 14 SOP WG (PAF) Nha Trang AB S. Vietnam 30Dec68; assigned variously Phan Rang, Tan Son Nhut, Tuy Hoa, Phu Cat, Da Nang S. Vietnam from 13Feb69 – MAP S. Vietnam 06Sep71, VNAF s/n: **525942**, 819 ASQ (53 TWG) Tan Son Nhut AB – Final Disposition: Unknown.

11130 / 52-5943 / C-119G-1-FA

Avble: 26Jan54; Acc: 18Feb54; Del: 19Mar54 to USAF – 483 TCWG (FEA) Ashiya AB Japan 19Mar54; bailment CAT (AMC) Taiwan 13Nov56 – 2347 RFC (CNR) Long Beach 25Jun58 – 452 TCWG (AFR) Long Beach 13Jun59; to March 01Oct60; 944 TCG (452 TCWG) (AFR) March 11Feb63 – MAP India 18Jul63 – Final Disposition: Unknown.

11131 / 52-5944 / C-119G-1-FA

Avble: 26Jan54; Acc: 18Feb54; Del: 23Feb54 to USAF – 313 TCWG (TAC) Sewart 23Feb54; depl Pope 20Apr54; depl Sumpter 19Jul54 – Fairchild (AMC) Hagerstown 10Aug54 – 483 TCWG (FEA) Ashiya AB Japan 24Sep54; to 319 AB WG (FEA) Andersen Guam 18Mar55 mnt; to 3960 AB WG (SAC) Andersen Guam 01Apr55 mnt; depl 316 TCWG (FEA) Ashiya AB Japan 01Jul55. Impacted a 6,100 ft mountain side 20 miles S Saijo on Shikoku Is. Japan 03Apr56 while on a cargo flight from Tachikawa AB to Ashiya AB Japan. Searchers sighted the wreckage a few days later, all 5 crew were killed – Final Disposition: Accident.

11132 / 52-5945 / C-119G-1-FA

Avble: 27Jan54; Acc: 18Feb54; Del: 25Feb54 to USAF – 313 TCWG (TAC) Sewart 25Feb54; depl Pope 19Apr54; depl Sumpter 19Jul54 – Fairchild (AMC) Hagerstown 54 – 483 TCWG (FEA) Ashiya AB Japan 17Sep54; depl 316 TCG (FEA) Ashiya AB Japan 01Jul55 – 2237 RFC (CNR) New Castle 03May57; to NAS Willow Grove 21Jul58 – Hayes (AMC) Birmingham 29Oct58 – 512 TCWG (AFR) NAS Willow Grove 07Jan59; 913 TCG (512 TCWG) (AFR) NAS Willow Grove 11Feb63 – Fairchild (LOG) St. Augustine 23Sep68, cvtd **AC-119K Stinger** 02Oct68 – 4410 CCT WG (TAC) Lockbourne 15Aug69 – 14 SOP WG (PAF) Phan Rang AB Japan 26Dec69; assigned variously Phu Cat, Da Nang S. Vietnam & Udorn, Nakhon Phanom Thailand from 01Feb70 – 313 ADH DV (PAF) Kadena AB Japan 24Aug71 – 56 SOP WG (PAF) Nakhon Phanom AB Thailand 09Oct71; assigned Da Nang 03Dec71 – MAP S. Vietnam 10Nov72, VNAF s/n: **525945**, 821 ASQ (53 TWG) Tan Son Nhut AB – Final Disposition: Unknown.

11133 / 52-5946 / C-119G-1-FA

Avble: 28Jan54; Acc: 23Feb54; Del: 02Mar54 to USAF – 313 TCWG (TAC) Sewart 02Mar54; depl Pope 20Apr54; depl Sumpter 19Jul54 – Fairchild (AMC) Hagerstown 17Aug54 – 483 TCWG (FEA) Ashiya AB Japan 17Sep54; depl 316 TCG (FEA) Ashiya AB Japan 23May55 – 2585 RFC (CNR) Miami 01Jun57 – 435 TCWG (AFR) Miami 19Dec58 **RECORDS INDECIPHERABLE** to Homestead 25Jul60; assigned 76 TCSq Homestead 01Oct61; 915 TCG (435 TCWG) (AFR) Homestead 13Feb63 – 926 TCG (446 TCWG) (AFR) NAS New Orleans 13Nov65 – 913 TAL GP (512 TCWG) (AFR) NAS Willow Grove 01Feb70 – Wfu to MASDC (LOG) Davis-Monthan 28Feb71 for storage, PCN: **CJ408** – DFI: 09Dec74 – Final Disposition: Scrapped.

Section Four—Service Histories

11134 / 52-5947 / C-119G-1-FA

Avble: 29Jan54; Acc: 19Feb54; Del: 02Mar54 to USAF – 313 TCWG (TAC) Sewart 02Mar54; depl Pope 19Apr54; depl Sumpter 19Jul54; depl Elmendorf 12Jan55 – 464 TCWG (TAC) Sewart 08Jun55 – bailment Fairchild (AMC) Hagerstown 18Jul56 – 314 TCWG (TAC) Sewart 19Aug56 – 2469 RFC (CNR) Scott 03May57; depls 2466 RFC (CNR) Bakalar 09Jul57, 18Aug57 & 12Dec57; cvtd **C-119J** 22Jul57 – 2466 RFC DT (CNR) Scott 14Dec57; to 5700 AB SQ (CNC) Albrook AFS Panama CZ 58 mnt – 73 TCSq (434 TCWG) (AFR) Scott 18Apr59 **RECORDS INDECIPHERABLE** 150 AML SQ (NJ-ANG) Newark 11Apr61 – **RECORDS MISSING 1961–1962** – 452 TCWG (AFR) March date unk – Wfu to 2704 ASD GP (LOG) Davis-Monthan 11Dec62 for storage – Fairchild (LOG) St. Augustine 31Dec63 – MAP Italy 06Mar64, AMI s/n: **MM52-5947** – Final Disposition: Scrapped.

11135 / 52-5948 / C-119G-1-FA

Avble: 01Feb54; Acc: 26Feb54; Del: 05Mar54 to USAF – 313 TCWG (TAC) Sewart 05Mar54; depl Pope 20Apr54; depl Sumpter 19Jul54 – Fairchild (AMC) Hagerstown 17Aug54 – 483 TCWG (FEA) Ashiya AB Japan 12Oct54; depl 316 TCWG (FEA) Ashiya AB Japan 01Jul55 – 2256 RFC (CNR) Niagara Falls 07Apr57 **RECORDS INDECIPHERABLE** 2237 RFC DT (CNR) Niagara Falls date unk – 328 TCSq (512 TCWG) (AFR) Niagara Falls 19Dec58; 914 TCG (512 TCWG) (AFR) Niagara Falls 11Feb63 – Wfu to MASDC (LOG) Davis-Monthan 23Nov70 for storage, PCN: **CJ364** – DFI: 21May71 – Kolar, Inc. Tucson Arizona 02Mar76 for scrap – Final Disposition: Scrapped.

11136 / 52-5949 / C-119G-1-FA

RECORDS MISSING – Avble: 03Feb54; Acc: 26Feb54; Del: 05Mar54 to USAF – 313 TCWG (TAC) Sewart 05Mar54, assigned 29 TCSq. Had a catastrophic engine failure after take-off from Sewart AFB Tennessee 11Jan55. Pilot: Ross N. Richards ordered an emergency bail out of 3 crew and the 34 paratroopers onboard. Richards and his co-pilot stayed at the controls until everyone had left the aircraft but were unable to bail out themselves and were killed on impact 5 miles NE Sewart AFB – Final Disposition: Accident.

11137 / 52-5950 / C-119G-1-FA

Avble: 03Feb54; Acc: 26Feb54; Del: 05Mar54 to USAF – 313 TCWG (TAC) Sewart 05Mar54; depl Pope 20Apr54; depl Sumpter 19Jul54 – Fairchild (AMC) Hagerstown 26Aug54 – 483 TCWG (FEA) Ashiya AB Japan 03Oct54; depl Akddoae? 03Aug55; depl 316 TCWG (FEA) Ashiya AB Japan 17Aug55 – SMAAR (AMC) McClellan 29Jun57 mnt – 1739 FRY SQ (MATS) Amarillo 25Jul57 – 2237 RFC (CNR) New Castle 24Jul57; to NAS Willow Grove 21Jul58 – 512 TCWG (AFR) NAS Willow Grove 01Dec58 **RECORDS INDECIPHERABLE** 913 TCG (512 TCWG) (AFR) NAS Willow Grove 11Feb63; depl 912 TCG (512 TCWG) (AFR) NAS Willow Grove 11Oct66 – Wfu to MASDC (LOG) Davis-Monthan 14Dec69 for storage, PCN: **CJ340** – DFI: 20Jan70 – Kolar, Inc. Tucson Arizona 17Feb76 for scrap – Final Disposition: Scrapped.

11138 / 52-5951 / C-119G-1-FA

Avble: 04Feb54; Acc: 26Feb54; Del: 05Mar54 to USAF – 313 TCWG (TAC) Sewart 05Mar54; depl Pope 20Apr54; depl Sumpter 19Jul54; depl Elmendorf 12Jan55 – 463 TCWG (TAC) Sewart (depl) 08Jun55 – 314 TCWG (TAC) Sewart 11Aug56; depl England Louisiana 05Nov56 – BMC (AMC) Birmingham 18Apr57 – 2472 RFC (CNR) Richards 08Jul57 – 442 TCWG (AFR) Richards 02Mar59 – 313 TCSq (349 TCWG) (AFR) Portland 01Jun61;

939 TCG (349 TCWG) (AFR) Portland 11Feb63 – 931 TAL GP (434 TCWG) (AFR) Bakalar 22Jun68 – 906 TAL GP (302 TCWG) (AFR) Clinton 16Feb69 – MAP Taiwan 05Jan70 – Final Disposition: Unknown.

11139 / 52-5952 / C-119G-1-FA

Avble: 08Feb54; Acc: 26Feb54; Del: 05Mar54 to USAF – 313 TCWG (TAC) Sewart 05Mar54; depl Pope 20Apr54; depl Sumpter 19Jul54; to 6606 AB WG (NEA) Goose Bay Canada 10May55 mnt – 463 TCWG (TAC) Sewart 08Jun55 – 314 TCWG (TAC) Sewart 12Aug56; depl England Louisiana 05Nov56 – BMC (AMC) Birmingham 22Apr57 – 2466 RFC (CNR) Bakalar 27Jul57; to 1001 AB WG (HQC) Andrews 03Jan59 mnt – 434 TCWG (AFR) Bakalar 19Apr59; 931 TCG (434 TCWG) (AFR) Bakalar 11Feb63; 930 TAL GP (434 TCWG) (AFR) Bakalar 25Apr68; depl 71 ACO SQ (1 ACO WG) (AFR) Lockbourne 20Jun68 – 906 TAL GP (302 TCWG) (AFR) Clinton 04Sep68 – MAP Taiwan 04Jan70 – Final Disposition: Unknown.

11140 / 52-5953 / C-119G-1-FA

Avble: 09Feb54; Acc: 26Feb54; Del: 05Mar54 to USAF – 313 TCWG (TAC) Sewart 05Mar54; depl Pope 20Apr54; depl Sumpter 19Jul54 – Fairchild (AMC) Hagerstown 30Aug54 – 483 TCWG (FEA) Ashiya AB Japan 54; to 1509 SUT SQ (MATS) Johnston 24Feb55 mnt; to 6486 AB WG (FEA) Hickham 01Apr55 mnt; depl 316 TCWG (FEA) Ashiya AB Japan 01Jul55; to 67 TR WG (FEA) Itami AB Japan 05Apr56 – 2237 RFC (CNR) New Castle 18Apr57; to NAS Willow Grove 21Jul58 – 512 TCWG (AFR) NAS Willow Grove 01Dec58; 913 TCG (512 TCWG) (AFR) NAS Willow Grove 11Feb63 – Wfu to MASDC (LOG) Davis-Monthan 14Dec69 for storage, PCN: **CJ341** – DFI: 20Jan70 – Kolar, Inc. Tucson Arizona 23Feb76 for scrap – Final Disposition: Scrapped.

11141 / 52-5954 / C-119G-1-FA

Avble: 10Feb54; Acc: 12Mar54; Del: 18Mar54 to USAF – 483 TCWG (FEA) Ashiya AB Japan 18Mar54 **RECORDS INDECIPHERABLE** → bailment CAT (AMC) Taiwan 28Feb56; depl Tachikawa AB Japan date unk – Wfu to 868 TM SQ (PAF) Tainan Taiwan 02Oct58 – MAP Taiwan 01Jul59 ← **RECORDS INDECIPHERABLE** Final Disposition: Unknown.

11142 / 52-9981 / C-119G-FA

Avble: 11Feb54; Acc: 12Mar54; Del: 18Mar54 to USAF – 483 TCWG (FEA) Ashiya AB Japan 18Mar54; bailment CAT (FEA) Taiwan 05Apr55 – 2346 RFC (CNR) Hamilton 24Jul58; to 124 CLM SQ (ID-ANG) Boise 16Aug58 mnt – 349 TCWG (AFR) Hamilton 29Nov58; 938 TCG (349 TCWG) (AFR) Hamilton 10Feb63 – 912 TCG (512 TCWG) (AFR) NAS Willow Grove 21Jun65; 913 TAL GP (512 TCWG) (AFR) NAS Willow Grove 05Jun68 – Wfu to MASDC (LOG) Davis-Monthan 19Dec69 for storage, PCN: **CJ345** – DFI: 20Jan70 – Kolar, Inc. Tucson Arizona 18Feb76 for scrap – Final Disposition: Scrapped.

11143 / 52-9982 / C-119G-FA

Avble: 12Feb54; Acc: 12Mar54; Del: 18Mar54 to USAF – 483 TCWG (FEA) Ashiya AB Japan 18Mar54; bailment CAT (FEA) Taiwan 16Jun55; depl Akddoae? 14Aug56 – 2346 RFC DT (CNR) Paine 19Apr58 – 97 TCSq (349 TCWG) (AFR) Paine 23Dec58; 941 TCG (349 TCWG) (AFR) Paine 19Jun63 – 924 TCG (446 TCWG) (AFR) Ellington 18Aug65 – 927 TAL GP (403 TCWG) (AFR) Selfridge 27Jan68 – Fairchild (LOG) St. Augustine 09Sep68, cvtd **AC-119K Stinger** 17Sep68 – 1 SOP WG (TAC) Lockbourne 18Jun69 – 4410

Section Four—Service Histories

CCT WG (TAC) Lockbourne 14Jul69; to 341 STM WG (SAC) Malmstrom 21Oct69 mnt – 14 SOP WG (PAF) Phan Rang AB S. Vietnam 27Dec69; assigned variously Phu Cat, Da Nang S. Vietnam & Udorn, Nakhon Phanom Thailand from 07Feb70; to 313 ADH DV (PAF) Kadena AB Japan 05Nov70 – SMAAR (LOG) Phan Rang S. Vietnam 23Aug71 – 56 SOP WG (PAF) Nakhon Phanom AB Thailand 26Aug71; assigned variously Da Nang S. Vietnam from 13Oct71 – MAP S. Vietnam 10Nov72, VNAF s/n: **529982**, 821 ASQ (53 TWG) Tan Son Nhut AB – Final Disposition: Unknown.

11144 / 52-6056 / C-119G-35-FA

Avble: 16Feb54; Acc: 11Mar54; Del: 16Mar54 to MDAP Italy, AMI s/n: **MM52-6056** – Final Disposition: Scrapped.

11145 / 52-6057 / C-119G-35-FA

Avble: 19Feb54; Acc: 11Mar54; Del: 16Mar54 to MDAP Italy, AMI s/n: **MM52-6057** – Final Disposition: Scrapped.

11146 / 52-6058 / C-119G-35-FA

Avble: 23Feb54; Acc: 12Mar54; Del: 18Mar54 to MDAP Belgium, BAF s/n: **CP-40**; BOC: 09May54 15 WG / 20 SQ Melsbroek AB Belgium, callsign: **OT-CBT**. Emergency landing due to engine failure Oct54. Minor ground collision with RAF C-130 (XV296) 15Oct68 – Wfu 30Dec71 and stored at Koksijde AB Belgium; SOC: Jul74 – Sold for scrap to J. Trappeniers 04Oct74, scrapped 1977-78 – Final Disposition: Scrapped.

11147 / 53-3136 / C-119G-36-FA

Avble: 11Mar54; Acc: 31Mar54; Del: 31Mar54 to USAF – Bailment Fairchild (AMC) Hagerstown 31Mar54, cvtd **JC-119G** 01Dec55; brief depl 464 TCWG (TAC) Pope 01Sep57 – Wfu to 2704 ASD GP (LOG) Davis-Monthan 04Feb64 for storage; cvtd **C-119G** – 942 TCG (452 TCWG) (AFR) March 16Mar64; 943 TCG (452 TCWG) (AFR) March 22May64 – Fairchild (LOG) St. Augustine 26Mar68, cvtd **AC-119G Shadow** 01Jul68 – 71 SOP SQ (1 ACO WG) (TAC) Lockbourne 08Aug68 – Fairchild (LOG) St. Augustine 13Nov68 – 14 SOP WG (PAF) Nha Trang AB S. Vietnam 10Dec68; to 3960 STR WG (SAC) Andersen Guam 13jan69; assigned variously Tan Son Nhut, Phan Rang, Tuy Hoa, Da Nang S. Vietnam from 29Jan69; to 18 Fighter WG (PAF) Kadena AB Japan 26Sep69 – MAP S. Vietnam 07Sep71, VNAF s/n: **533136**, 819 ASQ (53 TWG) Tan Son Nhut AB – Final Disposition: Unknown.

11148 / 53-3137 / C-119G-36-FA

Avble: 12Mar54; Acc: 29Mar54; Del: 02Apr54 to USAF – 464 TCWG (TAC) Lawson 02Apr54; depl Seymour 22Apr54; to Pope 13Sep54; depl Rhein-Main AB W. Germany 19Oct54; depl Aberdeen 05May56 – 2466 RFC (CNR) Bakalar 19Aug57; depl 2471 RFC (CNR) O'Hare 05Aug57; depl 2473 RFC (CNR) Gen Mitchell 27Sep57 – 434 TCWG (AFR) Bakalar 19Apr59; 931 TCG (434 TCWG) (AFR) Bakalar 11Feb63; 930 TAL GP (434 TCWG) (AFR) Bakalar 25Apr68; depl 71 SOP SQ (1 ACO WG) (TAC) Lockbourne 17Jun68 – 928 TAL GP (403 TCWG) (AFR) O'Hare 03Sep68 – Wfu to MASDC (LOG) Davis-Monthan 23Jan71 for storage, PCN: **CJ385** – DFI: 31Mar76 – Dross Metals Inc. Tucson Arizona 13Sep79 for scrap – Final Disposition: Scrapped.

11149 / 53-3138 / C-119G-36-FA

Avble: 16Mar54; Acc: 29Mar54; Del: 02Apr54 to USAF – 464 TCWG (TAC) Lawson 02Apr54; depl Seymour 22Apr54; to Pope 13Sep54; depl Rhein-Main AB W. Germany 19Oct54; depl Aberdeen 05May56 – 2472 RFC (CNR) Richards 09Jul57; to DT Tinker 09May58; depl DT Davis 20Jul58 – Hayes (AMC) Birmingham 12Mar59 – 305 TCSq (442 TCWG) (AFR) Tinker 20Apr59 – 65 TCSq (403 TCWG) (AFR) Davis 02Mar61; 929 TCG (403 TCWG) (AFR) Davis 11Feb63 – Fairchild (LOG) St. Augustine 16May63 – MAP India 08Jun63 – Final Disposition: Unknown.

11150 / 53-3139 / C-119G-36-FA

Avble: 16Mar54; Acc: 31Mar54; Del: 07Apr54 to USAF – 464 TCWG (TAC) Lawson 07Apr54; depl Seymour 22Apr54; to Pope 13Sep54; depl Rhein-Main AB W. Germany 19Oct54; depl Aberdeen 05May56 – 2469 RFC (CNR) Scott 04Jul57; depl 2466 RFC (CNR) Bakalar 18Aug57 – 2466 RFC DT (CNR) Scott 01Dec57 – 73 TCSq (434 TCWG) (AFR) Scott 18Apr59; 932 TCG (434 TCWG) (AFR) Scott 11Feb63 – 910 TCG (459 TCWG) (AFR) Youngstown 08Oct66 – Fairchild (LOG) St. Augustine 05Dec67 – MAP S. Vietnam 28Mar68, VNAF s/n: **533139**, 413 TSQ (53 TWG) Tan Son Nhut AB – Final Disposition: Unknown.

11151 / 53-3140 / C-119G-36-FA

Avble: 18Mar54; Acc: 31Mar54; Del: 07Apr54 to USAF – 464 TCWG (TAC) Lawson 07Apr54; depl Seymour 22Apr54; to Pope 13Sep54; depl Rhein-Main AB W. Germany 19Oct54; depl Aberdeen 05May56 – 2469 RFC (CNR) Scott 10Jun57; depl 2466 RFC (CNR) Bakalar 18Aug57 – 2466 RFC DT (CNR) Scott 01Dec57 – 73 TCSq (434 TCWG) (AFR) Scott 18Apr59; to 825 COS GP (SAC) Little Rock 28Aug59; 932 TCG (434 TCWG) (AFR) Scott 11Feb63 – 910 TCG (459 TCWG) (AFR) Youngstown 08Oct66 – Fairchild (LOG) St. Augustine 19Oct67 – MAP S. Vietnam 08Mar68, VNAF s/n: **533140**, 413 TSQ (53 TWG) Tan Son Nhut AB – Taiwan 73, RoCAF s/n: **3140** – Final Disposition: Unknown.

11152 / 53-3141 / C-119G-36-FA

Avble: 19Mar54; Acc: 31Mar54; Del: 07Apr54 to USAF – 464 TCWG (TAC) Lawson 07Apr54; depl Seymour 22Apr54; to Pope 13Sep54; depl Rhein-Main AB W. Germany 19Oct54; depl Aberdeen 05May56; to 3605 NTN WG (TC) Ellington 01Sep56 mnt – Hayes (AMC) Birmingham 15Jul57 – 2465 RFC (CNR) Minneapolis-St. Paul 17Oct57 – 2473 RFC DT (CNR) Minneapolis-St. Paul 01Dec57; to 2465 AB GP (CNC) Minneapolis-St. Paul 20Jun58 mnt – 96 TCSq (440 TCWG) (AFR) Minneapolis-St. Paul 19Dec59; 934 TCG (440 TCWG) (AFR) Minneapolis-St. Paul 11Feb63 – 906 TAL GP (302 TCWG) (AFR) Clinton 22Nov69; 907 TAL GP (302 TCWG) (AFR) Clinton 01Mar72 – MAP S. Vietnam 08Jul72, VNAF s/n: **533141**, 413 TSQ (53 TWG) Tan Son Nhut AB – Final Disposition: Unknown.

11153 / 53-3142 / C-119G-36-FA

Avble: 23Mar54; Acc: 07Apr54; Del: 15Apr54 to USAF – 483 TCWG (FEA) Ashiya AB Japan 15Apr54; bailment CAT (FEA) Taiwan 22Feb55 – 2346 RFC (CNR) Hamilton 07Jul58 – 349 TCWG (AFR) Hamilton 23Dec58; 938 TCG (349 TCWG) (AFR) Hamilton 10Feb63 – Fairchild (AMC) Hagerstown 22Jun66 – 913 TCG (512 TCWG) (AFR) NAS Willow Grove cvtd **YC-119K** 14Apr67 – bailment Fairchild (AMC) Hagerstown 01Jun67 – 907 TAL GP (302 TCWG) (AFR) Clinton 09Apr70 – WRAAR (LOG) Clinton 13Sep70 – AFT Center (SYS) Edwards 28Jul71 – Wfu to MASDC (LOG) Davis-Monthan 16Mar72 for storage, PCN: **CJ419** – DFI: 26Dec73 – Final Disposition: Scrapped.

11154 / 53-3143 / C-119G-36-FA

Avble: 26Mar54; Acc: 08Apr54; Del: 16Apr54 to USAF – 483 TCWG (FEA) Ashiya AB Japan 16Apr54; to 6319 AB WG (FEA) Andersen Guam 16Jul54; to 21 TCSq (483 TCWG) Tachikawa AB Japan 18Sep56 – Wfu to 868 TM SQ (PAF) Tainan Taiwan 08Oct58 – MAP Taiwan 01Jul59, RoCAF s/n: **3143** – Final Disposition: Unknown.

11155 / 53-3144 / C-119G-36-FA

Avble: 31Mar54; Acc: 15Apr54; Del: 23Apr54 to USAF – 483 TCWG (FEA) Ashiya AB Japan 23Apr54 – 2346 RFC DT (CNR) Portland 19Apr58 – 313 TCSq (349 TCWG) (AFR) Portland 23Dec58; 939 TCG (349 TCWG) (AFR) Portland 63 – 922 TAL GP (433 TCWG) (AFR) Kelly 13Jun68 – Wfu to MASDC (LOG) Davis-Monthan 28Jan71 for storage – Hayes (LOG) Dothan 18Jun71 mnt – 143 SOP GP (ANG) Theodore F. Green 09Sep71; cvtd **C-119L** 04Mar73 – Wfu to MASDC (LOG) Davis-Monthan 19Nov73 for storage – DFI: 31Mar76 – BofS Dross Metals, Inc. Tucson Arizona 12Sep80; aircraft actually acquired 15Dec78 and sold to William Waara (Michigan Aerial Applicator) Dothan Alabama 04Oct79 reg **N37484**, re-reg **N8512N** but ntu, N37484 retained – BofS Arbor Air (William Waara d/b/a Arbor Air) Ann Arbor Michigan 10Apr80 – BofS El Marc Air Columbus Ohio 13Apr81 – BofS Bud's Flying Service Ltd. Rising City Nebraska 04Oct82 – BofS Louis P. Minkoff Ypsilanti Michigan 24Dec82, this BofS was disputed by Bud's and a court case later ruled that Bud's Flying Service were the rightful owners of the aircraft – BofS Texas Aerial Applicators, Inc. Laredo Texas 04Aug83 – BofS American Air Freight Co. Laredo Texas 21Jan86; proposed export to Transporte De Carga Aerea Especializada Y Servicios of Mexico May87 but the sale never went through – Central Air Service, Inc. of Tucson Arizona circa 1987, no BofS document – BofS USAF Museum WPAFB Ohio 05Nov87 for static display at Hurlburt Field Memorial Air Park Florida. Apparently flew to Hurlburt in Sep87 from Texas without radios onboard and no flight plan filed. When the aircraft was picked up on radar DEA aircraft intercepted believing it to be a drug smuggler. Presently, correctly marked as s/n "33144" in a livery representing an AC-119G Shadow gunship, a role in which this airframe never actually served – Reg N37484 canx 12Sep12 – Final Disposition: Extant.

11156 / 53-3145 / C-119G-36-FA

Avble: 05Apr54; Acc: 13Apr54; Del: 20Apr54 to USAF – 483 TCWG (FEA) Ashiya AB Japan 20Apr54 – 2347 RFC (CNR) Long Beach 04Feb58; briefly to AASBR (AMC) Davis-Monthan 26Mar58 for storage – 452 TCWG (AFR) Long Beach 13Jun59; to March 01Oct60; 942 TCG (452 TCWG) (AFR) March 11Feb63; 943 TCG (452 TCWG) (AFR) March 31Mar65 – Fairchild (LOG) St. Augustine 17May68, cvtd **AC-119G Shadow** 01Jul68 – AFT Center (SYS) Edwards 15Sep68 – Fairchild (LOG) St. Augustine 29Jan69 – 1 SOP WG (TAC) Lockbourne 06Mar69 – 4410 CCT WG (TAC) Lockbourne 14Jul69 – Hayes (LOG) Dothan 07Feb71 – 14 SOP WG (PAF) Phan Rang AB S. Vietnam 27Feb71; assigned variously Tan Son Nhut S. Vietnam from 23Apr71 – MAP S. Vietnam 26Aug71, VNAF s/n: **533145**, 819 ASQ (53 TWG) Tan Son Nhut AB – Final Disposition: Unknown.

11157 / 53-3146 / C-119G-36-FA

Avble: 07Apr54; Acc: 19Apr54; Del: 27Apr54 to USAF – 483 TCWG (FEA) Ashiya AB Japan 27Apr54 – 2346 RFC DT (CNR) McClellan 01Jun58 – 314 TCSq (349 TCWG) (AFR) McClellan 23Dec58; 940 TCG (349 TCWG) (AFR) McClellan 10Feb63 – 925 TCG (446 TCWG) (AFR) Ellington 11Sep65 – Fairchild (LOG) St. Augustine 05Oct67 – MAP S. Vietnam 27Jan68, VNAF s/n: **533146**, 413 TSQ (53 TWG) Tan Son Nhut AB. Destroyed on the ground at Tan Son Nhut AB S. Vietnam 18Feb68 by a Communist mortar attack, no fatalities – Final Disposition: Combat Loss.

11158 / 53-3147 / C-119G-36-FA

Avble: 09Apr54; Acc: 26Apr54; Del: 04May54 to USAF – 483 TCWG (FEA) Ashiya AB Japan 04May54 2346 RFC (CNR) Hamilton 19Jul58 – 349 TCWG (AFR) Hamilton 23Dec58; 938 TCG (349 TCWG) (AFR) Hamilton 10Feb63 – 944 TCG (452 TCWG) (AFR) March 22Jun66 – 930 TAL GP (434 TCWG) (AFR) Bakalar 24Jan68; depl 71 ACO SQ (1 ACO WG) (TAC) Lockbourne 20Jun68 – 907 TAL GP (302 TCWG) (AFR) Clinton 04Sep68; to Lockbourne 24Jul71 – MAP S. Vietnam 03Jul72, VNAF s/n: **533147**, 413 TSQ (53 TWG) Tan Son Nhut AB – Final Disposition: Unknown.

11159 / 53-3148 / C-119G-36-FA

Avble: 13Apr54; Acc: 26Apr54; Del: 03May54 to USAF – 483 TCWG (FEA) Ashiya AB Japan 03May54 – 2347 RFC (CNR) Long Beach 11May58; to DT Hill 01Apr59 – 733 TCSq (452 TCWG) (AFR) Hill 03Apr59; 945 TCG (452 TCWG) (AFR) Hill 17Jan63 – 910 TCG (459 TCWG) (AFR) Youngstown 18Sep66 – Fairchild (LOG) St. Augustine 02Nov67 – MAP S. Vietnam 19Mar68, VNAF s/n: **533148**, 413 TSQ (53 TWG) Tan Son Nhut AB – 377 ABS WG (PAF) Tan Son Nhut S. Vietnam 15Jan73 – 405 Fighter WG (PAF) Clark AB The Philippines 20Jan73 for disposal – MAP Taiwan 25May73 – Final Disposition: Unknown.

11160 / 53-3149 / C-119G-36-FA

Avble: 15Apr54; Acc: 28Apr54; Del: 03May54 to USAF – 483 TCWG (FEA) Ashiya AB Japan 03May54; bailment CAT (AMC) Taiwan 18Nov55; to 21 TCSq (483 TCWG) Tachikawa AB Japan 18Sep56 – Wfu to 868 TM SQ (PAF) Tainan Taiwan 14Oct58 – MAP Taiwan 01Jul59 – Final Disposition: Unknown.

11161 / 53-3150 / C-119G-36-FA

Avble: 20Apr54; Acc: 29Apr54; Del: 06May54 to USAF – 483 TCWG (FEA) Ashiya AB Japan 06May54. Lost power in the right engine at take-off from Ashiya AB Japan 30Mar56. The pilot applied emergency braking but overshot the runway and came to rest in a grassy area. The 7 crew were uninjured but the aircraft was damaged beyond repair when the u/c collapsed and completely crumpled the fuselage underbelly – Final Disposition: Accident.

11162 / 53-3151 / C-119G-36-FA

Avble: 21Apr54; Acc: 04May54; Del: 10May54 to USAF – 483 TCWG (FEA) Ashiya AB Japan 10Mar54; bailment CAT (FEA) Taiwan 23Sep55 – 2347 RFC (CNR) Long Beach 11Jun58; to DT Hill 16Jan59 – 733 TCSq (452 TCWG) (AFR) Hill 03Apr59; 945 TCG (452 TCWG) (AFR) Hill 17Jan63 – Fairchild (LOG) St. Augustine 16May66 – 907 TCG (302 TCWG) (AFR) Clinton 28Jun66 – Wfu to MASDC (LOG) Davis-Monthan 31Aug70 for storage, PCN: **CJ354** – REC: 16Sep70 – DFI: 21May71 – Kolar, Inc. Tucson Arizona 05Apr76 for scrap – Final Disposition: Scrapped.

11163 / 53-3152 / C-119G-36-FA

Avble: 26Apr54; Acc: 05May54; Del: 12May54 to USAF – 483 TCWG (FEA) Ashiya AB Japan 12May54 – 2347 RFC (CNR) Long Beach 05Sep58 – 452 TCWG (AFR) Long Beach 13Jun59; to March 01Oct60; 942 TCG (452 TCWG) (AFR) March 11Feb63; 943 TCG

(452 TCWG) (AFR) March 31Mar65 – Wfu to MASDC (LOG) Davis-Monthan 10Feb69 for storage, PCN: **CJ281** – DFI: 09Dec74 – Final Disposition: Scrapped.

11164 / 53-3153 / C-119G-36-FA

Avble: 28Apr54; Acc: 14May54; Del: 19May54 to USAF – MIDAR (AMC) Olmsted 21May54 – 483 TCWG (FEA) Ashiya AB Japan 24Jul54; depl 374 TCWG (FEA) Tachikawa AB Japan 27Mar55; to 21 TCSq (483 TCWG) Tachikawa AB Japan 18Sep56 – Wfu to 868 TM SQ (PAF) Tainan Taiwan 08Oct58 – MAP Taiwan 01Jul59 – Final Disposition: Unknown.

11165 / 53-3154 / C-119G-36-FA

Avble: 30Apr54; Acc: 18May54; Del: 25May54 to USAF – MIDAR (AMC) Olmsted 25May54 – 483 TCWG (FEA) Ashiya AB Japan 05Oct54 – 4440 ADE GP (TAC) Langley 27Jan59 – 512 TCWG (AFR) NAS Willow Grove 15Feb59; depl 732 TCSq (94 TCWG) (AFR) Grenier 14Nov59; 913 TCG (512 TCWG) (AFR) NAS Willow Grove 11Feb63 – Fairchild (LOG) St. Augustine 08Aug68, cvtd **AC-119K Stinger** 12Aug68 – 1 SOP WG (TAC) Lockbourne 25Mar69 – 4410 CCT WG (TAC) Lockbourne 14Jul69 – 14 SOP WG (PAF) Phan Rang AB S. Vietnam 27Dec69; assigned variously Phu Cat, Da Nang, Tan Son Nhut S. Vietnam & Udorn, Nakhon Phanom Thailand from 07Feb70 – 56 SOP WG (PAF) Nakhon Phanom AB Thailand 23Aug71; to China Airlines (LOG) Taipei Taiwan 15Sep71 mnt; assigned variously Da Nang S. Vietnam from 12Nov71 – MAP S. Vietnam 10Nov72, VNAF s/n: **533154**, 821 ASQ (53 TWG) Tan Son Nhut AB – Final Disposition: Unknown.

11166 / 53-3155 / C-119G-36-FA

Avble: 05May54; Acc: 21May54; Del: 27May54 to USAF – MIDAR (AMC) Olmsted 27May54 – 483 TCWG (FEA) Ashiya AB Japan 27Jul54; bailment CAT (FEA) Taiwan 02Jul55; to 21 TCSq (483 TCWG) Tachikawa AB Japan 18Sep56 – Wfu to 868 TM SQ (PAF) Tainan Taiwan 02Oct58 – MAP Taiwan 01Jul59 – Final Disposition: Unknown.

11167 / 53-3156 / C-119G-36-FA

Avble: 07May54; Acc: 21May54; Del: 27May54 to USAF – MIDAR (AMC) Olmsted 27May54 – 483 TCWG (FEA) Ashiya AB Japan 23Jul54 – 2346 RFC (CNR) Hamilton 09Aug58; to DT McClellan 22Aug58 – 314 TCSq (349 TCWG) (AFR) McClellan 23Dec58; 940 TCG (349 TCWG) (AFR) McClellan 10Feb63 – 904 TCG (514 TCWG) (AFR) Stewart 26May65; 903 TAL GP (514 TCWG) (AFR) McGuire 25Feb67 – Fairchild (LOG) St. Augustine 08Aug68, cvtd **AC-119K Stinger** 12Aug68 – 1 SOP WG (TAC) Lockbourne 08Apr69 – Fairchild (LOG) St. Augustine 24May69 – 4410 CCT WG (TAC) Lockbourne 08Aug69 – 14 SOP WG (PAF) Phan Rang AB S. Vietnam 03Nov69; assigned variously Da Nang S. Vietnam from 17Dec69. Ran out of fuel and landed short while on approach to Da Nang AB S. Vietnam 19Feb70. 10 crew survived with minor injuries but the aircraft was damaged beyond repair – Final Disposition: Accident.

11168 / 53-3157 / C-119G-36-FA

Avble: 10May54; Acc: 25May54; Del: 02Jun54 to USAF – MIDAR (AMC) Olmsted 02Jun54 – 483 TCWG (FEA) Ashiya AB Japan 02Aug54 – 2346 RFC (CNR) Hamilton 31Jul58 – 349 TCWG (AFR) Hamilton 23Dec58; 938 TCG (349 TCWG) (AFR) Hamilton 10Feb63; depl 931 TCG (434 TCWG) (AFR) Bakalar 13Jul63 – 927 TCG (403 TCWG) (AFR) Selfridge 05Mar66 – Fairchild (LOG) St. Augustine 23Oct67 – MAP S. Vietnam 25Mar68, VNAF s/n: **533157**, 413 TSQ (53 TWG) Tan Son Nhut AB – 377 ABS WG (PAF) Tan Son

Nhut S. Vietnam 06Feb73 – 405 Fighter WG (PAF) Clark AB The Philippines 11Feb73 for disposal – DFI: 05Jun73 – Final Disposition: Scrapped.

11169 / 53-3158 / C-119G-36-FA

Avble: 12May54; Acc: 28May54; Del: 07Jun54 to USAF – MIDAR (AMC) Olmsted 07Jun54 – 483 TCWG (FEA) Ashiya AB Japan 11Aug54 – 2347 RFC (CNR) Long Beach 14Aug58 – 452 TCWG (AFR) Long Beach 13Jun59; to March 01Oct60; 942 TCG (452 TCWG) (AFR) March 11Feb63; 943 TCG (452 TCWG) (AFR) March 31Mar65 – Wfu to MASDC (LOG) Davis-Monthan 10Feb69 for storage – DFI: 26Dec73 – Final Disposition: Scrapped.

11170 / 53-3159 / C-119G-36-FA

Avble: 17May54; Acc: 28May54; Del: 10Jun54 to USAF – MIDAR (AMC) Olmsted 10Jun54 – 483 TCWG (FEA) Ashiya AB Japan 12Sep54 – 2347 RFC (CNR) Long Beach 20Feb58; to DT Hill 21Nov58 – 733 TCSq (452 TCWG) (AFR) Hill 03Apr59; 945 TCG (452 TCWG) (AFR) Hill 17Jan63 – 907 TCG (302 TCWG) (AFR) Clinton 15Apr66. Stalled and crashed in a field shortly after take-off from Clinton County AFB Ohio 09Aug68, 31 on board consisted of the 4 crew, 7 military and 20 civilian personnel on a flight to Otis AFB Massachusetts. 25 survived but 6 personnel from the 907 TCG were killed: A1C Paul L. Ruschau, A1C Michael L. Wilford, TSgt William B. Hansford III, Sgt Ernest L. Arehart, Sgt Richard N. Hall and Sgt David A. Husinga. IARC lists accident as 26Sep68 – Final Disposition: Accident.

11171 / 53-3160 / C-119G-36-FA

Avble: 18May54; Acc: 04Jun54; Del: 18Jun54 to USAF – MIDAR (AMC) Olmsted 18Jun54 – 483 TCWG (FEA) Ashiya AB Japan 11Aug54 – 2347 RFC (CNR) Long Beach 11Feb58 – 2346 RFC DT (CNR) McClellan 17Feb58 – 314 TCSq (349 TCWG) (AFR) McClellan 23Dec58; 940 TCG (349 TCWG) (AFR) McClellan 10Feb63 – 904 TCG (514 TCWG) (AFR) Stewart 02Apr65; depl McGuire 25May66; 903 TAL GP (514 TCWG) (AFR) McGuire 28Jan67; depl Myrtle Beach 16Jul67 – 928 TAL GP (403 TCWG) (AFR) O'Hare 13Jan71 – Fairchild (LOG) Crestview 15Jan71 – MAP Ethiopia 20Aug71, ETAF s/n: **917** – Wfu 86. Noted 90s stored at Debre Zeyit AB Ethiopia – Final Disposition: Scrapped.

11172 / 53-3161 / C-119G-36-FA

Avble: 20May54; Acc: 09Jun54; Del: 18Jun54 to USAF – MIDAR (AMC) Olmsted 18Jun54 – 483 TCWG (FEA) Ashiya AB Japan 30Aug54; bailment CAT (FEA) Taiwan 30May55 – 2242 RFC (CNR) Selfridge 28Mar58 – 2347 RFC DT (CNR) Hill 11Jun58 – Hayes (AMC) Birmingham 06Apr59 – 733 TCSq (452 TCWG) (AFR) Hill 23May59; 945 TCG (452 TCWG) (AFR) Hill 17Jan63 – 910 TCG (459 TCWG) (AFR) Youngstown 18Sep66 – Fairchild (LOG) St. Augustine 02Nov67 – MAP S. Vietnam 12Feb68, VNAF s/n: **533161**, 413 TSQ (53 TWG) Tan Son Nhut AB – 377 ABS WG (PAF) Tan Son Nhut 12Jan73 – 405 Fighter WG (PAF) Clark AB The Philippines 20Jan73 for disposal – MAP Taiwan 26May73 – Final Disposition: Unknown.

11173 / 53-3162 / C-119G-36-FA

Avble: 24May54; Acc: 11Jun54; Del: 18Jun54 to USAF – MIDAR (AMC) Olmsted 18Jun54 – 483 TCWG (FEA) Ashiya AB Japan 01Sep54 – 2346 RFC (CNR) Hamilton 08Aug58 – 349 TCWG (AFR) Hamilton 23Dec58; 938 TCG (349 TCWG) (AFR) Hamilton 10Feb63 – 906 TCG (302 TCWG) (AFR) Clinton 05Mar66; 907 TCG (302 TCWG) (AFR) Clinton 15Apr66 – Wfu to MASDC (LOG) Davis-Monthan 31Aug70 for storage,

PCN: **CJ356** – DFI: 16Sep70 – Kolar, Inc. Tucson Arizona 19Feb76 for scrap – Final Disposition: Scrapped.

11174 / 53-3163 / C-119G-36-FA

Avble: 27May54; Acc: 11Jun54; Del: 18Jun54 to USAF – MIDAR (AMC) Olmsted 18Jun54 – 483 TCWG (FEA) Ashiya AB Japan 07Sep54; bailment CAT (FEA) Taiwan 26May55; to 21 TCSq (483 TCWG) Tachikawa AB Japan 18Sep56 – Wfu to 868 TM SQ (PAF) Tainan Taiwan 10Oct58 – MAP Taiwan 01Jul59 – Final Disposition: Unknown.

11175 / 53-3164 / C-119G-36-FA

Avble: 02Jun54; Acc: 15Jun54; Del: 24Jun54 to USAF – MIDAR (AMC) Olmsted 24Jun54 – 483 TCWG (FEA) Ashiya AB Japan 28Sep54; bailment CAT (FEA) Taiwan 17Apr55; to 21 TCSq (483 TCWG) Tachikawa AB Japan 18Sep56 – Wfu to 868 TM SQ (PAF) Tainan Taiwan 06Oct58 – MAP Taiwan 01Jul59, RoCAF s/n: **3164**. Listed as damaged beyond repair 18Jul72 – Final Disposition: Accident.

11176 / 53-3165 / C-119G-36-FA

Avble: 04Jun54; Acc: 17Jun54; Del: 24Jun54 to USAF – MIDAR (AMC) Olmsted 24Jun54 – 483 TCWG (FEA) Ashiya AB Japan 18Sep54; to 21 TCSq (483 TCWG) Tachikawa AB Japan 18Sep56 – Wfu to 868 TM SQ (PAF) Tainan Taiwan 08Oct58 – MAP Taiwan 01Jul59, RoCAF s/n: **3165** – Final Disposition: Unknown.

11177 / 53-3166 / C-119G-36-FA

Avble: 08Jun54; Acc: 30Jun54; Del: 07Jul54 to USAF – MIDAR (AMC) Olmsted 07Jul54 – 483 TCWG (FEA) Ashiya AB Japan 23Oct54; depl 316 TCWG (FEA) Ashiya AB Japan 22Apr55; depl (316 TCWG) Akddoae? 25Aug55 – 2347 RFC (CNR) Long Beach 22Feb58 – 2346 RFC DT (CNR) Portland 14Mar58 – 313 TCSq (349 TCWG) (AFR) Portland 23Dec58; 939 TCG (349 TCWG) (AFR) Portland 10Feb63 – 943 TAL GP (452 TCWG) (AFR) March 18Apr68 – 922 TAL GP (433 TCWG) (AFR) Kelly 28Mar69 – 906 SOP GP (302 TCWG) (AFR) Clinton 06Nov70; to Lockbourne 24Jul71; 907 TAL GP (302 TCWG) (AFR) Lockbourne 01Mar72 – Wfu to MASDC (LOG) Davis-Monthan 04Mar73 for storage, PCN: **CJ294** (listed as assigned MASDC 26Mar69 matches PCN No. but IARC does not record it?) – DFI: 09Dec74 – Final Disposition: Scrapped.

11178 / 53-3167 / C-119G-36-FA

Avble: 10Jun54; Acc: 30Jun54; Del: 09Jul54 to USAF – MIDAR (AMC) Olmsted 09Jul54 – 483 TCWG (FEA) Ashiya AB Japan 18Sep54; bailment CAT (FEA) Taiwan 14May55 – 2347 RFC (CNR) Long Beach 04May58; to DT Hill 25Oct58 – 733 TCSq (452 TCWG) (AFR) Hill 03Apr59; 945 TCG (452 TCWG) (AFR) Hill 17Jan63 – 930 TCG (434 TCWG) (AFR) Bakalar 05Aug66 – 922 TCG (433 TCWG) (AFR) Kelly 21Sep66 – Fairchild (LOG) St. Augustine 10Oct67; to 3960 STA WG (SAC) Andersen Guam 24Feb68 – MAP S. Vietnam 25Mar68, VNAF s/n: **533167**, 413 TSQ (53 TWG) Tan Son Nhut AB – Final Disposition: Unknown.

11179 / 53-3168 / C-119G-36-FA

Avble: 14Jun54; Acc: 30Jun54; Del: 12Jul54 to USAF – MIDAR (AMC) Olmsted 12Jul54 – 483 TCWG (FEA) Ashiya AB Japan 23Sep54 – 2347 RFC (CNR) Long Beach 11Feb58 – 2346 RFC (CNR) Hamilton 16Feb58 – 349 TCWG (AFR) Hamilton 23Dec58; to 78 MSU GP (ADC) Hamilton 25Mar59 mnt; 938 TCG (349 TCWG) (AFR)

Hamilton 10Feb63 – 907 TCG (302 TCWG) (AFR) Clinton 18Apr66 – Wfu to MASDC (LOG) Davis-Monthan 31Aug70 for storage, PCN: **CJ355** – DFI: 16Sep70 – Kolar, Inc. Tucson Arizona 19Feb76 for scrap – Final Disposition: Scrapped.

11180 / 53-3169 / C-119G-36-FA

Avble: 15Jun54; Acc: 27Jul54; Del: 29Jul54 to USAF – MIDAR (AMC) Olmsted 29Jul54 – 483 TCWG (FEA) Ashiya AB Japan 24Sep54; to 21 TCSq (483 TCWG) Tachikawa AB Japan 18Sep56 – Wfu to 868 TM SQ (PAF) Tainan Taiwan 02Oct58 – MAP Taiwan 01Jul59, RoCAF s/n: **3169** – Final Disposition: Unknown.

11181 / 53-3170 / C-119G-36-FA

Avble: 18Jun54; Acc: 27Jul54; Del: 29Jul54 to USAF – MIDAR (AMC) Olmsted 29Jul54 – 483 TCWG (FEA) Ashiya AB Japan 30Sep54; depl 18 Fighter-Bomber WG (FEA) Kadena AB Japan 08Sep55 – 2346 RFC (CNR) Hamilton 24Jul58 – 349 TCWG (AFR) Hamilton 23Dec58; briefly to 313 TCSq (349 TCWG) Portland 20Mar61; 938 TCG (349 TCWG) (AFR) Hamilton 10Feb63 – 914 TCG (512 TCWG) (AFR) Niagara Falls 22Jan66 – Fairchild (LOG) St. Augustine 29Mar68, cvtd **AC-119G Shadow** 01Jul68 – 71 SOP SQ (1 ACO WG) (TAC) Lockbourne 20Sep68 – Fairchild at WRAAR (LOG) St. Augustine 26Dec68 – 14 SOP WG (PAF) Nha Trang AB S. Vietnam 29Jan69; assigned variously Tan Son Nhut, Phan Rang, Tuy Hoa, Phu Cat, Da Nang S. Vietnam from 07Mar69; to 313 ADH DV (PAF) Kadena AB Japan 29Nov70 – MAP S. Vietnam 08Sep71, VNAF s/n: **533170**, 819 ASQ (53 TWG) Tan Son Nhut AB – Final Disposition: Unknown.

11182 / 53-3171 / C-119G-36-FA

Avble: 22Jun54; Acc: 27Jul54; Del: 29Jul54 to USAF – MIDAR (AMC) Olmsted 29Jul54 – 483 TCWG (FEA) Ashiya AB Japan 08Oct54; to 21 TCSq (483 TCWG) Tachikawa AB Japan 18Sep56 – Wfu to 868 TM SQ (PAF) Tainan Taiwan 58 – MAP Taiwan 01Jul59, RoCAF s/n: **3171** – Final Disposition: Unknown.

11183 / 53-3172 / C-119G-36-FA

Avble: 25Jun54; Acc: 27Jul54; Del: 29Jul54 to USAF – MIDAR (AMC) Olmsted 29Jul54 – 483 TCWG (FEA) Ashiya AB Japan 07Oct54; to 21 TCSq (483 TCWG) Tachikawa AB Japan 18Sep56 – Wfu to 868 TM SQ (PAF) Tainan Taiwan 18Oct58 – MAP Taiwan 01Jul59, RoCAF s/n: **3172** – Final Disposition: Unknown.

11184 / 53-3173 / C-119G-36-FA

Avble: 29Jun54; Acc: 27Jul54; Del: 30Jul54 to USAF – MIDAR (AMC) Olmsted 30Jul54 – 483 TCWG (FEA) Ashiya AB Japan 08Oct54; bailment CAT (FEA) Taiwan 03May55 – 2347 RFC (CNR) Long Beach 01Sep58; to DT Hill 26Feb59 – 733 TCSq (452 TCWG) (AFR) Hill 03Apr59; 945 TCG (452 TCWG) (AFR) Hill 17Jan63 – 907 TCG (302 TCWG) (AFR) Clinton 03Jun66; to Lockbourne 24Jul71 – MAP S. Vietnam 31Jul72, VNAF s/n: **533173**, 413 TSQ (53 TWG) Tan Son Nhut AB – Final Disposition: Unknown.

11185 / 53-3174 / C-119G-36-FA

Avble: 01Jul54; Acc: 28Jul54; Del: 29Jul54 to USAF – MIDAR (AMC) Olmsted 29Jul54 – 483 TCWG (FEA) Ashiya AB Japan 12Oct54 – 2347 RFC (CNR) Long Beach 01Jul58 – 2346 RFC DT (CNR) Portland 17Jul58 – 313 TCSq (349 TCWG) (AFR) Portland 23Dec58; 939 TCG (349 TCWG) (AFR) Portland 63 – 943 TAL GP (452 TCWG) (AFR)

March 18Apr68 – 922 TAL GP (433 TCWG) (AFR) Kelly 28Mar69 – Wfu to MASDC (LOG) Davis-Monthan 26Jan71 for storage, PCN: **CJ387** – Hayes (LOG) Dothan 05Jun71 – 143 SOP GP (ANG) Theodore F. Green 17Aug71; cvtd **C-119L** 16Apr73 – Wfu to MASDC (LOG) Davis-Monthan 09Jun75 for storage – DFI: 31Mar76 – Dross Metals, Inc. Tucson Arizona 1980 reg **N55795**, no BofS – Juan Perez Miami Florida & Emmett B. Williams (Williams Auto Parts) Lavonia Georgia 1980, no BofS; noted at South Bimini Bahamas 01Jan80; noted at Port-au-Prince Haiti 15Feb80. An FAA Notice of Seizure was issued to the owners in Nov80 for making illegal flights without correct airworthiness and registration certificates. The notice also states that on 20Nov80 the C-119 departed Kissimmee Florida and has not been sighted since, nfd – Reg N55795 canx 29Nov10 – Final Disposition: Unknown.

11186 / 53-3175 / C-119G-36-FA

Avble: 06Jul54; Acc: 29Jul54; Del: 30Jul54 to USAF – MIDAR (AMC) Olmsted 30Jul54 – 483 TCWG (FEA) Ashiya AB Japan 27Oct54; depl 319 AB WG (FEA) Andersen Guam 19Mar55; to 3960 AB WG (SAC) Andersen Guam 01Apr55 mnt; depl 316 TCWG (FEA) Ashiya AB Japan 11Apr55 – 2347 RFC (CNR) Long Beach 29Aug58 – 452 TCWG (AFR) Long Beach 13Jun59; to March 01Oct60; 942 TCG (452 TCWG) (AFR) March 11Feb63; 944 TCG (452 TCWG) (AFR) March 31Mar65 – Fairchild (LOG) St. Augustine 12Oct67 – MAP S. Vietnam 06Feb68, VNAF s/n: **533175**, 413 TSQ (53 TWG) Tan Son Nhut AB – 377 ABS WG (PAF) Tan Son Nhut 07Jan73 – 405 Fighter WG (PAF) Clark AB The Philippines 08Jan73 for disposal – DFI: 05Jun73 – Final Disposition: Scrapped.

11187 / 53-3176 / C-119G-36-FA

Avble: 09Jul54; Acc: 30Jul54; Del: 18Aug54 to USAF – MIDAR (AMC) Olmsted 18Aug54 – 483 TCWG (FEA) Ashiya AB Japan 23Oct54; to 21 TCSq (483 TCWG) Tachikawa AB Japan 18Sep56 – Wfu to 868 TM SQ (PAF) Tainan Taiwan 18Oct58 – MAP Taiwan 01Jul59, RoCAF s/n: **3176** – Final Disposition: Unknown.

11188 / 53-3177 / C-119G-36-FA

Avble: 13Jul54; Acc: 30Jul54; Del: 18Aug54 to USAF – MIDAR (AMC) Olmsted 18Aug54 – 483 TCWG (FEA) Ashiya AB Japan 28Oct54; to 1500 FDM SQ (MATS) Hickham 01Mar55 mnt; to 6486 AB WG (FEA) Hickham 01Apr55 mnt; depl 316 TCG (FEA) Ashiya AB Japan 01Jul55; to 21 TCSq (483 TCWG) Tachikawa AB Japan 12Mar57 – Wfu to 868 TM SQ (PAF) Tainan Taiwan 02Oct58 – MAP Taiwan 01Jul59, RoCAF s/n: **3177** – Final Disposition: Unknown.

11189 / 53-3178 / C-119G-36-FA

Avble: 14Jul54; Acc: 19Aug54; Del: 25Aug54 to USAF – 464 TCWG (TAC) Lawson 25Aug54; to Pope 07Sep54; depl North Aux AF 06Jun55; depl Aberdeen 05May56 – 65 TCSq (403 TCWG) (AFR) Davis 26Nov58; 929 TCG (403 TCWG) (AFR) Davis 11Feb63 – 922 TCG (433 TCWG) (AFR) Kelly 18Oct65 – Fairchild (LOG) St. Augustine 29Apr68, cvtd **AC-119G Shadow** 01Jul68 – 71 SOP SQ (1 ACO WG) (TAC) Lockbourne 03Oct68 – Fairchild at WRAAR (LOG) St. Augustine 04Nov68 – 14 SOP WG (PAF) Nha Trang AB S. Vietnam 06Dec68; assigned variously Tan Son Nhut, Tuy Hoa, Phan Rang, Phu Cat S. Vietnam from 30Mar69; to 313 ADH DV (PAF) Kadena AB Japan 01Jun71 – MAP S. Vietnam 21Aug71, VNAF s/n: **533178**, 819 ASQ (53 TWG) Tan Son Nhut AB – Final Disposition: Unknown.

11190 / 53-3179 / C-119G-36-FA

Avble: 19Jul54; Acc: 31Aug54; Del: 27Sep54 to USAF – 317 TCWG (AFE) Neubiberg AB W. Germany 27Sep54; to 3110 MAI GP (AMC) RAF Burtonwood England 04May55 mnt; to Évreux-Fauville AB France 02Jun56; depl 465 TCWG (AFE) Évreux-Fauville AB France 01Jul56 – 2347 RFC (CNR) Long Beach 03Feb58 – 452 TCWG (AFR) Long Beach 13Jun59; to March 01Oct60; 942 TCG (452 TCWG) (AFR) March 01Jun63; 944 TCG (452 TCWG) (AFR) March 31Mar65; 943 TAL GP (452 TCWG) (AFR) March 23Jan68 – 934 TAL GP (440 TCWG) (AFR) Minneapolis-St. Paul 08Apr69 – 906 TAL GP (302 TCWG) (AFR) Clinton 15Nov69; to Lockbourne 24Jul71 – Wfu to MASDC (LOG) Davis-Monthan 13Mar72 for storage, PCN: **CJ418** – DFI: 09Dec74 – Final Disposition: Scrapped.

11191 / 53-3180 / C-119G-36-FA

Avble: 22Jul54; Acc: 31Aug54; Del: 28Sep54 to USAF – 317 TCWG (AFE) Neubiberg AB W. Germany 28Sep54; to 7373 MAI GP (AFE) Chateauroux AB France 18Mar55 mnt; depl 86 Fighter WG (AFE) Landstuhl AB W. Germany 20Feb56; to Évreux-Fauville AB France 11Jun56; depl 465 TCWG (AFE) Évreux-Fauville AB France 01Jul56; to 3110 MAI GP (AMC) RAF Burtonwood England 31Aug56 mnt – 2472 RFC DT (CNR) Davis 02Apr58 – 65 TCSq (403 TCWG) (AFR) Davis 18Oct58; to 810 AB GP (SAC) Biggs 10Nov58 mnt; 929 TCG (403 TCWG) (AFR) Davis 11Feb63 – 922 TCG (433 TCWG) (AFR) Kelly 18Oct65 – Wfu to MASDC (LOG) Davis-Monthan 02Dec70 for storage, PCN: **CJ377**; to Hayes (LOG) Dothan 26Jun71 & 28Jun72; PCN: **CJ437** 20Jul72 – 907 TAL GP (302 TCWG) (AFR) Lockbourne 27Jul72 – Wfu to MASDC (LOG) Davis-Monthan 04Mar73 for storage – DFI: 31Mar76, noted derelict at MASDC Oct79 in SEA camouflage with nose FM radio mast – Final Disposition: Scrapped.

11192 / 53-3181 / C-119G-36-FA

Avble: 26Jul54; Acc: 31Aug54; Del: 04Oct54 to USAF – 317 TCWG (AFE) Neubiberg AB W. Germany 04Oct54; to 3919 AB GP (SAC) RAF Fairford England 12Aug55 mnt; to Évreux-Fauville AB France 13Jun56; depl 465 TCWG (AFE) Évreux-Fauville AB France 01Jul56; to 3110 MAI GP (AMC) RAF Burtonwood England 06Jul56 mnt – 2472 RFC DT (CNR) Tinker 21Mar58; briefly to DT Davis 23Jul58 – 305 TCSq (442 TCWG) (AFR) Tinker 19Mar59 – 512 TCWG (AFR) NAS Willow Grove 10Feb61; 913 TCG (512 TCWG) (AFR) NAS Willow Grove 11Feb63 – 907 TAL GP (302 TCWG) (AFR) Clinton 21Dec69; to Lockbourne 24Jul71; stored MASDC (LOG) Davis-Monthan 10Jun72 – Fairchild (LOG) Bob Sikes (Crestview) 22Feb73 – Wfu to MASDC (LOG) Davis-Monthan cvtd **RC-119L** 29Aug73 for experimental duties until 07Dec73, PCN: **CJ448** – DFI: 31Mar76, noted burned on fire dump at MASDC 1978 – Final Disposition: Scrapped.

11193 / 53-3182 / C-119G-36-FA

Avble: 29Jul54; Acc: 31Aug54; Del: 08Oct54 to USAF – 317 TCWG (AFE) Neubiberg AB W. Germany 08Oct54; to 3110 MAI GP (AMC) RAF Burtonwood England 26Mar55 mnt; to Évreux-Fauville AB France 19Jun56; depl 465 TCWG (AFE) Évreux-Fauville AB France 01Jul56 – 2237 RFC (CNR) New Castle 20Feb58; to NAS Willow Grove 21Jul58 – Hayes (AMC) Birmingham 17Nov58 – 512 TCWG (AFR) NAS Willow Grove 03Feb59; 913 TCG (512 TCWG) (AFR) NAS Willow Grove 11Feb63 – 906 TAL GP (302 TCWG) (AFR) Clinton 25Dec69; to Lockbourne 24Jul71 – Wfu to MASDC (LOG) Davis-Monthan 25Sep71 for storage, PCN: **CJ415** – DFI: 31Mar76 – Final Disposition: Scrapped.

11194 / 53-3183 / C-119G-36-FA

Avble: 06Aug54; Acc: 31Aug54; Del: 07Oct54 to USAF – 317 TCWG (AFE) Neubiberg AB W. Germany 07Oct54; to 3110 MAI GP (AMC) RAF Burtonwood England 16Mar56 mnt; to Évreux-Fauville AB France 24Jun56; depl 465 TCWG (AFE) Évreux-Fauville AB France 01Jul56 – 2346 RFC DT (CNR) Paine 24Apr58; to 9 AB GP (SAC) Mt. Home 13Aug58 mnt – 97 TCSq (349 TCWG) (AFR) Paine 23Dec58; 941 TCG (349 TCWG) (AFR) Paine 63 – 925 TCG (446 TCWG) (AFR) Ellington 30Aug65; 924 TAL GP (446 TCWG) (AFR) Ellington 14Jan68 – 922 TAL GP (433 TCWG) (AFR) Kelly 15May68 – Wfu to MASDC (LOG) Davis-Monthan 09Dec70 for storage, PCN: **CJ383** – Hayes (LOG) Dothan 01Jul71 – 924 TAL GP (446 TCWG) (AFR) Ellington 19Jul72 – 907 TAL GP (302 TCWG) (AFR) Lockbourne 10Aug72 – Wfu to MASDC (LOG) Davis-Monthan 17Jan73 for storage – DFI: 09Dec74 – Final Disposition: Scrapped.

11195 / 53-3184 / C-119G-36-FA

Avble: 06Aug54; Acc: 31Aug54; Del: 12Oct54 to USAF – 317 TCWG (AFE) Neubiberg AB W. Germany 12Oct54; to Évreux-Fauville AB France 13Jun56; depl 465 TCWG (AFE) Évreux-Fauville AB France 01Jul56; to 3110 MAI GP (AMC) RAF Burtonwood England 11Jul56 mnt – 2472 RFC DT (CNR) Davis 10Apr58 – 65 TCSq (403 TCWG) (AFR) Davis 18Sep58; 929 TCG (403 TCWG) (AFR) Davis 11Feb63 – 922 TCG (433 TCWG) (AFR) Kelly 18Oct65 – 129 SOP GP (ANG) Hayward 03Apr71 – Hayes (LOG) Dothan 27May71 – 143 SOP GP (ANG) Theodore F. Green 27Jul71; cvtd **C-119L** 11Feb73 – Wfu to MASDC (LOG) Davis-Monthan 19May75 for storage – DFI: 14Dec76 – Final Disposition: Scrapped.

11196 / 53-3185 / C-119G-36-FA

Avble: 20Aug54; Acc: 31Aug54; Del: 18Oct54 to USAF – 317 TCWG (AFE) Neubiberg AB W. Germany 18Oct54; brief depl 60TCWG (AFE) Rhein-Main AB W. Germany 24May55; to 3110 MAI GP (AMC) RAF Burtonwood England 01Mar56 mnt; to Évreux-Fauville AB France 20Jun56; depl 465 TCWG (AFE) Évreux-Fauville AB France 01Jul56 – 2234 RFC (CNR) Hanscom 04Mar58 – 94 TCWG (AFR) Hanscom 19Mar59; 901 TCG (94 TCWG) (AFR) Hanscom 31May63 – 940 TCG (349 TCWG) (AFR) McClellan 15May64 – 901 TCG (94 TCG) (AFR) Hanscom 01Mar66 – Fairchild (LOG) Crestview 20Sep66 – 910 TCG (459 TCWG) (AFR) Youngstown 28Oct66 – 906 TAL GP (302 TCWG) (AFR) Clinton 15Oct69; to Lockbourne 24Jul71; 907 TAL GP (302 TCWG) (AFR) Lockbourne 28Sep71 – MAP S. Vietnam 03Jul72, VNAF s/n: **533185**, 413 TSQ (53 TWG) Tan Son Nhut AB – Final Disposition: Unknown.

11197 / 53-3186 / C-119G-36-FA

Avble: 25Aug54; Acc: 31Aug54; Del: 18Oct54 to USAF – 317 TCWG (AFE) Neubiberg AB W. Germany 18Oct54; to Évreux-Fauville AB France 15Jun56; depl 465 TCWG (AFE) Évreux-Fauville AB France 01Jul56; to 3110 MAI GP (AMC) RAF Burtonwood England 20Aug56 mnt – 2346 RFC DT (CNR) Paine 22Apr58; to 124 CLM SQ (ID-ANG) Boise 16Aug58 mnt – Hayes (AMC) Birmingham 30Oct58 – 97 TCSq (349 TCWG) (AFR) Paine 22Jan59; 941 TCG (349 TCWG) (AFR) Paine 63 – 924 TCG (446 TCWG) (AFR) Ellington 65 – 922 TAL GP (433 TCWG) (AFR) Kelly 10May68 – 129 SOP GP (ANG) Hayward 03Apr71 – Hayes (LOG) Dothan 23May71 – 143 SOP GP (ANG) Theodore F. Green 13Jul71; cvtd **C-119L** 18Jul73 – Wfu to MASDC (LOG) Davis-Monthan 18Jun75 for storage – DFI: 14Dec76 – Final Disposition: Scrapped.

11198 / 53-3187 / C-119G-36-FA

Avble: 27Aug54; Acc: 29Sep54; Del: 18Oct54 to USAF – 317 TCWG (AFE) Neubiberg AB W. Germany 18Oct54; to 3919 AB GP (SAC) RAF Fairford England 12Aug55 mnt; to Évreux-Fauville AB France 14Jun56; depl 465 TCWG (AFE) Évreux-Fauville AB France 01Jul56; to 3110 MAI GP (AMC) RAF Burtonwood England 27Jul56 mnt – 2472 RFC DT (CNR) Davis 16Mar58; to DT Tinker 26Mar58 – 305 TCSq (442 TCWG) (AFR) Tinker 19Mar59 – 512 TCWG (AFR) NAS Willow Grove 10Feb61; 912 TCG (512 TCWG) (AFR) NAS Willow Grove 11Feb63 – Fairchild (LOG) St. Augustine 12Apr68, cvtd **AC-119K Stinger** 01Jul68 – AFT Center (SYS) Edwards 19Jun69 – Fairchild (LOG) St. Augustine 28Feb70 – 4410 CCT WG (TAC) Lockbourne 23Jun70 – 1 SOP WG (TAC) Eglin 07Jul71 – 56 SOP WG (PAF) Nakhon Phanom AB Thailand 03Oct72; to Da Nang S. Vietnam 20Dec72 – 6498 ABS WG (PAF) Da Nang AB S. Vietnam to MAP S. Vietnam 31Jan73, VNAF s/n: **533187**, 821 ASQ (53 TWG) Tan Son Nhut AB – Final Disposition: Unknown.

11199 / 53-3188 / C-119G-36-FA

Avble: 02Sep54; Acc: 29Sep54; Del: 18Oct54 to USAF – 317 TCWG (AFE) Neubiberg AB W. Germany 18Oct54; to Évreux-Fauville AB France 19Jun56; depl 465 TCWG (AFE) Évreux-Fauville AB France 01Jul56; to 3110 MAI GP (AMC) RAF Burtonwood England 06Nov56 mnt – 2472 RFC DT (CNR) Davis 18Mar58; to DT Tinker 27Mar58 – 305 TCSq (442 TCWG) (AFR) Tinker 19Mar59; brief depl 435 TCWG (AFR) Miami 07Jun60 – 64 TCSq (403 TCWG) (AFR) O'Hare 27Feb61; 928 TCG (403 TCWG) (AFR) O'Hare 11Feb63 – Fairchild (LOG) Crestview 14Jan71 – 906 SOP GP (302 TCWG) (AFR) Clinton 20Jun71 – Fairchild (LOG) Crestview 09Jul71 – MAP Ethiopia 11Jul71, ETAF s/n: **915** – Wfu 86. Noted 90s stored at Debre Zeyit AB Ethiopia – Final Disposition: Scrapped.

11200 / 53-3189 / C-119G-36-FA

Avble: 08Sep54; Acc: 30Sep54; Del: 20Oct54 to USAF – 317 TCWG (AFE) Neubiberg AB W. Germany 20Oct54; to 3110 MAI GP (AMC) RAF Burtonwood England 12Mar56 mnt; to Évreux-Fauville AB France 22Jun56; depl 465 TCWG (AFE) Évreux-Fauville AB France 01Jul56; to 3150 MAI GP (AMC) Nouasseur AB, Morocco 12Jul56 mnt; to 3110 MAI GP (AMC) RAF Burtonwood England 23May57 mnt; to 3131 MAI GP (AMC) Chateauroux AB France 17Dec57 mnt – 60 TCWG (AFE) Dreux AB France 21May58 – 7305 AB GP (322 ADV) (AFE) Dreux AB France 24Sep58; 10 TCSq (322 ADV) (AFE) Dreux AB France 12Dec60 – 7305 CAM SQ (AFE) Dreux AB France 08Jan61 – 4440 ADE GP (TAC) Langley 24Mar61 – 96 TCSq (440 TCWG) (AFR) Minneapolis-St. Paul 30Mar61; 934 TCG (440 TCWG) (AFR) Minneapolis-St. Paul 11Feb63 – Fairchild (LOG) St. Augustine 24Jun68, cvtd **AC-119G Shadow** 01Jul68 – 71 SOP SQ (1 ACO WG) (TAC) Lockbourne 24Nov68 – 14 SOP WG (PAF) Nha Trang AB S. Vietnam 06Dec68; assigned variously Tan Son Nhut, Tuy Hoa, Phan Rang, Phu Cat S. Vietnam from 01Feb69; to 18 Fighter WG (PAF) Kadena AB Japan 08Nov69; to 313 ADH DV (PAF) Kadena AB Japan 04Feb71 – MAP S. Vietnam 18Aug71, VNAF s/n: **533189**, 819 ASQ (53 TWG) Tan Son Nhut AB – Final Disposition: Unknown.

11201 / 53-3190 / C-119G-36-FA

Avble: 09Sep54; Acc: 30Sep54; Del: 18Oct54 to USAF – 317 TCWG (AFE) Neubiberg AB W. Germany 18Oct54; to 7373 MAI GP (AFE) Chateauroux AB France 19Feb55 mnt; to 3110 MAI GP (AMC) RAF Burtonwood England 14Mar56 mnt; to Évreux-Fauville AB France 14Jun56; depl 465 TCWG (AFE) Évreux-Fauville AB France 01Jul56 – 2346 RFC DT (CNR) Paine 25Apr58 – 97 TCSq (349 TCWG) (AFR) Paine 23Dec58; 941 TCG (349 TCWG) (AFR)

Paine 63 – 904 TCG (514 TCWG) (AFR) Stewart 04Jun65; 903 TAL GP (514 TCWG) (AFR) McGuire 10May67; depl Myrtle Beach 16Jul67 – 928 TAL GP (403 TCWG) (AFR) O'Hare 07Aug68 – Wfu to MASDC (LOG) Davis-Monthan 17Dec70 for storage, PCN: **CJ384** – DFI: 21May71 – Kolar, Inc. Tucson Arizona 08Mar76 for scrap – Final Disposition: Scrapped.

11202 / 53-3191 / C-119G-36-FA

Avble: 13Sep54; Acc: 30Sep54; Del: 20Oct54 to USAF – 317 TCWG (AFE) Neubiberg AB W. Germany 20Oct54; to 7280 MAI GP (AFE) Nouasseur AB, Morocco 08Mar55 mnt; depl 7100 SUT WG (AFE) Wiesbaden AB W. Germany 23Nov55; to 3110 MAI GP (AMC) RAF Burtonwood England 22Mar56 mnt; to Évreux-Fauville AB France 15Jun56; depl 465 TCWG (AFE) Évreux-Fauville AB France 01Jul56 – 2473 RFC (CNR) Gen Mitchell 29Mar58 – 440 TCWG (AFR) Gen Mitchell 19Dec58; 933 TCG (440 TCWG) (AFR) Gen Mitchell 31May63 – Wfu to MASDC (LOG) Davis-Monthan 13Feb71 for storage, PCN: **CJ403** – DFI: 09Dec74 – Final Disposition: Scrapped.

11203 / 53-3192 / C-119G-36-FA

Avble: 22Sep54; Acc: 30Sep54; Del: 29Nov54 to USAF – 317 TCWG (AFE) Neubiberg AB W. Germany 29Nov54; to 1604 MAT SQ (MATS) Kindley 01Dec54 mnt; to Évreux-Fauville AB France 11Jun56; depl 465 TCWG (AFE) Évreux-Fauville AB France 01Jul56; to 3110 MAI GP (AMC) RAF Burtonwood England 25Oct56 mnt – 60 TCWG (AFE) Dreux AB France 15May58 – SABBE (AMC) Brussels Belgium 27Jun58 mnt – 7305 AB GP (322 ADV) (AFE) Dreux AB France 07Feb59; 11 TCSq (322 ADV) (AFE) Dreux AB France 12Dec60 – 7305 CAM SQ (AFE) Dreux AB France 08Jan61 – 4440 ADE GP (TAC) Langley 10Feb61 – 512 TCWG (AFR) NAS Willow Grove 17Feb61; 912 TCG (512 TCWG) (AFR) NAS Willow Grove 11Feb63; 913 TCG (512 TCWG) (AFE) NAS Willow Grove 31May63 – Fairchild (LOG) St. Augustine 17Jun68, cvtd **AC-119G Shadow** 01Jul68 – 71 SOP SQ (1 ACO WG) (TAC) Lockbourne 21Nov68 – 14 SOP WG (PAF) Nha Trang AB S. Vietnam 05Dec68; assigned variously Tan Son Nhut, Tuy Hoa, Phan Rang, Phu Cat S. Vietnam from 23Feb69; to 313 ADH DV (PAF) Kadena AB Japan 19Dec70 – MAP S. Vietnam 19Aug71, VNAF s/n: **533192**, 819 ASQ (53 TWG) Tan Son Nhut AB – Final Disposition: Unknown.

11204 / 53-3193 / C-119G-36-FA

Avble: 20Sep54; Acc: 30Sep54; Del: 18Oct54 to USAF – 317 TCWG (AFE) Neubiberg AB W. Germany 18Oct54; to 3110 MAI GP (AMC) RAF Burtonwood England 05Mar56 mnt; depl 465 TCWG (AFE) Évreux-Fauville AB France 16Feb57; to Évreux-Fauville AB France 08Jul57 – 2472 RFC DT (CNR) Davis 07Apr58 – 65 TCSq (403 TCWG) (AFR) Davis 18Sep58; 929 TCG (403 TCWG) (AFR) Davis 11Feb63 – 922 TCG (433 TCWG) (AFR) Kelly 18Oct65 – 129 SOP GP (ANG) Hayward 03Apr71 – Assigned variously Hayes (LOG) Dothan & General Dynamics (LOG) Theodore F. Green from 23May71 – 143 SOP GP (ANG) Theodore F. Green 24Dec71; cvtd **C-119L** 25Apr73 – Wfu to MASDC (LOG) Davis-Monthan 23Jun75 for storage – DFI: 14Dec76 – Final Disposition: Scrapped.

11205 / 53-3194 / C-119G-36-FA

Avble: 21Sep54; Acc: 30Sep54; Del: 09Nov54 to USAF – 317 TCWG (AFE) Neubiberg AB W. Germany 09Nov54; to 3110 MAI GP (AMC) RAF Burtonwood England 10Aug56 mnt; depl 465 TCWG (AFE) Évreux-Fauville AB France 07May57; to Évreux-Fauville AB France 08Jul57 – 2472 RFC DT (CNR) Davis 12Mar58; to DT Tinker 21Mar58 – 305 TCSq (442 TCWG) (AFR) Tinker 19Mar59 – 96 TCSq (440 TCWG) (AFR) Minneapolis-St. Paul

27Jan61; 934 TCG (440 TCWG) (AFR) Minneapolis-St. Paul 11Feb63 – 906 TAL GP (302 TCWG) (AFR) Clinton 22Nov69; 907 TAL GP (302 TCWG) (AFR) Clinton 25Nov69; to Lockbourne 24Jul71 – MAP S. Vietnam 23Jun72, VNAF s/n: **533194**, 413 TSQ (53 TWG) Tan Son Nhut AB – 405 Fighter WG (PAF) Clark AB The Philippines 08Mar73 for disposal – DFI: 05Jun73 – Final Disposition: Scrapped.

11206 / 52-5922 / C-119G-1-FA

Avble: 22Sep54; Acc: 30Sep54; Del: 27Oct54 to USAF – 317 TCWG (AFE) Neubiberg AB W. Germany 27Oct54; to 7559 MAI GP (FEA) RAF Burtonwood England 30May55 mnt; to 3110 MAI GP (AMC) RAF Burtonwood England 02Apr56 mnt; to Évreux-Fauville AB France 12Jun56; depl 465 TCWG (FEA) Évreux-Fauville AB France 01Jul56 – 2237 RFC DT (CNR) Niagara Falls 30Apr58 – 328 TCSq (512 TCWG) (AFR) Niagara Falls 19Dec58; 914 TCG (512 TCWG) (AFR) Niagara Falls 11Feb63 – Wfu to MASDC (LOG) Davis-Monthan 07Dec70 for storage, PCN: **CJ381** – DFI: 21May71 – Kolar, Inc. Tucson Arizona 06Feb76 for scrap – Final Disposition: Scrapped.

11207 / 52-5923 / C-119G-1-FA

Avble: 28Sep54; Acc: 12Oct54; Del: 18Oct54 to USAF – 317 TCWG (AFE) Neubiberg AB W. Germany 18Oct54 – **RECORDS MISSING 1954–1956** – to Évreux-Fauville AB France 12Jun56; depl 465 TCWG (FEA) Évreux-Fauville AB France 01Jul56; to 3150 MAI GP (AMC) Nouasseur AB, Morocco 08Sep56 mnt; to 3110 MAI GP (AMC) RAF Burtonwood England 24Sep56 mnt – 2346 RFC DT (CNR) Portland Apr58 – Hayes (AMC) Birmingham 24Nov58 – 313 TCSq (349 TCWG) (AFR) Portland 19Feb59; 939 TCG (349 TCWG) (AFR) Portland 63 – 908 TAL GP (435 TCWG) (AFR) Brookley 06May68 – 926 TAL GP (446 TCWG) (AFR) NAS New Orleans 12Feb69 – MAP Taiwan 08Apr69 – Final Disposition: Unknown.

11208 / 53-3195 / C-119G-36-FA

Avble: 01Oct54; Acc: 19Oct54; Del: 09Nov54 to USAF – 317 TCWG (AFE) Neubiberg AB W. Germany 09Nov54; depl 60 TCWG (AFE) Rhein-Main AB W. Germany 15Mar55; to 3110 MAI GP (AMC) RAF Burtonwood England 31Aug56 mnt; depl 48 Fighter-Bomber WG (AFE) Chaumont AB France 20Mar57; depl 465 TCWG (AFE) Évreux-Fauville AB France 10Apr57; to Évreux-Fauville AB France 08Jul57 – 2347 RFC (CNR) Long Beach 31Jan58; to CAR Center (ARD) Hanscom 04Feb58 rec canx; to Hayes (AMC) Birmingham 03Jun58 repairs – 452 TCWG (AFR) Long Beach 13Jun59; to 3560 PTN WG (ATC) Webb 21Oct59 mnt; to March 01Oct60; 944 TCG (452 TCWG) (AFR) March 11Feb63. Crashed along the ridge of Pallett Mountain California 14.4 miles W Wrightwood California 30Sep66 while on a night training flight out of March AFB, the weather at the time was stormy. The 4 AFR crew killed were: Maj Elvin Estes (pilot), Capt Norman Gassman, Capt Raymond Miller and SSgt Roger DuCharime. Significant pieces of wreckage are still at the crash site today – Final Disposition: Accident.

11209 / 53-3196 / C-119G-36-FA

Avble: 04Oct54; Acc: 21Oct54; Del: 27Oct54 to USAF – 317 TCWG (AFE) Neubiberg AB W. Germany 27Oct54; to 3131 MAI GP (AMC) Chateauroux AB France 05Jan56 mnt; to 3110 MAI GP (AMC) RAF Burtonwood England 29Sep56 mnt; depl 465 TCWG (AFE) Évreux-Fauville AB France 10Apr57; to Évreux-Fauville AB France 08Jul57 – 2347 RFC (CNR) Long Beach 11Feb58 – 2346 RFC DT (CNR) Portland 01Mar58 – Hayes (AMC) Birmingham 24Nov58 – 313 TCSq (349 TCWG) (AFR) Portland 19Mar59; 939 TCG (349

TCWG) (AFR) Portland 63 – 931 TAL GP (434 TCWG) (AFR) Bakalar 23Jun68 – 906 TAL GP (302 TCWG) (AFR) Clinton 02Apr69; to Lockbourne 24Jul71; 907 TAL GP (302 TCWG) (AFR) Lockbourne 28Sep71; wfu to MASDC (LOG) Davis-Monthan 10Jun72 for storage, reinstated 26Jun72 – Wfu to MASDC (LOG) Davis-Monthan 04Feb73 for storage, PCN: **CJ443** – DFI: 26Dec73 – Final Disposition: Scrapped.

11210 / 53-3197 / C-119G-36-FA

Avble: 06Oct54; Acc: 26Oct54; Del: 03Nov54 to USAF – 317 TCWG (AFE) Neubiberg AB W. Germany 03Nov54; to 7373 MAI GP (AFE) Chateauroux AB France 19Feb55 mnt; to 3110 MAI GP (AMC) RAF Burtonwood England 17Sep56 mnt; depl 465 TCWG (AFE) Évreux-Fauville AB France 10Apr57; to Évreux-Fauville AB France 08Jul57 – 2347 RFC (CNR) Long Beach 23Jan58 – 452 TCWG (AFR) Long Beach 13Jun59; to March 01Oct69; 942 TCG (452 TCWG) (AFR) March 11Feb63; 944 TCG (452 TCWG) (AFR) March 31Mar65 – 930 TAL GP (434 TCWG) (AFR) Bakalar 24Jan68; 931 TAL GP (434 TCWG) (AFR) Bakalar 25Apr68 – Fairchild (LOG) St. Augustine 20Jun68, cvtd **AC-119K Stinger** 01Jul68 – 4413 CCT SQ (1 ACO WG) (TAC) Lockbourne 08Jan69 – 1 SOP WG (TAC) Lockbourne 17Feb69 – 4410 CCT WG (TAC) Lockbourne 08Aug69 – WRAAR (LOG) Lockbourne 23Jun71 – Hayes (LOG) Dothan 09Jul71 – 1 SOP WG (TAC) Eglin 15Jul71; to Hurlburt 30Oct72 – MAP S. Vietnam 06Nov72, VNAF s/n: **533197**, 821 ASQ (53 TWG) Tan Son Nhut AB – Final Disposition: Unknown.

11211 / 53-3198 / C-119G-36-FA

Avble: 11Oct54; Acc: 22Oct54; Del: 01Nov54 to USAF – 317 TCWG (AFE) Neubiberg AB W. Germany 01Nov54; to 3110 MAI GP (AMC) RAF Burtonwood England 07Sep56 mnt; depl 465 TCWG (AFE) Évreux-Fauville AB France 10Apr57; to Évreux-Fauville AB France 08Jul57 – 2347 RFC (CNR) Long Beach 23Jan58 – 2346 RFC (CNR) Hamilton 30Jan58; to DT Paine 08Sep58 – Hayes (AMC) Birmingham 02Dec58 – 97 TCSq (349 TCWG) (AFR) Paine 25Feb59; 941 TCG (349 TCWG) (AFR) Paine 63 – 924 TCG (446 TCWG) (AFR) Ellington 02Oct65 – Fairchild (LOG) St. Augustine 02Nov67 – MAP S. Vietnam 27Jan68, VNAF s/n: **533198**, 413 TSQ (53 TWG) Tan Son Nhut AB – 377 ABS WG (PAF) Tan Son Nhut AB S. Vietnam 11Jan73 – 405 Fighter WG (PAF) Clark AB The Philippines 18Jan73 for disposal – MAP Taiwan 25May73 – Final Disposition: Unknown.

11212 / 53-3199 / C-119G-36-FA

Avble: 13Oct54; Acc: 27Oct54; Del: 06Nov54 to USAF – 317 TCWG (AFE) Neubiberg AB W. Germany 06Nov54; to 3110 MAI GP (AMC) RAF Burtonwood England 27Jun56 mnt; depl 465 TCWG (AFE) Évreux-Fauville AB France 10Apr57; to Évreux-Fauville AB France 08Jul57 – 2347 RFC (CNR) Long Beach 28Jan58 – 452 TCWG (AFR) Long Beach 13Jun59; to March 01Oct60; 942 TCG (452 TCWG) (AFR) March 11Feb63 – Fairchild (LOG) St. Augustine 05Jun63 – MAP India 11Jul63 – Final Disposition: Unknown.

11213 / 53-3200 / C-119G-36-FA

Avble: 14Oct54; Acc: 27Oct54; Del: 10Nov54 to USAF – 317 TCWG (AFE) Neubiberg AB W. Germany 10Nov54; to 3110 MAI GP (AMC) RAF Burtonwood England 27Mar56 mnt; to 3131 MAI GP (AMC) Chateauroux AB France 16Jun56 mnt; depl 465 TCWG (AFE) Évreux-Fauville AB France 10Apr57; to Évreux-Fauville AB France 08Jul57 – 60 TCWG (AFE) Dreux AB France 02Jun58 – SABBE (AMC) Brussels Belgium 03Jun58 mnt – 7305 AB GP (322 ADV) (AFE) Dreux AB France 10Nov58 – United Nations (UN), reg **UNO-101**, mission to The Congo 03Aug60 – MAP Italy 61, AMI s/n: **MM53-3200** – Public display at Pisa International Airport Pisa Italy – Final Disposition: Extant.

11214 / 53-3201 / C-119G-36-FA

Avble: 20Oct54; Acc: 27Oct54; Del: 12Nov54 to USAF – 317 TCWG (AFE) Neubiberg AB W. Germany 12Nov54; depl 465 TCWG (AFE) Évreux-Fauville AB France 10Apr57; to Évreux-Fauville AB France 08Jul57 – 2347 RFC (CNR) Long Beach 23Jan58 – 452 TCWG (AFR) Long Beach 13Jun59; to March 01Oct60; 944 TCG (452 TCWG) (AFR) March 11Feb63 – 930 TAL GP (434 TCWG) (AFR) Bakalar 19Jan68; depl 71 ACO SQ (1 ACO WG) (TAC) Lockbourne 20Jun68; 931 TAL GP (434 TCWG) (AFR) Bakalar 06Sep68 – 906 TAL GP (302 TCWG) (AFR) Clinton 19Mar69; to Lockbourne 24Jul71; 907 TAL GP (302 TCWG) (AFR) Lockbourne 28Sep71 – Wfu to MASDC (LOG) Davis-Monthan 04Mar73 for storage – DFI: 31Mar76 – Final Disposition: Scrapped.

11215 / 53-3202 / C-119G-36-FA

Avble: 21Oct54; Acc: 29Oct54; Del: 12Nov54 to USAF – 317 TCWG (AFE) Neubiberg AB W. Germany 12Nov54; to 3110 MAI GP (AMC) RAF Burtonwood England 19Jul56 mnt; depl 465 TCWG (AFE) Évreux-Fauville AB France 10Apr57; to Évreux-Fauville AB France 25Jul57 – 2347 RFC (CNR) Long Beach 28Jan58 – 2346 RFC DT (CNR) Paine 16Feb58 – 97 TCSq (349 TCWG) (AFR) Paine 23Dec58; 941 TCG (349 TCWG) (AFR) Paine 63 – 924 TCG (446 TCWG) (AFR) Ellington 09Aug65 – Fairchild (LOG) St. Augustine 06Nov67 – MAP S. Vietnam 14Mar68, VNAF s/n: **533202**, 413 TSQ (53 TWG) Tan Son Nhut AB – 377 ABS WG (PAF) Tan Son Nhut S. Vietnam 05Jan73 – 405 Fighter WG (PAF) Clark AB The Philippines 14Feb73 for disposal – DFI: 05Jun73 – Final Disposition: Scrapped.

11216 / 53-3203 / C-119G-36-FA

Avble: 22Oct54; Acc: 10Nov54; Del: 29Nov54 to USAF – 317 TCWG (AFE) Neubiberg AB W. Germany 29Nov54; to 3110 MAI GP (AMC) RAF Burtonwood England 20Mar56 mnt; depl 465 TCWG (AFE) Évreux-Fauville AB France 10Apr57; to Évreux-Fauville AB France 07Sep57 – 2347 RFC (CNR) Long Beach 13Feb58 – 2346 RFC DT (CNR) Paine 01Mar58 – 97 TCSq (349 TCWG) (AFR) Paine 23Dec58; 941 TCG (349 TCWG) (AFR) Paine 63 – 925 TCG (446 TCWG) (AFR) Ellington 02Oct65 – Fairchild (LOG) St. Augustine 20Nov67 – MAP S. Vietnam 29Mar68, VNAF s/n: **533203**, 413 TSQ (53 TWG) Tan Son Nhut AB – 377 ABS WG (PAF) Tan Son Nhut S. Vietnam 06Feb73 – 405 Fighter WG (PAF) Clark AB The Philippines 22Feb73 for disposal – DFI: 05Jun73 – Final Disposition: Scrapped.

11217 / 53-3204 / C-119G-36-FA

Avble: 27Oct54; Acc: 10Nov54; Del: 12Nov54 to USAF – 317 TCWG (AFE) Neubiberg AB W. Germany 12Nov54; to 3110 MAI GP (AMC) RAF Burtonwood England 07Mar56 mnt; depl 465 TCWG (AFE) Évreux-Fauville AB France 10Apr57; to Évreux-Fauville AB France 08Jul57 – 2347 RFC (CNR) Long Beach 24Jan58 – 2346 RFC DT (CNR) McClellan 07Feb58 – 314 TCSq (349 TCWG) (AFR) McClellan 23Dec58; 940 TCG (349 TCWG) (AFR) McClellan 10Feb63 – 904 TCG (514 TCWG) (AFR) Stewart 02Apr65; 903 TCG (514 TCWG) (AFR) McGuire 03Jun66; depl Myrtle Beach 16Jul67 – 914 TAL GP (512 TCWG) (AFR) Niagara Falls 07Jun68 – Wfu to MASDC (LOG) Davis-Monthan 01Feb71 for storage, PCN: **CJ396** – DFI: 31Mar76 – Final Disposition: Scrapped.

11218 / 53-3205 / C-119G-36-FA

Avble: 05Nov54; Acc: 16Nov54; Del: 30Nov54 to USAF – 317 TCWG (AFE) Neubiberg AB W. Germany 30Nov54; to 3110 MAI GP (AMC) RAF Burtonwood England 15Mar56 mnt; depl 465 TCWG (AFE) Évreux-Fauville AB France 10Apr57; to 7207 AB SQ (AFE) Aviano AB Italy 14Jun57; to Évreux-Fauville AB France 08Jul57 – 2347 RFC (CNR) Long Beach 11Feb58 – 2346 RFC (CNR) Hamilton 17Feb58 – 349 TCWG (AFR) Hamilton 29Dec58; 938 TCG (349 TCWG) (AFR) Hamilton 31May63; 939 TAL GP (349 TCWG) (AFR) Portland 31May67 – Fairchild (LOG) St. Augustine 13Mar68, cvtd **AC-119G Shadow** 01Jul68 – 71 SOP SQ (1 ACO WG) (TAC) Lockbourne 07Aug68 – Fairchild (LOG) St. Augustine 19Dec68 – 14 SOP WG (PAF) Nha Trang AB S. Vietnam 26Jan69; assigned variously Tuy Hoa, Tan Son Nhut, Phan Rang, Phu Cat S. Vietnam from 06Aug69 – MAP S. Vietnam 17Aug71, VNAF s/n: **533205**, 819 ASQ (53 TWG) Tan Son Nhut AB – Final Disposition: Unknown.

11219 / 53-3206 / C-119G-36-FA

Avble: 04Nov54; Acc: 18Nov54; Del: 29Nov54 to USAF – 317 TCWG (AFE) Neubiberg AB W. Germany 29Nov54; to 3110 MAI GP (AMC) RAF Burtonwood England 21Mar56 mnt; depl 465 TCWG (AFE) Évreux-Fauville AB France 10Apr57; to 7206 SUT GP (AFE) Athens Greece 08Jul57; to Évreux-Fauville AB France 12Jul57 – 2347 RFC (CNR) Long Beach 28Jan58 – 2346 RFC (CNR) Hamilton 16Feb58 – Hayes (AMC) Birmingham 30Oct58 – 349 TCWG (AFR) Hamilton 14Jan59; 938 TCG (349 TCWG) (AFR) Hamilton 10Feb63; 939 TCG (349 TCWG) (AFR) Portland 22Jun66 – 908 TAL GP (435 TCWG) (AFR) Brookley 06May68 – 933 TAL GP (440 TCWG) (AFR) Gen Mitchell 24Feb69 – Wfu to MASDC (LOG) Davis-Monthan 28Jan71 for storage – Hayes (LOG) Dothan 29Jun71 – 143 SOP GP (ANG) Theodore F. Green 21Sep71; cvtd **C-119L** 30May73 – Wfu to MASDC (LOG) Davis-Monthan 11Jun75 for storage – DFI: 31Mar76 – BofS Starbird, Inc. Seattle Washington 13Nov78 reg **N90268**; original reg **N4999K** but ntu; carried seafoods in Alaska. Upon landing at King Salmon Airport Alaska 05Jul80 the port engine fire warning light came on followed by the port wingtip bursting into flame. The two crew abandoned the aircraft and although the fire didn't spread the C-119 was later declared as damaged beyond repair. Both engines, other usable parts and equipment were later salvaged before it was scrapped – Reg N90268 canx 20Nov14 – Final Disposition: Accident.

11220 / 53-3207 / C-119G-36-FA

Avble: 09Nov54; Acc: 22Nov54; Del: 30Nov54 to USAF – 4501 SUT SQ (TAC) Donaldson 30Nov54; brief depl 464 TCWG (TAC) Pope 28Mar56 – 2347 RFC (CNR) Long Beach 15Jan58 – 452 TCWG (AFR) Long Beach 13Jun59; to March 01Oct60; assigned 733 TCSq Hill 31Jul62; 945 TCG (452 TCWG) (AFR) Hill 17Jab63 – MAP Taiwan 22Apr66, RoCAF s/n: **3129** – Final Disposition: Unknown.

11221 / 53-3208 / C-119G-36-FA

Avble: 12Nov54; Acc: 30Nov54; Del: 07Dec54 to USAF – 3750 TTN WG (TC) Sheppard 07Dec54 for ground instruction; to museum / school status 30Jun58 – 3750 MSU GP (ATC) Sheppard 01Feb62; redes **UC-119G** 30Jun62; redes **GC-119G** 30Nov62 – DFI: 09Feb65 – Final Disposition: Scrapped.

11222 / 53-3209 / C-119G-36-FA

Avble: 15Nov54; Acc: 30Nov54; Del: 03Dec54 to USAF – 6614 TSM SQ (NEA) Ernest Harmon Canada 03Dec54; to 6604 AB WG (NEA) Pepperrell Canada 21Jan55; to 6607 AB

WG (NEA) Thule AB Greenland 12Jun55; to 6606 AB WG (NEA) Goose Bay Canada 13Jul56, 10Jun56, 20Oct56; to 6605 AB WG (NEA) Ernest Harmon Canada 27Feb56 – MIDAR (AMC) Olmsted 04Mar57 – 4087 TSM SQ (SAC) Ernest Harmon Canada 08May57 – 2237 RFC DT (CNR) Niagara Falls 04May58 – 328 TCSq (512 TCWG) (AFR) Niagara Falls 19Dec58 – 4440 ADE GP (TAC) Langley 27Jun60 – MAP India 11Jul60 – Final Disposition: Unknown.

11223 / 53-3210 / C-119G-36-FA

Avble: 17Nov54; Acc: 30Nov54; Del: 10Dec54 to USAF – 6614 TSM SQ (NEA) Ernest Harmon Canada 14Dec54; to 6606 AB WG (NEA) Goose Bay Canada 03Jan55, 20May55, 28May55; to 6605 AB WG (NEA) Ernest Harmon Canada 03Apr56 – 4087 TSM SQ (SAC) Ernest Harmon Canada 01Apr57; to 4737 AB WG (ADC) Pepperrell Canada 06Feb58 mnt – 2472 RFC DT (CNR) Davis 11Apr58; 2472 RFC (CNR) Richards 16Apr58 – 65 TCSq (403 TCWG) (AFR) Davis 18Sep58 – 4440 ADE GP (TAC) Langley 14Jun60 – MAP India 26Jun60, IAF s/n: **BK510** – Final Disposition: Unknown.

11224 / 53-4639 / C-119G-36-FA

Avble: 19Nov54; Acc: 30Nov54; Del: 17Dec54 to MAP India, IAF s/n: **IK443** – Final Disposition: Unknown.

11225 / 53-4640 / C-119G-36-FA

Avble: 22Nov54; Acc: 16Dec54; Del: 17Dec54 to MAP India, IAF s/n: **IK444** – Cvtd to saloon at Agra Air Base Uttar Pradesh India, named *Shatrujeet Bar* – Final Disposition: Extant.

11226 / 53-4641 / C-119G-36-FA

Avble: 24Nov54; Acc: 14Dec54; Del: 17Dec54 to MAP India, IAF s/n: **IK445** – Final Disposition: Unknown.

11227 / 53-3211 / C-119G-36-FA

Avble: 30Nov54; Acc: 21Dec54; Del: 10Jan55 to USAF – 6614 TSM SQ (NEA) Ernest Harmon Canada 11Jan55; to 6606 AB WG (NEA) Goose Bay Canada 06Mar55, 09May55, 01Jun55; to 2500 AB WG (CNC) Mitchel 22May56 mnt – 4087 TSM SQ (SAC) Ernest Harmon Canada 01Apr57 – 2472 RFC DT (CNR) Davis 28Mar58 – 65 TCSq (403 TCWG) (AFR) Davis 18Sep58; 929 TCG (403 TCWG) (AFR) Davis 11Feb63 – Fairchild (LOG) S. Augustine 23Aug65 – 922 TCG (433 TCWG) (AFR) Kelly 04Oct65 – Fairchild (LOG) St. Augustine 01Jul68, cvtd **AC-119K Stinger** 26Jul68 – 1 SOP WG (TAC) Lockbourne 04Jun69 – 4410 CCT WG (TAC) Lockbourne 14Jul69 – 14 SOP WG (PAF) Phan Rang AB S. Vietnam 18Nov69; assigned variously Da Nang S. Vietnam & Nakhon Phanom Thailand from 10Dec69; to 313 ADH DV (PAF) Kadena AB Japan 31Aug70 – China Airlines (LOG) Taipei Taiwan 26Jul71 mnt – 56 SOP WG (PAF) Nakhon Phanom AB Thailand 09Oct71; to 432 TR WG (PAF) Udorn AB Thailand 22Oct71; assigned Da Nang S. Vietnam 06Jul72 – WRAAR (LOG) Nakhon Phanom AB Thailand 11Sep72 – 6498 ABS WG (PAF) Da Nang S. Vietnam 26Dec72 – MAP S. Vietnam 31Jan73, VNAF s/n: **533211**, 821 ASQ (53 TWG) Tan Son Nhut AB – Final Disposition: Unknown.

11228 / 53-3212 / C-119G-36-FA

Avble: 02Dec54; Acc: 21Dec54; Del: 12Jan55 to USAF – 463 TCWG (TAC) Ardmore 12Jan55 – MIDAR (AMC) Olmsted 27Aug56 – 2235 RFC (CNR) Grenier 13Apr57 – 732

TCSq (94 TCWG) (AFR) Grenier 19Mar59 – 4440 ADE GP (TAC) Langley 13Jun60 – MAP India 26Jun60 – Final Disposition: Unknown.

11229 / 53-3213 / C-119G-36-FA

Avble: 07Dec54; Acc: 30Dec54; Del: 28Feb55 to USAF – Bailment Fairchild (AMC) Hagerstown 28Feb55 – 464 TCWG (TAC) Pope 08Apr55; depl North Aux AF 03Jun55; bailment Fairchild (AMC) Hagerstown 25Jul55; depl Aberdeen 05May56; to 3800 AB WG (AU) Maxwell 12Aug56 mnt – 83 Fighter WG (TAC) Seymour 20Nov56; cvtd **C-119J** 01Aug57 – Hayes (AMC) Birmingham 23Oct57 – 2347 RFC (CNR) Long Beach 04Feb58 – 452 TCWG (AFR) Long Beach 13Jun59; to March 01Oct60 – 102 AML SQ (NY-ANG) New York ANGB 24Mar61 – 106 AML GP (ANG) New York ANGB 28Feb63 – Wfu to 2704 ASD GP (LOG) Davis-Monthan 24Apr63 for storage – DFI: 11May67 – Final Disposition: Scrapped.

11230 / 53-3214 / C-119G-36-FA

Avble: 09Dec54; Acc: 22Dec54; Del: 28Dec54 to USAF – 4501 SUT SQ (TAC) Donaldson 28Dec54. Landing mishap at Spartanburg MAP South Carolina 06Dec55 serious enough to damage the aircraft beyond repair, no fatalities – DFI: 20Dec55 – Final Disposition: Accident.

11231 / 53-3215 / C-119G-36-FA

Avble: 13Dec54; Acc: 28Dec54; Del: 05Jan55 to USAF – 314 TCWG (TAC) Sewart 05Jan55; depl England Louisiana 05Nov56 – 2230 RFC (CNR) NAS New York 12Sep57 – 2466 RFC (CNR) Bakalar 04Oct57 – 2242 RFC(CNR) Selfridge 17Oct57 – 403 TCWG (AFR) Selfridge 13Nov58 – 927 TCG (403 TCWG) (AFR) Selfridge 10Feb63 – MAP India 03Jul63 – Final Disposition: Unknown.

11232 / 53-3216 / C-119G-36-FA

Avble: 15Dec54; Acc: 29Dec54; Del: 12Jan55 to USAF – 463 TCWG (TAC) Ardmore 12Jan55; depls North Aux AF 07Apr55, 05May56; depl 464 TCWG (TAC) Pope 19Dec56 – Hayes (AMC) Birmingham 01Jul57 – 2230 RFC (CNR) NAS New York 20Sep57 – 2466 RFC (CNR) Bakalar 04Oct57 – 2242 RFC (CNR) Selfridge 18Oct57 – 403 TCWG (AFR) Selfridge 20Nov58; 927 TCG (403 TCWG) (AFR) Selfridge 10Feb63 – 129 SOP GP (ANG) Hayward 14Feb69; cvtd **C-119L** 05Mar73 – Wfu to MASDC (LOG) Davis-Monthan 07Dec73 for storage – DFI: 31Mar76 – BofS Dross Metals, Inc. Tucson Arizona 29Nov79 – BofS J.D. Gifford & Associates, Inc. Anchorage Alaska 28Sep80; ARAp 13Jan81 reg **N8504Y**; equipment removal to lighten airframe 28May81; inverter upgrade 31Aug81; cvtd **C-119L Jet-Pak** (J34-WE) 26Apr82 – Alaska Commercial Fishing and Agriculture Bank Anchorage Alaska 17Sep84 through repossession from J.D. Gifford; ARAp 01Feb85 – BofS Stebbins & Ambler Air Transport Anchorage Alaska 09Feb85. Subsequent history unk but thought to have been involved in some sort of minor accident and bku, listed in FAA records as "destroyed/scrapped." Reg N8504Y canx 21Feb04. Final Disposition: Unknown.

11233 / 53-3217 / C-119G-36-FA

Avble: 20Dec54; Acc: 29Dec54; Del: 14Jan55 to USAF – 463 TCWG (TAC) Ardmore 16Jan55; depl North Aux AF 09Apr55 – Hayes (AMC) Birmingham 03Jul57 – 2230 RFC (CNR) NAS New York 19Sep57 – 2466 RFC (CNR) Bakalar 04Oct57 – 2242 RFC (CNR) Selfridge 18Oct57 – 403 TCWG (AFR) Selfridge 13Nov58 – 4440 ADE GP (TAC) Langley 15Jun60 – MAP India 10Jul60 – Final Disposition: Unknown.

11234 / 53-3218 / C-119G-36-FA

Avble: 23Dec54; Acc: 17Jan55; Del: 20Jan55 to USAF – 60 TCWG (AFE) Rhein-Main AB W. Germany 20Jan55; to 6605 AB GP (NEA) Ernest Harmon AB Canada 22Jan55 mnt; to Dreux AB France 12Sep55; to 3110 MAI GP (AMC) RAF Burtonwood England 21Aug56 mnt – 7305 AB GP (322 ADV) (AFE) Dreux AB France 24Sep58; 11 TCSq (322 ADV) (AFE) Dreux AB France 12Dec60 – 4440 ADE GP (TAC) Langley 19Jan61 – 452 TCWG (AFR) March 27Jan61; 944 TCG (452 TCWG) (AFR) March 11Feb63 – Fairchild (LOG) St. Augustine 12Oct67 – MAP S. Vietnam 25Mar68, VNAF s/n: **533218**, 413 TSQ (53 TWG) Tan Son Nhut AB – 377 ABS WG (PAF) Tan Son Nhut S. Vietnam 11Jan73 – 405 Fighter WG (PAF) Clark AB The Philippines 22Jan73 for disposal – DFI: 05Jun73 – Final Disposition: Scrapped.

11235 / 53-3219 / C-119G-36-FA

Avble: 28Dec54; Acc: 18Jan55; Del: 03Feb55 to USAF – 60 TCWG (AFE) Rhein-Main AB W. Germany 03Feb55; to Dreux AB France Sep55; to 3110 MAI GP (AMC) RAF Burtonwood England 21Aug56 mnt – SABBE (AMC) Brussels Belgium 20Aug58 mnt – 7305 AB GP (322 ADV) (AFE) Dreux AB France 24Sep58 – United Nations (UN), reg **UNO-102**, mission to The Congo 03Aug60 – MAP Italy 61, AMI s/n: **MM53-3219** – Became a gate guardian at Pisa-San Guisto AB Italy – Ditellandia Air Park Castel Volturno Italy 1994–2004, removed and scrapped after the owner's death – Final Disposition: Scrapped.

11236 / 53-3220 / C-119G-36-FA

Avble: 29Dec54; Acc: 21Jan55; Del: 24Jan55 to USAF – 60 TCWG (AFE) Rhein-Main AB W. Germany 24Jan55; to Dreux AB France 23Sep55; to 3110 MAI GP (AMC) RAF Burtonwood England 02May56 mnt – 7305 AB GP (322 ADV) (AFE) Dreux AB France 24Sep58; 11 TCSq (322 ADV) (AFE) Dreux AB France 12Dec60 – 4440 ADE GP (TAC) Langley 11Jan61 – 78 TCSq (435 TCWG) (AFR) Bates 15Jan61 – 357 TCSq (302 TCWG) (AFR) Bates 08May61; 908 TCG (302 TCWG) (AFR) Bates 11Feb63 – 908 TCG (435 TCWG) (AFR) Bates 18Mar63; to Brookley 30Oct64; 915 TCG (435 TCWG) (AFR) Homestead 07Jan66; 908 TAL GP (435 TCWG) (AFR) Brookley 01Jul67 – Fairchild (LOG) St. Augustine 17Jul68 – MAP S. Vietnam 28Jul68, VNAF s/n: **533220**, 413 TSQ (53 TWG) Tan Son Nhut AB – 377 ABS WG (PAF) Tan Son Nhut S. Vietnam 10Jan73 – 405 Fighter WG (PAF) Clark AB The Philippines 16Jan73 for disposal – DFI: 29May73 – Final Disposition: Scrapped.

11237 / 53-3221 / C-119G-36-FA

Avble: 03Jan55; Acc: 21Jan55; Del: 03Feb55 to USAF – 60 TCWG (AFE) Rhein-Main AB W. Germany 03Feb55; to Dreux AB France 23Sep55; to 7310 AB GP (AFE) Rhein-Main AB W. Germany 07Dec55; to 3110 MAI GP (AMC) RAF Burtonwood England Jul56 mnt. Had a prop overspeed after take-off from Athens IAP Greece 01Jul57. While coming round for an emergency landing the pilot was forced to abort and belly-land the C-119 onto soft ground adjacent the runway as a DC-3 had been cleared to enter the runway in front of the Boxcar, there were no fatalities. The C-119 wreck was subsequently broken up for spares on site – Final Disposition: Accident.

11238 / 53-3222 / C-119G-36-FA

Avble: 06Jan55; Acc: 25Jan55; Del: 07Feb55 to USAF – 60 TCWG (AFE) Rhein-Main AB W. Germany 07Feb55. A group of 9 C-119s departed Stuttgart-Echterdingen AP W. Germany 11Aug55 for a training mission all with a full complement of army soldiers from the 499 Engineer Combat Battalion onboard. Near Altensteig W. Germany at 4,000 ft

C-119 53-3222 developed trouble in the left engine and while trying to leave the formation collided with C-119 53-7841. Both aircraft suffered a catastrophic loss of control and crashed killing all onboard. 53-3222 lost 5 crew and 15 passengers with 53-7841 5 crew and 41 passengers for a total of 66 souls killed making this the worst accident for the C-119 in USAF service. The incident became known as the Altensteig Mid–Air Collision and was the worst air accident in Europe at the time – Final Disposition: Accident.

11239 / 53-7826 / C-119G-36-FA

Avble: 10Jan55; Acc: 26Jan55; Del: 04Feb55 to USAF – 60 TCWG (AFE) Rhein-Main AB W. Germany 04Feb55; to Dreux AB France 12Sep55; to 3110 MAI GP (AMC) RAF Burtonwood England 09Apr56 mnt – 7305 AB GP (322 ADV) (AFE) Dreux AB France 24Sep58; 11 TCSq (322 ADV) (AFE) Dreux AB France 12Dec60 – 4440 ADE GP (TAC) Langley 09Jan61 – 78 TCSq (435 TCWG) (AFR) Bates 15Jan61 – 357 TCSq (302 TCWG) (AFR) Bates 08May61; 908 TCG (302 TCWG) (AFR) Bates 11Feb63 – 908 TCG (435 TCWG) (AFR) Bates 18Mar63; to Brookley 30Oct64 – Fairchild (LOG) St. Augustine 15Jul68, cvtd **AC-119K Stinger** 26Jul68 – 1 SOP WG (TAC) Lockbourne 24May69 – 4410 CCT WG (TAC) Lockbourne 14Jul69 – 14 SOP WG (PAF) Phan Rang AB S. Vietnam 21Oct69; assigned variously Da Nang, Phu Cat S. Vietnam & Nakhon Phanom Thailand from 29Nov69 – 56 SOP WG (PAF) Nakhon Phanom AB Thailand 23Aug71; assigned variously Da Nang S. Vietnam from 27Aug71. Assigned a daylight mission on 02May72 to destroy a U.S. airdrop of ammunition intended for the defenders of An Loc S. Vietnam which instead had fallen into enemy territory. While circling at 4,700 ft the gunship was fired at by an NVA 37mm anti-aircraft gun, several rounds of which struck the starboard wing setting it on fire. The starboard jet engine was blown off and starboard the engine failed with the u/c strut falling out of its bay creating uncontrollable drag. 7 of the 10 crew bailed out before the aircraft crashed 5 miles from An Loc, the survivors were rescued by helicopter over the next 4 hours requiring significant suppressing fire from supporting aircraft. Killed in Action were: Capt Terence F. Courtney (pilot), Capt David R. Slagle (sensor operator) and SSgt Kenneth R. Brown (illuminator operator). Rescued survivors were: Lt Col "Tash" Tashioglou (sensor operator), 1Lt Larry Barbee (sensor operator), 1Lt Jim Barkalow (co-pilot), SSgt "Yogi" Bare (flight engineer), SSgt Dale Iman (gunner?), SSgt "Ski" Sledzinski (gunner?) and A1C Richard "Craig" Corbett (gunner). All further AC-119 daylight missions were canceled after this mission – Final Disposition: Combat Loss.

11240 / 53-7827 / C-119G-36-FA

Avble: 18Jan55; Acc: 27Jan55; Del: 09Feb55 to USAF – 60 TCWG (AFE) Rhein-Main AB W. Germany 09Feb55; to Dreux AB France 01Oct55; to 3110 MAI GP (AMC) RAF Burtonwood England 26Oct56 mnt; to 3131 MAI GP (AMC) Chateauroux AB France 03Sep57 – 7305 AB GP (322 ADV) (AFE) Dreux AB France 24Sep58; 10 TCSq (322 ADV) (AFE) Dreux AB France 12Dec60 – 7305 CAM SQ (322 ADV) (AFE) Dreux AB France 08Jan61 – 4440 ADE GP (TAC) Langley 22Mar61 – 440 TCWG (AFR) Gen Mitchell 29Mar61; 933 TCG (440 TCWG) (AFR) Gen Mitchell 11Feb63 – MAP India 09Jul63 – Final Disposition: Unknown.

11241 / 53-4642 / C-119G-36-FA

Avble: 18Jan55; Acc: 31Jan55; Del: 04Feb55 to MAP India, IAF s/n: **IK446** – Final Disposition: Unknown.

11242 / 53-4643 / C-119G-36-FA

Avble: 26Jan55; Acc: 28Jan55; Del: 04Feb55 to MAP India, IAF s/n: **IK447** – Final Disposition: Unknown.

11243 / 53-4644 / C-119G-36-FA

Avble: 21Jan55; Acc: 08Feb55; Del: 15Feb55 to MAP India, IAF s/n: **IK448** – Final Disposition: Unknown.

11244 / 53-4645 / C-119G-36-FA

Avble: 24Jan55; Acc: 17Feb55; Del: 18Feb55 to MAP India, IAF s/n: **IK449** – Final Disposition: Unknown.

11245 / 53-7828 / C-119G-36-FA

Avble: 26Jan55; Acc: 15Feb55; Del: 16Feb55 to USAF – 60 TCWG (AFE) Rhein-Main AB W. Germany 16Feb55; to Dreux AB France 01Oct55; to 3110 MAI GP (AMC) RAF Burtonwood England 30Jul56 mnt; depl 48 Fighter-Bomber WG (AFE) Chaumont AB France 11Feb58 – 7305 AB GP (322 ADV) (AFE) Dreux AB France 24Sep58 – United Nations (UN), reg **UNO-103**, mission to The Congo 03Aug60 – MAP Italy 61, AMI s/n: **MM53-7828** – Final Disposition: Scrapped.

11246 / 53-7829 / C-119G-36-FA

Avble: 28Jan55; Acc: 17Feb55; Del: 24Feb55 to USAF – 60 TCWG (AFE) Rhein-Main AB W. Germany 24Feb55; to Dreux AB France 23Sep55; to 7310 AB GP (AFE) Rhein-Main AB W. Germany 12Jan56; to 3131 MAI GP (AMC) Chateauroux AB France 03Mar56; to 3110 MAI GP (AMC) RAF Burtonwood England 04Apr56 mnt; to 7425 AB GP (AFE) Hahn AB W. Germany 13Mar58 – 7305 AB GP (322 ADV) (AFE) Dreux AB France 24Sep58; to 7425 SUT GP (AFE) Hahn AB W. Germany 17Mar59; to 7310 SUT GP (AFE) Rhein-Main AB W. Germany 31Oct59 – 317 AB GP (AFE) Évreux-Fauville AB, France 17Jul60 – MAP Belgium, BAF s/n: **CP-41**; BOC: 17Jul60 15 WG / 20 SQ Melsbroek AB Belgium, callsign: **OT-CEA** – 15 WG / 40 SQ 01Dec61. Bird strike at take-off 28Jul72 – Wfu 02Jul73 and stored at Koksijde AB Belgium; SOC: 22Oct75 – To civil company International Engine Parts, scrapped 1977–78 – Final Disposition: Scrapped.

11247 / 53-7830 / C-119G-36-FA

Avble: 01Feb55; Acc: 25Feb55; Del: 04Mar55 to USAF – 60 TCWG (AFE) Rhein-Main AB W. Germany 04Mar55; to Dreux AB France 12Sep55; to 3110 MAI GP (AMC) RAF Burtonwood England 10Aug56 mnt – 7305 AB GP (322 ADV) (AFE) Dreux AB France 24Sep58; 11 TCSq (322 ADV) (AFE) Dreux AB France 12Dec60 – 4440 ADE GP (TAC) Langley 17Jan61 – 78 TCSq (435 TCWG) (AFR) Bates 29Jan61 – 357 TCSq (302 TCWG) (AFR) Bates 08May61; 908 TCG (302 TCWG) (AFR) Bates 11Feb63 – 908 TCG (435 TCWG) (AFR) Bates 18Mar63; to Brookley 30Oct64 – Fairchild (LOG) St. Augustine 15Jul68, cvtd **AC-119K Stinger** 26Jul68 – 1 SOP WG (TAC) Lockbourne 06Jun69 – 4410 CCT WG (TAC) Lockbourne 14Jul69 – 14 SOP WG (PAF) Phan Rang S. Vietnam 27Dec69; assigned variously Phu Cat, Da Nang, Tan Son Nhut S. Vietnam & Udorn, Nakhon Phanom Thailand from 15Feb70 – SMAAR (LOG) Phan Rang AB S. Vietnam 23Aug71 – 56 SOP WG (PAF) Nakhon Phanom AB Thailand 28Aug71; assigned variously Da Nang S. Vietnam from 03Sep71 – MAP S. Vietnam 10Nov72, VNAF s/n: **537830**, 821 ASQ (53 TWG) Tan Son Nhut AB – Final Disposition: Unknown.

11248 / 53-7831 / C-119G-36-FA

Avble: 14Feb55; Acc: 22Feb55; Del: 28Feb55 to USAF – 60 TCWG (AFE) Rhein-Main AB W. Germany 01Mar55; to 1604 MAT SQ (MATS) Kindley 13Mar55 mnt; to Dreux AB France 12Sep55; to 3110 MAI GP (AMC) RAF Burtonwood England 56 mnt **RECORDS INDECIPHERABLE** 7305 AB GP (322 ADV) (AFE) Dreux AB France 24Sep58; 11 TCSq (322 ADV) (AFE) Dreux AB France 12Dec60 – 4440 ADE GP (TAC) Langley 07Jan61 – 78 TCSq (435 TCWG) (AFR) Bates 12Jan61 – 357 TCSq (302 TCWG) (AFR) Bates 08May61; 908 TCG (302 TCWG) (AFR) Bates 11Feb63 – 908 TCG (435 TCWG) (AFR) Bates 18Mar63; to Brookley 30Oct64 – Fairchild (LOG) St. Augustine 13Jun68, cvtd **AC-119K Stinger** 01Jul68 – 4413 CCT SQ (1 ACO WG) (TAC) Lockbourne 19Dec68 – 1 SOP WG (TAC) Lockbourne 17Feb69 – Fairchild (LOG) St. Augustine 17Jun69 – 4410 CCT WG (TAC) Lockbourne 20Aug69 – 1 SOP WG (TAC) Eglin 07Jul71 – WRAAR (LOG) Robins 28Oct72 – 6498 ABS WG (PAF) Da Nang AB S. Vietnam 20Dec72 – MAP S. Vietnam 31Jan73, VNAF s/n: **537831**, 821 ASQ (53 TWG) Tan Son Nhut AB – Final Disposition: Unknown.

11249 / 53-7832 / C-119G-36-FA

Avble: 08Feb55; Acc: 28Feb55; Del: 07Mar55 to USAF – 60 TCWG (AFE) Rhein-Main AB W. Germany 07Mar55; to Dreux AB France 01Oct55; to 3110 MAI GP (AMC) RAF Burtonwood England 28May56 mnt – 7305 AB GP (322 ADV) (AFE) Dreux AB France 24Sep58; 10 TCSq (322 ADV) (AFE) Dreux AB France 12Dec60 – 7305 CAM SQ (AFE) Dreux AB France 08Jan61 – 4440 ADE GP (TAC) Langley 09Feb61 – 512 TCWG (AFR) NAS Willow Grove 17Feb61; 913 TCG (512 TCWG) (AFR) NAS Willow Grove 11Feb63 – 130 SOP GP (ANG) Kanawha 20Jan69 – Wfu to 3345 MSU GP (ATC) Chanute 11Jun69 as excess – DFI: 12Jun69 – Final Disposition: Scrapped.

11250 / 53-7833 / C-119G-36-FA

Avble: 10Feb55; Acc: 24Feb55; Del: 15Mar55 to USAF – 60 TCWG (AFE) Rhein-Main AB W. Germany 15Mar55; to Dreux AB France 23Sep55; to 3110 MAI GP (AMC) RAF Burtonwood England 03Apr56 mnt; to 3131 MAI GP (AFE) Chateauroux AB France 57 mnt; depl 36 Fighter WG (AFE) Bitburg AB W. Germany 12Jul57 – 7305 AB GP (322 ADV) (AFE) Dreux AB France 24Sep58; 11 TCSq (322 ADV) (AFE) Dreux AB France 12Dec60 – 7305 CAM SQ (AFE) Dreux AB France 08Jan61 – 4440 ADE GP (TAC) Langley 01Feb61 – 314 TCSq (349 TCWG) (AFR) McClellan 11Feb61; 940 TCG (349 TCWG) (AFR) McClellan 10Feb63 – 904 TCG (514 TCWG) (AFR) Stewart 12May65; depl (CNC) Boston ANG 21May66; 903 TCG (514 TCWG) (AFR) McGuire 31May67; depl Myrtle Beach 16Jul67 – Fairchild (LOG) St. Augustine 20Feb68, cvtd **AC-119G Shadow** 13Jun68 – 4413 CCT SQ (1 ACO WG) (TAC) Lockbourne 16Jun68 – 1 SOP WG (TAC) Lockbourne 17Feb69 – 4410 CCT WG (TAC) Lockbourne 14Jul69 – Hayes (LOG) Dothan 04Mar71 – 14 SOP WG (PAF) Phan Rang AB S. Vietnam 16Apr71; assigned variously Tan Son Nhut S. Vietnam from 01Jul71 – MAP S. Vietnam 30Aug71, VNAF s/n: **537833**, 819 ASQ (53 TWG) Tan Son Nhut AB – Final Disposition: Unknown.

11251 / 53-7834 / C-119G-36-FA

Avble: 14Feb55; Acc: 28Feb55; Del: 15Mar55 to USAF – 60 TCWG (AFE) Rhein-Main AB W. Germany 15Mar55; to Dreux AB France 01Oct55; to 3110 MAI GP (AMC) RAF Burtonwood England 18Apr56 mnt – 7305 AB GP (322 ADV) (AFE) Dreux AB France 19Aug58; depl 10 TR WG (AFE) Spangdah AB W. Germany 15Jan59; 10 TCSq (322 ADV) (AFE) Dreux AB France 12Dec60 – 7305 CAM SQ (AFE) Dreux AB France 08Jan61 – 4440 ADE GP

(TAC) Langley 21Mar61 – 440 TCWG (AFR) Gen Mitchell 25Mar61; 933 TCG (440 TCWG) (AFR) Gen Mitchell 11Feb63 – MAP India 08Jul63 – Final Disposition: Unknown.

11252 / 53-7835 / C-119G-36-FA

Avble: 17Feb55; Acc: 28Feb55; Del: 19Mar55 to USAF – 60 TCWG (AFE) Rhein-Main AB W. Germany 19Mar55; to Dreux AB France 23Sep55; to 3110 MAI GP (AMC) RAF Burtonwood England 14Apr56 mnt; to 7030 SUT GP (AFE) Ramstein AB W. Germany 28Jan58 – 7305 AB GP (322 ADV) (AFE) Dreux AB France 24Sep58; 11 TCSq (322 ADV) (AFE) Dreux AB France 12Dec60 – 4440 ADE GP (TAC) Langley 30Jan61 – 314 TCSq (349 TCWG) (AFR) McClellan 13Feb61; 940 TCG (349 TCWG) (AFR) McClellan 10Feb63 – 925 TCG (446 TCWG) (AFR) Ellington 03Sep65 – Fairchild (LOG) St. Augustine 18Jul67 – MAP Ethiopia 18Sep70, ETAF s/n: **914**, nfd – Final Disposition: Unknown.

11253 / 53-7836 / C-119G-36-FA

Avble: 23Feb55; Acc: 22Mar55; Del: 28Mar55 to USAF – 60 TCWG (AFE) Rhein-Main AB W. Germany 28Mar55; to Dreux AB France 01Oct55; to 3110 MAI GP (AMC) RAF Burtonwood England 24Aug56 mnt; depl 66 TR WG (AFE) Sembach AB W. Germany 08Nov57 – 7305 AB GP (322 ADV) (AFE) Dreux AB France 24Sep58; 10 TCSq (322 ADV) (AFE) Dreux AB France 12Dec60 – 4440 ADE GP (TAC) Langley 23Jan61 – 452 TCWG (AFR) March 03Feb61; 943 TCG (452 TCWG) (AFR) March 11Feb63 – 129 SOP GP (ANG) Hayward 27Dec68; cvtd **C-119L** 07Nov73 – Wfu to MASDC (LOG) Davis-Monthan 28Jan75 for storage, PCN: **CJ452** – DFI: 31Mar76 – BofS Dross Metals, Inc. Tucson Arizona 29Nov79 – BofS J.D. Gifford & Associates, Inc. Anchorage Alaska 28Sep80; ARAp 13Jan81 reg **N8504Z**; major radio, nav, flight, instrument upgrade and control surfaces recovered 25May82 – BofS Gerald C. Ball Anchorage Alaska 22Feb85 – BofS Northern Pacific Transport, Inc. Anchorage Alaska 29Aug86 – BofS Alaska Aircraft Leasing, Inc. Anchorage Alaska 01Apr87; nav upgrade 06Apr89; cvtd to **C-119L Jet-Pak** (J34-WE) 02Aug89 – The First National Bank of Alaska Anchorage Alaska 27Apr92 through repossession from Alaska Aircraft Leasing; ARAp 12Aug92 – BofS Hawkins & Powers Aviation, Inc. Greybull Wyoming 12Aug92 – BofS Anchorage Flight, Inc. Anchorage Alaska 01Sep92; avionics and GPS upgrade 01Jul93; bulk fuel system installed 30Jul93 – BofS Everts Air Fuel, Inc. Fairbanks Alaska 31Aug94; wfu and currently stored at Fairbanks Intl. Airport with nose art *Know Fear* – Final Disposition: Extant.

11254 / 53-7837 / C-119G-36-FA

Avble: 24Feb55; Acc: 24Mar55; Del: 28Mar55 to USAF – 60 TCWG (AFE) Rhein-Main AB W. Germany 28Mar55; to Dreux AB France 23Sep55; to 7310 AB GP (AFE) Rhein-Main AB W. Germany 29Dec55; to 3110 MAI GP (AMC) RAF Burtonwood England 23Oct56 mnt – 7305 AB GP (322 ADV) (AFE) Dreux AB France 23Sep58; 11 TCSq (322 ADV) (AFE) Dreux AB France 12Dec60 – 4440 ADE GP (TAC) Langley 23Jan61; to 1605 AB WG (MATS) Lajes Azores 06Jan61 mnt – 78 TCSq (435 TCWG) (AFR) Bates 14Jan61 – 357 TCSq (302 TCWG) (AFR) Bates 08May61; 908 TCG (302 TCWG) (AFR) Bates 11Feb63 – 908 TCG (435 TCWG) (AFR) Bates 18Mar63; to Brookley 30Oct64 – 130 SOP GP (ANG) Kanawha 06Feb69; cvtd **C-119L** 28Mar73 – Wfu to MASDC (LOG) Davis-Monthan 10Jul75 for storage, PCN: **CJ468** – DFI: 14Dec76 – Final Disposition: Scrapped.

11255 / 53-7838 / C-119G-36-FA

Avble: 01Mar55; Acc: 24Mar55; Del: 31Mar55 to USAF – 60 TCWG (AFE) Rhein-Main AB W. Germany 31Mar55; to Dreux AB France 23Sep55; to 3110 MAI GP (AMC) RAF Burtonwood England 02Jul56 mnt; to 3131 MAI GP (AMC) Chateauroux AB France 57 mnt – 7305 AB GP (322 ADV) (AFE) Dreux AB France 24Sep58; to 3131 MAI GP (AMC) Chateauroux AB France 31Sep58; depl 48 Fighter WG (AFE) Chaumont AB France 05Feb59; 11 TCSq (322 ADV) (AFE) Dreux AB France 12Dec60 – 7305 CAM SQ (AFE) Dreux AB France 08Jan61 – 4440 ADE GP (TAC) Langley 16Mar61 – 65 TCSq (403 TCWG) (AFR) Davis 17Mar61; 929 TCG (403 TCWG) (AFR) Davis 11Feb63 – Fairchild (LOG) St. Augustine 20Sep65 – 922 TCG (433 TCWG) (AFR) Kelly 26Oct65. Crashed 7 miles W Kelly AFB Texas 09Sep66 after take-off for a local training flight. 6 crew onboard with 2 fatalities, nfd – Final Disposition: Accident.

11256 / 53-7839 / C-119G-36-FA

Avble: 09Mar55; Acc: 23Mar55; Del: 28Mar55 to USAF – 60 TCWG (AFE) Rhein-Main AB W. Germany 28Mar55; to Dreux AB France 12Sep55; depl 388 Fighter-Bomber WG (AFE) Étain-Rouvres AB France 29Feb56; to 3110 MAI GP (AMC) RAF Burtonwood England 07Nov56 mnt – 7305 AB GP (322 ADV) (AFE) Dreux AB France 24Sep58; 11 TCSq (322 ADV) (AFE) Dreux AB France 12Dec60 – 7305 CAM SQ (AFE) Dreux AB France 08Jan61 – 4440 ADE GP (TAC) Langley 16Jan61; to 4500 AB WG (TAC) Langley 27Jan61 – 78 TCSq (435 TCWG) (AFR) Bates 09Feb61 – 357 TCSq (302 TCWG) (AFR) Bates 08May61; 908 TCG (302 TCWG) (AFR) Bates 11Feb63 – 908 TCG (435 TCWG) (AFR) Bates 18Mar63; to Brookley 30Oct64; depl 915 TCG (435 TCWG) (AFR) Homestead 04Feb66 – Fairchild (LOG) St. Augustine 01Apr68, cvtd **AC-119K Stinger** 01Jul68 – 4413 CCT SQ (1 ACO WG) (TAC) Lockbourne 06Nov68 – 1 SOP WG (TAC) Lockbourne 17Feb69 – 4410 CCT WG (TAC) Lockbourne 24Jul69 – 1 SOP WG (TAC) Eglin 08Jul71 – WRAAR (LOG) Robins 28Oct72 – 6498 ABS WG (PAF) Da Nang AB S. Vietnam 15Dec72 – MAP S. Vietnam 31Jan73, VNAF s/n: **537839**, 821 ASQ (53 TWG) Tan Son Nhut AB. While on a VNAF training flight out of Da Nang AB S. Vietnam 01Mar73, low visibility in fog and fuel starvation waiting to land caused the 5 USAF flight instructors and 8 VNAF trainees to bail out over the South China Sea. All 13 were rescued the next day but one VNAF crewman was drowned when his parachute became caught in a boat propeller, the abandoned aircraft came down in the sea – Final Disposition: Accident.

11257 / 53-7840 / C-119G-36-FA

Avble: 09Mar55; Acc: 30Mar55; Del: 11Apr55 to USAF – 60 TCWG (AFE) Rhein-Main AB W. Germany 11Apr55; to Dreux AB France 01Oct55; to 3110 MAI GP (AMC) RAF Burtonwood England 01Nov56 mnt **RECORDS INDECIPHERABLE** to 3131 MAI GP (AMC) Chateauroux AB France 25Sep57 – 7305 AB GP (322 ADV) (AFE) Dreux AB France 24Sep58; 10 TCSq (322 ADV) (AFE) Dreux AB France 12Dec60 – 7305 CAM SQ (AFE) Dreux AB France 08Jan61 – 4440 ADE GP (TAC) Langley 27Mar61 – 96 TCSq (440 TCWG) (AFR) Minneapolis-St. Paul 02Apr61; 934 TCG (440 TCWG) (AFR) Minneapolis-St. Paul 11Feb63 – 906 TAL GP (302 TCWG) (AFR) Clinton 08Nov69; 907 TAL GP (302 TCWG) (AFR) Clinton 12Nov69; to Lockbourne 24Jul71 – MAP S. Vietnam 01Jul72, VNAF s/n: **537840**, 413 TSQ (53 TWG) Tan Son Nhut AB – Final Disposition: Unknown.

11258 / 53-7841 / C-119G-36-FA

Avble: 10Mar55; Acc: 28Mar55; Del: 06Apr55 to USAF – 60 TCWG (AFE) Rhein-Main AB W. Germany 06Apr55. A group of 9 C-119s departed Stuttgart-Echterdingen AP W. Germany 11Aug55 for a training mission all with a full complement of army soldiers from the 499 Engineer Combat Battalion onboard. Near Altensteig W. Germany at 4,000 ft C-119 53-3222 developed trouble in the left engine and while trying to leave the formation collided with C-119 53-7841. Both aircraft suffered a catastrophic loss of control and crashed killing all onboard. 53-3222 lost 5 crew and 15 passengers with 53-7841 5 crew and 41 passengers for a total of 66 souls killed making it the worst accident for the C-119 in USAF service. The incident became known as the Altensteig Mid–Air Collision and was the worst air accident in Europe at the time – Final Disposition: Accident.

11259 / 53-7842 / C-119G-36-FA

Avble: 16Mar55; Acc: 29Mar55; Del: 21Apr55 to USAF – 60 TCWG (AFE) Rhein-Main AB W. Germany 21Apr55; to Dreux AB France 01Oct55; to 3110 MAI GP (AMC) RAF Burtonwood England 03Apr56 mnt – 7305 AB GP (322 ADV) (AFE) Dreux AB France 24Sep58; 10 TCSq (322 ADV) (AFE) Dreux AB France 12Dec60 – 7305 CAM SQ (AFE) Dreux AB France 08Jan61 – 4440 ADE GP (TAC) Langley 01Feb61 – 314 TCSq (349 TCWG)(AFR) McClellan 12Feb61; 940 TCG (349 TCWG) (AFR) McClellan 10Feb63 – 925 TCG (446 TCWG) (AFR) Ellington 11Sep65 – 906 TAL GP (302 TCWG) (AFR) Clinton 20Jan68; to Lockbourne 24Jul71 – 907 TAL GP (302 TCWG) (AFR) Lockbourne 01Mar72 – MAP S. Vietnam 01Jul72, VNAF s/n: **537842**, 413 TSQ (53 TWG) Tan Son Nhut AB – 405 Fighter WG (PAF) Clark AB The Philippines 09Mar73 for disposal – DFI: 05Jun73 – Final Disposition: Scrapped.

11260 / 53-7843 / C-119G-36-FA

Avble: 17Mar55; Acc: 30Mar55; Del: 11Apr55 to USAF – 60 TCWG (AFE) Rhein-Main AB W. Germany 11Apr55; to Dreux AB France 23Sep55; to 3110 MAI GP (AMC) RAF Burtonwood England 26Apr56 mnt – 7305 AB GP (322 ADV) (AFE) Dreux AB France 24Sep58; to 3131 ARP SQ (AMC) Chateauroux AB France 27Jun59 mnt – 317 AB GP (AFE) Évreux-Fauville AB, France 17Jul60 – MAP Belgium, BAF s/n: **CP-42**; BOC: 17Jul60 15 WG / 20 SQ Melsbroek AB Belgium, callsign: **OT-CEB**, to SIAI Marchetti for mnt 11Jul61 – 15 WG / 40 SQ 14Dec61. Minor in-flight incident 20Aug64 – Wfu 02May72 and stored at Koksijde AB Belgium; SOC: 22Oct75 – To civil company International Engine Parts, scrapped 1977-78 – Final Disposition: Scrapped.

11261 / 53-7844 / C-119G-36-FA

Avble: 21Mar55; Acc: 31Mar55; Del: 27Apr55 to USAF – 60 TCWG (AFE) Rhein-Main AB W. Germany 27Apr55; to Dreux AB France 12Sep55; to 7373 MAI GP (AFE) Chateauroux AB France 22Dec55 mnt; to 3131 MAI GP (AMC) Chateauroux AB France 01Jan56 mnt; to 3110 MAI GP (AMC) RAF Burtonwood England 27Sep56 mnt – 7305 AB GP (322 ADV) (AFE) Dreux AB France 24Sep58; 11 TCSq (322 ADV) (AFE) Dreux AB France 12Dec60 – 7305 CAM SQ (AFE) Dreux AB France 08Jan61 – 4440 ADE GP (TAC) Langley 28Feb61 – 434 TCWG (AFR) Bakalar 09Mar61; 930 TCG (434 TCWG) (AFR) Bakalar 11Feb63 – Wfu to Fairchild (LOG) St. Augustine 06Jul67 – MAP Ethiopia 25Aug70, ETAF s/n: **910**; cvtd **C-119K** – Wfu 86. Last noted derelict at Asmara AFB Eritrea – Final Disposition: Extant.

11262 / 53-4646 / C-119G-36-FA

Avble: 23Mar55; Acc: 31Mar55; Del: 31Aug55 to MAP India, IAF s/n: **IK450** – Preserved at the Indian Air Force Museum at Palam AFB New Delhi India – Final Disposition: Extant.

11263 / 53-4647 / C-119G-36-FA

Avble: 24Mar55; Acc: 31Mar55; Del: 31Aug55 to MAP India, IAF s/n: **IK451** – Final Disposition: Unknown.

11264 / 53-4648 / C-119G-36-FA

Avble: 28Mar55; Acc: 11Apr55; Del: 31Aug55 to MAP India, IAF s/n: **IK452** – Final Disposition: Unknown.

11265 / 53-4649 / C-119G-36-FA

Avble: 31Mar55; Acc: 13Apr55; Del: 31Aug55 to MAP India, IAF s/n: **IK453** – Final Disposition: Unknown.

11266 / 53-7845 / C-119G-36-FA

Avble: 04Apr55; Acc: 19Apr55; Del: 29Apr55 to USAF – 60 TCWG (AFE) Rhein-Main AB W. Germany 29Apr55; to Dreux AB France 12Sep55; to 3110 MAI GP (AMC) RAF Burtonwood England 05Aug56 mnt **RECORDS INDECIPHERABLE** 7305 AB GP (322 ADV) (AFE) Dreux AB France 24Sep58 – United Nations (UN), reg **UNO-104**, mission to The Congo 03Aug60 – MAP Italy 61, AMI s/n: **MM53-7845** – Final Disposition: Scrapped.

11267 / 53-7846 / C-119G-36-FA

Avble: 06Apr55; Acc: 22Apr55; Del: 29Apr55 to USAF – 60 TCWG (AFE) Rhein-Main AB W. Germany 29Apr55; to Dreux AB France 23Sep55; to 3110 MAI GP (AMC) RAF Burtonwood England 27Jul56 mnt – 7305 AB GP (322 ADV) (AFE) Dreux AB France 24Sep58; 11 TCSq (322 ADV) (AFE) Dreux AB France 12Dec60 – 7305 CAM SQ (AFE) Dreux AB France 08Jan61 – 4440 ADE GP (TAC) Langley 23Feb61 – 313 TCSq (349 TCWG) (AFR) Portland 06Mar61; 939 TCG (349 TCWG) (AFR) Portland 07May63; depl RCM WG (CNC) Portland 13Jun63 – MAP Taiwan 21Mar66 – Final Disposition: Unknown.

11268 / 53-7847 / C-119G-36-FA

Avble: 08Apr55; Acc: 21Apr55; Del: 29Apr55 to USAF – 60 TCWG (AFE) Rhein-Main AB W. Germany 29Apr55; to Dreux AB France 23Sep55; to 3110 MAI GP (AMC) RAF Burtonwood England 19Apr56 mnt – 7305 AB GP (322 ADV) (AFE) Dreux AB France 24Sep58; 11 TCSq (322 ADV) (AFE) Dreux AB France 12Dec60 – 7305 CAM SQ (AFE) Dreux AB France 08Jan61 – 4440 ADE GP (TAC) Langley 20Feb61 – 349 TCWG (AFR) Hamilton 22Apr61; 938 TCG (349 TCWG) (AFR) Hamilton 10Feb63 – 912 TCG (512 TCWG) (AFR) NAS Willow Grove 23Jan66 – MAP Taiwan 20Apr66, RoCAF s/n: **3131** – Final Disposition: Unknown.

11269 / 53-7848 / C-119G-36-FA

Avble: 11Apr55; Acc: 26Apr55; Del: 09May55 to USAF – 60 TCWG (AFE) Rhein-Main AB W. Germany 06May55; to Dreux AB France 23Sep55; to 3110 MAI GP (AMC) RAF Burtonwood England 16May56 mnt – 7305 AB GP (322 ADV) (AFE) Dreux AB France 24Sep58; 11 TCSq (322 ADV) (AFE) Dreux AB France 12Dec60 – 7305 CAM SQ (AFE) Dreux AB France 08Jan61 – 4440 ADE GP (TAC) Langley 17Mar61 – 65 TCSq (403 TCWG) (AFR) Davis 17Mar61; 929 TCG (403 TCWG) (AFR) Davis 11Feb63 – Fairchild (LOG) St.

Augustine 18Oct65 – 922 TCG (433 TCWG) (AFR) Kelly 30Nov65 – Fairchild (LOG) St. Augustine 28Mar68, cvtd **AC-119G Shadow** 01Jul68 – 71 SOP SQ (1 SOP WG) (TAC) Lockbourne 19Sep68 – Fairchild (LOG) St. Augustine 30Dec68 – 14 SOP WG (PAF) Nha Trang AB S. Vietnam 04Feb69; assigned variously Tan Son Nhut, Phan Rang, Tuy Hoa, Phu Cat, Da Nang S. Vietnam from 06Aug69; to 313 ADH DV (PAF) Kadena AB Japan 14Jan71 – MAP S. Vietnam 02Sep71, VNAF s/n: **537848**, 819 ASQ (53 TWG) Tan Son Nhut AB – Final Disposition: Unknown.

11270 / 53-7849 / C-119G-36-FA

Avble: 13Apr55; Acc: 27Apr55; Del: 10May55 to USAF – 60 TCWG (AFE) Rhein-Main AB W. Germany 10May55; to Dreux AB France 12Sep55; to 3110 MAI GP (AMC) RAF Burtonwood England 24Apr56 mnt – 7305 AB GP (322 ADV) (AFE) Dreux AB France 24Sep58; to 7100 SUT WG (AFE) Wiesbaden AB W. Germany 16Mar59; 11 TCSq (322 ADV) (AFE) Dreux AB France 12Dec60 – 7305 CAM SQ (AFE) Dreux AB France 08Jan61 – 4440 ADE GP (TAC) Langley 17Feb61 – 94 TCWG (AFR) Hanscom 26Feb61; depl 1 ACO GP (TAC) Howard Panama CZ (depl) 28Feb63; 901 TCG (94 TCWG) (AFR) Hanscom 31May63 – 910 TCG (459 TCWG) (AFR) Youngstown 03Dec66 – 130 SOP GP (ANG) Kanawha 02Jan69; to General Dynamics (LOG) Kanawha 31Jan73, cvtd **C-119L** 12Feb73 – Wfu to MASDC (LOG) Davis-Monthan 27Sep75 for storage, PCN: **CJ470** – DFI: 31Mar76 – Dross Metals, Inc. Tucson Arizona 15Dec78 – Lambeth Aircraft Corp. Phoenix Arizona 1979 reg **N1040E** ARAp 04May79. Suffered a serious engine failure that escalated into a fatal crash while attempting a forced belly landing in a field near Casa Grande Arizona 08Jul79. Crew members David D. Dwake (29yrs) and Sammy A. Middleton (32yrs) died soon after the crash. Howard B. Scruggs (37yrs) and Colombian national Luis Torrenegra survived but both later died in hospital. It was noted there were 30 55-gallon drums of aviation fuel onboard indicating the aircraft was attempting a lengthy flight, destination unknown – Reg N1040E canx 08Feb11 – Final Disposition: Accident.

11271 / 53-7850 / C-119G-36-FA

Avble: 20Apr55; Acc: 28Apr55; Del: 09May55 to USAF – 60 TCWG (AFE) Rhein-Main AB W. Germany 09May55; to Dreux AB France 23Sep55; to 3110 MAI GP (AMC) RAF Burtonwood England 14Sep56 mnt – 7305 AB GP (322 ADV) (AFE) Dreux AB France 24Sep58; 11 TCSq (322 ADV) (AFE) Dreux AB France 12Dec60 – 4440 ADE GP (TAC) Langley 12Jan61; to 1605 AB WG (MATS) Lajes Azores 13Jan61 mnt; to 4081 COS GP (SAC) Ernest Harmon Canada 23Jan61 mnt – 78 TCSq (435 TCWG) (AFR) Bates 27Mar61 – 357 TCSq (302 TCWG) (AFR) Bates 08May61; 908 TCG (302 TCWG) (AFR) Bates 11Feb63 – 908 TCG (435 TCWG) (AFR) Bates 18Mar63; to Brookley 02Oct64 – Fairchild (LOG) St. Augustine 01Aug68, cvtd **AC-119K Stinger** 07Aug68 – 1 SOP WG (TAC) Lockbourne 21Mar69 – 4410 CCT WG (TAC) Lockbourne 14Jul69 – 14 SOP WG (PAF) Phan Rang AB S. Vietnam 21Oct69; assigned variously Phu Cat, Da Nang, Tan Son Nhut S. Vietnam & Udorn, Nakhon Phanom Thailand from 11Feb70 – 56 SOP WG (PAF) Da Nang AB S. Vietnam 24Aug71; assigned variously Nakhon Phanom Thailand & Phan Rang S. Vietnam from 25Aug71 – MAP S. Vietnam 10Nov72, VNAF s/n: **537850**, 821 ASQ (53 TWG) Tan Son Nhut AB. Located at Ho Chi Minh City IAP Vietnam in derelict condition in the early 2000s, apparently stored for museum restoration and display – Final Disposition: Extant.

11272 / 53-7851 / C-119G-36-FA

Avble: 20Apr55; Acc: 28Apr55; Del: 16May55 to USAF – 60 TCWG (AFE) Rhein-Main AB W. Germany 16May55; to Dreux AB France 23Sep55; to 3110 MAI GP (AMC) RAF Burtonwood England 22Apr56 mnt – 7305 AB GP (322 ADV) (AFE) Dreux AB France 24Sep58; 11 TCSq (322 ADV) (AFE) Dreux AB France 12Dec60 – 7305 CAM SQ (AFE) Dreux AB France 08Jan61 – 4440 ADE GP (TAC) Langley 16Mar61 – 64 TCSq (403 TCWG) (AFR) O'Hare 24Mar61; 928 TCG (403 TCWG) (AFR) O'Hare 11Feb63; depl 1 ACO GP (TAC) Howard Panama CZ (depl) 28Feb63 – Fairchild (LOG) St. Augustine 20Apr68, cvtd **AC-119G Shadow** 01Jul68 – 71 SOP SQ (1 SOP WG) (TAC) Lockbourne 28Aug68 – Fairchild (LOG) St. Augustine 12Dec68 – 14 SOP WG (PAF) Nha Trang AB S. Vietnam 16Jan69; assigned variously Phan Rang, Tan Son Nhut, Tuy Hoa, Phu Cat S. Vietnam from 06Apr69; to SMAAR (LOG) Chu Lai S. Vietnam 20Nov69 mnt – MAP S. Vietnam 01Sep71, VNAF s/n: **537851**, 819 ASQ (53 TWG) Tan Son Nhut AB – Final Disposition: Unknown.

11273 / 53-7852 / C-119G-36-FA

Avble: 21Apr55; Acc: 28Apr55; Del: 12May55 to USAF – 60 TCWG (AFE) Rhein-Main AB W. Germany 12May55; to Dreux AB France Sep55; to 7373 MAI GP (AFE) Chateauroux AB France 26Aug55 mnt; to 3131 MAI GP (AMC) Chateauroux AB France 01Jan56 mnt; to 3110 MAI GP (AMC) RAF Burtonwood England 21Sep56 mnt – 7305 AB GP (322 ADV) (AFE) Dreux AB France 24Sep58; to 7425 SUT GP (AFE) Hahn AB W. Germany 26Jan59; 11 TCSq (322 ADV) (AFE) Dreux AB France 12Dec60 – 7305 CAM SQ (AFE) Dreux AB France 08Jan61 – 4440 ADE GP (TAC) Langley 09Mar61 – 64 TCSq (403 TCWG) (AFR) O'Hare 14Mar61; 928 TCG (403 TCWG) (AFR) O'Hare 11Feb63 – Fairchild (LOG) St. Augustine 20Apr68, cvtd **AC-119G Shadow** 01Jul68 – 71 SOP SQ (1 ACO WG) (TAC) Lockbourne 28Aug68 – Fairchild (LOG) St. Augustine 04Dec68 – 14 SOP WG (PAF) Nha Trang AB S. Vietnam 01Jan69; assigned variously Phan Rang, Tan Son Nhut, Tuy Hoa, Phu Cat, Da Nang S. Vietnam from 20Feb69 – MAP S. Vietnam 30Aug71, VNAF s/n: **537852**, 819 ASQ (53 TWG) Tan Son Nhut AB – Final Disposition: Unknown.

11274 / 53-7853 / C-119G-36-FA

Avble: 21Apr55; Acc: 28Apr55; Del: 12May55 to USAF – 60 TCWG (AFE) Rhein-Main AB W. Germany 12May55; to Dreux AB France 01Oct55; to 3110 MAI GP (AMC) RAF Burtonwood England 10Apr56 mnt – 7305 AB GP (322 ADV) (AFE) Dreux AB France 24Sep58; 10 TCSq (322 ADV) (AFE) Dreux AB France 12Dec60 – 7305 CAM SQ (AFE) Dreux AB France 08Jan61 – 4440 ADE GP (TAC) Langley 16Feb61 – 328 TCSq (512 TCWG) (AFR) Niagara Falls 20Feb61; 914 TCG (512 TCWG) (AFR) Niagara Falls 11Feb63 – 143 SOP GP (ANG) Theodore F. Green 07Apr71; depl 914 TAL GP (512 TCWG) (AFR) Niagara Falls 11May71; multiple assignments Hayes (LOG) Dothan & General Dynamics (LOG) Theodore F. Green from 08Nov71; cvtd **C-119L** 20Feb73 – Wfu to MASDC (LOG) Davis-Monthan 30Jun75 for storage, PCN: **CJ467** – DFI: 14Dec76 – Final Disposition: Scrapped.

11275 / 53-7854 / C-119G-36-FA

Avble: 25Apr55; Acc: 28Apr55; Del: 10May55 to USAF – 60 TCWG (AFE) Rhein-Main AB W. Germany 10May55; to Dreux AB France 23Sep55; to 3110 MAI GP (AMC) RAF Burtonwood England 29Oct56 mnt – 7305 AB GP (322 ADV) (AFE) Dreux AB France 24Sep58; 11 TCSq (322 ADV) (AFE) Dreux AB France 12Dec60 – 7305 CAM SQ (AFE) Dreux AB France 08Jan61 – 4440 ADE GP (TAC) Langley 06Mar61 – 73 TCSq (434 TCWG) (AFR) Scott 10Mar61; 932 TCG (434 TCWG) (AFR) Scott 02Jun63 – 910 TCG (459 TCWG) (AFR)

Youngstown 05Nov66 – Fairchild (LOG) St. Augustine 22Aug68, cvtd **AC-119K Stinger** 30Aug68 – 1 SOP WG (TAC) Lockbourne 12Apr69 – 4410 CCT WG (TAC) Lockbourne 14Jul69 – 14 SOP WG (PAF) Phan Rang AB S. Vietnam 18Nov69; assigned variously Da Nang, Tan Son Nhut S. Vietnam & Nakhon Phanom Thailand from 03Jan70 – 56 SOP WG (PAF) Da Nang AB S. Vietnam 24Aug71; assigned variously Nakhon Phanom Thailand from 04Sep71; to 313 ADH DV (PAF) Kadena AB Japan 09Sep71. Destroyed on the ground at Bien Hoa AB S. Vietnam 23Aug72 when the base came under fire from a rocket attack, the remains were abandoned at the site – Final Disposition: Combat Loss.

11276 / 53-7855 / C-119G-36-FA

Avble: 27Apr55; Acc: 12May55; Del: 19May55 to USAF – 314 TCWG (TAC) Sewart 20May55 – Hayes (AMC) Birmingham 26Jul57 – 2466 RFC (CNR) Bakalar 17Oct57 – 102 AML SQ (NY-ANG) New York ANGB 18Nov58; cvtd **C-119J** 31Jan59; to 3800 AB WG (AU) Maxwell 25Nov60 mnt; cvtd **MC-119J** 08Jun61; cvtd **C-119J** 20Jun61 – Wfu to 2704 ASD GP (LOG) Davis-Monthan 27Mar63 for storage – Fairchild (LOG) St. Augustine 29Jan67 – 4603 AB GP (ADC) Stewart 28Mar67 – 4676 AB GP (ADC) Richards 28Nov69 – 4650 COS SQ (ADC) Richards 30Jun70 – Wfu to MASDC (LOG) Davis-Monthan 11Jul72 for storage, PCN: **CJ432** – DFI: 09Dec74 – Final Disposition: Scrapped.

11277 / 53-7856 / C-119G-36-FA

Avble: 29Apr55; Acc: 12May55; Del: 20May55 to USAF – 314 TCWG (TAC) Sewart 19May55 – Hayes (AMC) Birmingham 28Aug57 – 2242 RFC (CNR) Selfridge 07Nov57 – 403 TCWG (AFR) Selfridge 13Nov58; 927 TCG (403 TCWG) (AFR) Selfridge 10Feb63 – Wfu to Fairchild (LOG) St. Augustine 17Jul67 – MAP Ethiopia 26Aug70, ETAF s/n: **911** – Wfu 1986. Noted 90s stored at Debre Zeyit AB Ethiopia – Final Disposition: Scrapped.

11278 / 53-7857 / C-119G-36-FA

Avble: 02May55; Acc: 12May55; Del: 24May55 to USAF – 313 TCG (TAC) Sewart 24May55 – 464 TCWG (TAC) Sewart 08Jun55 – 314 TCWG (TAC) Sewart 10Aug55; depl Dreux AB France 21Oct55 – 2230 RFC (CNR) NAS New York 17Aug57 – 2234 RFC (CNR) Hanscom 13Oct57 – Hayes (AMC) Birmingham 02Feb59 mnt – 94 TCWG (AFR) Hanscom 19Mar59; to 2500 AB GP (CNC) Mitchel 14Jul59 mnt; depl 452 TCWG (AFR) March 61; 901 TCG (94 TCWG) (AFR) Hanscom 31May63 – Fairchild (LOG) St. Augustine 06Sep66 – MAP Morocco 03Dec66, RMAF s/n: **CN-AMK**, 1 AT SQ – Final Disposition: Unknown.

11279 / 53-7858 / C-119G-36-FA

Avble: 10May55; Acc: 12May55; Del: 21May55 to USAF – 313 TCG (TAC) Sewart 21May55 – 463 TCWG (TAC) Sewart 08Jun55 – 314 TCWG (TAC) Sewart 55; depl Dreux AB France 01Oct55 – Hayes (AMC) Birmingham 07Aug57 – 2473 RFC (CNR) Gen Mitchell 28Oct57 – 440 TCWG (AFR) Gen Mitchell 19Dec58; 933 TCG (440 TCWG) (AFR) Gen Mitchell 11Feb63 – Wfu to MASDC (LOG) Davis-Monthan 29Jan71 for storage – 143 SOP GP (ANG) Theodore F. Green 05Jun71; multiple assignments Hayes (LOG) Dothan & General Dynamics (LOG) Theodore F. Green from 31Mar72; cvtd **C-119L** 27Feb73 – Wfu to MASDC (LOG) Davis-Monthan 18Jun75 for storage, PCN: **CJ463** – DFI: 01Jul77 – Final Disposition: Scrapped.

11280 / 53-7859 / C-119G-36-FA

Avble: 09May55; Acc: 12May55; Del: 24May55 to USAF – 313 TCG (TAC) Sewart 24May55 – 314 TCWG (TAC) Sewart 08Jun55; depl England Louisiana 05Nov56 – 2466

RFC (CNR) Bakalar 22Jun57; depl 2472 RFC (CNR) Richards 05Aug57; to 4238 AB GP (SAC) Barksdale 27Aug58 mnt – 434 TCWG (AFR) Bakalar 19Apr59 – 4440 ADE GP (TAC) Langley 28Jun60 – MAP India 10Jul60 – Final Disposition: Unknown.

11281 / 53-7860 / C-119G-36-FA

Avble: 11May55; Acc: 27May55; Del: 06Jun55 to USAF – 374 TCWG (FEA) Tachikawa AB Japan 06Jun55 – 421 ARM SQ (FEA) Yokota AB Japan 15Jul56 – 6007 CMP GP (FEA) Yokoya AB Japan 27Jul56; assigned 6021 RC SQ (PAF) Yokota AB Japan 01Jul57; assigned 6091 RC SQ (PAF) Yokota AB Japan 01Jul57 – 421 RFB SQ (PAF) Yokota AB Japan 17Jan58 – 6139 AB GP (PAF) Misawa AB Japan 11Jul58; to 8 Fighter WG (PAF) Itazuke AB Japan 05Jun59 for storage – MAP India 01Jun60 – Final Disposition: Unknown.

11282 / 53-7861 / C-119G-36-FA

Avble: 16May55; Acc: 31May55; Del: 14Sep55 to USAF – 421 ARM SQ (FEA) Yokota AB Japan 05Oct55 – **RECORDS MISSING 1955–1958** – 6102 AB WG (PAF) Yokota AB Japan 24Jul58 – 4440 ADE GP (TAC) Langley 01Sep60 – 452 TCWG (AFR) Long Beach 05Sep60; to March 01Oct60; 944 TCG (452 TCWG) (AFR) March 11Feb63 – Fairchild (LOG) St. Augustine 24Mar66 – MAP Morocco 22Jun66, RMAF s/n: **CN-AMG**, 1 AT SQ. Crashed into a mountain near Cabra Cordoba Spain 14Mar69, 6 crew and 1 civilian passenger killed. The aircraft was enroute from Paris France to Morocco – Final Disposition: Accident.

11283 / 53-7862 / C-119G-36-FA

Avble: 17May55; Acc: 27May55; Del: 06Jun55 to USAF – 374 TCWG (FEA) Tachikawa AB Japan 06Jun55 **RECORDS INDECIPHERABLE →** 421 ARM SQ (FEA) Yokota AB Japan 06Jun56; to unit unk (FEA) Johnson AB Japan 02Mar56; to 3960 AB WG (SAC) Andersen Guam 21Jul56 mnt – to 6041 AB GP (PAF) Johnson AB Japan 25Oct57 ← **RECORDS INDECIPHERABLE** 6143 AB GP (PAF) Itazuke AB Japan 29Apr58 – 6102 AB WG (PAF) Yokota AB Japan 12Jul60 – 4440 ADE GP (TAC) Langley 01Sep60 – 452 TCWG (AFR) Long Beach 05Sep60; to March 01Oct60; 943 TCG (452 TCWG) (AFR) March 11Feb63; depl 942 TCG (452 TCWG) (AFR) March 10May65 – Fairchild (LOG) St. Augustine 24Mar66 – MAP Morocco 22Jun66, RMAF s/n: **CN-AMH**, 1 AT SQ. Stored at Marrakesh Menara Airport Morocco – Final Disposition: Extant.

11284 / 53-7863 / C-119G-36-FA

Avble: 20May55; Acc: 31May55; Del: 09Jun55 to USAF – 374 TCWG (FEA) Tachikawa AB Japan 06Jun55 – 421 ARM SQ (FEA) Yokota AB Japan 09Jun56 – 6102 AB WG (PAF) Yokota AB Japan 16Jul58 – MAP India 01Jun60 – Final Disposition: Unknown.

11285 / 53-4650 / C-119G-36-FA

Avble: 23May55; Acc: 27May55; Del: 14Sep55 to MAP India, IAF s/n: **IK454** – Final Disposition: Unknown.

11286 / 53-4651 / C-119G-36-FA

Avble: 27May55; Acc: 31May55; Del: 10Sep55 to MAP India, IAF s/n: **IK455** – Final Disposition: Unknown.

11287 / 53-4652 / C-119G-36-FA

Avble: 31May55; Acc: 24Jun55; Del: 14Sep55 to MAP India, IAF s/n: **IK456** – Final Disposition: Unknown.

11288 / 53-4653 / C-119G-36-FA

Avble: 03Jun55; Acc: 29Jun55; Del: 10Sep55 to MAP India, IAF s/n: **IK457** – Final Disposition: Unknown.

11289 / 53-7864 / C-119G-36-FA

Avble: 07Jun55; Acc: 24Jun55; Del: 16Sep55 to USAF – 483 TCWG (FEA) Ashiya AB Japan 16Sep55 – 2347 RFC (CNR) Long Beach 03Sep58; to DT Hill 01Apr59 – 733 TCSq (452 TCWG) (AFR) Hill 03Apr59; 945 TCG (452 TCWG) (AFR) Hill 17Jan63 – MAP Taiwan 21May66 – Final Disposition: Unknown.

11290 / 53-7865 / C-119G-36-FA

Avble: 13Jun55; Acc: 28Jun55; Del: 23Sep55 to USAF – SOMAR (AMC) Clark AB The Philippines 13Nov55 – 2347 RFC (CNR) Long Beach 03Jan58 – 452 TCWG (AFR) Long Beach 13Jun59; to March 01Oct60; 943 TCG (452 TCWG) (AFR) March 11Feb63 – 129 SOP GP (ANG) Hayward 27Dec68; cvtd **C-119L** 09Mar73 – Wfu to MASDC (LOG) Davis-Monthan 19Feb75 for storage, PCN: **CJ454** – DFI: 28Jul76 – Final Disposition: Scrapped.

11291 / 53-7866 / C-119G-36-FA

Avble: 16Jun55; Acc: 29Jun55; Del: 26Sep55 to USAF – FELFR (FEA) FEAMCOM 26Sep55 – AMPAR (AMC) FEAMCOM 01Oct55. Crashed into a field 2.5 miles S Tachikawa AB Japan 01Jan56 while on a guided approach to that air base, crew were not injured – DFI: 18Apr56 – Final Disposition: Accident.

11292 / 53-7867 / C-119G-36-FA

Avble: 15Jun55; Acc: 30Jun55; Del: 21Sep55 to USAF – 3 Bomb WG (FEA) Johnson AB Japan 21Sep55; briefly to Joplin Missouri 21Sep55 – 6041 AB GP (PAF) Johnson AB Japan 25Oct57 – MAP India 01Jun60 – Final Disposition: Unknown.

11293 / 53-7868 / C-119G-36-FA

Avble: 20Jun55; Acc: 30Jun55; Del: 17Oct55 to USAF – 8 Fighter WG (FEA) Itazuke AB Japan 17Oct55 – 6143 AB GP (PAF) Itazuke AB Japan 01Oct57 – MAP India 01Jun60 – Final Disposition: Unknown.

11294 / 53-7869 / C-119G-36-FA

Avble: 22Jun55; Acc: 30Jun55; Del: 26Oct55 to USAF – 18 Fighter WG (FEA) Kadena AB Japan 26Oct55 – 6313 AB WG (PAF) Kadena AB Japan 01Oct57; 8 Fighter WG (PAF) Itazuke AB Japan 10Jun59 for storage – 4440 ADE GP (TAC) Langley 10Sep60 – 349 TCWG (AFR) Hamilton 17Sep60; assigned 314 TCSq McClellan 20Sep60; assigned 97 TCSq Paine 07Nov62; 941 TCG (349 TCWG) (AFR) Paine 63 – 904 TCG (514 TCWG) (AFR) Stewart 65; 903 TCG (514 TCWG) (AFR) McGuire 26May66; depl Myrtle Beach 20Jul67 – 914 TAL GP (512 TCWG) (AFR) Niagara Falls 08May68 – Wfu to MASDC (LOG) Davis-Monthan 07Dec70 for storage, PCN: **CJ382** – DFI: 21May71 – Kolar, Inc. Tucson Arizona 30Mar76 for scrap – Final Disposition: Scrapped.

11295 / 53-7870 / C-119G-36-FA

Avble: 27Jun55; Acc: 13Jul55; Del: 25Oct55 to USAF – 49 Fighter WG (FEA) Misawa AB Japan 25Oct55; to 54 Fighter GP (ADC) Greater Pittsburgh 26Oct55 mnt – 6139 AB GP (PAF) Misawa AB Japan 15Oct57; to 6143 AB GP (PAF) Itazuke AB Japan 03Jan60; to 6313 AB WG (PAF) Kadena AB Japan 03May60 – MAP India 01Jun60 – Final Disposition: Unknown.

11296 / 53-7871 / C-119G-36-FA

Avble: 29Jun55; Acc: 14Jul55; Del: 04Nov55 to USAF – 58 Fighter WG (FEA) Osan-Ni AB S. Korea 05Nov55 – 18 Fighter WG (FEA) Kadena AB Japan 28Feb56 – 6313 AB WG (PAF) Kadena AB Japan 01Oct57; to 8 Fighter WG (PAF) Itazuke AB Japan 06Jun59 for storage; to 6041 AB GP (PAF) Johnson AB Japan 24Jul59 for storage – 4440 ADE GP (TAC) Langley 10Sep60 – 403 TCWG (AFR) Selfridge 20Sep60; 927 TCG (403 TCWG) (AFR) Selfridge 10Feb63 – Fairchild (LOG) St. Augustine 07Apr69 – MAP Morocco 22Jun66, RMAF s/n: **CN-AMI**, 1 AT SQ – Final Disposition: Unknown.

11297 / 53-7872 / C-119G-36-FA

Avble: 05Jul55; Acc: 18Jul55; Del: 31Oct55 to USAF – 58 Fighter WG (FEA) Osan-Ni AB S. Korea 01Nov55 – 67 TR WG (FEA) Itami AB Japan 30Dec55 – 6102 AB WG (PAF) Yokota AB Japan 01Jul57 – MAP India 01Jun60 – Final Disposition: Unknown.

11298 / 53-7873 / C-119G-36-FA

Avble: 14Jul55; Acc: 25Jul55; Del: 10Oct55 to USAF – 8 Fighter WG (FEA) Itazuke AB Japan 10Oct55 – 6143 AB GP (PAF) Itazuke AB Japan 01Oct57 – 6102 AB WG (PAF) Yokota AB Japan 06Mar60 – 4440 ADE GP (TAC) Langley 02Sep60 – 434 TCWG (AFR) Bakalar 08Sep60; 931 TCG (434 TCWG) (AFR) Bakalar 11Feb63 – Fairchild (LOG) St. Augustine 11Oct67 – MAP S. Vietnam 25Mar68, VNAF s/n: **537873**, 413 TSQ (53 TWG) Tan Son Nhut AB – Final Disposition: Unknown.

11299 / 53-4654 / C-119G-36-FA

Avble: 19Jul55; Acc: 26Jul55; Del: 06Oct55 to MAP India, IAF s/n: **IK458** – Final Disposition: Unknown.

11300 / 53-4655 / C-119G-36-FA

Avble: 21Jul55; Acc: 29Jul55; Del: 06Oct55 to MAP India, IAF s/n: **IK459**. During a mnt flight an engine lost power where upon the crew attempted a forced landing in a stream in a mountainous area near Chushul India 19Sep60. The aircraft struck a boulder on landing killing 3 of the 9 onboard including both pilots. Some sources list the crash date as 01Jul61 – Final Disposition: Accident.

11301 / 53-4656 / C-119G-36-FA

Avble: 27Jul55; Acc: 17Aug55; Del: 12Oct55 to MAP India 11Oct55, IAF s/n: **IK460** – Final Disposition: Unknown.

11302 / 53-4657 / C-119G-36-FA

Avble: 01Aug55; Acc: 17Aug55; Del: 13Oct55 to MAP India 12Oct55, IAF s/n: **IK461** – Final Disposition: Unknown.

11303 / 53-7874 / C-119G-36-FA

Avble: 05Aug55; Acc: 25Aug55; Del: 07Nov55 to USAF – 18 Fighter WG (FEA) Kadena AB Japan 07Nov55 – 6313 AB WG (PAF) Kadena AB Japan 01Oct57 – MAP India 01Jun60 – Final Disposition: Unknown.

11304 / 53-7875 / C-119G-36-FA

Avble: 11Aug55; Acc: 25Aug55; Del: 27Sep55 to USAF – 483 TCWG (FEA) Ashiya AB Japan 26Sep55; depl 316 TCG (FEA) Ashiya AB Japan 19Mar56; to 21 TCSq (483 TCWG) (FEA) Tachikawa AB Japan 57 – Wfu to 868 TM SQ (PAF) Tainan Taiwan 18Oct58 – MAP Taiwan 01Jul59, RoCAF s/n: **3175** – Final Disposition: Unknown.

11305 / 53-7876 / C-119G-36-FA

Avble: 16Aug55; Acc: 25Aug55; Del: 19Sep55 to USAF – MADDP (FEA) Clark AB The Philippines 19Sep55 – SOMAR (AMC) Clark AB The Philippines 19Sep55 – 316 TCG (FEA) Ashiya AB Japan 14Nov55 – 483 TCWG (FEA) Ashiya AB Japan 02Apr56 – 2346 RFC (CNR) Hamilton 19Jul58 – 349 TCWG (AFR) Hamilton 23Dec58; 938 TCG (349 TCWG) (AFR) Hamilton 10Feb63 – 933 TCG (440 TCWG) (AFR) Gen Mitchell 26Jun65 – MAP Taiwan 21Apr66 – Final Disposition: Unknown.

11306 / 53-7877 / C-119G-36-FA

Avble: 18Aug55; Acc: 30Aug55; Del: 19Sep55 to USAF – FELFR (FEA) Clark AB The Philippines 19Sep55 – MADDP (FEA) Clark AB The Philippines 21Sep55 – SOMAR (AMC) Clark AB The Philippines 01Oct55 – 2347 RFC (CNR) Long Beach 04Jul58 – 452 TCWG (AFR) Long Beach 13Jun59; to March 01Oct60; 944 TCG (452 TCWG) (AFR) March 11Feb63; 943 TAL GP (452 TCWG) (AFR) March 23Jan68 – Fairchild (LOG) St. Augustine 25Feb68, cvtd **AC-119K Stinger** 01Jul68 – 4410 CCT WG (TAC) Lockbourne 25Oct69 – 1 SOP WG (TAC) Eglin 08Jul71 – 56 SOP WG (PAF) Nakhon Phanom AB Thailand 03Oct72 – MAP S. Vietnam 31Oct72, VNAF s/n: **537877**, 821 ASQ (53 TWG) Tan Son Nhut AB – Final Disposition: Unknown.

11307 / 53-7878 / C-119G-36-FA

Avble: 25Aug55; Acc: 08Sep55; Del: 03Oct55 to USAF – 3 Bomb WG (FEA) Johnson AB Japan 03Oct55 – 6041 AB GP (PAF) Johnson AB Japan 01Oct57 – 6102 AB WG (PAF) Yokota AB Japan 01Jul60 – 4440 ADE GP (TAC) Langley 03Sep60 – Fairchild (AMC) St. Augustine 14Nov60 mnt – 442 TCWG (AFR) Richards 22Dec60 – 97 TCSq (349 TCWG) (AFR) Paine 27May61; 941 TCG (349 TCWG) (AFR) Paine 63 – 925 TCG (446 TCWG) (AFR) Ellington 11Sep65 – 907 TAL GP (302 TCWG) (AFR) Clinton 20Jan68; to Lockbourne 24Jul71 – Wfu to MASDC (LOG) Davis-Monthan 04Mar73 for storage, PCN: **CJ447** – DFI: 31Mar76 – Final Disposition: Scrapped.

11308 / 53-7879 / C-119G-36-FA

Avble: 01Sep55; Acc: 16Sep55; Del: 12Oct55 to USAF – 49 Fighter-Bomber WG (FEA) Misawa AB Japan 11Oct55; depl 483 TCWG (FEA) Ashiya AB Japan 16Jul56 – 6139 AB GP (PAF) Misawa AB Japan 15Oct57 – 6102 AB WG (PAF) Yokota AB Japan 30Jun60 – 4440 ADE GP (TAC) Langley 02Sep60 – 440 TCWG (AFR) Gen Mitchell 08Sep60; 933 TCG (440 TCWG) (AFR) Gen Mitchell 11Feb63 – Fairchild (LOG) St. Augustine 16Aug68, cvtd **AC-119K Stinger** 30Aug68 – 1 SOP WG (TAC) Lockbourne 15Apr69 – 4410 CCT WG (TAC) Lockbourne 14Jul69 – 14 SOP WG (PAF) Phan Rang AB S. Vietnam 21Oct69;

assigned variously Da Nang, Phu Cat, Tan Son Nhut S. Vietnam & Udorn, Nakhon Phanom Thailand from 01Dec69; to 313 ADH DV (PAF) Kadena AB Japan 22Jul71 – 56 SOP WG (PAF) Nakhon Phanom AB Thailand 23Aug71; assigned variously Da Nang S. Vietnam from 18Sep71 – MAP S. Vietnam 10Nov72, VNAF s/n: **537879**, 821 ASQ (53 TWG) Tan Son Nhut AB – Final Disposition: Unknown.

11309 / 53-4658 / C-119G-36-FA

Avble: 06Sep55; Acc: 21Sep55; Del: 06Oct55 to MAP India, IAF s/n: **IK462** – Final Disposition: Unknown.

11310 / 53-4659 / C-119G-36-FA

Avble: 09Sep55; Acc: 29Sep55; Del: 06Oct55 to MAP India, IAF s/n: **IK463** – Final Disposition: Unknown.

11311 / 53-4660 / C-119G-36-FA

Avble: 15Sep55; Acc: 29Sep55; Del: 11Oct55 to MAP India, IAF s/n: **IK464** – Final Disposition: Unknown.

11312 / 53-4661 / C-119G-36-FA

Avble: 21Sep55; Acc: 30Sep55; Del: 12Oct55; to MAP India, IAF s/n: **IK465** – Final Disposition: Unknown.

11313 / 53-4662 / C-119G-36-FA

Avble: 26Sep55; Acc: 14Oct55; Del: 21Oct55 to MAP India, IAF s/n: **IK466** – Final Disposition: Unknown.

11314 / 53-7880 / C-119G-36-FA

Avble: 29Sep55; Acc: 18Oct55; Del: 01Nov55 to USAF – 67 TR WG (FEA) Itami AB Japan 01Nov55; depl 58 Fighter-Bomber WG (FEA) Osan-Ni AB S. Korea 30Dec55 – 3 Bomb WG (FEA) Johnson AB Japan 14Feb56; to 6200 AB WG (FEA) Clark AB The Philippines 08May56 – 6041 AB GP (PAF) Johnson AB Japan 25Oct57 – 4440 ADE GP (TAC) Langley 30Aug60 – 94 TCWG (AFR) Hanscom 07Sep60; 901 TCG (94 TCWG) (AFR) Hanscom 31May63 – Fairchild (LOG) St. Augustine 06Sep66 – MAP Morocco 03Dec66, RMAF s/n: **CN-AMJ**, 1 AT SQ – Final Disposition: Unknown.

11315 / 53-7881 / C-119G-36-FA

Avble: 06Oct55; Acc: 28Oct55; Del: 10Nov55 to USAF – 67 TR WG (FEA) Itami AB Japan 10Nov55 – 6102 AB WG (PAF) Yokota AB Japan 01Jul57 – MAP India 01Jun60 – Final Disposition: Unknown.

11316 / 53-7882 / C-119G-36-FA

Avble: 18Oct55; Acc: 31Oct55; Del: 14Nov55 to USAF – 421 ARM WG (FEA) Yokota AB Japan 01Nov55 – 6313 AB WG (PAF) Kadena AB Japan 03Jul58 – 4440 ADE GP (TAC) Langley 60 – 512 TCWG (AFR) NAS Willow Grove 19Sep60; 913 TCG (512 TCWG) (AFR) NAS Willow Grove 11Feb63 – 907 TAL GP (302 TCWG) (AFR) Clinton 21Dec69; to Lockbourne 24Jul71 – Wfu to MASDC (LOG) Davis-Monthan 28Dec72 for storage, PCN: **CJ438** – DFI: 31Mar76 – Final Disposition: Scrapped.

11317 / 53-7883 / C-119G-36-FA

Avble: 26Oct55; Acc: 09Nov55; Del: 17Nov55 to USAF – 3 Bomb WG (FEA) Johnson AB Japan 17Nov55 – 6041 AB GP (PAF) Johnson AB Japan 25Oct57 – 6102 AB WG (PAF) Yokota AB Japan 30Jun60 – 4440 ADE GP (TAC) Langley 03Sep60 – 435 TCWG (AFR) Homestead 13Sep60; assigned 78 TCSq Bates 01Oct60 – 357 TCSq (302 TCWG) (AFR) Bates 08May61; 908 TCG (302 TCWG) (AFR) Bates 11Feb63 – 908 TCG (435 TCWG) (AFR) Bates 18Mar63; to Brookley 30Oct64; depl 915 TCG (435 TCWG) (AFR) Homestead 01Dec65 – Fairchild (LOG) St. Augustine 01Aug68, cvtd **AC-119K Stinger** 07Aug68 – 1 SOP WG (TAC) Lockbourne 11Mar69 – 4410 CCT WG (TAC) Lockbourne 14Jul69 – 14 SOP WG (PAF) Phan Rang AB S. Vietnam 05Dec69; assigned variously Da Nang, Phu Cat, Tan Son Nhut S. Vietnam & Udorn, Nakhon Phanom Thailand from 07Dec69 – 56 SOP WG (PAF) Nakhon Phanom AB Thailand 17Oct71; assigned variously Da Nang S. Vietnam from 25Oct71 – MAP S. Vietnam 10Nov72, VNAF s/n: **537883**, 821 ASQ (53 TWG) Tan Son Nhut AB – Final Disposition: Unknown.

11318 / 53-7884 / C-119G-36-FA

Avble: 08Nov55; Acc: 14Nov55; Del: 05Dec55 to USAF – 483 TCWG (FEA) Ashiya AB Japan 05Dec55 del in error – AMPAR (AMC) FEAMCOM 25Jan56 – NMPAR (AMC) Tachikawa AB Japan 15Jun56 – SMPAR (AMC) Clark AB The Philippines 23Jun56 – 2346 RFC DT (CNR) Paine 19May58 – 97TCSq (349 TCWG) (AFR) Paine 58; 941 TCG (349 TCWG) (AFR) Paine 63 – 924 TCG (446 TCWG) (AFR) Ellington 09Aug65; 926 TAL GP (446 TCWG) (AFR) NAS New Orleans 11Jan68 – 130 SOP GP (ANG) Kanawha 08Jan69; to General Dynamics (LOG) Kanawha 05Mar73, cvtd **C-119L** 14Mar73 – Wfu to MASDC (LOG) Davis-Monthan 10Jun75 for storage, PCN: **CJ461** – DFI: 28Jul76 – BofS Dross Metals, Inc. Tucson Arizona 21Sep79; aircraft actually acquired 15Dec78 – BofS to William Waara (Michigan Aerial Applicator) Dothan Alabama 04Oct79 reg **N37483** but sale process appears to have stalled – Second BofS to Dross Metals, Inc. Tucson Arizona 12Sep80; second BofS to Michigan Aerial Applicator date unk, re-reg **N8512K** but ntu, N37483 retained – BofS Arbor Air (William Waara d/b/a Arbor Air) Ann Arbor Michigan 10Apr80 – BofS El Marc Air Columbus Nebraska 13Apr81 – BofS Jim Cottington [sic] (Jim Cottingham) address unk 22Sep81. Crashed under mysterious circumstances 9 miles NE Ciudad Victoria Mexico 01Oct81. N37483 left Laredo Intl. Airport on 30Sep81 for a cargo flight to Mexico with Howard F. (Jim) Cottingham (48yrs) at the controls and his son Thomas W. Cottingham (25yrs) assisting on the flight. A radio message was received from a panicked Cottingham stating that they were having Center-of-Gravity problems and that the cargo was being jettisoned, but nothing further was heard from the flight. A witness at the destination airstrip stated the C-119 circled the field for a long period before one of the engines "exploded" and caught fire. It then came down in a maize field clipping a power line before impacting. Jim Cottingham was killed on impact but his son Tommy survived only to succumb to injuries soon after being admitted to a local hospital. A post-crash inspection of the wreckage revealed the right engine propeller to have been stopped prior to impact and what appeared to be a spray of bullet holes along one side of the fuselage, whether this contributed to the crash is unknown – Reg N37483 canx 27Nov12 – Final Disposition: Accident.

KMC-101 / 51-8098 / C-119F-KM

First Kaiser C-119F built at Willow Run. Christened the *Robert L. Williams* in a ceremony 08Mar52, in honor of a local soldier killed in the Korean War. Avble: 28Mar52; Acc:

Section Four—Service Histories   385

29May52; Del: 18Nov52 to USAF **RECORDS INDECIPHERABLE** → 6510 AB WG (ARD) Edwards 08Nov52 – 6511 PDT GP (ARD) NAS El Centro 16Feb54 ← **RECORDS INDECIPHERABLE** to BMC (AMC) Birmingham 01Mar56, cvtd **C-119G** 20Apr56; cvtd **JC-119G** 31May56; to SBNAR (AMC) Norton 07Jan57 mnt – AFT Center (ARD) NAS El Centro (depl) 29Mar57; to SBNAR (AMC) Norton 29Mar57 mnt; cvtd **C-119G** 01Oct57 – 2346 RFC DT (CNR) McClellan 02Sep58 – Hayes (AMC) Birmingham 15Dec58 – 314 TCSq (349 TCWG) (AFR) McClellan 25Mar59; 940 TCG (349 TCWG) (AFR) McClellan 10Feb63 – 904 TCG (514 TCWG) (AFR) Stewart 13May65; 903 TCG (514 TCWG) (AFR) McGuire 02Jun66; depl Myrtle Beach South Carolina 16Jul67 – 928 TAL GP (403 TCWG) (AFR) O'Hare 08May68 – Wfu to MASDC (LOG) Davis-Monthan 05Nov68 for storage, PCN: **CJ264** – DFI: 26Dec73 – Final Disposition: Scrapped.

KMC-102 / 51-8099 / C-119F-KM

Avble: 10Apr52; Acc: 03May52; Del: 23Jun52 to USAF **RECORDS INDECIPHERABLE** → 516 TCWG (TAC) Memphis 23Jun52 – 463 TCWG (TAC) Memphis 53; to Ardmore 01Sep53 ← **RECORDS INDECIPHERABLE** 4405 OP SQ (TAC) Langley 19Apr55; depl 2589 RFC (CNR) Dobbins 01Nov55; to Fairchild (AMC) Hagerstown 08May56, cvtd **C-119G** 31May56; depl 405 Fighter WG (TAC) Langley 10Oct56 – 405 Fighter WG (TAC) Langley 13Feb57 – 836 AB GP (TAC) Langley 08Oct57 – 2242 RFC (CNR) Selfridge 14Apr58 – 403 TCWG (AFR) Selfridge 13Nov58; 927 TCG (403 TCWG) (AFR) Selfridge 10Feb63 – 913 TAL GP (512 TCWG) (AFR) NAS Willow Grove 08Jul69 – 907 TAL GP (302 TCWG) (AFR) Clinton 20Dec69 – MAP Taiwan 28Jan70, RoCAF s/n: **3142** – Final Disposition: Unknown.

KMC-103 / 51-8100 / C-119F-KM

Avble: 29Apr52; Acc: 30Jun52; Del: 08Jul52 to USAF **RECORDS INDECIPHERABLE** → 516 TCWG (TAC) Memphis 08Jul52 – 463 TCWG (TAC) Memphis 53; to Ardmore 01Sep53; depl Charleston 24Feb54 ← **RECORDS INDECIPHERABLE** to Fairchild (AMC) Hagerstown 17Jan55, cvtd **C-119G** 15Feb55; depls North Aux AF 11Apr55 & 05May56 – BMC (AMC) Birmingham 14Dec56 – 2466 RFC (CNR) Bakalar 09Mar57 – 434 TCWG (AFR) Bakalar 19Apr59; 930 TCG (434 TCWG) (AFR) Bakalar 11Feb63; depl 71 ACO SQ (1 ACO WG) (TAC) Lockbourne 20Jun68 – 913 TAL GP (512 TCWG) (AFR) NAS Willow Grove 05Sep68 – Wfu to MASDC (LOG) Davis-Monthan 06Nov68 for storage, PCN: **CJ268** – DFI: 31Mar76 – Final Disposition: Scrapped.

KMC-104 / 51-8101 / C-119F-KM

Avble: 29May52; Acc: 30Jun52; Del: 09Jul52 **RECORDS INDECIPHERABLE** → units unk possibly retained by Kaiser – 316 TCG (TAC) Sewart 07Feb53; depl Pope 20Apr54 – 314 TCWG (TAC) Sewart 15Nov54 ← **RECORDS INDECIPHERABLE** 456 TCWG (TAC) Charleston 24May55; to Shiroi AB Japan 10Nov55; depl 463 TCWG (TAC) Ardmore 18Apr56; to Ardmore 17May56 – 419 TCG (TAC) Ardmore 09Jul56 – 3345 TTA WG (TC) Chanute 31Aug56 – BMC (AMC) Birmingham 11Dec56, cvtd **C-119G** 01Mar57 – 2466 RFC (CNR) Bakalar 24Mar57 – 434 TCWG (AFR) Bakalar 19Apr59 – 440 TCWG (AFR) Gen Mitchell 01Nov61; 933 TCG (440 TCWG) (AFR) Gen Mitchell 11Feb63; depl 1 ACO WG (TAC) Lockbourne 10May68; depl 4413 CCT SQ (1 ACO WG) (TAC) Lockbourne 10May68 – Wfu to MASDC (LOG) Davis-Monthan 31Jan71 for storage, PCN: **CJ392** – DFI: 31Mar76 – Final Disposition: Scrapped.

KMC-105 / 51-8102 / C-119F-KM

Avble: 05May52; Acc: 30Jul52; Del: 06Aug52 **RECORDS INDECIPHERABLE** → 316 TCG (TAC) Sewart 05Aug52; depl Pope 19Apr54; depl Sumpter Alabama 24Aug54 – BMC (AMC) Birmingham 20Sep54 – 314 TCWG (TAC) Sewart 15Nov54 ← **RECORDS INDECIPHERABLE** to Fairchild (AMC) Hagerstown 31Jan55, cvtd **C-119G** 25Feb55 – 63 TCWG (TAC) Donaldson 14Apr55; depl 464 TCWG (TAC) Pope 12Apr55 – 4501 SUT SQ (TAC) Donaldson 04May55 – 2237 RFC (CNR) New Castle 27Sep57; to NAS Willow Grove 21Jul58 – Hayes (AMC) Birmingham 18Nov58 – 512 TCWG (AFR) NAS Willow Grove 10Mar59; 913 TCG (512 TCWG) (AFR) NAS Willow Grove 11Feb63 – 907 TAL GP (302 TCWG) (AFR) Clinton 21Dec69 – MAP Taiwan 10Jan70 – Final Disposition: Unknown.

KMC-106 / 51-8103 / C-119F-KM

Avble: 27Jun52; Acc: 28Jul52; Del: 02Aug52 **RECORDS INDECIPHERABLE** → 316 TCG (TAC) Sewart 02Aug52; depl Pope 19Apr54; depl Sumpter Alabama 24Aug54 – BMC (AMC) Birmingham 06Oct54 – 314 TCWG (TAC) Sewart 28Dec54 ← **RECORDS INDECIPHERABLE** 450 Fighter WG (TAC) Foster 18Mar55; to 3310 TTA WG (TC) Scott 07Oct55 mnt; to 3635 CCT WG (TC) Stead 10Nov55 mnt; BMC (AMC) Birmingham 25Apr56, cvtd **C-119G** 15Jun56; depl 363 TR WG (TAC) Shaw 14Nov56; to SPW Center (ARD) Kirtland 27Aug57 mnt – 831 AB WG (TAC) George 20Jul58 – 94 TCWG (AFR) Hanscom 18May59; 902 TCG (94 TCWG) (AFR) Grenier 31May63 – 926 TCG (446 TCWG) (AFR) NAS New Orleans 12Nov65 – Wfu to MASDC (LOG) Davis-Monthan 23Nov69 for storage, PCN: **CJ322** – DFI: 20Jan70 – Kolar, Inc. Tucson Arizona 29Jan76 for scrap – Final Disposition: Scrapped.

KMC-107 / 51-8104 / C-119F-KM

Avble: 08Jul52; Acc: 31Jul52; Del: 06Aug52 **RECORDS INDECIPHERABLE** → 316 TCG (TAC) Sewart 06Aug52; depl Pope 19Apr54; depl Sumpter Alabama 24Aug54 – 314 TCWG (TAC) Sewart 15Nov54; depl Elmendorf 11Jan55 ← **RECORDS INDECIPHERABLE** Wfu to 3345 TTN WG (TC) Chanute 09Jul55 for ground instruction – DFI: 26Feb58 – Final Disposition: Scrapped.

KMC-108 / 51-8105 / C-119F-KM

Avble: 11Aug52; Acc: 25Aug52; Del: 30Sep52 **RECORDS INDECIPHERABLE** → 6614 TSP SQ (NEA) Earnest Harmon Canada 30Sep52 – MIDAR (AMC) Olmsted date unk – 456 TCWG (TAC) Charleston 06Mar54 ← **RECORDS INDECIPHERABLE** CAR Center (ARD) Hanscom 12Apr55 test support – AFA Center (ARD) Eglin 03May55 test support – 4925 Test GP (ARD) Kirtland 03Oct55 test support; to 4900 AB GP (ARD) Kirtland 20Oct55 test support – 3510 CCT WG (TC) Randolph 17Nov55 – Fairchild (AMC) Hagerstown 06Jan56, cvtd **C-119G** 07Feb56 – 479 Fighter WG (TAC) George 22Feb56 – Hayes (AMC) Birmingham 23Sep57 – 831 AB GP (TAC) George 20Dec57; to 4510 CCT WG (TAC) Luke 19Oct58 – 77 TCSq (435 TCWG) (AFR) Donaldson 05Aug59. Both engines failed while on a training flight out of Donaldson AFB 27Feb60, the 4 crew bailed out and the abandoned C-119 came down narrowly missing a local high school building. Being a Saturday the school was empty anyway – Final Disposition: Accident.

KMC-109 / 51-8106 / C-119F-1-KM

Avble: 28Jul52; Acc: 29Aug52; Del: 04Dec52 – 6622 TSP SQ (NEA) Ernest Harmon Canada 04Dec52 – 6603 AB GP (NEA) Goose Bay Canada 21Jan53 – 6614 TSP SQ (NEA) Ernest Harmon Canada 53 – MIDAR (AMC) Olmsted 02Dec53 – 456 TCWG (TAC) Charleston

05Mar54 – CAR Center (ARD) Hanscom 18Mar55; to BMC (AMC) Birmingham 04Jan56, cvtd **C-119G** 05Mar56 – 2235 RFC (CNR) Grenier 03Jun57 – 2234 RFC (CNR) Hanscom 13Feb58; depl 512 TCWG (AFR) NAS Willow Grove 05Mar59 – 94 TCWG (AFR) Hanscom 19Mar59; 901 TCG (94 TCWG) (AFR) Hanscom 08Mar63; Fairchild (LOG) St. Augustine cvtd **C-119C** 11May64; cvtd **C-119G** 18Aug64 – 907 TCG (302 TCWG) (AFR) Clinton 23Jul66 – MAP Taiwan 05Jan70, RoCAF s/n: **3190** – Presently displayed at the Republic of China Air Force Museum Gangshan AB Taiwan – Final Disposition: Extant.

KMC-110 / 51-8107 / C-119F-KM

Avble: 13Aug52; Acc: 30Aug52; Del: 18Nov52 – 516 TCWG (TAC) Memphis 20Nov52 – 463 TCWG (TAC) Memphis 16Jan53; to Ardmore 17Aug53; depl Gray Texas 06May54 – 4405 OP SQ (TAC) Langley 09May55; to BMC (AMC) Birmingham 20Apr56, cvtd **C-119G** 11Jun56 – 405 Fighter WG (TAC) Langley 22Feb57 – 836 AB GP (TAC) Langley 08Oct57 – 2347 RFC (CNR) Long Beach 11May58 – 452 TCWG (AFR) Long Beach 13Jun59; to March 01Oct60; 943 TCG (452 TCWG) (AFR) March 11Feb63 – 908 TCG (435 TCWG) (AFR) Brookley 23Jul66 – 910 TAL GP (459 TCWG) (AFR) Youngstown 20Feb69 – Wfu to MASDC (LOG) Davis-Monthan 23Nov69 for storage, PCN: **CJ324** – DFI: 20Jan70 – Kolar, Inc. Tucson Arizona 30Jan76 for scrap – Final Disposition: Scrapped.

KMC-111 / 51-8108 / C-119F-KM

Avble: 25Aug52; Acc: 24Nov52; Del: 06Dec52 – 516 TCWG (TAC) Memphis 06Dec52 – 463 TCWG (TAC) Memphis 16Jan53; to Ardmore 17Aug53; depl 366 Fighter WG (TAC) Alexandria 05Nov54; to 62 TCWG (TAC) Larson 10Jan56 mnt; to Fairchild (AMC) Hagerstown 24Jan56, cvtd **C-119G** 16Feb56; depl North Aux AF 02Apr56 – 464 TCWG (TAC) Pope 26Sep56 – Hayes (AMC) Birmingham 18Nov58 – 97 TCSq (349 TCWG) (AFR) Paine 02Mar59; to 2347 AB GP (CNC) Long Beach 12Jul60 mnt; 941 TCG (349 TCWG) (AFR) Paine 63 – 925 TCG (446 TCWG) (AFR) Ellington 30Aug65; 926 TAL GP (446 TCWG) (AFR) NAS New Orleans 16Jan68 – Wfu to MASDC (LOG) Davis-Monthan 01Nov68 for storage, PCN: **CJ262** – DFI: 31Mar76 – Final Disposition: Scrapped.

KMC-112 / 51-8109 / C-119F-KM

Avble: 20Sep52; Acc: 26Nov52; Del: 04Dec52 – 316 TCG (TAC) Sewart 04Dec52 – BMC (AMC) Birmingham 54 – 314 TCWG (TAC) Sewart 24Nov54 – 450 Fighter WG (TAC) Foster 15Mar55; to BMC (AMC) Birmingham 01Sep55, cvtd **C-119G** 27Dec55 – 2473 RFC (CNR) Gen Mitchell 12Dec57 – Hayes (AMC) Birmingham 01Dec58 – 440 TCWG (AFR) Gen Mitchell 23Mar59; 933 TCG (440 TCWG) (AFR) Gen Mitchell 11Feb63 – Wfu to MASDC (LOG) Davis-Monthan 04Nov68 for storage, PCN: **CJ265** – DFI: 09Dec74 – Final Disposition: Scrapped.

KMC-113 / 51-8110 / C-119F-KM

Avble: 23Oct52; Acc: 15Jan53; Del: 24Jan53 – 514 TCWG (TAC) Mitchel 25Jan53 – 313 TCWG (TAC) Mitchel 01Feb53; to 4415 AB GP (TAC) Pope date unk; depl 62 TCWG (TAC) Larson 02Apr54; depl 465 TCWG (TAC) Mitchel date unk – 62 TCWG (TAC) Larson 07Jul54 – 463 TCWG (TAC) Ardmore 29Dec54 – 366 Fighter WG (TAC) Alexandria 22Apr55; to 805 AB GP (SAC) Barksdale 23Jan56 mnt; to Fairchild (AMC) Hagerstown 26Mar56, cvtd **C-119G** 12Apr56 – 834 AB GP (TAC) England 25Sep57 – 27 AB GP (TAC) Bergstrom 06Mar58 – 4510 CCT WG (TAC) Luke 09Dec58; to 3560 PTN WG (ATC) Webb 03Feb59 mnt; to 836 AB GP (TAC) Langley 05Apr59 – 732 TCSq (94 TCWG) (AFR)

Grenier 08Jul59; 902 TCG (94 TCWG) (AFR) Grenier 11Feb63 – 906 TCG (302 TCWG) (AFR) Clinton 19Nov65 – MAP Taiwan 01Dec69 – Final Disposition: Unknown.

KMC-114 / 51-8111 / C-119F-KM

Avble: 24Sep52; Acc: 26Nov52; Del: 10Dec52 to USAF – 316 TCG (TAC) Sewart 10Dec52 – BMC (AMC) Birmingham 04Oct54 – 314 TCWG (TAC) Sewart 20Dec54 – 450 Fighter WG (TAC) Foster 12Mar55; to 3415 TTN WG (TC) Lowry 12May55 mnt; depl 479 Fighter WG (TAC) George 14Dec55; to BMC (AMC) Birmingham 05Mar56, cvtd **C-119G** 26Apr56; to 3600 CCT WG (TC) Luke 29Oct56 mnt; depl 366 Fighter WG (TAC) England 18Apr57 – 27 AB GP (TAC) Bergstrom 08Apr58 – 4510 CCT WG (TAC) Luke 29Jan59 – 435 TCWG (AFR) Miami 20Jul59; to Homestead 25Jul60; assigned 76 TCSq Homestead 01Oct61; 915 TCG (435 TCWG) (AFR) Homestead 17Jan63 – 924 TCG (446 TCWG) (AFR) Ellington 09Aug65 – 906 TCG (302 TCWG) (AFR) Clinton 15Dec65 – 910 TAL GP (459 TCWG) (AFR) Youngstown 06Nov67 – Wfu to MASDC (LOG) Davis-Monthan 06Dec69 for storage, PCN: **CJ335** – DFI: 20Jan70 – Kolar, Inc. Tucson Arizona 29Jan76 for scrap – Final Disposition: Scrapped.

KMC-115 / 51-8112 / C-119F-KM

Avble: 03Dec52; Acc: 13Jan53; Del: 24Jan53 to USAF – 514 TCWG (TAC) Mitchel 25Jan53 – 313 TCWG (TAC) Mitchel 01Feb53; to 809 AB GP (SAC) MacDill 08Aug53 mnt; to Sewart 02Oct53 – 463 TCWG (TAC) Ardmore 01Apr54; depl Charleston 23Apr54; depl Gray Texas 06May54; to Fairchild (AMC) Hagerstown 18Jan55, cvtd **C-119G** 15Feb55; depl 464 TCWG (TAC) Pope 15Mar55; depl Laurinburg-Maxton Airport, North Carolina 02Aug55 – 2242 RFC (CNR) Selfridge 57 – 403 TCWG (AFR) Selfridge 13Nov58 – 4082 STR WG (SAC) Goose Bay Canada 13Sep62. Listed as accident 21Sep62, nfd – Final Disposition: Accident.

KMC-116 / 51-8113 / C-119F-KM

Avble: 03Dec52; Acc: 31Dec52; Del: 07Jan53 to USAF **RECORDS INDECIPHERABLE** → 514 TCWG (TAC) Mitchel 07Jan53 – 313 TCWG (TAC) Mitchel 53; to 4401 SUT WG (ARD) unk 53; depl 465 TCWG (TAC) Mitchel 53; to Sewart 02Oct53 – 6520 Test WG (ARD) Hanscom 11Jun54 ← **RECORDS INDECIPHERABLE** CAR Center (ARD) Hanscom Jan55 – 405 Fighter WG (TAC) Langley 01Mar55 mnt – 456 TCWG (TAC) Charleston 17Mar55; to Shiroi AB Japan 10Nov55; to Ardmore 05May56 – OGDAR (AMC) Hill 10Sep56 – BMC (AMC) Birmingham 22Oct56, cvtd **C-119G** 06Apr57 – 3499 MBT WG (TC) Chanute 18Apr57 – 150 AML SQ (NJ-ANG) Newark cvtd **C-119J** 01Mar58 – Fairchild (AMC) St. Augustine 31Jul62 – 512 TCWG (AFR) NAS Willow Grove 07Sep62 – Wfu to 2704 ASD GP (LOG) Davis-Monthan 22Jan63 for storage – Fairchild (LOG) St. Augustine 15Nov63 – MAP Italy 31Jan64 (USAF records list as lost to ground accident), AMI s/n: **MM51-8113** – Final Disposition: Scrapped.

KMC-117 / 51-8114 / C-119F-KM

Avble: 06Dec52; Acc: 19Jan53; Del: 26Jan53 to USAF **RECORDS INDECIPHERABLE** → 514 TCWG (TAC) Mitchel 25Jan53 – 313 TCWG (TAC) Mitchel 01Feb53; to Sewart 02Oct53; depl Pope 27Apr54; depl Sumpter Alabama 19Jul54 – 456 TCWG (TAC) Charleston 21Dec54 ← **RECORDS INDECIPHERABLE** to Shiroi AB Japan 10Nov55; to Ardmore 05May56 – ODGAR (AMC) Hill 10Sep56 – BMC (AMC) Birmingham 30Oct56, cvtd **C-119G** 05Jul56 – 2237 RFC (CNR) New Castle cvtd **C-119J** 14Aug57; to NAS Willow Grove 21Jul58; depl 2234 RFC (CNR) Hanscom 10Aug58 – 512 TCWG (AFR) NAS Willow Grove 01Dec58 – 94 TCWG (AFR) Hanscom 20May59 – 78 TCSq (435 TCWG) (AFR)

Bates Nov59; 77 TCSq (435 TCWG) (AFR) Donaldson 11Nov60; 435 TCWG (AFR) Homestead 03Jan61 – 156 AML SQ (NC-ANG) Douglas 26Feb61 – 183 AML SQ (MS-ANG) Hawkins 27May61 – 76 TCSq (435 TCWG) (AFR) Homestead 25Jul62 – Wfu to 2704 ASD GP (LOG) Davis-Monthan 29Nov62 for storage – REC: 27Jul64 – Final Disposition: Scrapped.

KMC-118 / 51-8115 / C-119F-KM

Avble: 24Dec52; Acc: 31Jan53; Del: 06Feb53 to USAF – 456 TCWG (TAC) Miami 06Feb53; to Charleston 15Aug53; to Shiroi AB Japan 10Nov55; to Ardmore 05May56 – ODGAR (AMC) Hill 10Sep56 – BMC (AMC) Birmingham 30Oct56, cvtd **C-119G** 13Jun57 – 2466 RFC (CNR) Bakalar 05Jul57; cvtd **C-119J** 22Jul57; depl 2469 RFC (CNR) Scott 20Aug57 – Fairchild (AMC) St. Augustine 04Aug58 – 6593 Test SQ (ARD) Edwards 19Sep58; to Hickham 01Dec58 – 4440 ADE GP (TAC) Langley 25Jul61 – 434 TCWG (AFR) Bakalar 28Jul61; 931 TCG (434 TCWG) (AFR) Bakalar 11Feb63 – Wfu to 2704 ASD GP (LOG) Davis-Monthan 27Dec63 for storage – REC: 27Jul64 – Final Disposition: Scrapped.

KMC-119 / 51-8116 / C-119F-KM

Avble: 09Jan53; Acc: 13Feb53; Del: 23Feb53 to USAF – 456 TCWG (TAC) Miami 23Feb53; to Charleston 15Aug53; to Shiroi AB Japan 10Nov55; to Ardmore 05May56 – OGDAR (AMC) Hill 10Sep56 – BMC (AMC) Birmingham 30Oct56, cvtd **C-119J** 09Aug57 – 1739 FRY SQ (MATS) Amarillo 30Aug57 – 2585 RFC (CNR) Miami 10Sep57; to 106 TPJ SQ (AL-ANG) Birmingham 18Jun58 mnt – 435 TCWG (AFR) Miami 19Dec58; depls 14 AF HQ (CNC) Robins 12Feb59 & 19Jun59; depl 512 TCWG (AFR) NAS Willow Grove 19Sep59; depl 94 TCWG (AFR) Hanscom 07Apr60; to 1604 AB WG (MATS) Kindley 10May60 mnt; assigned 78 TCSq (435 TCWG) (AFR) Bates 30Jun60; 435 TCWG (AFR) Homestead 13Nov60 – 156 AML SQ (NC-ANG) Douglas 26Feb61 – 183 AML SQ (MS-ANG) Hawkins 15May61 – 145 ATR SQ (OH-ANG) Akron-Canton 21Aug61; to Clinton 31Jul62 – 328 TCSq (512 TCWG) (AFR) Niagara Falls 30Aug62 – Wfu to 2704 ASD GP (LOG) Davis-Monthan 09Mar64 for storage – REC: 27Jul64 – Final Disposition: Scrapped.

KMC-120 / 51-8117 / C-119F-KM

Avble: 01Dec52; Acc: 19Jan53; Del: 24Jan53 to USAF – 514 TCWG (TAC) Mitchel 25Jan53 – 313 TCWG (TAC) Mitchel 01Feb53; to Sewart 02Oct53; depl Pope 19Apr54; depl Sumpter Alabama 19Jul54; to Fairchild (AMC) Hagerstown 12Dec54, cvtd **C-119G** 11Jan55 – 314 TCWG (TAC) Sewart 08Jun55 – BMC (AMC) Birmingham 21Feb57 – 2471 RFC (CNR) O'Hare 29May57; depl 2466 RFC (CNR) Bakalar 14Jun57 – 2473 RFC DT (CNR) O'Hare 01Dec57 – 2242 RFC DT (CNR) O'Hare 01Apr58 – Hayes (AMC) Birmingham 21Nov58 – 64 TCSq (403 TCWG) (AFR) O'Hare 06Mar59; 928 TCG (403 TCWG) (AFR) O'Hare 11Feb63 – MAP Taiwan 01Dec69 – Final Disposition: Unknown.

KMC-121 / 51-8118 / C-119F-KM

Avble: 17Nov52; Acc: 26Nov52; Del: 06Dec52 to USAF **RECORDS INDECIPHERABLE** → 316 TCG (TAC) Sewart 06Dec52; depl Pope 23Apr54 ← **RECORDS INDECIPHERABLE** 314 TCWG (TAC) Sewart 15Nov54 – Wfu to 3345 TTA WG (TC) Chanute 07Jul55 for ground instruction – DFI: 27Feb58 – Final Disposition: Scrapped.

KMC-122 / 51-8119 / C-119F-KM

Avble: 05Jan53; Acc: 31Jan53; Del: 06Feb53 to USAF **RECORDS INDECIPHERABLE** → 456 TCWG (TAC) Miami 06Feb53; to Charleston 15Aug53 ← **RECORDS**

INDECIPHERABLE to Shiroi AB Japan 10Nov55; to Ardmore 05May56 – OGDAR (AMC) Hill 10Sep56 – BMC (AMC) Birmingham 22Oct56, cvtd **C-119G** 19Jul57 – 2230 RFC (CNR) NAS New York cvtd **C-119J** 21Aug57 – 2234 RFC (CNR) Hanscom 11Oct57 – 94 TCWG (AFR) Hanscom 19Mar59 – 78 TCSq (435 TCWG) (AFR) Bates 25May60 – Hayes (AMC) Birmingham 07Jul60 – 187 AML SQ (WY-ANG) Cheyenne 01Apr61 – 183 AML SQ (MS-ANG) Hawkins 26Mar61 – 357 TCSq (435 TCWG) (AFR) Bates 12Jul62 – Wfu to PGC Center (SYS) Eglin 27Dec62 for storage – Wfu to 2704 ASD GP (LOG) Davis-Monthan 11Feb63 for storage – 902 TCG (94 TCWG) (AFR) Grenier 29Apr64 – REC: 13Aug64 – Final Disposition: Scrapped.

KMC-123 / 51-8120 / C-119F-KM

Avble: 18Dec52; Acc: 16Jan53; Del: 24Jan53 to USAF – 514 TCWG (TAC) Mitchel 25Jan53 – 313 TCWG (TAC) Mitchel 01Feb53; depl 465 TCWG (TAC) Mitchel 53; to Sewart 02Oct53 – 62 TCWG (TAC) Larson 02Apr54 – 450 Fighter WG (TAC) Foster 29Apr55; to Fairchild (AMC) Hagerstown 26Mar56, cvtd **C-119G** 16Apr56 – 506 Fighter WG (TAC) Tinker 23Jul58 – 4520 CCT WG (TAC) Nellis 18Dec58 – 512 TCWG (AFR) NAS Willow Grove 07Jun59; 913 TCG (512 TCWG) (AFR) NAS Willow Grove 11Feb63 – 907 TAL GP (302 TCWG) (AFR) Clinton 20Dec69 – MAP Taiwan 10Jan70, RoCAF s/n: **3192** – Presently displayed in Kinmen National Park Kinmen County Taiwan – Final Disposition: Extant.

KMC-124 / 51-8121 / C-119F-KM

Avble: 15Jan53; Acc: 13Feb53; Del: 27Feb53 to USAF – 456 TCWG (TAC) Miami 27Feb53; to Charleston 15Aug53; to Shiroi AB Japan 10Nov55; to Ardmore 05May56 – OGDAR (AMC) Hill 10Sep56 – BMC (AMC) Birmingham 13Nov56, cvtd **C-119G** 16Jul57 – 2471 RFC (CNR) O'Hare cvtd **C-119J** 06Aug57 – 2473 RFC DT (CNR) O'Hare 01Dec57 – 2242 RFC DT (CNR) O'Hare 01Apr58 – 102 AML SQ (NY-ANG) New York ANGB 03Oct58 – 64 TCSq (403 TCWG) (AFR) O'Hare 26Oct62 – Wfu to 2704 ASD GP (LOG) Davis-Monthan 04Jan63 for storage – Fairchild (LOG) St. Augustine 30Oct63 – MAP Italy 15Jan64 (USAF list as ground accident), AMI s/n: **MM51-8121**. Stored dismantled at a factory in Maserada Sul Piave Italy up to Oct16, owner unk – Final Disposition: Extant.

KMC-125 / 51-8122 / C-119F-KM

Avble: 20Jan53; Acc: 27Feb53; Del: 06Mar53 to USAF – 456 TCWG (TAC) Miami 06Mar53; to Charleston 15Aug53; depl 479 Fighter WG (TAC) George 18May54; to Shiroi AB Japan 10Nov55; to Ardmore 05May56 – OGDAR (AMC) Hill 10Sep56 – BMC (AMC) Birmingham 13Nov56, cvtd **C-119G** 18Jul57 – 2256 RFC (CNR) Niagara Falls cvtd **C-119J** 08Aug57 – 2242 RFC DT (CNR) Niagara Falls 01Dec57 – 2237 RFC DT (CNR) Niagara Falls 01Apr58 – 102 AML SQ (NY-ANG) New York ANGB 14Oct58. Crashed in Jamaica Bay off New York-Floyd Bennett Field New York 26Nov58 during a radar directed landing approach in heavy rain and fog, 3 of the 4 crew killed – Final Disposition: Accident.

KMC-126 / 51-8123 / C-119F-KM

Avble: 27Jan53; Acc: 27Feb53; Del: 07Mar53 to USAF – 456 TCWG (TAC) Miami 07Mar53; to Charleston 15Aug53; depl Langley 25Oct54; to Shiroi AB Japan 10Nov55; to Ardmore 05May56 – OGDAR (AMC) Hill 10Sep56 – BMC (AMC) Birmingham 30Oct56, cvtd **C-119G** 16May57 – 3345 TTA WG (TC) Chanute 29May57; cvtd **C-119J** 24Jul57 – 140 AML SQ (PA-ANG) Gen Spaatz 02May58; to Olmsted 01Apr61 – 64 TCSq (403 TCWG) (AFR) O'Hare 05Oct62 – Wfu to 2704 ASD GP (LOG) Davis-Monthan 04Jan63 for storage – REC: 27Jul64 – Final Disposition: Scrapped.

KMC-127 / 51-8124 / C-119F-KM

Avble: 29Jan53; Acc: 27Feb53; Del: 10Mar53 to USAF – 456 TCWG (TAC) Miami 10Mar53; to Charleston 15Aug53; to Shiroi AB Japan 10Nov55; to Ardmore 05May56 – OGDAR (AMC) Hill 10Sep56 – BMC (AMC) Birmingham 05Nov56, cvtd **C-119G** 10Apr57 – 3499 MBT WG (TC) Chanute 26Apr57; to 3345 TTA WG (TC) Chanute cvtd **C-119J** 12Sep57 – 183 AML SQ (MS-ANG) Hawkins 11Feb58; cvtd **MC-119J** 15Jul61 – 147 AML SQ (ANG) Greater Pittsburgh 04May62; cvtd **C-119J** 28Feb63 – Wfu to 2704 ASD GP (LOG) Davis-Monthan 21Aug63 for storage – Fairchild (LOG) Crestview 08Oct66 – 4603 AB GP (ADC) Stewart 06Jan67 – Wfu to MASDC (LOG) Davis-Monthan 20Nov69 for storage, PCN: **CJ319** – DFI: 20Jan70 – Final Disposition: Scrapped.

KMC-128 / 51-8125 / C-119F-KM

Avble: 03Feb53; Acc: 28Feb53; Del: 06Mar53 to USAF – 456 TCWG (TAC) Miami 06Mar53; to Charleston 15Aug53; to Shiroi AB Japan 10Nov55; to Ardmore 05May56 – OGDAR (AMC) Hill 10Sep56 – BMC (AMC) Birmingham 24Oct56, cvtd **C-119G** 13May57 – 3499 MBT WG (TC) Chanute 23May57; to 3345 TTA WG (TC) Chanute 21Oct57 – 150 AML SQ (NJ-ANG) Newark cvtd **C-119J** 08Mar58 – Wfu to 2704 ASD GP (LOG) Davis-Monthan 13Jan63 for storage – Fairchild (LOG) St. Augustine 06Nov63 – MAP Italy 21Nov63, AMI s/n: **MM51-8125** – Final Disposition: Scrapped.

KMC-129 / 51-8126 / C-119F-KM

Avble: 09Feb53; Acc: 27Feb53; Del: 07Mar53 to USAF – 456 TCWG (TAC) Miami 07Mar53; to Charleston 15Aug53; to Shiroi AB Japan 10Nov55; to Ardmore 05May56 – OGDAR (AMC) Hill 10Sep56 – BMC (AMC) Birmingham 13Nov56, cvtd **C-119G** 17Apr57 – 3499 MBT WG (TC) Chanute 03May57; to 3345 TTA WG (TC) Chanute 25Jun57; cvtd **C-119J** 26Dec57; to 3345 TTA WG (TC) Chanute 17Dec57 – 150 AML SQ (NJ-ANG) Newark 28Feb58 – 328 TCSq (512 TCWG) (AFR) Niagara Falls 12Oct62 – Wfu to 2704 ASD GP (LOG) Davis-Monthan 09Mar64 for storage – REC: 27Jul64 – Final Disposition: Scrapped.

KMC-130 / 51-8127 / C-119F-KM

Avble: 12Feb53; Acc: 12Mar53; Del: 20Mar53 to USAF – 456 TCWG (TAC) Miami 20Mar53; to Charleston 15Aug53; to Shiroi AB Japan 10Nov55; to Ardmore 05May56 – OGDAR (AMC) Hill 10Sep56 – BMC (AMC) Birmingham 05Nov56, cvtd **C-119G** 09May57 – 3499 MBT WG (TC) Chanute 28May57; to 3565 NAT WG (TC) James Connally cvtd **C-119J** 03Aug57; to 3345 TTA WG (TC) Chanute 27Aug57. Had an engine failure after take-off from Tinker AFB Oklahoma 24Oct57. The aircraft came down in a field within a suburb of Oklahoma City killing all 4 crew onboard – Final Disposition: Accident.

KMC-131 / 51-8128 / C-119F-KM

Avble: 18Feb53; Acc: 24Mar53; Del: 03Apr53 to USAF – 456 TCWG (TAC) Miami 03Apr53; to Charleston 15Aug53; to Shiroi AB Japan 10Nov55; to Ardmore 05May56 – OGDAR (AMC) Hill 10Sep56 – BMC (AMC) Birmingham 05Nov56, cvtd **C-119G** 07Jun57 – 2472 RFC (CNR) Richards 26Jun57; cvtd **C-119J** 31Jul57; to DT Davis 02Mar58 & 17May58; to 463 TCWG (TAC) Ardmore 28May58 mnt – 442 TCWG (AFR) Richards 02Mar59 – 187 AML SQ (WY-ANG) Cheyenne 04Mar61 – 140 AML SQ (PA-ANG) Olmsted 10May61 – Wfu to 2704 ASD GP (LOG) Davis-Monthan 03Dec62 for storage – Fairchild (LOG) St. Augustine date unk – MAP Italy 06Feb64, AMI s/n: **MM51-8128** – Final Disposition: Scrapped.

KMC-132 / 51-8129 / C-119F-KM

Avble: 19Feb53; Acc: 31Mar53; Del: 09Apr53 to USAF – 456 TCWG (TAC) Miami 09Apr53; to Charleston 15Aug53; to Shiroi AB Japan 10Nov55; to Ardmore 05May56 – OGDAR (AMC) Hill 10Sep56 – BMC (AMC) Birmingham 13Nov56, cvtd **C-119G** 17Apr57 – 3499 MBT WG (TC) Chanute 03May57; to 3345 TTA WG (TC) Chanute 11Oct57 – 145 AML SQ (OH-ANG) Akron-Canton cvtd **C-119J** 08Mar58; to Clinton 31Jul62 – 64 TCSq (403 TCWG) (AFR) O'Hare 08Sep62 – Wfu to 2704 ASD GP (LOG) Davis-Monthan 04Jan63 for storage – REC: 27Jul64 – Final Disposition: Scrapped.

KMC-133 / 51-8130 / C-119F-KM

Avble: 23Feb53; Acc: 21Mar53; Del: 27Mar53 to USAF – 456 TCWG (TAC) Miami 27Mar53; to Charleston 15Aug53; depl 63 TCWG (TAC) Charleston 09Nov54; to Shiroi AB Japan 10Nov55; to Ardmore 05May56 – OGDAR (AMC) Hill 10Sep56 – BMC (AMC) Birmingham 24Oct56, cvtd **C-119J** 14Aug57 – 2465 RFC (CNR) Minneapolis-St. Paul 11Sep57 – 2473 RFC DT (CNR) Minneapolis-St. Paul 01Dec57 – 96 TCSq (440 TCWG) (AFR) Minneapolis-St. Paul 19Dec58 – 140 AML SQ (PA-ANG) Olmsted 23Mar61 – 96 TCSq (440 TCWG) (AFR) Minneapolis-St. Paul 18Sep62 – Wfu to 2704 ASD GP (LOG) Davis-Monthan 16Jan63 for storage – Fairchild (LOG) St. Augustine 08Nov63 – MAP Italy 21Nov63, AMI s/n: **MM51-8130**; cvtd **EC-119J** 1975 – Final Disposition: Scrapped.

KMC-134 / 51-8131 / C-119F-KM

Avble: 27Feb53; Acc: 31Mar53; Del: 08Apr53 to USAF – 456 TCWG (TAC) Miami 08Apr53; to Charleston 15Aug53; depl Langley 25Oct54; to Shiroi AB Japan 10Nov55; to Ardmore 05May56 – OGDAR (AMC) Hill 10Sep56 – BMC (AMC) Birmingham 24Oct56, cvtd **C-119G** 14May57 – 3345 TTA WG (TC) Chanute 29May57; cvtd **C-119J** 24Jul57 – 140 AML SQ (PA-ANG) Gen Spaatz 02May58; to Olmsted 01Apr61 – 147 AML SQ (ANG) Greater Pittsburgh 31Dec62 – Wfu to 2704 ASD GP (LOG) Davis-Monthan 30Oct63 for storage – REC: 27Jul64 – Final Disposition: Scrapped.

KMC-135 / 51-8132 / C-119F-KM

Avble: 03Mar53; Acc: 31Mar53; Del: 28Apr53 to USAF – 456 TCWG (TAC) Miami 28Apr53; to Charleston 15Aug53; to Shiroi AB Japan 10Nov55; to Ardmore 05May56 – OGDAR (AMC) Hill 10Sep56 – BMC (AMC) Birmingham 09Nov56, cvtd **C-119G** 20Jun57 – 2466 RFC (CNR) Bakalar 16Jul57; cvtd **C-119J** 22Jul57 – 2347 RFC (CNR) Long Beach 15Jun58 – 452 TCWG (AFR) Long Beach 13Jun59; to March 01Oct60 – SBNAR (AMC) Norton 01Mar61 – REC: 27Jun61 – Final Disposition: Scrapped.

KMC-136 / 51-8133 / C-119F-KM

Avble: 06Mar53; Acc: 31Mar53; Del: 08Apr53 to USAF – 456 TCWG (TAC) Miami 28Apr53; to Charleston 15Aug53; bailment BMC (AMC) Birmingham 07Dec53 mnt. Damaged beyond repair in a ground taxi accident at Birmingham Airport Alabama 17Apr54. 2 of the 3 civilian employees from Hayes Aircraft Corp. sustained minor injuries after the aircraft lost control during a high speed taxi run – REC: 16Jun54 – DFI: 01Aug54 – Final Disposition: Accident.

KMC-137 / 51-8134 / C-119F-KM

Avble: 11Mar53; Acc: 31Mar53; Del: 08Apr53 to USAF **RECORDS INDECIPHERABLE** → 456 TCWG (TAC) Miami 08Apr53; to Charleston 15Aug53 ← **RECORDS**

INDECIPHERABLE to 3510 CCT WG (TC) Randolph 02Mar55 mnt; to Shiroi AB Japan 10Nov55; to Ardmore 05May56 – OGDAR (AMC) Hill 10Sep56 – BMC (AMC) Birmingham 02Nov56, cvtd **C-119G** 24May57 – 3700 MIT WG (TC) Lackland 12Jun57 – 3499 FTG WG (TC) Chanute cvtd **C-119J** 28Dec57 – 145 AML SQ (OH-ANG) Akron-Canton 27Mar58 – Fairchild (AMC) St. Augustine 09Jun61 mnt – 147 AML SQ (PA-ANG) Greater Pittsburgh 25Jul61 – Wfu to 2704 ASD GP (LOG) Davis-Monthan 17Dec62 for storage – REC: 27Jul64 – Final Disposition: Scrapped.

KMC-138 / 51-8135 / C-119F-15-KM

Avble: 16Mar53; Acc: 24Apr53; Del: 05May53 to USAF **RECORDS INDECIPHERABLE** → AFM Center (ARD) Patrick 05May53 ← **RECORDS INDECIPHERABLE** 366 Fighter WG (TAC) England 18Jan55; to Fairchild (AMC) Hagerstown 02Apr55, cvtd **C-119G** 30Apr55 – 2230 RFC (CNR) NAS New York 16Aug57 – 2466 RFC (CNR) Bakalar 11Oct57 – 2242 RFC (CNR) Selfridge 27Oct57 – AEMCO (AMC) Oakland 25Jul58 mnt – 403 TCWG (AFR) Selfridge 16Dec58; 927 TCG (403 TCWG) (AFR) Selfridge 11Feb63 – 930 SOP GP (434 TCWG) (AFR) Bakalar 01Jul69; to Grissom 12Jan70 – Wfu to MASDC (LOG) Davis-Monthan 02Sep70 for storage, PCN: **CJ361** – DFI: 24Sep70 – Kolar, Inc. Tucson Arizona 23Feb76 for scrap – Final Disposition: Scrapped.

KMC-139 / 51-8136 / C-119F-KM

Avble: 18Mar53; Acc: 24Apr53; Del: 07May53 to USAF **RECORDS INDECIPHERABLE** → AFM Center (ARD) Patrick 07May53 ← **RECORDS INDECIPHERABLE** 3510 CCT WG (TC) Randolph 03Nov55; to Fairchild (AMC) Hagerstown 28Dec55, cvtd **C-119G** 30Jan56 – 312 Fighter-Bomber WG (TAC) Clovis 11Apr56 – 832 AB GP (TAC) Cannon 08Oct57 – 2235 RFC (CNR) Grenier 17Nov57 – 2234 RFC (CNR) Hanscom 20Nov57 – 94 TCWG (AFR) Hanscom 19Mar59; to 2500 AB GP (CNC) Mitchel 11May59 mnt; 901 TCG (94 TCWG) (AFR) Hanscom 31May63 – 932 TCG (434 TCWG) (AFR) Scott 27May66 – 912 TCG (512 TCWG) (AFR) NAS Willow Grove 11Dec66 – 934 TAL GP (440 TCWG) (AFR) Minneapolis-St. Paul 15Aug68 – MAP Taiwan 08Apr69, RoCAF s/n: **3152** – Presently stored at Pingtung AB Taiwan – Final Disposition: Extant.

KMC-140 / 51-8137 / C-119F-KM

Avble: 21Mar53; Acc: 28Apr53; Del: 06May53 to USAF – 456 TCWG (TAC) Miami 06May53; to Charleston 15Aug53; to Shiroi AB Japan 10Nov55; to Ardmore 05May56 – OGDAR (AMC) Hill 10Sep56 – BMC (AMC) Birmingham 18Oct56, cvtd **C-119G** 06Feb57 – 3595 CCT WG (TC) Nellis 26Feb57; cvtd **C-119J** 27Jul57 – 183 AML SQ (MS-ANG) Hawkins 01Mar58 – 403 TCWG (AFR) Selfridge 08Sep62 – Wfu to 2704 ASD GP (LOG) Davis-Monthan 09Jan63 for storage – REC: 27Jul64 – Final Disposition: Scrapped.

KMC-141 / 51-8138 / C-119F-KM

Avble: 27Mar53; Acc: 29Apr53; Del: 06May53 to USAF – 456 TCWG (TAC) Miami 06May53; to Charleston 15Aug53; to Shiroi AB Japan 10Nov55; to Ardmore 05May56 – OGDAR (AMC) Hill 10Sep56 – BMC (AMC) Birmingham 07Nov56, cvtd **C-119G** 03Jun57 – 3700 MIT WG (TC) Lackland 25Jun57 – 183 AML SQ (MS-ANG) Hawkins cvtd **C-119J** 14Nov57 – 314 TCSq (349 TCWG) (AFR) McClellan 29Aug62 – Wfu to 2704 ASD GP (LOG) Davis-Monthan 04Jan63 for storage – REC: 27Jul64 – Final Disposition: Scrapped.

KMC-142 / 51-8139 / C-119F-KM

Avble: 01Apr53; Acc: 30Apr53; Del: 07May53 to USAF – 456 TCWG (TAC) Miami 07May53; to Charleston 15Aug53 – CAR Center (ARD) Hanscom 15Mar55; to 2845 AB GP (AMC) Griffiss 13Oct55 mnt; to RDP DP (AMC) Griffiss 29Feb56 mnt; to BMC (AMC) Birmingham 28Jun56, cvtd **C-119G** 17Oct56 – 2469 RFC (CNR) Scott 09May57; depl 2466 RFC (CNR) Bakalar 18Aug57 – 2466 RFC DT (CNR) Scott 01Dec57 – 73 TCSq (434 TCWG) (AFR) Scott 18Apr59 – 452 TCWG (AFR) March 09Dec61; 944 TCG (452 TCWG) (AFR) March 11Feb63; depl 945 TCG (452 TCWG) (AFR) Hill 21Apr66; 943 TAL GP (452 TCWG) (AFR) March 23Jan68 – Wfu to MASDC (LOG) Davis-Monthan 06Nov68 for storage, PCN: **CJ269** – DFI: 26Dec73 – Final Disposition: Scrapped.

KMC-143 / 51-8140 / C-119F-KM

Avble: 07Apr53; Acc: 30Apr53; Del: 07May53 to USAF **RECORDS INDECIPHERABLE →** 456 TCWG (TAC) Miami 07May53; to Charleston 15Aug53; depl Langley 19Oct54 **← RECORDS INDECIPHERABLE** to Shiroi AB Japan 10Nov55; to Ardmore 05May56 – OGDAR (AMC) Hill 10Sep56 – BMC (AMC) Birmingham 05Nov56, cvtd **C-119G** 21May57 – 3345 TTA WG (TC) Chanute 03Jun57; cvtd **C-119J** 24Jul57; to 5 AB GP (SAC) Travis 14Feb58 mnt – 183 AML SQ (MS-ANG) Hawkins 26Apr58 – 65 TCSq (403 TCWG) (AFR) Davis 31Jul62 – Wfu to 2704 ASD GP (LOG) Davis-Monthan 28Dec63 for storage – Fairchild (LOG) St. Augustine 19Oct66 – 4603 AB GP (ADC) Stewart 09Jan67 – Wfu to MASDC (LOG) Davis-Monthan 10Nov69 for storage, PCN: **CJ314** – DFI: 20Jan70 – MAP Italy 70?; AMI s/n: **MM51-8140**; cvtd **VC-119J** 1970? Noted stored at Pisa Italy 1980 – Final Disposition: Scrapped.

KMC-144 / 51-8141 / C-119F-KM

Avble: 10Apr53; Acc: 30Apr53; Del: 07May53 to USAF – 456 TCWG (TAC) Miami 07May53; to Charleston 15Aug53 **RECORDS INDECIPHERABLE** to Shiroi AB Japan 10Nov55; to Ardmore 05May56 – OGDAR (AMC) Hill 10Sep56 – BMC (AMC) Birmingham 05Nov56, cvtd **C-119G** 20Apr57 – 3499 MBT WG (TC) Chanute 01May57; depl 3345 TTA WG (TC) Chanute 25Jun57 – 183 AML SQ (MS-ANG) Hawkins cvtd **C-119J** 11Feb58 – 349 TCWG (AFR) Hamilton 31Aug62 – Wfu to 2704 ASD GP (LOG) Davis-Monthan 28Dec62 for storage – Fairchild (AMC) St. Augustine 19Oct66 – 4603 AB GP (ADC) Stewart 12Jan67 – 78 Fighter WG (ADC) Hamilton 17Jan69 – Wfu to MASDC (LOG) Davis-Monthan 12Nov69 for storage, PCN: **CJ317** – DFI: 20Jan70 – Final Disposition: Scrapped.

KMC-145 / 51-8142 / C-119F-KM

Avble: 01May53; Acc: 16May53; Del: 21May53 to USAF – AFM Center (ARD) Patrick 21May53 – 366 Fighter WG (TAC) England 20Aug55. Crashed after take-off 3.1 miles N England AFB Louisiana due to engine failure 06Sep55. 4 fatalities from 6 crew onboard – Final Disposition: Accident.

KMC-146 / 51-8143 / C-119F-KM

Avble: 06May53; Acc: 19May53; Del: 25May53 to USAF – AFM Center (ARD) Patrick 25May53 – 479 Fighter WG (TAC) George 31Aug55; to Fairchild (AMC) Hagerstown 21Mar56, cvtd **C-119G** 18Apr56. Damaged beyond repair while parked at George AFB California 05Jan57 when a Lockheed T-33A flying over the base caught fire and crashed destroying the C-119 and damaging another. Both pilots in the T-33A were killed – Final Disposition: Accident.

KMC-147 / 51-8144 / C-119F-KM

Avble: 07May53; Acc: 22May53; Del: 25May53 to USAF – 456 TCWG (TAC) Miami 25May53; to Charleston 15Aug53; to Shiroi AB Japan 10Nov55; to Ardmore 05May56 – OGDAR (AMC) Hill 10Sep56 – BMC (AMC) Birmingham 05Nov56, cvtd **C-119G** 02Jul57 – 2237 RFC (CNR) New Castle 22Jul57; cvtd **C-119J** 02Aug57; to NAS Willow Grove 21Jul58 – 102 AML SQ (NY-ANG) New York ANGB 10Sep58 – 732 TCSq (94 TCWG) (AFR) Grenier 28Nov62 – Wfu to 2704 ASD GP (LOG) Davis-Monthan 63 for storage – Fairchild (AMC) St. Augustine 16Nov63 – MAP Italy 31Jan64; AMI s/n: **MM51-8144**; cvtd **VC-119J** – Final Disposition: Scrapped.

KMC-148 / 51-8145 / C-119F-KM

Avble: 12May53; Acc: 26May53; Del: 28May53 to USAF – 313 TCWG (TAC) Mitchel 28May53; to Stewart 02Oct53; to Fairchild (AMC) Hagerstown 15Dec54, cvtd **C-119G** 18Jan55 – 464 TCWG (TAC) Sewart (depl) 08Jun55 – 314 TCWG (TAC) Sewart 10Aug55 – BMC (AMC) Birmingham 25Jul57 – 2237 RFC (CNR) New Castle 04Oct57; depl 5700 AB GP (CAC) Albrook 11Jan58; to NAS Willow Grove 21Jul57 – 512 TCWG (AFR) NAS Willow Grove 01Dec58; to 2500 AB GP (CNC) Mitchel 06Feb59 mnt; 912 TCG (512 TCWG) (AFR) NAS Willow Grove 11Feb63 – Fairchild (LOG) Fairchild-Stratos 12Apr68 mnt – 906 SOP GP (302 TCWG) (AFR) Clinton 10Feb70 – Wfu to MASDC (LOG) Davis-Monthan 31Aug70 for storage, PCN: **CJ353** – DFI: 16Sep70 – Kolar, Inc. Tucson Arizona 20Feb76 for scrap – Final Disposition: Scrapped.

KMC-149 / 51-8146 / C-119F-KM

Avble: 14May53; Acc: 28May53; Del: 01Jun53 to USAF – 313 TCWG (TAC) Mitchel 01Jun53; to Stewart 02Oct53; depl Sumpter Alabama 19Jul54 – 3600 CCT WG (TC) Luke 27May55, USAF *Thunderbirds* support aircraft; to 3510 CCT WG (TC) Randolph 26Jun55, 04Aug55 & 19Oct55; bailment Fairchild (AMC) Hagerstown 19Dec55, cvtd **C-119G** 23Jan56 – 3595 CCT WG (TC) Nellis 04Jun56 – 2585 RFC (CNR) Miami 08Apr57; to 464 TCWG (TAC) Pope 23Jan58 mnt; to 2584 AB SQ (CNC) Memphis 23Sep58 mnt – Hayes (AMC) Birmingham 25Nov58 mnt – 435 TCWG (AFR) Miami 02Apr59; to unk unit 59 mnt; to Homestead 25Jul60; assigned 76 TCSq Homestead 09Mar62; 915 TCG (435 TCWG) (AFR) Homestead 17Jan63 – 926 TCG (446 TCWG) (AFR) NAS New Orleans 13Nov65 – Wfu to MASDC (LOG) Davis-Monthan 30Nov69 for storage, PCN: **CJ325** – DFI: 20Jan70 – Kolar, Inc. Tucson Arizona 17Feb76 for scrap – Final Disposition: Scrapped.

KMC-150 / 51-8147 / C-119F-KM

Avble: 16May53; Acc: 28May53; Del: 01Jun53 to USAF – 313 TCWG (TAC) Mitchel 01Jun53; to Stewart 02Oct53 – 63 TCWG (TAC) Donaldson 02Apr54; bailment Fairchild (AMC) Hagerstown 17Jan55, cvtd **C-119G** 14Feb55; depl 464 TCWG (TAC) Pope 05Apr56 – 4501 SUT SQ (TAC) Donaldson 26Apr56; depl 314 TCWG (TAC) Sewart 26May56; to 3605 AOT WG (TC) Ellington 26Jun56 mnt – 2237 RFC (CNR) New Castle 28Sep57; to NAS Willow Grove 21Jul58 – 512 TCWG (AFR) NAS Willow Grove 01Dec58; 913 TCG (512 TCWG) (AFR) NAS Willow Grove 11Feb63 – 907 TAL GP (302 TCWG) (AFR) Clinton 20Dec69 – MAP Taiwan 09Jan70, RoCAF s/n: **3198** – Final Disposition: Unknown.

KMC-151 / 51-8148 / C-119F-KM

Avble: 19May53; Acc: 29May53; Del: 04Jun53 to USAF – 313 TCWG (TAC) Mitchel 04Jun53; to Stewart 02Oct53 – 63 TCWG (TAC) Donaldson 02Apr54; bailment Fairchild

(AMC) Hagerstown 03Jan55, cvtd **C-119G** 31Jan55 – 463 TCWG (TAC) Ardmore 12Feb55; depl 314 TCWG (TAC) Sewart 01Feb56; depl North Aux AF 02Apr56 – BMC (AMC) Birmingham 09Jan57 – 479 Fighter WG (TAC) George 02Apr57 – 831 AB GP (TAC) George 08Oct57 – 512 TCWG (AFR) NAS Willow Grove 08Jul58; 912 TCG (512 TCWG) (AFR) NAS Willow Grove 11Feb63; 913 TAL GP (512 TCWG) (AFR) NAS Willow Grove 24Jul68 – MAP Taiwan 21Apr69, RoCAF s/n: **3159** – Final Disposition: Unknown.

KMC-152 / 51-8149 / C-119F-KM

Avble: 21May53; Acc: 05Jun53; Del: 11Jun53 to USAF – AFM Center (ARD) Patrick 11Jun53 – 3510 CCT WG (TC) Randolph 30Sep55; bailment Fairchild (AMC) Hagerstown 14Dec55, cvtd **C-119G** 24Jan56. Damaged beyond repair in accident 12Feb56, nfd – Final Disposition: Accident.

KMC-153 / 51-8150 / C-119F-KM

Avble: 25May53; Acc: 08Jun53; Del: 15Jun53 to USAF – AFM Center (ARD) Patrick 15Jun53 – 3510 CCT WG (TC) Randolph 22Nov55 – 3345 TTA WG (TC) Chanute 10Feb56; to Fairchild (AMC) Hagerstown 06Mar56, cvtd **C-119G** 02Apr56 – 3595 CCT WG (TC) Nellis 22Aug56 – 2472 RFC (CNR) Richards 15Aug57 – 2465 RFC (CNR) Minneapolis-St. Paul 30Aug57 – 2473 RFC DT (CNR) Minneapolis-St. Paul 01Dec57 – 96 TCSq (440 TCWG) (AFR) Minneapolis-St. Paul 19Dec58; to 2465 AB GP (CNC) Minneapolis-St. Paul 10Jan59 mnt; 934 TCG (440 TCWG) (AFR) Minneapolis-St. Paul 11Feb63 – 906 TAL GP (302 TCWG) (AFR) Clinton 12Jan70 – MAP Taiwan 25Feb70, RoCAF s/n: **3204** – Presently stored at Pingtung AB Taiwan – Final Disposition: Extant.

KMC-154 / 51-8151 / C-119F-KM

Avble: 28May53; Acc: 11Jun53; Del: 12Jun53 to USAF – 313 TCWG (TAC) Mitchel 12Jun53; to Sewart 02Oct53. Landed short of runway at Eleuthera Aux AFB The Bahamas 06Jan54 causing serious damage to aircraft. Pilot: Robert S. Hermanson, no fatalities – AFM Center (ARD) Patrick for salvage and repairs – REC: 16Jun54 – DFI: 08Jul54 – Final Disposition: Accident.

KMC-155 / 51-8152 / C-119F-KM

Avble: 29May53; Acc: 13Jun53; Del: 18Jun53 to USAF – 456 TCWG (TAC) Miami 17Jun53; to Charleston 15Aug53; to Shiroi AB Japan 10Nov55; to Ardmore 05May56 – OGDAR (AMC) Hill 10Sep56 – BMC (AMC) Birmingham 09Nov56, cvtd **C-119G** 05Jul57 – 2237 RFC (CNR) New Castle cvtd **C-119J** 08Aug57; to NAS Willow Grove 21Jul58 – 102 AML SQ (NY-ANG) New York ANGB 23Aug58 – Wfu to 2704 ASD GP (LOG) Davis-Monthan 15May63 for storage – Fairchild (LOG) St. Augustine 05Dec63 – MAP Italy 19Feb64 (USAF list as lost to ground accident), AMI s/n: **MM51-8152** – Final Disposition: Scrapped.

KMC-156 / 51-8153 / C-119F-KM

Avble: 11Jun53; Acc: 25Jun53; Del: 01Jul53 to USAF – 456 TCWG (TAC) Miami 01Jul53; to Charleston 15Aug53. Impacted a ridge on Mt. Mission Point 12.5 miles NW of destination Burbank Airport California while on an instrument approach through fog 20Apr54, all 7 crew onboard were killed – DFI: 27May54 – Final Disposition: Accident.

KMC-157 / 51-8154 / C-119F-KM

Avble: 05Jun53; Acc: 22Jun53; Del: 26Jun53 to USAF – 456 TCWG (TAC) Miami 26Jun53; to Charleston 15Aug53; to Shiroi AB Japan 10Nov55; to Ardmore 05May56 – OGDAR (AMC) Hill 10Sep56 – BMC (AMC) Birmingham 05Nov56, cvtd **C-119G** 24Apr57 – 3499 MBT WG (TC) Chanute 14May57; cvtd **C-119J** 31Aug57 – 145 AML SQ (OH-ANG) Akron-Canton 27Mar58; to Clinton 14Jun62 – 512 TCWG (AFR) NAS Willow Grove 30Aug62 – Wfu to 2704 ASD GP (LOG) Davis-Monthan 22Jan63 for storage – Fairchild (AMC) St. Augustine 04Nov63 – MAP Italy 05Nov63, AMI s/n: **MM51-8154**. Damaged beyond repair due to an engine problem at Cisterna di Latina Italy 26Jun69, no fatalities – Final Disposition: Accident.

KMC-158 / 51-8155 / C-119F-KM

Avble: 17Jun53; Acc: 30Jun53; Del: 01Jul53 to USAF – 456 TCWG (TAC) Miami 01Jul53; to Charleston 15Aug53; to 4901 SUT WG (ARD) Kirtland 53 mnt; to Shiroi AB Japan 10Nov55; to Ardmore 05May56 – OGDAR (AMC) Hill 10Sep56 – BMC (AMC) Birmingham 05Nov56, cvtd **C-119G** 06May57 – 3499 MBT WG (TC) Chanute 17May57; cvtd **C-119J** 27Aug57; to 2750 AB WG (AMC) WPAFB 27Aug57 mnt – 145 AML SQ (OH-ANG) Akron-Canton 08Mar58; cvtd **MC-119J** 01Jul62; to Clinton 01Jul61 – Fairchild (AMC) St. Augustine 01Jun62 – 147 AML SQ (ANG) Greater Pittsburgh 19Jul62; cvtd **C-119J** 28Feb63 – Wfu to 2704 ASD GP (LOG) Davis-Monthan 21Aug63 for storage – REC: 27Jul64 – Final Disposition: Scrapped.

KMC-159 / 51-8156 / C-119F-KM

Avble: 19Jun53; Acc: 30Jun53; Del: 06Jul53 to USAF – 456 TCWG (TAC) Miami 06Jul53; to Charleston 15Aug53; depl 479 Fighter WG (TAC) George 18May54; to Shiroi AB Japan 10Nov55; to Ardmore 05May56 – OGDAR (AMC) Hill 10Sep56 – BMC (AMC) Birmingham 09Nov56, cvtd **C-119G** 24Jun57 – 2466 RFC (CNR) Bakalar 12Jul57; cvtd **C-119J** 22Jul57; depl 2471 RFC (CNR) O'Hare 12Aug57 – 434 TCWG (AFR) Bakalar 19Apr59 – 140 AML SQ (PA-ANG) Olmsted 29Mar61 – 440 TCWG (AFR) Gen Mitchell 27Sep62 – Wfu to 2704 ASD GP (LOG) Davis-Monthan date unk for storage – Fairchild (LOG) St. Augustine 01Dec63 – MAP Italy 14Feb64 (USAF list as lost to ground accident), AMI s/n: **MM51-8156** – Final Disposition: Scrapped.

KMC-160 / 51-8157 / C-119F-KM

Avble: 23Jun53; Acc: 21Jul53; Del: 31Jul53 to USAF – 456 TCWG (TAC) Miami 06Jul53; to Charleston 15Aug53; to Shiroi AB Japan 10Nov55; to 325 Fighter GP (ADC) McChord 13Mar56 mnt; to Ardmore 05May56 – OGDAR (AMC) Hill 10Sep56 – BMC (AMC) Birmingham 05Nov56, cvtd **C-119G** 07May57 – 3499 MBT WG (TC) Chanute 25May57; cvtd **C-119J** 01Jan58 – 140 AML SQ (PA-ANG) Gen Spaatz 27Mar58; to Olmsted 01Feb61; cvtd **MC-119J** 13Jun61 – 147 AML SQ (ANG) Greater Pittsburgh cvtd **C-119J** 15Dec62 – Wfu to 2704 ASD GP (LOG) Davis-Monthan 18Sep63 for storage – Fairchild (LOG) St. Augustine 26Oct66 – 78 Fighter WG (ADC) Hamilton 15Jan67 – Wfu to MASDC (LOG) Davis-Monthan 12Nov69 for storage, PCN: **CJ318** – DFI: 20Jan70 – Final Disposition: Scrapped.

KMC-161 / 51-8158 / C-119F-KM

Avble: 29Jun53; Acc: 27Jul53; Del: 01Aug53 to USAF – 456 TCWG (TAC) Miami 01Aug53; to Charleston 15Aug53; depl Langley 20Oct54; to Shiroi AB Japan 10Nov55; to Ardmore 05May56 – OGDAR (AMC) Hill 10Sep56 – BMC (AMC) Birmingham 09Nov56, cvtd **C-119G** 18Jun57 – 3595 CCT WG (TC) Nellis 29Jun57; cvtd **C-119J** 27Jul57 – 150 AML SQ (NJ-ANG) Newark 01Mar58; cvtd **MC-119J** 30Jun61 – Wfu to 2704 ASD GP (LOG) Davis-Monthan cvtd **C-119J** 14Feb63 – Fairchild (LOG) St. Augustine 01Dec63 – MAP Italy 14Feb64, AMI s/n: **MM51-8158**; cvtd **VC-119J** – Final Disposition: Scrapped.

KMC-162 / 51-8159 / C-119F-KM

Avble: 02Jul53; Acc: 27Jul53; Del: 03Aug53 to USAF – 456 TCWG (TAC) Miami 02Aug53; to Charleston 15Aug53; depl Langley 20Oct54; to Shiroi AB Japan 10Nov55; to Ardmore 05May56 – OGDAR (AMC) Hill 10Sep56 – BMC (AMC) Birmingham 18Oct56, cvtd **C-119G** 26Mar57 – 3499 MBT WG (TC) Chanute 08Apr57 – 140 AML SQ (PA-ANG) Gen Spaatz cvtd **C-119J** 17Apr58; to Olmsted 01Feb61 – Wfu to 2704 ASD GP (LOG) Davis-Monthan 03Dec62 for storage – Fairchild (LOG) St. Augustine 21Nov63 – MAP Italy 06Feb64 as a spares airframe – Final Disposition: Scrapped.

KMC-163 / 51-8160 / C-119F-KM

Avble: 20Jul53; Acc: 11Aug53; Del: 12Aug53 to USAF – 313 TCWG (TAC) Mitchel 13Aug53; to Sewart 02Oct53 – 63 TCWG (TAC) Donaldson 25Mar54; bailment Fairchild (AMC) Hagerstown 24Jan55, cvtd **C-119G** 21Feb55; depl 464 TCWG (TAC) Pope 03Apr55 – 63 AB GP (MATS) Donaldson 01Jul57 – 2465 RFC (CNR) Minneapolis-St. Paul 16Sep57 – 2473 RFC DT (CNR) Minneapolis-St. Paul 01Dec57 – 2347 RFC (CNR) Long Beach 24Jun58; to 78 MSU GP (ADC) Hamilton 24Aug58 mnt – 452 TCWG (AFR) Long Beach 13Jun59; to March 01Oct60; 943 TCG (452 TCWG) (AFR) March 11Feb63; 942 TCG (452 TCWG) (AFR) March 31Mar65 – 924 TCG (446 TCWG) (AFR) Ellington 12Aug65; depl 915 TCG (435 TCWG) (AFR) Homestead 29Nov65; 925 TAL GP (446 TCWG) (AFR) Ellington 06Dec67 – 927 TAL GP (403 TCWG) (AFR) Selfridge 26Jan68 – MAP Taiwan 06Mar69 – Final Disposition: Unknown.

KMC-164 / 51-8161 / C-119F-KM

Avble: 29Jul53; Acc: 19Aug53; Del: 21Aug53 to USAF – 313 TCWG (TAC) Mitchel 20Aug53; to Sewart 02Oct53; depl Sumpter Alabama 19Jul54; bailment Fairchild (AMC) Hagerstown 31Jan55, cvtd **C-119G** 25Feb55; depl 464 TCWG (TAC) Pope 17Mar55 – 464 TCWG (TAC) Sewart (depl) 08Jun55 – 314 TCWG (TAC) Sewart 09Aug55; depl Dreux AB France 21Oct55 – 2469 RFC (CNR) Scott 14Sep57 – 2466 RFC DT (CNR) Scott 01Dec57 – 2242 RFC (CNR) Selfridge 01Apr58 – 2347 RFC (CNR) Long Beach 27Jun58 – 452 TCWG (AFR) Long Beach 13Jun59; to March 01Oct60; 942 TCG (452 TCWG) (AFR) March 11Feb63 – 924 TCG (446 TCWG) (AFR) Ellington 02Aug65; 925 TCG (446 TCWG) (AFR) Ellington 11Sep65; 924 TAL GP (446 TCWG) (AFR) Ellington 14Jan68 – 908 TAL GP (435 TCWG) (AFR) Brookley 25May68 – MAP Taiwan 03Feb69 – Final Disposition: Unknown.

KMC-165 / 51-8162 / C-119F-KM

Avble: 07Aug53; Acc: 26Aug53; Del: 27Aug53 to USAF – 313 TCWG (TAC) Mitchel 26Aug53; to Sewart 02Oct53; depl Pope 19Apr54; depl Sumpter Alabama 19Jul54; bailment Fairchild (AMC) Hagerstown 07Dec54, cvtd **C-119G** 03Jan55 – 464 TCWG (TAC) Sewart (depl) 08Jun55 (listed as 463 TCWG but likely a typo) – 314 TCWG (TAC) Sewart 17Aug55; to 1608 FDM SQ (MATS) Charleston 01Nov55 mnt – BMC (AMC) Birmingham 26Feb57 – 2471 RFC (CNR) O'Hare 20May57 – 2466 RFC (CNR) Bakalar 14Jun57 – 2471 RFC (CNR) O'Hare 16Jul57 – 2466 RFC (CNR) Bakalar 21Jul57 – 2471 RFC (CNR) O'Hare 28Jul57 – 2473 RFC DT (CNR) O'Hare 01Dec57 – 2242 RFC DT (CNR) O'Hare 01Apr58 – 64 TCSq

(403 TCWG) (AFR) O'Hare 20Dec58 **RECORDS INDECIPHERABLE** 928 TCG (403 TCWG) (AFR) O'Hare 11Feb63 – Wfu to MASDC (LOG) Davis-Monthan 23Apr69 for storage, PCN: **CJ301** – DFI: 09Dec74 – Final Disposition: Scrapped.

KMC-166 / 51-8163 / C-119F-KM

Avble: 15Jul53; Acc: 30Jul53; Del: 05Aug53 to USAF – 456 TCWG (TAC) Miami 05Aug53; to Charleston 15Aug53. Lost control due to an engine failure 15.6 miles W Fort Bragg North Carolina 17Nov53. The C-119 was part of a 12 ship formation dropping paratroops over Holland Drop Zone near Fort Bragg when the incident occurred. Due to the engine loss, the aircraft slipped out of formation and collided with another C-119F (51–8122), this sending the aircraft plummeting into the massed paratroops still descending on their chutes. 4 crew and 1 passenger (a doctor from Fort Bragg) were killed when the C-119 impacted the ground, plus 10 paratroops who had just bailed out of the C-119 formation. C-119F 51-8122 made an emergency landing – DFI: 22Dec53 – Final Disposition: Accident.

KMC-167 / 51-8164 / C-119F-KM

Avble: 13Jul53; Acc: 31Jul53; Del: 07Aug53 to USAF – 456 TCWG (TAC) Miami 08Aug53; to Charleston 15Aug53; to 803 AB GP (SAC) Davis-Monthan 02Jul54 mnt; to Shiroi AB Japan 10Nov55; to Ardmore 05May56 – OGDAR (AMC) Hill 10Sep56 – BMC (AMC) Birmingham 02Nov56, cvtd **C-119G** 23May57 – 3700 MIT WG (TC) Lackland 13Jun57 – 183 AML SQ (MS-ANG) Hawkins cvtd **C-119J** 15Nov57; cvtd **MC-119J** 15Aug61 – 147 AML SQ (ANG) Greater Pittsburgh 10May62; cvtd **C-119J** 28Feb63 – Wfu to 2704 ASD GP (LOG) Davis-Monthan 09Mar64 for storage – REC: 27Jul64 – Final Disposition: Scrapped.

KMC-168 / 51-8165 / C-119F-KM

Avble: 22Jul53; Acc: 07Aug53; Del: 08Aug53 to USAF – 456 TCWG (TAC) Miami 12Aug53; to Charleston 15Aug53. Suffered a catastrophic engine fire after take-off from Charleston AFB South Carolina 23Aug55. While attempting to bring the aircraft around for a landing a wingtip clipped a tree causing the plane to crash into a residential suburb 1.7 miles from the base with the area erupting into flames from the gasoline. 5 of the 11 onboard the C-119 were killed with the 6 survivors being passengers in the cargo hold. 4 civilians in the Liberty Park housing area were killed with another 5 injured – Final Disposition: Accident.

KMC-169 / 51-8166 / C-119F-KM

Avble: 27Jul53; Acc: 18Aug53; Del: 25Aug53 to USAF – 313 TCWG (TAC) Mitchel 25Aug53; to Sewart 02Oct53; depl Pope 20Apr54; depl Sumpter Alabama 19Jul54 – 63 TCWG (TAC) Donaldson 28Sep54 – 366 Fighter WG (TAC) England 05Aug55; to Fairchild (AMC) Hagerstown 02Apr56, cvtd **C-119G** 30Apr56 – 834 AB GP (TAC) England 25Sep57 – 506 Fighter WG (TAC) Tinker 04Apr58; to SPW Center (ARD) Kirtland 01Jul58 mnt – 4520 CCT WG (TAC) Nellis 11Dec58 – 512 TCWG (AFR) NAS Willow Grove 14Oct59; 912 TCG (512 TCWG) (AFR) NAS Willow Grove 11Feb63; depl 913 TCG (512 TCWG) (AFR) NAS Willow Grove 07Sep66 – 927 TAL GP (403 TCWG) (AFR) Selfridge 19Aug68 – MAP Taiwan 06Mar69 – Final Disposition: Unknown.

KMC-170 / 51-8167 / C-119F-KM

Avble: 31Jul53; Acc: 25Aug53; Del: 26Aug53 to USAF – 313 TCWG (TAC) Mitchel 26Aug53; to Sewart 02Oct53 – 6520 Test WG (ARD) Hanscom 11Jun54 – CAR Center (ARD) Hanscom 01Jan55 – 456 TCWG (TAC) Charleston 09Mar55; to Shiroi AB Japan

10Nov55; to Ardmore 05May56 – OGDAR (AMC) Hill 10Sep56 – BMC (AMC) Birmingham 05Nov56; cvtd **C-119G** 07Jun57 – 3700 MIT WG (TC) Lackland 27Jun57 – 3499 FTG WG (TC) Chanute cvtd **C-119J** 27Nov57 – 140 AML SQ (PA-ANG) Gen Spaatz 10Apr58; to Olmsted 01Apr61; cvtd **MC-119J** 01Jun61 – 147 AML SQ (ANG) Greater Pittsburgh cvtd **C-119J** 14Dec62 – Wfu to 2704 ASD GP (LOG) Davis-Monthan 29Oct63 for storage – REC: 27Jul64 – Final Disposition: Scrapped.

KMC-171 / 51-8168 / C-119F-KM

Avble: 05Aug53; Acc: 25Aug53; Del: 26Aug53 to USAF – 313 TCWG (TAC) Mitchel 26Aug53; to Sewart 02Oct53; depl Sumpter Alabama 19Jul54; bailment Fairchild (AMC) Hagerstown 24Jan55, cvtd **C-119G** 18Feb55 – 464 TCWG (TAC) Sewart (depl) 08Jun55 – 314 TCWG (TAC) Sewart 11Aug55 – Hayes (AMC) Birmingham 21Aug57 – 2466 RFC (CNR) Bakalar 06Nov57 – 434 TCWG (AFR) Bakalar 19Apr59; 930 TCG (434 TCWG) (AFR) Bakalar 11Feb63; depl 71 ACO SQ (1 ACO WG) (TAC) Lockbourne 21Jun68; depl 931 TAL GP (434 TCWG) (AFR) Bakalar 28Sep68 – Wfu to MASDC (LOG) Davis-Monthan 01Dec69 for storage, PCN: **CJ331** – DFI: 20Jan70 – Kolar, Inc. Tucson Arizona 19Feb76 for scrap – Final Disposition: Scrapped.

KF-172 / 53-8069 / C-119G-FA

Avble: 04Mar54; Acc: 26Mar54; Del: 01Apr54 to USAF – 464 TCWG (TAC) Lawson 01Apr54; depl Seymour 15Apr54; to Pope 13Sep54; depl Rhein-Main AB W. Germany 19Oct54; depl Aberdeen 05May56 – 2585 RFC DT (CNR) Donaldson 10Nov58 – 77 TCSq (435 TCWG) (AFR) Donaldson 19Dec58 – 434 TCWG (AFR) Bakalar 11Jan61; 930 TCG (434 TCWG) (AFR) Bakalar 11Feb63 – Fairchild (LOG) St. Augustine 28Dec67, cvtd **AC-119G Shadow** 12Sep68 – 71 SOP SQ (1 ACO WG) (TAC) Lockbourne 29Nov68 – 14 SOP WG (PAF) Nha Trang AB S. Vietnam 06Dec68; assigned variously Tan Son Nhut, Phan Rang, Tuy Hoa S. Vietnam from 26Jan69; to 18 Fighter WG (PAF) Kadena AB Japan 16Oct69; to China Airlines (LOG) Taipei Taiwan 19Feb71 mnt – MAP S. Vietnam 20Aug71, VNAF s/n: **538069**, 819 ASQ (53 TWG) Tan Son Nhut AB – Final Disposition: Unknown.

KF-173 / 53-8070 / C-119G-FA

Avble: 04Mar54; Acc: 30Mar54; Del: 02Apr54 to USAF – 3510 FTN WG (TC) Randolph 02Apr54 – **RECORDS INDECIPHERABLE** → 3499 MBA WG (TC) Chanute 10Feb56 – 2585 RFC DT (CNR) Orlando 57 ← **RECORDS INDECIPHERABLE** 2473 RFC (CNR) Gen Mitchell 25Oct57 – 440 TCWG (AFR) Gen Mitchell 19Dec58 – 4440 ADE GP (TAC) Langley 15Jun60 – MAP India 30Jun60 – Final Disposition: Unknown.

KF-174 / 53-8071 / C-119G-FA

Avble: 04Mar54; Acc: 26Mar54; Del: 01Apr54 to USAF – 464 TCWG (TAC) Lawson 01Apr54; depl Seymour 22Apr54; to Pope 13Sep54; depl Rhein-Main AB W. Germany 19Oct54; depl Aberdeen 05May56 – 2346 RFC (CNR) Hamilton 29Jul58 – 349 TCWG (AFR) Hamilton 23Dec58; 938 TCG (349 TCWG) (AFR) Hamilton 10Feb63 – 907 TCG (302 TCWG) (AFR) Clinton 18Apr66; to Lockbourne 24Jul71 – Wfu to MASDC (LOG) Davis-Monthan 10Jun72 for storage – 907 TAL GP (302 TCWG) (AFR) Lockbourne 24Jun72 – Wfu to MASDC (LOG) Davis-Monthan 28Dec72 for storage, PCN: **CJ439** – DFI: 31Mar76 – noted bku Southwestern Alloys Corp. Tucson Arizona 2010 – Final Disposition: Scrapped.

KF-175 / 53-8072 / C-119G-FA

Avble: 08Mar54; Acc: 29Mar54; Del: 02Apr54 to USAF – 3510 FTN WG (TC) Randolph 02Apr54 – 312 Fighter-Bomber WG (TAC) Clovis 17Apr56 **RECORDS INDECIPHERABLE** → to unit unk (SAC) 57 mnt ← **RECORDS INDECIPHERABLE** 832 AB GP (TAC) Cannon 08Oct57 – 2347 RFC (CNR) Long Beach 29Jan58; to DT Hill 07Mar58 – 733 TCSq (452 TCWG) (AFR) Hill 03Apr59. Crashed near Hill AFB Utah 20Jan62. The C-119 took-off in the morning to practice instrument approaches using the runway at Hill AFB for touch-and-go passes. After one of these passes the aircraft was seen to climb out then veer to the left at about 600 ft and impact the ground near the base, all 5 crew onboard were killed – Final Disposition: Accident.

KF-176 / 53-8073 / C-119G-FA

Avble: 10Mar54; Acc: 31Mar54; Del: 02Apr54 to USAF – 3510 FTN WG (TC) Randolph 02Apr54 – 3499 MBA WG (TC) Chanute 10Feb56; to 3550 CCT WG (TC) Moody 06Aug56 **RECORDS INDECIPHERABLE** → Hayes (AMC) Birmingham 57 ← **RECORDS INDECIPHERABLE** 1739 FRY SQ (MATS) Amarillo 10Aug57 – 2585 RFC (CNR) Miami 27Aug57 – 435 TCWG (AFR) Miami 19Dec58; depl 512 TCWG (AFR) NAS Willow Grove 27Feb59; to 14 AFH HQ (CNC) Robins 03May59; to Homestead 25Jul60; assigned 76 TCSq Homestead 01Oct61; 915 TCG (435 TCWG) (AFR) Homestead 17Jan63 – 906 TCG (302 TCWG) (AFR) Clinton 20Nov65 – 130 SOP GP (ANG) Kanawha 02Jan69; to General Dynamics (LOG) Kanawha 12Apr73, cvtd **C-119L** 23Apr73 – Wfu to MASDC (LOG) Davis-Monthan 27Sep75 for storage, PCN: **CJ471** – DFI: 31Mar76 – BofS Starbird, Inc. Seattle Washington 13Nov78 reg **N9027K**; electrical upgrade 30Aug81 – Alaska Commercial Fishing and Agriculture Bank Anchorage Alaska 31Aug84, aircraft repossessed from Starbird, Inc.; ARAp 01Feb85 – BofS Northern Pacific Transport, Inc. Anchorage Alaska 09Dec85; lease to International Seafoods of Alaska, Inc. Jun86; lease to Arndt Brothers, Inc. Aug86 – BofS Alaska Aircraft Leasing, Inc. Anchorage Alaska 01Apr87; avionics upgrade 29Jun87 – The First National Bank of Anchorage 27Apr92 as a repossession from Alaska Aircraft Leasing – BofS Anchorage Flight, Inc. Anchorage Alaska 27Aug92 – BofS Classic Air Transport, Inc. Ridgewood New Jersey 01Aug93 – BofS Brooks Fuel Fairbanks Alaska 12Sep96 – BofS Alaska Aviation Museum Anchorage Alaska 15Nov17 for static display – Final Disposition: Extant.

KF-177 / 53-8074 / C-119G-FA

Avble: 10Mar54; Acc: 31Mar54; Del: 02Apr54 to USAF – 3510 FTN WG (TC) Randolph 02Apr54 – 3499 MBA WG (TC) Chanute 13Mar56 **RECORDS INDECIPHERABLE** → unit unk (CNR) 57 ← **RECORDS INDECIPHERABLE** 2473 RFC (CNR) Gen Mitchell 16Oct57 – 440 TCWG (AFR) Gen Mitchell 19Dec58; 933 TCG (440 TCWG) (AFR) Gen Mitchell 11Feb63 – 129 SOP GP (ANG) Hayward 27Dec68; cvtd **C-119L** 24May73 – Wfu to MASDC (LOG) Davis-Monthan 27Mar75 for storage, PCN: **CJ457** – DFI: 15Jul77 – Air Pro / Western Intl. Scrapyard Tucson 1990 for scrap – Howard E. Jenkins (Flying J Ranch Airpark) Pima Arizona 13Mar90. Displayed at Flying J Ranch Airport – Final Disposition: Extant.

KF-178 / 53-8075 / C-119G-FA

Avble: 25Mar54; Acc: 20Apr54; Del: 26Apr54 to USAF – 3510 FTN WG (TC) Randolph 26Apr54; to 2750 AB WG (AMC) WPAFB 24Jul54 mnt – 3499 MBT WG (TC) Chanute 04Feb56 **RECORDS INDECIPHERABLE** → 2472 RFC (CNR) Richards 57 ← **RECORDS INDECIPHERABLE** 442 TCWG (AFR) Richards 02Mar59; to 1 MSU GP

(ADC) Selfridge 05Mar59 mnt – 440 TCWG (AFR) Gen Mitchell 15Apr61; 933 TCG (440 TCWG) (AFR) Gen Mitchell 11Feb63 – Wfu to MASDC (LOG) Davis-Monthan 27Nov70 for storage, PCN: **CJ371** – DFI: 21May71 – Kolar, Inc. Tucson Arizona 09Mar76 for scrap – Final Disposition: Scrapped.

KF-179 / 53-8076 / C-119G-FA

Avble: 29Mar54; Acc: 14Apr54; Del: 19Apr54 to USAF – 3510 FTN WG (TC) Randolph 19Apr54 – 3499 MBT WG (TC) Chanute 15Feb56 **RECORDS INDECIPHERABLE** → 2472 RFC (CNR) Richards 57 ← **RECORDS INDECIPHERABLE** to 328 CLM SQ (ADC) Richards 13Feb58 mnt – 442 TCWG (AFR) Richards 02Mar59 – 435 TCWG (AFR) Homestead 18Apr61; assigned 76 TCSq Homestead 01Oct61; 915 TCG (435 TCWG) (AFR) Homestead 17Jan63 – 926 TCG (446 TCWG) (AFR) NAS New Orleans 08Nov65 – 129 SOP GP (ANG) Hayward 15Feb69; cvtd **C-119L** 17May73 – Wfu to MASDC (LOG) Davis-Monthan 01Jan75 for storage, PCN: **CJ456** – DFI: 31Mar76 – BofS Dross Metals, Inc. Tucson Arizona 29Nov79 – BofS J.D. Gifford & Associates, Inc. Anchorage Alaska 28Sep80; ARAp 13Jan81 reg **N8505A**; major radio, nav, flight instrument upgrade, control surfaces recovered and equipment removal to lighten airframe 15Apr82; horizontal stabilizer repair 21Mar83 – Alaska Commercial Fishing and Agriculture Bank Anchorage Alaska 31Aug84 through repossession from J.D. Gifford; ARAp 01Feb85 – BofS Northern Pacific Transport, Inc. Anchorage Alaska 09Dec85; lease to International Seafoods, Inc. 27Jun86; lease to Tate & Co. 09Sep86 – BofS Alaska Aircraft Leasing, Inc. Anchorage Alaska 01Apr87; nav upgrade 29Jun87; further lease to International Seafoods, Inc. 1990 – The First National Bank of Alaska Anchorage Alaska 27Apr92 through repossession from Alaska Aircraft Leasing; ARAp 12Aug92 – BofS Hawkins & Powers Aviation, Inc. Greybull Wyoming 12Aug92, stored – BofS D&G, Inc. Greybull Wyoming 01Jan93 – BofS The Pride Capital Group, LLC Deerfield Illinois 07Sep05 – BofS Harold Sheppard, Jr., d/b/a Sheppard Trucking Riverton Wyoming 2006; ARAp 19Feb07. Presently derelict at Greybull Wyoming along with several other C-119 and piston-engined aircraft – Final Disposition: Extant.

KF-180 / 53-8077 / C-119G-FA

Avble: 01Apr54; Acc: 19Apr54; Del: 26Apr54 to USAF – 3510 FTN WG (TC) Randolph 26Apr54 – 479 Fighter WG (TAC) George 03Apr56; to 810 AB GP (SAC) Biggs 14Apr56 mnt; 831 AB GP (TAC) George 08Oct57; to 325 Fighter GP (ADC) McChord 28Jan58 mnt; to unit unk (TAC) Langley 58 – 94 TCWG (AFR) Hanscom 04Aug59; 901 TCG (94 TCWG) (AFR) Hanscom 31May63 – Fairchild (LOG) St. Augustine 21Nov66 – 910 TCG (459 TCWG) (AFR) Youngstown 29Dec66 – 934 TAL GP (440 TCWG) (AFR) Minneapolis-St. Paul 13Sep69 – 906 TAL GP (302 TCWG) (AFR) Clinton 15Nov69; to Lockbourne 24Jul71; 907 TAL GP (302 TCWG) (AFR) Lockbourne 01Mar72 – MAP S. Vietnam 31Jul72, VNAF s/n: **538077**, 413 TSQ (53 TWG) Tan Son Nhut AB – Final Disposition: Unknown.

KF-181 / 53-8078 / C-119G-FA

Avble: 06Apr54; Acc: 20Apr54; Del: 26Apr54 to USAF – 3510 FTN WG (TC) Randolph 26Apr54 – 3499 MBT WG (TC) Chanute 15Feb56 **RECORDS INDECIPHERABLE** → 2473 RFC (CNR) Gen Mitchell 57 ← **RECORDS INDECIPHERABLE** 440 TCWG (AFR) Gen Mitchell 19Dec58; to 1604 AB WG (MATS) Kindley 18Mar59 mnt; to 5040 CLM GP (AAC) Elmendorf 10Feb61; 933 TCG (440 TCWG) (AFR) Gen Mitchell 11Feb63 – Fairchild (LOG) St. Augustine 03Jun63 – MAP India 10Jul63 – Final Disposition: Unknown.

KF-182 / 53-8079 / C-119G-FA

Avble: 08Apr54; Acc: 20Apr54; Del: 26Apr54 to USAF – 3510 FTN WG (TC) Randolph 26Apr54 – 3499 MBA WG (TC) Chanute 27Feb56 **RECORDS INDECIPHERABLE** → 2585 RFC DT (CNR) Orlando 57 ← **RECORDS INDECIPHERABLE** Hayes (AMC) Birmingham 26Aug57 – 2466 RFC (CNR) Bakalar 12Nov57 – 2473 RFC (CNR) Gen Mitchell 15Nov57 – AFT Center (ARD) Edwards 25Nov58 – 440 TCWG (AFR) Gen Mitchell 15Jan59 – 4440 ADE GP (TAC) Langley 28Jun60 – MAP India 10Jul60 – Final Disposition: Unknown.

KF-183 / 53-8080 / C-119G-FA

Avble: 14Apr54; Acc: 28Apr54; Del: 04May54 to USAF – 3510 FTN WG (TC) Randolph 04May54; depl 3550 FTN WG (TC) Moody 04Jun54; depl 3535 AOT WG (TC) Mather 16Oct55 – 312 Fighter-Bomber WG (TAC) Clovis 04May56 – 832 AB GP (TAC) Cannon 08Oct57 – 2234 RFC (CNR) Hanscom 22Nov57 – Hayes (AMC) Birmingham 07Jan59 – 94 TCWG (AFR) Hanscom 31Mar59; 901 TCG (94 TCWG) (AFR) Hanscom 31May63 – 907 TCG (302 TCWG) (AFR) Clinton 23Jul66; to Lockbourne 24Jul71 – MAP S. Vietnam 08Sep71, VNAF s/n: **538080**, 413 TSQ (53 TWG) Tan Son Nhut AB – VPAF Apr75, 918 Transport Regiment – Final Disposition: Unknown.

KF-184 / 53-8081 / C-119G-FA

Avble: 19Apr54; Acc: 30Apr54; Del: 08May54 to USAF – 3510 FTN WG (TC) Randolph 07May54 – 3415 TTA WG (TC) Lowry 23Feb56 – 3499 MBT WG (TC) Chanute 27Sep56 **RECORDS INDECIPHERABLE** → to unit unk (TC) Chanute 57; 2472 RFC (CNR) Richards 57 ← **RECORDS INDECIPHERABLE** 442 TCWG (AFR) Richards 02Mar59; to 2578 AB GP (CNC) Ellington 01Jul59 – 4440 ADE GP (TAC) Langley 15Jun60 – MAP India 30Jun60 – Final Disposition: Unknown.

KF-185 / 53-8082 / C-119G-FA

Avble: 23Apr54; Acc: 13May54; Del: 20May54 to USAF – 3510 FTN WG (TC) Randolph 20May54 – 3415 TTA WG (TC) Lowry 15Feb56 **RECORDS INDECIPHERABLE** → Hayes (AMC) Birmingham 56 – 3499 BMT WG (TC) Chanute 56 – 2473 RFC (CNR) Gen Mitchell 57 ← **RECORDS INDECIPHERABLE** 440 TCWG (AFR) Gen Mitchell 29Dec58; assigned 96 TCSq Minneapolis-St. Paul 14Jun60 – 4440 ADE GP (TAC) Langley 28Jun60 – MAP India 10Jul60 – Final Disposition: Unknown.

KF-186 / 53-8083 / C-119G-FA

Avble: 28Apr54; Acc: 17May54; Del: 20May54 to USAF – 3510 FTN WG (TC) Randolph 20May54; to 3800 AB WG (AU) Maxwell 10Jul54 mnt; to 3600 CCT WG (TC) Luke 04Aug55 – 3415 TTA WG (TC) Lowry 04Feb56; to 2750 AB WG (AMC) WPAFB 27Apr56 mnt **RECORDS INDECIPHERABLE** → 3700 MIT WG (TC) Lackland 56 – 2585 RFC (CNR) Miami 57 ← **RECORDS INDECIPHERABLE** 435 TCWG (AFR) Miami 19Dec58; depl 512 TCWG (AFR) NAS Willow Grove 27Feb59; to Homestead 25Jul60; assigned 76 TCSq Homestead 01Oct61; 915 TCG (435 TCWG) (AFR) Homestead 17Jan63 – 906 TCG (302 TCWG) (AFR) Clinton 26Nov65 – 130 SOP GP (ANG) Kanawha 16Jan69; to General Dynamics (LOG) Kanawha 19Jan73, cvtd **C-119L** 22Feb73 – Wfu to MASDC (LOG) Davis-Monthan 27Aug75 for storage, PCN: **CJ469** – DFI: 15Jul77 – Final Disposition: Scrapped.

KF-187 / 53-8084 / C-119G-FA

Avble: 29Apr54; Acc: 28May54; Del: 08Jun54 to USAF – 3510 FTN WG (TC) Randolph 08Jun54 – 4405 OP SQ (TAC) Langley 12Mar56; to 3310 TTA WG (TC) Scott 18May56 – 836 AB GP (TAC) Langley 08Oct57 – 27 AB GP (TAC) Bergstrom 14May58 – 4510 CCT WG (TAC) Luke 16Feb59 – 512 TCWG (AFR) NAS Willow Grove 09Jun59; 912 TCG (512 TCWG) (AFR) NAS Willow Grove 11Feb63; 913 TAL GP (512 TCWG) (AFR) NAS Willow Grove 06May68 – 130 SOP GP (ANG) Kanawha 28Jan69; to General Dynamics (LOG) Kanawha 13Feb73, cvtd **C-119L** 26Feb73 – Wfu to 314 TAL WG (MAC) Little Rock 26Sep75; assigned museum / school status 03Nov75. Presently displayed at Little Rock AFB Arkansas – Final Disposition: Extant.

KF-188 / 53-8085 / C-119G-FA

Avble: 04May54; Acc: 24May54; Del: 02Jun54 to USAF – 3510 FTN WG (TC) Randolph 02Jun54 – 3700 MIT WG (TC) Lackland 08Feb56 – 2585 RFC (CNR) Miami 57; to 1501 TSH WG (MATS) Travis 02Dec58 mnt – 435 TCWG (AFR) Miami 19Dec58; to Homestead 25Jul60; assigned 78 TCSq Bates 10Nov60 – 357 TCSq (302 TCWG) (AFR) Bates 08May61; 908 TCG (302 TCWG) (AFR) Bates 11Feb63 – 908 TCG (435 TCWG) (AFR) Bates 18Mar63 – Fairchild (LOG) St. Augustine 15Mar66 – MAP Taiwan 21Apr66, RoCAF s/n: **3133** – Final Disposition: Unknown.

KF-189 / 53-8086 / C-119G-FA

Avble: 06May54; Acc: 26May54; Del: 02Jun54 to USAF – 3510 FTN WG (TC) Randolph 02Jun54 – 3700 MIT WG (TC) Lackland 21Feb56 **RECORDS INDECIPHERABLE** → 2235 RFC (CNR) Grenier 57 ← **RECORDS INDECIPHERABLE** 732 TCSq (94 TCWG) (AFR) Grenier 19Mar59 – Fairchild (LOG) St. Augustine 06May63 – MAP India 27Jun63 – Final Disposition: Unknown.

KF-190 / 53-8087 / C-119G-FA

Avble: 11May54; Acc: 28May54; Del: 08Jun54 to USAF – 3510 FTN WG (TC) Randolph 08Jun54 – 483 TCWG (FEA) Ashiya AB Japan 23Mar56 – 2347 RFC (CNR) Long Beach 14Jul58 – 2346 RFC DT (CNR) McClellan 23Jul58 – 314 TCSq (349 TCWG) (AFR) McClellan 23Dec58; to 3615 PTN WG (ATC) Craig 03Jun59 mnt; 940 TCG (349 TCWG) (AFR) McClellan 10Feb63 – 904 TCG (514 TCWG) (AFR) Stewart 13May65; 903 TCG (514 TCWG) (AFR) McGuire 03Jun66; depl Myrtle Beach 16Jul67 – 914 TAL GP (512 TCWG) (AFR) Niagara Falls 07Jun68 – 130 SOP GP (ANG) Kanawha 23Jan69; to General Dynamics (LOG) Kanawha 02Mar73, cvtd **C-119L** 22Mar73 – Wfu 16Jul74 to museum / school status. Presently displayed at the 82 Airborne Division War Memorial Museum Fort Bragg North Carolina – Final Disposition: Extant.

KF-191 / 53-8088 / C-119G-FA

Avble: 13May54; Acc: 27May54; Del: 08Jun54 to USAF – 3510 FTN WG (TC) Randolph 08Jun54 – Bailment ALDIN (AMC) Indianapolis Indiana 13Apr56 – 2466 RFC (CNR) Bakalar 07Jun57 – 434 TCWG (AFR) Bakalar 19Apr59; 930 TCG (434 TCWG) (AFR) 11Feb63 – Fairchild (LOG) St. Augustine 20Nov67 – MAP S. Vietnam 24Mar68, VNAF s/n: **538088**, 413 TSQ (53 TWG) Tan Son Nhut AB – 405 Fighter WG (PAF) Clark AB The Philippines 10Jan73 for disposal – DFI: 05Jun73 – Final Disposition: Scrapped.

KF-192 / 53-8089 / C-119G-FA

Avble: 19May54; Acc: 10Jun54; Del: 18Jun54 to USAF – 3510 FTN WG (TC) Randolph 18Jun54; depl 483 TCWG (FEA) Ashiya AB Japan 26Mar56 – 483 TCWG (FEA) Ashiya AB Japan 14Jun56 – 2346 RFC DT (CNR) Portland 17Apr58 – 313 TCSq (349 TCWG) (AFR) Portland 23Dec58; 314 TCSq (349 TCWG) (AFR) McClellan 25Nov59; 940 TCG (349 TCWG) (AFR) McClellan 10Feb63 – 904 TCG (514 TCWG) (AFR) Stewart 13May65; 903 TCG (514 TCWG) (AFR) McGuire 31May67; depl Myrtle Beach 16Jul67 – Fairchild (LOG) St. Augustine 20Feb68, cvtd **AC-119G Shadow** 03Jun68 – 4413 CCT SQ (1 ACO WG) (TAC) Eglin 09Jun68; to Lockbourne 03Jul68 – 1 SOP WG (TAC) Lockbourne 18Feb69 – 4410 CCT WG (TAC) Lockbourne 14Jul69 – 14 SOP WG (PAF) Phan Rang AB S. Vietnam 01Feb71; to 43 STR WG (SAC) Andersen Guam 05Mar71; to 313 ADH DV (PAF) Kadena AB Japan 28Mar71; assigned variously Tan Son Nhut S. Vietnam from 02Aug71 – MAP S. Vietnam 28Aug71, VNAF s/n: **538089**, 819 ASQ (53 TWG) Tan Son Nhut AB – Final Disposition: Unknown.

KF-193 / 53-8090 / C-119G-FA

Avble: 24May54; Acc: 10Jun54; Del: 18Jun54 to USAF – 3510 FTN WG (TC) Randolph 18Jun54; to 63 TCWG (TAC) Donaldson 25Feb56 mnt – 363 TR WG (TAC) Shaw 29May56 – Hayes (AMC) Birmingham 26Sep57 – 2347 RFC (CNR) Long Beach 10Jan58; to DT Hill 25Jan58; to 325 MSU GP (ADC) McChord 10Sep58 mnt – 733 TCSq (452 TCWG) (AFR) Hill 03Apr59; 945 TCG (452 TCWG) (AFR) Hill 17Jan63 – MAP Taiwan 23Apr66 – Final Disposition: Unknown.

KF-194 / 53-8091 / C-119G-FA

Avble: 25May54; Acc: 11Jun54; Del: 19Jun54 to USAF – 3510 FTN WG (TC) Randolph 18Jun54 – 3700 MIT WG (TC) Lackland 06Feb56; depl Sheppard 18Sep56 – 2472 RFC (CNR) Richards 17Jun57 – 442 TCWG (AFR) Richards 02Mar59 – 328 TCSq (512 TCWG) (AFR) Niagara Falls 30Apr61; 914 TCG (512 TCWG) (AFR) Niagara Falls 11Feb63 – MAP Taiwan 20Apr66 – Final Disposition: Unknown.

KF-195 / 53-8092 / C-119G-FA

Avble: 01Jun54; Acc: 17Jun54; Del: 25Jun54 to USAF – 3510 TFN WG (TC) Randolph 25Jun54 – 479 Fighter WG (TAC) George 29Mar56; depl 366 Fighter WG (TAC) England 11Apr56 – 831 AB GP (TAC) George 08Oct57 – 512 TCWG (AFR) NAS Willow Grove 23Sep59 – 4440 ADE GP (TAC) Langley 27Jun60 – MAP India 10Jul60 – Final Disposition: Unknown.

KF-196 / 53-8093 / C-119G-50-FA

Avble: 03Jun54; Acc: 21Jun54; Del: 25Jun54 to USAF – 3510 FTN WG (TC) Randolph 24Jun54; film appearance *Battle Hymn* (1957) filmed Apr56 – 312 Fighter-Bomber WG (TAC) Clovis 11May56 – 832 AB GP (TAC) Cannon 08Oct57 – Hayes (AMC) Birmingham 18Oct57 – 2347 RFC (CNR) Long Beach 29Jan58; to DT Hill 07Feb58 – 733 TCSq (452 TCWG) (AFR) Hill 03Apr59; 945 TCG (452 TCWG) (AFR) Hill 17Jan63 – MAP Taiwan 21May66, RoCAF s/n: **3141** – Final Disposition: Unknown.

KF-197 / 53-8094 / C-119G-FA

Avble: 08Jun54; Acc: 25Jun54; Del: 02Jul54 to USAF – 3510 FTN WG (TC) Randolph 02Jul54 – 323 Fighter-Bomber WG (TAC) Bunker Hill 17Apr56 **RECORDS INDECIPHERABLE** → unit unk Langley 57 ← **RECORDS INDECIPHERABLE** 2230 RFC (CNR)

NAS New York 06Aug57 – 2234 RFC (CNR) Hanscom 05Oct57 – Hayes (AMC) Birmingham 19Dec58 – 94 TCWG (AFR) Hanscom 04Apr59 – 4440 ADE GP (TAC) Langley 15Jun60 – MAP India 26Jun60 – Final Disposition: Unknown.

KF-198 / 53-8095 / C-119G-FA

Avble: 11Jun54; Acc: 30Jun54; Del: 07Jul54 to USAF – 3510 FTN WG (TC) Randolph 07Jul54 – 323 Fighter-Bomber WG (TAC) Bunker Hill 24Apr56 – 2230 RFC (CNR) NAS New York 09Sep57 – 2234 RFC (CNR) Hanscom 09Oct57 – 94 TCWG (AFR) Hanscom 19Mar59; 901 TCG (94 TCWG) (AFR) Hanscom 31May63 – Fairchild (LOG) St. Augustine 06Sep66 – MAP Morocco 25Nov66, RMAF s/n: **CN-AML**, 1 AT SQ – Final Disposition: Unknown.

KF-199 / 53-8096 / C-119G-FA

Avble: 16Jun54; Acc: 28Jul54; Del: 04Aug54 to USAF – 1739 FRY SQ (MATS) Amarillo 04Aug54; to 3310 TTA WG (TC) Scott 02 Oct56 mnt; 3310 TTA WG briefly to Scituate Massachusetts 23Nov56; to 3750 TTWG (TC) Sheppard 14Dec56 mnt – 2347 RFC (CNR) Long Beach 29Jan58; to DT Hill 13Feb58 – 733 TCSq (452 TCWG) (AFR) Hill 03Apr59; 945 TCG (452 TCWG) (AFR) Hill 17Jan63 – MAP Taiwan 21May66 – Final Disposition: Unknown.

KF-200 / 53-8097 / C-119G-FA

Avble: 21Jun54; Acc: 28Jul54; Del: 04Aug54 to USAF – 1739 FRY SQ (MATS) Amarillo 04Aug54; to 1611 FDM SQ (MATS) McGuire 06Aug56 mnt – 2237 RFC (CNR) New Castle 19Feb58; to NAS Willow Grove 21Jul58 – 512 TCWG (AFR) NAS Willow Grove 01Dec58 – 4440 ADE GP (TAC) Langley 27Jun60 – MAP India 10Jul60 – Final Disposition: Unknown.

KF-201 / 53-8098 / C-119G-FA

Avble: 29Jun54; Acc: 28Jul54; Del: 02Aug54 to USAF – 313 TCG (TAC) Sumpter (depl) 02Aug54; to Sewart 19Aug54; depl Elmendorf 12Jan55 – 314 TCWG (TAC) Sewart 08Jun55; depl England Louisiana 05Nov56 – Hayes (AMC) Birmingham 57 – 2471 RFC (CNR) O'Hare 18Jul57; depl 2466 RFC (CNR) Bakalar 21Jul57; cvtd **C-119J** 22Jul57 – 2473 RFC DT (CNR) O'Hare 01Dec57 – 2242 RFC DT (CNR) O'Hare 01Apr58 – 64 TCSq (403 TCWG) (AFR) O'Hare 20Dec58 – 102 AML SQ (NY-ANG) New York ANGB 27Apr61 – 106 AML GP (ANG) New York ANGB 28Feb63 – Wfu to 2704 ASD GP (LOG) Davis-Monthan 04May63 for storage – Fairchild (LOG) St. Augustine 17Dec63 – MAP Italy 03Mar64, AMI s/n: **MM53-8098** – Final Disposition: Scrapped.

KF-202 / 53-8099 / C-119G-FA

Avble: 28Jun54; Acc: 29Jul54; Del: 02Aug54 to USAF – 313 TCG (TAC) Sumpter (depl) 02Aug54; to Sewart 19Aug54; depl Elmendorf 12Jan55 – 314 TCWG (TAC) Sewart 01Mar55; depl 313 TCWG (TAC) Sewart 01Jun55; depl 463 TCWG (TAC) Sewart 08Jun55; depl England Louisiana 05Nov56 – Hayes (AMC) Birmingham 13Aug57 – 2465 RFC (CNR) Minneapolis-St. Paul 31Oct57 – 2473 RFC DT (CNR) Minneapolis-St. Paul 01Dec57 – 96 TCSq (440 TCWG) (AFR) Minneapolis-St. Paul 19Dec59; 934 TCG (440 TCWG) (AFR) Minneapolis-St. Paul 11Feb63 – 906 TAL GP (302 TCWG) (AFR) Clinton 08Nov69; to Lockbourne 24Jul71; 907 TAL GP (302 TCWG) (AFR) Lockbourne 28Sep71 – Wfu to MASDC (LOG) Davis-Monthan 10Jun72 for storage – 907 TAL GP (302 TCWG) (AFR) Lockbourne 24Jun72; depl 906 TAL GP (302 TCWG) (AFR) Lockbourne 03Jul72 – Wfu to MASDC (LOG) Davis-Monthan 17Jan73 for storage – DFI: 09Dec74 – Final Disposition: Scrapped.

KF-203 / 53-8100 / C-119G-FA

Avble: 30Jun54; Acc: 29Jul54; Del: 02Aug54 to USAF – 313 TCG (TAC) Sumpter (depl) 02Aug54; to Sewart 19Aug54 – 314 TCWG (TAC) Sewart 01Mar55; depl 464 TCWG (TAC) Pope 09Feb56; depl England Louisiana 05Nov56 – Hayes (AMC) Birmingham 19Jul57 mnt – 2472 RFC (CNR) Richards 15Oct57 – 2347 RFC DT (CNR) Hill 14Jun58 – 2347 RFC (CNR) Long Beach 17Oct58 – SBNAR (AMC) Norton 06Feb59 mnt; declared excess 26Jun59 – REC: 30Jun59 – Final Disposition: Scrapped.

KF-204 / 53-8101 / C-119G-FA

Avble: 08Jul54; Acc: 30Jul54; Del: 10Aug54 to USAF – 313 TCG (TAC) Sumpter (depl) 10Aug54; to Sewart 19Aug54; depl Elmendorf 12Jan55 – 464 TCWG (TAC) Sewart 08Jun55 – 314 TCWG (TAC) Sewart 09Aug55 – 4501 SUT SQ (TAC) Donaldson 03Feb56 – Hayes (AMC) Birmingham 11Jul57 mnt – 2472 RFC (CNR) Richards 07Oct57; cvtd **C-119J** 31Jan59 – Hayes (AMC) Birmingham 16Mar59 mnt – 442 TCWG (AFR) Richards 16Mar59 – 137 AML SQ (NY-ANG) Westchester 04Mar61 – 102 AML SQ (NY-ANG) New York ANGB 15Mar61 – Wfu to 2704 ASD GP (LOG) Davis-Monthan 06Apr63 for storage – Fairchild (LOG) St. Augustine 15Dec66 – 78 Fighter WG (ADC) Hamilton 16Feb67 – 4676 AB GP (ADC) Richards 23Nov69 – 4650 COS GP (ADC) Richards 30Jun70; depl 460 Fighter SQ (ADC) Grand Forks 23Aug71 – Wfu to MASDC (LOG) Davis-Monthan 29Jun72 for storage, PCN: **CJ426** – DFI: 21Mar73 – Final Disposition: Scrapped.

KF-205 / 53-8102 / C-119G-FA

Avble: 09Jul54; Acc: 30Jul54; Del: 10Aug54 to USAF – 313 TCG (TAC) Sumpter (depl) 10Aug54; to Sewart 19Aug54. Struck Shin Hook Ridge about 5 miles SE New Hope Alabama 29Dec54 while enroute from Sewart AFB Tennessee to Brookley AFB Alabama. Killed in the crash were all 4 crew and 5 USAF staff passengers seated in the cargo hold: Capt Leslie D. Forguson (pilot); 1Lt Jay. B. Border; 2Lt Gerry M. Hall (passenger); 2Lt Charles Hawkins; 2Lt William T. Troy, Jr.; A1C Lawrence J. Foley; A2C Leon M. McKay (passenger); A2C Richard W. Miller; A3C Robert A. Shoemaker (passenger). Incredibly two passengers survived the crash when the clamshell doors were blown off and were sucked out the back: Airmen Robert Johnson and A3C Michael P. Kinane were later rescued – Final Disposition: Accident.

KF-206 / 53-8103 / C-119G-FA

Avble: 21Jul54; Acc: 20Aug54; Del: 27Aug54 to USAF – 464 TCWG (TAC) Lawson 27Aug54; to Pope 18Sep54; depl Aberdeen 05May56 – 461 BTA WG (TAC) Blytheville 08Nov56; cvtd **C-119J** 05Aug57 – 150 AML SQ (NJ-ANG) Newark 12Feb58; cvtd **MC-119J** 04Feb61 – 147 AML SQ (ANG) Greater Pittsburgh cvtd **C-119J** 16Nov62 – Wfu to MASDC (LOG) Davis-Monthan 30May63 for storage – Fairchild (LOG) St. Augustine 30Oct63 – MAP Italy 15Jan64, AMI s/n: **MM53-8103**; cvtd **VC-119J** – Final Disposition: Scrapped.

KF-207 / 53-8104 / C-119G-FA

Avble: 20Jul54; Acc: 31Aug54; Del: 29Sep54 to USAF – 317 TCWG (AFE) Neubiberg AB W. Germany 29Sep54; to 3110 MAI GP (AMC) RAF Burtonwood England 08May56 mnt; depl 465 TCWG (FEA) Évreux-Fauville AB France 12Apr57; to Évreux-Fauville AB France 08Jul57 – 2347 RFC (CNR) Long Beach 28Jan58; to DT Hill 18Oct58 – 733 TCSq (452 TCWG) (AFR) Hill 03Apr59; 945 TCG (452 TCWG) (AFR) Hill 17Jan63 – 907 TCG

(302 TCWG) (AFR) Clinton 03Jun66; to Lockbourne 24Jul71 – MAP S. Vietnam 03Jul72, VNAF s/n: **538104**, 413 TSQ (53 TWG) Tan Son Nhut AB – 405 Fighter WG (PAF) Clark AB The Philippines 11Mar73 for disposal – DFI: 05Jun73 – Final Disposition: Scrapped.

KF-208 / 53-8105 / C-119G-FA

Avble: 22Jul54; Acc: 31Aug54; Del: 28Sep54 to USAF – 317 TCWG (AFE) Neubiberg AB W. Germany 28Sep54; to 3110 MAI GP (AMC) RAF Burtonwood England 03Oct56 mnt; to Évreux-Fauville AB France 08Jul57 – 2347 RFC (CNR) Long Beach 23Jan58 – 2346 RFC DT (CNR) Portland 05Jul58 – 313 TCSq (349 TCWG) (AFR) Portland 22Dec58; 939 TCG (349 TCWG) (AFR) Portland 07May63; depl RCM SQ (CNC) Portland 13Jun63 – 943 TAL GP (452 TCWG) (AFR) March 24May68 – 922 TAL GP (433 TCWG) (AFR) Kelly 28Mar69 – 906 SOP GP (302 TCWG) (AFR) Clinton 06Nov70 – MAP Taiwan 27Nov70 – Final Disposition: Unknown.

KF-209 / 53-8106 / C-119G-FA

Avble: 28Jul54; Acc: 31Aug54; Del: 26Oct54 to USAF – 317 TCWG (AFE) Neubiberg AB W. Germany 26Oct54; to 3110 MAI GP (AMC) RAF Burtonwood England 08May56 mnt; to unit unk (AFE) 57; to Évreux-Fauville AB France 08Jul57 – 2347 RFC (CNR) Long Beach 01Feb58; to DT Hill 07Mar58 – Hayes (AMC) Birmingham 24Feb59 mnt – 733 TCSq (452 TCWG) (AFR) Hill 07Apr59; 945 TCG (452 TCWG) (AFR) Hill 17Jan63 – 907 TCG (302 TCWG) (AFR) Clinton 15Apr66; to Lockbourne 24Jul71 – MAP S. Vietnam 01Jul72, VNAF s/n: **538106**, 413 TSQ (53 TWG) Tan Son Nhut AB – Final Disposition: Unknown.

KF-210 / 53-8107 / C-119G-FA

Avble: 30Jul54; Acc: 31Aug54; Del: 20Oct54 to USAF – 317 TCWG (AFE) Neubiberg AB W. Germany 20Oct54; to 3110 MAI GP (AMC) RAF Burtonwood England 25Feb56 mnt; to Évreux-Fauville AB France 08Jul57 – 2347 RFC (CNR) Long Beach 28Jan58; to DT Hill 19Apr58. Had an engine failure enroute from Fort Bridger Wyoming to Hill AFB Utah 14Nov58. The 8 crew onboard bailed out over Wyoming but the C-119 on autopilot flew another 150 miles crashing near Swan Valley Idaho. 4 of the crew that bailed out were rescued, 2 were found dead and 2 remain missing – Final Disposition: Accident.

KF-211 / 53-8108 / C-119G-FA

Avble: 18Aug54; Acc: 31Aug54; Del: 20Oct54 to USAF – 317 TCWG (AFE) Neubiberg AB W. Germany 20Oct54; to 3110 MAI GP (AMC) RAF Burtonwood England 21Sep56 mnt; to Évreux-Fauville AB France 08Jul57 – 2242 RFC (CNR) Selfridge 31Mar58 – 403 TCWG (AFR) Selfridge 12Nov58; 927 TCG (403 TCWG) (AFR) Selfridge 10Feb63; 928 TAL GP (403 TCWG) (AFR) O'Hare 02Apr69 – Fairchild (LOG) Crestview 15Jan71 – MAP Ethiopia 06Aug71, ETAF s/n: **916** – Wfu 86. Noted 90s stored at Debre Zeyit AB Ethiopia – Final Disposition: Scrapped.

KF-212 / 53-8109 / C-119G-FA

Avble: 26Aug54; Acc: 31Aug54; Del: 05Oct54 to USAF – 317 TCWG (AFE) Neubiberg AB W. Germany 05Oct54; to 3110 MAI GP (AMC) RAF Burtonwood England 23Aug56 mnt; to Évreux-Fauville AB France 08Jul57 – 2234 RFC (CNR) Hanscom 21Feb58; to 4060 AB GP (SAC) Dow 28Mar58 mnt – 94 TCWG (AFR) Hanscom 19Mar59; 901 TCG (94 TCWG) (AFR) Hanscom 31May63 – 907 TCG (302 TCWG) (AFR) Clinton 16Jul66; to Lockbourne 24Jul71 – MAP S. Vietnam 08Jul72, VNAF s/n: **538109**, 413 TSQ (53 TWG) Tan Son Nhut AB – Final Disposition: Unknown.

KF-213 / 53-8110 / C-119G-FA

Avble: 30Aug54; Acc: 29Sep54; Del: 26Oct54 to USAF – 317 TCWG (AFE) Neubiberg AB W. Germany 26Oct54; to 3110 MAI GP (AMC) RAF Burtonwood England 16Jan56 mnt; depl 465 TCWG (AFE) Évreux-Fauville AB France 57; to Évreux-Fauville AB France 08Jul57 – 2347 RFC (CNR) Long Beach 14Feb58 – 2346 RFC (CNR) Hamilton 01Mar58 – 349 TCWG (AFR) Hamilton 23Dec58; to 831 AB GP (TAC) George 28Mar59 mnt; 938 TCG (349 TCWG) (AFR) Hamilton 10Feb63 – Fairchild (LOG) St. Augustine 13Dec65 – 933 TCG (440 TCWG) (AFR) Gen Mitchell 02Feb66 – Wfu to MASDC (LOG) Davis-Monthan 29Jan71 for storage, PCN: **CJ397** – DFI: 31Mar76 – Final Disposition: Scrapped.

KF-214 / 53-8111 / C-119G-FA

Avble: 03Sep54; Acc: 30Sep54; Del: 26Oct54 to USAF – 317 TCWG (AFE) Neubiberg AB W. Germany 26Oct54; to 3110 MAI GP (AMC) RAF Burtonwood England 56 mnt; depl 465 TCWG (AFE) Évreux-Fauville AB France 57; to Évreux-Fauville AB France 08Jul57 – 2346 RFC (CNR) Hamilton 09Apr58; to DT Paine 14Apr58 – 97 TCSq (349 TCWG) (AFR) Paine 23Dec58; 941 TCG (349 TCWG) (AFR) Paine 63 – 924 TCG (446 TCWG) (AFR) Ellington 09Aug65 – 922 TAL GP (433 TCWG) (AFR) Kelly 16Jan68 – Wfu to MASDC (LOG) Davis-Monthan 28Jan71 for storage – DFI: 26Dec73 – Final Disposition: Scrapped.

KF-215 / 53-8112 / C-119G-FA

Avble: 08Sep54; Acc: 30Sep54; Del: 20Oct54 to USAF – 317 TCWG (AFE) Neubiberg AB W. Germany 20Oct54 **RECORDS INDECIPHERABLE** → to 3110 MAI GP (AMC) RAF Burtonwood England 56 mnt; depl 465 TCWG (AFE) Évreux-Fauville AB France 57; to Évreux-Fauville AB France 08Jul57 ← **RECORDS INDECIPHERABLE** 2472 RFC DT (CNR) Davis 12Mar58; to DT Tinker 26Mar58; to 4750 ADF WG (ADC) Vincent 05Mar59 mnt – 305 TCSq (442 TCWG) (AFR) Tinker 19Mar59 – 313 TCSq (349 TCWG) (AFR) Portland 03Mar61; 939 TCG (349 TCWG) (AFR) Portland 63 – 931 TAL GP (434 TCWG) (AFR) Bakalar 24Jun68 – 906 TAL GP (302 TCWG) (AFR) Clinton 16Feb69; to Lockbourne 24Jul71; 907 TAL GP (302 TCWG) (AFR) Lockbourne 01Mar72 – MAP S. Vietnam 16Jul72, VNAF s/n: **538112**, 413 TSQ (53 TWG) Tan Son Nhut AB – Final Disposition: Unknown.

KF-216 / 53-8113 / C-119G-FA

Avble: 10Sep54; Acc: 30Sep54; Del: 18Oct54 to USAF – 317 TCWG (AFE) Neubiberg AB W. Germany 18Oct54; to 6605 AB WG (NEA) Ernest Harmon Canada 26Oct54 mnt **RECORDS INDECIPHERABLE** → to 3110 MAI GP (AMC) RAF Burtonwood England 56 mnt; depl 465 TCWG (AFE) Évreux-Fauville AB France 57; to Évreux-Fauville AB France 08Jul57 ← **RECORDS INDECIPHERABLE** 2473 RFC (CNR) Gen Mitchell 03Apr58 – 440 TCWG (AFR) Gen Mitchell 58; 933 TCG (440 TCWG) (AFR) Gen Mitchell 11Feb63 – Wfu to MASDC (LOG) Davis-Monthan 12Feb71 for storage – DFI: 09Dec74 – Final Disposition: Scrapped.

KF-217 / 53-8114 / C-119G-FA

Avble: 15Sep54; Acc: 30Sep54; Del: 18Oct54 to USAF – 317 TCWG (AFE) Neubiberg AB W. Germany 18Oct54 – **RECORDS MISSING 1955–1958** – 2346 RFC DT (CNR) Portland 18Apr58 – 313 TCSq (349 TCWG) (AFR) Portland 58; 939 TCG (349 TCWG) (AFR) Portland 63 – Fairchild (LOG) St. Augustine 23Feb68, cvtd **AC-119G Shadow** 18Jun68 – 4413 CCT SQ (1 ACO WG) (TAC) Lockbourne 27Jun68 – Fairchild (LOG) St. Augustine 21Jan69 – 1 SOP WG (TAC) Lockbourne 20Feb69 – 4410 CCT WG (TAC) Lockbourne

14Jul69 – 14 SOP WG (PAF) Phan Rang AB S. Vietnam 01Feb71; assigned variously Tan Son Nhut S. Vietnam from 18Mar70 – MAP S. Vietnam 06Sep71, VNAF s/n: **538114**, 819 ASQ (53 TWG) Tan Son Nhut AB – Final Disposition: Unknown.

KF-218 / 53-8115 / C-119G-FA

Avble: 20Sep54; Acc: 30Sep54; Del: 03Nov54 to USAF – 317 TCWG (AFE) Neubiberg AB W. Germany 18Oct54 **RECORDS INDECIPHERABLE** → to 3110 MAI GP (AMC) RAF Burtonwood England 56 mnt; depl 465 TCWG (AFE) Évreux-Fauville AB France 57; to Évreux-Fauville AB France 08Jul57 ← **RECORDS INDECIPHERABLE** 2472 RFC DT (CNR) Tinker 22Mar58 – Hayes (AMC) Birmingham 17Mar59 – 305 TCSq (442 TCWG) (AFR) Tinker 07May59 – 313 TCSq (349 TCWG) (AFR) Portland 61; 939 TCG (349 TCWG) (AFR) Portland 63 – Fairchild (LOG) St. Augustine 04Mar68, cvtd **AC-119G Shadow** 14Jul68 – 4413 CCT SQ (1 ACO WG) (TAC) Lockbourne 15Jul68 – 1 SOP WG (TAC) Lockbourne 27Feb69 – 4410 CCT WG (TAC) Lockbourne 14Jul69 – Hayes (LOG) Dothan 25Jan71 – 14 SOP WG (PAF) Phan Rang AB S. Vietnam 08Mar71; assigned variously Tan Son Nhut S. Vietnam from 11May71 – MAP S. Vietnam 03Sep71, VNAF s/n: **538115**, 819 ASQ (53 TWG) Tan Son Nhut AB – Final Disposition: Unknown.

KF-219 / 53-8116 / C-119G-FA

Avble: 22Sep54; Acc: 30Sep54; Del: 18Oct54 to USAF – 317 TCWG (AFE) Neubiberg AB W. Germany 18Oct54; to 3110 MAI GP (AMC) RAF Burtonwood England 01Mar56 mnt; depl 465 TCWG (AFE) Évreux-Fauville AB France 57; to Évreux-Fauville AB France 08Jul57 – 2237 RFC (CNR) New Castle 04Mar58; to NAS Willow Grove 20Jul58 – 512 TCWG (AFR) NAS Willow Grove 01Dec58; 912 TCG (512 TCWG) (AFR) NAS Willow Grove 11Feb63 – 397 Bomb WG (SAC) Dow 13Jun65. Listed as lost to accident 04Jun65 when the crew made an unspecified bail out 4 miles NE Lewiston Maine the aircraft then crashing. Pilot: Maj Samuel R. Foster, no fatalities – REC: 12Jul65 – Final Disposition: Accident.

KF-220 / 53-8117 / C-119G-FA

Avble: 29Sep54; Acc: 17Oct54; Del: 03Nov54 to USAF – 317 TCWG (AFE) Neubiberg AB W. Germany 03 Nov54; to 3110 MAI GP (AMC) RAF Burtonwood England 10Mar56 mnt; depl 465 TCWG (AFE) Évreux-Fauville AB France 57; to Évreux-Fauville AB France 08Jul57 – 2234 RFC (CNR) Hanscom 28Feb58 – 94 TCWG (AFR) Hanscom 19Mar59; 901 TCG (94 TCWG) (AFR) Hanscom 31May63 – 910 TCG (459 TCWG) (AFR) Youngstown 03Dec66 – 934 TAL GP (440 TCWG) (AFR) Minneapolis-St. Paul 13Sep69 – 906 TAL GP (302 TCWG) (AFR) Clinton 15Nov69; to Lockbourne 24Jul71; 907 TAL GP (302 TCWG) (AFR) Lockbourne 01Mar72 – MAP S. Vietnam 31Jul72, VNAF s/n: **538117**, 413 TSQ (53 TWG) Tan Son Nhut AB – Final Disposition: Unknown.

KF-221 / 53-8118 / C-119G-FA

Avble: 01Oct54; Acc: 18Oct54; Del: 01Nov54 to USAF – 317 TCWG (AFE) Neubiberg AB W. Germany 01Nov54. Unspecified ground accident 21Jan55 at Chateauroux AB France. Pilot: George G. Lynn, no fatalities; to 7373 MAI GP (AFE) Chateauroux AB France 21Jan55 for assessment – SAL: 13Apr55 – DFI: 18Jul55 – Final Disposition: Accident.

KF-222 / 53-8119 / C-119G-FA

Avble: 05Oct54; Acc: 20Oct54; Del: 01Nov54 to USAF – 317 TCWG (AFE) Neubiberg AB W. Germany 01Oct54 **RECORDS INDECIPHERABLE** → to 7280 MAI GP (AFE)

Nouasseur AB Morocco 22Feb55 mnt; to 3110 MAI GP (AMC) RAF Burtonwood England 56 mnt; depl 465 TCWG (AFE) Évreux-Fauville AB France 57; to Évreux-Fauville AB France 11Jul57 ← **RECORDS INDECIPHERABLE** 2472 RFC DT (CNR) Davis 12Mar58 – 65 TCSq (403 TCWG) (AFR) Davis 18Sep58; 929 TCG (403 TCWG) (AFR) Davis 11Feb63 – 922 TCG (433 TCWG) (AFR) Kelly 18Oct65 – 906 SOP GP (302 TCWG) (AFR) Clinton 06Nov70 – MAP Taiwan 04Jan71, RoCAF s/n: **3215** – Damaged beyond repair in an unspecified accident 27Jan89 – Final Disposition: Accident.

KF-223 / 53-8120 / C-119G-FA

Avble: 07Oct54; Acc: 25Oct54; Del: 09Nov54 to USAF – 317 TCWG (AFE) Neubiberg AB W. Germany 09Nov54; depl 60 TCWG (AFE) Rhein-Main AB W. Germany 05Feb55 **RECORDS INDECIPHERABLE** → to 3110 MAI GP (AMC) RAF Burtonwood England 17Aug56 mnt; depl 465 TCWG (AFE) Évreux-Fauville AB France 57; to Évreux-Fauville AB France 57 ← **RECORDS INDECIPHERABLE** 2346 RFC (CNR) Hamilton 29Apr58 – 831 AB GP (TAC) George 18Dec58 mnt – 349 TCWG (AFR) Hamilton 28Jan59; 938 TCG (349 TCWG) (AFR) Hamilton 10Feb63 – 934 TCG (440 TCWG) (AFR) Minneapolis-St. Paul 22Jan66 – MAP Taiwan 21Apr66 – Final Disposition: Unknown.

KF-224 / 53-8121 / C-119G-FA

Avble: 13Oct54; Acc: 26Oct54; Del: 08Nov54 to USAF – 317 TCWG (AFE) Neubiberg AB W. Germany 08Nov54 **RECORDS INDECIPHERABLE** → to 3110 MAI GP (AMC) RAF Burtonwood England 56 mnt; depl 465 TCWG (AFE) Évreux-Fauville AB France 57; to Évreux-Fauville AB France 08Jul57 ← **RECORDS INDECIPHERABLE** 2234 RFC (CNR) Hanscom 02Mar58 – 94 TCWG (AFR) Hanscom 19Mar59; 901 TCG (94 TCWG) (AFR) Hanscom 63 – 910 TCG (459 TCWG) (AFR) Youngstown 03Dec66 – Fairchild (LOG) St. Augustine 03Sep68, cvtd **AC-119K Stinger** 10Sep68 – 1 SOP WG (TAC) Lockbourne 18May69 – 4410 CCT WG (TAC) Lockbourne 14Jul69 – 14 SOP WG (PAF) Phan Rang AB S. Vietnam 21Oct69; assigned variously Da Nang, Phu Cat, Tan Son Nhut S. Vietnam & Nakhon Phanom Thailand from 29Nov69; to 313 ADH DV (PAF) Kadena AB Japan 02Aug71 – SMAAR (LOG) Phan Rang AB S. Vietnam 08Sep71 – 56 SOP WG (PAF) Nakhon Phanom AB Thailand 11Sep71; assigned variously Da Nang S. Vietnam from 12Oct71 – MAP S. Vietnam 10Nov72, VNAF s/n: **538121**, 821 ASQ (53 TWG) Tan Son Nhut AB – Final Disposition: Unknown.

KF-225 / 53-8122 / C-119G-FA

Avble: 15Oct54; Acc: 28Oct54; Del: 12Nov54 to USAF – 317 TCWG (AFE) Neubiberg AB W. Germany 12Nov54 **RECORDS INDECIPHERABLE** → to 3110 MAI GP (AMC) RAF Burtonwood England 56 mnt; to 7207 AB SQ (AFE) Aviano AB Italy 57; depl 465 TCSq (AFE) Évreux-Fauville AB France 57; to Évreux-Fauville AB France 08Jul57 ← **RECORDS INDECIPHERABLE** 2472 RFC DT (CNR) Davis 18Mar58; to DT Tinker 21Mar58 – 305 TCSq (442 TCWG) (AFR) Tinker 19Mar59; to 1405 AB WG (MATS) Scott 30Apr59 mnt – 96 TCSq (440 TCWG) (AFR) Minneapolis-St. Paul 27Jan61; 934 TCG (440 TCWG) (AFR) Minneapolis-St. Paul 11Feb63 – 906 TAL GP (302 TCWG) (AFR) Clinton 08Nov69; to Lockbourne 24Jul71; 907 TAL GP (302 TCWG) (AFR) Lockbourne 28Sep71 – Wfu to MASDC (LOG) Davis-Monthan 05Feb73 for storage – DFI: 26Dec73 – Final Disposition: Scrapped.

KF-226 / 53-8123 / C-119G-FA

Avble: 20Oct54; Acc: 28Oct54; Del: 12Nov54 to USAF – 317 TCWG (AFE) Neubiberg AB W. Germany 12Nov54; to 3110 MAI GP (AMC) RAF Burtonwood England 07May56 mnt; depl 465 TCWG (AFE) Évreux-Fauville AB France 57; to Évreux-Fauville AB France 08Jul57 – 2584 RFC (CNR) Memphis 21Feb58 – 2237 RFC (CNR) New Castle 11Mar58; to NAS Willow Grove 21Jul57 – 512 TCWG (AFR) NAS Willow Grove 01Dec58; 912 TCG (512 TCWG) (AFR) NAS Willow Grove 11Feb63 – Fairchild (LOG) St. Augustine 20Feb68, cvtd **AC-119G Shadow** 13Jun68 – 4413 CCT SQ (1 ACO WG) (TAC) Lockbourne 19Jun68 – 1 SOP WG (TAC) Lockbourne 17Feb69 – 4410 CCT WG (TAC) Lockbourne 14Jul69 – Hayes (LOG) Dothan 23Feb71 – 14 SOP WG (PAF) Phan Rang AB S. Vietnam 09Apr71; assigned variously Tan Son Nhut S. Vietnam from 30May71 – MAP S. Vietnam 23Aug71, VNAF s/n: **538123**, 819 ASQ (53 TWG) Tan Son Nhut AB – Final Disposition: Unknown.

KF-227 / 53-8124 / C-119G-FA

Avble: 26Oct54; Acc: 10Nov54; Del: 21Nov54 to USAF – 317 TCWG (AFE) Neubiberg AB W. Germany 21Nov54; to 3110 MAI GP (AFE) RAF Burtonwood England 04Mar55 mnt; depl 465 TCWG (AFE) Evreux-Fauville AB France 15Mar57; to Évreux-Fauville AB France 08Jul57 – 2347 RFC (CNR) Long Beach 04Feb58 – 2346 RFC (CNR) Hamilton 16Feb58; to DT McClellan 25Jun58 – 314 TCSq (349 TCWG) (AFR) McClellan 23Dec58; 940 TCG (349 TCWG) (AFR) McClellan 10Feb63 – 925 TCG (446 TCWG) (AFR) Ellington 02Sep65 – Fairchild (LOG) St. Augustine 20Nov67 – MAP S. Vietnam 22Mar68, VNAF s/n: **538124**, 413 TSQ (53 TWG) Tan Son Nhut AB – Final Disposition: Unknown.

KF-228 / 53-8125 / C-119G-FA

Avble: 28Oct54; Acc: 19Nov54; Del: 03Dec54 to USAF – 317 TCWG (AFE) Neubiberg AB W. Germany 03Dec54 **RECORDS INDECIPHERABLE** → depl 406 Fighter WG (AFE) RAF Manston England 20Feb55; to 3110 MAI GP (AMC) RAF Burtonwood England 17Sep56 mnt; depl 465 TCWG (AFE) Évreux-Fauville AB France 15Mar57; to Évreux-Fauville AB France 08Jul57 ← **RECORDS INDECIPHERABLE** 2242 RFC (CNR) Selfridge 26Mar58 – 403 TCWG (AFR) Selfridge 20Nov58; depl 434 TCWG (AFR) Bakalar 14Dec60; 927 TCG (403 TCWG) (AFR) Selfridge 10Feb63; 928 TAL GP (403 TCWG) (AFR) O'Hare 02Apr69 – Wfu to MASDC (LOG) Davis-Monthan 23Jan71 for storage / project, PCN: **CJ386**; to Hayes (LOG) Dothan 29Jun71 & 03Jul72 for project. Noted at MASDC in SEA three-tone camouflage, not the usual AFR livery – DFI: 31Mar76 – Final Disposition: Scrapped.

KF-229 / 53-8126 / C-119G-FA

Avble: 03Nov54; Acc: 18Nov54; Del: 29Nov54 to USAF – 317 TCWG (AFE) Neubiberg AB W. Germany 29Nov54; to 3110 MAI GP (AMC) RAF Burtonwood England 56 & 25Jul57 mnt; depl 465 TCWG (AFE) Évreux-Fauville AB France 57; to Évreux-Fauville AB France 08Jul57 – 2346 RFC DT (CNR) Paine 20Apr58 – 97 TCSq (349 TCWG) (AFR) Paine 58; 941 TCG (349 TCWG) (AFR) Paine 63 – 904 TCG (514 TCWG) (AFR) Stewart 04Jun65 – 903 TCG (514 TCWG) (AFR) Westover (depl) 06Apr66; to McGuire 26May67; depl Myrtle Beach 16Jul67 – 914 TAL GP (512 TCWG) (AFR) Niagara Falls 07Aug68 – 143 SOP GP (ANG) Theodore F. Green 30Mar71; multiple assignments Hayes (LOG) Dothan & General Dynamics (LOG) Theodore F. Green from 17Jun71; cvtd **C-119L** 15Mar73 – Wfu to MASDC (LOG) Davis-Monthan 23Jun75 for storage, PCN: **CJ466** – DFI: 15Jul77 – Final Disposition: Scrapped.

KF-230 / 53-8127 / C-119G-FA

Avble: 05Nov54; Acc: 30Nov54; Del: 07Dec54 to USAF – 3510 CCT WG (TC) Randolph 07Dec54; depl 3600 CCT WG (TC) Luke 31Oct55 – 3345 TTA WG (TC) Chanute 04Feb56 – 2585 RFC (CNR) Miami 19Aug57 – 435 TCWG (AFR) Miami 58; to Homestead 25Jul60; assigned 76 TCSq Homestead 01Oct61; 915 TCG (435 TCWG) (AFR) Homestead 17Jan63 – 926 TCG (446 TCWG) (AFR) NAS New Orleans 28Nov65 – 129 SOP GP (ANG) Hayward 15Feb69; cvtd **C-119L** 22Jun73 – Wfu to MASDC (LOG) Davis-Monthan 27Mar75 for storage, PCN: **CJ458** – DFI: 31Mar76 – BofS Starbird, Inc. Seattle Washington 20Nov78 reg **N90269**; original reg **N4999N** but ntu. Lease to Gifford Aviation, Inc. in Alaska for the carriage of seafoods. Crashed on take-off along a beach outside Big Creek Alaska 06Jul79. During the take-off run the aircraft was unable to gain enough airspeed to make lift-off, it appears the pilot tried to veer right to gain better ground but this caused a sideways slide. The take-off was continued until the aircraft crashed in the surf in the Bering Sea off the coast of Big Creek. The two pilots and three film crew onboard, making a PBS documentary, survived with injuries and were flown to Anchorage for care. The wreckage was pulled from the sea and left derelict in the sand dunes – Reg N90269 canx 05May15 – Final Disposition: Accident.

KF-231 / 53-8128 / C-119G-FA

Avble: 10Nov54; Acc: 30Nov54; Del: 07Dec54 to USAF – 3510 CCT WG (TC) Randolph 07Dec54 **RECORDS INDECIPHERABLE** → 4405 OP SQ (TAC) Langley 12Mar56 – 405 Fighter WG (TAC) Langley 57 ← **RECORDS INDECIPHERABLE** 836 AB GP (TAC) 08Oct57 – Hayes (AMC) Birmingham 16Oct57 – 2347 RFC (CNR) Long Beach 24Jan58; to DT Hill Oct58 – 733 TCSq (452 TCWG) (AFR) Hill 03Apr59 – 945 TCG (452 TCWG) (AFR) Hill 17Jan63 – Fairchild (LOG) St. Augustine 13Jun66 – MAP Brazil 17Oct66, FAB s/n: **2312** as attrition replacement, No. 2/1 Grupo de Transporte de Tropa (2 SQ / 1 GTT) Aerea dos Afonsos AB Rio de Janeiro. Crashed at Fazenda Santo Antonio near Luziania Goias 21Dec73 due to engine failure, nfd – Final Disposition: Accident.

KF-232 / 53-8129 / C-119G-FA

Avble: 16Nov54; Acc: 30Nov54; Del: 07Dec54 to USAF – 3510 CCT WG (TC) Randolph 07Dec54 – 3345 TTA WG (TC) Chanute 27Feb56 **RECORDS INDECIPHERABLE** → to units unk (TC) 56-57 – 2585 RFC (CNR) Miami 57 ← **RECORDS INDECIPHERABLE** 435 TCWG (AFR) Miami 58; to 95 COS GP (SAC) Biggs 24Feb59 mnt – 4440 ADE GP (TAC) Langley 16Jun60 – MAP India 26Jun60 – Final Disposition: Unknown.

KF-233 / 53-8130 / C-119G-FA

Avble: 18Nov54; Acc: 30Nov54; Del: 07Dec54 to USAF – 3750 TTN WG (TC) Sheppard 07Dec54 for ground instruction use; to museum / school status 20May58 – 3750 MSU GP (ATC) Sheppard redes **UC-119G** 30Jun62; redes **GC-119G** 30Nov62 – REC: 11Jun65 – Final Disposition: Scrapped.

KF-234 / 53-8131 / C-119G-FA

Avble: 26Nov54; Acc: 21Dec54; Del: 05Jan55 to USAF – 464 TCWG (TAC) Pope 05Jan55; to WRDCN (ARD) WPAFB 20Jul55 mnt; depl Aberdeen 05May56 **RECORDS INDECIPHERABLE** → 2346 RFC (CNR) Hamilton Jul58 – units unk 58 – 314 TCSq (349 TCWG) (AFR) McClellan 58 ← **RECORDS INDECIPHERABLE** 940 TCG (349 TCWG) (AFR) McClellan 10Feb63 – 904 TCG (514 TCWG) (AFR) Stewart 02Apr65; 903 TCG (514

TCWG) (AFR) McGuire 23Sep66; depl Myrtle Beach 16Jul67 – 927 TAL GP (403 TCWG) (AFR) Selfridge 13Jun68 – Fairchild (LOG) St. Augustine 21Jun68, cvtd **AC-119G Shadow** 01Jul68 – 71 SOP SQ (1 ACO WG) (TAC) Lockbourne 04Oct68 – 4413 CCT SQ (1 ACO WG) (TAC) Lockbourne 31Dec68 – 1 SOP WG (TAC) Lockbourne 17Feb69 – 4410 CCT WG (TAC) Lockbourne 14Jul69 – Hayes (LOG) Dothan 05Feb71 – 14 SOP WG (PAF) Phan Rang AB S. Vietnam 31Mar71; assigned variously Tab Son Nhut S. Vietnam from 14Jun71 – MAP S. Vietnam 24Aug71, VNAF s/n: **538131**, 819 ASQ (53 TWG) Tan Son Nhut AB – Final Disposition: Unknown.

KF-235 / 53-8132 / C-119G-FA

Avble: 01Dec54; Acc: 22Dec54; Del: 30Dec54 to USAF – 4501 SUT SQ (TAC) Donaldson 30Dec54; depl 464 TCWG (TAC) Pope 09Apr56 – 2347 RFC (CNR) Long Beach 19Jan58; to DT Hill 24Jun58 – Hayes (AMC) Birmingham 12Mar59 – 733 TCSq (452 TCWG) (AFR) Hill 20Apr59; 945 TCG (452 TCWG) (AFR) Hill 17Jan63 – MAP Taiwan 21May66, RoCAF s/n: **3139** – Final Disposition: Unknown.

KF-236 / 53-8133 / C-119G-FA

Avble: 03Dec54; Acc: 23Dec54; Del: 30Dec54 to USAF – 313 TCWG (TAC) Sewart 30Dec54; depl Elmendorf 12Jan55 – 314 TCWG (TAC) Sewart 08Jun55; depl England Louisiana 05Nov56 – Hayes (AMC) Birmingham 19Aug57 – 2235 RFC (CNR) Grenier 01Nov57 – 2234 RFC (CNR) Hanscom 20Nov57 – 94 TCWG (AFR) Hanscom 19Mar59; 901 TCG (94 TCWG) (AFR) Hanscom 31May63 – 910 TCG (459 TCWG) (AFR) Youngstown 27Oct66 – Fairchild (LOG) St. Augustine 04Oct67 – MAP S. Vietnam 25Mar68, VNAF s/n: **538133**, 413 TSQ (53 TWG) Tan Son Nhut AB – 405 Fighter WG (PAF) Clark AB The Philippines 04Jan73 for disposal – DFI: 05Jun73 – Final Disposition: Scrapped.

KF-237 / 53-8134 / C-119G-FA

Avble: 08Dec54; Acc: 28Dec54; Del: 04Jan55 to USAF – 313 TCWG (TAC) Sewart 04Jan55; depl Elmendorf 12Jan55; depl 463 TCWG (TAC) Sewart 08Jun55 – 314 TCWG (TAC) Sewart 55; depl England Louisiana 05Nov56 – Hayes (AMC) Birmingham 23Jul57 – 2472 RFC (CNR) Richards 02Oct57 – 442 TCWG (AFR) Richards 02Mar59 – 4440 ADE GP (TAC) Langley 28Jun60 – MAP India 10Jul60 – Final Disposition: Unknown.

KF-238 / 53-8135 / C-119G-FA

Avble: 10Dec54; Acc: 28Dec54; Del: 05Jan55 to USAF – 314 TCWG (TAC) Sewart 05Jan55 – Hayes (AMC) Birmingham 18Jul57 – 2472 RFC (CNR) Richards 02Oct57 – 442 TCWG (AFR) Richards 02Mar59 – 4440 ADE GP (TAC) Langley 28Jun60 – MAP India 10Jul60 – Final Disposition: Unknown.

KF-239 / 53-8136 / C-119G-FA

Avble: 15Dec54; Acc: 30Dec54; Del: 05Jan55 to USAF – 464 TCWG (TAC) Pope 05Jan55; depl Aberdeen 05May56; to 2750 AB WG (AMC) WPAFB 27Jul56 mnt – 2585 RFC DT (CNR) Donaldson 04Oct58 – 77 TCSq (435 TCWG) (AFR) Donaldson 58 – 4440 ADE GP (TAC) Langley 15Jun60 – MAP India 26Jun60, IAF s/n: **BK515** – Final Disposition: Unknown.

KF-240 / 53-8137 / C-119G-FA

Avble: 17Dec54; Acc: 29Dec54; Del: 05Jan55 to USAF – 464 TCWG (TAC) Pope 05Jan55; depl Aberdeen 05May56 – 2472 RFC DT (CNR) Tinker 58 – 305 TCSq (442

TCWG) (AFR) Tinker 19Mar59; to 445 TCA DT (CNC) Memphis 11Jan60 mnt; to 4520 CCT WG (TAC) Nellis 21Jan60 mnt – 4440 ADE GP (TAC) Langley 15Jun60 – MAP India 26Jun60 – final Disposition: Unknown.

KF-241 / 53-8138 / C-119G-FA

Avble: 22Dec54; Acc: 17Jan55; Del: 21Jan55 to USAF – 60 TCWG (AFE) Rhein-Main AB W. Germany 21Jan55; to Dreux AB France 01Oct55; to 3110 MAI GP (AMC) RAF Burtonwood England 30Apr56 mnt; to unit unk (AMC) 57; depl 50 Fighter-Bomber WG (AFE) Toul-Rosières AB France 27Nov57 – 7305 AB GP (322 ADV) (AFE) Dreux AB France 24Sep58; to 3131 ARP SQ (AMC) Chateauroux AB France 16Sep59 mnt – 317 AB GP (AFE) Évreux-Fauville AB France 60 – MAP Belgium17Jul60, BAF s/n: **CP-43**; BOC: 17Jul60 15 WG / 20 SQ Melsbroek AB Belgium, callsign: **OT-CEC**; to SIAI Marchetti for mnt 28Jun61 – 15 WG / 40 SQ 20Jan62. While parked was hit by a taxiing C-119G s/n: CP-9 25May67 – Wfu 13Dec71 and stored at Koksijde AB Belgium; SOC: 22Oct75 – To civil company International Engine Parts, scrapped 1977-78 – Final Disposition: Scrapped.

KF-242 / 53-8139 / C-119G-FA

Avble: 27Dec54; Acc: 17Jan55; Del: 20Jan55 to USAF – 60 TCWG (AFE) Rhein-Main AB W. Germany 20Jan55; to Dreux AB France 23Sep55; to 3110 MAI GP (AMC) RAF Burtonwood England 29Jun56 mnt; to 7425 AB GP (AFE) Hahn AB W. Germany 25Mar58 – 7305 AB GP (322 ADV) (AFE) Dreux AB France 24Sep58; 11 TCSq (322 ADV) (AFE) Dreux AB France 12Dec60 – 4440 ADE GP (TAC) Langley 03Jan61 – 78 TCSq (435 TCWG) (AFR) Bates 10Jan61; 435 TCWG (AFR) Homestead 19Apr61; assigned 76TCSq Homestead 01Oct61; 915 TCG (435 TCWG) (AFR) Homestead 17Jan63 – 906 TCG (302 TCWG) (AFR) Clinton 14Dec65; to Lockbourne 24Jul71 – MAP S. Vietnam 08Sep71, VNAF s/n: **538139**, 413 TSQ (53 TWG) Tan Son Nhut AB – Final Disposition: Unknown.

KF-243 / 53-8140 / C-119G-FA

Avble: 31Dec54; Acc: 31Jan55; Del: 31Mar55 to USAF – 60 TCWG (AFE) Rhein-Main AB W. Germany 31Mar55 – 1631 MAI SQ (MATS) Prestwick Scotland 06Apr55 – SABBE (AMC) Brussels Belgium 16Jan57 mnt – 60 TCWG (AFE) Dreux AB France 11Oct57 – 7305 AB GP (322 ADV) (AFE) Dreux AB France 24Sep58; 10 TCSq (322 ADV) (AFE) Dreux AB France 12Dec60; 7305 CAM SQ (AFE) Dreux AB France 08Jan61 – 4440 ADE GP (TAC) Langley 14Feb61 – 328 TCSq (512 TCWG) (AFR) Niagara Falls 20Feb61. Caught fire after take-off from Harlingen AFB Texas 22Jun61, the 4 crew bailed out and parachuted to the ground, the aircraft came down in the Kings Ranch area of Texas. It was found that one of the engines had exploded – Final Disposition: Accident.

KF-244 / 53-8141 / C-119G-FA

Avble: 04Jan55; Acc: 21Jan55; Del: 26Jan55 to USAF – 60 TCWG (AFE) Rhein-Main AB W. Germany 26Jan55; to Dreux AB France 01Oct55; to 3110 MAI GP (AMC) RAF Burtonwood England 56 mnt; to 7305 AB GP (322 ADV) (AFE) Dreux AB France 24Sep58 – 317 AB GP (AFE) Évreux-Fauville AB France 60 – MAP Belgium 17Jul60, BAF s/n: **CP-44**; BOC: 17Jul60 15 WG / 20 SQ Melsbroek AB Belgium, callsign: **OT-CED** – 15 WG / 40 SQ 15Feb61; to SIAI Marchetti for mnt 14Jun61 – Wfu 29Dec72 and stored at Koksijde AB Belgium; SOC: 22Oct75 – To civil company International Engine Parts, scrapped 1977-78 – Final Disposition: Scrapped.

KF-245 / 53-8142 / C-119G-FA

Avble: 07Jan55; Acc: 25Jan55; Del: 04Feb55 to USAF – 60 TCWG (AFE) Rhein-Main AB W. Germany 04Feb55; to Dreux AB France 01Oct55; to 3110 MAI GP (AMC) RAF Burtonwood England 56 mnt – 7305 AB GP (322 ADV) (AFE) Dreux AB France 24Sep58; 10 TCSq (322 ADV) (AFE) Dreux AB France 12Dec60 – 4440 ADE GP (TAC) Langley 25Jan61 – 452 TCWG (AFR) March 05Feb61; 942 TCG (452 TCWG) (AFR) March 01Jun63; 943 TCG (452 TCWG) (AFR) March 31Mar65 – 129 SOP GP (ANG) Hayward 15Jan69; cvtd **C-119L** 18Apr73; depl Aviano AB Italy 03Oct74 – Wfu to MASDC (LOG) Davis-Monthan 22Jan75 for storage, PCN: **CJ451** – DFI: 31Mar76 – BofS Dross Metals, Inc. Tucson Arizona 10Jul79 – BofS J.D. Gifford & Associates, Inc. Anchorage Alaska 28Sep80; ARAp 13Jan81 reg **N8504X**; major radio, nav, flight instrument upgrade and equipment removal to lighten airframe 29May81; electric emergency pump installed and control surfaces recovered 01Jun81; inverter upgrade 30Aug81; cvtd **C-119L Jet-Pak** (J34-WE) 13May82 – Alaska Commercial Fishing and Agriculture Bank Anchorage Alaska 31Aug84 through repossession from J.D. Gifford; ARAp 01Feb85 – BofS Northern Pacific Transport, Inc. Anchorage Alaska 12Apr85; lease to International Seafoods, Inc. 27Jun86; lease to Tate & Co. 09Sep86 – BofS Alaska Aircraft Leasing, Inc. Anchorage Alaska 01Apr87; nav upgrade 15May87. While on approach to Shageluk Airport Alaska 13May87 the aircraft was struck by a substantial downdraft which caused the C-119 to touchdown short of the runway. The force of impact collapsed the main u/c damaging the aircraft beyond economic repair, the two crew and three passengers were uninjured. The cargo hold was full of heavy building supplies which was a factor in the impact. The fuselage was later purchased for $1 by local resident Rudy Hamilton who converted it into a workshop. Reg N8504X canx 19Jun13 – Final Disposition: Accident.

KF-246 / 53-8143 / C-119G-FA

Avble: 13Jan55; Acc: 28Jan55; Del: 09Feb55 to USAF – 60 TCWG (AFE) Rhein-Main AB W. Germany 09Feb55; to Dreux AB France 01Oct55; to 3110 MAI GP (AMC) RAF Burtonwood England 56 mnt; to 3970 AB GP (SAC) Torrejón AB Spain 12Feb58 mnt – 7305 AB GP (322 ADV) (AFE) Dreux AB France 24Sep58 – 317 AB GP (AFE) Évreux-Fauville AB France 60 – MAP Belgium 17Jul60, BAF s/n: **CP-45**; BOC: 17Jul60 15 WG / 20 SQ Melsbroek AB Belgium, callsign: **OT-CEE**; to SIAI Marchetti for mnt 13Jun61 – 15 WG / 40 SQ 31May63. Aircraft struck by a stray British mortar round while flying over the Sennelager Range while on approach to RAF Gutersloh West Germany 26Jun63. Pilot: Lt Col Herman Kreps, 5 other crew and 42 paratroops onboard. The fatally wounded aircraft crashed near Detmold West Germany killing all onboard except for 9 paratroops who were able to bail out before impact. Final Disposition: Accident.

KF-247 / 53-8144 / C-119G-FA

Avble: 18Jan55; Acc: 31Jan55; Del: 14Feb55 to USAF – 60 TCWG (AFE) Rhein-Main AB W. Germany 14Feb55; to Dreux AB France 23Sep55; to 3110 MAI GP (AMC) RAF Burtonwood England 12Jul56 mnt – 7305 AB GP (322 ADV) (AFE) Dreux AB France 24Sep58; 10 TCSq (322 ADV) (AFE) Dreux AB France 12Dec60 – 7305 CAM SQ (AFE) Dreux AB France 08Jan61 – 4440 ADE GP (TAC) Langley 24Feb61 – 434 TCWG (AFR) Bakalar 04Mar61 – 357 TCSq (302 TCWG) (AFR) Bates 19Oct61; 908 TCG (302 TCWG) (AFR) Bates 11Feb63 – Fairchild (LOG) St. Augustine 08Mar63 – 908 TCG (435 TCWG) (AFR) Bates 16Apr63 – Fairchild (LOG) St. Augustine 02May63 – MAP India 28May63 – Final Disposition: Unknown.

Section Four—Service Histories

KF-248 / 53-8145 / C-119G-FA

Avble: 31Jan55; Acc: 15Feb55; Del: 25May55 to USAF – 60 TCWG (AFE) Rhein-Main AB W. Germany 25May55; to Dreux AB France 13Sep55; to 3110 MAI GP (AMC) RAF Burtonwood England 56 mnt – 7305 AB GP (322 ADV) (AFE) Dreux AB France 24Sep58; 10 TCSq (322 ADV) (AFE) Dreux AB France 12Dec60 – 7305 CAM SQ (AFE) Dreux AB France 08Jan61 – 4440 ADE GP (TAC) Langley 08Mar61 – 73 TCSq (434 TCWG) (AFR) Scott 15Mar61; 932 TCG (434 TCWG) (AFR) Scott 11Feb63 – 910 TCG (459 TCWG) (AFR) Youngstown 08Oct66 – Fairchild (LOG) St. Augustine 07May68, cvtd **AC-119K Stinger** 01Jul68 – 4413 CCT SQ (1 ACO WG) (TAC) Lockbourne 22Nov68 – 1 SOP WG (TAC) Lockbourne 17Feb69 – Fairchild (LOG) St. Augustine 13Jun69 – 4410 CCT WG (TAC) Lockbourne 10Aug69 – WRAAR (LOG) Lockbourne 23Jun71 mnt – Hayes (LOG) Dothan 24Jun71 – 1 SOP WG (TAC) Eglin 24Sep71 – WRAAR (LOG) Robins 28Oct72 – 6498 ABS WG (PAF) Da Nang S. Vietnam 15Dec72 – MAP S. Vietnam 31Jan73, VNAF s/n: **538145**, 821 ASQ (53 TWG) Tan Son Nhut AB – Final Disposition: Unknown.

KF-249 / 53-8146 / C-119G-FA

Avble: 02Feb55; Acc: 23Feb55; Del: 28Feb55 to USAF – 60 TCWG (AFE) Rhein-Main AB W. Germany 28Feb55; to Dreux AB France 01Oct55; to 3110 MAI GP (AMC) RAF Burtonwood England 21Apr56 mnt – 7305 AB GP (322 ADV) (AFE) Dreux AB France 24Sep58 – United Nations (UN), reg **UNO-105**, mission to The Congo 06Aug60 – MAP Italy 61, AMI s/n: **MM53-8146**; cvtd **EC-119G** 1976 – Preserved at Piana delle Orme Museum Italy – Final Disposition: Extant.

KF-250 / 53-8147 / C-119G-FA

Avble: 07Feb55; Acc: 24Feb55; Del: 07Mar55 to USAF – 60 TCWG (AFE) Rhein-Main AB W. Germany 07Mar55; to Dreux AB France 01Oct55; depl 7310 AB GP (AFE) Rhein-Main AB W. Germany 13Feb56; to 3110 MAI GP (AMC) RAF Burtonwood England 19Apr56 mnt – 7305 AB GP (322 ADV) (AFE) Dreux AB France 24Sep58; 10 TCSq (322 ADV) (AFE) Dreux AB France 12Dec60 – 7305 CAM SQ (AFE) Dreux AB France 08Jan61 – 4440 ADE GP (TAC) Langley 29Mar61 – 403 TCWG (AFR) Selfridge 05Apr61; 927 TCG (403 TCWG) (AFR) Selfridge 10Feb63 – Fairchild (LOG) St. Augustine 04Dec67 – MAP S. Vietnam 26Mar68, VNAF s/n: **538147**, 413 TSQ (53 TWG) Tan Son Nhut AB – 377 ABS WG (PAF) Tan Son Nhut S. Vietnam 05Jan73 – 405 Fighter WG (PAF) Clark AB The Philippines 06Jan73 for disposal – MAP Taiwan 26May73 – Final Disposition: Unknown.

KF-251 / 53-8148 / C-119G-FA

Avble: 09Feb55; Acc: 28Feb55; Del: 15Mar55 to USAF – 60 TCWG (AFE) Rhein-Main AB W. Germany 19Mar55; to Dreux AB France 12Sep55; depl 7310 AB GP (AFE) Rhein-Main AB W. Germany 28Mar56; to 3110 MAI GP (AMC) RAF Burtonwood England Nov56 mnt – 7305 AB GP (322 ADV) (AFE) Dreux AB France 24Sep58; 10 TCSq (322 ADV) (AFE) Dreux AB France 12Dec60 – 4440 ADE GP (TAC) Langley 27Jan61 – 452 TCWG (AFR) March 08Feb61; 944 TCG (452 TCWG) (AFR) March 11Feb63 – 931 TAL GP (434 TCWG) (AFR) Bakalar 19Jan68 – Fairchild (LOG) St. Augustine 15Jul68, cvtd **AC-119K Stinger** 26Jul68 – 1 SOP WG (TAC) Lockbourne 05Mar69 – 4410 CCT WG (TAC) Lockbourne 14Jul69 – 14 SOP WG (PAF) Phan Rang AB S. Vietnam 18Nov69; assigned variously Da Nang, Tan Son Nhut S. Vietnam & Udorn, Nakhon Phanom Thailand from 27Dec69 – CADTT (LOG) Taipei Taiwan 12Aug71 – 56 SOP WG (PAF) Nakhon Phanom AB Thailand 25Nov71; assigned variously Da Nang S. Vietnam from 15Mar72 –

MAP S. Vietnam 10Nov72, VNAF s/n: **538148**, 821 ASQ (53 TWG) Tan Son Nhut AB – Final Disposition: Unknown.

KF-252 / 53-8149 / C-119G-FA

Avble: 16Feb55; Acc: 28Feb55; Del: 10Mar55 to USAF – 60 TCWG (AFE) Rhein-Main AB W. Germany 10Mar55; to 1604 MAT SQ (MATS) Kindley 12Mar55 mnt; to Dreux AB France 12Sep55; to 3110 MAI GP (AMC) RAF Burtonwood England 03May56 mnt – 7305 AB GP (322 ADV) (AFE) Dreux AB France 24Sep58; to 3970 AB GP (SAC) Torrejón AB Spain 29Jun60 mnt; 10 TCSq (322 ADV) (AFE) Dreux AB France 12Dec60 – 7305 CAM SQ (AFE) Dreux AB France 08Jan61 – 4440 ADE GP (TAC) Langley 01Mar61 – 434 TCWG (AFR) Bakalar 09Mar61; 931 TCG (434 TCWG) (AFR) Bakalar 11Feb63 – 130 SOP GP (ANG) Kanawha 05Feb69; to General Dynamics (LOG) Kanawha 27Mar73, cvtd **C-119L** 19Apr73 – Wfu to MASDC (LOG) Davis-Monthan 27Sep75 for storage, PCN: **CJ472** – DFI: 14Jul77 – Final Disposition: Scrapped.

KF-253 / 53-8150 / C-119G-FA

Avble: 18Feb55; Acc: 28Feb55; Del: 10Mar55 to USAF – 60 TCWG (AFE) Rhein-Main AB W. Germany 10Mar55; to 1604 MAT SQ (MATS) Kindley 12Mar55 mnt; to Dreux AB France 01Oct55; to 3110 MAI GP (AMC) RAF Burtonwood England 56 mnt – SABBE (AMC) Brussels Belgium 27Jun58 mnt – 7305 AB GP (322 ADV) (AFE) Dreux AB France Nov58; 10 TCSq (322 ADV) (AFE) Dreux AB France 12Dec60 – 4440 ADE GP (TAC) Langley 20Jan61 – 452 TCWG (AFR) March 29Jan61; 943 TCG (452 TCWG) (AFR) March 11Feb63 – 129 SOP GP (ANG) Hayward 16Jan69; cvtd **C-119L** 04May73 – Wfu to MASDC (LOG) Davis-Monthan 04Feb75 for storage, PCN: **CJ453** – DFI: 31Mar76 – BofS Dross Metals, Inc. Tucson Arizona 29Nov79 reg **N37636** – BofS Raven International, Inc. Huntington West Virginia 30May80; noted at Oshkosh airshow Jul80; ARAp 18Nov80 – BofS D.M.I. Aviation, Inc. Tucson Arizona 01Mar82; used in a Ford truck commercial Jun82 involving dropping vehicles from the back of the aircraft over El Mirage California with up to ten parachutists also involved, well known film stunt pilot Art Scholl was coordinating the shoot; noted stored Ryan Field Arizona 1983; film appearance *Spies Like Us* (1985) – BofS Mike Ivers Yakutat Alaska 18Apr86 – BofS David Brady Cartersville Georgia 03Mar89; restored with USAF ANG livery, named *Georgia Box* making several airshow appearances – BofS Hawkins & Powers Aviation, Inc. Greybull Wyoming 28Dec90; ARAp 22Apr91 – BofS D&G, Inc. Greybull Wyoming 01Jan93; stored in outdoors becoming derelict – BofS The Pride Capital Group, LLC Deerfield Illinois 07Sep05 for auction – BofS Harold Sheppard, Jr., d/b/a Sheppard Trucking Riverton Wyoming 24Aug06; ARAp 19Feb07. Presently the aircraft remains derelict at Greybull Wyoming along with several other C-119 and piston-engined aircraft – Final Disposition: Extant.

KF-254 / 53-8151 / C-119G-FA

Avble: 02Mar55; Acc: 23Mar55; Del: 19May55 to USAF – 60 TCWG (AFE) Rhein-Main AB W. Germany 19May55; to Dreux AB France 23Sep55; to 3110 MAI GP (AMC) RAF Burtonwood England 18Apr56 mnt – 7305 AB GP (322 ADV) (AFE) Dreux AB France 24Sep58 – 317 AB GP (AFE) Évreux-Fauville AB France 60 – MAP Belgium 17Jul60, BAF s/n: **CP-46**; BOC: 17Jul60 15 WG / 20 SQ Melsbroek AB Belgium, callsign: **OT-CEH** – 15 WG / 40 SQ 24Feb61; to SIAI Marchetti for mnt 16Jun61. Minor landing incident 22May63 – Wfu 27Apr72 and stored at Koksijde AB Belgium; SOC: 12Dec74, to

Royal Army Museum for display – Transferred to the Brussels Air Museum Belgium by road 26Jan80 and restored for indoor display – Final Disposition: Extant.

KF-255 / 53-8152 / C-119G-FA

Avble: 04Mar55; Acc: 25Mar55; Del: 31Mar55 to USAF – 60 TCWG (AFE) Rhein-Main AB W. Germany 31Mar55; to Dreux AB France 23Sep55; to 3110 MAI GP (AMC) RAF Burtonwood England 56 mnt – 7305 AB GP (322 ADV) (AFE) Dreux AB France 24Sep58. Made a forced landing on a beach 1.4 miles SE Botricello Italy 05Mar60 after experiencing a runaway propeller. 14 passengers and the radio operator bailed out over Botricello. The 3 remaining crew made a superb landing on the beach intact but the sloping sands caused the aircraft to swing into the surf. The crew were able to escape from the aircraft and wade to shore. The heavy tides that followed overnight broke the C-119 in half causing it to be damaged beyond repair – Final Disposition: Accident.

KF-256 / 53-8153 / C-119G-FA

Avble: 16Mar55; Acc: 30Mar55; Del: 11Apr55 to USAF – 60 TCWG (AFE) Rhein-Main AB W. Germany 11Apr55; to Dreux AB France 01Oct55; to 7310 AB GP (AFE) Rhein-Main AB W. Germany 10Feb56; to 3110 MAI GP (AMC) RAF Burtonwood England 30Apr56 mnt; depl 50 Fighter-Bomber WG (AFE) Toul-Rosières AB France56 – 7305 AB GP (322 ADV) (AFE) Dreux AB France 24Sep58; 10 TCSq (322 ADV) (AFE) Dreux AB France 12Dec60 – 7305 CAM SQ (AFE) Dreux AB France 08Jan61 – 4440 ADE GP (TAC) Langley 13Mar61 – 64 TCSq (403 TCWG) (AFR) O'Hare 14Mar61; 928 TCG (403 TCWG) (AFR) O'Hare 11Feb63 – 129 SOP GP (ANG) Hayward 15Feb69; cvtd **C-119L** 11Jun73 – depl Aviano AB Italy 03Oct74 – Wfu to MASDC (LOG) Davis-Monthan 26Mar75 for storage, PCN: **CJ455** – DFI: 31Mar76 – BofS Dross Metals, Inc. Tucson Arizona 18Sep79 – BofS J.D. Gifford & Associates, Inc. Anchorage Alaska 28Sep80; ARAp 13Jan81 reg **N8504W**; rudder surfaces recovered 23May81; major radio, nav, flight instrument upgrade 24May81; equipment removal to lighten airframe 28May81; hydraulic pump installed 01Jun81. The right main gear shear pin failed causing an u/c collapse after landing at Dahl Creek Airport Alaska 07Sep81. The resulting damage completely crumpled the main fuselage belly almost snapping it in half behind the flight-deck. No fatalities. Reg N8504W canx 05Sep14 – Final Disposition: Accident.

KF-257 / 53-8154 / C-119G-FA

Avble: 23Mar55; Acc: 31Mar55; Del: 19Apr55 to USAF – 60 TCWG (AFE) Rhein-Main AB W. Germany 19Apr55; to Dreux AB France 01Oct55; to 3110 MAI GP (AMC) RAF Burtonwood England 56 mnt – 7305 AB GP (322 ADV) (AFE) Dreux AB France 24Sep58; to 3131 MAI GP (AMC) Chateauroux AB France 58 mnt; 10 TCSq (322 ADV) (AFE) Dreux AB France 12Dec60 – 7305 CAM SQ (AFE) Dreux AB France 08Jan61 – 4440 ADE GP (TAC) Langley 03Mar61 – 434 TCWG (AFR) Bakalar 10Mar61; 930 TCG (434 TCWG) (AFR) Bakalar 11Feb63; depl 71 ACO SQ (1 ACO WG) (TAC) Lockbourne 20Jun68; 931 TAL GP (434 TCWG) (AFR) Bakalar 16Sep68 – 910 TAL GP (459 TCWG) (AFR) Youngstown 25Feb69 – 130 SOP GP (ANG) Kanawha 28May69; cvtd **C-119L** 30Jan73 – Wfu to MASDC (LOG) Davis-Monthan 27Sep75 for storage, PCN: **CJ473** – DFI: 31Mar76 – BofS Starbird, Inc. Seattle Washington 13Nov78 reg **N90267**; used for crew training Jun79; carried seafoods in Alaska; electrical upgrade 30Aug81 – Alaska Commercial Fishing and Agriculture Bank Anchorage Alaska 31Aug84, aircraft repossessed from Starbird, Inc.; ARAp 01Feb85 – BofS Stebbins & Ambler Air Transport Anchorage Alaska 28May85. N90267 appears to have had some sort of minor accident making the airframe unairworthy and subsequently seems to have been bku for use as a spares source. N90267 de-reg 16Aug88. The aircraft was then repossessed again by Alaska Commercial Fishing and Agriculture Bank 29Nov89 but only as a derelict with missing parts – Hawkins & Powers Aviation, Inc. Greybull Wyoming as major parts circa early 1990s – Sold to Ed Rachanski Nevada 2006 who restored the front half of the fuselage as a mobile museum marked as: "927" (52–5927) – Donated Palm Springs Air Museum California as a display item – Final Disposition: Extant.

KF-258 / 53-8155 / C-119G-FA

Avble: 30Mar55; Acc: 11Apr55; Del: 27Apr55 to USAF – 60 TCWG (AFE) Rhein-Main AB W. Germany 27Apr55; to Dreux AB France 01Oct55; to 3110 MAI GP (AMC) RAF Burtonwood England 04Oct56 mnt; to 7310 AB GP (AFE) Rhein-Main AB W. Germany 57 – 7305 AB GP (322 ADV) (AFE) Dreux AB France 24Sep58; 10 TCSq (322 ADV) (AFE) Dreux AB France 12Dec60 – 7305 CAM SQ (AFE) Dreux AB France 08Jan61 – 4440 ADE GP (TAC) Langley 07Feb61 – 732 TCSq (94 TCWG) (AFR) Grenier 14Feb61; 902 TCG (94 TCWG) (AFR) Grenier 11Feb63 – 926 TCG (446 TCWG) (AFR) NAS New Orleans 15Nov65 – Fairchild (LOG) St. Augustine 02May68, cvtd **AC-119G Shadow** 01Jul68 – 71 ACO SQ (1 ACO WG) (TAC) Lockbourne 28Aug68 – Fairchild (LOG) St. Augustine 02Dec68 – 14 SOP WG (PAF) Nha Trang AB S. Vietnam 31Dec68; assigned variously Tan Son Nhut, Phan Rang, Tuy Hoa S. Vietnam from 05Feb69. Crashed due to engine failure after take-off 1.9 miles from Tan Son Nhut AB S. Vietnam 28Apr70 while enroute for a combat mission. The 6 crew killed were: Maj Meredith G. Anderson; 1Lt Charles M. Knowles; 1Lt Thomas L. Lubbers; MSgt Joseph C. Jeszeck; SSgt Robert F. Fage and Sgt Michael J. Vangelisti. 2 survivors were: Maj Robert Bokern and SSgt Allen Chandler – Final Disposition: Accident.

KF-259 / 53-8156 / C-119G-FA

Avble: 05Apr55; Acc: 22Apr55; Del: 29Apr55 to USAF – 60 TCWG (AFE) Rhein-Main AB W. Germany 29Apr55; to Dreux AB France 12Sep55; to 3110 MAI GP (AMC) RAF Burtonwood England 18May56 mnt – 7305 AB GP (322 ADV) (AFE) Dreux AB France 24Sep58; 10 TCSq (322 ADV) (AFE) Dreux AB France 12Dec60 – 7305 CAM SQ (AFE) Dreux AB France 08Jan61 – 4440 ADE GP (TAC) Langley 06Feb61 – 732 TCSq (94 TCWG) (AFR) Grenier 14Feb61; 902 TCG (94 TCWG) (AFR) Grenier 08Mar63 – 906 TCG (302 TCWG) (AFR) Clinton 19Nov65; to Lockbourne 24Jul71 – MAP S. Vietnam 02Sep71, VNAF s/n: **538156**, 413 TSQ (53 TWG) Tan Son Nhut AB – Final Disposition: Unknown.

Appendix I:
USAF & USMC Tail Codes

USAF

317 SOP SQ	**AH**	1 ACO WG, Lockbourne AFB, Ohio 1 ACO WG, Eglin AFB, Florida	1968–1972
17 SOP SQ	**EF**	14 SOP WG, Phan Rang AB, S. Vietnam	1969–1971
18 SOP SQ	**EH**	14 SOP WG, Phan Rang AB, S. Vietnam	1969–1971
71 SOP SQ	**IC**	1 ACO WG, Lockbourne AFB, Ohio	1968–1969
4413 CCT SQ	**IH**	4410 CCT WG, Lockbourne AFB, Ohio	1968–1969

USN

VR-24	**JM**	NAS Port Lyautey, Morocco	1960–1962

USMC

VMR-153	**AC**	MCAS Cherry Point, North Carolina	1952–1959
	BC	Detachment, NAS Port Lyautey, Morocco	1956–1957
VMR-252	**LH**	MCAS Cherry Point, North Carolina	1950–1957
	BH	MCAS Cherry Point, North Carolina	1957–1961
VMR-253	**AD**	MCAS El Toro, California	1952–1953
		MCAF Itami, Japan	1953–1955
		MCAF Iwakuni, Japan	1955–1958
	QD	MCAF Iwakuni, Japan	1958–1961
VMR-352	**QB**	MCAS El Toro, California	1959–1961
VMR-353	**MZ**	MCAS Miami, Florida	1953–1957
	DZ	MCAS Miami, Florida	1957–1959
		MCAS Cherry Point, North Carolina	1959–1963
VMGR-152	**BZ**	MCAF Iwakuni, Japan	1962

USMC (Reserve)

From 1958 to 1972 all USMCR aircraft used a numbered Station Code that denoted the operating base as opposed to the unit assignment.

VMR-216	**7T**	NAS Seattle, Washington State	1962–1968
	5T	NAS Seattle, Washington State	1968–1970
		NAS Whidbey Is., Washington State	1970–1972
	MV	NAS Whidbey Is., Washington State	1972

VMR-222	7Y	NAS Grosse Ile, Michigan	1961–1969
VMR-234	7E	NAS Minneapolis, Minnesota, to NAS Twin Cities from 1963	1961–1968
	5E	NAS Twin Cities, Minnesota	1968–1970
	5V	NAS Glenview, Illinois	1970–1972
	QH	NAS Glenview, Illinois	1972–1975

Appendix II:
C-119 Retirement Bases & Disposals

AASBR / 2704 ASD GP / MASDC, Davis-Monthan AFB, Arizona

Summary of Total Acquisitions

USAF Final Retirements:	456
USAF Temp Assignments:	7
Norway (via USAF) Retirements	7
USMC Final Retirements:	29
Total:	**499**

Summary of Final Disposals

USAF Scrapped:	408*
USAF Temp Assignments:	7
USAF to Civil (with reg):	25
USAF to Civil (un-reg):	3
MAP Ethiopia:	2
MAP Italy:	25
USMC Scrapped:	14*
USMC to Civil (with reg):	9
USMC to Civil (un-reg):	6
Total:	**499**

* *Includes disposals to local scrap dealers.*

C-119B (cvtd C-119C): 48-0320, 48-0322, 48-0326, 48-0328, 48-0329, 48-0331, 48-0332, 48-0337, 48-0340, 48-0341, 48-0343, 48-0344, 48-0347, 48-0349, 48-0351, 48-0352, 48-0354, 48-0355, 49-0101, 49-0104, 49-0106, 49-0107, 49-0108, 49-0109, 49-0111, 49-0112, 49-0113, 49-0115, 49-0116, 49-0117, 49-0118. ***Batch Total:*** 031

C-119C: 49-0119, 49-0121, 49-0122, 49-0124, 49-0125, 49-0127, 49-0129, 49-0131, 49-0132, 49-0133, 49-0135, 49-0141, 49-0142, 49-0143, 49-0144, 49-0151, 49-0154, 49-0156, 49-0157, 49-0159, 49-0164, 49-0165, 49-0167, 49-0170, 49-0174, 49-0176, 49-0177, 49-0178, 49-0179, 49-0184, 49-0185, 49-0188, 49-0191, 49-0193, 49-0194, 49-0196, 49-0197, 49-0198, 49-0199. ***Batch Total:*** 039

50-0119, 50-0120, 50-0121, 50-0122, 50-0123, 50-0124, 50-0125, 50-0126, 50-0129, 50-0130, 50-0131, 50-0132, 50-0134, 50-0136, 50-0139, 50-0140, 50-0141, 50-0142, 50-0143, 50-0144, 50-0146, 50-0148, 50-0150, 50-0151, 50-0152, 50-0154, 50-0155,

50-0156, 50-0157, 50-0158, 50-0161, 50-0165, 50-0166, 50-0167, 50-0168, 50-0169, 50-0170, 50-0171. *Batch Total:* 038

51-2532, 51-2533, 51-2534, 51-2536, 51-2537, 51-2538, 51-2539, 51-2540, 51-2541, 51-2543, 51-2545, 51-2547, 51-2550, 51-2553, 51-2554, 51-2555, 51-2557, 51-2558, 51-2559, 51-2561, 51-2562, 51-2563, 51-2564, 51-2565, 51-2568, 51-2571, 51-2572, 51-2576, 51-2578, 51-2579, 51-2580, 51-2581, 51-2582, 51-2583, 51-2584. *Batch Total:* 035

51-2587, 51-2588, 51-2589, 51-2591, 51-2592, 51-2593, 51-2594, 51-2595, 51-2596, 51-2597, 51-2598, 51-2599, 51-2600, 51-2601, 51-2603, 51-2604, 51-2605, 51-2606, 51-2607, 51-2608, 51-2610, 51-2612, 51-2613, 51-2614, 51-2615, 51-2617, 51-2618, 51-2619, 51-2620, 51-2622, 51-2623, 51-2625, 51-2626, 51-2627, 51-2628, 51-2629, 51-2630, 51-2631, 51-2632, 51-2633, 51-2634, 51-2636, 51-2637, 51-2638, 51-2641, 51-2642, 51-2643, 51-2644, 51-2646, 51-2647, 51-2649, 51-2650, 51-2651, 51-2652, 51-2654, 51-2655, 51-2656, 51-2657, 51-2658, 51-2660, 51-2661. *Batch Total:* 061

51-8233, 51-8236, 51-8237, 51-8238, 51-8239, 51-8240, 51-8242, 51-8243, 51-8245, 51-8246, 51-8247, 51-8248, 51-8249, 51-8250, 51-8251, 51-8252, 51-8253, 51-8254, 51-8255, 51-8256, 51-8257, 51-8258, 51-8259, 51-8260, 51-8261, 51-8263, 51-8264, 51-8265, 51-8266, 51-8267, 51-8268, 51-8269, 51-8270, 51-8271, 51-8272. *Batch Total:* 035

C-119F (incl. cvtd C-119G): 51-2662, 51-2663, 51-2664, 51-2666, 51-2668, 51-2670, 51-2672, 51-2673, 51-2674, 51-2676, 51-2678, 51-2682, 51-2684, 51-2685, 51-2686, 51-2692, 51-2693, 51-2697, 51-2698, 51-2699, 51-2702, 51-2705, 51-2708, 51-2712, 51-2713, 51-2714. *Batch Total:* 026

51-7968, 51-7969, 51-7972, 51-7975, 51-7976, 51-7982, 51-7986, 51-7987, 51-7988, 51-7991, 51-7997, 51-7998, 51-7999, 51-8001, 51-8003, 51-8006, 51-8009, 51-8010, 51-8012, 51-8014, 51-8022, 51-8023, 51-8029, 51-8030, 51-8033, 51-8035, 51-8036, 51-8038, 51-8039, 51-8040, 51-8041, 51-8042, 51-8043, 51-8045, 51-8046, 51-8049, 51-8050, 51-8052. *Batch Total:* 038

51-8098, 51-8100, 51-8101, 51-8103, 51-8107, 51-8108, 51-8109, 51-8111, 51-8113, 51-8114, 51-8115, 51-8116, 51-8119, 51-8121, 51-8123, 51-8124, 51-8125, 51-8126, 51-8128, 51-8129, 51-8130, 51-8131, 51-8134, 51-8135, 51-8137, 51-8138, 51-8139, 51-8140, 51-8141, 51-8144, 51-8145, 51-8146, 51-8152, 51-8154, 51-8155, 51-8156, 51-8157, 51-8158, 51-8159, 51-8162, 51-8164, 51-8167, 51-8168. *Batch Total:* 043

C-119F (former R4Q-2): 131662, 131664, 131669, 131670, 131673, 131675, 131676, 131677, 131678, 131679, 131680, 131683, 131687, 131688, 131689, 131690, 131691, 131694, 131695, 131698, 131699, 131700, 131701, 131702, 131704, 131706, 131708, 131715, 131717. *Batch Total:* 029

C-119G: 51-8053, 51-8054, 51-8055, 51-8059, 51-8061, 51-8062, 51-8063, 51-8069, 51-8072, 51-8073, 51-8078, 51-8081, 51-8082, 51-8087, 51-8088, 51-8089, 51-8093, 51-8096, 51-8097. *Batch Total:* 019

52-5840, 52-5845, 52-5847, 52-5849, 52-5851, 52-5862, 52-5863, 52-5866, 52-5868, 52-5871, 52-5872, 52-5873, 52-5874, 52-5876, 52-5877, 52-5880, 52-5883, 52-5884, 52-5890, 52-5895, 52-5896, 52-5897, 52-5899, 52-5900, 52-5901, 52-5903, 52-5906, 52-5913, 52-5914, 52-5915, 52-5917, 52-5918, 52-5931, 52-5932, 52-5933, 52-5934, 52-5939, 52-5946, 52-5947, 52-5948, 52-5950, 52-5953, 52-5922. *Batch Total:* 043

52-9981. *Batch Total:* 001

53-3136, 53-3137, 53-3142, 53-3144, 53-3145, 53-3151, 53-3152, 53-3158, 53-3162, 53-3166, 53-3168, 53-3174, 53-3179, 53-3180, 53-3181, 53-3182, 53-3183, 53-3184, 53-3186,

Appendix II: C-119 Retirement Bases & Disposals 425

C-119G s/n 52-5915 (11094) at MASDC, Arizona, in October 1979 is typical of the hundreds of Boxcar derelicts awaiting scrapping at that time (Richard Vandervord).

C-119G s/n 53-3180 (11191) at MASDC, Arizona, in October 1979, oddly with SEA camouflage and additional aerial fittings indicating possible test use for Vietnam War projects (Richard Vandervord).

53-3190, 53-3191, 53-3193, 53-3196, 53-3201, 53-3204, 53-3206, 53-3213, 53-3216. *Batch Total:* 028
    53-7836, 53-7837, 53-7849, 53-7853, 53-7855, 53-7858, 53-7865, 53-7869, 53-7878, 53-7882, 53-7884. *Batch Total:* 011
    53-8071, 53-8073, 53-8074, 53-8075, 53-8076, 53-8083, 53-8098, 53-8099, 53-8101, 53-8103, 53-8110, 53-8111, 53-8113, 53-8122, 53-8125, 53-8126, 53-8127, 53-8142, 53-8149, 53-8150, 53-8153, 53-8154. *Batch Total:* 022

NAF Litchfield Park, Arizona

    R4Q-1: 124324, 124326, 124327, 124328, 124329, 124330, 126574, 126575, 126576, 126577, 126578, 126580, 126581, 126582, 128723, 128724, 128725, 128727, 128728, 128729, 128730, 128731, 128732, 128733, 128734, 128735, 128736, 128737, 128738, 128739, 128740, 128742, 128743: *Batch Total:* 033

    R4Q-2: 131665, 131666, 131667, 131668, 131671, 131672, 131674, 131682, 131684, 131685, 131686, 131691, 131692, 131693, 131695, 131696, 131705, 131707, 131708, 131709, 131710, 131711, 131712, 131713, 131714, 131719. *Batch Total:* 026

USAF Disposals Other Than MASDC / MAP Operators

C-119B & C-119C

48-0319	Scrapped Sheppard AFB
48-0321	Scrapped WPAFB
48-0327	Scrapped O'Hare ARS
48-0333	Scrapped Sheppard AFB
48-0334	Destroyed for Tests
48-0335	Scrapped Greater Pittsburgh ARS
49-0110	Scrapped Pope AFB
49-0120	Scrapped WPAFB
49-0136	Scrapped Amarillo AFB
49-0139	Scrapped Amarillo AFB
49-0140	Scrapped Kelly AFB
49-0146	Scrapped Greater Pittsburgh ARS
49-0147	Scrapped Tinker AFB
49-0148	Scrapped Sheppard AFB
49-0155	Scrapped Amarillo AFB
49-0158	Destroyed for Tests
49-0162	Scrapped Amarillo AFB
49-0181	Destroyed for Tests
49-0182	Scrapped Amarillo AFB
50-0128	Museum / School Status
50-0133	Museum / School Status-Scrapped
50-0160	TL Code-Transferred out of USAF
51-2535	Scrapped WPAFB
51-2548	Scrapped Kelly AFB
51-2556	Scrapped Clark AB The Philippines
51-2566	Museum / School Status
51-2567	Museum / School Status
51-2569	Scrapped Ellington AFB
51-2573	Museum / School Status-Scrapped
51-2574	Scrapped Brookley AFB
51-2575	Scrapped Bakalar AFB
51-2577	TL Code-Transferred out of USAF
51-2609	Scrapped Malmstrom AFB
51-2616	Museum / School Status-Destroyed
51-2659	Scrapped Ellington AFB
51-8244	X Code-Commercial civil sale
51-8262	Scrapped Andrews AFB

C-119F (incl. cvtd C-119G)

51-2675	Museum / School Status
51-2717	Scrapped Chanute AFB
51-7983	Ex-VNAF, scrapped Clark AB, The Philippines
51-8024	Museum / School Status
51-8037	Museum / School Status
51-8104	Scrapped Chanute AFB
51-8118	Scrapped Chanute AFB

Appendix II: C-119 Retirement Bases & Disposals

51-8119   Scrapped Grenier AFB
51-8132   Scrapped (SBNAR) Norton AFB

131718   Scrapped Hayes Dothan

C-119G

52-5846   Museum / School Status-Sold civil operator
52-5850   Museum / School Status

53-3157   Ex-VNAF, scrapped Clark AB, The Philippines
53-3175   Ex-VNAF, scrapped Clark AB, The Philippines
53-3194   Ex-VNAF, scrapped Clark AB, The Philippines
53-3200   United Nations, s/n: UNO-101
53-3202   Ex-VNAF, scrapped Clark AB, The Philippines
53-3203   Ex-VNAF, scrapped Clark AB, The Philippines
53-3208   Scrapped Sheppard AFB
53-3218   Ex-VNAF, scrapped Clark AB, The Philippines
53-3219   United Nations, s/n: UNO-102
53-3220   Ex-VNAF, scrapped Clark AB, The Philippines

53-7828   United Nations, s/n: UNO-103
53-7832   Scrapped Chanute AFB
53-7842   Ex-VNAF, scrapped Clark AB, The Philippines
53-7845   United Nations, s/n: UNO-104

53-8084   Museum / School Status
53-8087   Museum / School Status
53-8088   Ex-VNAF, scrapped Clark AB, The Philippines
53-8100   Scrapped SBNAR, Norton AFB
53-8104   Ex-VNAF, scrapped Clark AB, The Philippines
53-8130   Scrapped Sheppard AFB
53-8133   Ex-VNAF, scrapped Clark AB, The Philippines
53-8146   United Nations, s/n: UNO-105

XC-120

48-330   Scrapped Eglin AFB

YC-119H

51-2585   Scrapped Sheppard AFB

USAF MAP Operator Disposals

C-119C: 49-0137 Morocco, 49-0171 Morocco, 49-0180 Morocco, 49-0183 Morocco, 49-0187 Morocco, 49-0190 Morocco. ***Batch Total:*** 006

C-119F (incl. cvtd C-119G): 51-2586 Taiwan, 51-2669 Taiwan, 51-2671 Taiwan, 51-2677 Taiwan, 51-2681 Taiwan, 51-2683 Taiwan, 51-2709 Taiwan, 51-2710 Taiwan, 51-2711 Taiwan, 51-2715 Taiwan, 51-2716 Taiwan. ***Batch Total:*** 011

51-8016 Taiwan, 51-7970 Taiwan, 51-7971 Taiwan, 51-7973 Taiwan, 51-7974 Taiwan, 51-7977 Taiwan, 51-7978 Taiwan, 51-7979 Taiwan, 51-7980 Taiwan, 51-7981 Taiwan, 51-7983 S. Vietnam, 51-7984 Taiwan, 51-7985 Taiwan, 51-7989 Taiwan, 51-7990 Taiwan, 51-7992 Taiwan, 51-7996 Taiwan, 51-8000 Taiwan, 51-8002 Taiwan, 51-8004 Taiwan, 51-8005 Taiwan, 51-8007 Taiwan, 51-8008 Taiwan, 51-8011 Taiwan, 51-8013

Taiwan, 51-8015 Taiwan, 51-8017 Taiwan, 51-8018 Taiwan, 51-8020 Taiwan, 51-8025 Taiwan, 51-8027 Taiwan, 51-8031 Taiwan, 51-8044 Taiwan, 51-8047 Taiwan, 51-8048 Taiwan. ***Batch Total:*** 035

51-8099 Taiwan, 51-8102 Taiwan, 51-8106 Taiwan, 51-8110 Taiwan, 51-8117 Taiwan, 51-8120 Taiwan, 51-8136 Taiwan, 51-8147 Taiwan, 51-8148 Taiwan, 51-8150 Taiwan, 51-8160 Taiwan, 51-8161 Taiwan, 51-8166 Taiwan. ***Batch Total:*** 013

C-119G: 51-8056 Taiwan, 51-8057 Taiwan, 51-8058 Taiwan, 51-8060 Taiwan, 51-8064 Brazil, 51-8065 Brazil, 51-8066 Brazil, 51-8067 Brazil, 51-8068 Brazil, 51-8070 Taiwan, 51-8071 Taiwan, 51-8074 Brazil, 51-8075 Brazil, 51-8076 Brazil, 51-8077 Brazil, 51-8079 Taiwan, 51-8080 Brazil, 51-8083 Taiwan, 51-8084 Brazil, 51-8091 Taiwan, 51-8092 Brazil, 51-8094 Taiwan. ***Batch Total:*** 022

52-5841 India, 52-5842 India, 52-5843 India, 52-5844 Taiwan, 52-5848 Taiwan, 52-5852 India, 52-5855 India, 52-5856 Taiwan, 52-5857 India, 52-5858 India, 52-5860 India, 52-5861 Taiwan, 52-5863 Jordan, 52-5864 S. Vietnam, 52-5865 India, 52-5867 India, 52-5869 Taiwan, 52-5870 Taiwan, 52-5875 Italy, 52-5878 India, 52-5880 Jordan, 52-5887 India, 52-5888 India, 52-5889 S. Vietnam, 52-5892 S. Vietnam, 52-5898 S. Vietnam, 52-5902 India, 52-5905 S. Vietnam, 52-5909 India, 52-5910 S. Vietnam, 52-5911 S. Vietnam, 52-5912 India, 52-5916 India, 52-5918 Jordan, 52-5920 India, 52-5924 Taiwan, 52-5925 S. Vietnam, 52-5926 S. Vietnam, 52-5927 S. Vietnam, 52-5928 Taiwan, 52-5929 Taiwan, 52-5936 Jordan, 52-5937 Taiwan, 52-5938 S. Vietnam, 52-5940 S. Vietnam, 52-5941 India, 52-5942 S. Vietnam, 52-5943 India, 52-5945 S. Vietnam, 52-5951 Taiwan, 52-5952 Taiwan, 52-5954 Taiwan, 52-5923 Taiwan. ***Batch Total:*** 053

52-9982 S. Vietnam. ***Batch Total:*** 001

53-3136 S. Vietnam, 53-3138 India, 53-3139 S. Vietnam, 53-3140 S. Vietnam then Taiwan, 53-3141 S. Vietnam, 53-3143 Taiwan, 53-3145 S. Vietnam, 53-3146 S. Vietnam, 53-3147 S. Vietnam, 53-3148 S. Vietnam then Taiwan, 53-3149 Taiwan, 53-3153 Taiwan, 53-3154 S. Vietnam, 53-3155 Taiwan, 53-3157 S. Vietnam, 53-3160 Ethiopia, 53-3161 S. Vietnam then Taiwan, 53-3163 Taiwan, 53-3164 Taiwan, 53-3165 Taiwan, 53-3167 S. Vietnam, 53-3169 Taiwan, 53-3170 S. Vietnam, 53-3171 Taiwan, 53-3172 Taiwan, 53-3173 S. Vietnam, 53-3175 S. Vietnam, 53-3176 Taiwan, 53-3177 Taiwan, 53-3178 S. Vietnam, 53-3185 S. Vietnam, 53-3187 S. Vietnam, 53-3188 Ethiopia, 53-3189 S. Vietnam, 53-3192 S. Vietnam, 53-3194 S. Vietnam, 53-3197 S. Vietnam, 53-3198 S. Vietnam then Taiwan, 53-3199 India, 53-3202 S. Vietnam, 53-3203 S. Vietnam, 53-3205 S. Vietnam, 53-3207 Taiwan, 53-3209 India, 53-3210 India, 53-3211 S. Vietnam, 53-3212 India, 53-3215 India, 53-3217 India, 53-3218 S. Vietnam, 53-3220 S. Vietnam. ***Batch Total:*** 051

53-7827 India, 53-7829 Belgium, 53-7830 S. Vietnam, 53-7831 S. Vietnam, 53-7833 S. Vietnam, 53-7834 India, 53-7835 Ethiopia, 53-7839 S. Vietnam, 53-7840 S. Vietnam, 53-7842 S. Vietnam, 53-7843 Belgium, 53-7844 Ethiopia, 53-7846 Taiwan, 53-7847 Taiwan, 53-7848 S. Vietnam, 53-7850 S. Vietnam, 53-7851 S. Vietnam 53-7852 S. Vietnam, 53-7856 Ethiopia, 53-7857 Morocco, 53-7859 India, 53-7860 India, 53-7861 Morocco, 53-7862 Morocco, 53-7863 India, 53-7864 Taiwan, 53-7867 India, 53-7868 India, 53-7870 India, 53-7871 Morocco, 53-7872 India, 53-7873 S. Vietnam, 53-7874 India, 53-7875 Taiwan, 53-7876 Taiwan, 53-7877 S. Vietnam, 53-7879 S. Vietnam, 53-7880 Morocco, 53-7881 India, 53-7883 S. Vietnam. ***Batch Total:*** 040

53-8069 S. Vietnam, 53-8070 India, 53-8077 S. Vietnam, 53-8078 India, 53-8079 India, 53-8080 S. Vietnam, 53-8081 India, 53-8082 India, 53-8085 Taiwan, 53-8086 India, 53-8088 S. Vietnam, 53-8089 S. Vietnam, 53-8090 Taiwan, 53-8091 Taiwan, 53-8092

Appendix II: C-119 Retirement Bases & Disposals

India, 53-8093 Taiwan, 53-8094 India, 53-8095 Morocco, 53-8096 Taiwan, 53-8097 India, 53-8104 S. Vietnam, 53-8105 Taiwan, 53-8106 S. Vietnam, 53-8108 Ethiopia, 53-8109 S. Vietnam, 53-8112 S. Vietnam, 53-8114 S. Vietnam, 53-8115 S. Vietnam, 53-8117 S. Vietnam, 53-8119 Taiwan, 53-8120 Taiwan, 53-8121 S. Vietnam, 53-8123 S. Vietnam, 53-8124 S. Vietnam, 53-8128 Brazil, 53-8129 India, 53-8131 S. Vietnam, 53-8132 Taiwan, 53-8133 S. Vietnam, 53-8134 India, 53-8135 India, 53-8136 India, 53-8137 India, 53-8138 Belgium, 53-8139 S. Vietnam, 53-8141 Belgium, 53-8143 Belgium, 53-8144 India, 53-8145 S. Vietnam, 53-8147 S. Vietnam then Taiwan, 53-8148 S. Vietnam, 53-8151 Belgium, 53-8156 S. Vietnam. **Batch Total:** 053

Appendix III:
Civil Registration Cross-Reference

G	**Great Britain**	
G-BLSW	10689 / 51-2700	C-119G

N	**United States of America**	
N131DM	10831 / BuNo. 131664	C-119F
N175ML	10844 / BuNo. 131677	C-119F
N1040E	11270 / 53-7849	C-119L
N1394N	10840 / BuNo. 131673	C-119F Jet-Pak
N13626	10836 / BuNo. 131669	C-119F
N13742	10431 / 49-194	C-119C Jet-Pak
N13743	10369 / 49-132	C-119C Jet-Pak
N13744	10436 / 49-199	C-119C Jet-Pak
N13745	10304 / 48-322	C-119C Jet-Pak
N13746	10334 / 48-352	C-119C Jet-Pak
N15501	10955 / (RCAF) 22130	C-119G-3E
N15502	10825 / (RCAF) 22114	C-119G
N15504	10774 / (RCAF) 22109	C-119G
N15505	10676 / (RCAF) 22101	C-119G
N15506	10736 / (RCAF) 22105	C-119G
N15507	10861 / (RCAF) 22117	C-119G
N15508	10993 / (RCAF) 22134	C-119G
N15509	10775 / (RCAF) 22110	C-119G-3E
N2700	10689 / 51-2700	C-119G
N383S	10992 / (RCAF) 22133	C-119G-3E
N3003	10737 / (RCAF) 22106	C-119G-3E
N3267U	10885 / BuNo. 131700	C-119F
N3558	10735 / (RCAF) 22104	C-119G
N3559	10870 / (RCAF) 22118	C-119G-3E
N3560	10957 / (RCAF) 22132	C-119G-3E
N3935	10824 / (RCAF) 22113	C-119G-3E
N37483	11318 / 53-7884	C-119L
N37484	11155 / 53-3144	C-119L
N37636	KF-253 / 53-8150	C-119L
N4234S	10875 / BuNo. 131690	C-119F
N4999K	11219 / 53-3206	C-119L	[ntu]
N4999N	KF-230 / 53-8127	C-119L	[ntu]
N48076	11005 / 52-5846	C-119G-3E
N49543	10844 / BuNo. 131677	C-119F

Appendix III: Civil Registration Cross-Reference 431

N5215R	10773 / (RCAF) 22108	C-119G	
N5216R	10956 / (RCAF) 22131	C-119G-3E	
N5217R	10860 / (RCAF) 22116	C-119G	
N55795	11185 / 53-3174	C-119L	
N7051U	10893 / BuNo. 131708	C-119F	
N8091	10906 / (RCAF) 22122	C-119G	
N8092	10678 / (RCAF) 22103	C-119G	
N8093	10776 / (RCAF) 22111	C-119G-3E	
N8094	10994 / (RCAF) 22135	C-119G	
N8501W	10880 / BuNo. 131695	C-119F	
N8504W	KF-256 / 53-8153	C-119L	
N8504X	KF-245 / 53-8142	C-119L Jet-Pak	
N8504Y	11232 / 53-3216	C-119L Jet-Pak	
N8504Z	11253 / 53-7836	C-119L Jet-Pak	
N8505A	KF-179 / 53-8076	C-119L	
N8512K	11318 / 53-7884	C-119L	[ntu]
N8512N	11155 / 53-3144	C-119L	[ntu]
N8682	10859 / (RCAF) 22115	C-119G-3E	
N8832	10907 / (RCAF) 22123	C-119G-3E	
N961S	10872 / (RCAF) 22120	C-119G-3E	
N962S	10905 / (RCAF) 22121	C-119G	
N963S	10954 / (RCAF) 22129	C-119G	
N964S	10823 / (RCAF) 22112	C-119G	
N965S	10943 / (RCAF) 22126	C-119G	
N966S	10738 / (RCAF) 22107	C-119G	
N9027K	KF-176 / 53-8073	C-119L	
N9955F	10370 / 49-133	C-119C	
N9956F	10811 / 51-8263	C-119C	
N9959F	10597 / 51-2608	C-119C	
N9960F	10607 / 51-2618	C-119C	
N9961F	10599 / 51-2610	C-119C	
N9966F	10594 / 51-2605	C-119C	
N90267	KF-257 / 53-8154	C-119L	
N90268	11219 / 53-3206	C-119L	
N90269	KF-230 / 53-8127	C-119L	
N99574	10855 / BuNo. 131688	C-119F	

3C-   **Equatorial Guinea**
3C-ABA   10689 / 51-2700   C-119G

5B-   **Republic of Cyprus**
5B-CFG   10885 / BuNo. 131700   C-119F

Appendix IV: Civil Airtankers

Tanker	Reg	msn / s/n	In Service	Notes
Aero Union Corp.				
E12	N13743	10369 / 49-132	1971–1975	To HVFS
E13	N13744	10436 / 49-199	1972–1976	To HVFS
C14	N13745	10304 / 48-322	1971–1975	To HVFS
85	N13742	10431 / 49-194	1970–1976	To HVFS
Hemet Valley Flying Service Co. (HVFS)				
81	N13743	10369 / 49-132	1975–1987	
82	N13745	10304 / 48-322	1975–1987	
86	N13744	10436 / 49-199	1976–1987	
87	N13746	10334 / 49-352	1976–1987	
88	N13742	10431 / 49-194	1976–1981	W/o 08Jul81
81	N13743	10369 / 49-132	1975–1987	
82	N13745	10304 / 48-322	1975–1987	
86	N13744	10436 / 49-199	1976–1987	
87	N13746	10334 / 49-352	1976–1987	
88	N13742	10431 / 49-194	1976–1981	W/o 08Jul81
Hawkins & Powers Aviation, Inc. (H&P)				
132	N8093	10776 / 22111	1979–1987	Re-numbered #**140** 1986.
133	N383S	10992 / 22133	1976–1979	W/o 08Jun79
134	N8832	10907 / 22123	1975–1985	W/o 17Aug85
135	N48076	11005 / 52-5846	1976–1987	W/o 16Sep87
136	N5216R	10956 / 22131	1975–1987	
137	N3559	10870 / 22118	1975–1987	
138	N8682	10859 / 22115	1972–1981	W/o 27Jun81
138	N15501	10955 / 22130	1981–1987	
139	N3935	10824 / 22113	1972–1987	
140	N3560	10957 / 22132	1973–1978	W/o 10Jun78
T&G Aviation, Inc.				
36	N15509	10775 / 22110	1976–1983	W/o 21Apr84

Appendix V: Aircraft Losses

C-119C s/n 50-138 (10456) would suffer a mid-air collision with C-119C s/n 50-135 on April 18, 1964, near Clinton County AFB, Ohio, resulting in all nine lives on "135" lost with eight of the ten onboard "138" also lost. Two survivors parachuted to the ground (USAF).

Date	Msn / s/n	Type / Operator	Country
28 Jun 1950	10371 / 49-134	C-119C / USAF	Texas USA
29 Jun 1950	10320 / 48-338	C-119B / USAF	Alabama USA
01 Sep 1950	10400 / 49-163	C-119C / USAF	Tennessee USA
13 Sep 1950	10335 / 48-353	C-119B / USAF	Japan
14 Sep 1950	10382 / 49-145	C-119C / USAF	Japan
25 Sep 1950	10306 / 48-324	C-119B / USAF	Japan
12 Oct 1950	10342 / 49-105	C-119B / USAF	Japan
02 Nov 1950	10426 / 49-189	C-119C / USAF	New York USA
09 Nov 1950	10403 / 49-166	C-119C / USAF	S. Korea
18 Dec 1950	10318 / 48-336	C-119B / USAF	N. Korea
07 Feb 1951	7002 / BuNo. 124325	R4Q-1 / USMC	Virginia USA
23 Mar 1951	10405 / 49-168	C-119C / USAF	S. Korea
29 Mar 1951	10327 / 48-345	C-119B / USAF	S. Korea
24 May 1951	10390 / 49-153	C-119C / USAF	Japan
03 Jun 1951	10332 / 48-350	C-119B / USAF	S. Korea
03 Jun 1951	10360 / 49-123	C-119C / USAF	S. Korea

Appendix V: Aircraft Losses

Date	Msn / s/n	Type / Operator	Country
04 Jun 1951	10429 / 49-192	C-119C / USAF	Tennessee USA
15 Jul 1951	10324 / 48-342	C-119B / USAF	Texas USA
27 Sep 1951	10406 / 49-169	C-119C / USAF	Japan
08 Dec 1951	10367 / 49-130	C-119C / USAF	Sea of Japan
04 Jan 1952	10387 / 49-150	C-119C / USAF	New Mexico USA
16 Jan 1952	10409 / 49-172	C-119C / USAF	S. Korea
29 Apr 1952	10410 / 49-173	C-119C / USAF	Japan
27 Jun 1952	7008 / BuNo. 124331	R4Q-1 / USMC	North Carolina USA
24 Jul 1952	10375 / 49-138	C-119C / USAF	Japan
25 Jul 1952	10340 / 49-103	C-119B / USAF	Japan
09 Oct 1952	10412 / 49-175	C-119C / USAF	Japan
07 Nov 1952	10518 / 51-2560	C-119C / USAF	Alaska USA
14 Nov 1952	10509 / 51-2551	C-119C / USAF	S. Korea
15 Nov 1952	10528 / 51-2570	C-119C / USAF	Alaska USA
17 Nov 1952	10579 / 51-2590	C-119CF / USAF	Montana USA
09 Feb 1953	10624 / 51-2635	C-119CF / USAF	South Carolina USA
10 Feb 1953	10445 / 50-127	C-119C / USAF	W. Germany
15 Feb 1953	10363 / 49-126	C-119C / USAF	Missouri USA
11 Mar 1953	10397 / 49-160	C-119C / USAF	S. Korea
15 May 1953	10783 / 51-8235	C-119CF / USAF	W. Germany
15 May 1953	10789 / 51-8241	C-119CF / USAF	W. Germany
17 May 1953	10830 / BuNo. 131663	R4Q-2 / USMC	Florida USA
23 Jun 1953	10398 / 49-161	C-119C / USAF	Japan
09 Aug 1953	10500 / 51-2542	C-119C / USAF	Libya
23 Aug 1953	10330 / 48-348	C-119B / USAF	Philippine Islands
17 Nov 1953	KMC-166 / 51-8163	C-119F / USAF	North Carolina USA
27 Nov 1953	10610 / 51-2621	C-119CF / USAF	France
09 Dec 1953	10980 / 51-8086	C-119G / USAF	Ohio USA
09 Feb 1954	10481 / 50-163	C-119C / USAF	Spain
26 Feb 1954	11061 / 52-5894	C-119G / USAF	Tennessee USA
06 Jan 1954	KMC-154 / 51-8151	C-119F / USAF	The Bahamas
11 Mar 1954	10504 / 51-2546	C-119C / French AF	N. Vietnam
13 Mar 1954	11012 / 52-5853	C-119G / USAF	Georgia USA
13 Mar 1954	11013 / 52-5854	C-119G / USAF	Georgia USA
13 Mar 1954	11046 / 52-5879	C-119G / USAF	Georgia USA
13 Mar 1954	11049 / 52-5882	C-119G / USAF	Georgia USA
13 Mar 1954	11060 / 52-5893	C-119G / USAF	Georgia USA
13 Mar 1954	11087 / 52-5908	C-119G / USAF	Georgia USA
19 Mar 1954	10732 / 51-7993	C-119F / USAF	Maryland USA
23 Mar 1954	10423 / 49-186	C-119C / French AF	N. Vietnam
30 Mar 1954	10668 / 51-2679	C-119F / USAF	North Carolina USA
17 Apr 1954	KMC-136 / 51-8133	C-119F / USAF	Alabama USA
20 Apr 1954	KMC-156 / 51-8153	C-119F / USAF	California USA
22 Apr 1954	11071 / 52-5904	C-119G / USAF	North Carolina USA
06 May 1954	10386 / 49-149	C-119C / French AF	Laos
04 Jun 1954	10548 / BuNo. 126579	R4Q-1 / USMC	S. Korea
20 Jun 1954	10821 / 51-8273	C-119CF / USAF	Belgium
06 Oct 1954	11018 / 52-5859	C-119G / USAF	North Carolina USA
20 Oct 1954	10510 / 51-2552	C-119C / USAF	S. Korea
06 Dec 1954	11100 / 52-5921	C-119G / USAF	Japan
29 Dec 1954	KF-205 / 53-8102	C-119G / USAF	Alabama USA
11 Jan 1955	11136 / 52-5949	C-119G / USAF	Tennessee USA
21 Jan 1955	KF-221 / 53-8118	C-119G / USAF	France

Appendix V: Aircraft Losses

Date	Msn / s/n	Type / Operator	Country
01 Mar 1955	11053 / 52-5886	C-119G / USAF	Japan
03 Mar 1955	10734 / 51-7995	C-119G / USAF	Maryland USA
08 Mar 1955	10351 / 48-114	C-119B / USAF	Pennsylvania USA
10 Mar 1955	10989 / 51-8095	C-119G / USAF	Alabama USA
01 Apr 1955	10654 / 51-2665	C-119F / USAF	Alabama USA
15 Apr 1955	10848 / BuNo. 131681	R4Q-2 / USMC	North Carolina USA
16 Apr 1955	10979 / 51-8085	C-119G / USAF	Japan
18 Apr 1955	10634 / 51-2645	C-119CF / USAF	France
18 Apr 1955	10637 / 51-2648	C-119CF / USAF	Corsica
11 Aug 1955	11238 / 53-3222	C-119G / USAF	W. Germany
11 Aug 1955	11258 / 53-7841	C-119G / USAF	W. Germany
23 Aug 1955	KMC-168 / 51-8165	C-119F / USAF	South Carolina USA
06 Sep 1955	KMC-145 / 51-8142	C-119F / USAF	Louisiana USA
09 Nov 1955	11058 / 52-5891	C-119G / USAF	Over Pacific Ocean
30 Nov 1955	11048 / 52-5881	C-119G / USAF	Japan
06 Dec 1955	11230 / 53-3214	C-119G / USAF	South Carolina USA
27 Dec 1955	10365 / 49-128	C-119C / USAF	Alaska USA
01 Jan 1956	11291 / 53-7866	C-119G / USAF	Japan
04 Feb 1956	10888 / BuNo. 131703	R4Q-2 / USMC	Florida USA
12 Feb 1956	10642 / 51-2653	C-119CF / USAF	France
12 Feb 1956	KMC-152 / 51-8149	C-119G / USAF	USA
19 Mar 1956	10908 / 22124	C-119G / RCAF	Canada
30 Mar 1956	11161 / 53-3150	C-119G / USAF	Japan
03 Apr 1956	11131 / 52-5944	C-119G / USAF	Japan
11 May 1956	10573 / BuNo. 128744	R4Q-1 / USMC	Japan
16 May 1956	10555 / BuNo. 128726	R4Q-1 / USMC	Japan
20 Jul 1956	11052 / 52-5885	C-119G / USAF	Over sea near Japan
27 Jul 1956	10477 / 50-159	C-119C / USAF	Wyoming USA
31 Aug 1956	10782 / 51-8234	C-119CF / USAF	Algeria
20 Sep 1956	10901 / BuNo. 131716	R4Q-2 / USMC	Florida USA
25 Oct 1956	10858 / 51-8032	C-119G / USAF	Canada
26 Oct 1956	10769 / 51-8026	C-119G / USAF	Pennsylvania USA
30 Nov 1956	10467 / 50-149	C-119C / USAF	W. Germany
05 Jan 1957	KMC-146 / 51-8143	C-119G / USAF	California USA
04 Mar 1957	10874 / 51-8034	C-119G / USAF	South Carolina USA
22 Apr 1957	11109 / 52-5930	C-119G / USAF	Japan
01 Jul 1957	11237 / 53-3221	C-119G / USAF	Greece
03 Jul 1957	10591 / 51-2602	C-119CF / USAF	France
24 Oct 1957	KMC-130 / 51-8127	C-119J / USAF	Oklahoma USA
13 Jan 1958	10942 / 22125	C-119G / RCAF	Canada
04 Feb 1958	10764 / 51-8021	C-119G / USAF	Illinois USA
07 Mar 1958	10570 / BuNo. 128741	R4Q-1 / USMC	Japan
27 Mar 1958	10432 / 49-0195	C-119C / USAF	Texas USA
16 Sep 1958	10339 / 49-0102	C-119C / USAF	USA
14 Nov 1958	KMC-125 / 51-8122	C-119J / ANG	New York USA
26 Nov 1958	KF-210 / 53-8107	C-119G / USAF	Idaho USA
13 Dec 1958	10945 / 22128	C-119G / RCAF	Italy
31 Dec 1958	10471 / 50-0153	C-119C / USAF	USA
03 Jan 1959	10613 / 51-2624	C-119CF / USAF	Indiana USA
05 Feb 1959	10328 / 48-0346	C-119C / USAF	California USA
06 Mar 1959	10882 / BuNo. 131697	R4Q-2 / USMC	North Carolina USA
07 Mar 1959	10929 / 51-8051	C-119J / USAF	Michigan USA
06 Jun 1959	10389 / 49-0152	C-119C / USAF	USA

Appendix V: Aircraft Losses

Date	Msn / s/n	Type / Operator	Country
20 Oct 1959	10305 / 48-0323	C-119C / USAF	Ohio USA
23 Dec 1959	unk	C-119G / IAF	India
27 Feb 1960	KMC-108 / 51-8105	C-119G / USAF	South Carolina USA
05 Mar 1960	KF-255 / 53-8152	C-119G / USAF	Italy
19 Jul 1960	11083 / CP-36	C-119G / BAF	The Congo
19 Sep 1960	11300 / IK459	C-119G / IAF	India
11 Aug 1960	10944 / 22127	C-119G / RCAF	Greenland
15 Jan 1961	10628 / 51-2639	C-119CF / USAF	Kentucky USA
02 Feb 1961	11076 / MM52-6037	C-119G / AMI	The Congo
15 Feb 1961	10869 / MM52-6011	C-119G / AMI	The Congo
22 Jun 1961	KF-243 / 53-8140	C-119G / USAF	Texas USA
17 Nov 1961	10911 / MM52-6014	C-119G / AMI	Tanzania
23 Nov 1961	10463 / 50-0145	C-119C / USAF	Canada
12 Dec 1961	10951 / CP-23	C-119G / BAF	Belgium
12 Dec 1961	11082 / CP-25	C-119G / BAF	Belgium
20 Jan 1962	KF-175 / 53-8072	C-119G / USAF	Utah USA
10 Mar 1962	10507 / 51-2549	C-119C / USAF	Louisiana USA
01 Apr 1962	10961 / 2302	C-119G / FAB	Brazil
25 Jun 1962	10482 / 50-0164	C-119C / USAF	Puerto Rico
28 Jul 1962	11174 (?) / 3163	C-119G / RoCAF	unk
21 Sep 1962	KMC-115 / 51-8112	C-119G / USAF	Canada
01 Nov 1962	10321 / 48-0339	C-119C / ANG	USA
01 Jan 1963	unk	C-119G / IAF	India
30 Mar 1963	10863 / MM52-6005	C-119G / AMI	Italy
08 May 1963	10733 / 51-7994	C-119G / USAF	Massachusetts USA
26 Jun 1963	KF-246 / CP-45	C-119G / BAF	W. Germany
09 Jul 1963	10307 / 48-0325	C-119C / USAF	USA
07 Oct 1963	10629 / 51-2640	C-119CF / USAF	USA
21 Oct 1963	unk / BK962	C-119G / IAF	India
30 Oct 1963	10455 / 50-0137	C-119C / USAF	USA
29 Jan 1964	10502 / 51-2544	C-119C / USAF	USA
04 Mar 1964	unk	C-119G / IAF	India
18 Apr 1964	10453 / 50-0135	C-119C / USAF	Ohio USA
18 Apr 1964	10456 / 50-0138	C-119C / USAF	Ohio USA
20 Apr 1964	11075 / MM52-6036	C-119G / AMI	Italy
11 May 1064	10984 / 51-8090	C-119G / USAF	Alabama USA
20 May 1964	10465 / 50-0147	C-119C / USAF	USA
30 Jul 1964	unk	C-119G / IAF	India
22 Oct 1964	11080 / MM52-6041	C-119G / AMI	France
14 Dec 1964	10924 / MM51-8046	C-119J / AMI	Italy
04 Jun 1965	KF-219 / 53-8116	C-119G / USAF	Maine USA
05 Jun 1965	10669 / 51-2680	C-119G / USAF	The Bahamas
06 Sep 1965	unk	C-119G / IAF	India
10 Sep 1965	unk	C-119G / IAF	India
14 Sep 1965	unk	C-119G / IAF	India
22 Oct 1965	11035 / CP-19	C-119G / BAF	W. Germany
08 Jan 1966	10600 / 51-2611	C-119CF / USAF	Pennsylvania USA
17 Apr 1966	10762 / 51-8019	C-119G / USAF	Illinois USA
25 Apr 1966	10480 / 50-0162	C-119C / USAF	USA
02 Aug 1966	11098 / 52-5919	C-119G / USAF	Florida USA
09 Sep 1966	11255 / 53-7838	C-119G / USAF	Texas USA
30 Sep 1966	11208 / 53-3195	C-119G / USAF	California USA
02 Mar 1967	10771 / 51-8028	C-119G / USAF	USA

Appendix V: Aircraft Losses

Date	Msn / s/n	Type / Operator	Country
18 Feb 1968	11157 / 533146	C-119G / VNAF	S. Vietnam
08 Aug 1968	10914 / 51-8036	C-119J / USAF	California USA
09 Aug 1968	11170 / 53-3159	C-119G / USAF	Ohio USA
06 Dec 1968	10684 / 12695	C-119G / RNAF	Norway
14 Dec 1968	10656 / 51-2667	C-119G / USAF	Puerto Rico
14 Mar 1969	11282 / CN-AMG	C-119G / RMAF	Spain
07 Apr 1969	unk / 3122	C-119G / RoCAF	unk
26 Jun 1969	KMC-157 / MM51-8154	C-119J / AMI	Italy
29 Jun 1969	unk	C-119G / IAF	India
11 Oct 1969	11074 / 52-5907	AC-119G / USAF	S. Vietnam
19 Feb 1970	11167 / 53-3156	AC-119K / USAF	S. Vietnam
25 Apr 1970	10948 / MM52-6018	C-119G / AMI	Italy
28 Apr 1970	KF-258 / 53-8155	AC-119G / USAF	S. Vietnam
06 Jun 1970	11114 / 52-5935	AC-119K / USAF	S. Vietnam
30 Oct 1970	unk / BK970	C-119G / IAF	India
10 Nov 1971	10683 / CP-14	C-119G / BAF	Belgium
10 Aug 1971	11160 (?) / 3149	C-119G / RoCAF	unk
20 Nov 1971	unk / 3124	C-119G / RoCAF	unk
14 Mar 1972	unk / 3117	C-119G / RoCAF	Taiwan
02 May 1972	11239 / 53-7826	AC-119K / USAF	S. Vietnam
06 May 1972	unk	C-119G / IAF	India
18 Jul 1972	11175 / 3164	C-119G / RoCAF	unk
23 Aug 1972	11275 / 53-7854	AC-119K / USAF	S. Vietnam
27 Oct 1972	11115 / 52-5936	C-119K / RJAF	France
01 Mar 1973	11256 / 537839	AC-119K / VNAF	S. Vietnam
21 Dec 1973	KF-231 / 2312	C-119G / FAB	Brazil
09 May 1974	unk / 3128	C-119G / RoCAF	Taiwan
09 May 1974	unk / 3166	C-119G / RoCAF	Taiwan
27 Jun 1974	10978 / 2307	C-119G / FAB	Brazil
29 Apr 1975	11056 / 525889	AC-119K / VNAF	S. Vietnam
26 Nov 1976	unk	C-119G / IAF	India
01 Jul 1977	unk / 3205	C-119G / RoCAF	unk
18 Aug 1977	unk / 3221	C-119G / RoCAF	unk
10 Jun 1978	10957 / N3560	C-119G / H&P Aviation	Wyoming USA
24 Jan 1979	11031 / MM52-6030	C-119G / AMI	Italy
08 Jun 1979	10992 / N383S	C-119G / H&P Aviation	California USA
06 Jul 1979	KF-230 / N90269	C-119L / Starbird	Alaska USA
08 Jul 1979	11270 / N1040E	C-119L / Lambeth Aircraft	Arizona USA
28 Aug 1979	unk / 3194	C-119G / RoCAF	Taiwan
03 Oct 1979	10605 / 51-2616	C-119CF / Bradley Air Mus.	Connecticut USA
23 Feb 1980	unk	C-119G / IAF	India
05 Jul 1980	11219 / N90268	C-119L / Starbird	Alaska USA
27 Jun 1981	10859 / N8682	C-119G / H&P Aviation	Alaska USA
08 Jul 1981	10431 / N13742	C-119C / HVFS	California USA
07 Sep 1981	KF-256 / N8504W	C-119L / J.D. Gifford & Assoc.	Alaska USA
01 Oct 1981	11318 / N37483	C-119G / Jim Cottingham	Mexico
07 Feb 1982	unk	C-119G / IAF	India
18 Apr 1983	unk / 3182	C-119G / RoCAF	unk
08 May 1983	10836 / N13626	C-119F / Supra Intl.	Alaska USA
06 Jun 1983	unk / 3197	C-119G / RoCAF	Taiwan
21 Apr 1984	10775 / N15509	C-119G / T&G	Alaska USA
17 Aug 1985	10907 / N8832	C-119G / H&P Aviation	Nevada USA
13 May 1987	KF-245 / N8504X	C-119L / Alaska Aircraft Leas.	Alaska USA

Date	Msn / s/n	Type / Operator	Country
16 Sep 1987	11005 / N48076	C-119G / H&P Aviation	California USA
27 Jan 1989	KF-222 / 3215	C-119G / RoCAF	unk
01 Jun 1996	10743 / 3181	C-119G / RoCAF	Taiwan

Appendix VI: Extant Airframes

C-119J s/n 51-8037 (10915) has been preserved at National Museum of the USAF since 1963 and was one of the nine Boxcars used for the Corona satellite recovery program in Hawaii (USAF).

RoCAF C-119G s/n 3158 (10710) is today preserved in a colorful livery at the China University of Science and Technology in Taiwan (Erik Sleutelberg).

Belgium
10690 / CP-10	Melsbroek Air Base Barracks, Brussels
10952 / CP-21	Fuselage only, current whereabouts unknown
KF-254 / CP-46	Royal Museum of the Armed Forces and Military History, Brussels

Brazil
10968 / FAB-2304	8th Parachutist Field Artillery Group, Rio de Janeiro
10970 / FAB-2305	Museu Aerospacial (MUSAL), Campos dos Afonos AB, Rio de Janeiro

Eritrea
11261 / 910	Asmara AFB / Intl. Airport, Asmara

Great Britain
10689 / 51-2700	Ex-civil N2700. Nose section only, current ownership unknown

India
11225 / IK444	Shatrujeet Officers' Mess, Agra
11262 / IK450	Indian Air Force Museum, Palam AFB, New Delhi
11286 / IK455	Stored Agra, unconfirmed
11310 / IK463	Stored Agra, unconfirmed

Italy
10950 / MM52-6020	Rivolto Air Base, Udine
11030 / MM52-6029	Museo Storico dell'Aeronauctica Militare, Vigna di Valle
11213 / MM53-3200	Pisa Intl. Airport, Pisa
KMC-124 / MM51-8121	Maserada Sul Piave
KF-249 / MM53-8146	Piana delle Orme Museum, Latina

South Korea
10739 / 3199	War Memorial of Korea, Seoul

Morocco
11283 / CN-AMH	Marrakesh Menara Airport, Marrakesh
10943 / CN-AMN	Kenitra Air Base, Kenitra, Morocco
2 unknown	Rabat-Sale Intl. Airport, Rabat, Morocco

Taiwan
10706 / 3202	Yuanzhiluxiuxian Park, Nanhua District
10710 / 3158	China University of Science and Technology, Hsinchu Campus
10724 / 3160	Xihujunji Park, Xihu, Changhua County
10964 / 3120	Pingtung AB, Pingtung
10965 / 3183	Junji Park, Changhua City
10973 / 3144	Kao Yuan Institute of Technology, Kaohsiung City
11003 / 3184	Junshi Park, Jiji City
KMC-109 / 3190	Republic of China Air Force Museum, Gangshan AB
KMC-123 / 3192	Kinmen National Park, Kinmen County
KMC-139 / 3152	Pingtung AB, Pingtung
KMC-153 / 3204	Pingtung AB, Pingtung

United States of America
10304 / 48-0322	Milestones of Flight Air Museum, Fox Field, California (Closed)
10334 / 48-0352	Air Mobility Command Museum, Dover AFB, Delaware
10369 / 49-0132	Pima Air & Space Museum, Tucson, Arizona
10394 / 49-0157	Pima Air & Space Museum, Tucson, Arizona
10436 / 49-0199	Castle Air Museum, Castle AFB, California

Appendix VI: Extant Airframes

10446 / 51-0128	82nd Airborne Division War Memorial Museum, Fort Bragg, North Carolina
10524 / 51-2566	Museum of Aviation, Robins AFB, Georgia
10525 / 51-2567	USAF History & Traditions Museum, Lackland AFB, Texas
10664 / 51-2675	U.S. Veterans Museum, Grandbury, Texas
10676 / (RCAF) 22101	Don F. Pratt Memorial Museum, Fort Campbell, Kentucky
10678 / (RCAF) 22103	National Warplane Museum, Geneseo, New York
10736 / (RCAF) 22105	Niagara Falls Air Reserve Station, New York
10737 / (RCAF) 22106	Atterbury-Bakalar Air Museum, Columbus, Indiana
10738 / (RCAF) 22107	Hill Aerospace Museum, Hill AFB, Utah
10767 / 51-8024	Strategic Air Command & Aerospace Museum, Ashland, Nebraska
10773 / (RCAF) 22108	Museum of Flight & Aerial Firefighting, Greybull, Wyoming
10776 / (RCAF) 22111	Hagerstown Aviation Museum, Hagerstown, Maryland
10824 / (RCAF) 22113	Museum of Flight & Aerial Firefighting, Greybull, Wyoming
10825 / (RCAF) 22114	Aerospace Museum of California, McClellan AFB, California
10840 / BuNo. 131673	Palmer Municipal Airport, Palmer, Alaska
10844 / BuNo. 131677	Mid Atlantic Air Museum, Reading, Pennsylvania
10846 / BuNo. 131679	Don F. Pratt Memorial Museum, Fort Campbell, Kentucky
10855 / BuNo. 131688	Pueblo Weisbrod Aircraft Museum, Pueblo, Colorado
10860 / (RCAF) 22116	National Infantry Museum, Fort Benning, Georgia
10870 / (RCAF) 22118	Air Mobility Command Museum, Dover AFB, Delaware
10880 / BuNo. 131695	Palmer Municipal Airport, Palmer, Alaska
10891 / BuNo. 131706	Southwestern Alloys Corp., Tucson, Arizona
10893 / BuNo. 131708	Flying Leatherneck Aviation Museum, MCAS Miramar, California
10906 / (RCAF) 22122	March Field Air Museum, March AFB, California
10915 / 51-8037	National Museum of the USAF, WPAFB Dayton, Ohio
10955 / (RCAF) 22130	Lauridsen Aviation Museum, Buckeye Municipal Airport, Arizona
10956 / (RCAF) 22131	Rolling Boxcar, Inc., Alaska
10993 / (RCAF) 22134	Travis AFB Heritage Center, Travis AFB, California
10994 / (RCAF) 22135	South Big Horn County Airport, Greybull, Wyoming
11009 / 52-5850	Grissom Air Museum, Grissom AFB, Indiana
11155 / 53-3144	Hurlburt Field Memorial Air Park, Hurlburt Field, Florida
11253 / 53-7836	Fairbanks Intl. Airport, Fairbanks, Alaska
KF-176 / 53-8073	Alaska Aviation Museum, Anchorage, Alaska
KF-177 / 53-8074	Flying J Ranch Airpark, Pima, Arizona
KF-179 / 53-8076	South Big Horn County Airport, Greybull, Wyoming
KF-187 / 53-8084	Little Rock AFB, Arkansas
KF-190 / 53-8087	82 Airborne Division War Memorial Museum, Fort Bragg, North Carolina
KF-253 / 53-8150	South Big Horn County Airport, Greybull, Wyoming
KF-257 / 53-8154	Palm Springs Air Museum, Palm Springs, California

Vietnam

11271 / 537850	Ho Chi Minh City Intl. Airport, Ho Chi Minh City

Known Wreck Sites & Rescued Wrecks

10431 / 49-0194	Los Padres National Forest, California, USA
10463 / 50-0145	24 miles South of Whitehorse, Yukon, Canada
10775 / (RCAF) 22110	Tobin Mine Creek Airstrip, Alaska, USA
10859 / (RCAF) 22115	25 miles S Bettles, Alaska, USA
KF-230 / 53-8127	Big Creek, Alaska, USA
KF-245 / 53-8142	Shageluk, Alaska, USA

Bibliography

Primary References

Federal Aviation Administration (FAA) N-number records. Aircraft Registration Branch, FAA, Oklahoma City, Oklahoma.
USAF Individual Aircraft Record Card (IARC) Reels & USAF Delivery Logs. The Air Force Historical Research Agency (USAFHRA), Maxwell AFB, Montgomery, Alabama.
USN Individual Aircraft Record Card (IARC) Reels (Partial). National Naval Aviation Museum (NNAM), NAS Pensacola, Florida.

Manuals, Technical Sheets, and Reports

FAA C-119 Type Certificates: A5NW; A6NW; A8NW; A21WE; A24WE; A32CE; A35CE.
RCAF C-119 Fleet Overhaul and Transfer Sheets.
USAF AC-119G Flight Manual. T.O. 1C-119(A)G-1, revised to January 1971.
USAF C-119B, C-119C & USN R4Q-1 Flight Manual. AN 01-115CCA-1, revised to July 1950.
USAF C-119B, C-119C & USN R4Q-1 Flight Manual. T.O. 1C-119B-1, revised to June 1956.
USAF C-119C / G / J & USN R4Q-2 Structural Repair Manual. T.O. 1C-119B-3, revised to April 1966.
USAF C-119C Maintenance Manual. T.O. 1C-119C-2-1, revised to August 1964.
USAF C-119F Flight Manual. AN 01-115CCB-1, revised to November 1953.
USAF C-119F & USN R4Q-2 Flight Manual. T.O. 1C-119F-1, revised to May 1956.
USAF C-119G Flight Manual. T.O. 1C-119G-1, revised to March 1956.
USAF C-119G / J Flight Manual. T.O. 1C-119G-1, revised to December 1966.
USAF / USN Accident Reports:
    C-119G s/n 52-5894 dated 26Feb54.
    C-119G s/n 53-8152 dated 05Mar60.
    R4Q-1 BuNo. 124325 dated 07Feb51.
United States Government Senate Subcommittee Hearing Reports:
    Aircraft Production Costs and Profits. February to March 1956.
    Contract Award of C-119 Cargo Planes by Air Force. June 1953.
    Review of the USFS Firefighting Aircraft Program. August 1993.

Books

Andrade, John M. *U.S. Military Aircraft Designations and Serials 1909–1979.* Midland Counties Publications, 1979.
Chakko, Gp Capt Jacob. *Memoirs.* Privately published, 1999.
Jane's All the World's Aircraft. Volume years 1945–1970. London: Haymarket Publishing Group.
Lloyd, Alwyn T. *Fairchild C-82 Packet and C-119 Flying Boxcar.* Aerofax (Midland Publishing), 2005.
Lunsford, Charles L. *Departure Message.* Great Unpublished, 2001.
Mann, Robert A. *Aircraft Record Cards of the United States Air Force.* McFarland, 2008.
Mikesh, Robert C. *Flying Dragons: The South Vietnamese Air Force.* Osprey, 1988.
Mitchell, Kent A. *Fairchild Aircraft 1926–1987.* Narkiewicz-Thompson, 1997.
Swanborough, Gordon, and Peter M. Bowers. *United States Military Aircraft Since 1909.* Putnam, 1989.
Yenne, Bill. *The History of the U.S. Air Force.* Bison Books, 1984.

Academic Publications, Articles, and Periodicals

Ballard, Lt Col Jack S. *The United States Air Force in Southeast Asia Development and Employment of Fixed-Wing Gunships 1962–1971*. Office of Air Force History. January 1974.
Cantwell, Gerald T. *Citizen Airmen: A History of the Air Force Reserve 1946–1994*. Air Force History and Museums Program. 1997.
Fletcher, Harry R. *Air Force Bases—Volume II*. Office of Air Force History, USAF. 1993.
Leeker, Dr. Joe F. *CAT in French Indochina*. University of Dallas. August 2015.
Mayne, Dr. Richard. "Keep Them Flying Part 1." *CAHS Journal*. Spring 2016.
Mayne, Dr. Richard. "Keep Them Flying Part 2." *CAHS Journal*. Summer 2016.
Mitchell, Kent A. "The C-119 Prototype—The One and Only Fairchild C-119A Cargo Plane." *AAHS Journal*. Winter 1993.
Mueller, Robert. *Air Force Bases—Volume I*. Office of Air Force History, USAF. 1989.
Ravenstein, Charles A. *Air Force Combat Wings Lineage and Honors History 1947–1977*. Office of Air Force History, USAF. 1984.
Van Nederveen, Capt Gilles K. *USAF Airlift into the Heart of Darkness, the Congo 1960–1978. Implications for Modern Air Mobility Planners*. Airpower Research Institute, USAF Air University. September 2001.
Williams, Nicholas M. "The People Beaters: Fairchild's R4Q / C-119F Packet in Navy and Marine Corps Service." *AAHS Journal*. Summer 1993.
Wilton, Dave. "USAF Tactical Airlift in Europe During the 1950s—Part 3." *Aeromilitaria*. Winter 2010.

Online References

www.ac119gunships.com—AC-119 technical data.
www.aerialvisuals.ca—C-119 historical data.
www.amarcexperience.com—C-119 retirements.
www.aviationarchaeology.com—accident data.
www.aviation-safety.net—accident data.
www.baaa-acro.com—accident data.
www.belgian-wings.be—BAF C-119 histories.
www.bharat-rakshak.com—IAF C-119 data.
www.caspir.warplane.com—RCAF C-119 data.
www.discover.dtic.mil—military and civil test reports.
www.faa.gov—civil histories.
www.goodall.com.au—C-119 airtanker histories.
www.joebaugher.com—serial numbers and history checks.
www.ruudleeuw.com—C-119 histories and data.
www.scramble.nl—foreign service serial numbers.
www.usafunithistory.com—USAF unit and wing data.
www.warbirdregistry.com—civil and museum locations.

Index

Numbers in ***bold italics*** indicate pages with illustrations

AC-119 gunships 12, 29–30, ***30***, 35–36, ***36***, ***37***, 59, 87
Aero Union Corp. 138
Air Defense Command (ADC) 62–66
Air Materiel Command (AMC) 66–68
Air National Guard (ANG) 34, 124–126
Aircraft International Associates 138–139

Belgian Air Force 41–42, ***42***
Brazilian Air Force 43, ***43***

C-82A series 9, ***9***
C-119: airframe structure 14–15, 18–19; airtankers ***12***, 38–40, 136, 138, 143, 144, ***150***, ***151***, 432; cargo capacity 15–16; crew numbers 15; design missions 15; electrical system 17; engines 16; fuel system 17; hydraulic system 17–18; oil system 17; production contracts 13–14; propellers 17; serial numbers 153–154; specifications (detailed) 13–17; tail codes 421–422; weights 16
C-119G-3E 39–40, ***40***
C-119J series 31–33, ***34***
C-119K series 34–35
C-119L series 37–38, ***37***
Chosin Reservoir, Battle of 11
Corona satellite recovery ***33***, ***62***, 69

Davis-Monthan AFB, Arizona 423–425
Dien Bien Phu, Battle of 10, 11, 46–47, 180
Dross Metals, Inc. 141–142

"Earthquake McGoon" 180–181
Ethiopian Air Force ***45***, 45–46

Fairchild Aircraft Company 13, 110
Flight of the Phoenix (film) 148–149, ***151***, 294, 296, 297, 310
French Air Force (Indochina) 46–48

Hagerstown Aviation Museum 143, ***150***
Hawkins & Powers Aviation, Inc. 143–144
Hemet Valley Flying Service Co. 144, ***151***

Indian Air Force 38, 48–50, ***49***
Italian Air Force 50–52, ***52***

Jet-Pak C-119 37–39, ***38***

Military Air Transport Service (MATS) 83–86
Moroccan Air Force 53, ***53***

Norwegian Air Force 54

Royal Canadian Air Force 44–45, ***45***
Royal Jordanian Air Force 52

South Vietnamese Air Force 132–134
Steward-Davis STOLmaster 39, ***39***, 318–319
Strategic Air Command 88–93

Tactical Air Command 93–103
Taiwanese Air Force 54–57, ***57***
Troop Carrier Groups / Wings ***10***, 58–59, 111

United States Air Forces (all namesakes) 10–11, 14, 58–61
USAF Museum Allocations 107–108
USAF Reserve 10–11, 110–124
United States Marine Corps 26–28, 126–131

XC-119A 19, ***20***, 154
XC-120 26, ***26***, 158

YC-119F 23, 228
YC-119H 25–26, ***25***, 228
YC-119K 34–35, 350